AF598623

THE SCIENCE of BIOLOGICAL SPECIMEN PREPARATION for MICROSCOPY

Proceedings of the 14th Pfefferkorn Conference, held August 6 to 11, 1995 at the Shrine of Our Lady of the Snows, Belleville, IL

Editors

M. Malecki and G.M. Roomans

Scanning Microscopy Supplement 10, 1996

Published By

Scanning Microscopy International

Post Office Box 66507

AMF O'Hare (Chicago), IL 60666, U.S.A.

swets 12/11/98

Scanning Microscopy International

This not-for-profit organization pursues the following goals.

(a). Promotion of advancement of science of scanned image microscopy methods and related microscopy and microanalysis techniques.

(b). Promotion of applications of these techniques in existing and new areas.

(c). Promotion of these techniques so that their users obtain the best information of the highest quality.

The organization attempts to fulfill its goals by sponsoring meetings and Pfefferkorn Conferences devoted to basic topics in scanning microscopy (see page vi for details); and by publishing (1) the quarterly journals Scanning Microscopy (previously Scanning Electron Microscopy), and Cells and Materials (2) the Proceedings of the Pfefferkorn Conferences (published as supplement to Scanning Microscopy from 1987); and (3) related publications.

Suggestions on other activities which may be undertaken to fulfill the above goals may be communicated to any one of the following persons:

John D. Fairing, President, Scanning Microscopy International
809 Westwood Drive, Ballwin, MO 63011 Phone: 314 227 8939

Godfried M. Roomans Vice President, and Editor-in-Chief, Scanning Microscopy
Box 571, Univ. Uppsala, S-75123 Uppsala, Sweden Phone: 46 18 4714114; FAX: 46 18 551120

Om Johari Secretary-Treasurer, Scanning Microscopy International
P.O. Box 66507, AMF O'Hare (Chicago), IL 60666 Phone: 847 524 6677; FAX: 847 985 6698

The topics covered in this supplement are of continuing interest and importance to our quarterly journal **Scanning Microscopy**. Contributions are invited. Pfefferkorn Conferences on topics related to physics and materials sciences included the 1st (1982), 3rd (1984), 5th (1986), 6th (1987), 8th (1989), and 10th (1991). Since 1987 the Proceedings of the Pfefferkorn Conferences are published as Scanning Microscopy Supplements.

The **16th Pfefferkorn Conference in 1998** (April 5-9, 1998, in Aberystwyth, Wales, UK) will be on Optimizing the Scanning Electron Microscope with Drs. Iolo ap Gwynn (Univ. Wales, Aberystwyth, UK) and Godfried M. Roomans (Univ. Uppsala, Sweden) as organizers.

Serial fee code: 0892-953X/93$5.00+.25.

For Additional Information about programs and publications of SMI please contact Om Johari or Godfried M. Roomans. Printed in United States of America

Fourteenth Pfefferkorn Conference on

The Science of Biological Specimen Preparation for Microscopy

held at Belleville, IL, during August 6-11, 1995

Foreword

The ultimate thrust for pursuing microscopy in life sciences is to gather information on functional structure of tissues, cells, and biomolecules. In practice, an important step in gathering that information is preparation of biological specimens so that they meet the technical limits of the microscopes. Each step in specimen preparation may not only help in facilitating microscopic imaging, but also can harm by inducing structural alterations. Therefore, for correct interpretation of images, thorough analysis of preparatory protocols is essential. Based upon this analysis, living cell phenomena can be deconvolved from the microscopic images.

The issues mentioned above were the driving force for organizing the 14th Pfefferkorn Conference on the Science of Biological Specimen Preparation for Microscopy in Belleville in August 1995. By bringing together experts from various areas of science related to microscopy, this conference was aimed at creating a forum for their presentations on thoroughly analyzed means of specimen preparation in life sciences. Discussions with the speakers which followed the presentations and often extended into long-after-midnight hours, constituted an important part of this conference.

An out-come of that conference is this book. The speakers at the conference constitute the majority of the authors. This team of authors was expanded for three reasons. Firstly, a few authors were unexpectedly prevented from attending the conference but submitted their contributions. Secondly, some speakers decided to rest their case with their oral presentation at the conference; thus leaving a gap in the spectrum of the procedures covered. Thirdly, during the editing of this book some important procedures were emerging, in particular in molecular biology, which were included to keep the book updated. All the papers were peer-reviewed. Discussions with the reviewers became a part of this book not only to keep the book in the spirit of the conference but also with the intention to challenge the opinions expressed by the authors and to bring controversies to the attention of the readers.

Finally, we would like to thank the authors and the reviewers for cooperation and patience throughout the editing process. We acknowledge with thanks strong encouragement, friendly support, and administrative help provided by Dr. Om Johari and Dr. Fairing of Scanning Microscopy International.

Marek Malecki — **University of Wisconsin, Madison**
Godfried M. Roomans — **University of Uppsala, Sweden**

Conference Sponsored by Scanning Microscopy International

John D. Fairing — President
Godfried M. Roomans — Vice-President
Om Johari — Secretary-Treasurer

Scanning Microscopy International is deeply grateful to the organizers for their efforts in bringing together experts in this field for the 14th Pfefferkorn Conference and putting together a successful meeting.

Om Johari, Conference Director

Scanning Microscopy Supplement 10, 1996 ISSN: 0892-953X
Proceedings of the Fourteenth Pfefferkorn Conference, Belleville, IL
Scanning Microscopy International, AMF O'Hare (Chicago), IL 60666, USA

The Science of Biological Specimen Preparation for Microscopy

Table of Contents

Pfefferkorn Conferences

Started in 1982, these conferences are dedicated to Late Professor Gerhard E. Pfefferkorn from Münster, Germany. The aim of this conference series is to cover basic topics related to scanning microscopy and related techniques.

The Twelfth Pfefferkorn Conference (held 1993 at Cambridge, U.K.), was devoted to *The Science of Biological Microanalysis* and the proceedings were published as **Scanning Microscopy Supplement 8**, 1994 (30 papers, 410 pages). **Price:** $65 (U.S.); $70 (elsewhere).

The Eleventh Pfefferkorn Conference (held 1992 at Amherst, Massachusetts, U.S.A.), was devoted to the *Physics of Generation and Detection of Signals Used for Microcharacterization* and the proceedings were published as **Scanning Microscopy Supplement 7**, 1993 (20 papers, 304 pages). **Price:** $65 (U.S.); $70 (elsewhere).

The Tenth Pfefferkorn Conference (held 1991 at Cambridge, U.K.), was devoted to *Signal and Image Processing in Microscopy and Microanalysis*, and the proceedings were published as **Scanning Microscopy Supplement 6**, 1992 (40 papers, 452 pages); editor: P.W. Hawkes. **Price:** $75 (U.S.); $80 (elsewhere).

The Ninth Pfefferkorn Conference (held August 1990, at Santa Cruz, California); Proceedings, published as **Scanning Microscopy Supplement 5**, 1991 (but bound in the same issue as Scanning Microscopy vol. 5, no. 4, 1991) was devoted to the *Science of Biological Specimen Preparation for Microscopy and Microanalysis* (10 papers, 114 pages). Editors: L. Edelmann and G.M. Roomans. **Price:** $52 (U.S.); $58 (elsewhere).

The Eighth Pfefferkorn Conference (held May 1989 at Park City, Utah); Proceedings, published as **Scanning Microscopy Supplement 4**, 1990, was devoted to the *Fundamental Electron and Ion Beam Interactions with Solids for Microscopy, Microanalysis and Microlithography* (21 papers, 370 pages). Editors: J. Schou, P. Kruit and D.E. Newbury. **Price:** $59 (U.S.); $64 (elsewhere).

The Seventh Pfefferkorn Conference (held September 1988 at the Univ. Guildford, Surrey, U.K.) Proceedings, published as **Scanning Microscopy Supplement 3, 1989**, were devoted to the *Science of Biological Specimen Preparation for Microscopy and Microanalysis* (29 new papers, 302 pages). Editors: R.M. Albrecht and R.L. Ornberg. **Price:** $57 (U.S.); $62 (elsewhere).

The Sixth Pfefferkorn Conference (held April 28 - May 2, 1987 at Niagara Falls, Canada) Proceedings, published as **Scanning Microscopy Supplement 2, 1988**, were devoted to *Image and Signal Processing in Electron Microscopy* (39 papers, 396 pages). Editors: P.W. Hawkes, F.P. Ottensmeyer, W.O. Saxton, and A. Rosenfeld. **Price:** $60 (U.S.); $65 (elsewhere).

The Fifth Pfefferkorn Conference (held October 1986 at Brüggen, Germany) Proceedings, published as **Scanning Microscopy Supplement 1, 1987**, were devoted to *Physical Aspects of Microscopic Characterization of Materials* (27 papers; 254 pages). Editors: J. Kirschner, K. Murata, and J.A. Venables. **Price:** $48 (U.S.); $52 (elsewhere).

Also available are the Proceedings of the Fourth Pfefferkorn Conference (1985): *The Science of Biological Specimen Preparation for Microscopy and Microanalysis 1985* [eds.: M. Müller, R.P. Becker, A. Boyde, and J.J. Wolosewick (ISBN 0-931288-37-1) **Price:** $46 (U.S.); $50 (elsewhere)]; Third Pfefferkorn Conference (1984): *Electron Optical Systems for Microscopy, Microanalysis and Microlithography* [eds.: J.J. Hren, F.A. Lenz, E. Munro and P.B. Sewell (ISBN 0-931288-34-7) **Price:** $44 (U.S.); $48 (elsewhere)]; Second Pfefferkorn Conference (1983) *The Science of Biological Specimen Preparation for Microscopy and Microanalysis* [eds.: J.-P. Revel, T. Barnard, and G.H. Haggis (ISBN 0-931288-32-0) **Price:** $40 (U.S.); $44 (elsewhere)]; and the First Pfefferkorn Conference (1982) *Electron Beam Interactions with Solids for Microscopy, Microanalysis and Microlithography* [eds.: D.F. Kyser, D.E. Newbury, H. Niedrig, and R. Shimizu. (ISBN 0-931288-30-4) **Price:** $51 (U.S.); $56 (elsewhere)].

The Proceedings of the Thirteenth Pfefferkorn Conference on *Cathodoluminescence* (held May 1994 in Niagara Falls), are in press.

The Sixteenth Pfefferkorn Conference will be held on "*Optimizing the Scanning Electron Microscope*" during April 5-9, 1998 at Aberystwyth, Wales, U.K.

For additional information contact: Dr. Om Johari, Scanning Microscopy International, P.O. Box 66507, Chicago (AMF O'Hare), IL 60666, USA (phone 847 524 6677; FAX: 847 985 6698).

Scanning Microscopy Supplement 10, 1996 (pages 1-16)
Scanning Microscopy International, Chicago (AMF O'Hare), IL 60666 USA

0892-953X/96$5.00+.25

PREPARATION OF PLASMID DNA IN TRANSFECTION COMPLEXES FOR FLUORESCENCE AND ELECTRON SPECTROSCOPIC IMAGING

Marek Malecki*

Molecular Biology Laboratory, University of Wisconsin at Madison and Integrated Microscopy Resource, National Institutes of Health Biotechnology Resource, Madison, WI

(Received for publication October 1, 1995 and in revised form June 18, 1996)

Abstract

The aim of this project was to develop procedures necessary to study mechanisms of receptor mediated gene transfer by means of integrated microscopy.

Plasmid DNA was incorporated into a transfection complex consisting of poly(L)lysine and transferrin to which the nuclear localization signal was conjugated. This complex was presented to cultured glioma cells. Preparation of the transfected DNA for imaging was pursued by two methods. In the first method tetramethylrhodamine, nanogold, and ferritin were linked through streptavidin to the biotinylated plasmid DNA. Trafficking of the fluorescent derivatives was studied in living cells with fluorescence microscopy. Then, selected cells were rapidly cryo-immobilized. Ultrastructural distribution of the transfected DNA was imaged with energy filtering transmission electron microscopy. In the second method, the unmodified transfected DNA was detected in cryo-immobilized cells by *in situ* polymerase chain reaction and *in situ* hybridization. For laser scanning fluorescence microscopy probes were labeled with tetramethylrhodamine. For ultrastructural analysis by electron spectroscopic imaging, probes containing incorporated digoxigenin were labeled with anti-digoxigenin boronated antibodies.

Based upon the developed procedures, it has been demonstrated that the presence of the nuclear localization signal in the transfection complex resulted in rapid nuclear import of the transfected DNA.

Key Words: Transfected DNA, receptor mediated gene transfer, polymerase chain reaction, *in situ* hybridization, immunolabeling, fluorescence microscopy, energy filtering transmission electron microscopy.

*Address for correspondence:
Marek Malecki
Molecular Biology Laboratory, University of Wisconsin
1675 Observatory Drive, Madison, WI 53706
Telephone Number: 608-263-8481
FAX Number: 608-265-4076
E-mail: malecki@macc.wisc.edu

Introduction

Intranuclear delivery of selected genes is an essential step toward their expression in selected cells (Anderson, 1992). Mechanisms of gene delivery through receptor mediated endocytosis are not yet recognized (Malecki *et al.*, 1995). The specific aim of this project was to develop specimen preparation techniques necessary for imaging of cellular trafficking pathways of the transfected DNA. Recognition of these mechanisms should help us in designing efficient methods of gene delivery.

Gene delivery through receptor mediated endocytosis using non-viral vectors is particularly important for targeted gene therapy *in vivo* (Wagner *et al.*, 1990; Zenke *et al.*, 1990; Curiel *et al.*, 1991). A low level of gene expression reported with this method was thought to be related to lysosomal degradation, endosomal arrest, or perinuclear entrapment (Wagner *et al.*, 1992; Chowdhury *et al.*, 1993; Harris *et al.* 1993; Malecki *et al.*, 1995).

Nuclear transport of the transfected plasmid DNA is a particularly important phase of gene transfer, yet very poorly documented. These concerns could be effectively addressed by revealing trafficking of the transfecting plasmid DNA as well as the cellular structures involved in this trafficking through the use of techniques of modern microscopy. The main advantage of this approach is that it allows pursuit of these studies on intact cells retaining integrity of their complex endocytotic pathways and signaling systems (Hopkins *et al.*, 1990). This approach is particularly powerful, if it involves integrated microscopy in which the images of the same cell from different microscopes are integrated into one comprehensive message concerning cellular functions (Malecki, 1992).

The project had two specific aims. The first, to determine a panel of reporter molecules for the recombinant plasmid DNA and the transfection complexes to reveal their intracellular trafficking. The second, to develop protocols for intracellular detection of small quantities of the unlabeled plasmid DNA by *in situ* hybridization and of the transfection complexes by immunolabeling for fluorescence microscopy (FM) and energy filtering transmission electron microscopy (EFTEM). For both parts of the project cryo-immobilization of the events for ultrastructural analysis was essential.

As to the first specific aim, fluorochromes tagged to the transfected DNA were considered the most suitable reporter molecules for detection of intracellular trafficking within living cells. Probes and antibodies coupled to fluorochromes were also helpful for evaluation of the labeled samples as a preliminary step before getting involved into far more labor-intensive studies on ultrastructure of frozen cells with electron microscopy. For these reasons, it is useful to summarize features of ideal fluorochromes. They should have high quantum efficiency to ensure good signal detection. They should be physiologically inert so their derivatives do not change physiological properties of probes. Products of their decay due to illumination should not be toxic for cells. Ideal reporter molecules should maintain stable emission. For multiple labeling, they should have significant Stoke's shifts and band separation to avoid over-lapping leading to bleed-through. In these respects, studies of basic features of various fluorochromes and derived biomolecules excited by one-, two-, and three-photon excitation are particularly important (Tsien and Waggoner, 1990; Lakowicz and Gryczynski, 1992; Gryczynski and Lakowicz, 1994; Szmacinski *et al.*, 1995). Shortcomings in meeting the requirements summarized above put additional requirements on the instrumentation. Indeed, an important factor contributing to recent improvements in imaging of fluorescent reporter molecules was also due to rapid progress in laser scanning fluorescence microscopy (Tsien and Bacskai, 1995; Denk *et al.*, 1995; Brakenhoff and Visscher, 1995). In particular, development of two-photon laser scanning fluorescence microscopy (Denk *et al.*, 1990) and improvements in real-time laser scanning confocal microscopy (Petran and Hadravsky, 1968; Draaijer and Houpt, 1988) carry much promise. In two-photon laser scanning fluorescence microscopy, photon density leading to effective excitation, thus fluorescent emission is limited to the focal plane only. As a result, effective signal is spatially confined while glare from above and below that plane is reduced. Moreover, fading of fluorochromes is limited to the volume brought to the excitation state. A volume from which free radicals are generated is limited only to that volume, therefore cell viability should be improved as compared to other imaging modes in which fluorescence is generated from the cell entire volume. In video-rate laser scanning microscopy, acousto-optical deflector (AOD) generates scans at the super-video rates up to 240 frames/sec. The point of interest for this project relies in the fact that these scanning rates create conditions for imaging of fast cellular phenomena.

As to the second specific aim, fluorescent emission from the plasmid DNA marked with fluorochromes has to be sufficiently strong to become detectable. This is achieved by incorporating high concentration of reporter molecules into the plasmid DNA perhaps leading to changes in its physiological properties, as was reported for biotin (Khine and Lingwood, 1994). In another approach, the concentration of the plasmid DNA presented to the cells is increased during transfection perhaps leading to cytotoxicity. Both approaches may lead to false results. In an alternative approach procedures of *in situ* hybridization (ISH) (Gall and Pardue, 1969; Pardue and Gall, 1969; John *et al.*, 1969) can be applied for detection of the transfected DNA (Malecki *et al.*, 1995). In ISH, weak reporter signals often create difficulties for their detection. In view of these difficulties, it was necessary to develop various amplification procedures whereby a low number of unlabeled copies could be detected. Protocols to carry on polymerase chain reactions in situ (Haase *et al.*, 1990; Nuovo *et al.* 1991a,b; Yap and McGee, 1991), to design more efficient probes (Konat *et al.*, 1991; Thiry 1995), and to enhance reporter signals (Hacker *et al.*, 1995; Malecki *et al.*, 1995) were developed. These protocols served as the starting point for their modification for the purpose of this project, whereby a sequence of the transfected DNA in cryo-immobilized and freeze-substituted cells is amplified by polymerase chain reaction in situ followed by in situ hybridization for imaging with laser scanning fluorescence microscopy.

For both specific aims, studies on the transfected DNA pathways in living cells can be greatly enriched by studies of the same cells with electron microscopy. For ultrastructural analysis of the plasmid DNA trafficking, it is worthwhile referring to experience gathered in studies on endocytosis, exocytosis, or nuclear transport. In these processes, molecules traffic through various cellular compartments (Rothman, 1994). In particular, it was recognized early on in studies on synaptic vesicles that chemical fixation used for preservation of ultrastructural studies resulted in generation of artifacts. This occurred because of a cross-linking agent which was diffusing from the cell surface and did not prevent synaptic vesicles from diffusing from the unfixed cellular interior toward the cell periphery. This led to imaging of spatial relations different from those that occurred in living cells and therefore to false interpretations of living cell phenomena. In a similar manner, cell permeabilization or homogenization procedures may very seriously alter spatial relationships. In response to this problem, various rapid cryo-immobilization procedures have been developed (Van Harreveld and Crowell, 1964; Heuser *et al.*, 1979; Rash, 1983; Ryan and Knoll, 1994). Anticipating similar problems with analysis of processes involved in gene delivery, the special cryo-immobilization equipment had to be designed. In particular, assembly of an optic system integrated into one apparatus with a rapid freezer allowed us imaging of a living cell followed by rapid cryo-immobilization of this cell for ultrastructural analysis (Malecki, 1992).

Moreover, an essential part of ultrastructural analysis is unambiguous identification of structures. For this purpose, the most frequently used markers in electron microscopy at the present time are antibodies conjugated to colloidal gold (Faulk and Taylor, 1971) and probes tagged with biotin or digoxigenin for subsequent labeling with colloidal gold marked antibodies (Hutchison *et al.*, 1982). Unfortunately, colloidal gold beads do not form covalent bonds with antibodies leading often to their detachment (Kramarcy and Sealock, 1990). To overcome this limitation gold clusters with functional groups have been designed (Hainfeld, 1988). Their sizes and formed covalent bonds determine accuracy and reliability of immunolabeling (Hainfeld and Furuya, 1992). In conventional transmission electron microscopy (CTEM) of cells, these clusters are often buried within grains of metal stains (e.g., Os, U, and Pb), therefore they require silver enhancement (Hacker *et al.*, 1995). On the contrary, larger diameters of colloidal gold beads compromise efficiency of labeling (Takizawa and Robinson, 1994). For ultrastructural identification, electron spectroscopic imaging (Ottensmeyer and Andrew, 1980; Ottensmeyer, 1984; Edgerton, 1993) offers an interesting alternative. Atoms of a selected element can be incorporated into the probes to be used as reporters for electron spectroscopic mapping. An important part of this procedure is selection of an element, which must be absent from the structures of interest. In this respect boronated antibodies offer a possibility for specific discrimination of their distribution (Barth *et al.*, 1986; Bendayan *et al.*, 1989; Kessels *et al.*, 1996; Malecki *et al.*, 1995). Moreover, new great opportunities open up when atoms of different elements are incorporated into different probes or antibodies to pursue studies on colocalization of labeled molecules (Malecki, 1995). An obtained map may then be superimposed onto the image of the cell ultrastructure captured through contrast tuning. This approach is applicable for derivatives of molecules presented to living cells as well as derivatives of probes and antibodies used to label molecules of interest in frozen cells. This method offers unambiguous location of reporter molecules at the ultrastructural level.

This project was focused on preparation of the plasmid DNA in the transfection complexes in order to image their cellular pathways with fluorescence microscopy and energy filtering transmission electron microscopy.

Materials and Methods

Cell cultures

Human cell lines were from American Type Culture Collection (ATCC, Rockville, MD). Cells of human glioblastoma cell line (ATCC HTB-14 U-87 MG) were grown in Eagle's Minimal Essential Medium (MEM) with non-essential amino acids, 1 mM pyruvate, Earle's balanced salt solution (BSS), 20 mM Hepes, 10% Donor Calf Serum (DCS) (Gibco, Gaithersburg, MD). $2x10^6$ cells plated into a 75 cm^2 cell culture flask (Corning, Midland, MI) reached confluence within 7 days. Cells of human glioblastoma cell line (ATCC HTB-15 U-118 MG) were grown in Dulbecco's modified Eagle's medium (DMEM) supplemented with 10% DCS. $2x10^6$ cells plated into a 75 cm^2 cell culture flask reached confluence within 7 days. For retaining stock cultures, cells were grown in polystyrene flasks (Corning). For imaging, cells were plated onto 3 mm diameter, polystyrene filmed carriers 24 hours before experiments and closed into the viewing chambers as described previously (Malecki, 1991). Cell viability was tested with the Live/Dead Euk system (Molecular Probes, Eugene, OR). The DNA of the cell nuclei was stained with bisbenzimide (Sigma, St. Louis, MO).

Plasmid DNA, Transfection complex, Transfection

The plasmids with the human cytomegalovirus (CMV) promoter (courtesy of Dr. Skowron, Primerix, Madison, WI) and the green fluorescent protein (GFP) (courtesy of Dr. Prasher, United States Department of Agriculture, Otis, MA and Clonetech, Palo Alto, CA) or β galactosidase (lacZ) (Promega, Madison, WI) coding sequences were used. For imaging in living cells, the plasmid DNA was photo-biotinylated (Sigma) prior to transfection (1:100-2000). Streptavidin was coupled with tetramethylrhodamine for imaging with laser scanning fluorescence microscopy (LSFM) at absorption 515nm / emission 590nm. Streptavidin was also coupled to ferritin for imaging at the iron edge at 708 eV or nanogold (Nanoprobes, Stony Brook, NY) at zero loss. Incorporation of reporter molecules was evaluated with EFTEM. Alternatively, the plasmid DNA had been labeled by nick translation (Maniatis *et al.*, 1975). The labeling solution contained 0.5 μg of the plasmid DNA, 10 μl of 10x nick translation buffer (0.5 M Tris, 0.1 M $MgCl_2$, 0.08 M 2-mercaptoethanol, 0.5 mg/ml bovine serum albumin (BSA) pH 7.5, 1 μl of stock solution of DNAase1 (2ng/ml), 1 μl of each stock solution of deoxynucleoside triphosphates (dNTPs) (100 μM) (Boehringer, Indianapolis, IN), 1 μl of biotin-dUTP (300 μM) (Boehringer, Indianapolis, IN), 32 μl water. To that solution 1 μl of Taq polymerase (5-10 units activity) had been added (Perkin Elmer, Foster City, CA). Reaction was carried for 1 hour. It was quenched by the addition of 4μl of 0.25 M EDTA, 2 μl of tRNA (10mg/ml), and 150 μl of Tris (10 mM), pH 7.5. The biotinylated DNA was conjugated with streptavidin tagged with tetramethylrhodamine (Molecular Probes), nanogold (Nanoprobes), or ferritin (Sigma). Unreacted dNTPs and reporter molecules were removed after each step of nick translation and reporter molecule incorporation by chromatography (Pharmacia, Piscataway, NJ). The plasmid DNA was spread on filmed grids (Sogo *et al.*, 1984). The samples were plunge-frozen and freeze-dried. Contrast was enhanced by low angle rotary shadowing with platinum and in BAF400 (Baltec, Balzers, Liechtenstein). The unlabeled plasmid DNA was detected in cryo-immobilized cells by *in situ* hybridization as described below.

Transfection complexes consisted of three components linked by covalent bonds through bifunctional linkers. Poly(L)lysine was linked to transferrin through the heterobifunctional linker N-succinimidyl-3-(2-pyridyldithio) propionate (SPDP) (Pierce, Rockford, IL) (1:1). The nuclear localization signal (NLS) was linked to transferrin (Sigma) through the heterobifunctional linker m-maleimidobenzoyl-N-hydroxysuccimide ester (MBS) (Pierce) (6:1) (Malecki and Skowron, 1995). The complexes of proteins were dialyzed against HEPES buffered saline pH 7.2 (HBS) between all linking steps. Transferrin was marked with fluorescein isothiocyanate (Sigma) for imaging with LSFM at absorption 495 nm / emission 519 nm.

The plasmid DNA was linked to the transfection complex 1h prior to transfection in HBS pH 7.2 (1:1) (Wagner *et al.*, 1990). The plasmid DNA within the transfection complexes was presented to the glioblastoma cells in HBS.

Cryo-immobilization

Rapid cryo-immobilization was pursued through the use of two instruments. The first was a prototype of the apparatus integrating into one unit: optic system and rapid freezer (Malecki, 1992). The second was an apparatus which was designed to pursue rapid cryo-immobilization of cells being imaged with a laser scanning fluorescence microscope equipped with a conventional specimen stage (Malecki, in preparation). Frozen samples were maintained in liquid nitrogen until further processing. Freezing rates were measured from thermocouples inserted into the specimen viewing chamber.

Polymerase chain reaction, *in situ* hybridization

Cryo-immobilized cells were freeze-substituted in methanol in three steps: -80°C, -35°C, and 0°C with each step lasting 24 hours in the freeze-substitution apparatus (Baltec, Balzers, Liechtenstein). The samples were rehydrated at 0°C through three methanol/water mixtures (3/1, 1/1, 1/3). 20 nucleotides long primers for amplification of the GFP coding sequence fragment were generated in a synthesizer (Applied Biosystems, Foster City, CA) with a sequence determined with Prime software (Genetic Computer Group, Madison, WI). An advantage of this approach was that the cell carriers were small enough to fit into small tubes of a conventional thermal cycler (Perkin Elmer, Foster City, CA). Therefore, there was no need for application of a larger cycler hosting slides and no need for using larger volumes of solutions. Moreover, the polymerase chain reaction (PCR) protocol could be closer to the liquid phase PCR (Mullis *et al.*, 1986) rather than to the *in situ* PCR (Nuovo *et al.*, 1991a) due to the same surface-to-volume ratios and temperature gradients. The PCR mixture consisted of: 50 mM KCl, 2 mM $MgCl_2$, 10 mM Tris-HCl, 0.01% gelatin, 0.1 mM dNTP (Boehringer), 10 pM primers. The cells on the 3 mm cell carriers were inserted into the small vials. The temperature was increased to 95°C and kept for at least 10 min. Then 5 U/μl of Taq polymerase (Perkin Elmer) was added. PCR amplification proceeded in 20 cycles each consisting of three steps: denaturation: at 95°C for 5 min the first, and 1 min all subsequent; annealing at 53°C for 2 min; and polymerization at 70°C for 3 min in all cycles and 10 min the last one. These conditions were established after running controls in which the primers were irrelevant or omitted altogether. The amplified sequences in PCR protocols and unamplified sequences in ISH protocols were hybridized with ssDNA probes marked with digoxigenin (Konat *et al.*, 1991) (1 ng/μl) in Tris-buffered salt solution (TSS) [0.3 M NaCl, 0.03 M Tris-HCl (2xTSS) pH 6.8] containing *E. coli* denatured DNA (200 ng/μl), at 67°C for 12 h. Unconjugated probes were rinsed off with 50% formamide in saline sodium citrate (SSC) [0.3 M NaCl, 0.03 M sodium citrate pH 7.0 (2xSSC)] and 2xSSC at 39°C. The probes were then labeled with monoclonal anti-digoxigenin antibodies (Boehringer) followed by secondary antibodies which were marked either with tetramethylrhodamine (Molecular Probes) or with boron (Barth *et al.*, 1986). Transferrin was labeled with monoclonal antibodies followed by secondary antibodies conjugated with ferritin (Singer, 1959; Rifkind *et al.*, 1960).

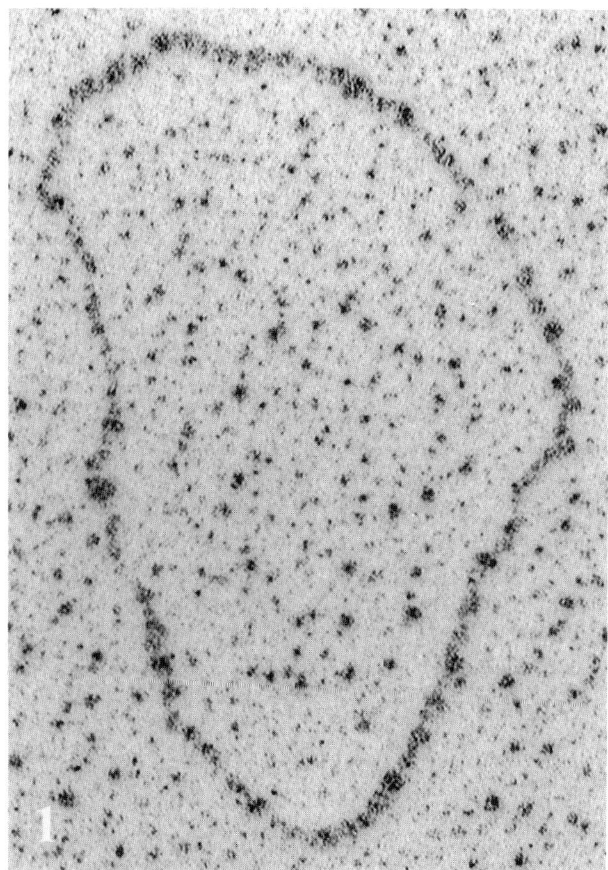

Figure 1. The plasmid DNA (GFP) had incorporated biotin labeled with streptavidin. After spreading on polystyrene films, the samples were frozen, freeze-dried, and low-angle rotary shadowed with Pt followed by supportive coating with C. Imaging was pursued with the EM 912 Omega operated at 120 kV with the filter set at O loss. The image was recorded with a slow-scan CCD. Horizontal field width 394 nm.

Fluorescence microscopy

For the labeled plasmids, imaging of steps in gene delivery by receptor mediated endocytosis was pursued on the living cells closed within the environmental chambers. For the unlabeled plasmids, cryo-immobilized cells after in situ hybridization and immunolabeling were mounted in glycerol containing HBS saline. Morphological analysis was performed with fluorescence microscopes. The optics, additionally equipped with the Xenon illuminator, of the imaging-freezing apparatus was described elsewhere (Malecki, 1992). Images on this system were captured on 6400ASA film (Kodak, Rochester, NY) or cooled charge-coupling device (CCD) camera (Photometrics, Tucson, AZ) 6400ISO equivalent. The Axiovert (Zeiss, Oberkochen, Germany) was equipped with the Enterprise argon ion (457 nm, 488 nm, 529 nm lines) and ultraviolet (UV) (364 nm line) lasers and Odyssey XL digital video-rate confocal laser scanning imaging system operated up to 240 frames/s under control of Intervision software (Noran, Middleton, WI). The Diaphot (Nikon, Garden City, NY) was equipped with the Microlase diode-pumped Nd:YLF solid state laser (1048 nm line), the pulse compressor (resulting pulses of approximately 300 fs at 120 MHz), and the BioRad MRC600 scanning system under control of Comos software (the system assembled at the Integrated Microscopy Resource, Madison, WI).

Images were deconvolved after their import to the Indy workstation (Silicon Graphics, Mountain View, CA) using recently developed algorithms (Avinash, 1995). Deconvolved images after conversion into the TIFF format were assembled into galleries in Photoshop (Adobe, Mountain View, CA).

Energy filtering transmission electron microscopy

The frozen cells containing the transfected plasmid DNA marked with reporter molecules were freeze-substituted and embedded in Lowicryl K4M or Epon 812 followed by cutting ultrathin serial sections (Malecki and Small, 1987). The unmodified transfected plasmid DNA in the frozen cells was detected after freeze-substitution and embedding followed by ultramicrotomy and etching of plastics for immunolabeling and *in situ* hybridization (Ris and Malecki, 1993). Ultrastructural analysis was pursued in energy filtering transmission electron microscopes. The EM912 (Zeiss) with the LaB_6 source was operated at 120kV. The energy filter was the in-column Omega. The microscope was equipped with the cryo-transfer system (Oxford Instruments, Oxford, UK). Images were acquired on the CCD (1kx1k) camera. Image acquisition and processing was pursued with ESIVision software (Soft Imaging Software, Golden, CO). The CM120 (Philips Electron Optics, Eindhoven, The Netherlands) with LaB_6 source was operated at 120kV. The energy filter was the post-column Gatan Imaging Filter (GIF) (Gatan, Warrendale, PA). The microscope was equipped with the cryo-transfer system (Gatan). Images were acquired and processed with Microscopist software (Gatan). For image analysis, files were imported, processed, and analyzed using ESIVision software (Soft Imaging Software).

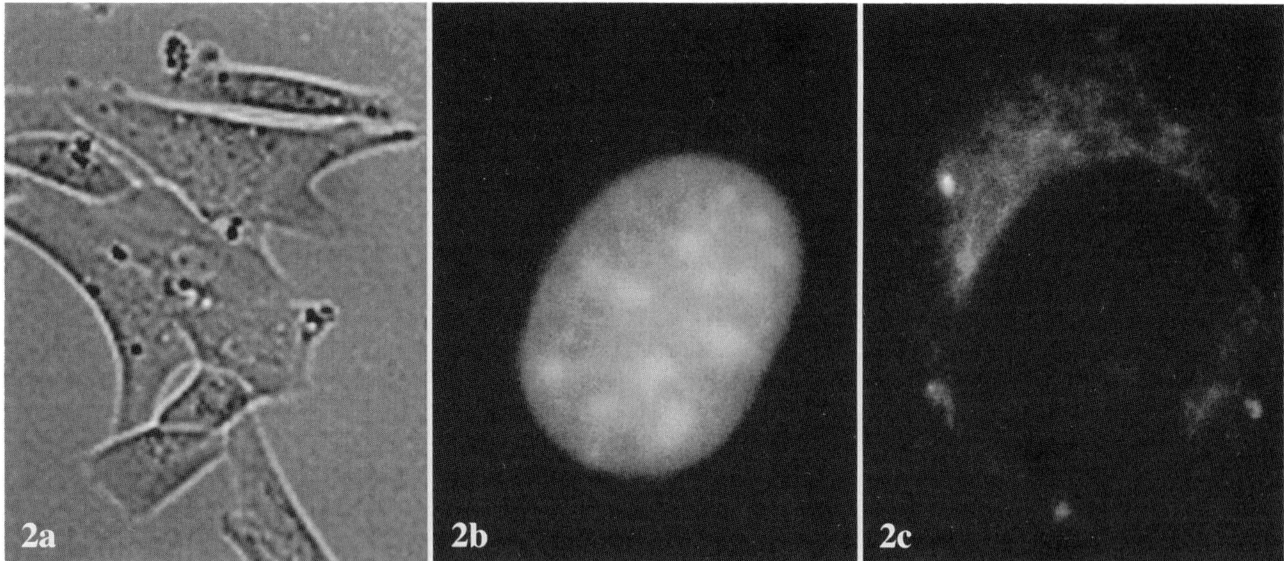

Figure 2. **(a)** The cells of the ATCC HTB 14 U87 MG line were grown on a polystyrene film in the viewing chamber. Their morphology is identical to that of cells grown in polystyrene flasks. Horizontal field width 55 μm. **(b)**. The nuclei of the ATCC HTB 14 U87 MG cells were stained with bisbenzimide. Imaging was pursued with the fluorescence optics of the apparatus allowing us rapid freezing of the cells after their recording (Malecki, 1992). The image was recorded on the 6400ASA Kodak film. Horizontal field width 5.5 μm. **(c)**. The same cell as in Fig. 2b. This cell was transfected with the plasmid DNA incorporated into the Tf-PLL transfection complex without NLS. The plasmid DNA was labeled with tetramethylrhodamine. The image was recorded 30 min after presentation of transfection complexes to the glioblastoma cells at 37°C. Around the nucleus, there is accumulation of the fluorescent transfected DNA. As the result of the depth of field and the point spread function, it is impossible to determine, if fluorescent patches near the center of this image are due to the fluorescence from within the nucleus or from beyond. Horizontal field width as in Fig. 2b.

Figure 3 (*on facing page*). **(a-f)** Three-dimensional distribution of the transfected DNA in the living ATCC HTB 14 U-87 MG cells. The cells were transfected with the plasmid DNA in the Tf-PLL-NLS complexes. The biotinylated plasmid DNA was marked TMR-streptavidin. Imaging was pursued with Odyssey XL Noran. Optical sections were collected starting from the cell base upwards and selected near the equator of the cell nucleus based upon referring to the nuclear DNA stained with bisbenzimide shown in Fig. 3 g-l. Clearly, threads of fluorescence are present within the nuclear space in Fig. 3b. This can be determined based upon the fact that fluorescence from the transfected DNA is spread only within some central optical sections. These fluorescent threads very sharp in Fig. 3b become fuzzy in neighboring sections in Fig. 3c, but disappear in more distant sections in Fig. 3a and Figs. 3e,f. Clearly, the threads of the fluorescent plasmid DNA extend within the nucleus. They may represent highlighted routes of the intranuclear distribution system. Horizontal field width of the single frame 5 μm. **(g-l)**. The nuclei of the cells presented in Figs. 3a-f stained with bisbenzimide. Fluorescent pattern of the nuclear DNA served as reference for distribution of the transfected DNA. Horizontal field width of the single frame is the same as in Fig. 3a-f.

Results

Preparation of the plasmid DNA in the transfection complexes by incorporating reporter molecules before transfection

The first part of this project was aimed at developing protocols to incorporate suitable reporter molecules into the plasmid DNA and the transfection complexes, so that they could be detected with FM and EFTEM. Incorporation of reporter molecules into the plasmid DNA was best studied with EFTEM as demonstrated in Fig. 1. Topography of the entire plasmid DNA circle was observed at the zero loss due to contrast generated after low angle rotary shadowing. The plasmid DNA:biotin:streptavidin ratios were necessary to optimize composition of the transfection complexes and to quantitate effectiveness of gene transfer protocols.

The transfected plasmid DNA which was modified with fluorochromes was first detected within living cells with epi-fluorescence microscopy (Fig. 2). To deter-

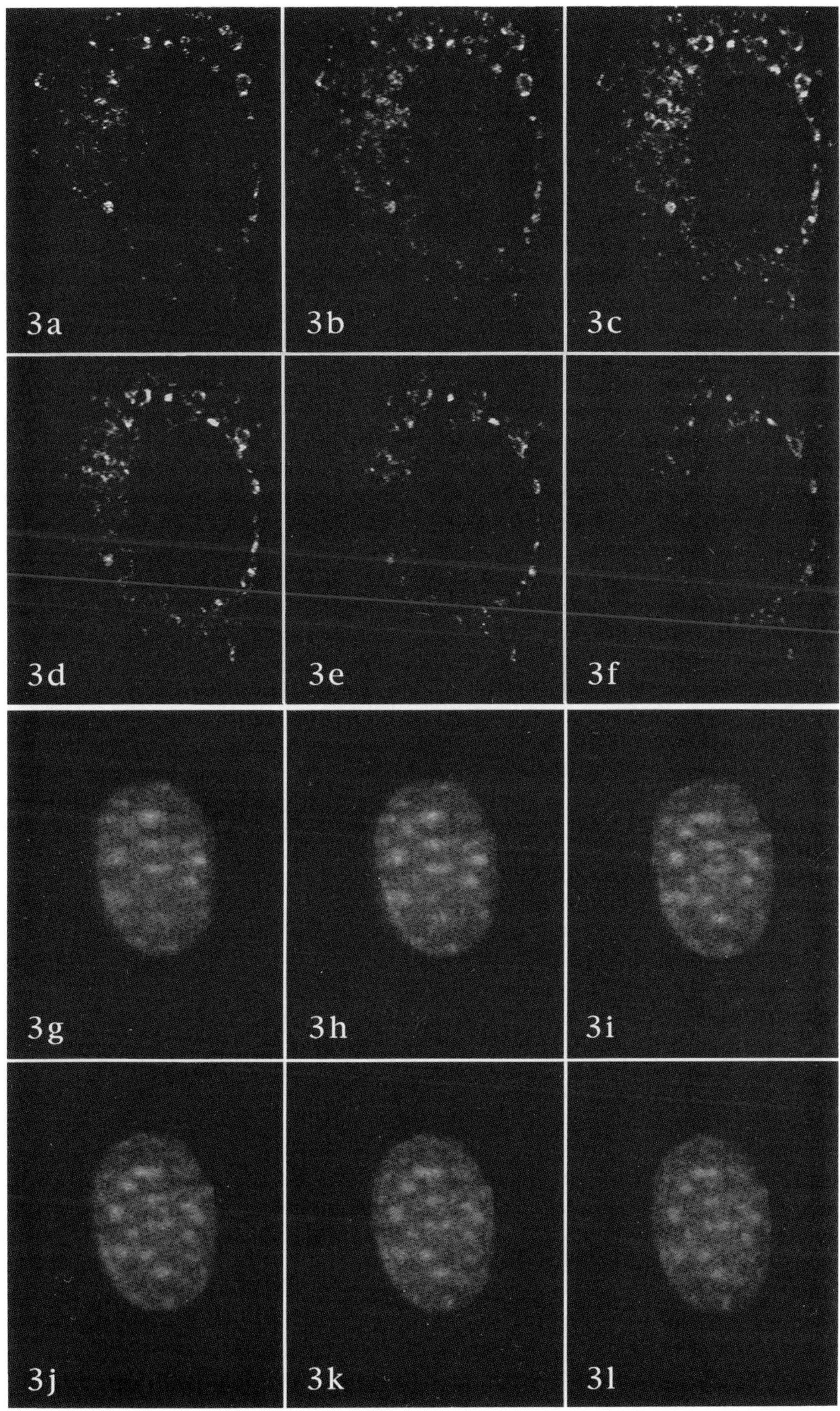
3a
3b
3c
3d
3e
3f
3g
3h
3i
3j
3k
3l

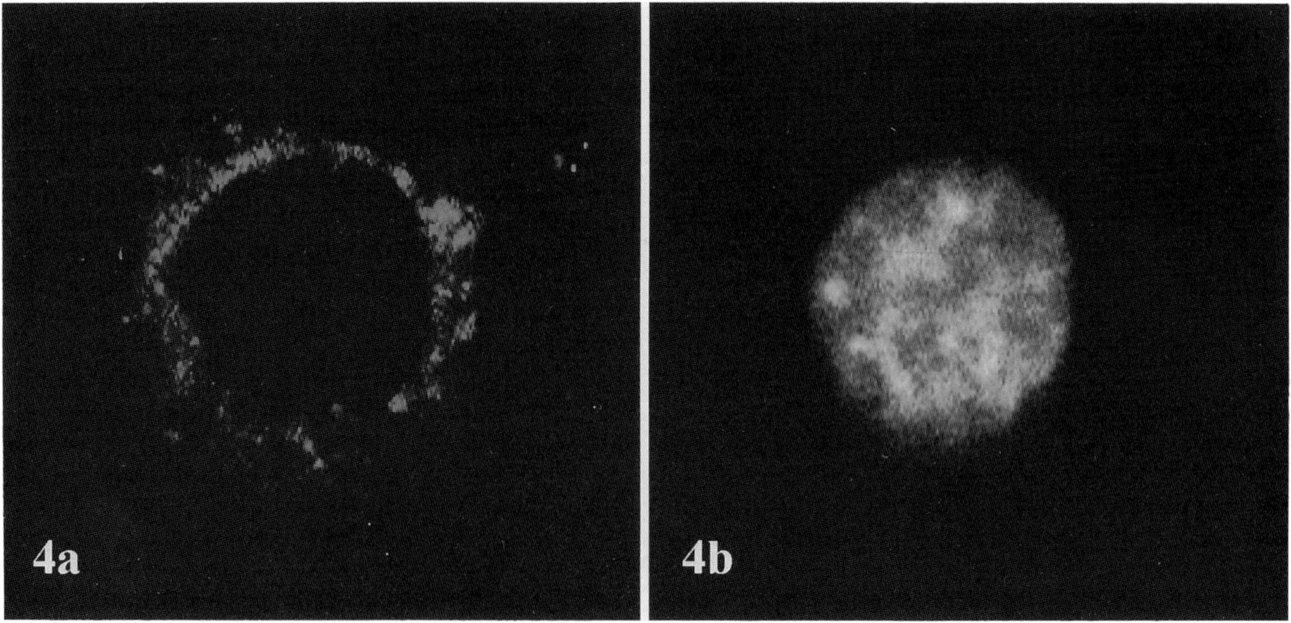

Figure 4. (**a**) Perinuclear distribution of the plasmid DNA marked with TMR. The image was recorded 1 hour after delivery of the plasmid at 37°C within the Tf-PLL complex into the ATCC HTB 15 U-118 MG cell. An optical section was taken near the perimeter of the nucleus as evaluated based upon fluorescence from the nuclear DNA stained with bisbenzimide presented in Fig. 4b. Inside the nuclear volume, there is no fluorescence from the fluorescent derivatives of the plasmid DNA. Images were obtained at the rate 1 frame/s with accumulation of 10 frames in the two-photon laser scanning fluorescence microscope assembled at the Integrated Microscopy Resource. Horizontal field width 8 μm. (**b**) The nucleus of the same cell as in Fig. 4a stained with bisbenzimide. Based upon the diameter, this optical section was selected at the perimeter of the nucleus. Both images were recorded in the same optic plane.

---

mine specificity of binding of the transfection complexes directed by transferrin to specific receptor, the transfection complexes were presented to glioma cells at 4°C. In this approach, the plasmid DNA within the transfection complex is present only on the cell surface since internalization did not occur at this temperature. Fluorescent labeling outlines the cell shape which can be verified with phase-contrast. Cell nuclei stained with bisbenzimide served as reference structures. Temperature increase to 37°C resulted in fast internalization of the transfection complexes. The transfection complexes soon accumulated in the perinuclear space. Based upon epi-fluorescence only, it was nearly impossible to determine exact relationship to the nuclear envelope, i.e., to determine if the plasmid DNA entered the nucleus. Preliminary attempts toward computer deconvolution of these images were not successful. These images were recorded with films or CCD cameras with sensitivity equivalent to 6400 ISO. Therefore, low levels of incorporation of fluorochromes were still sufficient to detect the transfected DNA. Moreover, low concentrations of the transfected plasmid DNA in the cells were detectable. In experiments performed entirely at 37°C, after initial attachment of the transfection complexes to the cell membrane manifested by the presence of fluorescence on the cell surface, within 5 min the transfection complexes were internalized similarly to the results described elsewhere (Grady *et al.*, 1994; Rizova *et al.*, 1994; Ghosh and Maxfield, 1995). Temporary suppression of FITC fluorescence indicated passage through an endosomal compartment. Considering this time-frame and taking into account that these events were happening in three dimensions (3D), application of a digital, video-rate laser scanning confocal microscope was essential. Moreover, deconvolution was necessary to reduce out-of-focus glare.

For evaluation of three-dimensional distribution of the quickly trafficking plasmid DNA video-rate laser scanning fluorescent microscopy, was applied (Fig. 3). Presence of a slit or a pinhole is seriously limiting intensity of detected signal in LSFM. Therefore, at least 10x higher concentrations of fluorochromes had to be incorporated into the plasmid DNA to ensure detectable

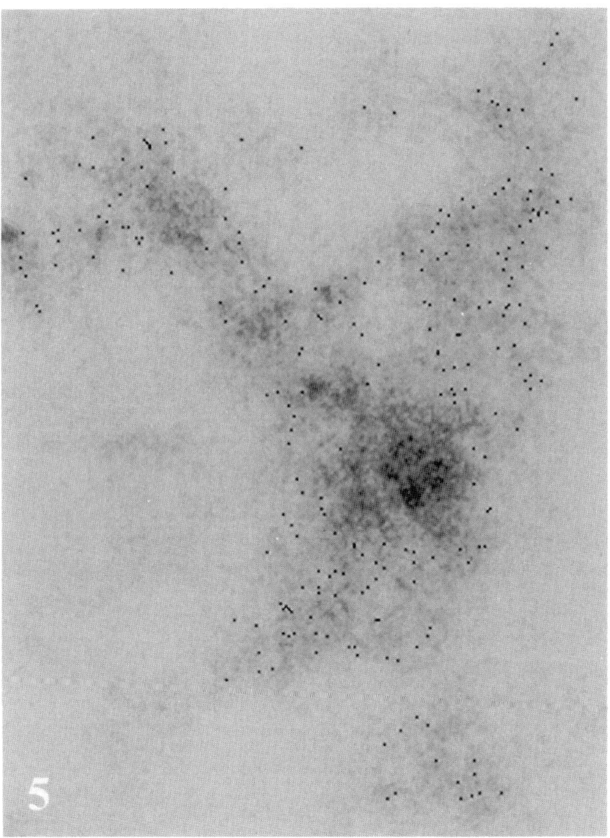

Figure 5. Ultrastructural localization of the transfected biotinylated DNA marked with streptavidin-nanogold in the ATCC HTB 14 U87 MG cell. This cell was transfected within the Tf-PLL-NLS transfection complex. 1 hour after transfection at 37°C, the cell was rapidly frozen, freeze-substituted, embedded, and sectioned. Patches of nanogold beads are present inside the cell nucleus. This pattern of distribution of the transfected DNA reminds that observed in living cells with LSFM e.g., Fig. 3a-i. Imaging was pursued in the Zeiss 912 Omega operated at 120 kV with the filter set at 0 eV energy loss and the slit at 20 eV energy window. Horizontal field width 360 nm.

----

signal with LSFM. The nuclear DNA stained with bisbenzimide was detected by using the UV laser, thus outlining the intranuclear space, i.e., the painted nuclear DNA served as a reference system which helped to verify intranuclear entry of the plasmid DNA. In this experiment the transfection complexes were coupled to NLS. Presence of NLS resulted in intranuclear import of the plasmid DNA. Confinement of glare from below and above the focal plane was achieved by restricting the point spread function range through deconvolution (Avinash, 1995). The video-rate performance of this confocal microscope was due to fast horizontal scanning utilizing an acousto-optical deflector (AOD). This approach helped to confirm the general time-frame of the transfection complexes trafficking with good temporal resolution (Malecki *et al.*, 1995). However, this approach had also drawbacks. Only one fluorochrome could be excited with one wave-length at a time, i.e., it was not possible to record multiple labeling within one focal plane during one scan. Furthermore, in this system an entire cell volume was exposed to laser irradiation. This led to bleaching of fluorochromes within the entire volume. Laser irradiation led to generation of free radicals from the entire cell volume and therefore to seriously reduced cell viability. Additionally, the final exposure was the product of multiplication of the number of optical sections in a series that have to be recorded for reconstruction of spatial relations by the number of series necessary for reconstruction of the events in time. Both factors contributed to accelerated bleaching and toxicity even further. Moreover, changes of the emission intensity due to fading and variations of concentration due to trafficking made quantitative analysis very complex. Drawbacks described above were reduced by increasing scan speeds leading to reduced dose-rates and by introducing neutral density filters to reduce illumination intensities.

An attempt has also been made toward imaging of the transfected DNA trafficking using two-photon fluorescence microscopy (Fig. 4). In the two-photon system, at the sufficient photon density in the beam waist the system was capable to induce excitation of fluorochromes. Distribution of the transfected plasmid DNA was based upon the derivatives with tetramethylrhodamine. This fluorochrome was excited at 523 nm with emission at 590 nm. Interestingly, distribution of the nuclear DNA painted with bisbenzimide was also visible. Bisbenzimide has the excitation peak at 364 nm with emission at 490nm. These observations indicate two-photon, but possibly also three-photon excitation. The images were acquired at the rate of 1 frame per second. At least twenty optical sections had to be acquired to consider reconstruction of three-dimensional distribution of the transfection complexes within the cells. Very weak signals required averaging of more than 10 frames. This summed up to exposure lasting about 200 seconds for evaluation of one time point. Higher acquisition rates were at the expense of either signal-to-noise ratio or resolution. Nevertheless, pulsed-laser had definite advantages. As it was clear from reconstructions of z-planes in experiments in which illu-

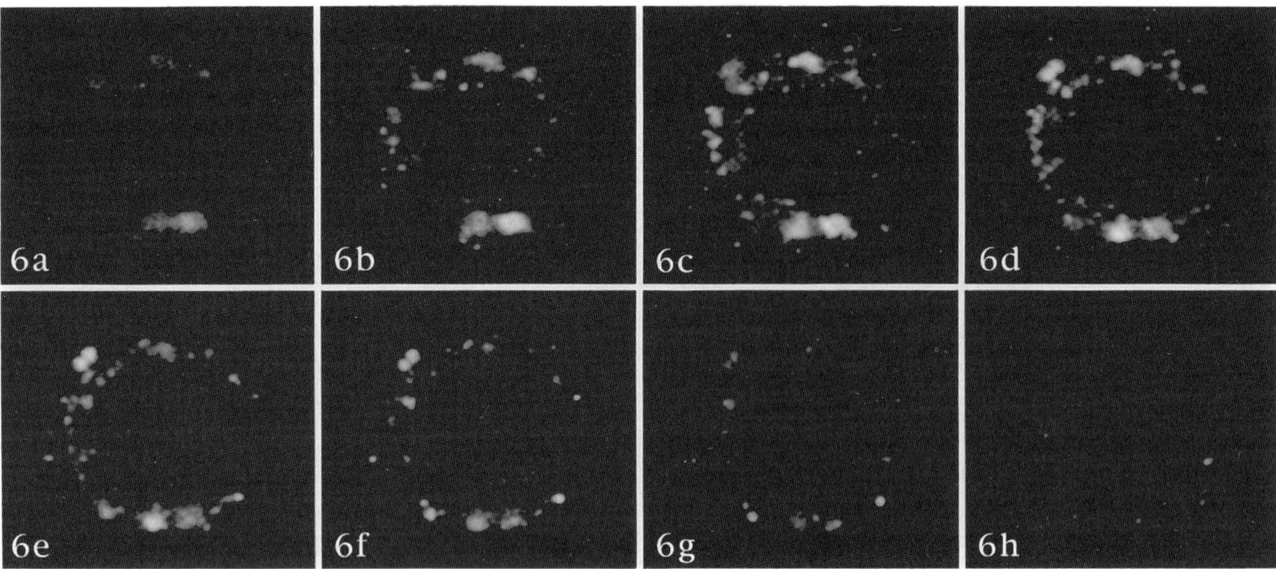

Figure 6. (a-i) Three-dimensional distribution of the transfected DNA within the ATCC HTB 14 U87 MG cell. This cell was cryo-immobilized 30 min after the non-viral vector was presented to the cells at 37°C. The sequence of the transfected DNA was amplified through polymerase chain reaction followed by *in situ* hybridization with the probes containing incorporated dUTP digoxigenin and labeling with the monoclonal anti-digoxigenin antibodies. The images were recorded using the Odyssey XL Noran. Inside the nucleus, there are focal accumulations of the probes as seen in Fig. 6c. Those bright spots are limited to that one optical section only as it is clear from comparison with sections in Fig. 6b and Fig. 6d. This pattern of labeling indicates intranuclear import of the transfected DNA. Similar distribution of the plasmid DNA was observed in living cells e.g., Figs. 2 and 3 or was revealed ultrastructurally in Fig. 5. Significant amounts of plasmid DNA are still present in the perinuclear space. Horizontal field width of the single frame 11 μm.

---

mination and recording of images was limited to one focal plane only, bleaching was limited to this thin slice only. Therefore, deconvolution was unnecessary. Excitation limited to a small volume resulted in the reduced fading of fluorochromes and reduced amounts of cytotoxic products of photo-decay. This allowed prolonged observations of the plasmid intracellular trafficking and an extended viability of the cells under investigation. Simultaneous detection of two probes by two- and three-photon excitation and acquisition of fluorescence from within the same focal plane allowed pursuit of the direct collocation studies. These features allowed us further insight into the trafficking events. After being internalized, the plasmid DNA was moving toward the perinuclear region and was accumulating there. At this phase, trafficking was slower than at early stages of the delivery (Malecki *et al.*, 1995). Evaluation of spatial relationships between the intranuclear compartment and the plasmid DNA was of the utmost importance. Here, it was also possible to confirm earlier observations that the plasmid DNA within the transfection complexes which did not contain NLS had been entrapped in the perinuclear region for up to 2 hours (Malecki and Skowron, 1995).

The plasmid DNA contained also nanogold tagged through streptavidin conjugated to biotin incorporated either through photo-biotinylation or nick translation. Therefore, the same samples which were recorded with FM were also suitable for ultrastructural analysis. Detailed information concerning trafficking of the plasmid DNA was obtained after rapid freezing, embedding, and ultramicrotomy for EFTEM (Fig. 5). In conventional transmission electron microscopy (CTEM) nanogold is too small to be detected within the cell ultrastructure directly. Instead, it requires silver enhancement. In EFTEM, filtration of inelastic scattering at the zero loss mode resulted in visualization of nanogold within the cellular ultrastructure very clearly. These beads pointed out the distribution of the plasmid DNA.

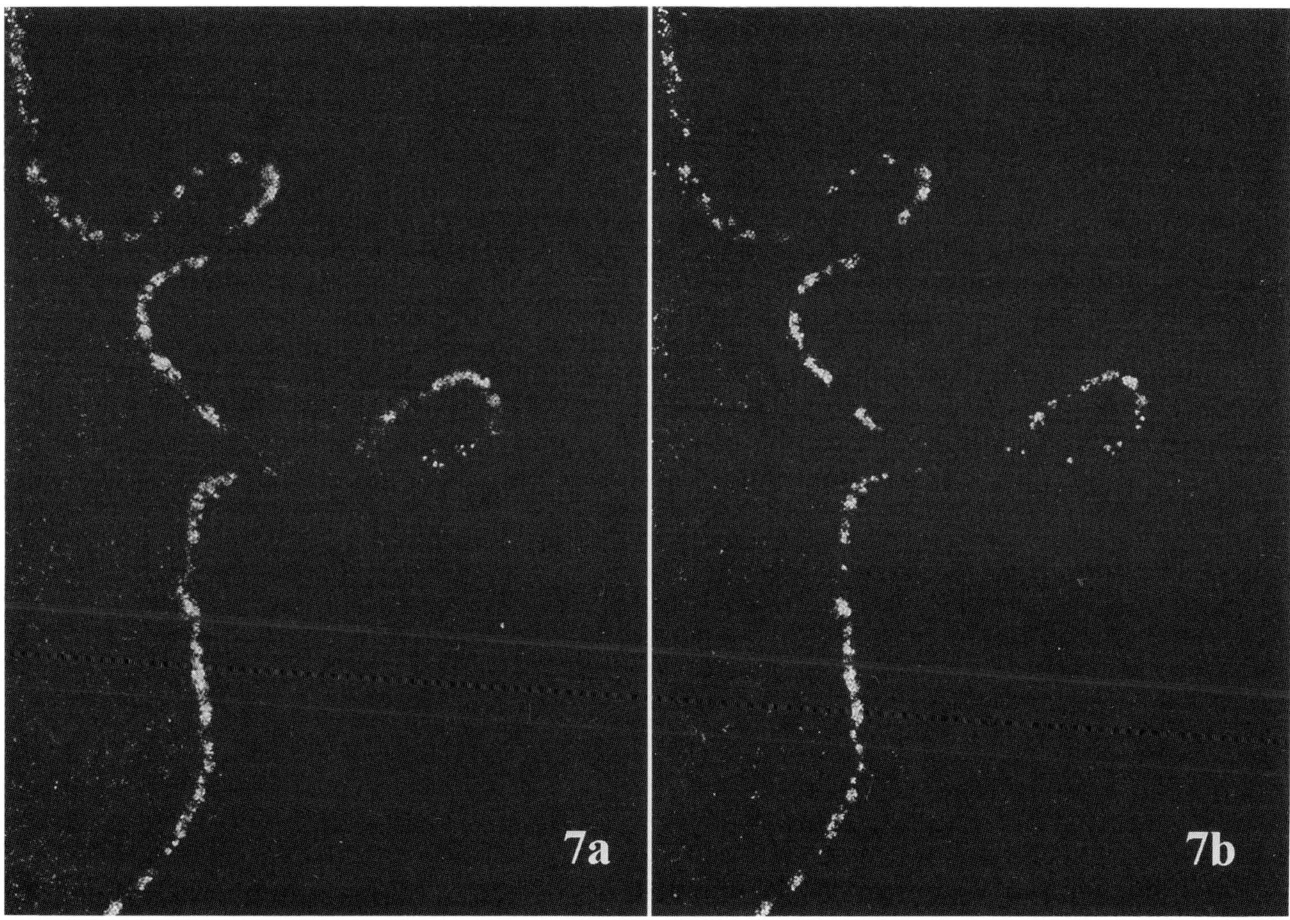

Figure 7. (**a**) The plasmid DNA with the PLL-Tf-NLS was presented to the ATCC HTB 15 U-118 MG cell at 4°C to prevent internalization. After 15 min the cell was rapidly cryo-immobilized. The plasmid DNA was *in situ* hybridized with the probes containing incorporated dUTP digoxigenin followed by labeling with boronated anti-digoxigenin antibodies. Imaging was pursued in the Zeiss 912 Omega with the filter set at 201 eV. The image was acquired on the slow-scan CCD camera and processed by the three-window method. Signal is restricted to the cell surface. Horizontal field width 1.7 μm. (**b**) The same area of the cell as in Fig. 7a. Transferrin was labeled with the specific antibodies followed by the secondary antibodies conjugated to ferritin. Imaging was pursued in the Zeiss 912 Omega with the filter set at 708 eV. The image was acquired and processed as described in Fig. 7a. Signal accumulation indicates patching. Horizontal field width as in Fig. 7a.

Preparation of the plasmid DNA in the transfection complexes by *in situ* hybridization and immunolabeling in cryo-immobilized cells

The second part of this project was aimed at developing procedures for cellular location of the unlabeled transfected plasmid DNA. Particularly challenging might be cases where the plasmid DNA would be present within cells in such a low number of copies that it might remain undetectable after labeling through in situ hybridization. For this part of the project cultured glioma cells were rapidly cryo-immobilized, followed by freeze-substitution, and rehydration. Sequences of interest were amplified by polymerase chain reaction followed by *in situ* hybridization with probes marked with tetramethyl-rhodamine (Fig. 6). Other factors also contributed to enhanced imaging qualities. Phototoxicity was not a major concern in this part of this the project, therefore much higher light intensities were applied leading to improved signal-to-noise ratios. Agents which could stabilize fluorochromes, reduce fading, or improve their quantum efficiency could be applied without risks of their potential toxicity. Practically, if the cells were properly sealed in the viewing chamber with a reference system, the same cell could be even imaged several times on the same or other microscopes. Patterns of labeling were similar to those obtained with fluorescent derivatives of the plasmid DNA in living cells. Particularly intriguing were interactions of the transfection complexes with the nuclear envelope. Incorporation of the nuclear localization signal to the transfection complexes resulted in their abundant presence within the nuclei. For unambiguous determination if the plasmid was truly within the intranuclear compartment deconvolution of the images was necessary. Analysis of images resulted in definite verification of signal within the nucleus, thus a proof for a successful entrance of the plasmid DNA. Signal from the plasmid DNA was detectable around the nuclear envelope. Within the nucleus, labeling was organized in a patchy pattern. Further ultrastructural analysis of these structures and their associations with the transfection complexes was far beyond resolution limits of light microscopy.

Ultrastructural analysis was pursued on ultrathin sections obtained after freeze-substitution and embedding of rapidly cryo-immobilized cells. The most significant difference in preparation of these samples compared to the samples for CTEM was absence of heavy metal stains in cell preparation protocols, i.e., heavy metals were not applied en bloc or on sections. In Fig. 7, there is an example of the glioma cell to which the transfection complexes were presented at 4°C. The plasmid DNA in sections was hybridized in situ with probes containing digoxigenin followed by labeling with boronated antibodies. Transferrin was labeled with monoclonal antibodies coupled to ferritin. The labeled sections were imaged with EFTEM. Net distribution of boron and iron was obtained through the three-window method. Distribution of the plasmid DNA determined based upon the signal for boron and distribution of the transfection complex based upon the signal for iron are restricted to the cell membrane. Collocation of both signals is apparent. Clearly, distribution of the transferrin being the targeting part of the transfection complex was restricted to the cell surface. This is the result of inhibition of all internalization processes at this temperature. If surface receptors were saturated with transferrin only as a control, then in situ hybridization resulted in absence signal from the plasmid DNA. This experiment indicated that binding of the transfection complexes to the cell surface was through transferrin specific receptors. Collocation of signals from the transfected DNA and the transferrin indicated that the plasmid DNA remained associated with the transfection complexes at this stage of transfection. There are only minimal differences in distribution, which may be referred to transfection complexes which do not contain the plasmid DNA. Multiple labeling should help us to determine in details cellular pathways for the transfection complexes carrying the transfected plasmid DNA. By energy loss tuning, contrast reversing, and image processing, it was possible to generate images resembling those which were obtained after conventional staining with heavy metals for CTEM. Therefore, it should allow us to pursue identification of molecules with far better accuracy than methods previously available. This approach should allow us to refer or even superimpose the information concerning distribution of reporter molecules and the information concerning ultrastructural organization of this cell.

Discussion

Fluorescent derivatives of the plasmid DNA allowed us to highlight trafficking pathways of transfection complexes in living cells with fluorescence microscopy (Malecki *et al.*, 1995). This approach was significantly strengthened by using laser scanning fluorescence microscopy often followed by deconvolution of images. It also helped to determine general spatial and temporal arrays of events during gene transfer. Further refinement of these results can be best achieved by using better fluorochromes. Fluorochromes currently available are quite sensitive to illumination and their fast bleaching is a major problem. This problem is even more severe in three-dimensional and time-resolved studies. In those studies acquisition of images would have to be repeated several times inducing bleaching. Moreover, detection of fluorochromes requires fluorescent emission of sufficiently high intensity. This can be reached either by local accumulation of fluorescent derivatives in a cell, e.g., transfected DNA in glioblastoma cells or incorporation of great number of fluorochromes into a single molecule, e.g., number of fluorochromes in the plasmid DNA. Both may compromise physiological behavior of that cell or that molecule. Therefore, development of stable fluorochromes with high quantum efficiency is still of high priority and worth efforts (Gryczynski and Lakowicz, 1995; Tsien and Waggoner, 1995). Efficient detection of fluorescent derivatives can also be improved by more sensitive detectors. In particular, new CCD cameras extending their sensitivity to detect intensities below 10^{-6} lux are particularly promising in fluorescence microscopy. Two other immediate applications can be foreseen. In the first, even more sensitive cameras, capable for photon-counting, may be able to detect low emissions from threshold excitations to be followed by further image enhancements and three-dimensional reconstructions. In the second, CCD cameras can be installed onto emission paths of LSFMs, thus to avoid light losses during descanning. Above described improvements have an ultimate goal to minimize intervention during preparation of the transfection complexes for imaging, thus to retain their physiological features, while to ensure accuracy of their detection.

Ultrastructural analysis was efficiently pursued based upon images of cryo-immobilized cells. These cells were rapidly frozen in the apparatus for their observation until freezing (Malecki, 1992). They were watched while closed in the viewing chambers in the suitable environment until the very moment of rapid freezing. Observation of these cells ensured that they were happy until that very moment. On the contrary, all the steps necessary to freeze cells using conventional freezers, e.g., transfers from cell culture dishes or mounting in a conventional freezer might very severely damage cell environment leading to severe cellular changes which might remain unnoticed until after long processing to CTEM. These risks were successfully avoided here. With the high freezing rate cellular organization was captured in the state close to that in a living cell, but now available for ultrastructural deliberation. Another advantage pouring from application of freezing techniques in this study was enhancement of accuracy of structure labeling. It was recognized earlier that chemical fixatives and heavy metals which might interfere with effectiveness of IL, ISH, or PCR. The procedures applied in this study did not include these harmful steps, but rather were based upon low temperature techniques. This led to efficient and accurate identification of ultrastructure.

Ultrastructural analysis of cellular distribution of the transfected DNA and other components of the transfection complexes in this project was primarily based upon electron spectroscopic imaging. Success of this approach was based not only on the advanced cryo-techniques and the energy filtering transmission electron microscope at the cutting edge of imaging technology, but also on preparation of cells and probes specifically for this instrument. Boronation of antibodies was based upon the protocol developed years ago for neutron therapy (Gilliand *et al.*, 1980; Barth *et al.*, 1986). In that application the highest number of atoms attainable within an antibody molecule was necessary for efficient neutron capture, but the molecule still had to retain its ability to recognize the epitope. In this project the least number of atoms, but concentrated as close as possible to the probe's binding site are more important. Therefore, further attempts in this direction are worth effort (Kessels *et al.*, 1996; Malecki, 1995). Investigations are required to determine that minimal number of atoms in derivatives which will be detectable in the microscope, i.e., minimum detectable mass. Moreover, for pursuit of electron spectroscopic imaging electron doses delivered to the specimens have to be sufficiently high to ensure detectable signal after filtration. Therefore, to endure high electron doses for electron spectroscopic imaging the transfected DNA has to be protected by embedments stable under the beam. Further attempts to develop resins better protecting specimens for electron spectroscopic imaging are on the way (Malecki, in preparation). Clearly, intensified CCD cameras and cryo-stages would further enhance chances for success of these attempts.

Finally, I would like to summarize briefly practical and immediate consequences of this work for gene transfer projects. Targeted gene delivery represents two major challenges for gene transfer protocols. The first challenge, a selected sequence has to be delivered into the cells of choice only. To meet this goal, a vector has to be developed that recognizes epitopes on the cells of choice only on the same manner as viruses do, i.e., with their own host cell preferences. Targeted gene delivery is the key factor for applying these results for *in vivo* gene therapy. The experimental design presented in this report is a step in this direction. By imaging fluorescent derivatives of the transfection complexes and/or labeling cellular receptors, it is possible to evaluate accuracy of targeting much faster than with other methods currently applied in gene therapy laboratories.

The other challenge is to ensure intranuclear delivery of the sequence of choice. Intranuclear gene delivery into postmitotic cells may carry a different set of problems than into proliferating cells. In postmitotic cells, e.g., neurons, the nuclear envelope creates a very sophisticated gate mechanism possibly associated with an intranuclear distribution system in which intranuclear channels, as described so far only in *Xenopus* oocytes (Ris and Malecki, 1993) and/or nuclear compartments as those described in HEp-2 cells (Thiry 1995) may play an essential role. In proliferating cells, e.g., glioblastomas, nuclear envelopes are present in the interphase, but they disintegrate during late prophase. Therefore, the transfected DNA may follow two different trafficking pathways leading to nuclear compartment. There, the transfected DNA may start to play an active role interacting with the nuclear DNA and associated structures, i.e., either it may integrate into the nuclear DNA, or it may become closed within the nuclear compartment for transient expression. Evaluation of the final destination of the transfected DNA either with newly designed gene transfer vectors or by applying new transfection protocols can be efficiently evaluated using the procedures described in this work.

Acknowledgments

This study was supported by the following awards: Grant BIR-9420056 from the National Science Foundation (Principal Investigator: Dr. Marek Malecki); Grant BIR-9522771 from the National Science Foundation (Principal Investigator: Dr. Marek Malecki), and Research Resource Grant RR-570 from the National Institutes of Health (Principal Investigator: Dr. John White). The author acknowledges with thanks comments as well as access to the instrumentation provided by Drs Ueli Aebi, Ralph Albrecht, Brian Andrews, Gopal Avinash, Robert Bremmel, Gary Case, Kirk Czymmek, Marion Greaser, Heinz Gross, James Hainfeld, Tom Kelly, Joseph Lakowicz, Wolfgang Probst, Hans Ris, Piotr Skowron, Jose Sogo, Waclaw Szybalski, and John White. The author thanks Raf Malecki for image processing and Dr. Joyce Sexton for editing the manuscript.

References

Anderson WF (1992) Human gene therapy. Science **256**: 808-813.

Avinash G (1995) New algorithm for simultaneous blur and image restoration in three-dimensional optical microscopy. J NIH Res **7**: 78.

Barth RF, Alam F, Soloway AH, Adams DM, Steplewski Z (1986) Boronated monoclonal antibody 17-1A for potential neutron capture therapy of colorectal cancer. Hybridoma **5**: S43.

Bendayan M, Barth RF, Gingras D, Londofio I, Robinson I, Alam F, Adams DM, Mattiazzi L (1989) Electron spectroscopic imaging for high resolution immunocytochemistry; use of boronated protein A. J Histochem Cytochem **37**: 573-580.

Chowdhury NR, Wu CH, Wu GY, Yerneni PC, Hays RM, Bommineni VR, Chowdhury JR (1993) Fate of DNA targeted to the liver by asialoglycoprotein receptor-mediated endocytosis in vivo. Prolonged persistence in cytoplasmic vesicles after partial hepatectomy. J Biol Chem **268**: 12265-11271.

Brakenhoff GJ, Visscher K (1995) Real-time stereo (3D) confocal microscopy. In: Handbook of Biological Confocal microscopy. Pawley JB (ed). Plenum Press, New York. pp 155-177.

Curiel DT, Agarwal S, Wagner E, Cotten M (1991) Adenovirus enhancement of transferrin-polylysine-mediated gene delivery. Proc Natl Acad Sci USA **88**: 8850-8854.

Denk W, Strickler JH, Webb WW (1990) Two-photon laser scanning fluorescence microscopy. Science **248**: 73-76.

Denk W, Strickler JH Webb WW (1995) Two-photon molecular excitation in laser scanning microscopy. In: Handbook of Biological Confocal Microscopy. Pawley JB (ed). Plenum Press, New York. pp 445-458.

Draaijer A, Houpt PM (1988) A standard video-rate confocal laser-scanning reflected and fluorescence microscope. Scanning **10**: 139-145.

Edgerton RF (1986) Electron Energy-Loss Spectroscopy in the Electron Microscope. Plenum Press, New York. pp 291-352.

Faulk WP, Taylor GM (1971) An immunocolloid method for the electron microscope. Immunocytochemistry **8**: 1081-1083.

Gall JG, Pardue ML (1969) Formation and detection of RNA-DNA hybrid molecules in cytological preparations. Proc Natl Acad Sci USA **63**: 378-383.

Ghosh RN, Maxfield FR (1995) Evidence for nonvectorial retrograde trafficking in the early endosomes in HEp2 cells. J Cell Biol **128**: 549-561.

Gilliand KT, Steplewski Z, Collier RJ, Mitchell KF, Chang TH, Koprowski H (1980) Antibody-directed cytotoxic agents: use of monoclonal antibody to direct the action of toxin A chains to colorectal carcinoma cells. Proc Natl Acad Sci USA **77**: 4539-4543.

Grady EF, Slice LW, Brant WO, Walsh JH, Payan DG, Bunnett NW (1994) Direct observation of endocytosis of gastrin releasing peptide and its receptor. J Biol Chem **270**: 4603-4611.

Gryczynski I, Lakowicz J (1994) Fluorescence intensity and anisotropy decays of the DNA stain Hoechst 33342 resulting from one-photon and two-photon excitation. J Fluorescence **4**: 331-336.

Haase AT, Retzel EF, Staskus KA (1990) Amplification and detection of lentiviral DNA inside cells. Proc Natl Acad Sci USA **87**: 4971-4975.

Hacker GW, Zehbe I, Hauser-Kronberger C, Gu J, Graf AH, Dietze O (1994) In situ detection of DNA and mRNA sequences by immunogold-silver staining (IGSS). Cell Vision **1**: 30-37.

Hainfeld JF (1988) Gold cluster-labelled antibodies. Nature **333**: 281-282.

Hainfeld JF, Furuya FR (1992) A 1.4nm gold cluster covalently attached to antibodies improves immunolabeling. J Histochem Cytochem **40**: 177-184.

Harris CE, Agarwal S, Hu P, Wagner E, Curiel DT (1993) Receptor-mediated gene transfer to airway epithelial cells in primary culture. Am J Resp Cell Mol Biol **9**: 441-447.

Heuser JE, Reese TS, Dennins MJ, Jan Y, Jan L, Evans L (1979) Synaptic vehicle exocytosis captured by quick freezing and correlated with quantal transmitter release. J Cell Biol **81**: 275-300.

Hopkins CR, Gibson A, Shipman M, Miller K (1990) Movement of internalized ligand - receptor complexes along a continuous endosomal reticulum. Nature **346**: 335-338.

Hutchison N, Langer-Safer P, Ward D, Hamkalo B (1982) In situ hybridization at the electron microscope level: hybrid detection by autoradiography and colloidal gold. J Cell Biol **95**: 609-618.

John HA, Birnstiel ML, Jones LW (1969) RNA-DNA hybrids at the cytological level. Nature **223**: 582-587.

Kessels MM, Qualman B, Klobasa F, Sierralta WD (1996) Immunocytochemistry by electron spectroscopic imaging using homogeneously boronated peptide. Cell Tissue Res **284**: 239-245.

Khine AA, Lingwood CA (1994) Capping and receptor-mediated endocytosis of cell-bound verotoxin (Shiga-like toxin). 1: Chemical identification of an amino acid in the B subunit necessary for efficient receptor glycolipid binding and cellular internalization. J Cell Physiol **161**: 319-332.

Konat G, Laszkiewicz I, Bednarczuk T, Kanoh M, Wiggins RC (1991) Generation of radioactive and non-radioactive ssDNA hybridization probes by polymerase chain reaction. J Meth Cell Mol Biol **3**: 64-68.

Kramarcy NR, Sealock R (1990) Commercial preparations of colloidal gold-antibody complexes frequently contain free antibody. J Histochem Cytochem **39**: 967-968.

Lakowicz J, Gryczynski I (1992) Fluorescence intensity and anisotropy decay of the 4', 6'-diamidino-2phenyindole-DNA complex resulting from one-photon and two-photon excitation. J Fluorescence **2**: 117-121.

Malecki M (1991) High voltage electron microscopy and low voltage scanning electron microscopy of human neoplasmic cells in culture. Scanning Microscopy Suppl **5**: S53-S73.

Malecki M (1992) Light microscopy of living cells correlative to high voltage electron microscopy and low voltage scanning electron microscopy of these cells cryo-whole mounts. Proc 50th Annual Meeting EMSA. San Francisco Press, San Francisco, CA. pp 566-567.

Malecki M (1995) Detection of transfected DNA by *in situ* polymerase chain reaction and *in situ* hybridization. J Histochem Cytochem **44**: 782-788.

Malecki M, Skowron P (1995) Nuclear localization signal facilitates intranuclear entry of plasmid DNA. Mol Biol Cell **6**(313a): 1823 (abstr).

Malecki M, Small JV (1987) Immunocytochemistry of contractile and cytoskeletal proteins in smooth muscle. Protoplasma **139**: 160-169.

Malecki M, Skowron P, Ris H (1995) Imaging of transgenes with laser scanning confocal microsocopy and field emission scanning electron microsocopy. Scanning Suppl 1: 16-17.

Maniatis TA, Jeffrey A, Kleid DG (1975) Nucleotide sequence of the rightward operator of phage λ. Proc Natl Acad Sci USA **72**: 1184-1189.

Mullis KB, Faloona FA, Scharf S, Saiki R, Horn G, Erlich H (1986) Specific enzymatic amplification of DNA in vitro: Polymerase chain reaction. Cold Spring Harbor Symp Quant Biol **51**: 263-283.

Nuovo GJ, MacConnell P, Forde A, Delvenne P (1991a) Detection of human papillomavirus DNA in formalin fixed tissues by in situ hybridization after amplification by PCR. Am J Pathol **139**: 847-854.

Nuovo GJ, Gallery F, MacConnell P, Becker J, Bloch W (1991b) An improved technique for the detection of DNA by in situ hybridization after in situ PCR. Am J Pathol **139**: 1239-1244.

Ottensmeyer WP, Andrew JW (1980) High resolution microanalysis of biological specimens by electron energy loss spectroscopy and electron spectroscopic imaging. J Ultrastruct Res **72**: 336-348.

Ottensmeyer WP (1984) Electron spectroscopy imaging: parallel enrgy filtering and microanalysis in the fixed beam electron microscope. J Ultrastruct Res **88**: 121-134.

Pardue ML, Gall JG (1969) Molecular hybridization of radioactive DNA to DNA of cytological preparations. Proc Natl Acad Sci USA **64**: 600-604.

Petran M, Hadravsky M (1968) Tandem-scanning reflected light microscope. J Opt Soc **58**: 661-664.

Rash ER (1983) The rapid-freeze technique in neurobiology. Trends Neurosci **6**: 208-212.

Ris H, Malecki M (1993) High resolution field emission scanning electron microscopye imaging of internal cell structure by epon extraction from sections. J Struct Biol **111**: 148-157.

Rizova H, Carayon P, Michel L, Barbier A, Lacheretz F, Dubertet L (1994) Internalization of surface HLA-DR molecules by human Langerhans cells. Cell Biol Toxicol **10**: 367-373.

Rothman JE (1994) Mechanisms of intracellular protein transport. Nature **372**: 55-63.

Ryan K, Knoll E (1994) The rapid-freeze technique in neurobiology. Trends Neurosci **6**: 208-212.

Sogo J, Lozano M, Salas M (1984) Binding of proteins to DNA. Nucl Acid Res **12**: 1943-1955.

Szmacinski H, Gryczynski I, Lakowicz J (1995) Three-photon induced fluorescence of the calcium probe Indo-1. Photochem Photobiol **62**: 804-808.

Takizawa T, Robinson JM (1994) Use of 1.4nm immunogold particles for immunocytochemistry on ultra-thin sections. J Histochem Cytochem **43**: 1615-1623.

Thiry M (1995) Nucleic acid compartmentalization within the cell nucleus by in situ transferase-immunogold techniques. Microsc Res Techn **31**: 4-21.

Tsien RY, Waggoner A (1990) Fluorophores for confocal microsocopy: Photophysics and photochemistry.

In: Handbook of Confocal Microsocopy. Pawley JB (ed). Plenum Press, New York. pp 169-178.

Tsien RY, Bacskai BJ (1995) Video-rate confocal microscopy. In: Handbook of Biological Confocal Microscopy. Pawley JB (ed). Plenum Press, pp 459-478.

Van Harreveld A, Crowell J (1964) Electron microscopy after rapid freezing on a metal surface and substitution fixation. Anat Rec **149**: 381-386.

Wagner E, Zenke M, Cotten M, Beug H, Birnstiel ML (1990) Transferrin-polycation conjugates as carriers for DNA uptake into cells. Proc Natl Acad Sci USA **87**: 3410-3414.

Wagner E, Plank C, Zatloukal K, Cotten M, Birnstiel ML (1992) Influenza virus hemagglutinin HA-2 N-terminal fusogenic peptides augment gene transfer by transferrin-polylysine-DNA complexes: toward a synthetic virus-like gene-transfer vehicle. Proc Natl Acad Sci USA **89**: 7934-7938.

Yap EPH, McGee JOD (1991) Slide PCR: DNA amplification from cell samples on microscopic glass slides. Nucleic Acid Res **19**: 15.

Discussion with Reviewers

G. Konat: How were the stoichiometric ratios for transfection complexes determined and optimized?
Author: The Tf:PLL:plasmid DNA ratio was based upon the results in which gene expression was reported in the result of gene transfer via receptor mediated endocytosis (Wagner *et al.* 1990). To determine NLS:Tf ratio, the data from the studies on intranuclear import of albumin coupled to NLS were applied (Kalderon *et al.*, 1984). These ratios were verified by measuring higher levels of lacZ and GFP expression after non-viral vectors conjugated with NLS were used for gene transfer via receptor mediated endocytosis (Malecki, Case, and Skowron, in preparation).

M. Thiry: Could you determine mechanisms involved in intranuclear delivery of the plasmid DNA?
Author: To determine these mechanisms, digitonin treated cells were used as a model system for studies of nuclear transport (Adam and Gerace, 1991). In this system, we were able to demonstrate perinuclear retention of the transfected DNA in the absence of NLS in transfection complexes (Malecki *et al.* 1995). Moreover, presence of NLS in transfection complex resulted in rapid nuclear import. Import was abolished by reduction of temperature and depletion of ATP. Based upon these data, active nuclear transport is supposed to play an essential role in intranuclear delivery.

Additional References

Adam SA, Gerace L (1991) Cytosolic proteins that specifically bind nuclear localization signals are receptors for nuclear import. Cell **66**: 837-847.

Kalderon D, Richardson WD, Markham AF, Smith AE (1984) Sequence requirements for nuclear location of simian virus 40 large T antigen. Nature **311**: 33-38.

Scanning Microscopy Supplement 10, 1996 (pages 17-26) 0892-953X/96$5.00+.25
Scanning Microscopy International, Chicago (AMF O'Hare), IL 60666 USA

SIMULTANEOUS IDENTIFICATION OF A SPECIFIC GENE PROTEIN PRODUCT AND TRANSCRIPT USING COMBINED IMMUNOCYTOCHEMISTRY AND *IN SITU* HYBRIDIZATION WITH NON-RADIOACTIVE PROBES

Gwen V. Childs*

Department of Anatomy and Neurosciences, University of Texas Medical Branch, Galveston, TX

(Received for publication March 10, 1996 and in revised form September 26, 1996)

Abstract

Simultaneous identification of messenger RNA (mRNA) and proteins in the same cells or tissues is a valuable tool to help the cell biologist evaluate the cell secretory cycle. Some cells may produce the mRNA and delay the production of the proteins. Alternatively, the proteins may be rapidly secreted. Other cells may produce both in sequence within the same time frame. Because of this difference, some cells can only be identified by their mRNA product. Others may have both products. This presentation describes a non-radioactive approach to the detection of both products with dual-peroxidase labeling protocols in use in this laboratory since 1983. The first detection system uses biotinylated cRNA probes or oligoprobes in *in situ* hybridization along with antisera to biotin to detect the hybrid. The detection system is amplified by 2-3 layers of anti-biotin, second antibody (made against the anti-biotin) and streptavidin conjugated to horseradish peroxidase. After the mRNA is detected with a blue-black substrate (nickel intensified diaminobenzidine), the antigens are detected with immunoperoxidase techniques and orange-amber substrate. The *in situ* hybridization protocol can also be used at the electron microscopic level. Trouble shooting and control protocols are also described. This approach has been shown to be valuable for detection of pituitary hormones, growth factors mRNAs and antigens.

Key Words: *In situ* hybridization, immunoperoxidase cytochemistry, messenger RNA (mRNA), cRNA, oligoprobe, avidin-biotin cytochemistry, pituitary hormones, electron microscopy

*Address for correspondence:
Gwen V. Childs
Department of Anatomy and Neurosciences
University of Texas Medical Branch, MRB 10-104
Galveston, TX 77555-1043
Telephone Number: 409-772-2101
FAX Number: 409-772-4687
E-mail: gvchilds@utmb.edu

Introduction

Hybridization refers to the reaction between two single-stranded nucleic acid molecules that bind by means of hydrogen bonding of complementary base pairs. Hybridization performed "*in situ*" refers to those techniques that allow binding and detection of hybrids in a cell or tissue section.

Initially, *in situ* hybridization was performed to detect deoxyribonucleic acid (DNA) targets or amplified ribosomal ribonucleic acid (RNA genes) within cell nuclei (Gall and Pardue, 1969; John *et al.*, 1969; Buorgiorno-Nardelli and Amaldi, 1970). Early researchers and clinicians have also mapped the location of genes within chromosomal preparations or nuclei (Pardue and Dawid, 1981; Fostel *et al.*, 1984). The early uses of *in situ* hybridization for cytoplasmic RNA involved detection of viral nucleic acid sequences within infected tissues (Brahic and Haase, 1978; Haase *et al.*, 1981).

Early investigators used radiolabeled complementary cDNA or cRNA probes for the hybridization reaction. Detection of the hybrids then involved autoradiography. However, during the past decade other detection systems have developed that can be as efficient as those with radioactive labels and take only a day or two to perform. Most of these involve the attachment of a signaling hapten molecule such as biotin or digoxygenin to the cDNA or cRNA probe. The detection systems for these molecules are enhanced by sandwich techniques that apply different layers of reactants, or enzyme reactions, or both. The techniques can be readily applied to the detection of both mRNA and antigens within the cells (by dual-labeling approaches).

The purpose of this presentation will be to describe a method for the detection of mRNAs and antigens with dual-labeling immunocytochemistry (Childs *et al.*, 1987, 1990, 1991a,b, 1994; Kaiser *et al.*, 1992; Lee *et al.*, 1993; Fan *et al.*, 1995). We use biotinylated complementary oligonucleotide or cRNA probes to hybridize with the cytoplasmic mRNAs. Then we detect the probes with anti-biotin and enhance the reaction by a sandwich technique that provides layers of biotin and streptavidin-peroxidase (McQuaid and Allan, 1992; Fan *et al.*, 1995). After the initial reaction, we can detect the pituitary antigens by classical immunoperoxidase cytochemistry with the use of a different colored substrate (Childs *et al.*, 1987, 1990, 1991a,b, 1994; Kaiser *et al.*, 1992; Lee *et al.*, 1993).

Materials and Methods

When we first detected mRNA hybrids in whole pituitary cells, we used the avidin biotin peroxidase complex technique (ABC *Elite*, Vector Laboratories, Burlingame, CA) (Childs *et al.*, 1987, 1990, 1991a,b, 1992a,b, 1994; Kaiser *et al.*, 1992; Lee *et al.*, 1993). We no longer use this complex to detect the biotinylated probes because we frequently encountered high background with ABC kits used from 1993 to the present. The early *Elite* kits did not have these background problems. Thus, we now use a new five-step immunolabeling protocol that detects biotin conjugated to the cRNA probes with anti-biotin. It was first described by McQuaid and Allan (1992). This protocol will be described in the following paragraphs.

Preparation of pituitary cells

To maximize our ability to detect mRNA and antigens, the protocol is applied to cells plated for 1-36 h on 13 mm glass coverslips (Thomas Scientific, Suresdesboro, NJ; Catalog number 6672A75) in 24 well trays (Fisher Scientific, Houston, TX; Catalog number 08-757-156). The protocol can also be used on frozen sections or paraffin sections. The probes used for the hybridization are either cRNA probes (Childs *et al.*, 1987, 1990, 1991a,b; Fan *et al.*, 1995) or complementary oligonucleotide probes at least 30 mer long (Childs *et al.*, 1992 a,b, 1994; Kaiser *et al.*, 1992; Lee *et al.*, 1993). They may be biotinylated by the Vector Photobiotin kit (Vector Laboratories) following kit instructions (Childs *et al.*, 1987; Wu *et al.*, 1991, 1992). More recently, we have ordered the oligonucleotide probes commercially prepared, with Biotin attached. Alternatively, we add Biotin-UTP to the cRNA probes produced during in vitro transcription (Fan *et al.*, 1995).

All fixation, washing, and handling methods are run under sterile conditions to prevent RNase contamination. The buffer or diluent solutions are made, Millipore filtered, and stored frozen in small aliquots. Controls include substitution of labeled sense sequences, or omission of the labeled probe. The protocol can be adapted for use with dispersed cells at the electron microscopic level.

Preparation of solutions

The following list includes the solutions needed for *in situ* hybridization. Table 1 provides information about some of the major vendors that supply these reagents.

(1) Phosphate buffered saline (PBS): 0.1 M phosphate buffer + 0.9% NaCl pH 7.2

(2) Triton X-100 (0.3%): 300 μl Triton X-100 (Sigma Chemical, St. Louis, MO); bring to 100 ml with 0.1 M PBS.

(3) EDTA (50 mM): add 1.46 g EDTA to 100 ml of 0.1 M Tris buffer (Sigma Chemical, Catalog number T-5030) (20 ml 0.5 M Tris + 80 ml Millipore-filtered water), pH 8.0.

(4) Para-formaldehyde (4%): 4.0 g para-formaldehyde in 100 ml 0.1 M PBS, pH 7.2.

(5) Acetic anhydride (0.25%) + 0.1M triethanolamine: 250 μl acetic anhydride; bring to 100 ml with Millipore filtered water, then add 1.856 g triethanolamine, pH 8.0 (Sigma Chemical)

(6) Deionized formamide (50%) in 2X Sodium Salt Citrate (SSC): make a 1:10 dilution from 20X SSC with Millipore-filtered water and add 1:1 formamide (Sigma Chemical, Catalog number F-7503)

(7) Sodium Salt Citrate Stock (SSC): make 20X SSC stock by adding 3 M sodium chloride + 0.3 M Sodium Citrate to Millipore filtered water. Make 4X solution with a 1:5 dilution from stock 20X SSC with Millipore filtered water

Another task is to prepare the *in situ* hybridization buffer. The components are added in sequence in the following instructions. It usually takes about a day and it can be prepared in advance. Most of these components come from Sigma Chemical, unless otherwise noted (see Table 1).

(1) Deionize formamide: add 1.5 g REXYN I-300 (Fisher Scientific, Catalog number R-208) to 100 ml formamide, stir for 45-60 min, filter through Whatman filter paper.

(2) Combine 100 ml deionized formamide + 100 ml 4X SSC (20 ml 20X SSC + 80 ml Millipore filtered water). The final concentration will be 50% formamide.

(3) Add 7.88 g Tris salt (0.25 M), pH 7.5, warm to about 37°C in 1.0 ml Millipore filtered water. Dissolve Tris.

(4) With sterile syringe and needle, add the following to the warm formamide-Tris solution: 20 g dextran sulfate; 0.5 g bovine serum albumin (RIA grade) (Sigma Chemical, Catalog number A-7638), 0.5 g Ficoll-400, 0.5 g Polyvinyl pyrrolidone-360, 1.0 g sodium pyrophosphate, 1.0 g sodium lauryl sulfate.

(5) Finally, add 200 μl of dissolved salmon sperm DNA (ssDNA) to 200 ml of hybridization buffer.

(6) Aliquot into 10 ml units and store frozen (-20°C) in 15 ml centrifuge tubes

Prehybridization protocol

The next series of steps involve preparing the cells for the hybridization. Several components are used to

Table 1. Vendors for *in situ* hybridization histochemistry, including Uniform Resource Locator (URL, Internet link)

Vendor	Address	Telephone number	Internet
DAKO Corporation	6392 Via Real, Carpenteria, CA 93013	800-235-5763	Dimensions has a web page devoted to DAKO Corporation (http://www2.multinet.net/dli/dako.htm) which has access to data sheets
Fisher Scientific	10700 Rockley Road, Houston, TX 77099	800-876-1900 800-766-7000	http://www.fisher1.com
JRH Biosciences	P.O.Box 14848, Lenexa, KS 66325	800-255-6032	
Sigma Chemical	P.O.Box 14508; St. Louis, MO 63178	800-325-3010	http://www.sigma.sial.com/sigma/sigma.html
Thomas Scientific	99 High Hill Road, P.O.Box 99, Suresdesboro, NJ 08085-0099	800-345-2100	
Vector Laboratories	30 Ingold Road, Burlingame, CA 94010	800-227-6666	http://www2.multinet.net/dli/vector.htm

Table 2. Internet Uniform Resource Locator (URL) links to courses, protocols and services in *in situ* hybridization

(1) Link to our protocols including the colored versions of Fig. 1 and other photographs. We will continue to update this page as we progress. URL: http://cellbio.utmb.edu/childs/in_situ.htm

(2) The Anderson Laboratory in situ protocol (URL: http://www.cco.caltech.e...r/htmls/Big_In_situ.html)describes non-radioactive methods for labeling frozen sections, whole mount embryos and cultured cells.

(3) This protocol (URL: http://cellbio.ucdavis.e...mePage/Tucker/WCRDB.html describes single cell RT-PCR and quantitative *in situ* hybridization for beginners.

(4) Dr. Jan Blancato (ONCOR, Inc.) gives a course in *in situ* hybridization. The URL is: http://www.cua.edu/www/catc/ish.htm

(5) Exon-Intron, Inc. runs a course in *in situ* hybridization. The URL is: http://www.dnatech.com/insitu.htm.

(6) Paul Hough at Brookhaven National Laboratory runs an advanced *in situ* hybridization course. The URL is http://www.cshl.org/meetings/96situ.htm.

---

aid penetration and several steps are used to prevent non-specific reactions. In our protocol, the pituitary cells are plated on glass coverslips and grown in Dulbecco's Modified Eagle's Medium (DMEM, JRH Biosciences, Lenexa, KS; Catalog number 56499-10L), and fixed in 2.5% glutaraldehyde (in 0.1 M phosphate buffer) for 30 min at room temperature. Then cells are washed for 1 h in 4 changes of phosphate buffer + 4.5% sucrose. We work under sterile conditions even during the fixation process and store the cells no longer than 1 week at 2-4°C. The steps are taken to prevent RNase contamination which would eliminate RNA from the tissue or cells.

On the day of the hybridization, we first rinse the fixed cells on coverlips with fresh, sterile 0.1 M phosphate buffered saline (PBS) for 5 min at room temperature, shaking. Then we treat in the following sequence of solutions. (All of the following chemicals but the p-formaldehyde came from Sigma Chemical). All steps are done at room temperature.

(1) To aid penetration of reagents, the cells are first treated with 0.3% Triton X-100 for 15 min at room temperature. This is followed by washing with 0.1M PBS (twice, 3 min each, while gently shaking on a side-to side shaker).

(2) To further improve penetration and remove associated proteins, we next treat with Proteinase K (1 μg/ml) for 15 min at room temperature.

(3) The cells are then stabilized by postfixation with 4% para-formaldehyde/0.1M PBS for 5 min at room temperature. This is followed by washing with 0.1 M PBS (twice, 3 min each, while gently shaking).

(4) To cover non-specific reactive sites, we next treat the cells with 0.25% acetic anhydride for 10 min while shaking.

(5) We then treat the cells with 50% formamide/2X SSC for 10 min while shaking the tray at room temperature. Then we continue the incubation at 37°C for 10 min. This solution prepares the cells for the hybridization buffer and helps break weak hydrogen bonds in the non-specific linkages between the complementary probe and the surrounding tissue. If there is a 2-3% mismatch in the complementary probe, the concentration of this solution should be lowered to 40-45% to prevent breaking bonds in the specific probe-mRNA hybrid. This lower concentration should also be used to make the hybridization buffer. (Note that the concentration in the buffer is 50% in the above formulation).

Alternative pre-hybridization technique

We are currently testing pepsin digestion in lieu of the proteinase K and have had successful results with our cells fixed in either glutaraldehyde or para formaldehyde. The Pepsin is purchased from Sigma (Catalog number PN6887) and is made up just before use. Two mg of pepsin are added per ml of 0.01 N HCl. It is stored at room temperature for no longer than two hours. It is added to the cells for 25 min at 37°C followed by an additional 10 min at room temperature.

Hybridization conditions

The cells are then placed in the hybridization solutions which is 300 ng/ml of biotinylated oligonucleotide or cRNA-probe diluted in hybridization buffer (see above). They are incubated for 12-15 h at 37°C. The temperature of hybridization may be varied according to the melting temperature of the probe. We have found that a range from 36 to 42 C works for most of our probes against gonadotropins (Childs *et al.*, 1987, 1992 a,b) or pro-opiomelanotropin mRNAs (Wu *et al.*, 1992).

Post-hybridization protocol

After the overnight hybridization the post-hybridization washing steps are followed. We also add an RNase treatment step if the probe used was a cRNA (Childs *et al.*, 1987; Fan *et al.*, 1995). The coverslips are first washed with 4X SSC (three times, 15 min each; this is done at room temperature with gentle shaking during the last 5 min). This wash is at low enough stringency (high salt concentration) to remove the excess unreacted probe without removing the probe that has specifically hybridized to the mRNA. If there is background with the reaction, higher stringency washes can be done. Also, one can raise the temperature of the washes. This involves continuing the washes with reduced concentrations of SSC (2X, 1X and 0.5X), 15 min each wash. However, we have found this to be unnecessary, especially with our oligonucleotide probes.

Detection of the biotin

After the post-hybridization washes, the detection protocol can be followed.
The detection protocol begins with a blocking step in a solution containing proteins. They cover non-specific sites and prevent background reactions, but they do not react with any of the components of the detection protocol. Therefore, we first block with 0.05 M Tris buffered saline containing 1% bovine serum albumin + 10% normal horse serum (Vector Laboratories) for 15 min at room temperature (pH 7.6). Then we detect the biotin on the hybrids with the following sequence:

(1) The cells are first treated with monoclonal anti-biotin (1:30) (Dako Corp.) for 30 min at 37°C. They are then washed with 0.05 M Tris buffered saline (twice, for 3 min each), pH 7.6).

(2) The cells are then incubated with biotinylated horse anti-mouse IgG (rat absorbed, Vector Laboratories), 1:100, for 10 min at room temperature. They are then washed with 0.05 M Tris buffered saline (twice, 3 min each). It is very important to use rat-absorbed anti-mouse IgG because most of these antisera will react with rat tissues, non-specifically.

(3) The cells are incubated with a second layer of monoclonal Anti-Biotin (1:100) for 30 min at 37°C. Coverslips are then washed with 0.05 M Tris buffered saline (twice, 3 min each).

(4) The cells are incubated with a second layer of biotinylated horse anti-mouse IgG (rat absorbed) 1:100 for 10 min at room temperature and washed with 0.05 M Tris buffered saline (twice, 3 min each).

(5) Finally, we incubate the cells with a 1:10 dilution of peroxidase conjugated streptavidin (Dako, Catalog number P-397) for 5 min at room temperature. The cells are then washed with 0.05 M Tris buffered saline (twice).

At this point, the nickel-intensified diaminobenzidine (DAB) is prepared by dissolving 0.45 g nickel ammonium sulfate in 30 ml 0.05 M acetate buffer, adding 1 diaminobenzidine tablet (Sigma Chemical, Catalog number D-5905). The solution is dissolved. Then we add 20 μl of 30% hydrogen peroxide. The solution is filtered on a Whatman filter paper and then added to the cells for 6 min. The reaction product is blue-black. This reaction cannot be improved by longer times. There is a limited life and time-span of this solution. Therefore, to increase signal, we vary concentrations of the probe or other components of the detection protocol (see

Figure 1. A cluster of pituitary cells, one of which is dual-labeled for FSH-beta mRNA (black, arrows) and LH-beta antigens (gray, L). Note that the label for mRNA is in a linear pattern. The gray label for the LH antigens detects them throughout the cell except on the nucleus. Bar = 10 μm. Reproduced with permission from Childs *et al.* (1994).

Discussion).

After the diaminobenzidine step, the coverslips are washed with 0.05 M acetate buffer (twice), dehydrated, dried and mounted to glass slides, cell side up with Permount. A second square coverslip is mounted over the cells. Note: This DAB reaction product is dissolved in water soluble solutions and therefore cannot be used with water soluble mounting media or glycerol. Therefore, use only organic solvent soluble mounting media with it. One can view the slides for a brief period mounted in glycerol, however eventually this will remove the reaction product.

After the mRNA is detected with the blue-black peroxidase substrate, the cell can be further labeled by immunocytochemistry for its protein or antigen content. For this we use the dual-labeling protocol with contrasting color peroxidase substrates.

Applications of this technique to electron microscopic preparations

We apply the above technique to the detection of mRNA at the electron microscopic level by leaving the dissociated pituitary cells in suspension (Childs *et al.*, 1990). Then, we spin the cells down at 900 rpm after each step. The gentle spinning is effective and creates a pellet which must be resuspended in the new solution. After the diaminobenzidine step, the cells are exposed to 1% osmium tetroxide for 30 min at refrigerator temperatures (2-4°C). This fixes the membrane lipids and enhances the peroxidase reaction. The cell pellets are then embedded in Epon according to traditional electron microscopic methods.

Dual-labeling for antigens

After biotinylated ligands, mRNA or a first antigen is detected we use contrasting colored substrates and immunocytochemistry to detect an antigen (e.g., pituitary hormone, growth factor, or *c-fos*.) The following shows an example of the protocol that detects the antigen. This allows us to identify the hormone content of the cell that expresses the mRNA. The basic dual-labeling techniques have been published since 1983 (Childs *et al.*, 1983). They are outlined below.

(1) Rinse coverslips with 0.05M Tris buffered saline (once).

(2) Block again with 0.05 M Tris buffered saline + 1% bovine serum albumin (Sigma Chemical, Catalog number A-7638) for 15 min at room temperature.

(3) Incubate with specific antisera to antigen in question (diluted 1:5000—1:50,000) for 30 min at 37°C. Wash coverslips with 0.05 M Tris buffered saline (three times).

(4) Incubate with biotinylated goat anti-rabbit IgG (Vector Laboratories, Catalog number BA-1000) (This is made with 25 μl stock Biotinylated-IgG + 25 μl normal goat serum in 2 ml of buffer). Incubate for 20 min at room temperature. Wash coverslips with 0.05 M Tris buffered saline (twice).

(5) Incubate with 1:200 peroxidase conjugated streptavidin (DAKO, Catalog number P-397) for 20 min at room temperature. Wash coverslips with 0.05 M Tris buffered saline (twice).

(6) To detect peroxidase, prepare orange-amber diaminobenzidine. Dissolve 1 Tris buffer tablet in 15 ml Millipore filtered water, add 1 diaminobenzidine tablet (Sigma Chemical, Catalog number D-5905) + 12 μl 30% hydrogen peroxide; filter on Whatman filter paper and use immediately. Apply to cells for 5-7 min at room temperature. Wash coverslips with Millipore filtered water (three times). Dehydrate, dry and mount on slides with Permount.

Note: diaminobenzidine is eventually washed out in water soluble mounting media, especially temporary mounting media like glycerol. Therefore, if another peroxidase substrate is used, that requires water soluble media, please store the slides dry and use glycerol ONLY for short periods of time. The best way to store the slides is to dehydrate the tissues and use permount.

Table 2 lists Internet Uniform Resource Locator

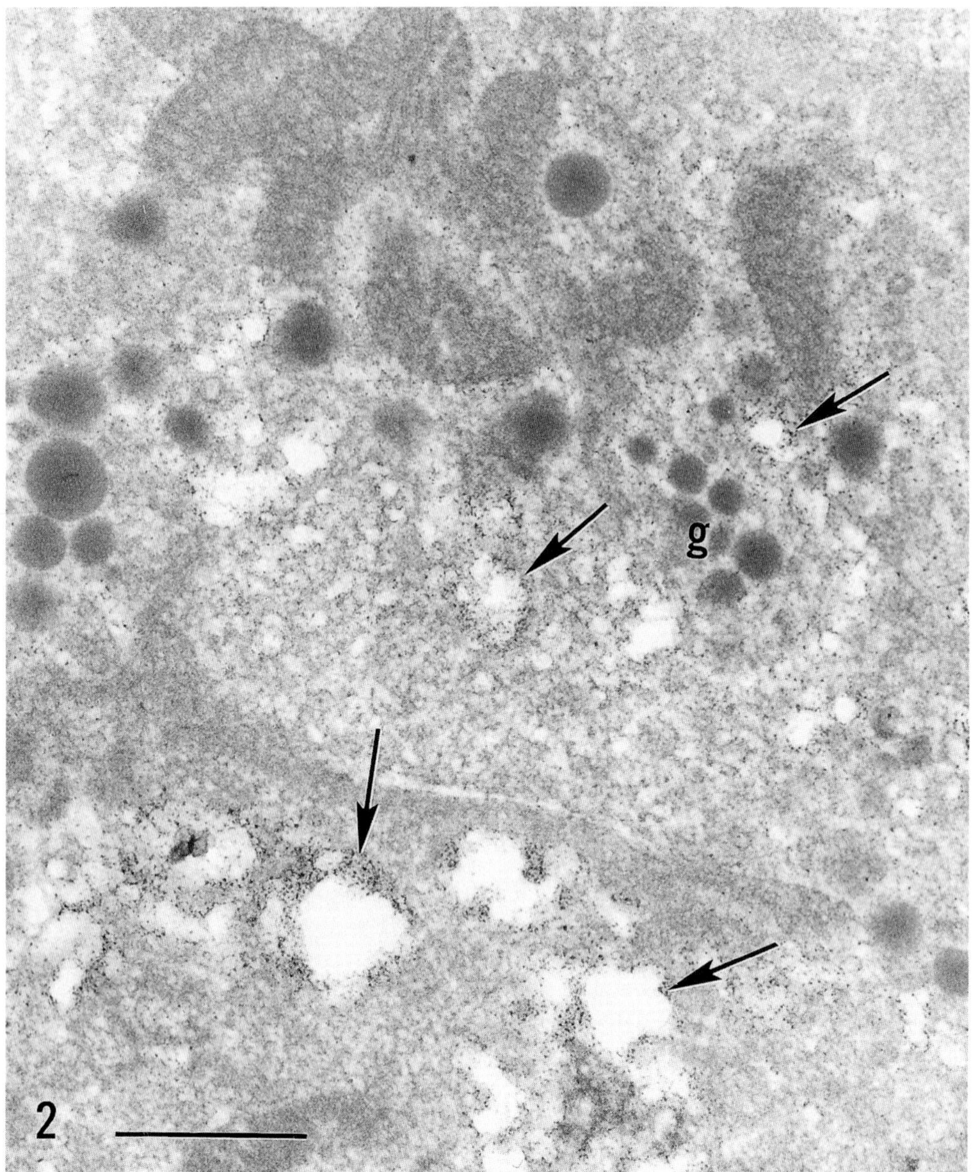

Figure 2. Label for LH-beta mRNA on the surface of rough endoplasmic reticulum detected with a biotinylated cRNA probe and avidin peroxidase (arrows). Note that the secretory granules (g) remain unlabeled. Bar = 0.25 μm.

(URL) links to courses, protocols and services in *in situ* hybridization.

Results and Discussion

The labeling for the mRNA is in dark patches or lines in the specific cell type. This can be confirmed by immunolabeling for the antigen in the same cell (Childs, 1987, 1991a,b, 1992 a,b, 1994; Kaiser *et al.*, 1992; Lee *et al.*, 1993; Fan *et al.*, 1995). However, there are clearly cells that contain only the antigen or only the mRNA (Childs *et al.*, 1994). Thus, we have discovered cells in different stages of their secretory cycle (Childs *et al.*, 1994).

Fig. 1 shows dual labeling for FSH mRNA and LH antigens (Childs *et al.*, 1994). The cell in this figure expresses FSH mRNA in a black linear pattern, and Luteinizing hormone (LH) antigens, labeled orange. In black and white, the orange label is gray and fills the cell.

Fig. 2 shows that, at the electron microscopic level, the label is associated with dilated profiles of rough endoplasmic reticulum (Childs *et al.*, 1990). This field shows labeling for Luteinizing hormone beta subunit mRNA on dilated rough endoplasmic reticulum (RER) in a pituitary gonadotrope. Studies from our laboratory

(Childs *et al.*, 1990) show similar patterns of labeling for Follicle stimulating hormone mRNA.

Troubleshooting this protocol

The label is easy to detect in cells because of its density and linear or patchy pattern. However, it should not be mistaken for the metallic deposits sometimes seen in the cells (pseudoperoxidase activity). The label in cells that do not have amber label should be blue-black or a purplish blue. A steel-gray background over all the cells indicates background problems that should be eliminated (see below).

The *in situ* hybridization fields should always be quantified. Furthermore comparisons should be made between single and dual-labeled preparations. One should count the percentages of labeled cells in fields exposed only to the single labeling protocol After one is confident of the percentages, one should then count the percentages of mRNA-bearing cells in the dual-labeling protocol. They should match. To facilitate this comparison, we have developed a Lotus 1,2,3 or Exel Spreadsheet that automatically calculates these percentages after we load in the raw counts (Childs *et al.*, 1991 a,b, 1994; Lee *et al.*, 1993). We also compare percentages of antigen bearing cells with those obtained through single labeling. These tests are vital to insure that the dual-labeling protocol does not interfere or add to the detected label for either mRNA or antigens. Controls run with the labeling protocol (see below) will also help test the specificity of the reaction.

The labeling density for the mRNA (sensitivity) can be improved by adding another layer of anti-biotin and streptavidin. However, one runs the risk of higher background (McQuaid and Allan, 1992). Also, tests of different temperatures of hybridization may show an optimal temperature that promotes labeling. Finally, one can vary the concentration of probe from 0.1-100 ng/ml for example and the reaction should increase with increasing concentration (up to a plateau point).

cRNA probes usually are more sensitive because they carry more biotin molecules/probe and they are longer. Yet, they are sticky and difficult to work with. They also require expertise in molecular biology techniques for their production. Oligonucleotide probes may carry only one biotin although they can be engineered to carry more. Their sensitivity may also vary with the size and number of biotin molecules. Some low abundance mRNAs may require cRNA probes for adequate detection. Nevertheless, for most of our studies, biotinylated oligonucleotide probes label expected populations of cells at concentrations of 1-100 ng/ml. One could also engineer several biotinylated probes directed against different parts of the mRNA transcript. This may increase sensitivity. All probes must react with unique sequences, however, or specificity will be compromised.

In our hands, this technique does not give high background. Usually the background is crystal clear and the unlabeled cells are difficult to see without the benefit of special optics. However, if there is background, it is usually seen as a steel gray (or light blue) deposit all over the field or cells. Also, it could be seen in dual-labeled fields as light amber label all over the cells. In that case, one might suspect the second labeling protocol. However, both could be at fault. And, both should be checked.

There are several ways to reduce background. First, one can remove one layer of the detection system for the mRNA (an anti-biotin and a anti-mouse IgG step) or dilute the components further. This should be done only if the signal is strong, however. Second, one can increase SSC washes, reducing the concentration to 0.1X. This low salt provides high stringency conditions needed to wash out the non-specifically bound probes that may cause the background. This is especially a problem with the sticky cRNA probes. Caution must be given, however, to the possibility that the washes will wash out the probe attached in the specific hybrids. Finally, the use of different concentrations of the detecting reagents for the biotin (anti-biotin and streptavidin-peroxidase) will reduce background. If the signal is strong, one can dilute both and achieve optimal results.

Controls

We have controlled for interference during the dual labeling by eliminating either the complementary RNA probe from the first sequence or the specific antibody from the second sequence. In each case, the resulting labeling reflects only that which had the complete sequence. Thus, we have proved that the first detection protocol does not add to the label in the second (and visa versa). Furthermore, we have absorbed the specific antibody in the second sequence with its specific antigens. This has eliminated all second reactions and only the blue-black label for the mRNA is visible. This not only proves the specificity of the labeling, it proves that the amber diaminobenzidine does not cause the blue-black label to change color. In studies since 1983, we have shown that the blue-black diaminobenzidine is stable, as long as the second reaction is not prolonged (Childs *et al.*, 1983). This is why the rapid ABC or DAKO immunoperoxidase kits are so valuable. If immunolabeling requires 48 hour incubation in antisera, one runs the risk of losing all detectable reaction product during the second incubation periods. Thus, immunoperoxidase kits with short incubation times should be chosen.

Summary

To summarize, this protocol has been used successfully for the detection of mRNAs in tissue sections (McQuaid and Allan, 1992) and cell cultures (Fan *et al.*, 1995). It is preferred over the direct avidin detection systems because of its sensitivity and the fact that some avidin-containing kits produce high background. The anti-biotin sandwich method provides flexibility and enhancement potential along with a streptavidin solution that works at neutral pH and gives low background reactions.

References

Brahic M, Haase AT (1978) Detection of viral sequences of low reiteration frequency by *in situ* hybridization. Proc Natl Acad Sci USA **75**: 6125-6127.

Buorgiorno-Nardelli S, Amaldi F (1970) Autoradiographic detection of molecular hybrids between rRNA and DNA in tissue sections. Nature **225**: 946-948

Childs GV, Naor Z, Hazum E, Tibolt R, Westlund KM, Hancock MB (1983) Cytochemical characterization of pituitary target cells for biotinylated gonadotropin releasing hormone. Peptides **4**: 549-555.

Childs GV, Lloyd J, Unabia, G, Gharib, SD, Weirman, ME, Chin, WW (1987) Detection of LH mRNA in individual gonadotropes after castration: use of a new in situ hybridization method with a photobiotinylated cRNA probe. Mol Endocrinol **1**: 926-932.

Childs GV, Unabia G, Weirman ME, Gharib SD Chin WW (1990) Castration induces time-dependent changes in the FSH -mRNA-containing gonadotrope cell population. Endocrinology **126**: 2205-2213.

Childs GV, Patterson J, Unabia G, Rougeau D, Wu P (1991a) Epidermal growth factor enhances ACTH secretion and expression of POMC mRNA by corticotropes in mixed and enriched cultures. Mol Cell Neurosci **2**: 235-243.

Childs GV, Taub K, Jones KE, Chin WW (1991b) Tri-iodothyronine receptor β-2 mRNA expression by somatotropes and thyrotropes: Effect of propylthiouracil-induced hypothyroidism in rats. Endocrinology **129**: 2767-2773.

Childs GV, Unabia G, Lloyd J (1992a) Recruitment and maturation of small subsets of luteinizing hormone (LH) gonadotropes during the estrous cycle. Endocrinology **130**: 335-345.

Childs GV, Unabia G, Lloyd JM (1992b) Maturation of FSH gonadotropes during the rat estrous cycle. Endocrinology **131**: 29-36.

Childs GV, Unabia G, Rougeau D (1994) Cells that express luteinizing hormone (LH) and follicle stimulating hormone (FSH) beta (β) subunit mRNAs during the estrous cycle: The major contributors contain LH, FSH and/or growth hormone. Endocrinology **134**: 990-997.

Fan X, Childs GV (1995) EGF and TGF mRNA and their receptors in the rat anterior pituitary: localization and regulation. Endocrinology **136**: 2284-2324.

Fostel J, Narayanswami S, Hamkalo B, Clarkson SG, Pardue ML (1984) Chromosomal location of a major tRNA gene cluster of *Xenopus laevis*. Chromosoma **90**: 254-260.

Gall JG, Pardue M (1969) Formation and detection of RNA-DNA hybrid molecules in cytological preparations. Proc Natl Acad Sci USA **63**: 378-383.

Haase AT, Venture P, Gibbs C, Touretellotte W (1981) Measles virus nucleotide sequences: Detection by hybridization *in situ*. Science **212**: 672-673.

John HA, Birnstiel ML, Jones KW (1969) RNA-DNA hybrids at the cytological level. Nature **223**: 582-587.

Kaiser U, Lee BL, Unabia G, Chin W, Childs GV (1992) Follistatin gene expression in gonadotropes and folliculostellate cells of diestrous rats. Endocrinology **130**: 3048-3056.

Lee BL, Unabia G, Childs G (1993) Expression of follistatin mRNA in somatotropes and mammotropes early in the estrous cycle. J Histochem Cytochem **41**: 955-960.

McQuaid S, Allan GM (1992) Detection protocols for biotinylated probes. Optimization using multistep techniques. J Histochem Cytochem **40**: 569-574.

Pardue ML, Dawid IB (1981) Chromosomal locations of two DNA segments that flank ribosomal insertion-like sequences in *Drosophila*: Flanking sequences are mobile elements. Chromosoma **83**: 29-43.

Wu PA, Childs GV (1990) Cold and novel environment stress affects AVP mRNA in the paraventricular nucleus, but not the supraoptic nucleus: an *in situ* hybridization study. Mol Cell Neurosci **1**: 233-249.

Wu PA, Childs GV (1991) Changes in rat pituitary POMC mRNA after exposure to cold or a novel environment detected by *in situ* hybridization. J Histochem Cytochem **39**: 843-852.

Discussion with Reviewers

Reviewer I: In many protocols, methanol treatment makes a variety of cultured cells sufficiently permeable to the probe. Your prehybridization protocol includes proteinase K treatment "to improve penetration (of probe) and remove associated proteins". Proteinase K is a very active, non-specific endopeptidase, at high concentration capable of degrading proteins into single amino-acids. Isn't the proteinase K treatment too hazardous for epitopes to be detected in the same samples, after the hybridization step?

Author: That is an excellent question. The original technique was for use with cRNA probes to detect LH-beta and FSH-beta mRNAs in rat pituitary cells. These large probes (300- 500 bp) required more intense deproteination. Therefore, we used the Proteinase K effectively to aid penetration. At the same time, we recognized the danger to the antigens to be detected in the immunolabeling protocol. Therefore, we tried fixation in 2% glutaraldehyde for 30 min at room temperature before any of the hybridization steps. This fixative gives much stronger cross-linking than any formalin fixative because the glutaraldehyde has two aldehyde groups that can cross link amino acids (formaldehyde has only one group).

We already knew that the anterior pituitary antigens were preserved well in glutaraldehyde-fixed tissues. Therefore, we were pleased when we discovered, in 1987, that the in situ hybridization protocol with proteinase K did not destroy any of the pituitary antigens used to date. In fact, the major result was increased efficiency in the reactions. All primary antisera dilutions were increased at least 2-fold. For example, in a recent experiment with Growth hormone detection, the optimal dilution of anti-GH after dual immunolabeling (two antigens) was 1:20,000-1:40,000. However, after *in situ* hybridization, the optimal dilution of anti-GH was 1:70,000-1:80,000. These antigens are stored in granules and dilated rough endoplasmic reticulum and this may also be a factor in their recovery. Cytoplasmic antigens may be more labile and require less intense deproteination.

Reviewer I: You have discussed reasons for high background associated with detection system for non-radiolabeled probes. However, regardless of the probe labeling method, radioactive or non-radioactive, lack of probe specificity remains one of them. In our hands (35S labeled riboprobes), the most effective way to improve signal to noise ratio, was to use a probe that in Northern analysis would detect a single band of a size matching the expected mRNA size. If this condition was fulfilled, the RN-ase treatment was found unnecessary. Do you think that evaluating specificity of biotin-labeled probes by Northern hybridization might be worth while?
Author: Yes, all of our biotin-labeled cRNA probes (LH, FSH, follistatin, EGF) have been evaluated by either Northern Hybridization or RNase protection assay for specificity. They detected a single band. They were used to assay pituitary mRNAs in collaborative studies.

Reviewer II: The author states that non-radioactive detection systems can be as sensitive as autoradiographic techniques with radiolabeled probes. How can one determine the sensitivity of the hybridization reaction with biotinylated probes?
Author: Actually, these techniques are not purely quantitative in that they do not allow one to detect amounts of DNA or mRNA in a cell. Thus, "efficient" is the operative word, rather than sensitivity. With our system, we use concentrations of cRNA probes for detection that are comparable to those used in autoradiographic detection protocols. If the same enzyme techniques are used in an *in situ* PCR reaction, the amplifying effects of the enzyme substrates allow detection of as little as one copy of mRNA per cell.

Reviewer II: How do various fixation conditions affect the sensitivity of the detection method. Is 2.5% glutaraldehyde used for both light and electron microscopy and/or when performing a dual labeling experiment? Doesn't glutaraldehyde fixation abolish or reduce the antigenicity of endogenous proteins when performing a dual labeling experiment for both mRNA and endogenous proteins.?
Author: We always run comparative studies with any new antigen or mRNA and in every case have found increased, or equal detection efficiency when we used glutaraldehyde. This means that we can use lower concentrations of cRNA probes to get the same signal. We have compared results in a variety of fixatives, including para-formaldehyde, Bouin's, 10% formalin and glutaraldehyde. We just completed another comparison between formalin and 2% glutaraldehyde and its effects on the detection of calcium channel mRNA in pituitary cells. The cells fixed in glutaraldehyde are much more intensely labeled. Furthermore, the peptide antigens are held in place by the stronger fixative and we have a much cleaner reaction in glutaraldehyde fixed tissues. Some of the larger glycoprotein antigens can be detected with much higher dilutions of antisera (1:125,000) if para-formaldehyde is used, however the cell fixation is not optimal. Thus, we choose glutaraldehyde and these same antigens are detected with 1:30,000-1:60,000 dilutions of their primary antisera. One trades efficiency (more dilute antisera) for optimal morphology. However, as stated above, we have never encountered loss of efficiency when detecting mRNA in glutaraldehyde-fixed tissues. Each new graduate student is asked to do the same comparison with a new mRNA and the conclusion has always been the same.

Reviewer II: What is the major advantage of using a biotinylated detection method? What is the major disadvantage of using a biotinylated detection method?
Author: The major advantage is the fact that one can achieve signal in a day, after as little as 2 hours of incubation in the cRNA probe. The longest protocols (with oligoprobes) are 1.5 days. It is also easily adapted

to a rapid immunolabeling technique, so in less than two days, one can detect the mRNA and the antigens. Also, no radioactive compounds are required to add to the growing waste. Finally, because the background is so low (absent), one need not view or depict the cells in dark field. Frequently, autoradiograms must be depicted in dark field, because the background grain level is so high in surrounding tissues. Finally, I have seldom seen autoradiograms where the labeling grains are well confined to the cells in question. It is impossible to detect regional labeling in such cells. Since we know that most cells are not filled with mRNA (100%), most of this is artifactual spread. In the case of the non-radioactive systems, one can also get this spread, however, it can be more readily controlled. In the case of peroxidase, the labeling can be confined to the region of the mRNA, where it belongs (Figure 2). There are no real disadvantages to using any non-radioactive detection system, since it can be adapted to In situ PCR for the low abundance mRNAs. If one has endogenous biotin, one can use the blocking kits supplied by Vector Laboratories. Also, there are blocking kits and reagents available for the enzymes in the detection systems.

Reviewer II: Assuming that the mRNA is associated with cytoplasmic polyribosomes and/or with RER, what is the advantage of detecting mRNA hybrids at the ultrastructural level? What is the major advantage of being able to detect mRNA at the ultrastructural level, i.e., by electron microscopy.

Author: In the pituitary, one cannot always differentiate the cell types accurately by morphology alone. Thus one needs a detection system that identifies the antigens within the granules and/or the mRNA on the rough endoplasmic reticulum. In the case of poorly differentiated pituitary cells, one might have only polyribosomes and few storage granules. Such cells can only be identified by their content of mRNA for the particular hormone. Such cells are abundant among developing gonadotropes, for example.

Reviewer II: How does one adapt this protocol for detection of mRNA in tissue sections? Are different fixation conditions used?

Author: We have used frozen pituitary sections with biotinylated oligoprobes for AVP. We rapidly froze the small tissue blocks and cut the sections with a cryotome. Then, we fixed them in para-formaldehyde vapors. They were then detected with an avidin detection system. In this case, my student (Dr. Ping Wu) preferred to use streptavidin alkaline phosphatase. This was quite successful and her papers are cited in the reference list. We are now working on adapting the streptavidin alkaline phosphatase technique to a dual-labeling protocol and have had some success with the use of immunoperoxidase as the detection system for the antigens. This protocol can be used as is for either cells in culture or frozen sections.

Reviewer II: Where does one block for endogenous biotin, peroxidase, and/or alkaline phosphatase in the method described.

Author: We have no endogenous biotin. We have been using biotin-avidin detection systems in the pituitary for over 15 years, and have never encountered any evidence of endogenous biotin. However, if it is a problem, one can use the blocking reagents supplied by Vector Laboratories. We also have been using peroxidase detection systems since 1972. We see pseudoperoxidase activity exhibited by red blood cells in our immunolabeling protocols. This can be eliminated by pretreatment with hydrogen peroxide. Interestingly, we never see this activity when running *in situ* hybridization protocols. Thus, something in our sequence must quench it. Our control fields are so crystal clear, it is difficult to find the cells without special optics. With regard to alkaline phosphatase, there are blocking kits and reagents available. However, our recent work has shown that they are unnecessary. A recent inadvertent proof of this happened when my technician accidentally left out the streptavidin alkaline phosphatase in the sequence that detected calcium channel mRNA. The cells were exposed to an extended time in the substrate solution. There was absolutely no reaction in these glutaraldehyde fixed cells. Finally, our current experiments have been run on cells plated on silane coated slides with two wells indented in each slide. One of the wells is always a control, the other receives the biotinylated cRNA probe. We wash these slides in coplin jars, in the same solutions. Each slide is exposed to the same substrate solution. Not only are the controls for the single labeling protocol completely clean (showing no endogenous alkaline phosphatase), but the controls for the dual-labeling protocol remain clean (showing no endogenous peroxidase from the second protocol). The control and experimental fields on the same slide are as different as night and day, even if the experimental fields contain many intensely labeled cells.

Scanning Microscopy Supplement 10, 1996 (pages 27-47)
Scanning Microscopy International, Chicago (AMF O'Hare), IL 60666 USA
0892-953X/96$5.00+.25

IN SITU HYBRIDIZATION, *IN SITU* TRANSCRIPTION, AND *IN SITU* POLYMERASE CHAIN REACTION

L.E. De Bault[1*] and J. Gu[2]

[1]Department of Pathology, University of Oklahoma Health Sciences Center, Oklahoma City, OK 73190
[2]Institute for Molecular Morphology, Mount Laurel, NJ 08054

(Received for publication October 1, 1996 and in revised form December 31, 1996)

Abstract

In situ hybridization, *in situ* transcription, and *in situ* polymerase chain reaction (PCR) are techniques used to detect DNA and RNA sequences within a cell or tissue structure. These three *in situ* methodologies employ the principles of recombinant DNA to form double-stranded hybrids of DNA-DNA, DNA-RNA, or RNA-RNA. The essence of *in situ* hybridization (ISH) is the hybridization of a labeled probe to a complementary target sequence, whereas *in situ* transcription (IST) is the synthesis of complementary DNA incorporating a label directly on the target DNA or RNA within a cell or tissue. In the case of *in situ* PCR (ISPCR), it is the repeated *in situ* duplication of both the sense and antisense strands of DNA to increase the number of copies of the target sequence. ISH, IST, and ISPCR each have their advantages and disadvantages. The purpose of this chapter is to address *in situ* considerations requried of these techniques, emphasizing tissue fixation, pre-hybridization steps, DNA probes, RNA probes, oligoprobes, and probe labeling. Five successfully used protocols are presented as examples. Any given nucleotide target sequence may have its own unique set of optimum conditions, thus requiring some adjustment in the hands of the user.

Key Words: *In situ* hybridization, *in situ* transcription, *in situ* polymerase chain reaction, molecular morphology, cytochemistry.

*Address for correspondence:
L.E. De Bault
Department of Pathology
P.O. Box 26901, BMSB 451
Oklahoma City, OK 73190
Telephone number: (405) 271-7300
FAX number: (405) 271-1107
E-Mail: L-DeBault@UOKHSC.Edu

Table of Contents

Introduction

In situ hybridization, *in situ* transcription, and *in situ* polymerase chain reaction (PCR) are techniques used to detect DNA and RNA sequences in chromosomes, cells or intact tissue sections. These three *in situ* methodologies employ the principles of recombinant DNA (rtDNA) as they are broadly defined [49], and rely on the powerful and widely used technique of nucleic acid hybridization which exploits the ability of complementary sequences in single-stranded DNAs and RNAs to pair with each other to form double-stranded hybrids of DNA-DNA, DNA-RNA, or RNA-RNA. Man-made sequences of nucleotide that specifically bond or hybridize to a target sequence in an A-T (or U) and C-G complementary fashion are called DNA or RNA probes. In the *in situ* hybridization technique, the probes are labeled with a radioisotope or chemically tagged for the detection and localization of the hybridized probes. The hybridized probes are then detected by either a direct or indirect method. In the *in situ* transcription technique, an unlabeled probe is used as a primer and labeled nucleotides are incorporated into the cDNA that is synthesized as the primer is extended on the target template. In the case of *in situ* PCR, the target sequences are bracketed by two primers and are amplified by making a large number of copies, and then the copies are detected or directly visualized.

The essence of *in situ* techniques

The essence of *in situ* hybridization (ISH) is the hybridization of a labeled probe to a complementary target sequence. In its simplest form, ISH is performed by separating the strands of double-stranded nucleic acids by denaturing, hybridizing a labeled probe to its complementary DNA or RNA in tissue sections or individual cells, washing away the unhybridized probe, and detecting the label on the bound probe. Protocols will be presented that will take advantage of simplicity and directness of ISH. The advantages and disadvantages of ISH and the need for controls will be discussed.

The essence of *in situ* transcription (IST) is the synthesis of complementary DNA (cDNA) within a cell. In its simplest form, IST is performed by annealing a specific unlabeled primer to its complementary mRNA in a tissue section or individual cell, washing away the unhybridized primer, and synthesizing cDNA in the presence of reverse transcriptase and deoxynucleotide triphosphates (dNTPs), some of which are labeled. Protocols will be presented that take advantage of the simplicity of IST, the use of Digoxigenin (DIG)-labeled dUTP and an immunochemical bridge. The advantages and disadvantages of IST and Immunogold Silver Staining (IGSS) will be discussed.

The essence of *in situ* PCR (ISPCR) is repeated *in situ* transcriptions of both the sense and antisense strands of DNA to increase the number of copies of the target sequence. In its simplest form, ISPCR is performed by denaturing double-stranded nucleic acids to single-strands, two primers bracketing a sequence of interest are hybridized to their targets, the sequence amplified with PCR, and the amplicons detected. The advantages of ISPCR and the need for controls will be discussed. For more detailed discussion of these techniques, readers are referred to the monographs and reviews published on the topics [14, 16, 58, 63, 75].

Historical background

ISH, IST, and ISPCR, like other scientific tools and the new knowledge they provide, were preceded by other key discoveries and innovations. From our perspective, there were about a dozen or so historic milestones spanning approximately two decades which made molecular biology possible and the application of ISH, IST, and ISPCR to tissues and cells *in situ*, practical. The 1969 work of Gall and Pardue [24], demonstrating the formation and detection of RNA-DNA hybrid molecules *in situ*, were proof-of-principle findings that made the ISH technique possible. In 1970, the discovery of RNA-dependent DNA polymerase (or reverse transcriptase) by two independent groups, Baltimore [5] and Temin and Mizutani [88], and the 1971 work of Kleppe *et al.* [51] demonstrating the replication of short synthetic DNA segments catalyzed by DNA polymerases, demonstrated the possibility of IST. But it was not until Langer *et al.* [59] enzymatically synthesized Biotin-labeled polynucleotides, followed by the work of Saiki *et al.* in 1985 [82] that achieved *in vitro* primer-mediated amplification of genomic DNA, that practical IST was made possible. The IST technique was first described and applied to tissue sections by Tecott *et al.* in 1988 [87] and was confirmed and applied to cells *in vitro* by Longley *et al.* in 1989 [67]. The field picked up steam with three landmark discoveries: i.e., the 1973 work of Cohen *et al.* [17] demonstrating the cloning DNA fragments into plasmid vectors, the 1975 work of Kohler and Milstein [57] demonstrating the development and *in vitro* production of monoclonal antibodies, and the 1976 work of Chien *et al.* [15] discovering thermostable DNA polymerase. This was later followed in 1987 by Kogan *et al.* [56] who introduced the use of thermostable DNA polymerase and made possible the practical use of the polymerase chain reaction (PCR). The introduction of liquid phase PCR and its automation dramatically improved the rate of progress for molecular biology [77]. The aforementioned milestones were essential advances for the advent of ISH, IST, and ISPCR, and were indirectly necessary for the proof-of-principles of these *in situ* techniques.

Scope of Chapter

Many names have been given to these techniques encompassing a wide range of protocols, but the principles remain the same. In this chapter, we give a concise and comprehensive review of the principles and procedures of these techniques and their major variations. As examples, certain protocols that have been tested by the authors are included at the end of this chapter.

In situ Conditions

General considerations

In vivo nucleotide sequences anneal and separate with a high degree of precision. *In vitro* their behavior also follows certain general rules. The factors that affect their behavior the most include temperature, salt concentration, length of the probes, percentage of C-G content, pH and concentrations of certain chemicals such as formamide. Changing the temperature or pH are the most commonly used methods to manipulate the annealing and denaturation of DNA or RNA [16].

Fixation

The most commonly used fixatives for *in situ* hybridization have been aldehyde-based fixatives such as 10% formalin or 4% paraformaldehyde [14]. The aldehyde fixatives cross link proteins, or proteins and nucleotides, and effectively preserve the nucleotide sequence. Fixatives with picric acid such as Bouin's fixative or heavy metals such as Zenker's solution should be avoided as damage of target nucleotide sequences may occur [27]. Fixation from three hours to overnight appears to be an acceptable range depending on the size and nature of the tissue and the strength of the fixation. Frozen sections can also be used.

Pre-hybridization steps

The aldehyde-fixed samples need to be digested to expose the targets, yet this digestion should be optimized to also preserve the morphology and retain the target sequences after hybridization. Proteinase K, trypsin, pronase and other proteases have all been used successfully. The digested sections should be washed thoroughly to remove or inactivate all the enzymes. A prehybridization step is recommended which should eliminate or reduce the background nonspecific reactions. This usually consists of a cocktail of the hybridization solution without the specific probe to exhaust possible nonspecific bonding sites prior to incubation of the specific probes. Additional pretreatment using levamisol, normal rabbit serum, etc. can also be performed to prevent the subsequent detection step causing any background. The samples are then equilibrated in a solution that is close to the hybridization condition [37] and proceed with the hybridization steps. We have also found that it may be beneficial to let the samples dry after the equilibrate washing and applying the hybridization cocktail directly on the dry samples. When nonspecific background staining is controlled, this usually gives stronger signals, presumably because of a better penetration of the probes to the tissue sample and a more controlled concentration of the salts in the immediate environment of the hybridization reaction.

DNA probes

The key in probe design and selection is that the probe sequence has to be specific to the target and not complementary to any other sequence. Double-stranded DNA probes were the initial approach developed for *in situ* hybridization because of their relative stability. They can, however, be destroyed by DNase which requires magnesium to function and can be inhibited by EDTA [80]. They are usually labeled by nick translation or random primer extension. These probes can be as long as a few thousand nucleotides or as short as 1 or 2 dozen nucleotides [80]. Double-stranded probes must be denatured before hybridization, which is generally achieved by increasing the temperature. The drawback of double-stranded probes is that the two strands tend to re-anneal to themselves during hybridization, thereby decreasing the detecting efficiency. Single-stranded DNA probes can also be produced by using the 13M cloning vectors [42, 72] and PCR [68], overcoming the disadvantage of probe re-annealing. Thus, they are more efficient at hybridizing to their targets than double-stranded probes [78].

RNA probes

A riboprobe is a single-stranded RNA probe produced by cloning man-made vectors containing promoters and an insert of cDNA fragment of interest at a known orientation [47, 71, 85]. The recombinant plasmid vectors can be grown and amplified in appropriate bacterial hosts, whereby the inserted cDNA grows in quantity. The inserted sequences are then transcribed with RNA polymerase into RNA probes. Radioisotope or other labeling molecules can be incorporated during transcription. By reversing the orientation of the inserted cDNA, both anti-sense and sense RNA probes can be produced. A "sense" probe refers to a sequence equivalent to the target mRNA and an "antisense" probe is complementary to the sequence and is used to hybridize with the target. Riboprobes are superior to DNA probes in that they are single-stranded and do not reanneal to themselves, resulting in a more efficient target detection [3, 18, 64, 80]. When hybridized to mRNA, it is more stable than DNA-RNA hybridization and can stand stringent hybridization conditions and washing [44]. Following hybridization, it is possible to destroy the unhybridized single-stranded probes with RNase, further reducing the background. A large amount of the same kind of probe can be generated. However, the preparation of riboprobes requires the availability of the appropriate cDNA, considerable experience with molecular biological skills to prepare and propagate the plasmids and transcripts, and an RNase-free environment for probe preparation and handling. It is more specific than the shorter oligoprobes. A good riboprobe can bind its complementary mRNA more tightly than the shorter oligoprobes allowing for more stringent washing, which can lead to lower backgrounds and a more specific detection of the target.

Oligoprobes

Oligoprobes (or oligoprimers) are small single-stranded probes that possess many advantages [43]. The simplicity in manufacturing those probes make them an easy choice for *in situ* hybridization. If the target sequence is known, oligoprobes can be synthesized with a DNA synthesizer or ordered from a molecular biologic reagent company at a reasonable cost. The length of an oligoprobe should be about 14-50 bases [80]. The sequence has to be unique to the target of interest so that unwanted hybridization will be minimized. The percent of G and C content is also important as this will affect the thermostability of the hybrid or the background labeling. Optimal conditions for individual probes and targets should be established empirically, particularly for oligoprobes, as the window for optimal hybridization is narrower than for longer probes.

Probe labeling

A variety of molecules have been used to label probes. Radioisotope labeling remains the most sensitive approach. ^{35}S and ^{33}P have been a good compromise for probe labeling as they do not require too long an exposure time for autoradiography and give relatively high resolution. These labeling methods are widely used because of their sensitivity and reliability [47, 64].

Non-radioisotope labeling has become increasingly popular and, with refinement of the technique, more sensitive [29, 58, 76]. These methods avoid the hazardous radioisotope handling and disposal, shorten the experimental protocol, and yield high resolution results that can be combined with other techniques such as immunohistochemistry. It is also suitable for *in situ* hybridization at the electron microscopic level when used in combination with immunogold labels. Commonly used labeling molecules include biotin, digoxigenin, alkaline phosphatase, hapten and fluorescein. Biotin enjoys the popularity because of the availability of a wide variety of detection kits. Digoxigenin is a plant molecule that can be recognized by specific antibodies, providing added specificity and sensitivity due to reduced nonspecific binding, and is preferred by many investigators. Fluorescein can yield very high signal-to-noise ratios and detecting sensitivities, and therefore has been widely used in chromosomal labeling, and often goes by the acronym FISH (Fluorescence *In Situ* Hybridization) [48, 69, 93].

Optimizing *in situ* Conditions

The search for optimal condition for hybridization is one of the key issues for successful experiments. It is important to determine the melting temperature at which 50% of the double-stranded sequences are denatured. Equations are available for different factors that govern the relationship between the two complementary sequences in the denaturing and reannealing processes. Generally for ideal hybridization, the reannealing temperature should be set to 20-25°C lower than the melting temperature to generate a stable hybridization product. To ensure specificity for shorter probes, the hybridization temperature may be adjusted at 12-17°C below the melting point. The optimal hybridization temperature should be established empirically in addition to being predicted from the formulas. A temperature range of 37-42°C is a good starting point for the hybridization step. This is derived from inclusion of 50% formamide in the hybridization solution which competes with bases for the formation of hydrogen bonds and thus lowers the temperature requirement. For a DNA sequence with 50% C-G contents in 0.4 M sodium chloride containing 50% formamide, the melting temperature is about 54.5°C, and the hybridization temperature is adjusted to 37-42°C. The concentration of a probe is also important. It is usually used at around 300 ng/ml with a range of 200-1500 ng/ml depending on the size of the probes. The hybridization kinetic is also affected by salt concentrations. At concentrations below 1.5 M, the higher the salt concentration, the higher the rate of hybridization. Polymers of high molecular weight, such as dextran sulfate and polyethylene glycol, are often added at a concentration of about 10% (v/w). Those molecules can create a networking phenomenon, taking up sufficient space in the solution to artificially increase the probe concentration and thus the hybridization rate by about tenfold without creating any noticeable side effects **[1,2]**. The washing solution should be designed to contain the appropriate salt content that will wash away unbound probes and probes in less stable, mismatched hybrids, and leave the perfectly-matched hybrids intact. To set up an *in situ* hybridization protocol for a new target, the best approach is to follow previously established procedures and then adjust the various conditions, one at a time, starting from the probe design, hybridization solution and incubation temperature. Changing probe design is often the most effective measure to make *in situ* hybridization work.

In situ Hybridization

Technically *in situ* hybridization is a well established method. The protocols are fairly straight forward. When the right probes are selected and the target sequences are preserved and accessible, the properly executed hybridization procedure should lead to clear and well defined signals with low background. The window of opportunity for achieving optimal reaction is fairly broad. Certain small mishaps in performing the technique would not lead to unusable results.

Advantages of ISH

One of the advantages of *in situ* hybridization, in comparison to *in situ* transcription and *in situ* PCR, is that the probes are hybridized to the target DNA or RNA sequences directly without target manipulation or amplification. What are amplified in the *in situ* hybridization protocols are the labeling molecules not the targets themselves. Therefore, there is little problem of product diffusion or amplicon back diffusion, sometimes a formidable problem for *in situ* PCR. In addition, when the signal amplification steps are constant, the amount of reporting signals are approximately proportional to that of the target, thus allowing semiquantitative assessment of the target DNA or RNA. When evaluated jointly with the results obtained with immunocytochemistry, this semiquantitative information can lead to meaningful interpretations of the cellular components under study. For example, they may indicate the balance or imbalance of the expression of a particular gene and the production of the coded protein. When the distributions and relative quantities of the gene expression and its protein are different, certain cellular events could be indicated [29, 86, 88].

Disadvantages of ISH

The disadvantage of *in situ* hybridization is that it is not as sensitive as *in situ* PCR, or *in situ* transcription. Nevertheless, with the newly reported, more powerful signal enhancement techniques such as the nanogold-silver immunostaining and computer enhanced fluorescence *in situ* hybridization, the detecting sensitivity of *in situ* hybridization has been improved markedly [35, 69, 78].

In situ Transcription

IST is initiated by the hybridization of a primer (usually a specific oligonucleotide) to target mRNA in a tissue section or cell preparation on a glass slide as is done in conventional ISH [18, 38, 50, 62, 63, 65, 66]. Transcription of the target mRNA is achieved by adding reverse transcriptase and labeled nucleotides which allow the synthesis of a labeled complementary DNA (cDNA). Since the synthesized cDNA remains associated with its mRNA template as a cDNA-mRNA hybrid, anatomical distribution and cellular localization are preserved, thus, it is one of the main advantages of IST.

In the original works [67, 87], radio labels were used and the localization of the labeled cDNA was detected by autoradiography. Many non-radioactive detection systems have been developed for ISH and can be applied to IST, i.e., fluorochrome-labeled dNTPs in fluorescence methods [6, 7, 52, 65, 66, 89, 93], biotin-labeled [26, 39, 50, 53, 55, 59, 65, 73, 74] or digoxigenin-labeled [19, 20, 36, 39, 79, 91, 92] dNTPs in immunochemical detection methods.

One of the first applications of the IST technique was to the localization of proopinmelanocortin (POMC) mRNA on fresh-frozen paraformaldehyde-fixed sections of rat pituitary using reverse transcriptase and radio-labeled ^{3}H-dCTP in the cDNA elongation process followed by autoradiography [87]. In another early study, IST was applied to localized alpha-2 domain of CD1a mRNA using an oligonucleotide primer specific to the target mRNA, incorporating ^{35}S-dCTP into cDNA followed by autoradiography [67]. IST has also been used to investigate the temporal expression of mRNAs in developing embryos [22, 23]. One of the most intriguing uses of IST was in the analysis of gene expression in live cells where mRNA in single live neurons *in vitro* was injected with primer, dNTPs, and reverse transcriptase via whole cell patch electrode. The contents of the cell was then harvested by suction back into the electrode, the electrode incubated *in vitro* where the cDNA is first synthesized by IST followed by replication of cDNA to many copies of amplified RNA (aRNA) [90]. The aRNA was assessed by Southern and Northern blot analysis. More recently IST has been used to localize γ-GTP mRNA in paraffin section from rat kidney [91] and from cell cultures of rat brain microvessels [92].

One technique closely related to IST for mRNA is termed, PRimed *in situ* labeling (PRINS), which was first used by a Danish group to localize chromosomal DNA [52-55], and was later confirmed by a group in Scotland [26, 73]. For clarity, the PRINS technique applied to DNA is referred to as PRINS-DNA. The PRINS-DNA procedure uses unlabeled DNA probes or oligonucleotides as the primer, and DNA polymerase (Klenow or *Taq*-I) and biotin or digoxigenin labeled dNTPs to synthesize labeled DNA *in situ*. The site of synthesis was detected by immunocytochemistry using fluorochromes as reporter molecules. A variation of PRINS-DNA can be used in the detection of mRNA [54,

74], and when applied to mRNA it is referred to as PRINS-mRNA. PRINS-mRNA is virtually the same as IST in that it uses an unlabeled primer complementary to a specific mRNA sequence, and reverse transcriptase and labeled dNTPs to synthesize a labeled cDNA. Thus all comments made about IST should also apply to PRINS-mRNA.

Advantages of IST

Since chain elongation is independent of the length of the primer, oligonucleotide primers induce as much (if not more) labeling of the target mRNA than a much longer pre-labeled probe, thus increasing the sensitivity of the method. Another advantage of IST is its application to the detection of minor sequence variations in RNA by the proper selection of the sequence; thus, the position of the primer can control whether or not there will be chain elongation. The application of IST to the detection of minor sequence variations should be superior to ISH in that it is well known that the last few nucleotides of the primer are crucial to the initiation of chain elongation [10]. A significant technical advantage of the method is that the probe (primer) is unlabeled and labeling occurs only secondarily to specific hybridization, and the unincorporated labeled nucleotide can be washed away easier, which results in a lower background. In addition, fewer procedural steps allow for a shorter cumulative incubation time resulting in less degradation in the tissue morphology which often accompanies ISH or ISPCR. Another advantage of the milder conditions of IST is that it allows for the detected mRNA in cell suspension intended for flow cytometry without clumping and disintegration of cells [4].

Disadvantages of IST

The main disadvantage of IST is that it requires a high copy number of target mRNA (or DNA). If the copy number is low, ISH may be the technique of choice. If the copy number is very low, then ISPCR is the technique of choice, and has been shown to be able to detect a single copy. See ISPCR below.

In situ PCR

In situ PCR combines the DNA amplifying power of liquid phase PCR with the localizing capability of *in situ* hybridization [28, 63]. First, it amplifies minute quantities of DNA or RNA fragments to millions or billions of identical copies at the site of the original template and then detects or visualizes the amplified signal *in situ*. This technique fills a technical gap and allows detection of low copy numbers of nucleotide sequences against the background of tissue structure, even detecting a single copy of DNA or RNA per cell [78]. ISPCR is particularly important in detecting latent virus infections and studying the pathogenesis of many viral and oncogenic diseases [28, 78].

In situ PCR can be performed in several different ways. It can directly incorporate labeling molecules into amplified products by using labeled primers or labeled free nucleotides in PCR. This way, all amplified sequences have the labeling molecule built into them for subsequent detection. The amplified sequence can also be detected by employing *in situ* hybridization with a labeled probe complementary to the amplified target. RNA can be reverse transcribed into cDNA and then amplified and detected. Several *in situ* PCR machine models are available on the market. Each has its advantages and limitations.

It is well recognized that the key steps in *in situ* PCR include tissue preparation, pretreatment, primer and probe design, washing and detection. However, the prevention of amplified product diffusion is perhaps the most important consideration in designing any *in situ* PCR protocol. This technique is relatively new and technically challenging. False positivity and negativity can occur easily and should be carefully checked with a number of control experiments, including positive and negative tissue samples, and omission of primer, probe or other key components, one at a time, in the amplification or detection mixtures. The ISPCR results should be compared with results obtained by liquid phase PCR using the same primers and probes to detect the same sequences on DNA or RNA extracts from the same tissue samples.

Advantages of ISPCR

The most obvious advantage of *in situ* PCR is, of course, its very high detecting sensitivity while retaining tissue morphology so that minute quantities of DNA or RNA can be visualized and correlated to the surrounding morphology. It was reported that it can detect down to a single copy of a DNA or RNA sequence in intact cells or on tissue sections. This makes it a very valuable tool for many purposes, especially the detection of early or latent viral infections or early genetic changes in oncogenesis. All these can be achieved with commonly used enzyme labeling methods and viewed with a transmitting light microscope. The detecting sensitivity of conventional *in situ* hybridization is not entirely clear, but is believed to need at least 20 copies for detecting and thus leaves a technical gap where *in situ* PCR finds most of its applications. A second advantage is, in theory, *in situ* PCR can be combined with other methods to demonstrate two targets simultaneously on the same tissue preparation. It can also be performed at the electron microscopic level, although the preliminary reports in this regard (mostly in abstract forms) showed very poor ultrastructure preservation. A third advantage, again in theory, is that direct *in situ* PCR can detect DNA or RNA targets without knowing the entire sequence by using a pair of primers flanking the two ends of the target. Overall, *in situ* PCR has a high detecting sensitivity that has made *in situ* PCR so popular.

Disadvantages of ISPCR

In comparison with *in situ* hybridization and other highly sensitive methods, *in situ* PCR has a number of disadvantages. First, it is technically challenging to set up reliable *in situ* PCR and, once set up, the technique is often not very stable. This is due primarily to the large number of critical steps in the protocol. These include adequate enzyme digestion, efficient amplification, prevention of amplicon diffusion, and reduction of background. It often takes longer and needs more controls to establish reliable *in situ* PCR protocols. The second disadvantage is that the tissue morphology is less than ideal. Because of digestion and, in particular, the harsh treatment by the thermal cycling, the morphology of the samples, although still recognizable, is often distorted and damaged. The third disadvantage is that the protocol of *in situ* PCR requires specially designed machines which makes it more expensive and less convenient to perform. Fourth, because of the very high detecting sensitivity, the less than desirable morphology and the potential problems associated with amplicon diffusion and inadequate digestion or washing, the results could be difficult to interpret. Extensive controls are sometimes called for and this might make the experiment lengthy and less manageable. Strict internal and external controls are needed for *in situ* PCR; however, optimal control kits are not currently available for this technique.

It should be mentioned that although *in situ* PCR is a very powerful and attractive technique, it is not the only one for detecting low copy number of nucleotide fragments. Other procedures such as FISH, nanogold-silver method, 3-SR technique, radioisotope labeled *in situ* hybridization, and autoradiography, etc., can also detect those sequences without some of the drawbacks of *in situ* PCR. Each of the methods has it strengths and weaknesses and should be considered before committing a given *in situ* PCR protocol.

Controls

As for immunohistochemistry, it is extremely important to perform adequate controls to verify the specificity of *in situ* hybridization [14, 16, 58, 63, 75]. For single-stranded probes, a common approach is to use a sense instead of antisense probe as a negative control and follow the exact same protocol. However, it has been reported [44], although rarely noted in the literature, that a small number of antisense sequences may be produced by the target cells, possibly by a mechanism of transcription regulation. Therefore, a sense probe may result in a reduced but still specific and positive labeling. We have observed this phenomenon and learned to interpret the sense probe results with caution [29]. Using an unrelated probe of the same length and C + G content is the best way to get around this problem. Positive controls may include a tissue preparation with a known quantity and distribution of the target sequence. Negative controls may include a piece of tissue that is known not to contain the target sequence. This can be created artificially by treating the tissue sections with RNase or DNase to destroy the target sequences. However, this needs to be performed carefully and followed by extensive washing. Any trace residue of the enzyme that finds its way into the real experiments may destroy the target or probe. Additional controls may include the omission of each of the key elements in the incubation cocktails. This is often effective in checking the ingredients that cause the false positivity or give high background. It is always advisable to perform Southern or Northern blot analysis on DNA or RNA extracts of the same target tissue side-by-side with *in situ* hybridization to verify the presence and quantity of the target sequences. The specificity of the visualization methods also needs to be checked including replacement of each of the key elements, particularly the primary antibody or the first linking reagent to the probe. Only after the key controls give the expected results can the observations with *in situ* hybridization be validated.

It should be noted that new techniques are emerging that can detect low copy numbers of DNA and RNA without going through the elaborate *in situ* amplification steps. The reporting signals can be amplified instead of, or in addition to, the amplification of the target sequences themselves.

EM *in situ* hybridization

In situ hybridization has been applied at the electron microscopical levels using electron-opaque labels [25]. The attempts of applying *in situ* PCR at the EM level have not been very successful because of the deterioration of the ultrastructure caused by the many PCR cycles, although signals have been reported to be visualized under the electron microscope.

Both preembedding and postembedding *in situ* hybridization can be performed on electron microscopic grids. The procedure is similar to that for the light microscope except that the protocols are adjusted to the EM conditions with much gentler digestion and washing. Colloidal gold remains to be the best labeling method at the electron microscope level. A double labeling with 2 differently sized gold particles can be performed on the same grid to demonstrate the RNA or DNA and its corresponding protein simultaneously, greatly facilitating the morphological elucidation of the subcellular regulatory mechanism of a particular gene and its product [25].

Technically, EM *in situ* hybridization is quite challenging. The tissue samples are very delicate and the optimal balance of the many treatments, washings and reactions need to be established empirically for each target sequence in its host tissue. For detailed protocols, readers are referred to a monograph on this topic edited by Morel [75].

Protocols

Recommended protocols of *in situ* hybridization

The following protocols are selected from many published procedures. They have been tested in our own laboratories and found to be reliable and reproducible. The first is for RNA-RNA detection using riboprobe to detect mRNA. It has been successfully used to detect c-myc, N-myc and L-myc gene expression in small cell lung cancers [31] and atrial natriuretic peptide gene expression in the heart [30, 32]. The second protocol uses oligoprobe that was labeled with FITC and detected with a specific antibody to FITC. We used this protocol to detect peptide mRNA including ANP, neuropeptide Y (NPY), insulin and glucagon. These protocols can be adapted to different purposes. They should be adjusted individually with particular probes, targets and tissues. The protocols rarely give optimal results by just copying and switching from one probe or target to another. Controls should always be performed with the experiments to assist in trouble-shooting and result interpretation. There is inconsistency in tissue preparation, digestion, pretreatment, and detection among the protocols. This was due to the fact that they were developed in different laboratories by different individuals. They have all worked well for their particular applications at the time of the studies.

Common materials and equipment. The equipment needed are incubation oven, adjustable precision pipettes, glassware, glass slides, coverslips, etc. The reagents and solutions needed are xylene, alcohol, dextran sulfate, saline sodium citrate (SSC) (20x, 5x, 2x, 1x), EDTA (1 mM), Tris-buffered saline (TBS) (50 mM Tris/HCl, 150 mM NaCl pH 7.6), alkaline phosphate substrate buffer (100 mM Tris/HCl, 50 mM $MgCl_2$, 100 mM NaCl pH 9.0), formamide ($HCONH_2$), Diethyl Pyrocarbonate (DEPC; Sigma, St. Louis, MO).

Protocol #1: *in situ* hybridization using biotin-labeled riboprobes on tissue sections.

1. Prepare tissue sections by dewaxing, rehydration, washing in DEPC-treated water, etc.
2. Digest tissue with Proteinase K, 5-30 μg/ml at 37°C for 10-30 min in humid chamber.
3. Wash in DEPC-treated water for 3x2 min.
4. Deactivate Proteinase K at 70-75°C for 1 min (optional) wash in double distilled water 5 mins.
5. Prepare prehybridization solution (for 1 ml):

50% Dextran sulfate	250 μl
20xSSC	149 μl
EDTA (1 mM)	120 μl
Herring Sperm DNA	33 μl
DEPC-treated water	448 μl

6. Apply prehybridization solution to tissue section and incubate at 42°C for 30 min.
7. Prepare hybridization solution with biotinylated probe.

Hybridization solution (H.S.) (for 1 ml):

50% Dextran sulfate	250 μl
EDTA (1 mM)	120 μl
20xSSC	100 μl
Formamide	450 μl
Herring Sperm DNA	33 μl
DEPC-treated water	47 μl

Mix well and add probe. Probe final concentration: 0.2 - 1.5 ng/μl.

8. Apply hybridization solution.
9. Incubate 3 hours to overnight at 42°C in humid chamber.
10. Wash in 5x SSC for 5 min.
11. Wash in 2x SSC for 5 min.
12. Proceed to detection with a biotin detection kit with sufficient background blockage.
13. Wash in distilled water (2x5 min).
14. Counterstain and mount.

Notes to Protocol #1

1. Radioisotope or digoxigenin-labeled in situ hybridization can be more sensitive than biotin-labeled in situ hybridization.

2. The probe in the hybridization solution (step 8) can be heated to 95°C for 5 mins before being applied to the tissue section to eliminate any self-annealing or secondary structure formation of the probes.

3. Up to step 10, the procedure should be performed in a RNase-free manner.

4. The hybridization mixture can be applied on dehydrated dry sections or on wet sections. If the latter, there should be as little liquid remaining on the slides as possible to avoid diluting the concentrations of the ingredients in the hybridization solution.

5. For hybridization solution of less than 40 μl per tissue section, a coverslip can be applied and sealed at the hybridization step (step 9) to prevent evaporation.

6. RNase may be used in post hybridization treatment to remove single-stranded probes, thereby reducing *background. This can be applied at 100 μg/ml RNase A and 1 unit/μl RNase T_1 for 30 mins at 37°C.*

7. Counterstain should not mask or overshadow the specific labeling.

Protocol #2: *in situ* hybridization using FITC-Labeled oligoprobes on tissue sections.

1. Prepare tissue section by dewaxing, rehydration, washing in DEPC-treated water, etc.
2. Digest tissue sections with Proteinase K at 5-30 μg/ml in 0.05 M Tris/HCl buffer pH 7.6 made with DEPC-treated water, and incubate for 10-30 min at 37°C.
3. Immerse in DEPC-treated, double-distilled water for 3 x 2 min.
4. Dehydrate in increasing grades of ethanol.
5. Air dry for 2 min.
6. Prepare hybridization solution with FITC-labeled probe.

Hybridization solution

30% Formamide
10% Dextran Sulfate
0.6 M NaCl
Mix well and add probe. Probe concentration: 0.2-1.5 ng/μl, optimized for probe and tissue sample selected.

7. Apply hybridization solution and coverslip.
8. Incubate 3 hours at 42°C in humid chamber.
9. Wash slides in TBS containing 0.1% Triton-X-100 for 3 x 3 min. Allow coverslips to drain off in the washing solution - Do not touch coverslips to remove. Dipping may be required to remove any remaining coverslips.
10. Place slides on incubation tray and cover sections with 100 μl of TBS, containing 3% bovine serum albumin, 0.1% Triton-X-100, 20% normal rabbit serum. Incubate for 10 min.
11. Tip off the blocking solution and add rabbit Fab anti FITC conjugated to alkaline phosphatase diluted 1:100 - 1:200 in TBS containing 3% bovine serum albumin and 0.1% Triton-X-100. Incubate for 30 min - 3 hours at 20°C.
12. Wash slides in TBS for 2 x 3 min.
13. Wash slides in alkaline phosphatase substrate buffer pH 9.0 for 5 min.
14. Place slides in humid chamber and demonstrate alkaline phosphatase activity by covering sections with the following solution (for 200 μl):

80 μl 5-Bromo-4-chloro-3-indolylphosphate (BCIP) - 50 mg/ml in dimethyl formamide
80 μl Nitro blue tetrazolium (NBT) - 75 mg/ml in 70% dimethyl formamide
10 μl 1 M levamisole
10 ml alkaline phosphatase substrate buffer

Color development may take from 1 hour to overnight.

15. Wash in running water for 5 min.
16. Counterstain and coverslip.

Notes to Protocol #2

1. In step #2, the enzyme activity can be stopped by optionally heating to 70°C for 1 min. Generelly the activity is stopped by dilution in the subsequent washings.

2. Up to step 9, the procedure should be performed in a RNase-free manner.

3. All the other notes described in Protocol #1 also apply to Protocol #2.

Recommended protocols of *in situ* transcription

Animals perfuse-fixed with 3% paraformaldehyde and 0.5% glutaraldehyde in 0.1M PBS (phosphae buffered saline), and the tissue embedded in paraffin by standard procedures. Under RNAse free conditions, 25-nt oligonucleotide probes complementary to target mRNA were hybridized 5 μm tissue sections, overnight. The primed mRNA was then transcribed *in situ* by incubation with a mixture of nucleotide precursors containing DIG-labeled dUTP, and AMV (Avian Myeloblastosis Virus) reverse transcriptase. After washing the *in situ* transcribed sections, the incorporated DIG was bridged with sheep anti-DIG IgG, and detected with 10nm gold conjugated rabbit anti-sheep IgG followed by silver enhancement. Controls consisted of the omission of the oligonucleotide probes, or the substitution of unrelated 25-nt oligonucleotides in the hybridization step.

IST has been applied to tissue, perfuse-fixed, paraffin-embedded, and handled to permit IST, immunocytochemistry, and enzymecytochemistry on sections all from the same paraffin block. This has been illustrated in detail by De Bault and Wang [19] using γ-Glutamyl Transpeptidase (γ-GTP) in rat kidney to demonstrate the localization of γ-GTP enzyme activity, γ-GTP protein, and γ-GTP mRNA.

Common materials and equipment: The equipment needed is: incubation oven, hot plate, ice bucket, adjustable precision pipettes, glassware, glass slides, coverslips, etc. The reagents and solutions needed are xylene, alcohol, sucrose, bovine serum albumin (BSA), dithiotheitol (DTT), gum arabic, triton X-100, $MgCl_2$, KCl, citric acid, sodium citrate, SSC (20x, 5x, 2x, 1x), EDTA (1 mM), TBS (50 mM Tris/HCl, 150 mM NaCl pH 7.6), RNase inhibitor, Formamide ($HCONH_2$), DEPC, silver lactate or silver acetate, ice, etc.

Protocol #3: *In situ* transcription using unlabeled oligoprimers and DIG-labeled dUTP on tissue sections.

1. Tissue was fixed in 3% paraformaldehyde + 0.5% glutaraldehyde + 2.5% sucrose in 0.1 M PBS, pH

7.4, followed by standard paraffin embedding.

2. Tissue sections were dewaxed, rehydrated in a graded ethanol series, and washed with 2xSSC (2xSSC solution consists of 0.15 M NaCl, 0.015 M sodium citrate, and adjusted to pH 7.0 with 1N HCl). DEPC-treated water was used to make all solutions and buffers.

3. Each slide is incubated with 20 μl of ISH mixture containing oligonucleotide primer(s) (covered with a glass coverslip) at 42°C overnight, followed by 30 min incubation at room temperature. The final mixture contained:

50% Formamide
4xSSC
0.02% bovine serum albumin (BSA)
5 mM DTT
0.6U RNase Inhibitor/μl
10μM Oligonucleotide (see notes #5 and #6 below for ISH solution)

4. Wash 2 times with 2xSSC and 2 times with 0.5xSSC, 15 min each at room temperature.

5. Hold slides in 0.5xSSC for 2 hr before proceeding with IST.

6. Each slide was incubated with 15 μl of IST mixture containing DIG-labeled dUTP (covered with a glass coverslip) at 37°C for 60 min followed by 45°C for 10 min. The final mixture contained:

50 mM Tris-HCl, pH 7.4
6 mM $MgCl_2$
40 mM KCl
5 mM DTT
0.02% BSA
0.1 mM DIG-DNA Labeling Mixture
0.6U RNase Inhibitor/μl
1U AMV Reverse Transcriptase/μl (see note #7 below for IST Solution

7. Wash 2 times with 2xSSC 30 min each at room temperature.

8. Wash 2 times with 0.05xSSC 1 hr each at 35°C.

9. Wash 3 times with 0.05 M TBS pH 7.4 containing 0.25% Triton X-100 10 min each at room temperature.

10. Slides were pre-incubated in 1% BSA in 0.05 M TBS for 15 min at room temperature.

11. Slides were incubated with 50 μl of 1° antibody mixture at +4°C overnight, followed by 1 hour incubation at room temperature. The final mixture contained Sheep Anti-Digoxigenin antibody diluted 1:50 in 1% BSA in 0.05 M TBS.

12. Wash 3 times with 0.05 M TBS 10 min each followed by wash with 0.02 M TBS pH 8.2 containing 0.25% Triton X-100 for 10 min at room temperature.

13. Incubate in 1% BSA in 0.02 M TBS pH 8.2 for 15 min at room temperature.

14. Incubate with 50 μl of 2° antibody mixture for 1 hr at room temperature. The final mixture contained 10nm gold conjugated Rabbit Anti-Sheep IgG diluted 1:10 in 1% BSA in 0.02 M TBS.

15. Wash once with 1% BSA in 0.02 M TBS, pH 8.2 for 10 min, followed by 3 washes of 0.05 M TBS 10 min each.

16. Wash with distilled water.

17. Incubate in Silver Enhancement mixture for about 20 min at 22°C. The final mixture contained [34, 40]:

5.5 mM silver lactate or silver acetate
77 mM hydroquinon
120 mM citric acid
80 mM sodium citrate
10% gum arabic (see note #10 below for Silver Enhancement Solution)

18. Wash with distilled water 5 times 2 min each.

19. Counter stain lightly with hematoxylin and eosin (H&E) (optional).

20. Mount in Permount® and coverslip.

Notes to Protocol #3:

1. All reagents used and steps in the perfusion and fixation procedure were performed at +4°C, unless otherwise stated [45, 61].

2. In the standard paraffin embedding, infiltration was by machine processing and included 2 changes of 95% Ethanol for 20 min each, 3 changes of 100% ethanol for 15, 20, and 30 min respectively, 2 changes of Xylene for 20 and 30 min respectively, all at 40°C, and 2 changes of paraffin for 45 min each at 57°C. The tissue is embedded in flat molds [70, 83]. Completion of the entire embedding process on the same day that the perfusion fixation is performed is important in maintaining maximum antigenicity and mRNA reactivity.

3. 5 μm paraffin sections were floated on a 0.1% DEPC-treated water bath at 42°C, and picked up on silanized slides [81]. It is ***important*** *to perform the in situ transcription and immunogold-silver staining or other detecting procedures immediately after cutting; cut sections stored for days or weeks lose reactivity and background often increases.*

4. All in situ reagents were prepared on ice and used at the indicated temperatures [46, 74, 87].

5. Oligonucleotides were synthesized on an Applied Biosystems, Inc. (Forest City, CA) Model 380B or 394A according to the β-cyanoethyl phosphoramidite chemistry method [12]. The newly synthesized oligo-nucleotides were purified by reverse phase high pressure liquid chromatography on a 4.6 x 250 mm C18 column (Rainin Instrument Co., Woburn, MA) The column was equilibrated with 0.02 M triethylammonium acetate, pH 7.0; and the elution was accomplished by a linear grad-ient of 5% to 30% acetonitrile in 12 min. This typically yields 100 pM/μl of 25-nt oligonucleotides [19].

6. Preparation of in situ Hybridization (ISH) Solutions for step #3: A minimum of 20 μl of final ISH working solution containing the oligoprimers is needed for each slide. The following stock and final working solutions are recommended:

Solution #1

Formulation	*Volumes*	*Final Concentration*
Deionized Formamide	*500 μl*	*50% Formamide*
20xSSC	*200 μl*	*4xSSC*
10mg/10 ml BSA	*200 μl*	*200μg BSA/ml*
38.5mg/5 ml DTT	*100 μl*	*5 mM DTT*

Total	*1 ml*	

Solution #2

Take 160 μl of solution #1 and add 2.5 μl of 40U/μl RNase Inhibitor (Boehringer Biochemica, Mannheim, FRG) to give a total of 100U in 162.5 μl. (Enough for 8 slides) [8-11].

Solution #3

Final working solutions: Makes 44 μl to 46 μl of ISH mixture containing Oligonucleotide. Enough for 2 slides when used at 20 μl/slide. Adjust volume for additional slides.

7. Preparation of in situ Transcription (IST) Solution: Prepare solution with cold reagents and hold on ice until used. 15 μl of final working IST solution is needed for each slide.

Solution #1

Formulation	*Volumes*	*Final Concentration*
0.05M Tris-HCl pH 6.5	*40 ml*	
$MgCl_2$.$6H_2O$ (MW:203.3)	*48 mg*	*6 mM*
KCl (MW:74.55)	*120 mg*	*40 mM*
DTT (MW:154.3)	*31 mg*	*5 mM*
BSA	*8 mg*	*200 μg/ml*

Total	*40 ml*	

Solution #2: Final Working Solution:

Solution #1	*140.0 μl*	
BM Digoxigenin-DNA Labeling Mix (10X)*	*15.0 μl*	*0.1mM/base*
BM RNase Inhibitor (40U/μl)	*2.5 μl*	*100U/163.5μl*
BM AMV Reverse Transcriptase (24U/μl)	*6.0 μl*	*150U/163.5μl*

Total	*163.5 μl*	

** = Boehringer Mannheim Biochemica [13, 21, 33, 41, 84, 94].*

8. In negative control slides the procedure was modified by: a) omitting the ISH Oligonucleotide step #3, b) omitting the DIG-DNA Labeling Mixture from IST step #6, c) omitting the AMV Reverse Transcriptase from IST step #6, or d) a combination of these omissions.

9. In the immunocytochemical detection of Digoxigenin all immunoreagents were diluted with 1% BSA in 0.05 M Tris buffer saline (TBS) pH 7.4 [60, 87, 95].

10. Preparation of Silver Enhancement Solutions: In the silver enhancement step, a 100 ml final working solution is prepared as follows:

Formulation	*Volumes*	*Final Concentration*
Solution #1:		
Citric Acid, monohydrate	*2.55 g*	*120 mM*
Sodium Citrate, dihydrate	*2.35 g*	*80 mM*
Distilled H_2O	*50 ml*	
Add 50% gum arabic (in H_2O)	*20 ml*	*10%*

Subtotal	*70 ml*	
Solution #2		
Hydroquinon	*0.85 g*	*77 mM*
Distilled H_2O	*15 ml*	

Subtotal	*15 ml*	
Solution #3		
Silver Lactate	*0.11 g*	*5.5 mM*
Distilled H_2O	*15 ml*	

Subtotal	*15 ml*	

Solution #4

Final Working Solution: First mix solutions #1 and #2, just before use add solution #3. The final working solution should be kept in the dark as much as possible, i.e., wrap working solution container and staining jars with aluminum foil.

Recommended protocols of *in situ* PCR

The published protocols for *in situ* PCR and *in situ* RT PCR are very different. Here we presented the protocols for detecting HIV RNA and DNA as an example to illustrate the techniques [96]. HIV ISPCR and HIV RT-ISPCR protocols take an average of three full working days and follow the same guidelines of other ISPCR methods. The same kits are used as for conventional PCR and RT-PCR. The technician's schedule must be planned in advance. An ISPCR laboratory should have a full-time and motivated staff.

Common materials and equipment. The equipment and reagents needed are identical to those for *in situ* hybridization plus reagents, enzymes and probes specified in the protocols. An *in situ* PCR machine is also required. For our purposes, we employed the *in situ* PCR Machine (PTC-100-16MS) for MJ Research Inc. (Watertown, MA). Other *in situ* PCR machines designed for slides can also be used.

Protocol #4: *In situ* polymerase chain reaction using HIV biotinylated probe on tissue sections.

1. Deparaffined tissue sections or cytological samples are digested with 30 μg/mL of proteinase K solution, inside a humid chamber at 37°C for 15 to 30 min.

2. Wash sections in double-distilled (dd) H_2O for 10 min twice.

3. Place slides on a block at 70-75°C for 1 min to inactivate the enzyme, followed by washing.

4. Prehybridize the sections inside a humid chamber at 42°C for 30 min using 50 μL of a solution composed of 12% dextran sulfate , 2x SSC, 0.12 mM EDTA and 0.33 mg/mL salmon sperm DNA.

5. Remove the excess prehybridization solution from each section and heat each of them to 75°C in a slide thermocycler (MJ Research).

6. Slowly add 50μL of PCR mixture to each section.

PCR Mixture:

	Volume	Final Concentration
ddH_2O	33.75 μL	
10x PCR Buffer II	5 μL	1x
25 mM $MgCl_2$	5 μL	2.5 mM
10 mM each dNTP	1 μL	200 μM each
25 mM each primer (SK38/39 or SK102/432)	1 μL	0.5 mM each
5 U/mL *Taq* DNA polymerase	0.25 μL	1.25 U/50 μL

7. Cover each section with a glass coverslip one at a time and completely seal the edges with an appropriate amount of transparent nail polish. Keep all the slides at 75°C until the PCR mixture has been added to the last one.

8. Start the thermocycling. The temperature parameters for ISPCR are 95°C 1.5 min (initial denaturing), 30 cycles of 95°C (denaturing) 30 s, 55°C (annealing) 45 s, and 72°C (extension) 30 s. This is followed by 72°C (final extension) 1.5 min, and storage at 4°C (soaking).

9. The nail polish is softened with acetone and the cover slips are carefully removed with a surgical blade. Immediately, the sections are washed in 5x SSC, 2x SSC and PBS for 5 min each.

10. Bake the sections in an oven at 60°C for 20 min.

11. One hundred microliters of hybridization solution are added to each section. Heat to 95°C, 5 min for denaturing. Avoid evaporation of the solution. Hybridize overnight in a humid chamber at 45°C.

Hybridization Solution
50% formamide
25% dextran sulfate
2x SSC
0.33 mg/mL salmon sperm DNA
HIV-1 biotinylated probe (SK19 or SK102 for SK38/39 and SK145/431 amplifications, respectively) at 250 to 400 pg/mL concentration.

12. Wash the sections with 5x SSC, 2x SSC and PBS for 5 min each.

13. Detection is performed using a kit for biotinylated probes (K600: DAKO, Carpenteria, CA) based on the linkage of streptavidin and biotinylated alkaline-phosphatase coupled to NBT/BCIP calorimetric reaction (blue color).

14. Once the detection is completed, wash the section in PBS for 5 min. If desired, slides may be slightly counter stained with Pyronin-Y, Nuclear Fast Red or Fast Green. These dyes should be dissolved in 2x SSC or PBS and not in ddH_2O.

15. Dry sections at 50°C in an oven and cover using permanent mounting media.

Notes to protocol # 4.

1. In step #1, the digestion time and temperature vary according to type of specimen and must be empirically verified.

2. In step # 7, avoid excessive nail polish; otherwise, slides will not fit into the compartments of the thermocycler. Use the pipet tip to adjust the cover slips. Bubbles trapped beneath the coverslip will usually come out by themselves during heating. Therefore, do not try to remove them by pressing the coverslip.

3. In step # 13, this reaction is carried out in darkness and monitored at about 15-min intervals under a light microscope, usually for no more than 1 h.

Protocol #5. Reverse transcriptase-*in situ* PCR on tissue sections. Method A. Reverse transcriptase-driven RT-ISPCR.

1. Deparaffined tissue sections or cytological specimens are digested with 30 μg/mL proteinase K solution inside a humid chamber at 30°C for 15 to 30 min. Digestion process may be varied (see above).

2. Sections are washed twice in ddH_2O for 10 min and heated to 70-75°C for 1 min to inactivate the enzyme.

3. Sections are pretreated with RNase-free DNase. Ten to twenty U/section incubating at 37°C for a minimum of 4 h. Overnight incubation is strongly advised to completely destroy the DNA.

4. The sections are extensively washed with several changes of ddH_2O for 20 min.

5. Ten microliters of RT mixture (GeneAmp RNA PCR Kit) are added to each section and incubated inside a humid chamber at room temperature for 15 min.

RT Mixture:

	Volume	Final Concentration
$MgCl_2$	2.0 μL	5 mM
10x Buffer II	1.0 μL	1x
ddH_2O	1.5 μL	-
dNTPs each	1.0 μL	1 mM
RNase inhibitor	0.5 μL	1 U/10 μL
Reverse transcriptase	0.5 μL	2.5 U/10 μL
Random hexamers	0.5 μL	2.5 μM
Total volume	10.0 μL	

6. Sections are incubated in a humid chamber at 42°C for 20 min.

7. Twenty microliters of prehybridization solution (same as used for HIV ISPCR) are added to each section.

8. Slides are placed in the slide thermocycler (MJ Research) set with one cycle above 95°C for 3 min (to inactivate the reverse transcriptase) and 5°C for 5 min.

9. Sections are incubated with the residual prehybridization solution in a humid chamber at 42°C for 20 min.

10. PCR mixture is added to the sections, 40 μL/section.

PCR Mixture

	Volume	Final Concentration
$MgCl_2$	2.0 μL	2 mM
10x PCR buffer	4.0 μL	1x
ddH_2O	37.75 μL	-
Taq DNA polymerase	0.25 μL	1.25 U/50 μL
Primer SK38	0.5 μL	0.25 μM
Primer SK39	0.5 μL	0.25 μM
Total volume	40 μL	

11. The amplification is performed in the slide thermocycler using the same parameters as for HIV ISPCR (see above).

12. Hybridization and detection are also performed as described in HIV ISPCR.

Method B. r*Tth* DNA polymerase-driven RT-ISPCR

1. Deparaffined tissue sections or cytological specimens are digested with 30μg/mL proteinase K solution inside a humid chamber at 37°C for 15 or 30 min.

2. Sections are washed twice in ddH_2O for 10 min and heated to 70-75°C for 1 min to inactivate enzyme.

3. RNase-free DNase pretreatment is performed as described above.

4. Sections are extensively washed in ddH_2O for 20 min.

5. Twenty microliters of RT mixture (Thermostable r*Tth* Reverse Transcriptase RNA PCR Kit; Perkin-Elmer) are added to each section and incubated inside a humid chamber at 70°C for 25 min.

RT Mixture

	Volume	Final Concentration
ddH_2O	11.5 μL	-
10x RT Buffer	2.0 μL	1x
$MnCl_2$	2.0 μL	1 mM
dNTPs each	0.4 μL	200 μM
r*Tth* DNA polymerase	2.0 μL	5 U/20 μL
"Downstream" primer SK39	1.0 μL	0.75 μM
Total volume	20.0 μL	

6. Next the humid chambers containing the slides are placed in a refrigerator at 4°C to stop the reaction.

7. Ten microliters of 12% dextran sulfate solution containing 1 mg of glycogen are added to each section.

8. Eighty microliters of PCR mixture are added to the sections.

PCR Mixture

	Volume	Final Concentration

ddH_2O	61.0 μL	-
10x Chelating buffer	8.0 μL	0.8x
$MgCl_2$	10.0 μL	2.5 mM
"Upstream" primer SK38	1.0 μL	0.25 μM
Total volume	80.0 μL	

9. Amplification is performed in the slide thermocycler (MJ Research) using the same parameters as for HIV ISPCR (see above).

10. Hybridization and detection are also performed as described for HIV ISPCR.

Notes to Protocol #5.

1. Radioisotope or digoxigenin-labeled in situ hybridization could be more sensitive than biotin-labeled in situ hybridization.

2. The probe in the hybridization solution (step 8) can be heated to 95°C for 5 min before being applied to the tissue section to eliminate any self-annealing or secondary-structure formation by the probes.

3. Up to step 10, procedure should be performed in an RNase-free manner.

4. The hybridization mixture can be applied to dehydrated dry sections or to wet sections. If the latter, there should be as little liquid remaining on the slides as possible to avoid disbursing concentrations of the ingredients in the hybridization solution.

5. For a hybridization solution of less than 40 μl per tissue section, a coverslip can be applied and sealed at the hybridization step (step 9) to prevent evaporation.

6. RNase may be used in posthybridization treatment to remove single-stranded probes, thereby reducing the background. The RNase can be applied at 100 μg of RNase A per ml and 1 U of RNase T_1 per μl for 30 min at 37°C.

7. Counterstain should not mask or overshadow the specific labeling.

Acknowledgments

The authors wish to thank Neera Agrawal, Nancyleigh Carson, Brett Levine, Michelle Forte and Howard Doughty for their technical assistance, Yanhui Chang, M.D., Ph.D. and Lan Su, D.M.D., Ph.D. for their critical comments and Gayle Enghund and Nelba Harris for typing the manuscript.

This work was supported in part by NIH grant NS-18775 to LED, and NIH grant HL-42975 and a grant from Patterson Foundation to JG.

References

1. Amasino RM (1986) Acceleration of nucleic acid hybridization rate by polyethylene glycol. Anal Biochem **152**: 304-307.

2. Anderson MLM, Young BD (1985) Quantitative filter hybridisation. In: Nucleic Acid Hybridisation: A Practical Approach. Hames BD, Higgins SJ (eds). IRL Press, Oxford, pp. 73-85.

3. Angerer RC, Cox KH, Angerer LM (1985) *In situ* hybridization to cellular RNAs. In: Genetic Engineering. Setlow JK, Hollaender A (eds). Plenum Press, New York, pp. 43-57,

4. Bains MA, Agarwal R, Pringle JH, Hutchinson RM, Lauder I (1993) Flow cytometric quantitation of sequence-specific mRNA in hemopoietic cell suspensions by primerinduced *in situ* (PRINS) fluorescent nucleotide labeling. Exptl Cell Res **208**: 321-326.

5. Baltimore D (1970) Viral RNA-dependent DNA polymerase. Nature **226**: 1209-1211.

6. Bauman JGJ, Bayer JA, van Dekken H (1990) Fluorescent in-situ hybridization to detect cellular RNA by flow cytometry and confocal microscopy. J Microsc **157**: 73-81.

7. Bayer JA, Baumann JGJ (1990) Flow cytometric detection of "b"-globulin mRNA in murine haemopoietic tissues using fluorescent *in situ* hybridization. Cytometry **11**: 132-143.

8. Blackburn P (1979) Ribonuclease inhibitor from human placenta: rapid purification and assay. J Biol Chem **254**: 12484-12487.

9. Blackburn P, Jailkhani BL (1979) Ribonuclease inhibitor from human placenta: interaction with derivatives of ribonuclease A. J Biol Chem **254**: 12488-12493.

10. Blackburn P, Wilson G, Moore S (1977) Ribonuclease inhibitor from human placenta. Purification and properties. J Biol Chem **252**: 5904-5910.

11. Bradford MM (1976) A rapid and sensitive method for the quantitation of microgram quantities of protein utilizing the principle of protein-dye binding. Anal Biochem **72**: 248-254.

12. Caruthers MH (1985) Gene systhesis machine: DNA chemistry and its uses. Science **230**: 281-285.

13. Chen EY, Seeburg PH (1985) Supercoil sequencing: a fast and simple method for sequencing plasmid DNA. DNA **4**: 165-170.

14. Chesselet MF (1991) *In situ* Hybridization Histochemistry. CRC Press, Boston.

15. Chien A, Edgar DB, Trela JM (1976) Deoxyribonucleic acid polymerase from the extreme thermophile *Thermus acquaticus*. J Bacteriol **127**: 1550-1557.

16. Choo KH (1996) *In Situ* Hybridization Protocols. Humana Press, Totowa, NJ.

17. Cohen SN, Chang ACY, Boyer H, Helling R (1973) Construction of biologically functional bacterial plasmids in vitro. Proc Natl Acad Sci USA **70**: 3240-3244.

18. Cox KH, DeLeon DV, Angerer LM, Angerer RC (1984) Detection of mRNAs in sea urchin embryos by *in situ* hybridization using asymmetric RNA probes. Dev Biol **101**: 485-502.

19. De Bault LE, Wang BL (1994) Localization of mRNA by *in situ* transcription and immunogold-silver staining. Cell Vision **1**: 67-70.

20. De Bault LE, Wang BL (1995) Oligonucleotide--primed *in situ* transcription and immunogold silver staining systems: Localization of mRNA in tissues and cells. In: *In Situ* Polymerase Chain Reaction and Related Technology. Gu J (ed). Eaton Publishing, Natick, Ma, pp. 99-114.

21. Feinberg AP, Vogelstein B (1983) A technique for radiolabeling DNA restriction endonuclease fragments to high specific activity. Anal Biochem **132**: 6-13.

22. Finnell RH, Bennett GD, Karras SB, Mohl VK (1988) Common hierarchies of susceptibility to the induction of neural tube defects in mouse embryos by valproic acid and its 4-propyl-4-pentennoic acid metabolite. Teratology **38**: 313-320.

23. Finnell RH, Moon SP, Abbott LC, Golden JA, Chernoff GF (1986) Stain differences in heat-induced neural tube defects in mice. Teratology **33**: 247-252.

24. Gall JG, Pardue ML (1969) Formation and detection of RNA-DNA hybrid molecules in cytological preparations. Proc Natl Acad Sci USA **63**: 378-383.

25. Gingras D, Bendayan M (1995) Colloidal gold electron microscopic *in situ* hybridization: Combination with immunocytochemistry for the study of insulin and amylase secretion. Cell Vision **2**: 218-225.

26. Gosden J, Hanratty D, Starling J, Fantes J, Mitchell A, Porteous D (1991) Oligonucleotide-primed *in situ* DNA synthesis (PRINS): a method for chromosome mapping, banding, and investigation of sequence organization. Cytogen Cell Gen **57**: 100-104.

27. Greer CE, Peterson SL, Kiviat NB, Manos MM (1991) PCR amplification from paraffin-embedded tissues. Effects of fixative and fixation time [see comments]. Am J Clin Pathol **95**: 117-124.

28. Gu J (1994) Principles and applications of *in situ* PCR. Cell Vision **1**: 8-19.

29. Gu J (1997) *In situ* hybridization as an adjunct method with immunohistochemistry. In: Manual of Clinical Laboratory Immunology. 5th ed. Gu J (ed). American Society for Microbiology Press, Washington, DC, pp. 414-421.

30. Gu J, D'Andrea M, Seethapathy M (1989) Atrial natriuretic peptide and its messenger ribonucleic acid in overloaded and overload-released ventricles of rat. Endocrinology **125**: 2066-2074.

31. Gu J, Linnoila RI, Seibel NL, Gazdar AF, Minna JD, Brooks BJ, Hollis GF, Kirsch IR (1988) A study of myc-related gene expression in small cell lung cancer by *in situ* hybridization. Am J Pathol **132**: 13-17.

32. Gu J, McGrath LB (1990) Localized endocrine conversion of ventricular cardiocytes in ventricular aneurysm. J Histochem Cytochem **38**: 1659-1668.

33. Gubler U, Hoffman BJ (1983) A simple and very efficient method for generating cDNA libraries. Gene **25**: 263-269.

34. Hacker GW, Grimelius L, Danscher G, Bernatzky G, Muss W, Adam H, Thurner J (1988) Silver acetate autometallography: An alternative enhancement technique for immunogold-silver staining (IGSS) and silver amplification of gold, silver, mercury and zinc in tissues. J Histotech **11**: 213-221.

35. Hacker GW, Zebbe I, Hainfeld J, Sallstrom J, Hauser-Kronberger C, Gruf AH, Su H, Dietze O, Bagasra O (1996) High-performance nanogold *in situ* hybridization and *in situ* PCR. Cell Vision **3**: 209-215.

36. Hacker GW, Zehbe I, Hauser-Kronberger C, Gu J, Graf AH, Dietze O (1994) *in situ* detection of DNA and mRNA sequences by immunogold-silver staining (IGSS). Cell Vision **1**: 30-37.

37. Harper ME, Marselle LM, Gallo RC, Wong-Staal F (1986) Detection of lymphocytes expressing human T-lymphotropic virus type III in lymph nodes and peripheral blood from infected individuals by *in situ* hybridization. Proc Natl Acad Sci USA **83**: 772-776.

38. Harrison PR, Conkie D, Affara N, Paul J (1974) *In situ* localization of globin messenger RNA formation. I. During mouse fetal liver development. J Cell Biol **63**: 402-413.

39. Hindkjar J, Koch J, Mogensen J, Pedersen S, Fischer H, Nygaard M, Junker S, Gregersen N, Kolvraa S, Bolund L (1991) *in situ* labelling of nucleic acids for gene mapping, diagnostics and functional cytogenetics. In: Advances in Molecular Genetics: Vol 4: Genome Analysis From Sequence to Function. Collins J, Driesel AJ (eds). Huthig Buch Verlag, Heidelberg, pp. 45-59.

40. Holgate CS, Jackson P, Cowen PN, Bird CC (1983) Immunogold-silver staining: New method of immunostaining with enhanced sensitivity. J Histochem Cytochem **31**: 938-944.

41. Houts GE, Miyagi M, Ellis C, Beard D, Beard JW (1979) Reverse transcriptase from avian myeloblastosis virus. J Virol **29**: 517-522.

42. Hu NT, Missings J (1982) The making of strand-specific M23 probes. Gene **17**: 271.(Abstract)

43. Itakura K, Rossi JJ, Wallace RB (1984) Synthesis and use of synthetic oligonucleotides. [Review]. Annu Rev Biochem **53**: 323-356.

44. Izant JG, Weintraub H (1984) Inhibition of thymidine kinase gene expression by anti-sense RNA: a molecular approach to genetic analysis. Cell **36**:

1007-1015.

45. Janssen K (1993) *In situ* hybridization and immunohistochemistry. In: Current Protocols in Molecular Biology. 7th ed. Janssen K (ed). Greene Publishing Associates, Inc. and John Wiley & Sons, Inc., U.S.A., pp. 14.1.4.

46. Janssen K (1993) *In situ* hybridization and immunohistochemistry. In: Current Protocols in Molecular Biology. 7th ed. Janssen K (ed). Greene Publishing Associates, Inc. and John Wiley & Sons, Inc., U.S.A., pp. 14.3.1-14.3.7.

47. Johnson MT, Johnson BA (1984) Efficient synthesis of high specific activity 35S-labelled human beta-globin pre-mRNA. BioTech **2**: 156 (Abstract).

48. Kearns WG, Pearson PL (1994) Fluorescent *in situ* hybridization using chromosome-specific DNA libraries. In: *In situ* Hybridization Protocols. Choo KGA (ed). Humana Press, Totowa, NJ, pp. 15-22.

49. Kendrew J (1994) The Encyclopedia of Molecular Biology. Blackwell Science, Inc., Cambridge, MA, p. 943.

50. Kievits T, Dauwerse JG, Wiegant J, Devilee P, Breuning MH, Cornelisse CJ, van Ommen GJB, Pearson PL (1990) Rapid subchromosomal localization of cosmids by nonradioactive *in situ* hybridization. Cytogen Cell Gen **53**: 134-136.

51. Kleppe K, Ohtsuka E, Kleppe R, Molineux I, and Khorana HG (1971) Studies on polynucleotides XCVI. Repair replication of short synthetic DNAs as catalyzed by DNA polymerases. J Mol Biol **56**: 341-361.

52. Koch J, Fischer H, Askholm H, Hindkjaer J, Pedersen S, Kolvraa S, Bolund L (1993) Identification of a supernumerary der(18) chromosome by a rational strategy for the cytogenetic typing of small marker chromosomes with chromosome-specific DNA probes. Clinical Genetics **43**: 200-203.

53. Koch J, Hindkjaer J, Mogensen J, Kolvraa S, Bolund L (1991) An improved method for chromosome--specific labeling of alpha satellite DNA *in situ* by using denatured double-stranded DNA probes as primers in a primed *in situ* labeling (PRINS) procedure. Genetic Analysis, Techniques & Applications **8**: 171-178.

54. Koch J, Mogensen J, Pedersen S, Fischer H, Hindkjaer J, Kolvraa S, Bolund L (1992) Fast one-step procedure for the detec-tion of nucleic acids *in situ* by primer-induced sequence-specific labeling with fluorescein-12dUTP. Cytogen Cell Gen **60**: 1-3.

55. Koch JE, Kolvraa S, Petersen KB, Gregersen N, Bolund L (1989) Oligonucleotide-priming methods for the chromosomespecific labelling of alpha satellite DNA *in situ*. Chromosoma **98**: 259-265.

56. Kogan SC, Doherty M, Gitschier J (1987) An improved method for prenatal diagnosis of genetic diseases by analysis of amplified DNA sequences. N Engl J Med **317**: 985-990.

57. Kohler G, Milstein C (1975) Continuous cultures of fused cells secreting antibody of predefined specificity. Nature (London) **256**: 495-497.

58. Kricka LJ (1992) Nonisotopic DNA Probe Techniques. Academic Press, San Diego, CA.

59. Langer PR, Waldrop AA, Ward DC (1981) Enzymatic synthesis of biotin-labeled polynucleotides: Novel nucleic acid affinity probes. Proc Natl Acad Sci USA **78**: 6633-6637.

60. Larsson LI (1989) Immunocytochemical detection systems. In: Immunocytochemis-try: Theory and Practice. 2nd ed. Larsson LI (ed). CRC Press, Inc., Boca Raton, Florida, pp. 77-146.

61. Larsson LI (1989) Fixation and tissue pretreatment. In: Immunocytochemistry: Theory and Practice. 2nd ed. Larsson LI (ed). CRC Press, Inc., Boca Raton, Florida, pp. 41-76.

62. Lawrence JB, Singer RH (1985) Quantitative analysis of *in situ* hybridization methods for the detection of actin gene expression. Nucleic Acids Research **13**: 1777-1799.

63. Lawrence JB, Taneja K, Singer RH (1989) Temporal resolution and sequential expression of muscle-specific genes revealed by *in situ* hybridization. Devel Biol **133**: 235-246.

64. Lewis ME, Baldino F Jr (1990) Probes for *in situ* hybridization histochemistry. In: *In situ* Hybridization Histochemistry. Chesselet MF (ed). CRC Press, Boca Raton, pp. 1-22.

65. Lichter P, Cremer T, Tang CJC, Watkins PC, Manuelidis L, and Ward DC (1988) Rapid detection of human chromosome 21 aberrations by *in situ* hybridization. Proc Natl Acad Sci USA **85**: 9664-9668.

66. Lichter P, Tang CJC, Call K, Hermanson G, Evans GA, Housman D, Ward DC (1990) High--resolution mapping of human chromosome 11 by *in situ* hybridization with cosmid clones. Science **247**: 64-69.

67. Longley J, Merchant MA, Kacinski BM (1989) *In situ* transcription and detection of CD1a mRNA in epidermal cells: an alternative to standard *in situ* hybridization techniques. J Invest Dermatol **93**: 432-435.

68. Lu LH, Gillett NA (1994) An optimized protocol for *in situ* hybridization using PCR-generated 33P-labeled riboprobes. Cell Vision **1**: 169-176.

69. Luke S, Belogolovkin V, Varkey JA, Ladoulis CT (1997) Fluorescence *in situ* hybridization (FISH). In: Analytical Morphology: Theory Applications and Protocols. Gu J (ed). Birkhouser Publishing, Boston, MA, pp. 139-174.

70. Luna LG (1986) Manual of Histologic Staining Methods of the Armed Forces Institute of Pathology. 3rd ed. McGraw-Hill Co, New York.

71. Melton DA, Krieg PA, Rebagliati MR, Maniatis T, Zinn K, Green MR (1984) Efficient *in vitro* synthesis of biologically active RNA and RNA hybridization probes from plasmids containing a bacteriophage SP6 promoter. Nucleic Acids Research **12**: 7035-7056.

72. Messing J (1983) New M13 vectors for cloning. Meth Enzymol **101**: 20-78.

73. Mitchell A, Jeppesen P, Hanratty D, Gosden J (1992) The organisation of repetitive DNA sequences on human chromosomes with respect to the kinetochore analysed using a combination of oligonucleotide primers and CREST anticentromere serum. Chromosoma **101**: 333-341.

74. Mogensen J, Kolvraa S, Hindkjaer J, Petersen S, Koch J, Nygaard M, Jensen T, Gregersen N, Junker S, and Bolund L (1991) Nonradioactive, sequence-specific detection of RNA *in situ* by primed *in situ* labeling (PRINS). Exptl Cell Res **196**: 92-98.

75. Morel G (1993) Hybridization Techniques for Electron Microscopy. CRC Press, Boca Raton, FL.

76. Mulder H, Ahren B, Sundler F (1995) Recent advances in localization and quantification of gene expression - a lesson from islet amyloid polypeptide. Cell Vision **2**: 94-103.

77. Mullis KB, Faloona F, Scharf SJ, Saiki RK, Horn GT, Erlick HA (1986) Specific enzymatic amplification of DNA in vitro. Cold Spring Harbor Symposium, Quantitative Biology **51**: 263-273.

78. Nuovo GJ (1994) *In situ* Hybridization: Protocols and Applications. Raven Press, New York.

79. Nuovo GJ, Gallery F, MacConnell P, Becker J, Bloch W (1991) An improved technique for the *in situ* detection of DNA after polymerase chain reaction amplification. A J Pathol **139**: 1239-1244.

80. Perrett C (1993) The molecular principles of *in situ* hybridization. In: Hybridization Techniques for Electron Microscopy. Morel G (ed). CRC Press, Boca Raton, FL, pp. 13-44.

81. Rentrop M, Knapp B, Winter H, Schweizer J (1986) Aminoalkylsilane-treated glass slides as support for *in situ* hybridization of keratin cDNAs to frozen tissue sections under varying fixation and pretreatment conditions. Histochem J **18**: 271-276.

82. Saiki RK, Scharf S, Faloona F, Mullis KB, Horn GT, Erlich HA, Arnheim N (1985) Enzymatic amplification of beta-globin genomic sequences and restriction site analysis for diagnosis of sickle cell anemia. Science **230**: 1350-1354.

83. Sheehan DC, Hrapchak BB (1980) Processing of tissue: dehydrants, clearing agents, and embedding media. In: Theory and Practice of Histotechnology. Sheehan DC, Hrapchak BB (eds). 2nd ed. Battelle Press, Columbus, pp. 59-78.

84. Shimomaye E, Salvato M (1989) Use of avian myeloblastosis virus reverse trans-criptase at high temperature for sequence analysis of highly structured RNA. Gene Analysis Techniques **6**: 25-28.

85. Su L (1993) Detection of keratin gene expression in normal and neoplastic epithelia by *in situ* hybridisation with specifically constructed riboprobes. Doctoral Thesis. University of London, UK.

86. Su L, Morgan PR (1994) Discrepancies between keratin mRNA and protein expression in mucosal and glandular epithelia by *in situ* hybridization and immunohistochemisty. Cell Vision **1**: 146-153.

87. Tecott LH, Barchas JD, Eberwine JH (1988) *In situ* transcription: specific synthesis of complementary DNA in fixed tissue sections. Science **240**: 1661-1664.

88. Temin HM, Mizutani S (1970) RNA-dependent DNA polymerase in virions of Rous sarcoma virus. Nature **226**: 1211-1213.

89. Trask BJ (1991) Fluorescence *in situ* hybridization: applications in cytogenetics and gene mapping. Trends in Genetics **7**: 149-154.

90. Van Gelder RN, von Zastrow ME, Yool A, Dement WC, Barchas JD, Eberwine JH (1990) Amplified RNA synthesized from limited quantities of heterogeneous cDNA. Proc Natl Acad Sci USA **87**: 1663-1667.

91. Wang BL, De Bault LE (1994) Co-localization of γ-glutamyl transpeptidase and its mRNA in paraffin embedded sections of rat kidney: A light microscopic study using enzymecytochemistry, immunocytochemistry, *in situ* transcription, and immunogold-silver staining systems. Cell Vision **1**: 122-130.

92. Wang BL, Grammas P, Ringer DP, Kong J, De Bault LE (1995) Localization of γ-glutamyl transpeptidase and its mRNA in a spontaneous tube-forming clone of endothelial cells from rat cerebral resistance-vessel endothelium. Cell Vision **2**: 196-202.

93. Ward DC, Menninger J, Lieman J, Desai T, Banks A, Boyle A, Bray-Ward P, Haaf T (1994) Integration of the physical, genetic and cytogenetic maps of human chromosomes: Implications for the development of diagnostic DNA probes. Cell Vision **1**: 61-66.

94. Weih F, Stewart AF, Schutz G (1988) A novel and rapid method to generate single stranded DNA probes for genomic footprinting. Nucleic Acids Research **16**: 1628.

95. Zehbe I, Hacker GW, Rylander E, Sallstrom J, Wilander E (1992) Detection of single HPV copies in SiHa cells by *in situ* polymerase chain reaction (*in situ* PCR) combined with immunoperoxidase and immunogold-silver staining (IGSS) techniques. Anticancer Research **12**: 2165-2168.

96. Zevallor EA, Bard E, Anderson VM, Choi TS, Gu J (1995) Conventional PCR, *in situ* PCR and reverse transcription *in situ* for HIV detection. In: *In situ*

Polymerase Chain Reaction and Related Technology. Gu J (ed). Birkhauser Publishing, Boston, MA, pp. 77-98.

Discussion with Reviewers

Reviewer I: Among different fixatives which one would you prefer for expression analyses of a particular gene in the tissue sections?
Authors: 10% formalin or 4% paraformaldehyde in 0.01 M PBS adjusted to pH 7.4.

Reviewer I: Would it be possible to amplify a particular genetic element from a tissue section which has been fixed in paraformaldehyde or formalin for longer period of time like 5-10 years?
Authors: It is possible to amplify targets in formalin or paraformaldehyde fixed archive tissue samples. One example was reported by Isaacson *et al.* (1994) who examined polio, measles, influenza and HTLV-1 in archival brain tissue samples with *in situ* RT-PCR and detected their RNA sequences in individual neurons, glia and vascular endothelial cells. Some of the paraffin tissue blocks were over 25 years old. However, it should be pointed out that the degree of success with archival material is dependent on the completeness of the fixation and the nature of the target sequence.

Reviewer I: You mentioned that the concentration of probes in the range of 200-1500 ng/ml depended on the size of the probe. How does the size of the probe relate to concentration, is it empirical or certain defined rules that govern the concentration and size of the probe?
Authors: The concentration in this case was given in ng/ml. At a given concentration, the larger the probe, the fewer the number probes available per ml of solution. Thus, the concentration should be increased proportionally with the increase in probe length to ensure that a sufficient number of probes are available for hybridization.

Reviewer II: The authors state that single stranded (ss) DNA probes are superior to double stranded (ds) DNA probes because they do not self-anneal. The hybridization signal's intensity ultimately reflects the number of reporter nucleotides on the probe-target complex. For a target of 8,000 base pairs (a small virus), a ds probe can easily be made that covers this entire region; this is much more difficult with the ss DNA probe. In my experience ds DNA probes usually give the best results; this is most notable with the ultimate small single stranded DNA probe, i.e, the oligoprobe. Do the authors have any direct experience showing that a ss DNA probe is superior to a ds DNA probe and, if so, would this be expected for large targets (>1000b)?
Authors: This depends on the size of the probe and the target and how the probes are labeled. When the single stranded and double stranded probes are the same size and identically labeled, single stranded probes are more efficient as they do not normally anneal to themselves. In the case of a large target where large single stranded probes are difficult to make, double stranded probes will be superior. On the other hand, longer probes will find less targets than shorter ones, thereby having decreased detecting sensitivity but giving a more specific signal.

Reviewer II: In the section on trouble-shooting for *in situ* hybridization, the authors state that probe design is a very important factor. In my experience, nick translation and random primer usually give good results for probe synthesis for any DNA template of >500 bp. Define probe design and explain its role in troubleshooting. In the case of no signal with *in situ* hybridization, I would recommend that the investigator use his/her labeling system and an alu DNA probe tem-plate, as the repetitive alu sequence is present as thousands of copies in mammalian cells. Do the authors agree with this strategy or do they recommend some other method to deter-mine the cause of a lack of signal with *in situ* hybridization that involves changing probe design?
Authors: Probe design in this context refers to the selection of a particular fragment of the sequence complementary to the target and the length of this sequence, rather than the way the probes are labeled. In our experience, when experimenting with a new target, repeated negative signals, even after modifying the hybridization conditions, call for a new probe to a different portion of the target. Certain probes work well and others do not, i.e., either too faint a signal or too much a background, even if, in theory, the new probe should work equally well. This may have to do with the degree of the uniqueness of the probe sequence and the way the target sequences are folded and embedded in the tissue sample, presenting different availability of the target sequences to the particular probes selected. The suggested alu DNA probe templet approach is a good method to check the labeling efficiency, the general hybridization conditions and target (alu sequences) availability, but give little indication to the suitability of the particular probe that gives weak or no signal. In addition to employing the alu sequence, shifting the target sequence or using a collection of different sequences to different or overlapping portions of the target should be tested. By changing the probe design, i.e., to use a different probe sequence, we have solved the problems of false negative hybridization in a number cases while other modifications produced little success.

M. Malecki: Do you perform liquid phase PCR as a

way to optimize annealing temperature for a specific target?

Reviewer I: Besides the critical parameters like Proteinase K, probes, and primers, do you think optimization of annealing temperature is also critical, and if it is so, how can one proceed with optimization of the system?

Authors: Optimization of annealing temperature is also very important, of course. This can be established by keeping the other factors constant and testing a number of temperature settings. It can also be established by liquid phase PCR with the same target and primers. Nevertheless, the temperature should not vary too much from target to target.

Reviewer II: Background signals with *in situ* hybridization are due to nonspecific sticking of the labeled probe to cellular proteins and/or nucleic acids. That is, the probe, which presumably enters all cells, will diffuse out unless it finds its complementary target or non-specifically sticks to cellular components. From a practical viewpoint, I would argue that the most likely causes of background are: a) posthybridization wash not stringent enough, where temperature is a key factor; b) probe concentration too high. The questions are: 1) why do the protocols not include a temperature for the post hybridization wash. 2) Most protocols do not include a prehybridization wash, which, in my experience, does not affect background. Have the authors done comparative studies of background with and without prehybridization?

Authors: Background signals with *in situ* hybridization are caused by nonspecific or less specific sticking of the labeled probes to cellular proteins and/or nucleic acids or may be trapped by cytoskeleton structures that in theory can be removed by critical washing. Temperature, stringency of the washing solution, washing duration and vigilance are all factors affecting the efficiency of probe removal from the tissue sample. Temperature is one of the factors that affect the outcome significantly. We usually raise the temperature up to 50°C if the initial washing does not remove most of the background staining. The optimal condition should be established empirically for each case, therefore no particular temperature is recommended. We have performed comparisons between protocols with and without prehybridization washes and found that prehybridization washing reduced background staining is some cases, presumably by blocking potential nonspecific bonding sites in the tissue sample. We also know that this prehybridization washing does not affect the efficacy of amplification nor does it increase background staining. We routinely perform prehybridization washing for both *in situ* hybridization and *in situ* PCR.

Reviewer II: With IST, one is making labeled cDNA inside the cell. Do the authors know of instances where this labeled product diffused out of the cell and into another cell type known to not have the target?

Authors: We know of no published case where the labeled cDNA product of IST diffused into another cell type not containing the target sequence. In IST, the cDNA is literally synthesized on the target sequence creating a double stranded nucleic acid duplex with little or no mismatched NTP pairs. Such double stranded duplexes are very stable and require strong denaturing conditions to separate the new cDNA from its complementary strand, i.e., high melting temperatures well in excess of the 45°C used in the last 10 min of the transcription step, high stringency conditions in excess of 2xSSC at room temperature plus 0.05xSSC at 35°C.

Reviewer II: The authors state that extensive washing of the protease and DNAse is needed with *in situ* PCR. However, in my experience, a 1 minute wash in water and a 1 minute wash in 100% ethanol is enough to eliminate these enzymes. Have the authors demonstrated any change in the *in situ* PCR signal relative to the length of wash after one or both of these enzymes?

Authors: You may be right, but we prefer to be on the safe side. When there are so many steps that can go wrong we want to be sure that the experiments are not ruined by easily avoidable mishaps at the beginning of the protocol. We also know that extensive washing does no harm to the targets and subsequent reactions.

Reviewer II: The authors claim that one can make millions of copies of the amplicon in the cell during *in situ* PCR. However, most investigators report a 200+ fold increase in copy number, and not a million fold increase. It would, on a theoretical basis, seem unlikely that one could amplify one million copies in a space of 10μ, given the extremely high concomitant amplicon concentration which is one of the limits on the extent of amplification in a 100μl reaction volume. What is the basis for the authors claim for such a marked increase in copy number during *in situ* PCR?

Authors: It is true that PCR on tissue sections or intact cells is not as efficient as liquid phase PCR due to the limited accessibility of the target and the interference of fixed tissue structures as well as the smaller amount of solution (we add about 50-100 μl of PCR solution to the slide depending on the size of the sample). But once a sequence is amplified, more than a few hundred copies will result within the next ten cycles. Nobody knows, even roughly, how many copies of amplicons there are after amplification *in situ*. It is difficult, if not impossible, to measure. By theory, the amplicons are free floating and they diffuse out freely following each cycle of amplification, therefore leaving rooms for new amplicons to form. We believe that the estimated 200 plus copies of the amplicons by others are not grounded. One thing appears to be true, i.e, there are not too many copies left *in situ* by the time the signals are examined under the microscope. Judging by the intensity and the size of the reporting signal, there should not be more than a few hundred copies of the ampli-cons remaining in their original location. Nevertheless, this does not mean that only those amplicons are produced by the PCR, as most of them have diffused or been washed away. There is no reasonable mechanism to keep the amplicons remaining *in situ*, except the possible long amplicon theory proposed by one of the authors (JG) and some possible trapping and network-forming phenomenon to hold the amplicons on site. We speculate that most amplicons diffuse to the supernatant easily.

Reviewer II: The authors state that back diffusion is a formidable problem with *in situ* PCR. Back diffusion would presumably be due to the large amount of labeled amplicon in the overlying solution sticking to cellular proteins and/or nucleic acids. This is equivalent to every *in situ* hybridization reaction, where one has a large amount of labeled DNA that can stick non-specifically to cellular components. This can be removed by a high stringent wash, owing to the weak force of these bonds versus the much stronger hydrogen bonds with 100% homology between probe and target. Explain why back diffusion (background) is any more of an issue with *in situ* PCR versus *in situ* hybridization.

Authors: The answer to this question is related to the issue discussed in a previous question ("In the section on troubleshooting ..."). In the situation of *in situ* PCR, the concentrations of the amplicons are much higher than that used for *in situ* hybridization, and increasingly so until a plateau is reached. This large amount of amplicons will diffuse to other parts of the tissue sample and may stick to them, semi-specifically or non-specifically by bonding to proteins, or simply be trapped and tangled at nonspecific sites. In addition, the variations in sizes of the amplicons may "stick" to non-specific sites more easily than the more uniformly sized probes in the solution of *in situ* hybridization. More-over, the tissue samples for *in situ* PCR tend to be more harshly treated than those used in straight *in situ* hybridization. For these reasons, the so-called back diffusion may be more readily seen during *in situ* PCR than during *in situ* hybridization. It should also be noted that back diffusion does not occur only with labeled amplicons, but unlabeled amplicons produced in the so-called indirect *in situ* PCR protocol can also cause nonspecific sticking and subsequently be picked up by hybridization step used in the detection process resulting in nonspecific staining.

Reviewer I: Between the two *in situ* PCR strate-gies, i.e., direct incorporation of labeled molecules into amplified products or indirect labeling like addition of labeled probe, which one is better and why? Would you please comment on this?

Authors: We prefer the indirect methods, as do most investigators. It has an added step to check the specificity. Since some nonspecific sequences are amplified in PCR, they will not be picked up by the specific probes in the hybridization step used in the detection system. It should also be pointed out that in the indirect method, nonspecific DNA repair, etc. will not affect the final results. This has been discussed extensively in a number of publications.

Reviewer II: An important measure of specificity with ISH, IST, or *in situ* PCR is afforded by knowing which cell type likely contains the target of interest. There is often striking localization of the target to specific cell types. An example would be parvoviral infections where the target (the nucleated red blood cell) is easily identified on morphologic grounds. According to published reports, with *in situ* hybridization and reverse transcriptase (RT) *in situ* PCR, the signal only localizes to the target cell, assuming a high stringent wash. If the wash is not stringent enough, then other cell types show signals with both methods. Explain why background should be any more of a problem with *in situ* PCR versus *in situ* hybridization, where an excess of probe far greater than can be synthesized during the cycling process is present, and why it is not evident in many published reports of viral infection and *in situ* PCR where the target cell is known?

Authors: This question has been partially answered in a previous question ("The authors state that back diffusion ...."). To start with, *in situ* PCR is usually employed only after the conventional *in situ* hybridization failed to detect any signal convin-cingly. The extremely high detecting sensitivity of *in situ* PCR leads the investigators to a new territory where more marginal positivity may be present and positive signals may show up in unexpected cell types. When those occur, extensive controls are called for and the new results may still not be black and white. These all make the interpretation of the *in situ* PCR and, in particular, RT *in situ* PCR results more difficult. For target mRNA, it is often a matter of difference in quantity among different cell types rather than a yes or no answer. We speculate that when the target locations are known, the authors are more confident to publish their results and the articles are more likely to be accepted by journal referees. The many, less than clear cut observations, even though they may be closer to the truth, were buried in the lab's notebooks and data bases.

M. Malecki: Would you be able to estimate the distances between the target sequences and the reporter molecules in ISH, IST, and ISPCR?
Reviewer III: Could you analyze differences in the distribution of reporter molecules around the target sequence labeled by IST and ISPCR as compared to ISH?
Authors: The distances between target sequences and reporter molecules in different methods depend on many factors, including the size of the probes, the size of the target, the type of tissue, the labeling methods, the signal detecting methods, the visualization approach, etc. It all depends on how much the targets have been amplified and retained and what projection range the reporter signal build-up has achieved by the particular detecting method. It can range from the immediate proximity as in IST, to a signal that can engulf the entire cell by one target sequence in IS PCR. In general, *in situ* hybridization gives more localized signals than the other two techniques.

Reviewer I: In the HIV ISPCR protocol, prehybridization is recommended before doing actual ISPCR. Does it provide certain (extra) advantage(s) for *in situ* PCR amplification?
Authors: We routinely perform prehybridization for *in situ* hybridization and find this step reducing background labeling. For indirect *in situ* PCR, prehybridization is also necessary. We perform this step before the PCR cycles rather than afterwards, i.e., directly before the subsequent *in situ* hybridization. This way, we avoided the prehybridization step between the PCR cycles and the hybridization step and believe that this is beneficial for preventing diffusion of or washing away the amplicons from their original sites.

Reviewer I: In the HIV ISPCR protocol, in step #9 acetone is recommended for softening of nail polish. Most people use absolute ethanol for this purpose. Do you think acetone is better over ethanol?
Authors: It is our preference. Both will work. We have not made a systematic comparison.

Additional Reference

Isaacson SH, Asher DM, Gibbs CJ, Gajdusek DC (1994) *In situ* RT-PCR amplification in archival brain tissue. Cell Vision 1: 84.

Scanning Microscopy Supplement 10, 1996 (Pages 49-55)
Scanning Microscopy International, Chicago (AMF O'Hare), IL 60666 USA

0892-953X/96$5.00+.25

PREPARATION OF SAMPLES FOR POLYMERASE CHAIN REACTION *IN SITU*

Gerard J. Nuovo*

MGN Medical Research Laboratories, Setauket, NY

(Received for publication October 1, 1996 and in revised form December 20, 1996)

Abstract

The purpose of this paper is to describe the key variables in sample and reagent preparation needed for successful polymerase chain reaction (PCR) *in situ*. Tissue or cell preparations should be fixed in a cross linking fixative, such as 10% buffered formalin, preferably from 15 to 48 hours. Tissues should be embedded in paraffin; cell preparations can be fixed when near confluence, then physically removed and processed. When possible three samples (4 μM tissue sections or 1-5000 cells) should be placed on silane coated glass slides. Digestion in pepsin (2 mg/ml) for 30 min is adequate for DNA detection by PCR *in situ* hybridization whereas optimal protease digestion time is variable and related to formalin fixation time for reverse transcriptase (RT) *in situ* PCR. RT *in situ* PCR requires an overnight digestion with DNase. The amplifying solution should contain 4.5 mM $MgCl_2$, 0.05% bovine serum albumin, and, for RNA analysis, the reporter nucleotide. A false positive signal would be evident with incorporation of the reporter nucleotide for DNA targets due to DNA repair; this can be avoided with frozen, fixed tissues and the hot start maneuver. Otherwise, one needs to use a labeled probe and a hybridization step to detect amplified DNA targets in paraffin embedded tissues.

Key Words: polymerase chain reaction *in situ*, *in situ* hybridization, formalin, fixative, hot start.

*Address for correspondence:
Gerard J. Nuovo
MGN Medical Research Laboratories
8 Huckleberry Lane
Setauket, NY 11733
Telephone Number: (516)-941-3183
FAX Number: (516)-941-3549
E-mail: gnuovomd@mem.po.com

Introduction

The purpose of this paper is to provide a concise, simple reference for sample and reagent preparation for those investigators who wish to do PCR *in situ*. This discussion will presuppose that the reader has minimal experience with cell and tissue preparation in a pathology laboratory. The basis of much of the information presented in this manuscript can be found in several publications [4, 6, 7, 8, 9, 10, 11]. Readers who wish a more in-depth discussion of the different variables important for PCR *in situ* are referred to a textbook [2]. Finally, a video presentation is available for those wishing to see more directly some of the technical, hands-on features of doing PCR *in situ* [3].

The key variables for PCR *in situ*

Introductory statements. From the standpoint of the technique itself and reproducibility of the data PCR *in situ*, in my opinion, is no more difficult than solution phase PCR or *in situ* hybridization. However, it does have a large disadvantage when compared to either of these other two techniques. Many people who do solution phase PCR usually have much expertise in molecular analyses but, often, do not have much experience working with intact tissue sections. Similarly, those who do *in situ* hybridization often have a strong background in histology but not as extensive an experience with other molecular-based methodologies. To achieve the maximum advantages with PCR *in situ*, one needs to have some knowledge of tissue preparation and histologic interpretation as well as basic concepts of molecular analyses in general and PCR in particular.

Definition of terms. There are many different approaches to amplifying DNA or cDNA inside an intact cell. One fundamental difference is whether one detects the amplified DNA or cDNA by incorporating a labeled reporter nucleotide into the amplicon as it is being synthesized or by use of a labeled probe. The following terms relate to this important point:

Polymerase chain reaction (PCR) *in situ*: A general term for amplifying DNA or cDNA inside an intact cell.

PCR *in situ* hybridization: Detection of the amplicon by doing a hybridization step after PCR using a labeled probe. This is best suited to detection of DNA targets in paraffin embedded tissue.

In situ PCR: Detection of the PCR product via a labeled nucleotide (or, less commonly, a labeled primer) that is incorporated into the amplicon during DNA synthesis. This can not be done for a specific DNA target in paraffin embedded tissues due to the ubiquitous presence of primer independent DNA synthesis in such material. *In situ* PCR is the basis, however, for RNA detection.

Reverse transcriptase (RT) *in situ* PCR: Detection of a specific cDNA target via a labeled nucleotide that is incorporated into the amplicon during amplification. A key step with RT *in situ* PCR is DNase digestion, which, after optimal protease digestion, eliminates the various DNA synthesis pathways which may be operative during *in situ* PCR.

Primer independent DNA synthesis: DNA synthesis operative during *in situ* PCR in cells or tissue that have been exposed to dry heat (e.g., 65°C for 15 min) prior to PCR. The DNA synthesis is due to repair of single stranded gaps. This DNA repair pathway is always operative in paraffin embedded tissues due to the obligatory heating step [2, 10].

Sample Preparation

The glass slide

Cell and tissue adherence is an important issue with PCR *in situ* as the repeated cyclings at elevated temperatures would tend to separate the sample from the slide. Fortunately, one can obtain near 100% adherence of the tissue or cell sample using silane coated slides. Organosilane (Sigma, St. Louis, MO), can be purchased separately and diluted to 2% in acetone for those wishing to pretreat the slides themselves. I recommend that one purchases slides already pretreated in silane, which are readily available and inexpensive from many commercial sources (e.g., ONCOR, Gaithersburg, MD or Enzo Diagnostics, Farmingdale, NY). In my experience, silane coated slides may be used years after they are prepared with no loss of tissue adherence. Other pretreatment regimes to improve tissue adherence are available (e.g., glue, SOBO, L-lysine). However, in my experience, these do not work as well as silane.

The fixative

Fixatives are used in the pathology lab to render native degradative enzymes inoperative and to harden tissues such that they can be cut into thin (1-6 μm sections) for histologic analysis. There are two classes of fixatives which comprise nearly all such material used in the surgical pathology laboratory - cross linking fixatives and denaturing fixatives. The most common cross linking fixative formulation is 10% buffered formalin. Over 95% of surgical specimens in most surgical pathology laboratories will be fixed in buffered formalin. Other cross linking fixatives include glutaraldehyde, routinely used for electron microscopy, and paraformaldehyde. Acetone and 95% ethanol are examples of denaturing type fixatives. They are rarely used in diagnostic pathology, although ethanol is by far the most common fixative used in diagnostic cytopathology. Another category of fixatives that one may encounter in diagnostic pathology include formalin based fixatives to which has been added either picric acid (Bouin's solution) or a heavy metal such as mercury (Zenker's solution).

The optimal fixative for PCR *in situ* is 10% buffered formalin. Most surgical specimens will be fixed from 8 to 24 hours. If one can control the fixation time, it is recommended to use 24-48 hours as this will allow for a broader range of optimal protease digestion times for RT *in situ* PCR, which will be discussed shortly. Do not use fixatives which contain picric acid or a heavy metal, as they cause extensive degradation of the DNA and preclude successful amplification [2, 5, 10]. One can obtain a signal with PCR *in situ* using ethanol or acetone fixation. Such fixatives have the advantage that a protease digestion step is not needed, as compared to the same tissue type fixed in formalin [9]. However, in my experience, cell localization, critical for interpretation with RT *in situ* PCR, is not as distinct with the denaturing fixatives. Further, the detection rate is less than 100% with *in situ* PCR and denaturing fixatives when analyzing for a target that is known to be present in all cell types. This should be contrasted with formalin fixed material where, after optimal protease digestion, the detection rate will be 100%. Under these conditions, if one analyzes the over lying amplifying solution, the amplicon will not be detectable as compared to ethanol or acetone fixed material, where the amplified product is readily detectable in the overlying solution [9].

Unfixed, cryostat sections are sometimes used in immunohistochemistry because certain epitopes may be destroyed after formalin fixation. If one wishes to use such sections with PCR *in situ*, it is recommended that the slides be fixed overnight in 10% buffered formalin prior to amplification. Indeed, one can do target specific direct incorporation hot start *in situ* PCR for a DNA target using such material. This is because such material lacks the primer independent DNA synthesis pathway as it was never exposed to dry heat [2, 10].

Cells versus tissue sections

Tissue sections should be processed in the routine manner of surgical pathology laboratories: 3-4 four μm sections placed on a given silane coated slide. The multiple sections on one slide allows for direct compari-

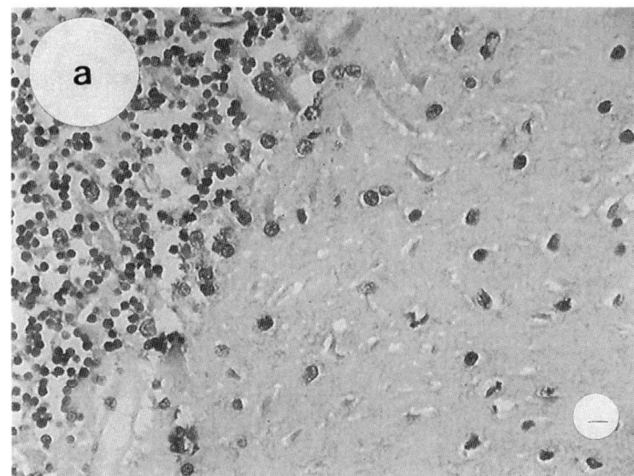

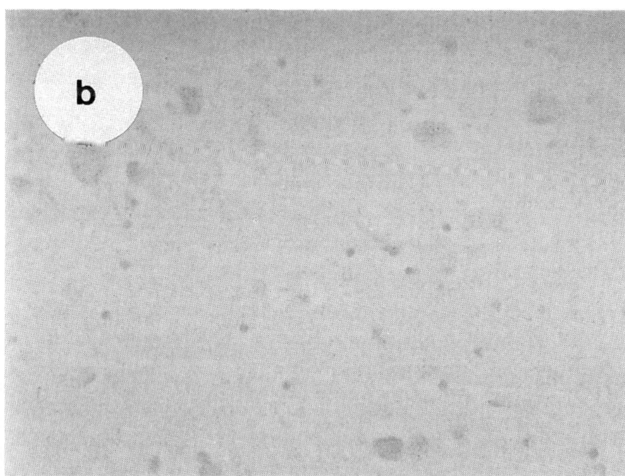

Figure 1. Determination of the optimal protease digestion time. This brain tissue was digested with pepsin (2 mg/ml) for 30, 60, and 90 min. At 90 min, an intense signal was evident in most cell types (**a**). This non specific signal was eliminated by overnight digestion in DNase (**b**), demonstrating that 90 min is the optimal protease digestion time for doing RT *in situ* PCR in this tissue section (see Fig. 4). The bar in each figure represents 10 μm.

1. Optimal protease digestion time

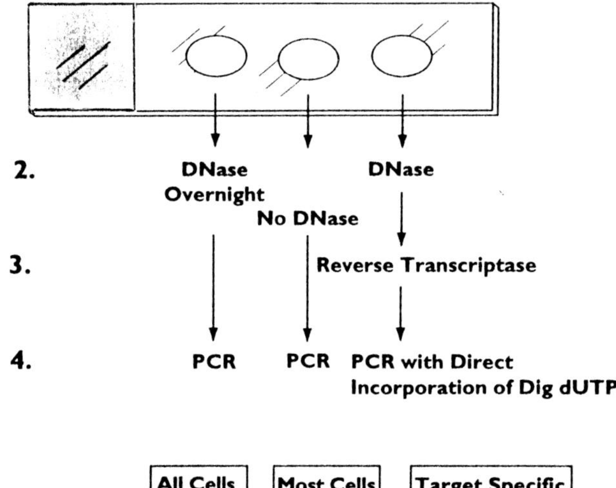

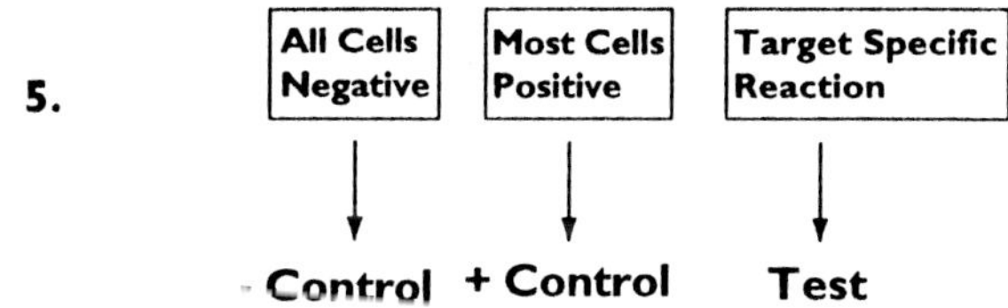

6. If the morphology is poor, then less protease
If the - control has + cells, then more protease

Figure 2. RT *in situ* PCR protocol. After determining the optimal protease digestion time, RT *in situ* PCR can be done by performing the positive control (no DNase), negative control (DNase, no RT or RT with irrelevant primers) and the test (DNase and RT) on the same glass slide.

son in the same cells of the critical positive and negative controls.

One has several options when working with cell preparations. If grown in a tissue culture plate, one can added sterile, silane coated slides (or coverslips) and have the cells grow on them directly. Alternatively, one can grow the cells to near confluence, remove the growth media, and then add an ample amount of 10% buffered formalin for 2-3 days. The cells can then be scraped off the plate, the solution placed in a conical tube, and the cells washed twice in diethylpyrocarbonate (DEPC) (RNase free) treated water. The cells can then be resuspended to a concentration of about 5000 cells per 50 μl, and three 50 μl "spots" can be placed on the silane slide and allowed to dry. For those doing RNA analysis, I recommend exposing the cells to dry heat at 60°C for 15 min to induce DNA repair; this allows for a stronger positive control, as discussed below. For those doing DNA analysis, do **not** expose the cells to dry heat if you wish to do target specific *in situ* PCR.

Protease digestion

If ones uses formalin fixed material, as recommended, then protease digestion will be needed to obtain success with PCR *in situ* [2, 9, 10, 11]. There are a variety of proteases that are routinely used in the diagnostic pathology laboratory. These include pepsin, trypsin, proteinase K, and pronase. It is recommended that one choose one of these proteases and use it exclusively to become familiar with its nuances.

I prefer pepsin (or trypsin) over proteinase K as it can be inactivated by increasing the pH to 8.3 and a

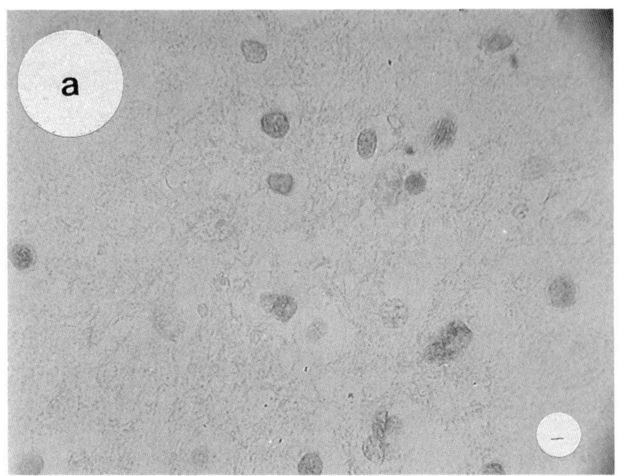

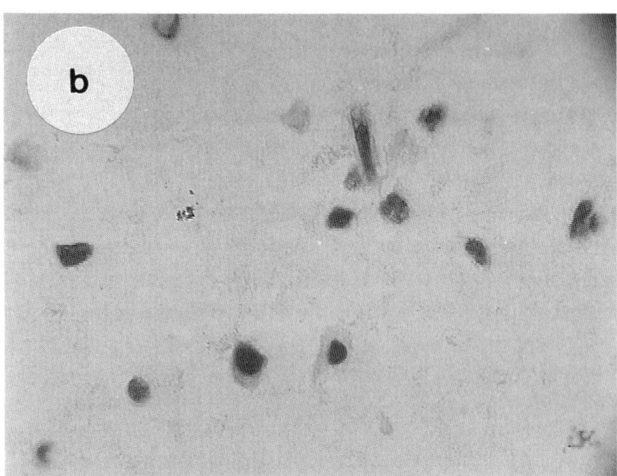

Figure 3. Persistence of non specific signal with DNase digestion after suboptimal protease digestion. This is the same brain tissue illustrated in Fig. 1. A signal is not evident with the no DNase positive control (**a**) if the protease digestion time is suboptimal, e.g., 30 min. However, note that a signal is evident under these conditions with the negative control (DNase, no RT) (**b**). The nuclear based signal in most cell types demonstrates that the signal is non specific.

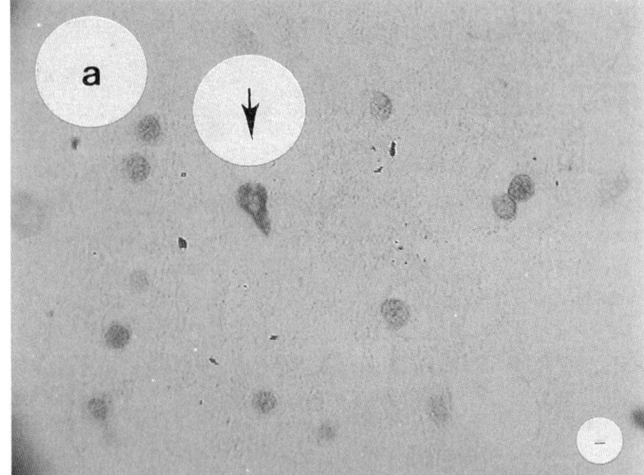

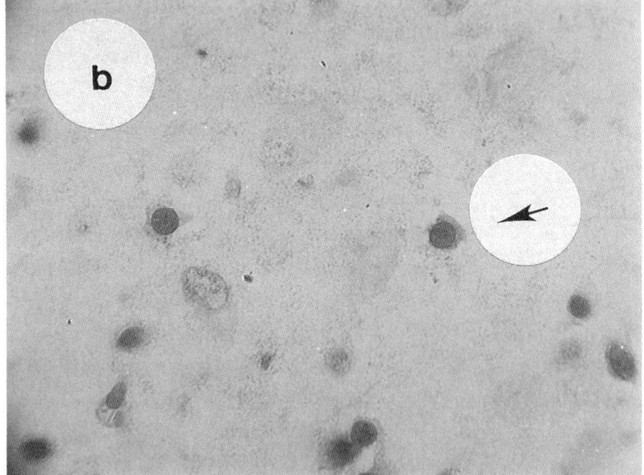

Figure 4. Specific localization of the signal with RT *in situ* PCR. This brain tissue is the same illustrated in Fig. 1. After optimal protease digestion, note that the signal for tumor necrosis factor mRNA localizes to the cytoplasm of a few cells (**a**). This is to be contrasted with the nuclear based signal for HIV-1 RNA (**b**) from this case of a person with severe AIDS dementia.

simple rinse. Also, pepsin is less likely to over digest the sample, which precludes successful PCR *in situ* (unpublished observations). If one does use proteinase K, do not dry heat inactivate as this will render the DNA unsuitable for subsequent hybridization [2].

To prepare the protease solution, take 20 mg of pepsin and add 9.5 ml DEPC water and 0.5 ml of 2 N HCl. The pepsin should dissolve immediately. Store 1 ml aliquots at -20°C for up to one week. Thaw at either room temperature or 37°C and use immediately after thawing.

The optimal protease digestion time for *in situ* PCR and PCR *in situ* hybridization is very similar for a wide range of tissues whether they have been fixed for a short period of time (e.g., 8 hours) or a long period of time (e.g., 1 week). In most of these instances, 20-30 min of digestion with pepsin at room temperature will allow for successful *in situ* PCR or PCR *in situ* hybridization [2]. It is worthwhile noting that 20-30 min of pepsin digestion is adequate for over 95% of routinely processed surgical pathology specimens analyzed by standard *in situ* hybridization [2]. To remove the protease, simply wash the solution off the glass slide using sterile water, wash in sterile water for another minute, then wash in 100% ethanol for 15 seconds, and allow to air dry.

The optimal protease digestion time is much less

straightforward for RT *in situ* PCR. The likely explanation relates to the need for DNase digestion with RT *in situ* PCR. That is, the function of the protease with RT *in situ* PCR is not only to permit entry of the DNA polymerase but also to render the entire genomic DNA susceptible to DNase digestion! Longer formalin fixation times likely increase the number of protein-DNA cross links which must be removed for DNase to degrade the DNA. This may explain the strong relationship between formalin fixation and protease digestion time for successful RT *in situ* PCR [2, 10].

To determine the optimal protease digestion time for a given sample for RT *in situ* PCR, it is recommended that one place 3 sections on a given silane coated glass slide and digest each a different time with pepsin. For routinely processed surgical pathology material, it is recommended that one start with 20 min, 40 min, and 60 min of pepsin digestion. Then, do *in situ* PCR. For paraffin embedded tissues, it will not matter which primer pair one uses as the primer independent DNA synthesis pathway will be operative. Whichever protease digestion times gives an intense signal in at least 50% of the cells will be the optimal for that particular specimen [2]. To corroborate that this is the optimal protease digestion time, one should digest 2 of the samples for that time, and digest one overnight in DNase (described shortly). The definitive proof of optimal protease digestion time is an intense signal in most cells that is completely eliminated by DNase digestion (Fig. 1). This simple statement is the foundation of RT *in situ* PCR (Fig. 2). The elimination of the genomic based DNA synthesis pathways by DNase digestion after optimal protease digestion will allow for target specific direct incorporation of the reporter nucleotide into the cDNA of interest. It cannot be stressed enough that suboptimal protease digestion will NOT permit for cDNA target specific incorporation during RT *in situ* PCR due to persistence of the genomic, nuclear based DNA synthesis pathways (Fig. 3) [2, 10]. To remove the protease for RT *in situ* PCR, simply wash the solution off the glass slide using DEPC water, wash in DEPC water for another minute, then wash in 100% ethanol for 15 sec, and allow to air dry. I do not follow routine RNase precautions until the protease digestion step, and then use gloves and RNase free materials until the RT step is completed.

DNase digestion

After optimal protease digestion, the genomic DNA can be degraded such that it can no longer serve as a template for DNA synthesis during PCR by DNase digestion. Although 7 hours of DNase digestion can accomplish this task, it is recommended that one do this step overnight [2]. The DNase solution can be prepared by diluting both the PCR buffer II (Perkin Elmer, Norwalk, CT) and the RNase free DNase (Boehringer Mannheim, Indianapolis, IN) 1:10 with DEPC water. To remove the DNase for RT *in situ* PCR, wash the solution off the glass slide using DEPC water, wash in DEPC water for another minute, then wash in 100% ethanol for 15 sec, and allow to air dry.

Reagent conditions

When preparing the amplifying solution for PCR *in situ* hybridization, one can use the same formulation of buffer, nucleotides, and primers as for solution phase PCR. However, one must adjust the concentration of $MgCl_2$ and taq polymerase for optimal PCR *in situ*; further, it may be advantageous to include bovine serum albumin (BSA) in the amplifying solution to block adsorption of the taq polymerase to the glass, thus raising the effective concentration of the enzyme [9]. Optimal magnesium concentration for PCR *in situ* is 4.5 mM. This increased concentration relative to solution phase PCR applies to a wide variety of primer pairs and targets [2, 9]. The optimal concentration of the taq polymerase is 25 U/50 μl of amplifying solution. One can use 5 U/50 μl of the taq polymerase if BSA is added to the amplifying solution [2, 9]. The recipe that is recommended for PCR *in situ* hybridization follows [2]:

5 μl of GeneAmp buffer (the reagents are from the PCR *in situ* kit from Perkin Elmer)
9 μl of $MgCl_2$ (25 mM stock)
8 μl dNTP (final concentration 200 μM)
1.5 μl 2% BSA
3 μl of primer pair (20 μM stock)
22.5 sterile water
1 μl Taq polymerase (hot start, see below)

The recipe that is recommended for RT *in situ* PCR (from the EZ rTth kit from Perkin Elmer) follows [2, 4]:

10 μl EZ buffer
1.6 μl each of dATP, dCTP, dGTP, dTTP (final concentration 200 μM)
1.6 μl of 2% bovine serum albumin
1.0 μl of RNasin
3.0 μl of primers 1 and 2 (20 μM stock) (for the negative control, use nonspecific primers or omit the primers)
0.6 μl digoxigenin dUTP (1 mM stock)
14.6 μl DEPC water
12.4 μl 10 mM $MnCl_2$
2.0 μl rTth

Hot start maneuver

There are several DNA synthesis pathways which may be operative during PCR, either in solution phase or *in situ*. Of course, the pathway specific for the DNA or cDNA target of interest is the one we wish to be operative. However, the primers may initiate DNA synthesis after annealing to non target DNA (mis priming) or to themselves (primer oligomerization). In either instance, a large amount of DNA may be synthesized which may interfere with target specific amplification. The other DNA synthesis pathway which may be operative during PCR *in situ* is primer independent and is induced by dry heating of the cells (7), which initiates the formation of single stranded DNA gaps. These gaps can serve as "surrogate" primers and initiate DNA synthesis during *in situ* PCR in any paraffin embedded tissue since such specimens are routinely heated during processing.

If one inhibits both mis priming and primer oligomerization during *in situ* PCR, target specific DNA synthesis is enhanced, even if the primer independent pathway is still operative [2, 6, 7]. Inhibition of the two primer dependent nonspecific DNA synthesis pathways is readily accomplished by realizing that their melting temperatures (Tms) will be much lower than for primer-target annealing, due to their lesser homology. That is, if one withholds either the taq polymerase or some essential reagent (such as magnesium) from the amplifying solution until the temperature of the reaction denatures the primer-non target and primer-primer hybrids, without denaturing the primer-target hybrids, then these primer dependent non specific DNA pathways can be blocked. This can be achieved by withholding the taq polymerase until 55°C; hence, the term hot start PCR [1, 2, 6, 7].

The hot start maneuver is needed for routine detection of 1 DNA target per cell using PCR *in situ* hybridization [2, 6, 7]. It is not needed for RT *in situ* PCR, as DNase digestion eliminates all the genomic-based DNA synthesis pathways.

One can either manually withhold the taq polymerase until the temperature of the cycler reaches 55°C, or add some chemical which inhibits primer annealing to non target DNA and other primers. The latter can be achieved with single stranded binding proteins [2, 7, 9] or antibodies against taq polymerase (unpublished observations). The antitaq polymerase antibodies prevents polymerase activity at room temperature; at >55°C, the antibodies are denatured and the taq polymerase can then function.

Cycling parameters

The recommended cycling parameters for PCR *in situ* hybridization and RT *in situ* PCR (using the one step rTth) are listed below. A key point to remember is that one is more likely to inactivate the taq polymerase with PCR *in situ* relative to solution phase PCR due to the much greater surface to volume ratio. Thus, the denaturing times and temperatures should be kept as short as possible.

The cycling parameters for RT *in situ* PCR are [2, 4]:

Incubate at 65°C for 30 min; denature at 94°C for 3 min; cycle at 60°C for 1.5 min, 94°C for 45 sec; do 20 cycles.

The cycling parameters for PCR *in situ* hybridization are [2]:

Incubate at 55°C for manual hot start; add taq polymerase, denature at 94°C for 3 min; cycle at 60°C for 1.5 min, 94°C for 45 sec; do 35 cycles, then do hybridization step.

References

[1] Chou Q, Russell M, Birch DE, Raymond J, Bloch W (1992) Prevention of pre-PCR mis-priming and primer dimerization improves low-copy-number amplifications. Nucl Acid Res **20**: 1717-1723.

[2] Nuovo GJ (1996). PCR *in situ* hybridization: Protocols and Applications, 3rd edition. Lippincott-Raven Press, New York, NY.

[3] Nuovo GJ (1996) Keys to successful *in situ* PCR (video). Lippincott-Raven Press, New York, NY.

[4] Nuovo GJ, Forde A (1995). An improved system for reverse transcriptase *in situ* PCR. J Histotech **18**: 295-299.

[5] Nuovo GJ, Silverstein SJ (1988) Comparison of formalin, buffered formalin, and Bouin's fixation on the detection of human papillomavirus DNA extracted from genital lesions. Lab Invest **59**: 720-724.

[6] Nuovo GJ, MacConnell P, Forde A, Delvenne P (1991) Detection of Human Papillomavirus DNA in formalin fixed tissues by *in situ* hybridization after amplification by the polymerase chain reaction. Am J Pathol **139**: 847-854.

[7] Nuovo GJ, Gallery F, MacConnell P, Becker J, Bloch W (1991) An improved technique for the detection of DNA by *in situ* hybridization after PCR-amplification. Am J Pathol **139**: 1239-1244.

[8] Nuovo GJ, Gorgone G, MacConnell P, Goravic P (1992) *In situ* localization of human and viral cDNAs after PCR-amplification. PCR Method Applic **2**: 117-123.

[9] Nuovo GJ, Gallery F, Hom R, MacConnell P, Bloch W (1993) Importance of different variables for optimizing *in situ* detection of PCR-amplified DNA. PCR Method Applic **2**: 305-312.

[10] Nuovo GJ, Gallery F, MacConnell P (1994) Analysis of non-specific DNA synthesis during *in situ* PCR. PCR Meth Applic **4**: 342-349.

[11] Yoo BJ, Selby MJ, Choe J, Suh BS, Choi SH, Joh JS, Nuovo GJ, Lee H, Houghton M, Han JH (1995). Transfection of a differentiated human hepatoma

cell line (Huh7) with in vitro transcribed hepatitis C Virus (HCV) RNA and establishment of a long term culture persistently infected with HCV. J Virol **69**: 32-38.

Discussion with Reviewers

M. Malecki: Have you noticed any differences in the distribution of the reporter molecules around the target sequence labeled by *in situ* PCR and PCR-*in situ* hybridization?
Author: No. With adequate protease digestion, DNA based signals are evident in the nucleus. RNA based signals vary from nuclear (HIV-1), subnuclear (peri-nucleolar or nucleolar for premRNAs), nuclear membrane (hepatitis C), and cytoplasmic (tumor necrosis factor (TNF) mRNA). However, with overprotease digestion, these various signals tend to become cytoplasmic. A signal on the cytoplasmic membrane usually reflects background; i.e., sticking of labeled primer oligomers from the amplifying solution onto the cell surface due to an inadequate post PCR wash.

Reviewer I: Would you be able to evaluate the ranges of distances between the target sequences and the reporter molecules in *in situ* PCR and PCR-*in situ* hybridization depending on amplification protocols?
Author: The reporter nucleotide is directly incorporated into the target sequence during *in situ* PCR; about 1 labeled nucleotide will be added for every 20 nucleotides of the amplicon. For PCR-*in situ* hybridization, the label is included on the probe. If the probe is made by nick translation or random priming, then the ratio of label to unlabeled nucleotides will be about the same as for *in situ* PCR. If terminal transferase is used, then from 1 to 5 labeled nucleotides will be present on the probe.

Reviewer II: In sample preparation sometimes epithelial cells like human endothelial cells need specific treatment like coating of the surface with different attachment factor. Do you think that silane coating will interfere with attachment factor if one is growing these cells on silane coated slides?
Author: No. My collaborators and I have obtained good results growing cells on silane coated slides.

Reviewer II: If a choice of fixatives is available which one would you recommend for tissue sections and which for established cell lines?
Author: I would strongly recommend 10% buffered formalin, 2-3 days, for each. This will allow one a broader optimal protease window for RT *in situ* PCR.

Reviewer II: In proteinase K digestion some researchers have mentioned appearance of small dots in the membrane peripheral area, have you experienced the same in your proteinase K digestion strategies?
Author: I do not rely on dots, presumably representing early breakdown of the cell membrane from over protease digestion. Rather, I rely on the primer independent signal to show me what is the optimal protease for RT *in situ* PCR. This is less important for PCR *in situ* hybridization, where 15-30 min of digestion with pepsin at 2 mg/ml will be adequate for most samples. I describe this in detail in reference [2]. Briefly, one can digest paraffin embedded (i.e., tissues exposed to heat) tissues for different times with a given protease, do *in situ* PCR or even isothermal DNA synthesis for 15-30 min, and determine optimal protease based on the strength of the signal.

Reviewer II: In definition of terms you mention *in situ* PCR as detection of the PCR product via a labeled nucleotide and less commonly by a labeled primer. Do you think that labeled nucleotide strategy is advantageous over a labeled probe?
Author: For RNA analysis - yes, for DNA analysis on paraffin embedded tissues, definitely no. The advantage is the time saved and the simplicity of not having to do a hybridization step. The DNase digestion after optimal protease digestion allows on to do target specific incorporation into the cDNA with RT *in situ* PCR. However, for DNA targets, the primer independent signal invariably present in paraffin embedded tissues necessitates detection by a probe step.

Reviewer II: In reagent conditions, you recommend the same formulation of buffer, nucleotides and primers as for solution phase PCR which is quite comprehensible, as far as concentration of these reagents is concerned you only take into consideration the $MgCl_2$ concentration and taq polymerase. What about nucleotides and primer concentration as we know that in the cellular milieu the conditions are not the same as in a tube. Would you recommend higher concentration of primers and probes for *in situ* PCR?
Author: We have varied the concentration of the primers up to 9 μM (about 10 times greater than the protocol we recommend) with no enhancement of the signal [8]. Indeed, the increased risk of primer oligomerization and, potentially, mispriming may be disadvantageous. We have not varied the nucleotide concentration. With regards to the probe, when using an oligoprobe and PCR-*in situ* hybridization, the concentration is a key factor in background and signal.

Scanning Microscopy Supplement 10, 1996 (pages 57-60)
Scanning Microscopy International, Chicago (AMF O'Hare), IL 60666 USA
0892-953X/96$5.00+.25

GENERATION OF HIGH EFFICIENCY ssDNA HYBRIDIZATION PROBES BY LINEAR POLYMERASE CHAIN REACTION (LPCR)

Gregory W. Konat

Department of Anatomy, West Virginia University School of Medicine, Morgantown, WV

(Received for publication October 2, 1995 and in revised form July 10, 1996)

Abstract

The polymerase chain reaction (PCR) methodology can be employed to produce DNA hybridization probes. The major advantages of this paradigm over other techniques include superior specific activity of the probes, the versatility of sequence selection, the ability to produce short probes, and the simplicity of the procedure. We have further improved the efficiency of PCR probes by generating single stranded (ssDNA) probes that do not reanneal with themselves in solution, and hence, their availability for the interaction with the complementary sequences of the target is profoundly increased. Protocols for ^{32}P-dCTP labeled and digoxigenin-dUTP labeled probes have been elaborated to maximize the incorporation rate of the label as well as to provide for the production of full-length probes. The ssDNA probes may be particularly suitable for nucleic acid detection in tissues by *in situ* hybridization.

Key Words: ssDNA hybridization probes, linear polymerase chain recation (PCR), digoxigenin-labeled probes, radiolabeled probes.

*Address for correspondence:
Gregory W. Konat
Department of Anatomy
West Virginia University School of Medicine
4052 Health Sciences North
P.O.Box 9128, Morgantown, WV 26506-9128.
Telephone number: (304) 293-2212
FAX number: (304) 293-8159
E-mail: Gkonat@wvu.edu

Introduction

The quality of hybridization probes is a decisive factor in the sequence-specific detection and/or quantitation of nucleic acids. The polymerase chain reaction (PCR) (Saiki *et al.*, 1985, 1988) provides a convenient technique to generate hybridization probes that in many aspects are superior to those prepared by other techniques (Konat *et al.*, 1994). Both internal labeling, i.e., the incorporation of labeled dNTPs into the extending strand, and end labeling, i.e., the extension of labeled primer can be performed. Protocols for both isotopic (Jansen and Ledly, 1989; Schowalter and Sommer, 1989; Bednarczuk *et al.*, 1991; Konat *et al.*, 1991; Blakeley and Carman, 1991) and nonradioactive variety (Day *et al.*, 1990; Lo *et al.*, 1988; Konat *et al.*, 1991; Liesack *et al.*, 1990; Lion and Haas, 1990; Lanzillo, 1990; Emanuel, 1991; Tabibzadeh *et al.*, 1991; Uchimura *et al.*, 1991) are available. The detection sensitivity of such probes is comparable to riboprobes (Schowalter and Sommer, 1989). Other major advantages of PCR include the versatility of sequence selection, the ability to produce discrete short probes, the requirement for only small amounts of even unpurified DNA template, and the simplicity of the procedure.

The efficiency of PCR probes can further be improved by generating single stranded (ssDNA) probes (Bednarczuk *et al.*, 1991; Konat *et al.*, 1991) that do not reanneal with themselves in solution, and hence, their availability for the interaction with the complementary sequences of the target is profoundly increased. For example, the use of ssDNA probes results in an approximately 8-fold increase in the signal intensity as compared to double stranded (dsDNA) probes as assessed by Northern blot analysis (Bednarczuk *et al.*, 1991; Fig. 1). The reaction conditions for the generation of ^{32}P-dCTP-labeled and digoxigenin-dUTP labeled probes have been elaborated to maximize the incorporation rate of the label as well as to provide for the production of full-length probes (Konat *et al.*, 1991). The ssDNA PCR technology that yields discrete, high efficiency probes may be particularly suitable for nucleic acid detection in tissues by *in situ* hybridization.

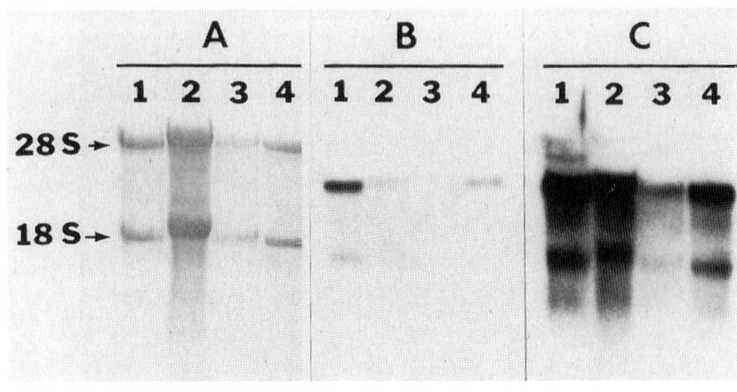

Figure 1: Hybridization efficiency of ssDNA versus dsDNA PCR probes. Total rat brain RNA was isolated by the acidic guanidinium thiocyanate-phenol-chloroform method (Chomczynski and Sacchi, 1987), size fractionated by electrophoresis on 1% agarose gel, vacuum transferred onto a Nytran membrane (Schleicher and Schuell, Keen, NH), and UV cross-linked. The membranes were stained in 0.02% methylene blue to visualize the RNA (panel A). The membranes were subsequently hybridized with radiolabeled probes, and washed at high stringency (0.2XSSC, 0.1% SDS, 60°C, 60 min). Both, the dsDNA probe (panel B) and the ssDNA probe (panel C) were generated using 250 μCi of α^{32}P-dCTP (Du Pont NEN, Boston, MA). The autoradiograms show Northern blots following 4 h exposure with an intensifying screen. The positions of 28S and 18S ribosomal RNA are the same in all three panels. The lanes 1 and 4 contained 8 and 3 μg, respectively, of total RNA from 3-week-old rat brain, while the lanes 2 and 3 contained 24 and 2 μg, respectively, of total RNA from 3-month-old rat brain. Densitometric quantitation revealed that ssDNA hybridization probes resulted in approximately 8 times higher signal intensities as compared to dsDNA probes.

Reproduced from Bednarczuk *et al.* (1991) by permission of Eaton Publishing Co., 154 E. Central St., Natick, MA 01760-3644 USA.

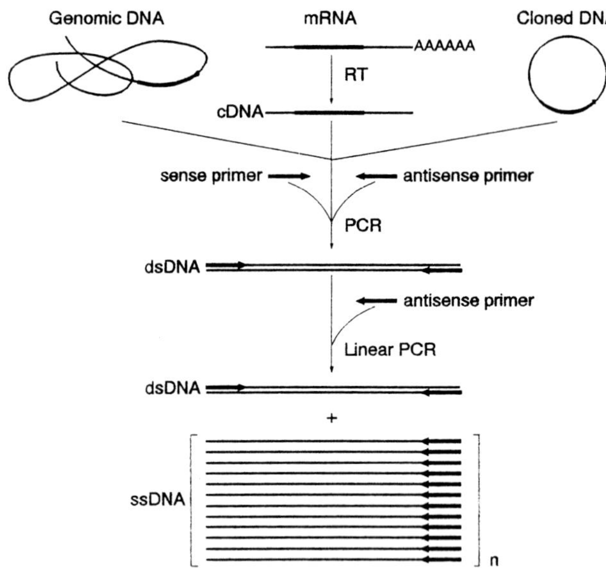

Figure 2: Flow diagram for generation of hybridization probes by linear PCR. In the initial step, the DNA fragment of interest is amplified either from reversely transcribed (RT) mRNA, from cloned DNA, or from genomic DNA by conventional PCR using a pair of specific primers. This double stranded (dsDNA) fragment is subsequently used as a template for the generation of single stranded DNA molecules (ssDNA) by linear PCR with only one, e.g., antisense primer. Because the ssDNA is arithmetically amplified, the number of ssDNA strands (n) generated on one molecule of dsDNA template corresponds to the number of PCR cycles. The labeling of ssDNA probes is attained by the incorporation of labeled nucleotides into the linear PCR reaction.

Materials and Methods

The strategy for the production of ssDNA probes is schematized in Fig. 2, and detailed protocols for the generation of ^{32}P-dCTP-labeled and digoxigenin-dUTP labeled probes (Konat *et al.*, 1991) are provided below.

Materials and equipment

GeneAmpT DNA Amplification Reagent Kit is obtained from Perkin Elmer Cetus (Norwalk, CT). ^{32}P-α-dCTP (>3,000 Ci/mmol) is from ICN Biochemicals (Irvine, CA). Digoxigenin-11-dUTP (DIG-dUTP) and Quick Spin G-50 columns are purchased from Boehringer Mannheim (Indianapolis, IN), and Elutip-d from Schleicher and Schuell (Keen, NH). Oligonucleotide primers (20-mers) are custom synthesized and HPLC-purified by Operon Technologies (Alameda, CA). All PCR amplifications are performed in a DNA Thermal Cycler (Perkin-Elmer Cetus, Norwalk, CT) in standard 0.5 ml eppendorf tubes. The reagents and equipment of comparable quality obtained from other commercial suppliers may also be used.

Preparation of dsDNA template

A DNA fragment is amplified by the conventional two-primer PCR from an appropriate DNA, e.g., reversely transcribed RNA, genomic digest, or cloned inserts (Fig. 2). The amplicon is subsequently isolated by 1% agarose gel electrophoresis and purified by Elutip-d elution using the manufacturer's protocol. The dsDNA template is quantitated spectrophotometrically at 260 nm.

Generation of radiolabeled ssDNA probes

The PCR amplification reaction contains in a total volume of 25 μl: 1X PCR buffer (10 mM Tris-HCl, pH 8.3, 50 mM KCl, 1.5 mM $MgCl_2$, 0.01% gelatin, pH 8.3), 20 μM of each dATP, dTTP and dGTP, approximately 25 ng of dsDNA template, 1 μM primer (antisense), 2 μM ^{32}P-dCTP (150 μCi), and 0.62 units of Taq polymerase. The mix is overlaid with approximately 25 μl of mineral oil, and processed through 30 cycles at the following profile: (1) denaturation for 2 min at 94°C, (2) annealing for 2 min at appropriate temperature (determined for a particular template/primer set), (3) extension for 5 min at 72°C.

Subsequently, the reaction mix is diluted with 25 μl water, and the probe is purified from unincorporated nucleotides by passing the mix through a Quick Spin G50 column. Although the best results are obtained with freshly prepared probes, the probe can be stored for several days at -20°C.

Generation of digoxigenin-labeled ssDNA probes

The PCR amplification reaction contains in a total volume of 100 μl: 1X PCR buffer (10 mM Tris-HCl, pH 8.3, 50 mM KCl, 1.5 mM $MgCl_2$, 0.01% gelatin, pH 8.3), 100 μM of each dATP, dCTP and dGTP, 90 μM dTTP, 10 μM DIG-dUTP, approximately 400 ng of dsDNA template, 3 μM primer (antisense), and 7.5 units of Taq polymerase. The mix is overlaid with approximately 25 μl of mineral oil, and processed through 45 cycles at the following profile: (1) denaturation for 1.5 min at 94°C, (2) annealing for 1.5 min at appropriate temperature (determined for a particular template/primer set), (3) extension for 2 min at 72°C.

Subsequently, the aqueous phase is mixed with 0.1 vol of 4 M LiCl and 3 vols of ethanol, and placed at -20°C for 30 min. The DNA is spun down (12,000 g, 20 min), and the pellet is washed with 100 μl 70% ethanol, air-dried, dissolved in TE buffer, and stored at -20°C.

Commentary

The identity of templates prepared from complex nucleic acid mixtures, i.e., genomic digest or total RNA should be confirmed by restriction and sequence analysis. The removal of unincorporated dNTPs and primers from dsDNA template is another critical step required for successful generation of high quality ssDNA probes. Although the procedure described in Materials and Methods yields highly purified templates, alternative methods can also be employed.

The ssDNA protocols employ a linear (arithmetic) amplification paradigm in which the production of probes is proportional to the number of thermal cycles, and to the amount of dsDNA template. Thus, relatively high concentrations of the template have to be used to maximize the generation of the probes. Excessive amounts of template, however, are undesirable as they result in higher proportion of unlabeled DNA, and consequently may reduce the intensity of hybridization signal through annealing with the labeled probe.

The experimental conditions for individual template/primer sets may require adjustment to optimize the quality (average length) and the quantity (yield) of the probes (Konat *et al.*, 1991). This pertains chiefly to the annealing temperature and magnesium concentration. Furthermore, some DNAs may require DMSO or other agents of choice as well as hot start to attain efficient amplification/labeling. In general, the optimal conditions found for the dsDNA template amplification should also be optimal for the generation of ssDNA probes.

In the radiolabeled protocol the reaction conditions are suboptimal in regard to dNTP, and especially, dCTP concentration. Consequently, the radiolabeling reaction requires a 5 min extension time to ensure high proportion of full-length probe. Under these conditions, approximately 50% yield of ^{32}P-dCTP incorporation is reached after 30 cycles of amplification (Konat *et al.*, 1991), which is comparable to the efficiency of incorporation observed for dsDNA probes (Jansen and Ledley, 1989). For some applications (e.g., *in situ* hybridization) ^{35}S or ^{3}H labeled nucleotides can be incorporated using the same protocol.

In the nonradioactive probe paradigm the template concentration is increased four fold as compared to the radiolabeling protocol to maximize the production of the ssDNA probe at high dNTP concentration. The optimal concentration of both Taq polymerase and the primer is three times higher, whereas the concentration of dNTPs is 50% lower than the standard concentrations for exponential (two primer) PCR (Saiki, 1989). Under these conditions, approximately 8 μg of full-length digoxigenin-labeled ssDNA probe (60% incorporation) can be generated in a 100 μl reaction using 45 cycles (Konat *et al.*, 1991). The specific labeling of these probes can be further augmented by increasing the ratio of DIG-dUTP to dTTP up to 1:2 (Emanuel, 1991). The DIG-labeled probes preserve their activity for at least one year when stored at -20°C (Lion and Haas, 1990).

The versatility of ssDNA technique combined with high efficiency of the probes will undoubtedly be valuable for the detection of nucleic acids in tissues. For example, short discrete probes specific to any segment of a message or a gene can be easily produced by selecting appropriate primers. Such discrete probes will be highly desirable for detection of differentially spliced messages, or genetic alleles. The simplicity of sequence selection will also help to overcome possible problems of disadvantageous secondary structure of certain regions of nucleic acids that may impede hybridization. Furthermore, if required, both sense and antisense probes can be conveniently generated from the same dsDNA template by extending one of the two primers.

References

Bednarczuk TA, Wiggins RC, Konat G (1991) Generation of high efficiency, single-stranded DNA hybridization probes by PCR. BioTechn **10**: 478.

Blakeley MS, Carman MD (1991) Generation of an S1 probe using arithmetic polymerase chain reaction. BioTechn **10**: 52-51.

Chomczynski P, Sacchi N (1987) Single-step method of RNA isolation by acidic guanidinium thiocyanate-phenol-chloroform extraction. Anal Biochem **162**: 156-159.

Day PJR, Bevan IS, Gurney SJ, Young LS, Walker MR (1990) Synthesis *in vitro* and application of biotinylated DNA probes for human papilloma virus type 16 by utilizing the polymerase chain reaction. Biochem J **267**: 119-123.

Emanuel JR (1991) Simple and efficient system for synthesis of non-radioactive nucleic acid hybridization probes using PCR. Nucleic Acid Res **19**: 2790.

Jansen R, Ledley FD (1989) Production of discrete high specific activity DNA probes using the polymerase chain reaction. Gene Anal Techn **6**: 79-83

Konat G, Laszkiewicz I, Bednarczuk TA, Kanoh M, Wiggins RC (1991) Generation of radioactive and nonradioactive ssDNA hybridization probes by polymerase chain reaction. Technique **3**: 64-68.

Konat GW, Laszkiewicz I, Grubinska B, Wiggins RC (1994) Generation of labeled DNA probes by PCR. In: PCR Technology: Current Innovations. Griffin HG, Griffin AM (eds) CRC Press Inc, pp. 37-42.

Lanzillo JJ (1990) Preparation of digoxigenin-labeled probes by the polymerase chain reaction. BioTechn **8**: 621-622.

Liesack W, Menke MAOH, Stackebrandt E (1990) Rapid generation of vector-free digoxigenin-dUTP labeled probes for nonradioactive hybridization using the polymerase chain reaction (PCR) method. System Appl Microbiol **13**: 255-256.

Lion T, Haas OA (1990) Nonradioactive labeling of probe with digoxigenin by polymerase chain reaction. Anal Biochem **188**: 335-337.

Lo YMD, Mehal WZ, Fleming KA (1988) Rapid production of vector-free biotinylated probes using the polymerase chain reaction. Nucleic Acid Res **16**: 8719.

Saiki RK (1989) The design and optimization of the PCR. In: PCR Technology: Principles and Applications for DNA Amplification. Erlich HA (ed) Stockton Press, New York, pp 7-16.

Saiki RK, Scharf S, Faloona F, Mullis KB, Horn GT, Erlich HA, Arnheim N (1985) Enzymatic amplification of β-globin genomic sequences and restriction site analysis for the diagnosis of sickle-cell anemia. Science **230**: 1350-1354.

Saiki RK, Gelfand DH, Stoffel S, Scharf SJ, Higuchi R, Horn GT, Mullis KB, Erlich HA (1988) Primer-directed enzymatic amplification of DNA with a thermostable DNA polymerase. Science **239**: 487-491.

Schowalter DB, Sommer SS (1989) The generation of radiolabeled DNA and RNA probes with polymerase chain reaction. Anal Biochem **177**: 90-94.

Tabibzadeh S, Bhat UG, Sun X (1991) Generation of nonradioactive bromodeoxyuridine labeled DNA probes by polymerase chain reaction. Nucleic Acid Res **19**: 2783.

Uchimura Y, Ishida H, Asada K, Mukai H, Kato I (1991) Nonradioactive labeling with chemically modified cytosine tails by polymerase chain reaction. Gene **108**: 103-108.

Discussion with Reviewers

M. Malecki: In your work you reported 8-fold increase in signal intensity generated with ssDNA as compared to the dsDNA. Have you noticed variability in signal intensity dependent on the reporter molecules used?

Author: The increased hybridization efficiency of ssDNA probes as compared to dsDNA probes is mostly attributed to their inability to reanneal in solution. The sequence of a probe itself should not affect the ratio of hybridization efficiencies, however, we only compared probes generated on the PLP-specific template.

M. Malecki: *Taq* polymerase is recommended in your protocol. Would you anticipate a higher fidelity through proof reading using, e.g., *Pfu*?

Author: A few misincorporated bases introduced *in vitro* by DNA polymerases would in most cases have a minimal effect on the stringency of hybridization. However, during the generation of radiolabeled probes when the concentration of dNTPs (and especially dCTP) is greatly reduced the infidelity of polymerases may increase significantly, and consequently reduce the stringency. *Pfu* polymerase could reduce the mutation rate by approximately 10-fold. However, it should also be considered that the integral 3'-(editing)-exonuclease activity of *Pfu* may increase at low dNTP concentrations, and hence, profoundly reduce the amount of ssDNA molecules, or their length.

Scanning Microscopy Supplement 10, 1996 (pages 61-71)
Scanning Microscopy International, Chicago (AMF O'Hare), IL 60666 USA
0892-953X/96$5.00+.25

NUCLEIC ACID DETECTION BY *IN SITU* MOLECULAR IMMUNOGOLD LABELING PROCEDURES

Marc Thiry

Laboratoire de Biologie Cellulaire et Tissulaire, Université de Liège, Liège, Belgium

(Received for publication October 30, 1995 and in revised form September 23, 1996)

Abstract

We have recently combined immunogold labeling procedures with molecular biology methods to pinpoint the precise locations of nucleic acids in biological material at the ultrastructural level. These new immunocytological approaches involve the incorporation of labeled nucleotides in the nucleic acids present at the surface of ultrathin sections prior to immunogold labeling. The antibodies used recognize a nucleoside analogue (bromodeoxyuridine) or a hapten (biotin) employed to label nucleotides. Examples of high-resolution detection include DNA or RNA present in different substructures of cell nuclei, and in particular, in adenovirus-induced intranuclear regions of HeLa cells. In addition to being highly sensitive and specific, these new methods offer the possibility of studying the spatial distribution of nucleic acids in very well preserved, readily recognizable structures.

Key Words: DNA, RNA, immunogold techniques, ultrastructure, electron microscopy.

*Address for correspondence:
Marc Thiry
Laboratoire de Biologie Cellulaire et Tissulaire
Université de Liège (Bât. L3), rue de Pitteurs, 20
B-4020 Liège, Belgium
Telephone Number: +32(4)366.51.60
FAX Number: +32(4)366.51.73
E-mail: mthiry@ulg.ac.be

Introduction

Different methodological approaches are available to distinguish DNA and RNA from other macromolecules present in biological material at the ultrastructural level (Bendayan, 1984; Gautier, 1976; Moyne, 1980). A powerful strategy for selectively labeling DNA and RNA molecules involves the use of antibodies raised against DNA or RNA (Scheer *et al.*, 1987; Thiry *et al.*, 1991). These techniques exploit the highly specific reaction between antigen and antibody, using labeled immune reagents to locate antigens *in situ*. Unlike autoradiography after thymidine or uridine uptake, moreover, they are less time-consuming. The *in situ* hybridization is a molecular biology technique that has been now successfully applied on ultrathin sections of different biological materials to locate and to map specific nucleic acid sequences (for review, see Morel *et al.*, 1995). To improve the resolution of the *in situ* hybridization technique considerably, a very sensitive immunogold detection procedure was used to detect hybrid molecules. Recently, we have developed *in situ* terminal deoxynucleotidyl transferase (TdT)-immunogold technique, for DNA (Thiry, 1992b) and the *in situ* polyadenylate nucleotidyl transferase (PnT)-immunogold technique, for RNA (Thiry, 1993b) detection, to pinpoint the precise location of nucleic acids on ultrathin sections.
The main goal of the present review is to compare these new approaches with the classical immunocytological approaches. Special emphasis is placed on the potential values and limitations of these two new, immunocytological techniques.

Principle of the *in situ* Transferase-Immunogold Techniques

In these new methods for *in situ* detection of nucleic acids, free DNA or RNA ends generated by sectioning are specifically elongated by transferases using labeled nucleotides. Subsequently, these modified nucleotides are visualized by immunogold electron microscopy. Thus, these techniques include two successive steps, the enzymatic reaction followed by an immunocytological

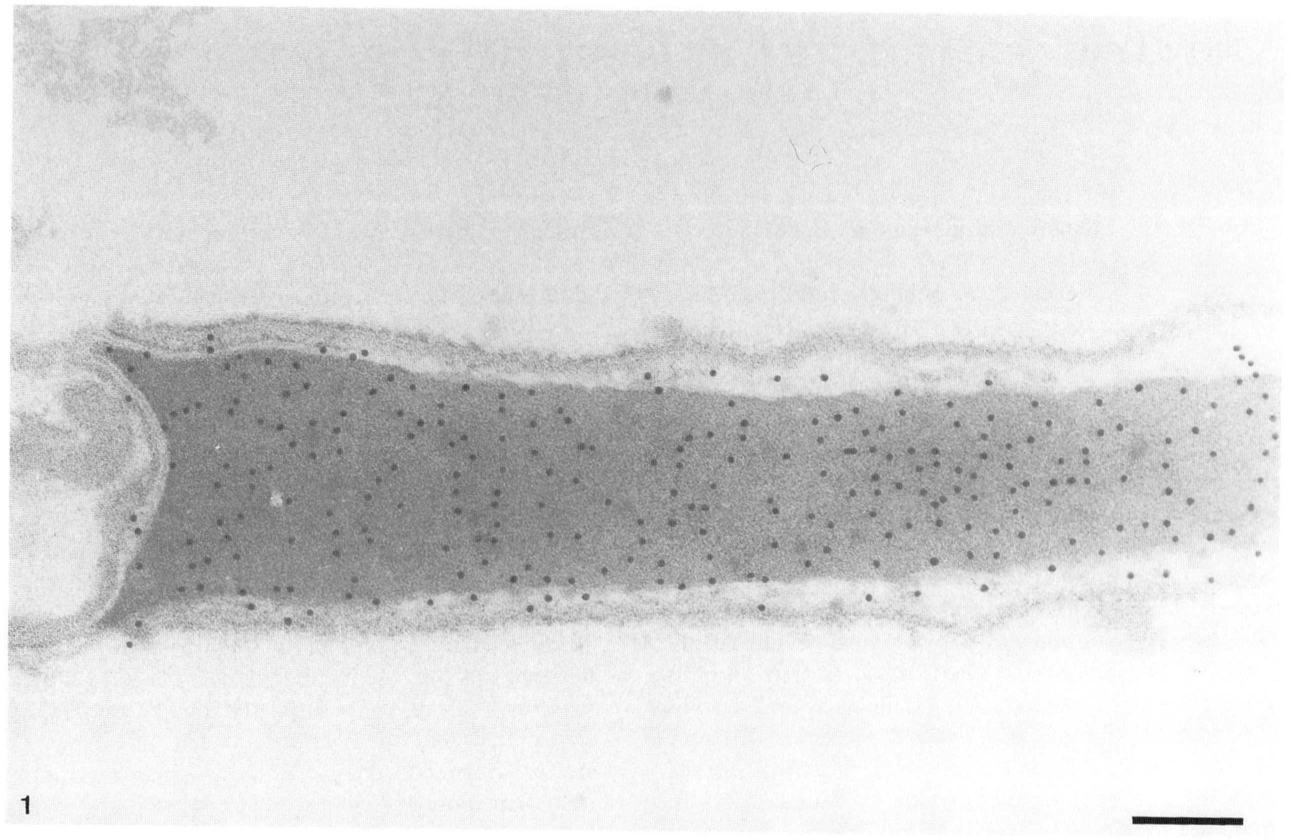

Figure 1. Immunodetection of BUdR triphosphates added by TdT on ultrathin sections in a portion of bull spermatozoa. Intense labeling is found over the condensed chromatin of the nucleus. Glutaraldehyde (1.6 %)/Epon/10 nm. Bar = 0.2 μm.

(*Figures 2 and 3 on facing page*)

Figure 2. Location of DNA on Epon sections in the intranuclear fibrillar spots of an Ad5-infected HeLa cell at an intermediate stage of nuclear transformation, by the *in situ* TdT-immunogold technique. Label is found over all the fibrillar, electron-dense spots (stars) located inside the labeled fibrillogranular network (N) near the viral ssDNA accumulation sites (A). No label is present over the interchromatin granule cluster (IG). Bar = 0.2 μm.

Figure 3. Location of DNA in different virus-induced intranuclear structures of an Epon-embedded HeLa cell at a late stage of nuclear transformation by the *in situ* TdT-immunogold technique. Label is densely distributed over the condensed host chromatin (C). Some viral nucleoids are also labeled (arrows). No label is visualized over the electron-dense amorphous inclusions (I) or the crystalloid (T). Except for some mitochondria (M), the cytoplasm (P) is also gold-free. Bar = 0.2 μm.

procedure. In the enzymatic reaction for identifying DNA, TdT and biotinylated deoxynucleotides or bromodeoxyuridine triphosphates are employed. TdT is an unusual DNA polymerase which catalyzes a template-independent addition of deoxyribonucleotide triphosphates to the 3'-OH ends of double-or single-stranded DNA (Bollum, 1974; Chang and Bollum, 1986). The labeled nucleotides added by TdT on ultrathin sections are then visualized by an indirect immunogold labeling technique involving either an anti-biotin antibody or a monoclonal anti-bromodeoxyuridine antibody and a secondary antibody coupled to colloidal gold.

PnT and biotinylated ATP are used to detect RNA. PnT, an *E. coli* enzyme catalyzes the addition of 5'-adenosine monophosphate to the 3'-OH end of single-stranded RNA (Sippel, 1973). All classes of RNA can be used as substrates (Sippel, 1973). The biotinylated ATP bound to the surface of ultrathin sections is then revealed by means of an anti-biotin antibody followed by a second, colloidal-gold-coupled antibody.

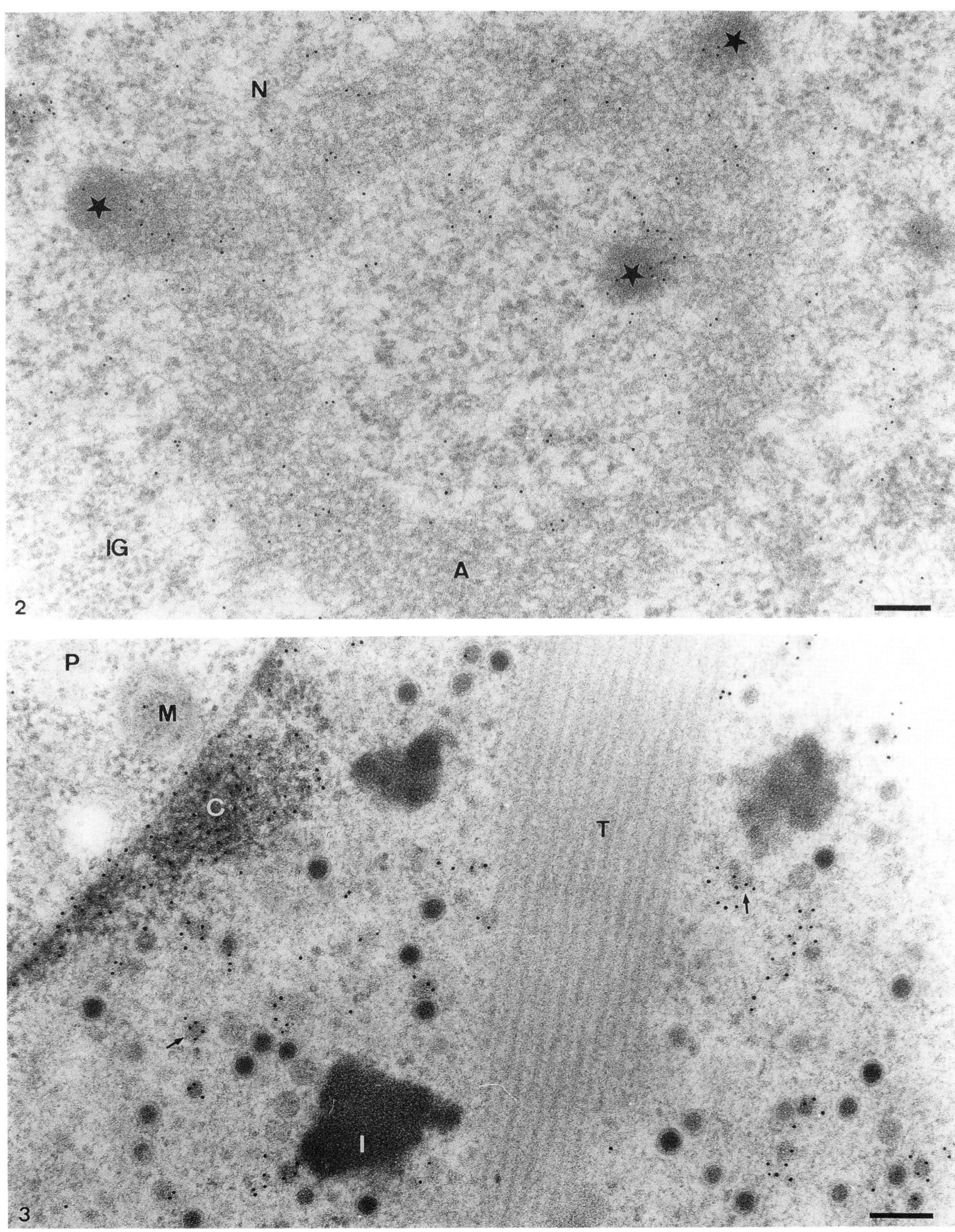
N
IG
A
2
P
M
C
T
I
3

Materials and Methods

TdT method

Cells or tissues are fixed for 30 min at 4°C in 2 % glutaraldehyde in 0.1 M Sorensen's buffer (pH 7.4), rinsed in buffer, dehydrated and embedded in Epon, Lowicryl K4M, and London Resin (LR) White.

Ultrathin sections are collected in platinum rings and stored on distilled water until used.

Ultrathin sections are:

- incubated for 10-30 min at 37°C at the surface of the following medium (100 mM Na cacodylate, 2 mM $MnCl_2$, 10 mM beta-mercaptoethanol, 50 μg/ml bovine serum albumin (BSA), 125 U/ml calf thymus TdT, and 20 μM 5-bromo-2-deoxyuridine triphosphate (pH 6.5-7),
- incubated for 10-30 min at 37°C in the same medium supplemented with 4 μM each of dCTP, dGTP, and dATP.

For labeling, the grids are:

- incubated by floating them face-down in a drop of phosphate buffered saline (PBS) (0.14 M NaCl, 0.006 M Na_2HPO_4, 0.004 M KH_2PO_4, pH 7.2) containing goat normal serum (GNS) diluted 1/30, and 1 % bovine serum albumin (BSA),
- washed with PBS containing 0.2 % BSA (pH 7.2),
- incubated for 4 hours with monoclonal anti-BUdR antibody (Becton Dickinson, Mt. View, CA) diluted 1/50 in PBS containing 0.2 % BSA (pH 7.2), and GNS diluted 1/50,
- rinsed four times with PBS containing 1 % BSA (pH 7.2),
- rinsed one time with PBS containing 0.2 % BSA (pH 8.2),
- incubated for 60 min at room temperature with goat anti-mouse IgG coupled to colloidal gold (5 nm diameter, diluted 1/40 in PBS, pH 8.2, containing 0.2 % BSA),
- washed four times with PBS containing 1 % BSA (pH 8.2),
- rinsed four times in deionized water,
- dried.

Sections are stained with uranyl acetate and lead citrate before examination in an electron microscope.

PnT method

Cells or tissues are fixed at 4°C for 60 min in 1.6 % glutaraldehyde in 0.1 M Sorensen's buffer (pH 7.4), dehydrated through graded acetone solutions, acetylated, then processed for embedding in Epon.

Ultrathin sections are either collected in platinum rings and stored on distilled water until used, or mounted on collodion-coated nickel grids.

Ultrathin sections are:

- incubated for 5 min at 37°C with 50 mM Tris HCl (pH 7.9), 10 mM beta-mercaptoethanol, 10 mM $MgCl_2$, 2.5 mM $MnCl_2$, 0.25 M NaCl, 1 mg/ml bovine serum albumin, 25 U/ml *E. coli* PnT and 0.2 mM biotinyl-17-ATP,
- rinsed five times in bidistilled water,
- incubated for 30 min in PBSB (34 mM NaCl, 0.7 mM KCl, 20 mM Na_2HPO_4, 0.4 mM KH_2PO_4, 1 % BSA, pH 7.2) containing normal rabbit serum (NRS) diluted 1/30,
- incubated for 60 min at room temperature with goat anti-biotin antibodies diluted 1/500 in PBSB containing NRS diluted 1/50,
- rinsed four times in PBSB (pH 7.2),
- rinsed in PBSB (pH 8.2),
- incubated for 60 min at room temperature with medium containing rabbit anti-goat IgG coupled to gold particles either 5 nm or 10 nm in diameter, respectively diluted 1/50 or 1/200 in PBSB (pH 8.2),
- rinsed four times in PBSB,
- rinsed four times in bidistilled water,
- dried and prestained with salts of uranium and lead.

The *in situ* PnT-immunogold procedure is frequently performed on both sides of ultrathin sections. Once labeled on one side, the sections are mounted on collodion-coated nickel grids and the procedure applied to the second face.

(*Figures 4 and 5 on facing page*)

Figure 4. Detection of RNA in tangentially sectioned nuclear envelope of acetylated Ehrlich tumor cells by the PnT-immunogold technique. Label is frequently found in the lumen of nuclear pores (P). In contrast, the condensed chromatin blocks are completely devoid of gold particles. T = cytoplasm, M = mitochondria. Bar = 0.2 μm.

Figure 5. Intranucleolar distribution of RNA in an acetylated Ehrlich tumor cell as visualized with the *in situ* PnT-immunogold method. Gold particles are present in all the three main regions of the nucleolus, including the fibrillar centers (FC). On the contrary, the condensed chromatin (C) associated with the nucleolus is completely devoid of label. Evident label is also detected over an interchromatin granule cluster (IG). D = dense fibrillar component, G = granular component, N = nucleolar canal. Bar = 0.2 μm.

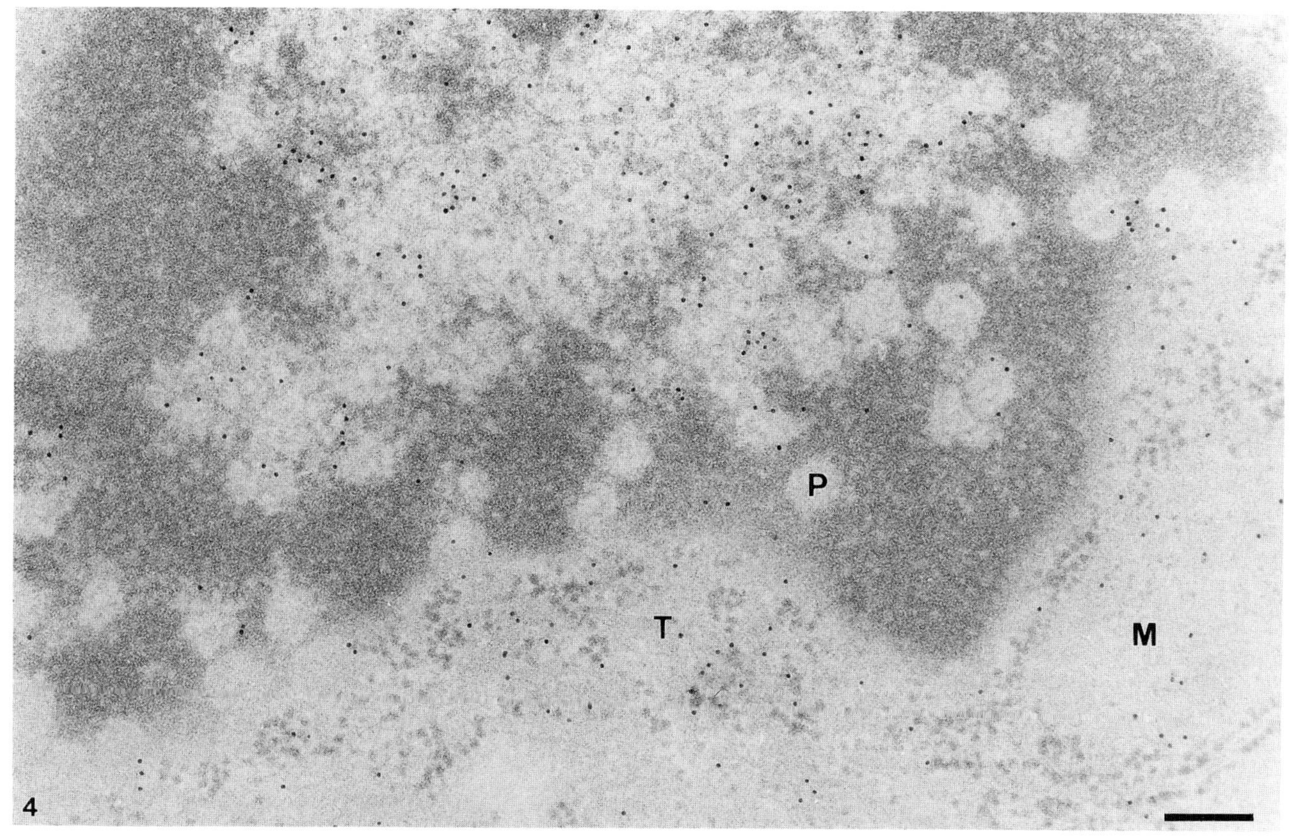
P
T
M
4

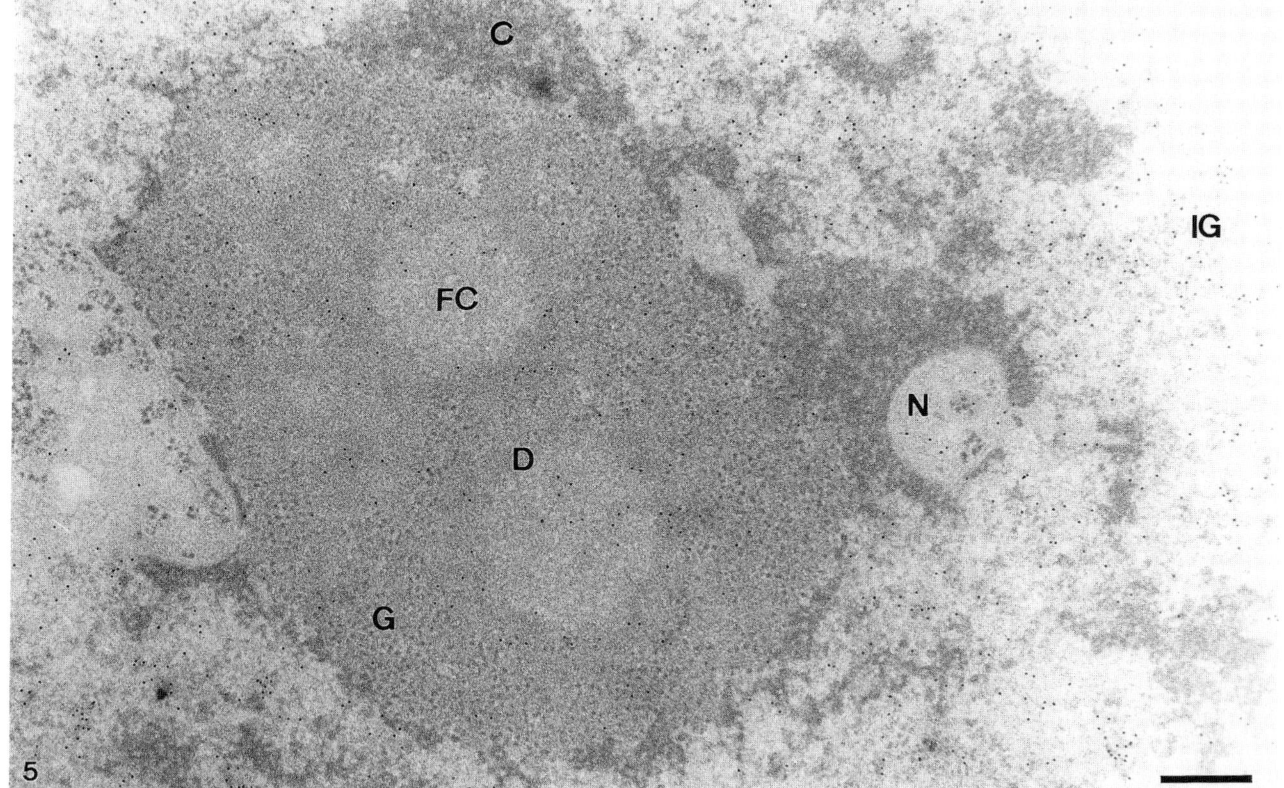
C
IG
FC
N
D
G
5

Potential Values and Limitations

The techniques are based on the highly specific enzyme-substrate reaction, on the one hand, and on the reaction between antigen and antibody, also highly specific, on the other hand. Consequently, highly specific results are obtained by these approaches. An exhaustive series of experimental controls has, moreover, confirmed this specificity (Thiry, 1992a,b, 1993b). As for any other immunogold technique, however, careful quantitative evaluation of the background signals is imperative (Thiry, 1993a,c, 1994; Thiry and Puvion-Dutilleul, 1995; Vandelaer *et al.*, 1993). Because gold particles are highly electron-dense with a perfect round shape and easily recognizable under electron microscope, the resolution of the labeling method is very high. As in the case of all postembedding gold methods, these immunogold labeling procedures are restricted to binding sites exposed at the surface of the sections (Thiry, 1992b). Immunoglobulin-gold complexes do not penetrate into resin sections (Bendayan and Stephens, 1984; Bendayan *et al.*, 1987; Stierhof and Schwarz, 1988; Stierhof *et al.*, 1986) thus revealing only those nucleic acids exposed by the cutting procedure. These constitute a minor but representative portion of the nucleic acids present in biological material sections. While being a limitation of these approaches, this also represents a significant advantage in that it allows for intracellular labeling of DNA and RNA molecules without limitations due to diffusion of reagents, permeation of membrane, and accessibility of nucleic acids. The successful labeling of various nucleic acid-containing structures, some containing very low amounts of nucleic acids (i.e. mitochondria, mycoplasmas, virus) (Thiry, 1992a,b, 1993, 1995a), demonstrates the high sensitivity of these approaches.

As compared with immunocytological techniques involving antinucleic acid antibodies these new approaches present, however, two essential advantages.

First, they show a more general distribution of nucleic acids in biological materials. The *in situ* transferase-immunogold techniques reveal with great precision the specific nucleic acid-containing structures in a wide variety of organisms, from prokaryotes to eukaryotes (Thiry, 1992a,b,c, 1993a,b,c, 1994, 1995a,b; Thiry *et al.*, 1993; Thiry and Puvion-Dutilleul, 1995; Vandelaer *et al.*, 1993). For instance, using the TdT method, label is not only found over condensed chromatin like the metaphase chromosomes or the spermatozoon nuclei (Fig. 1), but also over DNA in a completely extended configuration, like the highly active genes present in the Balbiani rings of *Chironomus tentans* salivary glands (Thiry, unpublished results). In adenovirus infected (Ad5) HeLa cells (Thiry and Puvion-Dutilleul, 1995) (Figs. 2 and 3), the labeling reveals not only the condensed host chromatin but also the different types of Ad5 DNA fibers, i.e., active Ad5 genomes, inactive free Ad5 genomes, inactive encapsidated Ad5 genomes and single-stranded Ad5 DNA, in different virus-induced substructures which appear in the centers of infected nuclei. Using the PnT procedure, label is visualized over mitochondria, some nucleoplasmic substructures such as interchromatin granule clusters and coiled bodies, and the three main components of the nucleolus (Figs. 4 and 5). Examples of sensitive, high resolution detection of nucleic acid-containing structures have also been reported with the use of antinucleic acid antibodies (Hansmann *et al.*, 1986; Hayashi-isiharu *et al.*, 1993; Mena *et al.*, 1994; Puvion-Dutilleul and Pichard, 1993; Scheer *et al.*, 1987; Thiry 1992c, 1993c, 1994; Thiry and Puvion-Dutilleul, 1995). However, it is important to remember that no universal antinucleic acid antibody exists. Each antibody recognizes a certain nucleic acid "region" (distinguished by its configuration, sequence, etc.) (Angelier *et al.*, 1986; Lubit *et al.*, 1976; Soyer-Gobillard *et al.*, 1990). In this respect, in adenovirus-infected HeLa cells, by means of the *in situ* TdT method, label was consitently revealed over round fibrillar spots (Thiry and Puvion-Dutilleul, 1995) (Fig. 2). The use of anti-DNA antibodies has not allowed the detection of DNA in this substructure (Puvion-Dutilleul and Pichard, 1993). Consequently, the antibodies to nucleic acids do not constitute an useful mean for studying the general distribution of nucleic acids in biological materials.

The second advantage of the approaches compared with the use of antinucleic acid antibodies is that they are compatible with various fixation and embedding conditions. These new approaches are applicable to

(*Figures 6 and 7 on facing page*)

Figure 6. Distribution of DNA in a portion of Ehrlich tumor cell after acetylation and EDTA regressive staining as revealed by the TdT-immunogold technique. Label is preferentially found over the completely bleached blocks of condensed chromatin (C). A few gold particles are also visualized in interchromatin spaces (S). However, the interchromatin granule (IG) cluster is totally unlabeled. T= cytoplasm. Bar = 0.2 μm.

Figure 7. Immunodetection of biotinylated ATP added by PnT on ultrathin sections of acetylated Ehrlich tumor cells. The gold particle concentration is particularly high over the nucleoplasmic spaces (S) between the heterochromatin blocks. In addition, intense label is also present in cytoplasmic ribosome-rich regions (T). No label occurs over the condensed chromatin (C) associated with the nuclear envelope (NE). Bar = 0.2 μm.

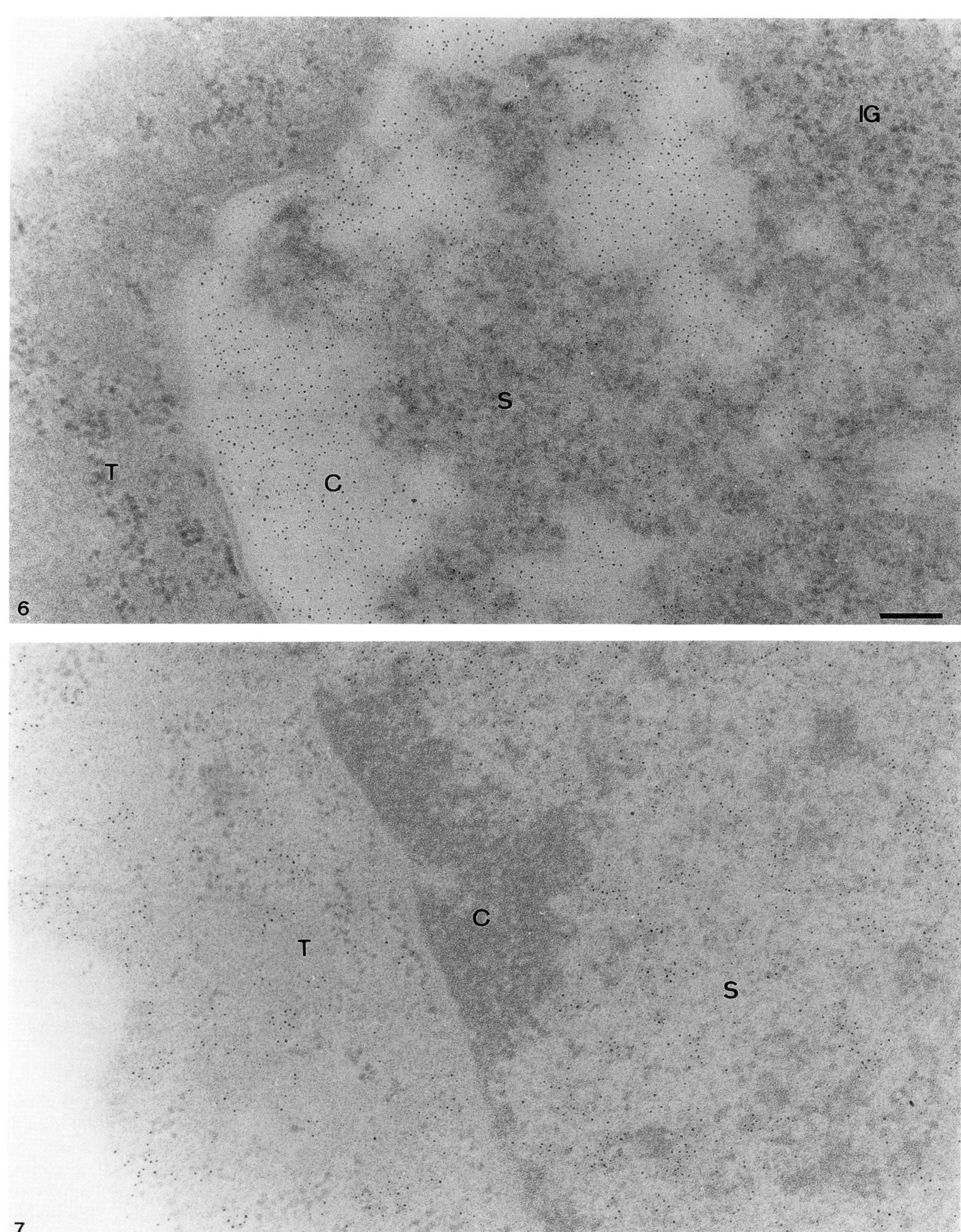
IG
S
T
C
6
C
T
S
7

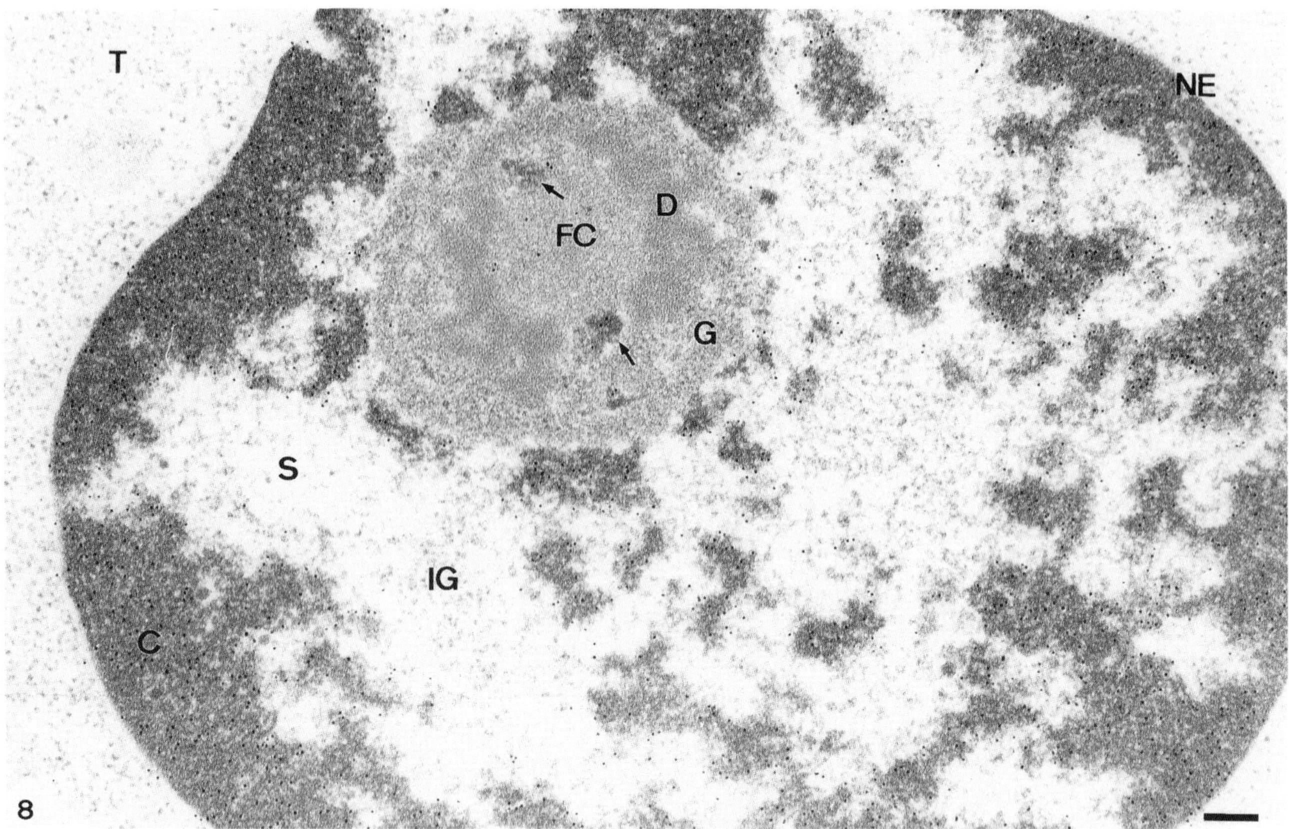

Figure 8. Immunodetection of BUdR triphosphates added by TdT on ultrathin sections of acetylated human resting T lymphocytes. The gold particle concentration is particularly high over the condensed chromatin (C) blocks associated with the nuclear envelope (NE) and the nucleolus. A few gold particles are also found in nucleoplasmic regions (S), however, interchromatin granule (IG) clusters are completely devoid of label. Inside the nucleolus, evident label is exclusively found over the intranucleolar clumps of condensed chromatin (arrows) and over the fibrillar centers (FC), whereas no significant label is detected over the dense fibrillar component (D) and the granular component (G). Except for some mitochondria, the cytoplasm (T) is also gold-free. Bar = 0.2 μm.

sections fixed with glutaraldehyde, formaldehyde, glutaraldehyde/formaldehyde and even glutaraldehyde/osmium tetroxide, whether embedded in Epon, Agar 100, Spurr, Lowicryl K4M or LR white (Thiry 1992a, b, 1993b; Thiry *et al.*, 1993). They can be further combined with cytochemical methods (Thiry 1992a,b, 1993b), such as acetylation or EDTA regressive staining, which intensify the contrast between some of the various nuclear compartments (Figs. 6-7). These methods thus offer the possibility of determining the precise location of nucleic acids in very well preserved, readily recognizable structures.

On the other hand, to respect a perfect compromise among preservation of fine cellular structure, retention of antigenic determinants, and optimal specificity, antibodies to nucleic acids are only applicable to biological material fixed and embedded in a way that does not always optimally preserve the structure to be studied. In this respect, to facilitate the identification of DNA positive sites within the mammalian cell nucleolus, the *in situ* TdT technique was applied on acetylated material (Thiry 1992a,b, 1993a,b; Thiry *et al.*, 1993; Vandelaer *et al.*, 1993), a cytochemical method allowing an excellent distinction between the various nucleolar components and in particular a high contrast of the condensed chromatin (Thiry and Goessens, 1986; Wassef *et al.*, 1979). Under these experimental conditions (Fig. 8), in a large variety of cell nucleoli, besides the presence of an intense label over the intranucleolar and perinucleolar condensed chromatin, DNA was consistently found in the fibrillar centers whereas no significant label was detected in the dense fibrillar component and the granular component. This analysis also revealed the presence of small clumps of condensed chromatin in close contact with the fibrillar centers and interrupting the layer of the dense fibrillar component.

Such observations cannot be achieved with antibodies to DNA because they are not compatible with acetylation conditions.

Conclusions

The immunocytological approaches represent powerful means of studying and identifying the nucleic acids in biological materials at the ultrastructural level.

The *in situ* transferase-immunogold techniques offer the possibility of studying the spatial distribution of nucleic acids in very well preserved, readily recognizable structures. While the antinucleic acid antibodies do not constitute an useful mean of studying the general distribution of nucleic acids in biological materials, their high elective specificity makes it possible to investigate the precise location and functional role of certain specific nucleic acid regions (i.e., Z-DNA, 5-methylcytosine). The use of antinucleic acid antibodies can therefore provide complementary informations on nucleic acids visualized by the *in situ* transferase-immunogold techniques.

Acknowledgments

The author is grateful to Prof. G. Goessens (University of Liège, Belgium) for encouraging discussions and for critical reading of the manuscript. He also acknowledges the skillful technical and secretarial assistance provided by Miss F. Skivée and Mrs S. Bodson. This work received financial support from the "Fonds de la Recherche Scientifique Médicale" (grant n°3.4512.90) and from the "Actions de Recherche Concertées" (grant n°91/95-152). Marc Thiry is researcher at the National Fund for Scientific Research (Belgium).

References

Angelier N, Bonnanfant-Jaïs M, Moreau N, Gounon P, Lavaud A (1986) DNA methylation and RNA transcriptional activity in amphibian lampbrush chromosomes. Chromosoma **94**: 169-182.

Bendayan M (1984) Enzyme-gold electron microscopic cytochemistry: A new affinity approach for the ultrastructural localization of macromolecules. J Electron Microsc Techn **1**: 349-372.

Bendayan M, Stephens H (1984) Double labelling cytochemistry applying the protein A-gold technique. In: Immunolabelling for Electron Microscopy. Polak J, Varndell I (eds). Elsevier, Amsterdam New York. pp 143-154.

Bendayan M, Nanci A, Kan F (1987) Effect of tissue processing on colloidal gold cytochemistry. J Histochem Cytochem **35**: 983-996.

Bollum FJ (1974) Terminal deoxynucleotidyl transferase. In: The Enzymes. Boyer PD (ed). Academic Press, New York. **10**: 145-171.

Chang LMS, Bollum FJ (1986) Molecular biology of terminal transferase. Crit Rev Biochem **21**: 27-52.

Gautier A (1976) Ultrastructural localization of DNA in ultrathin tissue sections. Int Rev Cyt **44**: 113-191.

Hansmann P, Falk H, Scheer U and Sitte P (1986) Ultrastructural localization of DNA in two cryptomonas species by use of a monoclonal DNA antibody. Eur J Cell Biol **42**: 152-160.

Hayashi-isiharu Y, Ueda K, Nonaka M (1993) Detection of DNA in the nucleoids of chloroplasts and mitochondria in Euglena gracilis by immunoelectron microscopy. J Cell Sci **105**: 1159-1164.

Lubit B W, Pham T D, Miller O J, Erlanger B F (1976) Localization of 5-methylcytosine in human metaphase chromosomes by immunoelectron microscopy. Cell **9**: 503-509.

Mena C G, Testillano P S, Gonzalez-Melendi P, Gorab E, Risueno MC (1994) Immunoelectron microscopy of RNA combined with nucleic acid cytochemistry in plant nucleoli. Exp Cell Res **212**: 393-408.

Morel G, Le Guellec D, Mertani H, Trembleau A (1995) *In situ* hybridization for electron microscopy. In: Visualization of Nucleic Acids. Morel G (ed). CRC Press, Boca Raton, pp 229-257.

Moyne G (1980) Methods in ultrastructural cytochemistry of the cell nucleus. Progr Histochem Cytochem **13**: 1-70.

Puvion-Dutilleul F, Pichard E (1993) Superiority of *in situ* hybridization over immunolabeling for detecting DNA on lowicryl sections: a study on adenovirus-infected cells. J Histochem Cytochem **41**: 1537-1546.

Scheer U, Messner K, Hazan R, Raska I, Hansmann P, Falk H, Spiess E, Franke W (1987) High sensitivity immunolocalization of double- and single-stranded DNA by a monoclonal antibody. Eur J Cell Biol **43**: 358-371.

Soyer-Gobillard M O, Geraud M L, Coulaud D, Barray M, Theveny B, Revet B, Delain E (1990) Location of B- and Z-DNA in the chromosomes of a primitive eukaryote Dinoflagellate. J Cell Biol **111**: 293-308.

Sippel AE (1973) Purification and characterization of adenosine triphosphate: ribonucleic acid adenyltransferase from Escherichia coli. Eur J Biochem **37**: 31-40.

Stierhof YD, Schwarz H (1988) *Immunoelectron-microscopy on cryosections section permeability to specific antibodies, protein A-gold complexes and ferritin conjugated IgGs.* Inst Phys Conf Ser **93**, **3**: 519-520.

Stierhof YD, Schwarz H, Frank H (1986) Transverse sectioning of plastic-embedded immunolabeled cryosections: Morphology and permeability to protein A-colloidal gold complexes. J Ultrastruct Res **97**: 187-196.

Thiry M (1992a) Ultrastructural detection of DNA within the nucleolus by sensitive molecular immunocytochemistry. Exp Cell Res **200**: 135-144.

Thiry M (1992b) Highly sensitive immunodetection of DNA on sections with exogenous terminal deoxynucleotidyl transferase and non-isotopic nucleotide analogues. J Histochem Cytochem **40**: 411-419.

Thiry M (1992c) New data concerning the functional organization of the mammalian cell nucleolus: detection of RNA and rRNA by *in situ* molecular immunocytochemistry. Nucleic Acids Res **20**: 6195-6200.

Thiry M (1993a) Ultrastructural distribution of DNA and RNA within the nucleolus of human Sertoli cells as seen by molecular immunocytochemistry. J Cell Sci **105**: 33-39.

Thiry M (1993b) Immunodetection of RNA on ultrathin sections incubated with polyadenylate nucleotidyl transferase. J Histochem Cytochem **41**: 657-665.

Thiry M (1993c) Differential location of nucleic acids within interchromatin granule clusters. Eur. J Cell Biol **62**: 259-269.

Thiry M (1994) Cytochemical and immunocytochemical study of coiled bodies in different cultured cell lines. Chromosoma **103**: 268-276.

Thiry M (1995a) Nucleic acid compartmentalization within the cell nucleus by *in situ* transferase-immunogold techniques. Microsc Res Techn **31**: 4-21.

Thiry M (1995b) Behavior of interchromatin granules during the cell cycle. Eur J Cell Biol **68**: 14-24.

Thiry M, Goessens G (1986) Ultrastructural study of the relationships between the various nucleolar components in Ehrlich tumour and HEp-2 cell nucleoli after acetylation. Exp Cell Res **164**: 232-242.

Thiry M, Goessens G (1992a) Where, within the nucleolus, are the rRNA genes located? Exp Cell Res **200**: 1-4.

Thiry M, Puvion-Dutilleul F (1995) Differential distribution of single-stranded DNA, double-stranded DNA, and RNA in adenovirus-induced intranuclear regions of HeLa cells. J Histochem Cytochem **43**: 749-759.

Thiry M, Scheer U, Goessens G (1991) Localization of nucleolar chromatin by immunocytochemistry and *in situ* hybridization at the electron microscopic level. Electron Microsc Rev **4**: 85-110.

Thiry M, Ploton D, Ménager M, Goessens G (1993) Ultrastructural distribution of DNA within the nucleolus of various animal cell lines or tissues revealed by terminal deoxynucleotidyl transferase. Cell Tiss Res **271**: 33-45.

Vandelaer M, Thiry M, Goessens G (1993) Ultrastructural distribution of DNA within the ring-shaped nucleolus of human resting T lymphocytes. Exp Cell Res **205**: 430-432.

Wassef M, Burglen J, Bernhard W (1979) A new method for visualization of preribosomal granules in the nucleolus after acetylation. Biol Cell **34**: 153-158.

Discussion with Reviewers

M. Malecki: What is your recommendation for the choice of embedment, i.e., giving the strongest labeling and the lowest background?

Author: The new techniques are compatible with various fixation and embedding conditions. The labeling intensity obtained with the TdT method does not appear to be affected by the fixation and embedding conditions. On the other hand, the *in situ* PnT-immunogold technique is compatible with fewer fixation and embedding conditions than the *in situ* TdT-immunogold technique. In particular, the labeling intensity obtained with the PnT method varies according to the fixation and embedding used. Best results were obtained on 1.6% glutaraldehyde-fixed and Epon-embedded cells.

M. Malecki: Did you attempt to use resinless sections or sucrose infused cryo-sections for derivatized analog incorporation/immuno-gold labeling?

Author: No, nevertheless, it is interesting to note that the TdT method is applicable to cell sections cryofixed, cryosubstituted in acetone, and embedded in Epon.

M. Malecki: Can you explain your preference for using gold conjugated antibodies against biotin rather than streptavidin-gold?

Author: We have not tested different procedures for revealing biotin. The use of anti-biotin antibodies was currently used in our laboratory and provided satisfactory results.

M. Malecki: What is your experience with other analogs in particular those with digoxigenin ?

Author: In our studies, we have commonly used two analogs: biotinylated nucleotides and bromodeoxyuridine triphosphates. In our hands, best results were obtained with the BudR system. Moreover, we have used digoxigenin-ATP as substrate in the PnT method. The labeled sites were subsequently revealed by an anti-digoxigenin antibody and antibodies coupled to colloidal gold. Under these conditions, results were very poor; only a few gold particles were present. These results seemed to indicate that digoxigenin-ATP is not efficiently incorporated into RNA by PnT.

Reviewer II: What is the specificity of TdT and/or PnT

labeling in preserved tissue, and how has it been checked ?

Author: The specificity of the TdT and/or PnT labeling was tested in several ways.

TdT: When TdT or BUdR triphosphate were omitted from the TdT medium, the ultrathin sections were devoid of label. No labeling was observed when BUdR triphosphate was replaced by BUdR monophosphate. When the sections of Lowicryl-embedded cells were incubated with trypsin, pronase, protease, or RNase, the labeling persisted; when DNA was specifically removed from the sections by treatment with DNaseI, the labeling was completely abolished.

PnT: When PnT or biotinylated ATP was omitted from the PnT medium, no labeling occurred. When acetylated cell sections were pre-incubated with RNase A at a high concentration (10 mg/ml), labeling was completely abolished. At a lower RNase A concentration, labeling was only reduced. Labeling was likewise weaker when acetylated cell sections were pre-incubated with RNase T2. No label was detected on acetylated cell sections pre-treated with RNase T2, followed by RNase A at a lower concentration. Preincubation with DNase I, on the other hand, did not prevent labeling. Finally, when the RNase T2 incubation was carried out after the PnT reaction, no labeling occurred. Finally, when the RNase T2 incubation was carried out after the PnT reaction, no labeling occurred.

The immunolabeling specificity was also tested. When the primary antibody was omitted, no labeling occurred. Gold lacking the antibody tag did not bind to the sections. All these experimental controls support the high specificity of these methods.

Scanning Microscopy Supplement 10, 1996 (pages 73-80) 0892-953X/96$5.00+.25
Scanning Microscopy International, Chicago (AMF O'Hare), IL 60666 USA

HYDRATION-SCANNING TUNNELING MICROSCOPY AS A RELIABLE METHOD FOR IMAGING BIOLOGICAL SPECIMENS AND HYDROPHILIC INSULATORS

M. Heim[1*], R. Eschrich[1] A. Hillebrand[1], H. F. Knapp[1], G. Cevc[2] and R. Guckenberger[1]

[1]Max-Planck-Institut für Biochemie, Abt. Molekulare Strukturbiologie, Martinsried,
[2]Technische Universität München, Klinikum Rechts der Isar, Medizinische Biophysik, München, Germany

(Received for publication October 17, 1995, and in revised form February 9, 1996)

Abstract

The recently discovered high lateral conductivity of molecularly thin adsorbed water films enables investigation of biological specimens, and even of surfaces of hydrophilic insulators by scanning tunneling microscopy (STM). Here we demonstrate the capabilities of this method, which we call hydration-STM (HSTM), with images of various specimens taken in humid atmosphere: We obtained images of a glass coverslip, collagen molecules, tobacco mosaic virus, lipid bilayers and cryosectioned bovine achilles tendon on mica. To elucidate the physical mechanism of this conduction phenomenon we recorded current-voltage curves on hydrated mica. This revealed a basically ohmic behavior of the I-V curves without a threshold voltage to activate the current transport and indicates that electrochemistry probably does not dominate the surface conductivity. We assume that the conduction mechanism is due to structuring of water at the surface.

Key Words: Surface conductivity, structured water, thin films, surface electrochemistry, glass, collagen, bovine achilles tendon, lipid layers.

*Address for correspondence:
Manfred Heim
Max-Planck-Institut für Biochemie
Abt. Molekulare Strukturbiologie
D-82151 Martinsried
Germany

Telephone number: 49-89-85782651
FAX number: 49-89-85782641
email: heim@alf.biochem.mpg.de

Introduction

The recently discovered high electrical conductivity of surface-adsorbed water molecules (Guckenberger *et al.*, 1994; Heim *et al.*, 1995) permits scanning tunneling microscopy (STM) investigation of biological samples and even hydrophilic insulators in humid air. Water films which are only a few Å thick adsorb to hydrophilic surfaces in humid air and act as a conductive coverage of the surface to be investigated. This conductivity is exploited for imaging with the scanning tunneling microscope, a method we term hydration-STM (HSTM).

A conductive metal skin with a minimum thickness of 1-2 nm, normally necessary for studying insulators with the STM, limits resolution substantially, whereas the conductive water has the advantage of being very thin (less than 0.2 nm on average can be sufficient). Moreover, the biological specimens in HSTM experiments are present in a semihydrated state, which can be important for the biological material not to denature.

For biological applications the atomic force microscope (AFM) is the most widely used probe microscope since it does not rely on sample conductivity and is capable of measuring various mechanical properties of the sample such as friction (Mate *et al.*, 1987), elasticity (Maivald *et al.*, 1991; Radmacher *et al.*, 1993), viscosity, etc. even in an aqueous environment. However, the AFM always exerts forces upon the studied objects which can induce conformational changes in the samples. For STM studies, sharper tips can easily be manufactured and the shorter decay length of tip-sample interaction, at least in principle, allows achievement of higher resolution with the STM (Rohrer, 1990). This implies that the STM is still attractive for investigation of soft materials. In this work we describe how HSTM can be easily applied for the imaging of various biological molecules or other hydrophilic surfaces. This is demonstrated with HSTM images of glass, collagen molecules, the tobacco mosaic virus, a lipid bilayer and a cryosection of bovine achilles tendon on mica.

Apart from the possibility of imaging insulators with the STM, the effect of the lateral conductivity of ultrathin water films is of fundamental interest itself since the physical background of the exceptionally high conductivity is not yet well understood. We hypothesize that electrical conductivity arises from a specifically ordered structure of the water molecules on the adsorbing surfaces. The influence of the surface forces could immobilize water molecules in such a way that they form a hydrogen-bond network which mediates charge transport. The surface adsorbed water would have ice-like rather than water-like properties. An indication of this was recently observed in images of surface-adsorbed water on mica, taken in humid air with the atomic force microscope (Hu *et al.*, 1995) (for a review on water structure on surfaces see Bockris and Jeng, 1990). The surface conductivity of water can probably be attributed to conformational changes in the water layer and, therefore, corresponds to well known phenomena like the increase of conductivity and the dielectric constant in frozen water (Israelachvili, 1992). However, the existence and the role of such thin layers of structured water with properties very different from bulk water is still in discussion (Israelachvili and Wennerström, 1996). We have recorded voltage-current curves on hydrated mica in order to derive more information about the properties of the conductivity of molecularly thin water films.

Experimental

All experiments were done with a home built STM described in ref. (Guckenberger *et al.*, 1988). One important feature of this device is the use of a very sensitive preamplifier (input stage, 20 GΩ feedback resistor, commercially available from Quintenz Hybridtechnik, Kramerstrasse 3, 82061 Neuried, Germany) which enables us to image with tunneling currents as low as 50 fA. The relative humidity in the humidity chamber is controlled by mixing a stream of dry nitrogen with a humid nitrogen stream which bubbles through deionized water (with 18 MΩcm resistivity from a Milli-Q plus system, Millipore Corporation, Bedford, MA). The tips used in our imaging experiments were electrochemically etched tungsten wires (Hacker *et al.*, 1992). All images were taken in the constant current mode.

The glass coverslip was cleaned first in a 2 % Hellmanex solution in an ultrasonic bath for 15 minutes and then washed in Milli-Q water at 30 °C for 15 minutes in the ultrasonic bath.

Collagen molecules were deposited on mica as follows: A small drop of collagen solution was placed on a piece of freshly cleaved mica and left to adsorb for about 1 min. The sample was then washed by placing it on top of a large drop of pure water for 5 seconds and the excess water was blotted off with a filter paper. The same procedure was also used to prepare tobacco mosaic virus (TMV) on mica. The initial TMV suspension contained 43 mg/ml TMV in sodium phosphate solution (0.1 M; pH 7.0). The suspension was diluted 9:1 with pure water prior to deposition on the mica substrate.

The lipid bilayer consisted of DPPE (1,2-dipalmitoylphosphatidylethanolamine) as the first layer on mica and DPPC (1,2-dipalmitoylphosphatidylcholine, both from Sygena, Liestal, Switzerland) as the top layer. Both lipids were prepared with the Langmuir-Blodgett technique in a film balance. The lateral pressure of the lipids in the film balance was maintained at 40 mN/m for 45 minutes prior to the deposition. After preparation of the lipid bilayer the sample was dried as described in ref. (Heim *et al.*, 1995).

The bovine achilles tendon was embedded in glucose and frozen in liquid nitrogen. The tendon then was sectioned with a microtome and mounted on mica. To remove glucose, the sample was washed with pure water, then dyed with methylene blue to facilitate recognition in the light microscope. All samples were fixed mechanically onto a sample holder made of stainless steel, which provides electrical contact via mechanical contact to the sample side facing the tip.

The current voltage curve was recorded with a gold tip in mechanical contact with the hydrated mica surface. In this experiment the ring-shaped sample holder with an inner diameter of 6 mm was covered with gold by evaporation. The preparation of gold tips is described in (Knapp, 1996).

Results and Discussion

HSTM images of insulators and biological specimens

In this work we describe the possibilities and limitations of the newly developed hydration scanning tunneling microscopy (HSTM) method.

Glass. The first example (Fig. 1) shows the HSTM image of a glass coverslip. At a relative humidity of 68 %, the glass surface is hydrated, and the wetting film is capable of laterally transporting the tunneling current from the spot beneath the tip to the sample holder. The glass surface investigated displays holes with a diameter of approximately 20 nm. Many smaller holes can also be seen on the glass surface with a diameter of about 2.5 - 8 nm. It is noteworthy that the image is recorded with a sample bias of -0.7 V which demonstrates that imaging at low voltages, and thus, not only in the field emission regime, is possible.

A HSTM image of a plasma cleaned glass coverslip presented in (Heim *et al.*, 1996) shows basically the same structure (with a slightly reduced resolution). The big as well as the small holes are visible on the plasma cleaned and the detergent/water cleaned coverslip. This shows that plasma cleaning does not influence the glass surface on the nanometer scale. The glass shown in Fig. 1 becomes hydrophilic enough for HSTM imaging after

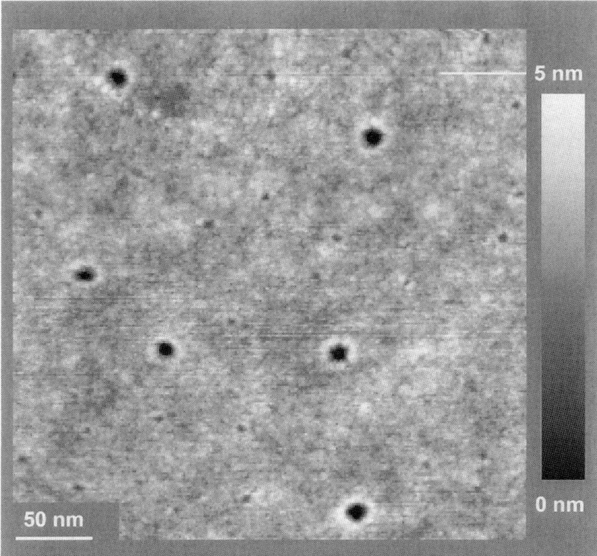

Figure 1: HSTM image of an uncoated glass coverslip taken at 0.7 V sample bias, a tunneling current of 1 pA, and at a relative humidity of 68 %. The surface adsorbed water molecules make the glass surface sufficiently conductive for STM imaging. The glass shows several holes which are about 20 nm wide and many small holes with diameters between 2.5 and 6 nm.

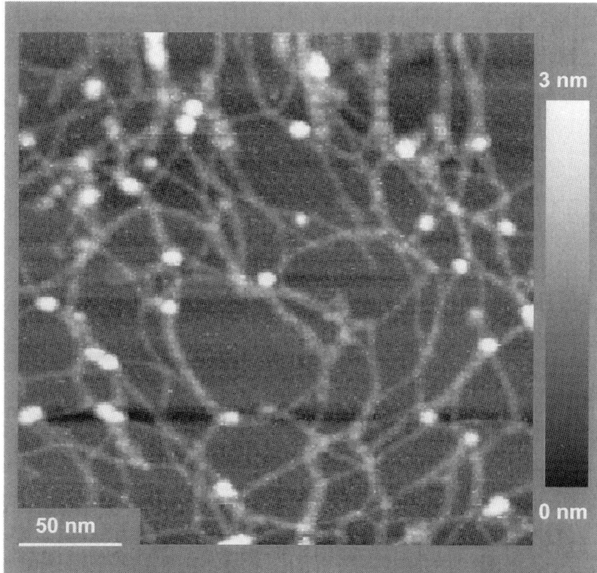

Figure 2: HSTM image of collagen IV molecules on mica. At 53 % relative humidity the uncoated sample was imaged with a sample bias of -4.5 V and 0.06 pA.

----

a conventional washing procedure, which implies that plasma cleaning is not the crucial factor for water adsorption in humid air. The important point is to thoroughly clean the surfaces in order to make them sufficiently hydrophilic.

Mica. Mica physisorbs many biological materials and is smooth on atomic scale which makes it one of the most widely used substrates for investigation of biological materials in scanning probe microscopy (SPM). Like glass, freshly cleaved mica adsorbs water molecules to its surface and therefore becomes conductive when exposed to humid air.

Collagen. Type IV collagen forms long (approximately 320 nm) triple-helical strands (approximately 1.5 nm in diameter) with a terminal globular domain. Individual collagen strands interconnect with each other either via this globular domain or via a rod-like segment at the opposite end (Timpl *et al.*, 1981). Fig. 2 shows the image of collagen IV molecules on mica imaged with the HSTM. The thinnest filaments visible in the image probably are single collagen molecules whereas the wider filaments are likely to consist of several molecules. The protrusions in the image can be attributed to the globular part of the collagen molecules, since such features appear exclusively on the filaments. The measured heights of the thinnest filaments are about 0.5 nm in contrast to their real height of 1.5 nm.

Similar to collagen, plasmid DNA can be prepared on mica and imaged with the HSTM method (Guckenberger *et al.*, 1994).

TMV. The described method also can be applied for biological specimens several nanometers thick such as tobacco mosaic virus (figure 3). The TMV virus consists of a RNA strand which is encapsulated in helically arranged proteins forming a 300 nm long cylinder. The diameter of the cylinder is 18 nm. Viruses which have lost their central RNA often break into several pieces and thus virus tubes much shorter than 300 nm can be found. The HSTM image delivers a TMV height of 15 nm which corresponds to the STM experiments of metal coated TMV (data not shown).

Lipid Bilayer. Fig. 4 shows the image of a lipid bilayer membrane on mica consisting of DPPE as the first layer and DPPC as the top layer prepared with Langmuir-Blodgett technique. This example shows that HSTM imaging works on films of organic molecules which are extended laterally over a large area. The surface consists of polar DPPC headgroups which adsorb water from the humid air and therefore produce a conductive surface. Defects are seen in the form of circular holes, which are about 6 nm deep, as expected for a DPPE/DPPC bilayer.

If only one lipid layer is transferred on mica, the surface becomes hydrophobic, since the hydrocarbon chains of the lipids then point towards the air. No water adsorbs on such a surface, and thus no surface conductivity is measured (Heim *et al.*, 1995). Examination of such specimens with the HSTM method is not possible.

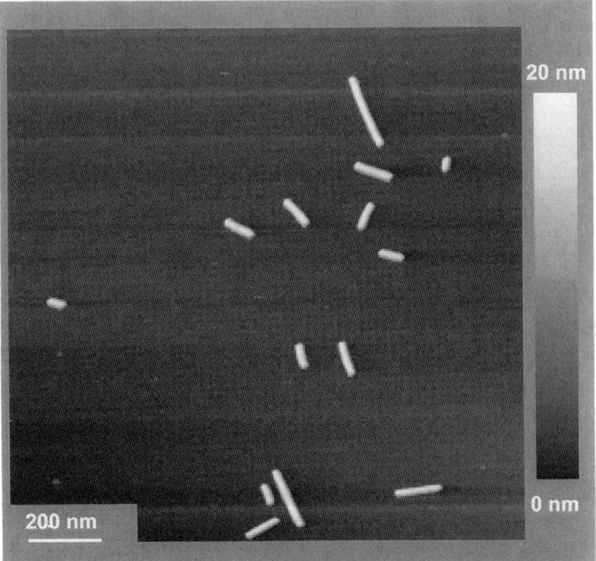

Figure 3: Uncoated tobacco mosaic virus on mica, imaged by HSTM at -7.2 V, 0.3 pA in the atmosphere of 72 % relative humidity. The height of the viruses measured with this HSTM method is 15 nm, which corresponds to the height value deduced from STM experiments with metal coated TMV.

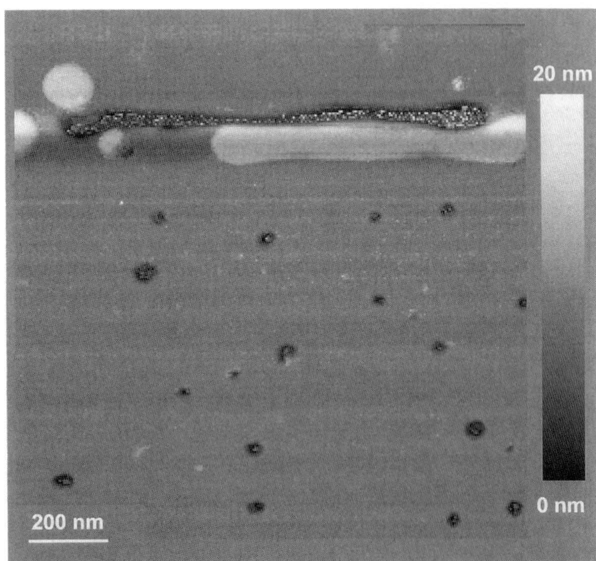

Figure 4: HSTM image of a lipid bilayer consisting of DPPE as the underlayer and DPPC as the top layer. The lipid layers were prepared with the Langmuir-Blodgett method (Heim *et al.*, 1995). The holes are about 6 nm deep, corresponding to the expected thickness of a DPPE/DPPC bilayer. In the upper part of the image, a piece of membrane is seen which has detached, and then came to lie near the rim of the hole. It is not clear if such a detaching of a membrane piece is tip induced or not. (Imaging parameters: +7.7 V sample bias, 0.25 pA, 50 % relative humidity.)

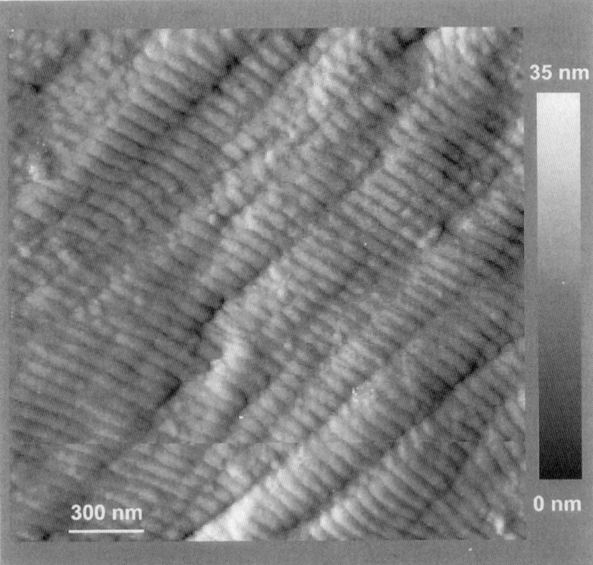

Figure 5: HSTM image of a cryosectioned bovine achilles tendon. (Imaging parameters: +7 V sample bias, 0.1 pA and 66 % relative humidity.) The sample is approximately 100 nm thick, extends over several millimeters in x-y direction, and its surface corrugation spans about 35 nm. In spite of its inhomogeneous surface topography it is hydrated uniformly enough to exhibit a sufficient conductivity for HSTM imaging. The surface reveals the d-band structure, well known from electron microscopical investigations.

Bovine tendon. Fig. 5 presents the image of a cryosectioned bovine achilles tendon which was prepared on a freshly cleaved mica surface. The well known d-band structure can be seen. Although the bovine tendon section is about 100 nm thick and shows a surface roughness of about 35 nm, it hydrates homogeneously enough to allow imaging with the STM at 66 % relative humidity. In this case, the current has to advance along the tendon surface over several millimeters until it reaches the smooth hydrated mica substrate or the conductive sample holder.

The results documented in this work show that STM on insulators is a straightforward procedure as long as the insulating surface is hydrophilic. Likewise, poorly conducting biological specimens can be imaged irrespective of their lateral and vertical size. Imaging is reproducibly possible without visible changes in the images after repeated scanning. However, during preparation, one has to ensure that the initial solution does not contain too many amphiphilic molecules since they can assemble into a monolayer covering the substrate and/or the sample surface, resulting in a hydrophobic and therefore, nonconductive system.

Independent of imaging, an aging effect of the surface-conductivity is observed on mica as well as on glass. The conductivity of mica decreases substantially within hours. This is probably due to contamination of the mica surface which reduces its surface hydrophilicity. Such contamination is directly visible in HSTM images of aged mica surfaces (several days after cleavage).

In contrast to thin filamentous biological specimens where the measured sample heights are mostly too small by a factor of up to 3 (shown for collagen and DNA), the HSTM method delivers true sample heights in the case of thick biological specimens, as demonstrated here with TMV and bovine achilles tendon. Sample heights of thin but laterally extended structures like mica steps (1 nm) are measured correctly with this method.

The imaging properties are as for all probe microscopy techniques highly dependent on the tip in use. In case of HSTM, some tips tend to build up a water meniscus between the hydrated surface and the tip. The build-up and break-down of a water meniscus leads to oscillations which prevent stable imaging (for details see (Heim *et al.*, 1996), for a discussion see (Fan and Bard, 1995) and (Guckenberger and Heim, 1995)).

Current-voltage relationship of thin water films on mica

The physical mechanism for the surprisingly high conductivity of surface adsorbed water is still unclear. To shed some light on this surface conduction phenomenon we measured the current-voltage relationship of thin water films on mica.

Fig. 6 shows a representative I-V map measured at 50 % relative humidity displaying a strictly linear behavior. Nothing indicative of conventional electrochemical reactions like peaks or a threshold voltage, can be discerned in the I-V map. Scan speeds between 2 V/s and 0.1 V/s gave identical linear behavior in the I-V map. (The plot shown was recorded with 0.1 V/s). The described experiment can also be done with tungsten tips and results in similar maps.

Only in a few cases do peaks appear in the I-V maps, for example. at higher voltages, under very humid conditions or after surface treatment (more detailed experiments are ongoing). This indicates that electrochemical reactions can contribute to the surface conductivity but the major factor of conductivity is a non-electrochemical conduction phenomenon. Recent experiments revealed that the charge transfer between the tip and the sample during imaging is based on the tunneling effect and not on electrochemistry in a tip-sample water bridge. This is indicated by the exponential increase of the current when approaching or retracting the tip from the sample (Guckenberger and Heim, 1995; Heim *et al.*, 1996). Thus, our results support the proposal that the conduction phenomenon is due to the surface-adsorbed water layer itself, rather than conventional ion movement.

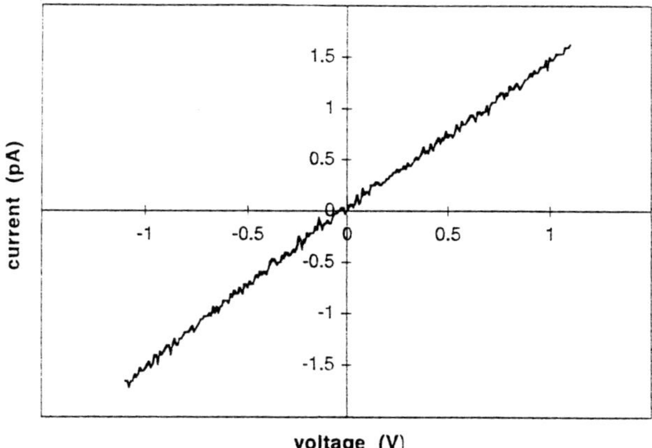

Figure 6: Voltage-current relationship on mica in air at 50 % relative humidity. The map was recorded with a gold tip in mechanical contact to the hydrated mica surface. The second electrode was a ring-shaped, gold-covered sample holder. The hysteresis between increasing and decreasing voltage scan is negligibly small. Therefore, only the curve of increasing sample bias is shown. The linear behavior indicates a non-electrochemical conduction phenomenon. The scan speed was 0.1 V/s.

Conclusions

The images presented in this work document that STM based on the conductivity of surface adsorbed water molecules, termed hydration-STM (HSTM), is a reliable method to image hydrophilic insulators and biological specimens, irrespective of their lateral and vertical size, or their surface roughness. Repeated imaging of the same sample areas can be done without changes in the image. The prerequisite for imaging with the hydration-STM method is that the specimens are hydrophilic, which is the case for most biological materials. The second condition for successful HSTM imaging is that the STM is able to work with low currents (below 1 pA). The specimens are exposed to humid conditions where they adsorb water from the ambient air. Thereby, the hydrated surface becomes sufficiently conductive for STM imaging.

The heights of the HSTM-imaged biological material are realistic for specimens several nanometer thick, whereas in case of thin filamentous samples, the measured heights are too low. In contrast, mica step-heights are measured correctly with HSTM.

The current-voltage relationship at 50 % relative

humidity revealed a strictly ohmic behavior of the thin water films on mica. Therefore, we conclude that a non-electrochemical conduction mechanism is responsible for the surface currents. This knowledge, and the previously published result that electronic tunneling takes place between tip and hydrated mica, leads us to assume that a specific structure of the surface water is responsible for the current transport in HSTM.

Acknowledgements

This work was supported by the Deutsche Forschungsgemeinschaft (SFB 266). Manfred Heim would like to thank the Technische Universität München for financial support. The authors thank H. Wiedemann for preparing the cryosectioned samples of the bovine achilles tendon, Joe Butler for providing tobacco mosaic virus and Mary Kania for critically reading the manuscript.

References

Bockris JOM, Jeng KT (1990) Water structure at interfaces: the present situation. Adv Colloid Interface Sci **33**: 1-54.

Fan F-RF, Bard AJ (1995) STM on wet insulators: electrochemistry or tunneling ? Technical comment. Science **270**: 1849-1851.

Guckenberger R, Kösslinger C, Gatz R, Breu H, Levai N, Baumeister W (1988) A scanning tunneling microscope (STM) for biological applications: design and performance. Ultramicroscopy **25**: 111-122.

Guckenberger R, Heim M, Cevc G, Knapp HF, Wiegräbe W, Hillebrand A (1994) Scanning tunneling microscopy of insulators and biological specimens based on lateral conductivity of ultrathin water films. Science **266**: 1538-1540.

Guckenberger R, Heim M (1995) STM on wet insulators: electrochemistry or tunneling ? Response to technical comment. Science **270**: 1851-1852.

Hacker B, Hillebrand A, Hartmann T, Guckenberger R (1992) Preparation and characterization of tips for scanning tunneling microscopy of biological specimens. Ultramicroscopy **42**: 1514-1518.

Heim M, Cevc G, Guckenberger R, Knapp HF, Wiegräbe W (1995) Lateral electrical conductivity of mica-supported lipid bilayer membranes measured by scanning tunneling microscopy. Biophys J **69**: 489-497.

Heim M, Eschrich R, Hillebrand A, Knapp HF, Cevc G, Guckenberger R (1996) STM based on the conductivity of surface adsorbed water. Charge transfer between tip and sample via electrochemistry in a water meniscus or via tunneling? J Vac Sci Tech B **14**:1498-1502.

Hu J, Xiao XD, Ogletree DF, Salmeron M (1995) Imaging the condensation and evaporation of molecularly thin films of water with nanometer resolution. Science **268**: 267-269.

Israelachvili J (1992) Special interactions: hydrogen-bonding, hydrophobic and hydrophilic interactions. In: Intermolecular and Surface Forces. Academic Press, London.

Israelachvili J, Wennerström H (1996) Role of hydration and water structure in biological and colloidal interactions. Nature **379**: 219-225.

Knapp HF (1996) Electrochemical etching of au tips. In: Procedures in Scanning Probe Microscopies. Part One: Instrumentation. Heckl W, Guckenberger R, eds, John Wiley and Sons, Chichester, in press.

Maivald P, Butt HJ, Gould SAC, Prater CB, Drake B, Gurley JA, Elings VB, Hansma PK (1991) Using force modulation to image surface elasticities with the atomic force microscope. Nanotechnology **2**: 103-106.

Mate CM, McClelland GM, Erlandsson R, Chiang S (1987) Atomic force microscopy studies of frictional forces and force effects in scanning tunneling microscopy. Phys Rev Lett **59**: 1942.

Radmacher M, Tillmann RW, Gaub HE (1993) Imaging viscoelasticity by force modulation with the atomic force microscope. Biophys J **64**: 735-742.

Rohrer H (1990) Scanning Tunneling Microscopy - Methods and Variations. Kluwer, Dordrecht, Netherlands. pp 1-25.

Timpl R, Wiedemann H, Van Delden V, Furthmayr H, Kühn K (1981) A network model for the organization of type IV collagen molecules in basement membranes. Eur J Biochem **120**: 203-211.

Discussion with Reviewers

M. Amrein: The authors ascribe the high surface conductivity of their samples to the ordering of the water layer at the interface to the hydrophilic solid. However, it is the hydrophobic surfaces that are known to increase the ordering of the water in the proximity of the interface. Is this not somewhat contradictory?

Authors: The ordering of the water molecules has been reported on hydrophobic as well as on hydrophilic substrates (Bockris and Jeng, 1990; Toney *et al.*, 1994). On a hydrophilic substrate ordered water molecules can be present on the surface in a thin film without surrounding bulk water. At the surface of a hydrophobic substrate, it is the surrounding bulk water (together with the hydrophobic surface) which enforces the ordering. Without surrounding bulk water the water molecules would flee the interface. We measured electrical conductivity in thin surface adsorbed water layers on a hydrophilic substrate. However, such a conduction phenomenon might also occur in ordered water layers in the vicinity of a hydrophobic surface. To measure the conductivity of ordered water on a hydrophobic substrate is more difficult since the surrounding bulk water will deliver a high background signal.

M. Amrein: It is widely believed (and proven in many cases) that upon air-drying, the surface tension of the water causes a protein sample to collapse. Do you expect your samples to be in a native state?

Authors: Since our sample preparation was very similar to an air drying procedure we expect that the surface tension of the water will exert forces to the samples. Another effect of drying any biological specimen is the loss of structural water. However, this surface water which contributes to the stability of biological macromolecules is preserved when imaging in humid atmosphere. Conditions of high humidity in some cases even suffice to preserve biochemical activity, as it is true for bacteriorhodopsin in purple membrane.

A. Quist: How close to the surface can you go without building up a water meniscus between the tip and the surface? You describe that a water meniscus is absent; can you relate the presence/absence of a water meniscus between tip and sample to the absence/presence of a specific structure of the surface adsorbed water?

Authors: To evaluate the distance between the tip and a hydrated sample surface is quite difficult and we can only indirectly estimate this distance. In experiments where the tip was approached, and subsequently retracted from a hydrated mica surface while recording the electrical current, we found that a water meniscus between tip and sample could be pulled to a length of only about 1 nm (Heim *et al.*, 1996). This indicates that the tip can approach the sample to a distance of probably 1 nm or less without the build-up of a water meniscus.

We can only speculate about the structure of the water in case of the presence of a water meniscus.

The water molecules are probably bound quite strongly to a hydrophilic substrate, decreasing the tendency of the surface adsorbed water to build up a water meniscus even when the tip approaches within molecular distances. The forces between substrate and water molecules would on the one hand, induce the ordering of the water (and thus induce the conductivity) and on the other hand, prevent the build-up of a water meniscus between the tip and surface.

A. Quist: The collagen is measured at one third of its true height. Is collagen (or other small molecules) "buried" partially in the water layer, resulting in a lower height measurement?

Authors: If the collagen molecules were partially buried in the water layer (without any flattening of the molecules itself) the measured height would be reduced by no more than the thickness of the water layer on the surrounding mica (less than 0.5 nm according to (Beaglehole *et al.*, 1991) and our measurements). The height of thin filamentous molecules like collagen or DNA is found to be reduced by more than 0.5 nm (which is found also in most AFM experiments). This indicates that the measured height artifacts cannot arise from the water layers alone. One possibility is that the conductivity on the collagen molecules is reduced, resulting in a decreased tip-sample distance on the collagen causing reduced height values of the molecules. Another reason for the reduced heights could be real flattening of the samples due to the influence of surface forces. This interpretation is supported by the observation that thin but laterally extended biological objects (e.g., the lipid layers) do not show such a dramatic height artifact. Such layers can not be so easily flattened since the lipid molecules are stabilized by neighboring molecules. However, it is difficult to believe that helical structures like DNA and collagen are flattened so dramatically. Further experiments are necessary to understand the background of the observed height artifacts.

A. Quist: Different images are taken with very different biases, (+ and -). Why? Is there a difference for each sample?

Authors : It is possible to image all samples at all biases with a good quality. While doing the experiments we often changed the bias as well as the current without finding preferences for the different samples. As a general tendency, imaging seems to be more stable at positive sample bias (Heim *et al.*, 1996).

A. Quist: Can any of the holes observed on glass be due to patchy adsorption of water on the surface?

Authors: The holes greater than ten nanometers in diameter and several nanometers deep, do not arise from patchy adsorbed water. If there were no holes at these places, the tip would approach several nanometers to the glass, hit the glass surface and break. Furthermore, these holes appear in AFM images of the samples as well as in STM images of metal coated coverslips.

The smaller (2 - 5 nm in diameter) holes, which appear to be only 1 - 2 nm deep are not visible in AFM images, and at least in principle, could arise from holes in the water layer. At such places the conductivity is reduced or even nonexistent leaving a hole in the HSTM image. Nevertheless, we think that the small holes are real topographical holes since we have never found such phenomena on other substrates like mica or thermally oxidized silicon wafers.

Additional References

Beaglehole D, Radlinska EZ, Ninham BW, Christenson HK (1991) Inadequacy of lifshitz theory for thin liquid films. Phys Rev Lett **66**: 2084-2087.

Toney MF, Howard JN, Richer J, Borges GL, Gordon JG, Melroy OR, Wiesler DG, Yee D, Sorensen LB (1994) Voltage-depending ordering of water molecules at an electrode-electrolyte interface. Nature **368**: 444-446.

Scanning Microscopy Supplement 10, 1996 (pages 81-96) 0892-953X/96$5.00+.25
Scanning Microscopy International, Chicago (AMF O'Hare), IL 60666 USA

IMAGING MOLECULAR STRUCTURE OF CHANNELS AND RECEPTORS WITH AN ATOMIC FORCE MICROSCOPE

Ratneshwar Lal

Neuroscience Research Institute, University of California, Santa Barbara, CA 93106,

(Received for publication October 10, 1995 and in revised form April 24, 1996)

Abstract

Biological membranes contain specialized protein macromolecules such as channels, pumps and receptors. Physiologically, membranes and their constituent macromolecules are the interface surfaces toward which most of the regulatory biochemical and other signals are directed. Yet very little is known about these surfaces. The structure of biological membranes has been analyzed primarily using imaging techniques that are limited in their resolution of surface topology. An atomic force microscope (AFM) developed by Binnig, Quate and Gerber, can image molecular structures on specimen surfaces with subnanometer resolution, under diverse environmental conditions. Also, AFM can manipulate surfaces with molecular precision: it can nanodissect, translocate, and reorganize molecules on surfaces. The surface topology has been imaged for several hydrated channels, pumps and receptors which were a) present in isolated native membranes, b) reconstituted in artificial membrane or, c) expressed in an appropriate expression system. These images, at molecular resolution, reveal exciting new findings about their architecture. AFM induced "force dissection" reveals surfaces which are commonly inaccessible. In whole cell studies, in addition to the molecular structure of membrane receptors and channels, correlative electrical and biochemical activities have been examined. Such study suggests a "single cell" experiment where the structure-function correlation of many cloned channels and receptors can be understood.

Key Words: Scanning force microscopy, Biomolecular imaging, Quaternary structure, Imaging receptors and channels, gap junctions, porins, acetylcholine receptor, vacuolar ATPase, bacteriorhodopsin, cholera toxin.

*Address for correspondence:
R. Lal
Neuroscience Research Institute,
University of California
Santa Barbara, CA 93106,
Telephone number: 805-893-2350
FAX number: 805-893-2005
E-mail: rlal@sbphy.physics.ucsb.edu

Introduction

Currently, molecular structures of channels and receptors are being studied primarily by electron microscopy, electron and X-ray diffractions, and infra-red spectroscopy. Molecular functions, on the other hand, are being examined using various biochemical, electrophysiological and molecular biological techniques. However, it is difficult, if not impossible, to combine both structure and function studies with these techniques, mainly because the techniques used for structural studies require an unfavorable operating environment and an extensive sample preparation which limits the functional states of biological macromolecules. Moreover, these techniques provide very little information about the surfaces of channels and receptors, the very sites of molecular interaction.

An atomic force microscope (AFM) (Binnig *et al.*, 1986) can image the three-dimensional (3D) surface structure of a wide variety of biological specimens (for reviews see Lal and John, 1994; Hansma and Hoh, 1994, Yang and Shao, 1995). AFM can image, at molecular resolution, native samples bathed in an aqueous medium. This provides the opportunity of observing biochemical and physiological processes in real time at molecular and often atomic level and thus can be used for direct molecular structure-function studies of membrane channels and receptors. In addition, AFM is uniquely qualified to provide topographical information of the surfaces to which most of the regulatory biochemical and other signals are directed. Other microscopical techniques such as the scanning electron microscope (SEM) can also view surfaces. However, unlike SEM, AFM can image in an aqueous environment and often at a greater resolution. AFM, due to its significantly greater signal-to-noise ratio, can provide an accurate measure of long-range packing such as crystallinity even for a small cluster of particles (for example, 20 x 20 particles).

AFM has been used to image a wide variety of individual biological macromolecules, whole cells, membranes and membrane-bound proteins, and a few interactive processes (for review see Lal and John, 1994). We have recently imaged ion channels in a)

native membranes such as gap junctions (Hoh *et al.*, 1991; Lal *et al.*, 1995b), b) reconstituted membranes such as porin channels (Lal *et al.*, 1993), and c) expressed in *Xenopus* oocyte such as acetylcholine receptors (AChR) (Lal and Yu, 1993). Our development of membrane *force dissection* (Hoh *et al.*, 1991, Lal *et al.*, 1995a) suggests that the commonly inaccessible regions can be imaged. Dynamic biochemical processes such as enzyme-substrate reactions, crystal growth and protein polymerization, and physico-chemical properties such as elasticity, viscosity and various chemical forces of biological specimen have been studied (Drake *et al.*, 1989; Durbin *et al.*, 1993; Florin *et al.*, 1994; Shroff *et al.*, 1995). In addition, an AFM can be used as a nanomanipulator (for nanodissection and for force-induced changes in molecular conformations; Hoh *et al.*, 1991; Lal *et al.*, 1995a; Muller *et al.*, 1995) as well as an exquisite sensor of molecular forces - hydrogen bonds, Van der Waals and electrostatic forces (for review, see Lal and John, 1994).

Principle of Operation

For a detailed treatise on the principle of operation, resolution, and limitations of AFM, see Lal and John, 1994.

Briefly, an AFM image consists of a series of parallel line contours obtained by scanning the surface of a specimen placed on a xyz translator across a molecularly sharp tip with a small (~ a nanoNewton (nN)) tracking force. The tip is mounted on a microcantilever with a small spring constant. As the tip is scanned, the net interaction force between the molecules on probe tip and sample surface deflects the tip and hence the cantilevered spring. The deflection is sensed by detecting the angular deflection of a laser beam reflected off the back of the mirrored cantilever, which is converted to electrical signals by photodetectors. An electronic feedback loop keeps the cantilever deflection and hence the tracking force constant by moving the sample up and down as the tip traces over the contours of the surface (Figure 1). An image is obtained by plotting the vertical motion (z) of the xyz translator and hence the sample height (z) as a function of specimen lateral position (xy).

Recent improvements in AFM imaging include the "Tapping mode imaging" (Hansma *et al.*, 1994a) and combined light fluorescence, laser confocal and atomic force microscopes (Hansma *et al.*, 1994b; Hillner *et al.*, 1995; Lal *et al.*, 1995a,b). In the tapping mode imaging, the lateral force of interaction is significantly reduced and a considerably greater lateral resolution has been achieved on imaging isolated biological macromolecules. The inverted microscope based dual fluorescence and force microscope (Figure 1) allows simultaneous multimodal imaging and the stationary sample stage facilitates the ease of combining other electrophysiological and biochemical techniques for correlational studies.

Two major factors for the AFM to be applicable to biological imaging are:

a) Resolution: With AFM, the images are formed by reconstructing the contour of the interaction forces between the probe and the specimen. By selecting a small imaging area and appropriate operating conditions, one can distinguish adjacent structures less than a nanometer apart as the signal-to-noise ratio in an AFM image is sufficiently large compared to EM and other techniques. For hard crystals such as calcite, atomic resolution is obtained (Ohnesorge and Binnig, 1993). For soft biological specimens such as a living cell, the resolution is reduced (for review, see Lal and John, 1994). In biological specimens with a high density of protein and limited mobility such as membranes with channels and receptors, however, the resolution could be comparable to that for a hard crystal. On many biological membranes including gap junctions, Na^+-K^+ ATPase, AChR, bacteriorhodopsin and porin channels, a subnanometer resolution has been obtained (see below).

b) Identity of the imaged structures; correlative studies: Though AFM could provide, with molecular resolution, surface information of crystalline as well as amorphous materials, it is often difficult to identify the nature of individual components, especially if the surface contains a heterologous population of structures. This is often the case with most of the biological membranes, except in favorable systems like purified native gap junctions (Lal *et al.*, 1995b), purple membrane bacteriorhodopsin (Butt *et al.*, 1990; Muller *et al.*, 1995) and reconstituted channels and receptors (porins and cholera toxins (Lal *et al.*, 1993; Mou *et al.*, 1995). The suitability of using the AFM for biological imaging will rely on an unambiguous identification of imaged structures. For mixed macromolecules, critical comparisons obtained from AFM and alternative/complementary techniques, such as structural probes of EM and X-ray diffraction, biochemical and immunological binding assays (Figures 2,3), pharmacological labeling and electrophysiological measurements will be essential (Lal *et al.*, 1995b; Lal and Yu, 1993; Kordylewski *et al.*, 1994; Bustamante *et al.*, 1995; Han *et al.*, 1995; Arakawa *et al.*, 1992; for review see Lal and John, 1994).

Kordylewski *et al.* (1994) have reported an elegant correlative study between AFM and the transmission electron microscopy (TEM). They examined freeze fractured replicas of adult rat atrial tissue by both TEM and AFM (Figure 2). The same replicas were analyzed and the same details were identified which allowed a critical comparison of surface topography by both

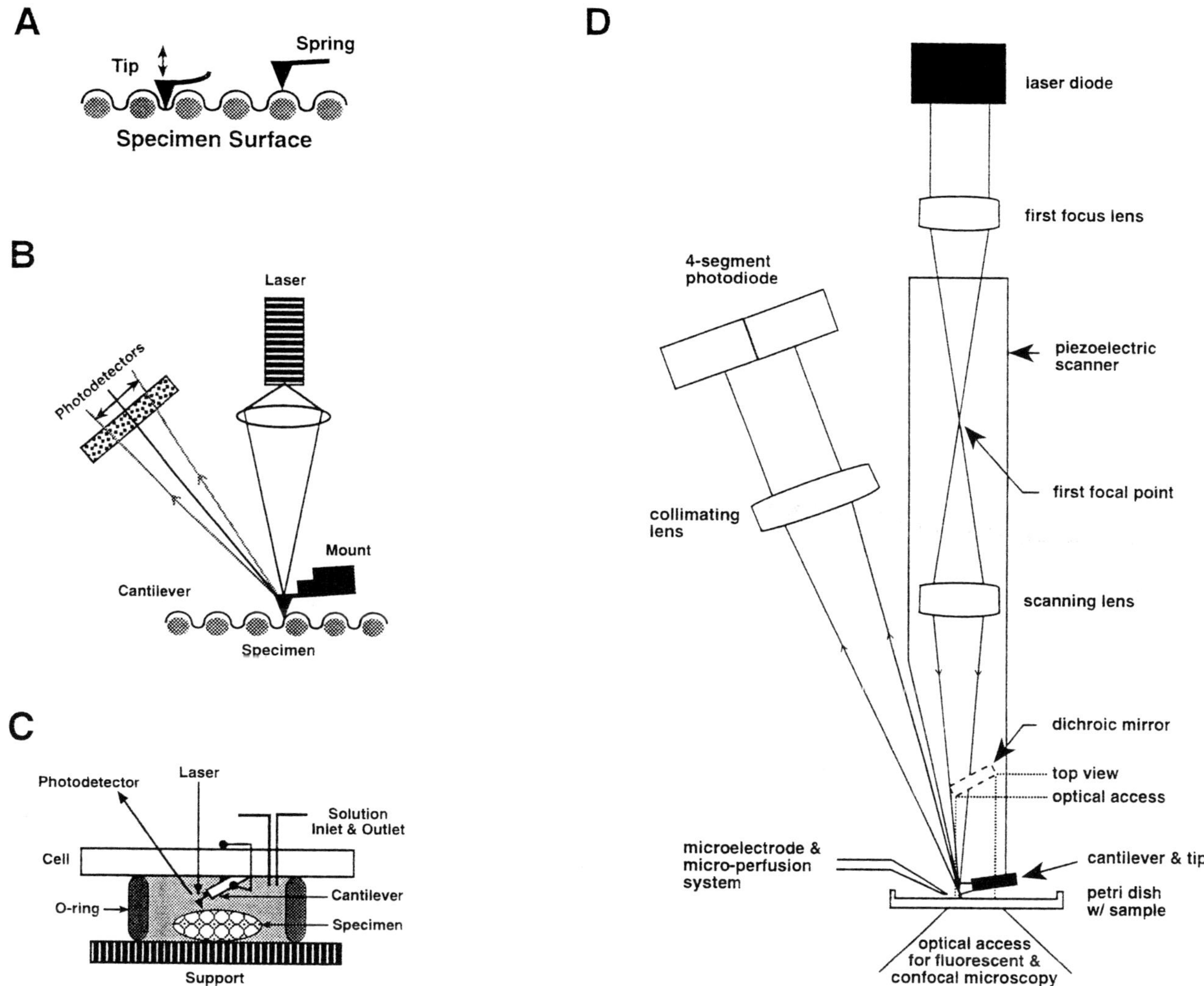

Figure 1: Schematics of operating principle of an AFM. **A**. Force sensor for an AFM. The force sensor is a spring deflection system consisting of a sharp tip micro fabricated onto a cantilever spring with a small spring constant. As the tip scans the specimen surface, the interaction forces between the tip and specimen deflect the tip and hence the cantilever. The deflection equivalent to the interactive forces (as a function of tip-surface separation) is used to create 3-D surface topography. **B**. An optical deflection detection system, which is the most commonly used. A cantilever deflection reflects the incident collimated laser beam. The reflected beam is incident asymmetrically on position sensitive photo detectors. The output of these detectors is used in a feedback loop to keep the deflection (and hence the force) constant while the tip scans the specimen. **C**. An AFM with a fluid cell. Both the cantilevered tip and the specimen are immersed in a fluid. The surface of the fluid cell is transparent and has two openings for fluid inlet and outlet. **D**. A schematic diagram of the optical component of a combined light fluorescence and force microscope. The first focal point is located inside the upper portion of the piezoelectric scanner. After the positions of the lenses are adjusted, the scanning focused spot accurately tracks the cantilever and the zero-deflection signal from the 4-segment photodiode is independent of position within the scan area. One of the key advantages of the new AFM is that there is optical access to the sample from above and below. Thus the new AFM can be combined with an optical microscope of high numerical aperture. The other key advantage is that since the sample is stationary during scanning and can be large, techniques for on-line perturbations and recordings can easily be incorporated. (For details, see Lal and John, 1994, and Hansma *et al.*, 1994b).

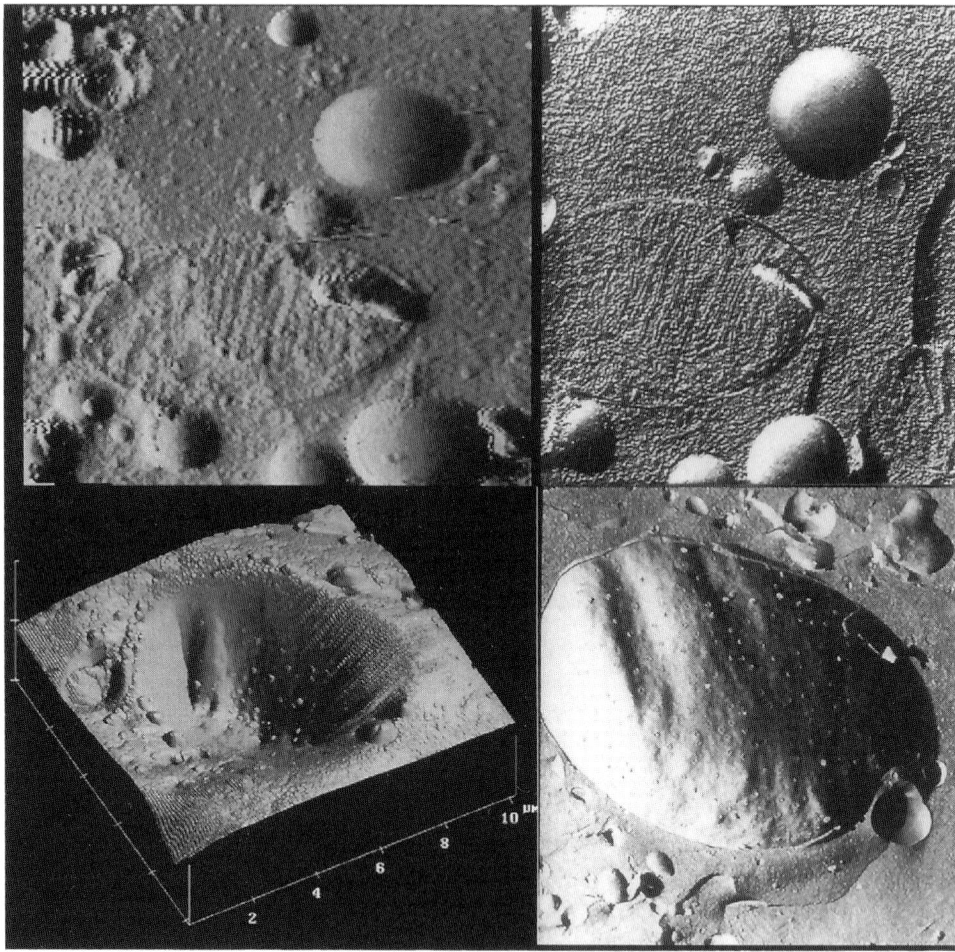

Figure 2: Comparison of the same structural details in a freeze-fracture replica of a rat cardiac tissue imaged with AFM (left) and TEM (right). Top: atrial granules and mitochondrion with cristae; bottom: a cell nucleus with pores in the nuclear envelope. (for details see Kordylewski *et al.*, 1994).

---

techniques. AFM images of large scale subcellular structures (nuclei, mitochondria, granules) correlated well with those from the TEM. However, AFM images of smaller features and surface textures appeared somewhat different from TEM images. This presumably reflects the difference in the surface sensitivity of the AFM vs TEM, as well as the nature of images in AFM (3D surface contour) and TEM (2D projection). In addition, AFM images provided new information about the replica itself. Unlike TEM, it was possible to examine both sides of the replica with AFM; the resolution on one side was significantly greater compared to the other side. It was also possible to obtain quantitative height information which is not readily available with TEM.

The simple design of AFM allows it to be integrated with other techniques, such as fluorescence and laser confocal microscopies (Hillner *et al.*, 1995; Lal *et al.*, 1995b; Figure 3). This can permit an independent verification with appropriately labeled markers: first use fluorescent signals to identify specific areas and then use atomic force imaging to obtain the ultrastructural details. We have imaged purified cardiac gap junctions that were labeled with anti-Cx43 antibody (Figure 3). All fluorescently labeled regions when imaged by AFM showed gap-junctional plaque-like features. In this preliminary study, as our goal was simply to locate all the membranes that were gap junction specific, we had used a considerably large amount of antibodies which covered the entire membrane, resulting in the net increase of ~ 5 nm in membrane height. Antibodies adsorbed to the glass substrate alone had a similar height variation. Thus, it is feasible to obtain molecular resolution structural information on membrane proteins present singly and in small clusters as long as they have detectable immunofluorescence signals.

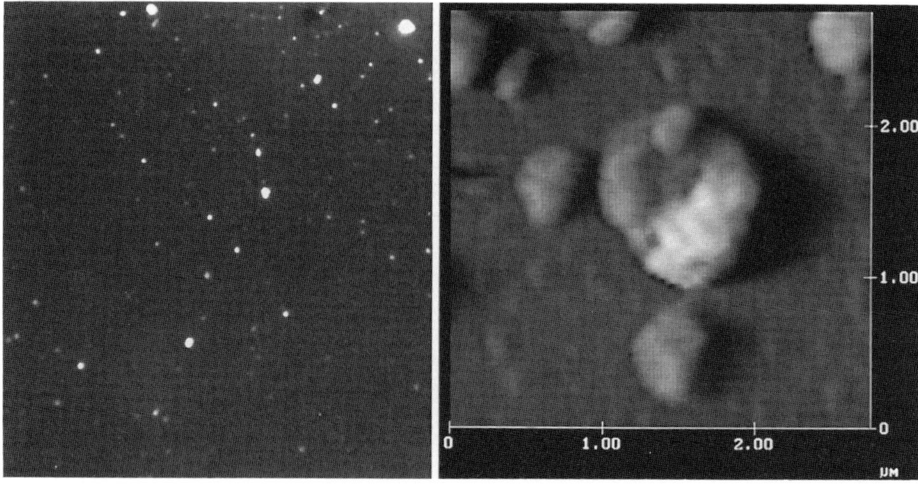

Figure 3: Left: Fluorescence image of anti-Cx43 labeled gap junctions. Right: AFM image of gap junctions in the same field of view which showed fluorescence labeling (for details see Lal *et al.*, 1995b).

Sample Preparation

The atomic force microscope can be used to image specimens in both aqueous and dry conditions, at ambient temperature and pressure. Imaging conditions influence the choice of substrate, the stability of the specimen with respect to its interaction with the probe and the preservation of the specimen with respect to its physiological or biochemical functions.

The physico-chemical characteristics of the sample and sample-support interactions determine or suggest ways under which it can be imaged. Problems encountered can merely be as simple as getting the sample to attach to the support. Techniques can include the drying down of samples and adsorption against specially prepared surfaces. Low sample-support interaction requires that low forces be used for imaging or the tip essentially sweeps the sample from the support. Originally, graphite (HOPG hydrophobic uncharged), mica (hydrophilic negatively charged) and glass were the supports most routinely used. These supports can also be modified chemically to adjust their hydropathicity, charge density, and their polarity. Today the repertoire has expanded greatly.

It is possible to modify or coat the support such that it can act as a ligand for the sample and thus orient the specimen in a defined way. It is also possible to use artificial systems to generate constraints where there were none before. One such elegant study is that of Yang *et al.* (1993) who incorporated the cholera toxin into synthetic phospholipid bilayers followed by covalent cross-linking. The pentameric structure of the B-subunit was clearly visible. This approach may prove to be widely applicable to other isolated membrane proteins.

Simple adsorption in the presence of appropriate counterions or cross-linking of the membrane fragments to a chemically modified substratum (such as covalent linking) is the most commonly used sample preparation for membrane-bound proteins Karrasch *et al.*, 1993; Yang *et al.*, 1993; Mou *et al.*, 1995). However, such method may not be suitable to observe structural changes *in situ*. For double bilayer membranes such as gap junctions and reconstituted vesicles and planer bilayers, they nevertheless, can leave the upper portions of the proteins free for conformational changes in response to on-line perturbations.

A novel sample preparation system for imaging living cells consists of having a suction pipette hold single cells. By this approach Hörber *et al.* (1992) showed the vaccinia virus leaving a kidney cell. Recently, Hörber *et al.* (1995) have adsorbed membrane patches to the tip of a glass patch pipette and imaged the surface morphology and its modification as a function of the applied force.

Representative Examples

Purified membranes

Gap junctions and hemi-junctions: Proteins that are naturally embedded within membranes and form 2-D crystalline arrays have been examined unfixed and under appropriate buffers. High resolution images of the extracellular face of liver gap junctions show a typical hexameric packing of gap junction channels (connexons?) with pore like indentations (Hoh *et al.*, 1991, 1993). Similar observations were made with cardiac gap junctions (Lal *et al.*, 1995b).

The gap junction channels consist of two hemichannels (connexons), one arising from each plasma membrane. Our AFM images present evidence of hemiplaques in purified heart gap junctions samples (Figure

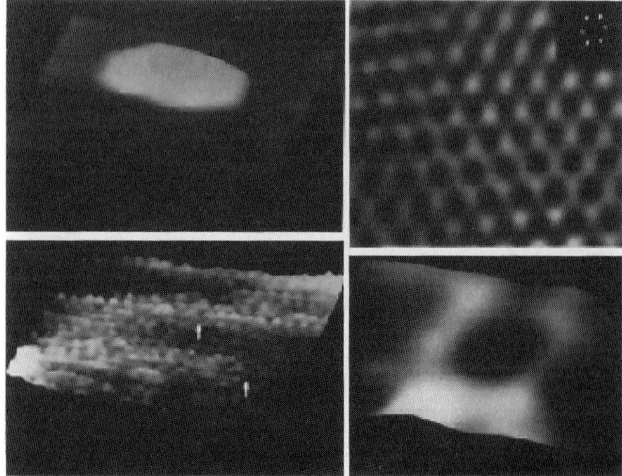

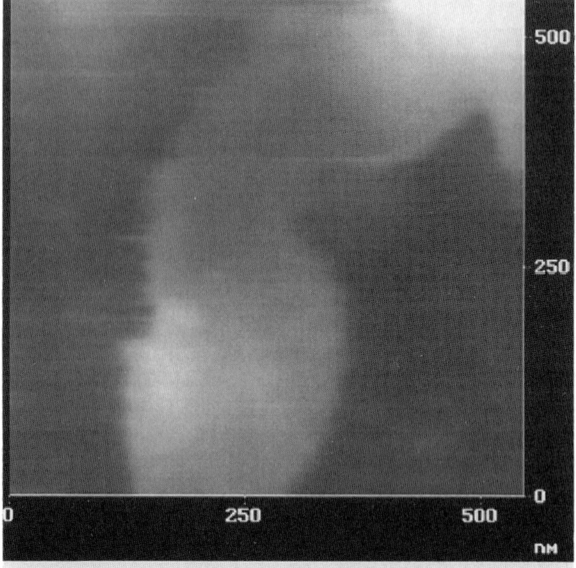

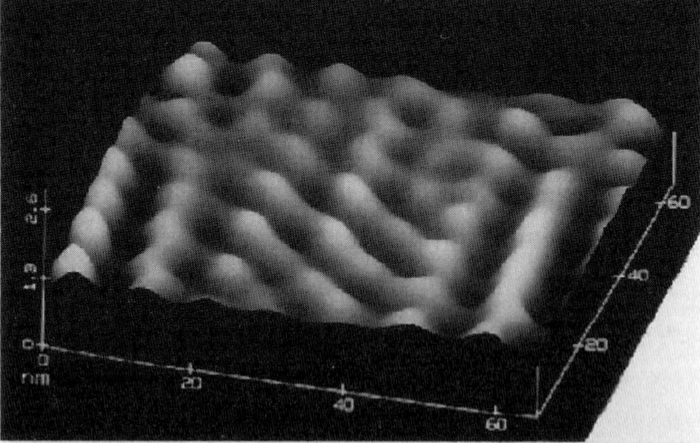

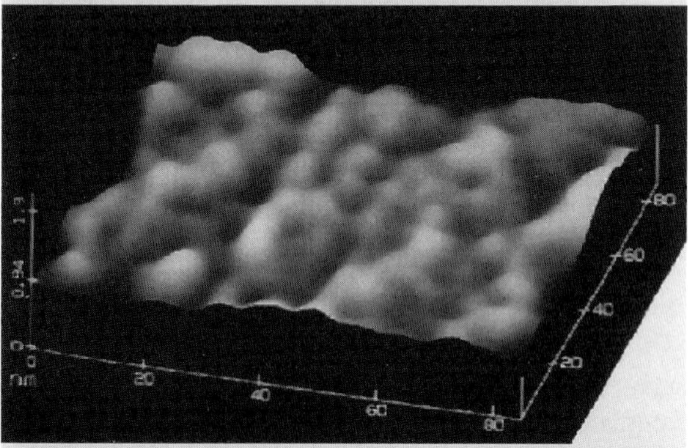

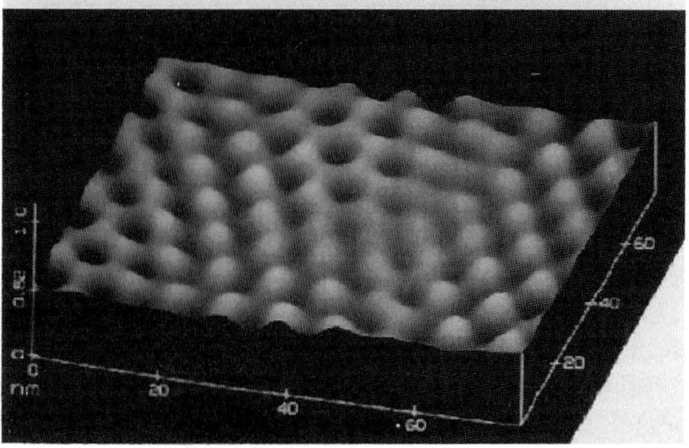

Figure 4: Images of hemi-gap junctions in 30% dextrose solution. Top left: Low magnification surface view image. The scan size is 0.9 x 0.9 μm^2. The thickness of membrane is about 9.6 nm. Bottom left: Medium resolution (scan size 130 x 130 nm^2) surface view image showing arrays of particles which presumably constitute connexons (arrows). The images in Figures A and B are unfiltered. Top right: Inverse Fourier transformed, top view image of another scan area showing a hexagonal long-range packing of connexons with a center to center spacing about 9.7 nm. Bottom right: High resolution (scan size 10 x 10 nm^2) surface view image of a single connexon. Subunits (1-1.5 nm diameter) and a funnel shaped central pore with the outer diameter of abour 2.5 nm are visible. Number of subunits (5 or 6) is difficult to distinguish as reported for liver gap junctions (Hoh *et al.*, 1993). For details see Lal *et al.* (1995b).

Figure 5 : AFM images of vacuolar proton pumps (V-H^+-ATPases). Purified acidosomes were imaged dry or in PBS. Top panel: membranes typical of the purified acidosome fraction. They appear to be collapsed vacuoles, typically <0.5 μm in diameter. Wet imaging. Second panel from the top: Inverse 2D FFT filtered image of the outer surface (presumably, cytoplasmic). A hexagonal long-range packing of proteins with central pore-like depressions is visible. The center-to-center spacing of these pores are ~ 12.8 nm. Dry imaging. Second panel from the bottom: Low pass filtered image of the integral portion of the acidosomes. The head-, and stalk- portions of the acidosomes were biochemically dissected. Structures with central pore-like depression enclosed by proteins with uneven protrusions are apparent. Wet imaging. Bottom panel: Inverse 2D FFT filtered image of the integral portion showing a hexagonal long-range packing. Wet imaging.

4) (Lal *et al.*, 1995b). These samples were characterized by SDS PAGE, thin section EM and negative staining. We have characterized these as hemiplaques based on their thickness (~7-11 nm), as compared to whole junctions (~ 22-25 nm); the presence of subunits with a ~10 nm center-to-center spacing; and the quasi-crystalline hexagonal subunit packing. These dimensions are in close agreement with those determined by high resolution EM. Hemi-plaques comprised ~17-18 % of all plaques imaged and were independent both of sample treatment (trypsinization, and glutaraldehyde fixation) and of imaging conditions (dry, PBS, and 30% dextrose solution). At higher resolution, we were able to delineate individual connexons and subunit structures, showing a central depression ~1.5-2.8 nm in diameter, perhaps representing a pore, and the surrounding 5-6 subunits with an overall diameter of ~6 nm.

It is noteworthy that recent electrophysiological, molecular and immunocytological studies indicate the presence of individual hemichannels in the plasma membrane, though their clustering in plaque-like structure is unclear. Hemichannels, in native plasma membranes, would have to remain closed during normal condition for the cell viability or permselectivity. The role of hemichannels in single cells could be related to the exchange of molecules between the cytoplasm and the extracellular region, regulation of cell volume and programmed cell death. Alternately, the putative hemi-plaques may simply be an intermediate stage of junction formation, or it may be the breakdown product.

Purple membrane: Purple membrane has been imaged with an AFM in Several experimental conditions. Molecular resolution images of purple membrane show a three fold symmetry suggesting a trimeric arrangement of bacteriorhodopsin molecules (Butt *et al.*, 1990; Muller *et al.*, 1995a,b). Subunit structure and a force-induced conformational changes have also been reported. These two membrane proteins have been extensively characterized by alternative techniques such as EM. The results from AFM studies are in remarkable agreement with those of negative stain EM images.

One advantage of using AFM for membrane imaging is that the membrane polarity (extracellular vs cytoplasmic surface) could be made unambiguously. For example, a double bilayer gap junction will always have a cytoplasmic surface facing upward. The other advantage is a more precise measurement of the membrane thickness. Such measurement was used to directly and easily distinguish single bilayer vs double bilayer gap junction structures, an observation further confirmed by Nanodissection of the upper layer of the gap junctions (Lal *et al.*, 1995b; Hoh *et al.*, 1991). For purple membranes the thickness measured was substrate dependent, an interesting and important observation that may reflect

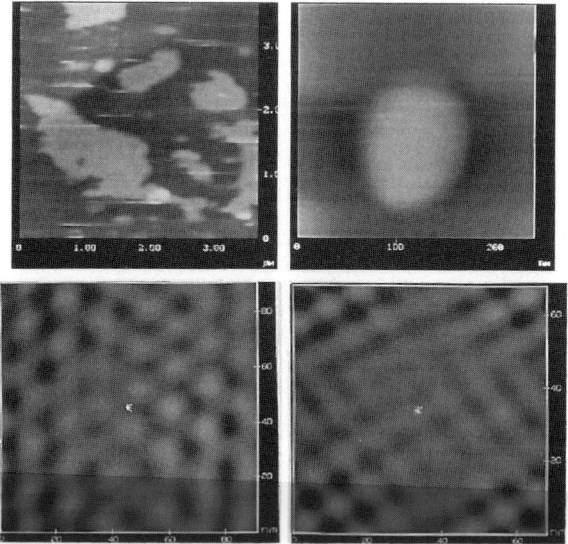

Figure 6: Images of DMPC vesicle morphology and reconstituted OmpF porins for different protein-to-lipid ratios. The protein concentration was kept constant and the lipid concentration was changed ($\mu g/\mu g$) such that the protein-to-lipid (P/L ratio) varied from 0.2 to 2.0. Top left: Vesicles with P/L ratio of 0.2. The vesicles are large and irregular. Top right: A vesicle with P/L ratio of 2.0. The vesicles are small and oval. Bottom left: Inverse 2-D FFT filtered image of porin channels on the upper surface of a vesicle with a P/L ratio of 0.2. Porins appear to be arranged in a long-range hexagonal order. Bottom right: Inverse 2-D FFT filtered image of porin channels on the upper surface of a vesicle with a P/L ratio of 2.0. Porins appear to be arranged in a long-range rectangular order.

as yet uncharacterized tip-sample-substrate interactions.

HPI layer: Karrasch *et al.* (1993) have imaged, in buffer, the hexagonally packed intermediate layer (HPI layer) of *D. radiodurans* covalently linked to derivatized glass. At low magnification, flat single membrane layers were visible. These layers consist of doughnut-shaped units which are packed in a hexagonal arrangement. Interestingly, HPI layers were not previously amenable to AFM imaging in buffer without covalent immobilization on a solid substrate. In the present study, the resolution was further improved during repetitive scanning of the same region, though it is unlikely that there was a force-induced rearrangement.

Na^+,K^+-ATPase: Paul *et al.* (1994) have obtained molecular resolution images of Na^+,K^+-ATPase present in purified canine kidney membranes. Imaging under dry condition but using the "tapping mode" of imaging, they reported a channel-like structure with a central pore in the middle of each macromolecule on the cytoplasmic

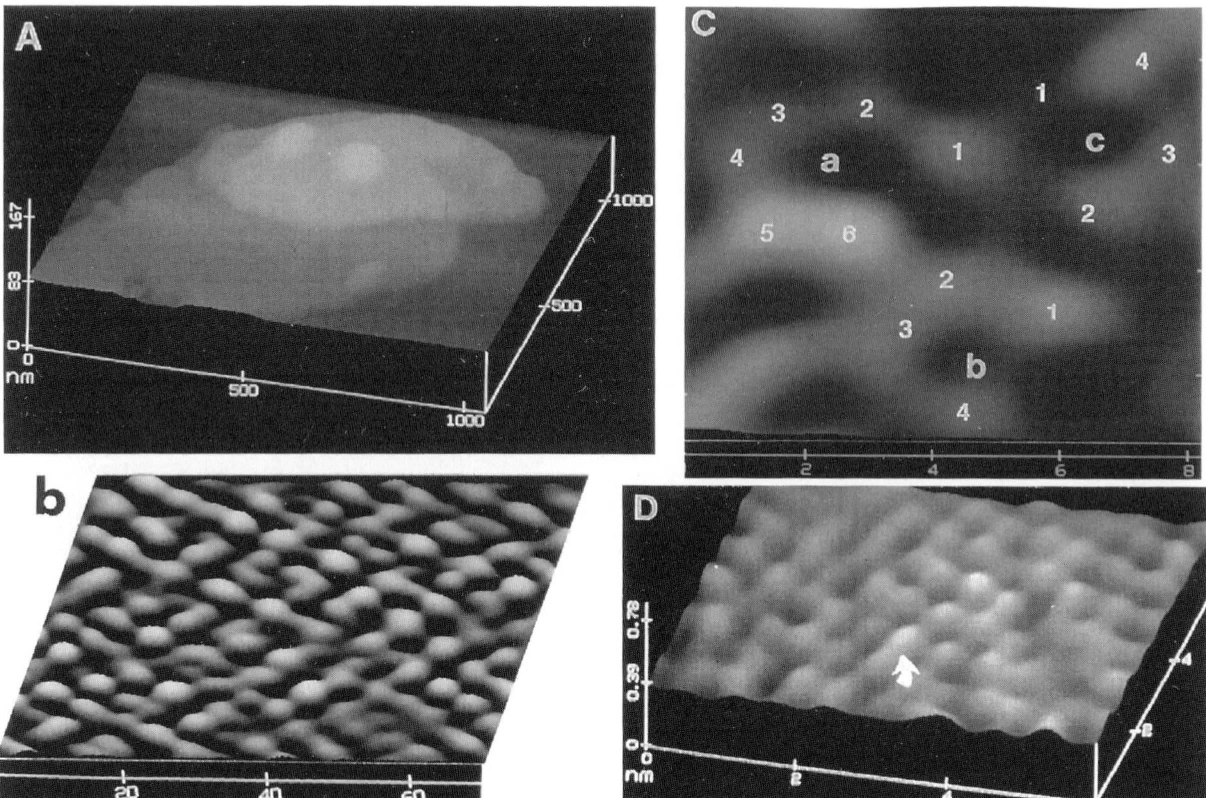

Figure 7: Reconstituted porin channels imaged with an AFM. **A**. Reconstituted DMPC lipid vesicles imaged in PBS. Note, the overlapping multilammelar vesicles. **B**. Inverse 2-D FFT filtered image of *Bordetella pertussis* porin channels reconstituted in DMPC vesicles and imaged under PBS with an AFM. The raw image showed identical pattern but was noisy. Note that the long-range packing is rectangular. Unit cell (~7.9 x 13.8 nm) shows two trimers. The lattice parameters are similar to that reported by EM study. **C**. Subnanometer resolution **unfiltered** image (in 3-D) showing the fine structures in the individual monomers. Note several "bead"-like structures and asymmetrically located pores (marked a,b,c). **D**. Subnanometer resolution images of reconstituted DMPC vesicles without any protein added. Only the filtered image is shown although a similar pattern was visible in the raw data. Individual phospholipid headgroups (~0.5 nm in size; arrowhead) appear to be arranged in a crystalline fashion. For details see Lal *et al.*, 1993.

---

face of the isolated membranes. The pore diameter ranged from 0.6 -2 nm for a range of different sample treatments. Additionally, in the regions where the protein macromolecules were absent, they were able to see individual lipid head structures with an orthorhombic lattice.

Vacuolar proton pumps (V-H^+-ATPase): Vacuolar proton pumps (V-H^+-ATPase) present in *Dictyostelium discoideum* have been imaged in our laboratory. The acidosomes were isolated and biochemically characterized by Nolta *et al.* (1991). They have previously reported (Nolta *et al.*, 1991) that these vacuolar proton pumps are made of eight subunits, two polypeptides constituting the integral portion while the other six polypeptides are peripheral. Based on the analogy with V-H^+-ATPases of other species, it was proposed that the head-portion consists of two polypeptides, each as three catalytic and three regulatory subunits, and four other polypeptides making the stalk. The V-H^+-ATPase appears to be distinct from either the F_1F_2-type or the E_1E_2-type ATPases.

Figure 5 shows AFM images of V-H^+-ATPase in purified acidosomes. Top panel shows a typical membrane which appeared to be a collapsed vacuole and resembles the previous EM images (Nolta *et al.*, 1991). We were unable to obtain molecular resolution on head-portion of isolated acidosomes while imaging in an aqueous medium. However, on fixed and air-dried membranes, we obtained a higher resolution (~6 nm) on the head-portion of the outer (presumably, cytoplasmic) surface. In some areas of membranes, V-H^+-ATPases were highly packed and as shown in the second panel from the top they were arranged in a hexagonal order with a center-to-center spacing of ~13 nm. Nolta

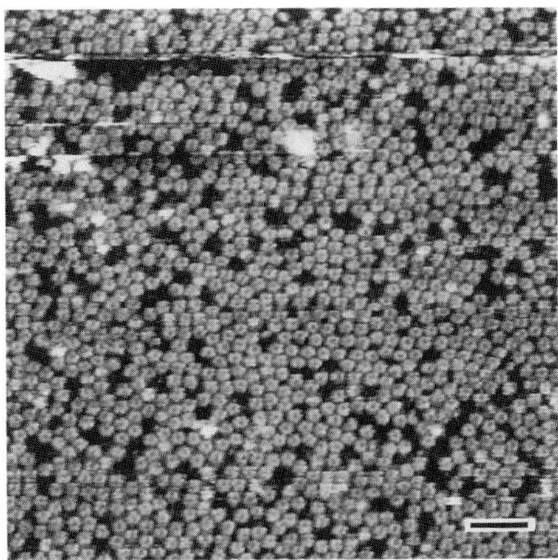

Figure 8: AFM images of individual cholera toxin B-oligomers which were bound to the gangliosides in a bilayer made of egg-PC by the vesicle fusion method. The full coverage of the bilayer is achieved with 10 mol% GM1. Left panel: medium resolution image of the reconstituted bilayer in the gel phase. Scale bar = 30 nm. Right panel: High resolution image. The image quality is comparable to that for the gel-phase image (left side) suggesting that the fluidity of the bilayer did not seriously reduce the resolution in AFM images. B-oligomers with both 6-fold and 5-fold symmetries are visible. Missing subunits are also visible in some B-oligomers. Scale bar = 10 nm. (For details, see Mou *et al.*, 1995).

---

et al. (1991) have previously reported from their EM studies, polygonal particles ~12-13 nm in size with a 3-fold rotational symmetry and suggested that they represent the head portion of the whole V-H^+-ATPase.

After biochemical dissection of the V-H^+-ATPase (Nolta *et al.*, 1991), we were able to obtain a higher resolution (~3 nm), while imaging in an aqueous medium, on the remaining integral portion of V-H^+-ATPase. The bottom two panels show arrangement of individual ATPases. The surface appears to have a fluctuation in height of subunits perhaps reflecting variable extra-membranous protrusions of subunits forming the integral portion. These ATPases were spaced ~13 nm apart. A highly processed image (bottom panel) shows their hexagonal packing. Further detailed studies will be required to determine the correspondence of each head portion to their integral counter part. Such study is feasible by first imaging the whole membrane (and thus imaging the head portion), force-dissecting the head and stalk portions and then imaging the integral portion.

Reconstituted channels and receptors

AFM imaging is not confined to native membranes only. Examples have already been given of using synthetic membranes to provide order to embedded proteins, a technique that may prove beneficial for imaging proteins that do not ordinarily form extended arrays. Membrane proteins forming 2-D arrays in artificial systems have also been imaged.

OmpF porin: Bacterial porins are one of the best studied channel forming membrane proteins. Recently, a number of these porins reconstituted as 2-D crystals into DMPC lipid vesicles were imaged in a fluid environment. For OmpF porins reconstituted in DMPC vesicles, Lal *et al.* (1993) reported that these vesicles assume flattened, double bilayer configurations and also provided support for the 2-D crystal porins (Figures 6,7) (Lal *et al.*, 1993). Often, multilammelar vesicles with incremental step membrane thicknesses were apparent. Molecular resolution images show the predicted trimer formation of porins: similar to that shown by X-ray crystallography and EM reconstruction. Their long range packing was dependent on the lipid-to-protein ratio. OmpF porins showed a mixed hexagonal and rectangular packing, the center to center spacing was consistent with the X-ray diffraction study and the mixed patterns correlate well with the given protein to lipid ratio (0.7).

We undertook a systematic study of the protein-to-lipid interaction and its role in vesicle formation and the regularity of protein packing. Results from our preliminary observation are shown in Figure 6. When the protein-to-lipid ratio was changed from 0.5 to 5.0 (by reducing the lipid concentration while keeping the protein concentration constant), the shape and size of the vesicles changed from very large and irregular to small and spherical. In addition, the packing of proteins changed from hexagonal to rectangular.

Recently, Schabert *et al.* (1995) were able to resolve fine-details of OmpF porins which were reconstituted in DMPC vesicles at a protein-to-lipid ratio of 2.0.

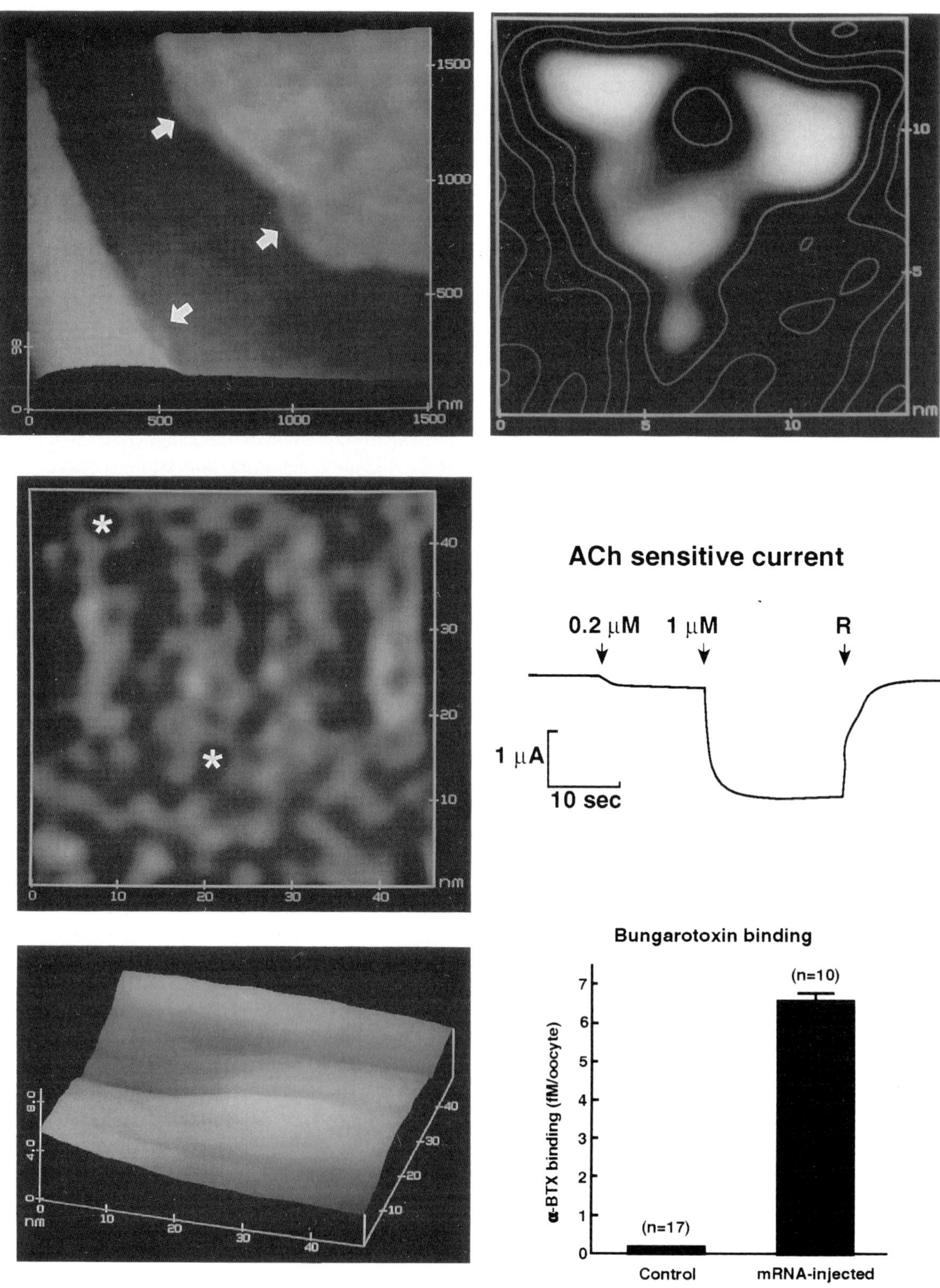

They were able to resolve rectangular unit cells (a = 13.5 nm and b = 8.2 nm) that comprise two trimers with central pore-like protrusions. Interestingly, they reported two conformations of these channels while imaging at a low force (0.1 nN) suggesting that AFM can be used to monitor conformational states of membrane proteins.

***Bacillus pertussis* porin**: Lal *et al.* (1993) were able to achieve molecular resolution on *B. pertussis* porins

Figure 9 on facing page

Figure 9: Molecular resolution AFM imaging of AChR expressed in *Xenopus* oocytes. After recording ACh sensitive electrical current in an oocyte, its vitelline membrane was manually removed for AFM imaging. The cell was cut into two halves and one half was imaged. Top left: Low magnification AFM image of the plasma membrane. A crack in the membrane is shown. The thickness appears to be ~13 nm, the predicted thickness of AChR spanning the plasma membrane. Middle left: High-resolution **low pass filtered** images of an oocyte expressing the mouse AChR channels, showing the irregular spacing of receptor clustering. Bottom left: High resolution image of a control oocyte not injected with any mRNA. Top right: Line graph of an image of a single AChR channel (~10.5 nm dia), showing the contour map of individual subunits and the central pore. The five subunits (~1-1.5 nm in dia) with a central pore like structure are shown. The protrusion of one unit is not apparent in the image. Middle right: ACh sensitive current measured in an oocyte after the AChR expression. The dose-dependent ACh sensitive current was reversible. Bottom right: α-bungarotoxin binding assay revealed a high level of AChR expression. The channel density calculated from AFM, electrophysiological, and pharmacological studies were comparable. For details see Lal and Yu, 1993.

----

which were packed in a quasi-rectangular order. These images also showed a lack of strong long range order, commonly observed for naturally occurring membranes and viewed by EM; indicating the molecular motion in a native environment and perhaps tip induced perturbations. In addition, molecular resolution surface topology of *B. pertussis* porins was obtained for the first time (Figure 7). The individual trimeric porins were visible. In a few cases on can see monomeric components that had pore like central depressions and several surrounding bead like protrusions, perhaps the ß-sheet folding of polypeptide.

Cholera-toxin B-oligomers: Yang *et al.* (1993), incorporated purified cholera toxin into synthetic phospholipid bilayers by covalent cross-linking. Imaging under an appropriate buffer they were able to obtain images of a pentameric structure at molecular resolution. The lateral dimension of individual subunits was comparable to that obtained by other methods, however the height of these units was significantly lower. It is not clear whether this reflects the partial embedding of protein in the lipid substrate and /or imaging force induced compression of the proteins.

Mou *et al.* (1995) have imaged isolated cholera toxin B-oligomers which were anchored to supported bilayers made of DPPC and POPG (Figure 8). Cholera toxin B-oligomers were seen to bind to the ganglioside in the fluid phase. Individual oligomers, their subunit organization, and few missing subunits were clearly resolved at a lateral resolution of ~1 nm. The image quality and resolution were comparable in both the fluid phase and the gel phase of the bilayer, suggesting that the bilayer fluidity did not introduce any significant perturbation during high resolution imaging. In addition, they were able to grow 2D arrays of B-oligomers directly on these model membranes without any special treatment. These studies suggest that isolated membrane proteins can be imaged with AFM without any significant chemical modifications such as cross-linking with the substrate.

Ca-ATPase: Lacapere *et al.* (1992) have imaged 3D crystals of Ca-ATPase which were reconstituted from 2D crystalline membranes isolated from sarcoplasmic reticulum vesicles. In their study, the lateral resolution was limited and they were not able to image crystal periodicity on the surface. However, the steps of membrane planes were measured which correspond to the unit cell spacings. While imaging under aqueous medium, they observed a significantly greater mobility on the surface, perhaps reflecting the intrinsic flexibility of layers of membrane sheets.

As the spectrum of AFM imaging is expanding, it is becoming clear that molecular resolution can be obtained on non-crystalline specimens in fluid medium. This opens a new avenue for the study of molecular structure of biological macromolecules such as ion channels and receptors that can be easily expressed in an appropriate expression system such as Xenopus oocyte, or simply isolated and anchored properly on a suitable substrate.

Whole cell plasma membrane and isolated organelles

AChR: Using the Xenopus oocyte expression system, nicotinic acetylcholine receptor (AChR) has recently been imaged (Lal and Yu, 1993) (Figure 9). Imaging whole and hemi oocytes whose vitelline membrane was removed showed a typical plasma membrane with thickness approaching 13 nm. This thickness (compared to a thickness of ~5.5 nm for a typical lipid bilayer) is consistent with the thickness of AChR spanning the plasma membrane which was predicted based on EM reconstruction (Unwin, 1993). Additionally, there was a noticeable step in the membrane indicating(?) the lipid leaflet interface, analogous perhaps, to images generated by freeze fracture EM. High resolution images on these membranes showed clusters of pentameric structures with a center to center spacing of 10-11 nm, close to the minimum center to center spacing of

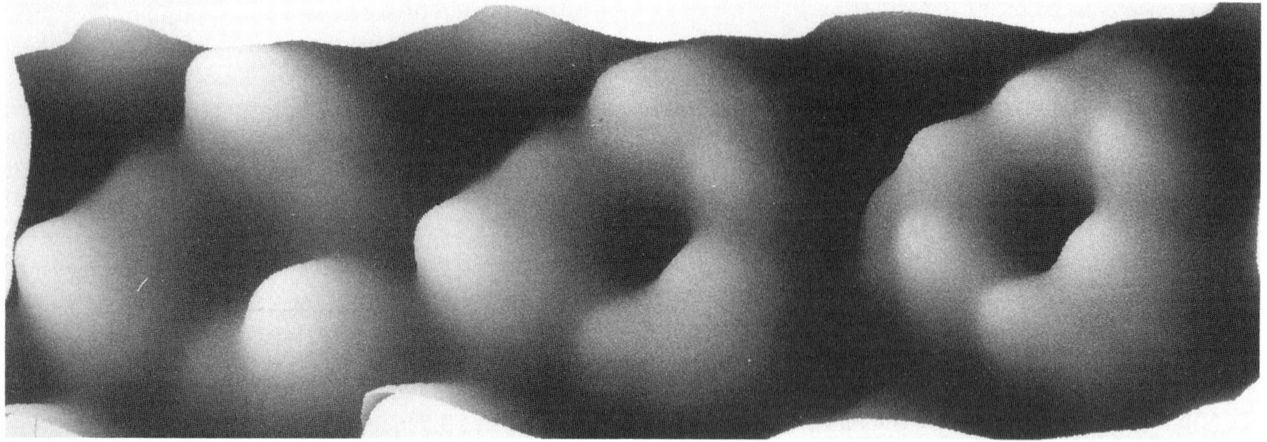

Figure 10: Force dependent conformational change of bacteriorhodopsin. Perceptive view (scale bar represents 2 nm) of the transition from native (left) to donut-shaped bacteriorhodopsin trimers (right). The central trimer is a composition of the left trimer recorded at 100 pN and the right trimer recorded at 300 pN. (For details, see Muller *et al.*, 1995).

about 9-10 nm for the 250 kD multisubunit AChR. The structure imaged shows three main protrusions, perhaps subunits from the membrane, and two minor protrusions, (perhaps other two subunits), and a pore like center depression. The angle between the two alpha subunits was ~128 degrees, in agreement with the reported value. The five subunit is consistent with that organization predicted from EM studies (Unwin, 1993). The height of these protrusions varied (0.7 ± 0.1 nm) but was significantly smaller than the previously reported value, which maybe accounted for by either tip induced flattening or drying of the sample.

Calcium channel: Haydon *et al.* (1994) have reported the localization of calcium channels on the calyx-type nerve terminal of the fixed chick ciliary ganglion in culture. These calcium channels were localized by imaging avidin-coated 30 nm gold particles which were incubated with biotinylated w-conotoxin GVIA. Calcium channels were clustered in low (~1 per μm^2) as well as high (55 per μm^2) density with an interchannel spacing of 40 nm. No molecular structure of individual calcium channels was reported possibly due to a low resolution commonly obtained while imaging intact cells. The interchannel spacing of 40 nm may reflect the spatial limitation due to the tagging with 30 nm gold particles. Also, individual calcium channels probably will be much smaller in diameter. Other membrane channels such as sodium and potassium channels and gap junctions are ~6-10 nm in diameter.

NPC: Isolated cellular organelles have been imaged with AFM. Oberleithner *et al.* (1994) have imaged nuclear pore complexes (NPC) in isolated and air dried nuclear surface of cultured kidney cells. These NPCs were ~134 nm in outer diameter with a central pore-like trough. NPCs were randomly distributed and their density increased from 7.4 to 9.8 per μm^2 in cells which were previously treated with aldosterone.

These studies of cells and cellular surfaces both fixed and unfixed, and imaged in dry or wet environment provides encouragement for undertaking studies for direct structure-function studies.

Structure-Function Studies

Significant advantages of AFM over conventional high resolution microscopy is that AFM allows imaging under an aqueous environment and AFM can be combined with other techniques for simultaneous or successive structure-function studies. Such potentials of AFM have just begun to be used for biological studies.

Lal and Yu (1993) reported a "single cell" experiment where electrical activity and surface images were obtained from *Xenopus* oocytes that were expressing AChR (Figure 9). Electrophysiological studies on the expressed oocytes revealed ACh sensitive current. Immunolabeling experiments on the expressing oocytes, by binding of AChR specific α-bungarotoxin, were also conducted in parallel. The receptor density calculated from AFM studies correlated well with the electrophysiological measurements and toxin binding, but the clustering of AChR was found to differ from the predicted uniform distribution of the expressed receptors as commonly assumed in electrophysiological studies. The density in the clusters was as high as seen in neuromuscular junctions. With further understanding and improvement in AFM imaging, it may not be long before when one can design whole cell-to-individual molecule experiment in a physiological environment.

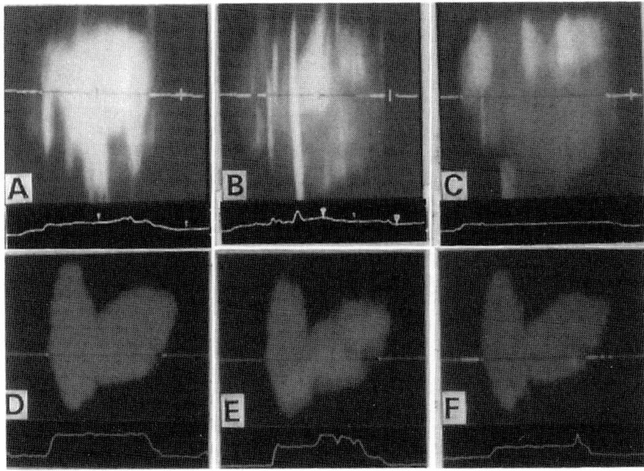

Figure 11: Comparison of force-dissection of heart (A, B, C) and liver (D, E, F) gap junctions. The top view images and thickness profiles along a line-cut perpendicular to the scan direction are shown for a series of scans at increasing forces of imaging. A, B, C: Heart gap junction. Imaging forces are 0.1 nN (A), 5.2 nN (B) and 11.6 nN (C), respectively. The thickness of the dissected single layer is reduced from ~21 nm to ~9 nm. The scan size is 875 x 875 nm^2. D, E, F: Liver gap junction. Imaging forces are 0.8 nN (D), 3.1 nN (E), and 10.1 nN (F), respectively. The thickness is reduced by half from ~15.5 nm to ~7 nm. The scan size is 1.5 x 1.5 μm. (for details, see Hoh *et al.*, 1991; Lal *et al.*, 1995b).

Bustamante *et al.* (1996) have reported correlative patch-clamp electrical recording and AFM of transcription factor IID (TFIID) interactions with the nuclear pore complex (NPC) isolated from kidney cells. The nuclei were fixed in glutaraldehyde after incubation with or without TFIID and ATP. Individual NPCs were imaged and a change in surface topology was apparent after treatment with TFIID as if the TFIID interaction was able to constrict the pore size. In a parallel set of experiments, they were able to show NPC unplugging accompanied by prolonged electrical current through the channel, perhaps reflecting the reopening of the channels.

Hörber *et al.* (1995) have combined AFM with patch-clamp technique in the same experiments and measured the electrical current in the excised membrane patches from *Xenopus* oocyte and attached to the patch pipette tip while imaging the surface topology with AFM. Notably, these membrane patches were significantly more stable and produced higher resolution images than imaging over a whole cell. Also, they were able to deform the membrane surface by applying pressure through the patch pipette and observed the lateral displacement of features. The resolution was limited to about 10 nm. Nevertheless, it shows the promises of direct structure-function studies of membrane structures and with further improvement it can be extended to molecular level.

Butt (1992) and Butt *et al.* (1993) have simultaneously imaged the 3-D structure of purple membrane, measured ion transport through the membrane, and examined the electrical properties of the membrane adsorbed onto a lipid monolayer. Dietz *et al.* (1991) have measured electric charge transfer through synthetic ultrafiltration membranes. And, Proksch *et al.* (1996) have simultaneously imaged the surface structure of nucleopore filters and measured electrical current passing across the filter through pores of different diameters. For these studies, they developed a combined scanning ion conductance microscope which can record electrical activity while imaging the 3D structure of various membranes.

Imaging force can be varied considerably during the experiments. Such feature has been used to nanomanipulate proteins and membrane structures (Florin *et al.*, 1994; Hoh *et al.*, 1991; Lal *et al.*, 1995b; Hörber *et al.*, 1995; Muller *et al.*, 1995a). Muller *et al.* (1995a) have elegantly reported two different conformations at the cytoplasmic surface of purple membranes (Figure 10). When the imaging force was reduced from ~300 picoNewton to 100 picoNewton, individual donut-shaped trimeric bacteriorhodopsin molecules transformed into units with three dominant protrusions.

Previously, Lal and his collaborators (Hoh *et al.*, 1991; Lal *et al.*, 1994,1995b) have shown that by a controlled increase in the imaging force it was possible to nanodissect the double bilayer gap junctions (Figure 11). Interestingly, it was also possible to obtain higher resolution on the membrane surface exposed after nanodissection (Hoh *et al.*, 1991,1993). Whether such enhanced resolution was due to some force-induced modulation or simply a function of the reduced extra-membranous mass of gap junction polypeptides (connexins) need to be resolved.

Conclusion

The simple design and invariance to the operating environment allows AFM to be integrated with other techniques for correlational studies. The integration of the AFM and fluorescence microscope is one such exciting development (Hansma *et al.*, 1994b, Lal *et al.*, 1995a,b). Schmitt *et al.* (1992) have studied molecular recognition reactions at receptor-substrate interfaces using fluorescence, plasmon surface polaritons and AFM. Toledo-Crow *et al.* (1992) have combined near-field differential scanning optical microscope with AFM. Emerging technique of AFM based NMR imaging

(Rugar *et al.*, 1993) provides the most promising avenue of 3-D structural analysis of individual macromolecules without the need of crystallization and related complications. New cryo-atomic force microscope, developed by Shao and his collaborators (Han *et al.*, 1995) will provide a significant impetus for high resolution and even atomic resolution imaging of channels and receptors, both isolated and those present in cell membranes.

The studies of hydrated channels and receptors show potential for direct structure-function studies of these biological macromolecules that can be properly anchored, reconstituted and or expressed in a suitable expression system. As these membranes can be imaged under an appropriate buffer, it may be possible to observe short- and long-term molecular interactions with various ligands (e.g., antibodies, toxins, drugs).

Acknowledgments

Vacuolar ATPase was prepared and characterized by Drs. Kathleen Nolta and Ted Steck at the University of Chicago. I thank them for providing acidosomes and for their participation during the preliminary work related to vacuolar ATPase. The porin channel reconstitution and the lipid-protein interaction were carried out by Drs. Hoein Kim and Michael Garavito. I thank them for their effort and encouragement. I thank Dr. Z. Shao for providing images of cholera-toxins and Dr. D.J. Muller for providing images of bateriorhodopsins. Supported by the Cottage Hospital-Digital Instruments Special Research Award and the Alzheimer's Disease Program, Department of Health, California.

References

Arakawa H, Umemura K, Ikai A (1992) Protein images obtained by STM, AFM and TEM. Nature **358**: 171-173.

Binnig G, Quate CF, Gerber C (1986) Atomic force microscope. Phys Rev Lett **56**: 930-933.

Bustamante JO, Liepins A, Prendergast RA, Oberleithner H (1996) Patch-clamp and atomic force microscopy demonstrate transcription factor IID (TFIID) interactions with the nuclear pore complex. J Membr Biol **146**: 263-72.

Butt HJ (1992) Measuring local surface charge densities in electrolyte solutions with a scanning force microscope. Biophys J **63**: 578-582.

Butt HJ, Downing KH, Hansma PK (1990) Imaging the membrane protein bacteriorhodopsin with the atomic force microscope. Biophys J **58**: 1473-1480.

Butt HJ, Seifert K, Fendeler K, Bamberg E (1993) Characterizing solid supported membranes with the atomic force microscope. Biophys J **64**: A14.

Dietz P, Hansma PK, Herrmann K-H, Inacker O, Lehmann HD (1991) Atomic force microscopy of synthetic ultrafiltration membranes in air and under water. Ultramicroscopy **35**: 155-159.

Drake B, Gould SC, Weisenhorn AL, Hansma HG, Hansma PK, Quate CF, Prater CB, Albrecht TR, Cannell DS (1989) Imaging crystals, polymers, and processes in water with the atomic force microscope. Science **243**: 1586-1589.

Durbin SD, Carlson WE, Saros MT (1993) In situ studies of protein crystal growth by atomic force microscopy. J Phys D Appl Phys **26**: B128-132.

Florin E-L, Moy VT, Gaub HE (1994) Adhesion forces between individual ligand-receptor pairs. Science **264**: 415-417.

Han W, Mou J, Yang J, Shao Z (1995) Cryo atomic force microscopy: A new approach for biological imaging at high resolution. Biochemistry **34**: 8216-8220.

Hansma HG, Hoh JH (1994) Biomolecular imaging with the atomic force microscope. Annu Rev Biophys Biomol Struct **23**: 115-128.

Hansma PK, Cleveland JP, Radmacher M, Walters DA, Hillner P, Bezanilla M, Fritz M, Vie D, Hansma HG, Prater CB, Massie J, Gurley J, Elings V (1994a) Tapping mode atomic force microscopy in liquids. Appl Phys Lett **64**: 1738-1742.

Hansma PK, Drake B, Grigg D, Prater CB, Yasher F, Gurley G, Elings V, Feinstein S, Lal R (1994b) A new, optical-lever based atomic force microscope. J Appl Phys **76**: 796-799.

Haydon PG, Henderson E, Stanley EF (1994) Localization of individual calcium channels at the release face of a presynaptic nerve terminal. Neuron **13**: 1275-1280.

Hillner PE, Walters DA, Lal R, Hansma HG, Hansma PK (1995) Combined atomic force and confocal laser scanning microscope. J Micro Soc Am **1**: 123-126.

Hoh JH, Lal R, John SA, Revel JP, Arnsdorf MF (1991) Atomic force microscopy and dissection of gap junctions. Science **253**: 1405-1408.

Hoh JH, Sosinsky GE, Revel J-P, Hansma PK (1993) Structure of the extra-cellular surface of gap junction by atomic force microscopy. Biophys J **65**: 149-163.

Hörber JKH, Haberle W, Ohnesorge F, Binnig G, Liebich HG, Czerny CZ, Mahnel H, Mayr A (1992) Investigation of living cells in the nanometer regime with the scanning force microscope. Scanning Microsc **6**: 919-930.

Hörber JKH, Mosbacher J, Haberle W, Rupersberg JP, Sakmann B (1995) A look at membrane patches with a scanning force microscope. Biophys J **68**: 1687-1693.

Karrasch S, Dolder M, Schabert F, Ramsden J, Engel E (1993) Covalent binding of biological samples to solid supports for scanning probe microscopy in buffer solution. Biophys J **65**: 2437-2446.

Kordylewski L, Saner D, Lal R (1994) Atomic force microscopy of freeze-fracture replicas of rat atrial tissue. J Microscopy **173**: 173-181.

Lacapere JJ, Stokes DL, Chatenay D (1992) Atomic force microscopy of 3-dimensional membrane-protein crystals - Ca-ATPase of sarcoplasmic reticulum. Biophys J **63**: 303-308.

Lal R, Kim H, Garavito RM, Arnsdorf MF (1993) Molecular resolution imaging of outer membrane channels reconstituted in an artificial bilayer. Am J Physiol **265**: C851-C856.

Lal R, Drake B, Blumberg D, Saner D, Hansma PK, Feinstein S (1995a) Imaging neurite outgrowth and cytoskeletal reorganization with an atomic force microscope: studies on PC12 and NIH3T3 cells in culture. Am J Physiol **269**: C275-C285.

Lal R, John SA, Laird DW, Arnsdorf MF (1995b) Heart gap junction preparations reveal hemiplaques by atomic force microscopy. Am J Physiol **268**: C968-C977.

Lal R, Yu L (1993) Molecular structure of cloned nicotinic AChR receptors expressed in *Xenopus* oocyte as revealed by atomic force microscopy. Proc Natl Acad Sci **90**: 7280-7284.

Lal R, John SA (1994) Biological application of atomic force microscopy. Am J Physiol **256**: C1-C21.

Mou J, Yang J, Shao Z (1995) Atomic force microscopy of cholera toxin B-oligomers bound to bilayers of biologically relevant lipids. J Mol Biol **248**: 507-512.

Muller DJ, Buldt G, Engel A (1995a) Force-induced conformational change of bacterio-rhodopsin. J Mol Biol **249**: 239-243.

Muller DJ, Shabert FA, Buldt G, Engel A (1995b) Imaging purple membranes in aqueous solutions at sub-nanometer resolution by atomic force microscopy. Biophys J **68**: 1681-1686.

Nolta KV, Padh H, Steck TL (1991) Acidosomes from *Dictyostelium*. Initial biochemical characterization. J Biol Chem **266**: 18318-18323.

Oberleithner H, Brinckmann E, Schwab A, Krohne G (1994) Imaging nuclear pores of aldosterone-sensitive kidney cells by atomic force microscopy. Proc Natl Acad Sci **91**: 9784-9788.

Ohnesorge F, Binnig G (1993) True atomic resolution by Atomic Force Microscopy through repulsive and attractive Forces. Science **260**: 1451-1456.

Paul JK, Nettikadan SR, Ganjeizadeh M, Yamaguchi M, Takeyasu K (1994) Molecular imaging of Na-K-ATPase in purified kidney membranes. FEBS Lett **346**: 289-294.

Proksch RA, Lal R, Hansma PK, Morse D, Stucky G (1996) Imaging the internal and external pore structures of membrane in fluid: tapping mode scanning ion conductance microscopy. Biophys J **71**: 2155-2157.

Rugar D, Yannoni CS, Sidles JA (1992) Mechanical detection of magnetic resonance. Nature **360**: 563-566.

Schabert FA, Henn C, Engel A (1995) Native *E coli* OmpF porin surfaces probed by atomic force microscopy. Science **268**: 92-94.

Schmitt FJ, Weisenhorn AL, Hansma PK, Knoll W (1992) Molecular recognition reactions at interfaces as seen by fluorescence, plasmon surface-polaritons and atomic force microscopy. Thin Solid Films **210**: 666-669.

Shroff S, Saner D, Lal R (1995) Dynamic micromechanical properties of contractile cells measured by atomic force microscopy. Am J Physiol **269**: C286-292.

Toledo-Crow R, Yang PC, Chen Y, Vaez-Iravani M (1992) Near-field differential scanning optical microscope with atomic force regulation. Appl Phys Lett **60**: 2957-2959.

Unwin N (1993) Nicotinic acetylcholine receptor at 9 A resolution. J Mol Biol **229**: 1101-1124.

Yang J, Tamm LK, Tillack TW, Shao Z (1993) New approach for atomic force microscopy of membrane proteins - the imaging of cholera toxin. J Mol Biol **229**: 286-290.

Yang J, Shao Z (1995) Recent advances in biological atomic force microscopy. Micron **26**: 35-49.

Discussion with Reviewers

R. Marchant: I think some attention should be given to various technical difficulties and imaging artifacts that are encountered when using AFM. Examples include thermal drift and friction effects that can effect the precision and reproducibility of measurements in all dimensions; the non-linear combination of the tip and specimen that cause significant broadening of lateral measurements, particularly for biomolecules that are taller than 1 nm; the effect of imaging forces on delicate biomolecules (an adverse example of "nano-dissection"); and the need for well characterized tips to account for anomalies seen in the images; drawbacks of the optical lever detection, sample preparation issues, and anomalies observed in images as a result of the feedback system, particularly at faster scanning rates.

Author: I agree with the reviewer that these points should be discussed in a balanced and broad review of the biological application of AFM. However, this manuscript is written primarily to provide the breadth and scope of imaging membrane channels and receptors. And as such, in addition to providing some new data (Figures 5 and 6), the available literature on imaging channels and receptors were summarized. For detailed treatise of the technical limitations and difficulties, readers can always refer to several current review articles such as Lal & John, 1994, Hansma & Hoh, 1995, Yang & Shao, 1995.

J. Zasadzinski: What limits the resolution with contact mode AFM? Will non-contact modes - tapping, adhesion etc. or lateral foerce provide higher resolution or other important data?

Author: The scope of the present manuscript is limited to providing examples of AFM applications to channels and receptors. The reviewer's question is better answered in the above mentioned review articles and other cross references.

J. Zasadzinski: What are the major benefits to AFM studies of membrane proteins over cryo-electron microscopy? Are there certain systems that just cannot be done with TEM that can be examined by AFM? Are there systems in which the AFM resolution is superior to the TEM?

Author: There are several advantages of AFM imaging over EM. (a) The signal to noise ratio in AFM is high enough that one can view micro-details in even single channel and receptors. For similar information in EM studies, one need to average information from a large number of channels and receptors which are in a crystalline order; (b) The AFM images contain a greater detail of the surface topography which is an essential information about membrane proteins which has a large portion not embedded in the lipid bilayer. For example, heart gap junctions has a large cytoplasmic domain which prevents any high resolution structural information.

J. Zasadzinski: How important is the environment or the substrate in the protein conformation? Is the problem with resolution more that the protein is moving over the time scales of the images, thereby blurring the images? What are the benefits to such kinetic information that obviously cannot be determined from fixed or frozen material?

Author: Again this problem is well discussed in the several recent review articles cited above. A major advantage of AFM imaging is that, since it images hydrated specimen, it allows for real-time dynamic studies of protein conformations. This is still a new domain of AFM application and need to be further developed. The temporal resolution in an AFM images depends on several factors including the image size, spatial resolution and protein mobility. In general, slow lateral movement of small proteins can be followed in real-time (Thomson *et al.*, 1996). At present time, the rotational motion of protein molecules are too fast to be tracked in AFM imaging.

Additional Reference

Thomson NH, Fritz M, Radmacher M, Cleveland JP, Schmidt CF, Hansma PK (1996) Protein tracking and observation of protein motion using atomic force microscope. Biophys J **70**: 2421-2431.

Scanning Microscopy Supplement 10, 1996 (pages 97-109)
Scanning Microscopy International, Chicago (AMF O'Hare), IL 60666 USA
0892-953X/96$5.00+.25

ATOMIC FORCE MICROSCOPY OF DNA, NUCLEOPROTEINS AND CELLULAR COMPLEXES: THE USE OF FUNCTIONALIZED SUBSTRATES

Yuri L. Lyubchenko[1,2*], Robert E. Blankenship[3], Alexander A. Gall[4], S. M Lindsay[2], Ottavio Thiemann[5], Larry Simpson[5] and Luda S. Shlyakhtenko[1,6]

Departments of [1]Microbiology, [2] Physics, [3] Chemistry and Biochemistry, Arizona State University, Tempe, AZ, [4]MicroProbe Corp., Bothell, WA, [5]Howard Hughes Medical Institute and Department of Biology, Microbiology and Immunology, University of California, Los Angeles, CA, [6]Molecular Imaging Corp., Tempe, AZ.

(Received for publication August 6, 1995 and in revised form May 8, 1996)

Abstract

Progress towards rapid and simple characterization of biomolecular samples by scanning probe microscopy is impeded mainly by limitations of the current approach to sample preparation. We are working on approaches based on chemical functionalization of mica. Treatment of mica with aminopropyltriethoxy silane (APTES) makes the surface positively charged (AP-mica) and able to hold DNA in place for imaging, even in water. We have shown that AP-mica is an appropriate substrate for numerous nucleoprotein complexes as well. The AFM images of the complex of DNA with RecA protein are stable and indicate a structural periodicity for this filament. AP-mica holds strongly such large DNA complexes as kinetoplast DNA (kDNA) and is an appropriate substrate for their imaging with AFM. We have further develop this approach for making hydrophobic substrates. Silylation of mica surface with hexamethyldisilazane (Me-mica) allowed us to get AFM images of chlorosomes, an antenna complex isolated from green photosynthetic bacteria. Me-mica may be converted into a positively charged substrate after treatment with water solutions of tetraethylammonium bromide or cetyltrimethylammonium bromide. These activated surfaces show high activity towards binding the DNA molecules.

Key Words: Scanning probe microscopy, AFM, SFM, mica, silylation, molecular imaging, DNA, SAM

*Address for correspondence:
Yuri L. Lyubchenko
Department of Microbiology, Arizona State University, Box 872701, Tempe, AZ 85287-2701
Phone number: (602) 965-8430
FAX number: (602) 965-0098
e-mail: lyubchenko@phyast.la.asu.edu

Introduction

Sample preparation is a crucial steps for any type of microscopy, and particularly for scanning probe microscopy (SPM). In this novel type of microscopy, which mostly include scanning tunneling microscopy (STM) and atomic force microscopy (AFM), a very sharp tip "reads" the surface profile [6, 7]. Therefore SPM imaging of molecular adsorbents requires their strong attachment to the surface in order to avoid resolution-limiting motion occasioned by the tip during scanning, which in extreme cases can result in sweeping the adsorbed molecules away from the surface. The effect of tip motion is reduced if the AFM instrument is operated in non-contact or tapping mode [see, e.g., 12, 15, 16], but even in this case the problem of sample preparation is an important one. The class of suitable substrates for AFM is rather narrow and mostly includes natural mineral mica, glass and silica surfaces. In a very limited number of cases biomolecular adsorbents bind to a bare substrate allowing their imaging with AFM. For example, images of chromatin [1] were obtained for samples prepared on cover glass. Membrane proteins can be imaged with very high resolution because on the mica surface they form a 2D array which stabilizes lateral movement of proteins during scanning [16, 18]. Note that the progress in AFM imaging of nucleic acids was made through the development of substrate preparation methods based predominantly on the polyelectrolyte character of DNA and RNA molecules (see reviews [11, 16, 25-27]). Vesenka, Bustamante and co-workers [11, 39] developed the method of ionic treatment of mica. In this approach the mica surface is treated with divalent cations (e.g. Mg^{2+}) to increase its affinity for DNA which is held in place strongly enough to permit reliable imaging by AFM. Dramatic improvements were obtained by using specially made tips and by imaging under propanol (as reviewed in [11, 15, 16]). Other di- or trivalent cations can be used for pre treatment of mica [36]. The presence of a multivalent cation helps spread the DNA and bind it to the mica. In addition to imaging of DNA, this method of sample preparation was successfully applied to studies of complexes of DNA with proteins [11, 12, 26]. Yang *et al.* [42] have developed an approach which is a modification of the well-known electron microscopic procedure. They deposited the DNA molecules onto a carbon coated mica substrate using cytochrome C spreading technique. Gold substrates can be activated by self-assembled monolayers (SAM) of thiols for reliable STM and AFM imaging of DNA [2, 3, 17]. Note that both physisorbed and chemisorbed SAM are used to make monolayer and multilayer structures with predetermined molecular architecture [38]. Non-treated cover glass can be used as the AFM substrate for binding chromatin, although the rinsing step (to remove non-bound material and salt components) should be done gently and the dried sample should be imaged immediately [1]. High resolution AFM images of chromatin were obtained by use of mica treated with spermidine [20, 43].

We [21-26] have worked out a procedure for mica modification (AP-mica) that allowed us to routinely perform visualization of DNA with AFM, achieving resolution as good as that of traditional electron microscopy (EM). The method is based on covalent attachment of aminopropyltriethoxy silane to the surface of mica, which makes it positively charged. DNA samples are deposited directly onto the modified mica (AP-mica) through self-adsorption in a *one-step procedure.* After that, DNA samples can be imaged directly with AFM. It is important to note that deposition of samples in our procedure can be performed under a wide range of environmental conditions (pH, ionic strength, temperature); this feature of AP-mica is an advantage for imaging of nucleoprotein complexes of different types [24-26]. Another important feature of AP-mica is its high stability; this often allows, for example, preparation of the substrate well in advance of deposition.

AP-mica, however, is only a good substrate for holding molecules bearing a negative charge. Biological objects have a broad spectrum of physical chemical characteristics; therefore, substrates with different affinities are required for strong attachment of various biological objects. Keeping in mind the positive results obtained with AP-mica, we investigated the use of chemical modification procedures to obtain surfaces with various other characteristics, such as hydrophobic and other positively charged surfaces. Hydrophobic surfaces were used for imaging of chlorosomes, an antenna complexes isolated from green photosynthetic bacteria. We have shown that a hydrophobic surface can be transformed into a positively charged one after treatment with salts carrying hydrophobic chains. These surfaces bind DNA molecules as well as AP-mica. Strongly positively charged surfaces (obtained after modifications with iodopropyltrimethoxy silane followed by surface activation with DABCO) were used for imaging of non-membrane proteins.

Materials and Methods

Mica modification procedures

AP-mica: Modification was performed in vapors of 30-aminopropyltriethoxy silane (APTES) following the procedure described elsewhere [5, 21, 25]. Briefly, freshly cleaved strips of mica were left in the APTES atmosphere created in 2l glass desiccator containing 30 μl of vacuum distilled APTES under ambient conditions for 2 hours.

Me-mica: 30 μl of methyltrichlorosilane or hexamethyldisilazane (Aldrich, Milwaukee, WI) in a small plastic vial was placed on the bottom of a 2l-desiccator. Freshly cleaved strips of mica were mounted at the top of the desiccator and left for 2 hours for modification after filling the system with argon. Positively charged mica was obtained by incubation of Me-mica in saturated water solutions of tetraethylammonium bromide (TEAB) or cetyltrimethylammonium bromide (CTAB) for 10 min. at room temperature.

IP-mica: Iodopropyltrimthoxy silane (IPTMS) was synthesized as described in [9]. 10 μl of IPTMS in a 1.5 ml plastic vial were placed on the bottom of a 100-ml glass bottle and freshly cleaved mica strips were mounted at the top of the glassware. The bottle was placed into a preheated oven (80°C) and mica modification allowed to proceed for 2 hours. N4-mica substrates were obtained by 30 min. incubation of IP-mica in 1 mM solution of diazabicyclo [2,2,2] octane (DABCO, Aldrich, Milwaukee, WI) for 30 min.

Chlorosome samples: *Chloroflexus aurantiacus* was grown photoheterotrophically in 1-liter batch cultures at 55°C as described in [29]. The cells were harvested by centrifugation after 3-4 days of growth and chlorosomes were isolated by the method of Gerola and Olson [14] using a 2M concentration of the chaotropic agent, NaSCN. After a 10-50% linear sucrose gradient ultracentrifugation, the chlorosome band was collected for spectroscopic studies.

Kinetoplast DNA (kDNA): Sample preparation was performed by a protocol described elsewhere [32, 33, 35]. Namely, kDNA was isolated from stationary phase *Leishmania tarentolae* (UC strain) cells grown in brain heart infusion medium (Difco Laboratories, Detroit, MI) supplemented with 10 mg/ml hemin. The cells were washed in SET buffer (150 mM NaCl, 100 mM EDTA, 10 mM Tris-HCl, pH 7.5), resuspended in SET at a cell density of 1×10^9 cells/ml and lysed with 0.5 mg/ml pronase and 3% Na sarcosinate for 1 h at 60°C . The genomic DNA was fragmented by passing the lysate through an 18 gauge needle at 25 psi. The kDNA networks, which are relatively resistant to shear forces, are pelleted by centrifugation in the SW28 rotor at 22,000 rpm for 1.5 hr at 4°C. The kDNA is resuspended in 10 mM Tris HCl (pH 7.9), 1 mM EDTA (TE buffer), and sedimented through a CsCl step gradient in an SW 28 rotor for 15 min at 20,000 rpm. The kDNA networks sedimented to the interface and were visualized by ethidium bromide staining. The band was recovered, the dye removed by n-butanol extraction, and the DNA concentrated with sec-butanol and dialyzed against TE buffer. The DNA was then extracted with phenol/chloroform, ethanol-precipitated and resuspended in TE. The network DNA was examined for integrity in the fluorescence microscope after staining with DAPI (1 mg/ml).

Double-stranded RNA samples from retrovirus containing dsRNA molecules of 11 different sizes were prepared as described elsewhere [21, 23].

Radiolabeled assay: Mica substrates were incubated for different times at room temperature in a solution of ^{32}P-labeled HpaII restriction fragments of pBR 322 plasmid DNA (New England BioLabs, Beverly, MA) at concentration of 0.2 ng/ml in 10 mM Tris-HCl buffer (pH 7.5), containing 20 mM NaCl, 5 mM EDTA. Ratios of bound to total DNA were calculated from the radioactivity of the rinsed mica plates and of the DNA solution before and after the binding procedure.

Apparatus: Imaging was carried out on a NanoScope II or a NanoScope III instruments (Digital Instruments, Inc., Santa Barbara, CA) in contact mode using commercial AFM cantilevers from Park Scientific Instruments (Sannyvale, CA) with a nominal spring constant 0.6 N/m. Imaging in solution was performed by using a fluid cell provided by Digital Instruments. Scanning rate was 2-3 Hz, typical for scanning over 1-10 μm areas .

Results

Our approach to substrate preparation for AFM is based on functionalization of silicon surfaces through a covalent attachment of a silane compound with a specific characteristic. The chemistry of silane coupling agents was founded by E. Plueddemann almost 50 years ago and is currently well understood [30]. A class of organofunctional silanes of the general formula X-Si-Y (**I**) (where Y is the organofunctional group and X is a hydrolyzable group) is widely used for bringing an organo-functional group Y to the surface. A general scheme for this reaction is as follows:

$$(SiOH)_n + X\text{-}Si\text{-}Y \rightarrow (SiO)_n\text{-}(X\text{-}Si\text{-}Y)_m$$

The functional groups (Y) are chosen for reactivity or compatibility with the biopolymer, while the hydrolyzable groups (X, $(CH_2O)_3$, for example) are responsible for covalent bonding of (**I**) to mineral surfaces $(SiOH)_n$. Most commercial coupling agents are supplied as alkoxysilanes (methoxy or ethoxysilanes).

Positively Charged AP-Mica: Imaging of kDNA

We used aminopropyltriethoxy silane (APTES) for preparing a positively charged mica surface (AP-mica, [5, 21, 25, 26]). This compound with $(EtO)_3$ as X and an aminopropyl moiety as Y attaches covalently to silicon surfaces at ambient conditions [21, 25] leaving the amino group on the surface. This group protonates in water solutions making the surface positively charged. Modification of mica in very low concentration solutions of APTES or in its vapors allowed us to obtain uniformly modified smooth AP-mica surfaces suitable for AFM imaging of DNA, RNA and bacteriophages in air and under water [21, 22, 24, 25]. Examples of images of lambda DNA and circular and linear DNA molecules coated with RecA protein are shown in Figure 1A and 1B respectively (RecA-DNA complexes were provided by Dr. A. Stasiak, Lausanne University, Switzerland). A periodic structure of RecA filament is resolved for complexes prepared on both linear and circular DNA templates. It turned out that AP-mica surface strongly holds these samples, allowing imaging in air, propanol and water (Lyubchenko *et al.*, in preparation).

We also investigated the possibility of using AP-mica as a substrate for AFM imaging of large DNA aggregates such as kinetoplast DNA (kDNA). kDNA, the mitochondria-type DNA in trypanosomes and related protozoan parasites, is unusual in both its structure and genetic function. kDNA is a giant network composed of several thousands minicircles and a few dozen of maxicircles, which are topologically interlocked (reviewed in [10]). Electron microscopic studies showed that sizes of isolated networks varied for different species between 4-6 μm for tripanosomids and 15-20 μm for *Crithidia* genus [10].

A 30 μl droplet of the kDNA sample (concentration 10 μg/ml in TE buffer (pH 7.5)) isolated from *Leishmania tarentolae* was applied onto AP-mica for 5 min. allowing the DNA molecules to adsorb, then the mica sample was rinsed with deionized water, dried in argon and used for AFM imaging. A large scale image (27 x 27 μm) obtained by using the J-scanner (maximum scanning area is 100 μm x 100 μm) is shown in Figure 2A. The majority of particles have a cap-like geometry ~3 μm in diameter with a bright rim (see, for example particle 1; cf. with EM images of kDNA particles with rims in recent paper [31]). These images are qualitatively similar to those obtained for the same sample with fluorescence microscopy [35]. A few particles with a

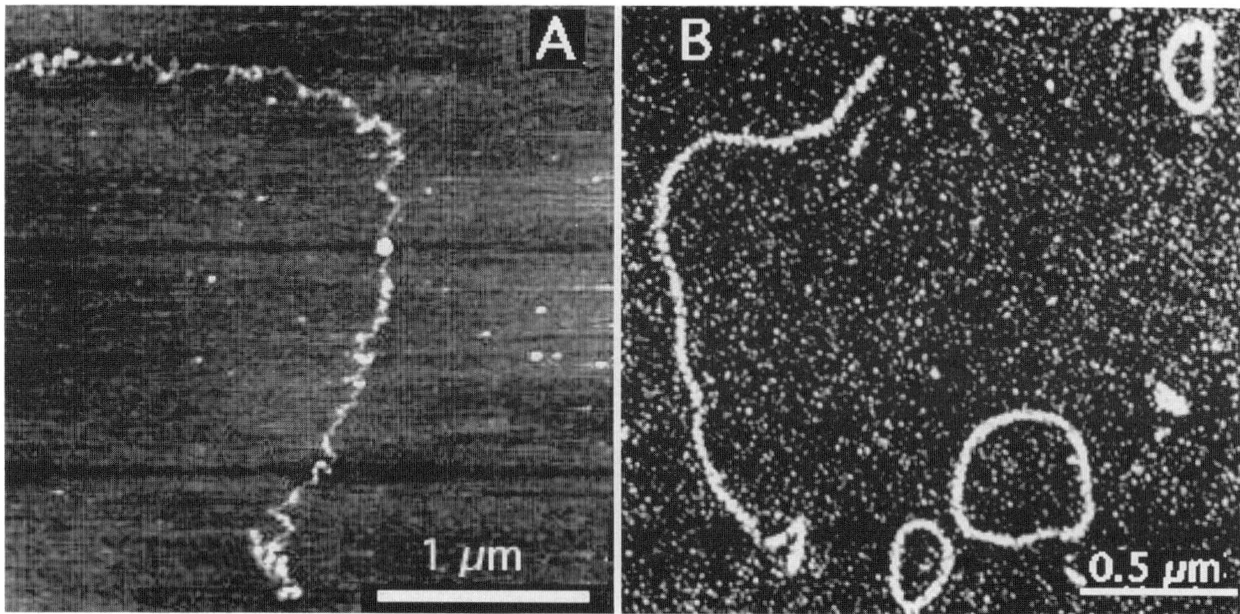

Figure 1: (**A**) AFM images of lambda DNA. DNA concentration is 0.01 μg/ml in buffer containing 0.1 mM Tris-HCl, pH7.6, 0.01 mM EDTA was deposited onto AP-mica according the procedure described elsewhere [21-23]. (**B**) AFM images of RecA-DNA complexes deposited onto AP-mica from the reaction mixture for RecA protein [26].

---

rim are more elongated (see particle #2) and in addition to these cap-like features particles like #3, without a rim can be found. Higher resolution images of particles were obtained using of a medium-size scanner (15 μm, D head). Imaging was done in propanol. Examples of images of individual particles are shown in Figure 2B, C and D. These particles have a clear-cut cap structure. The background on these images is due to presence in the sample of kDNA of small fragments of kDNA particles as interlocked multimers of mini circles. All particles are bound strongly to the substrate, there is no movement of particles or parts of them during repetitive scanning. This can be seen from Figures 2E and F which are consecutive zoomed images in the middle of particle in D around the rim. These images show that the network is rather dense, and it is difficult to resolve the structure of the network. The main problem is the small size of minicircles for this type of kDNA: 0.29 μm in their length or 92 nm in diameter for perfect circles [10]. Even fragments of kDNA particles are rather tight associates of DNA minicircles (data not shown). Individual DNA strands in the kDNA particles are resolved on EM and AFM images of kDNA from *C. fasciculata* ; note that this type of kDNA has DNA minicircles three times larger than our sample [10]. The study of kDNA from *L. tarentolae* with high resolution and in aqueous solutions is in progress.

Preparation of Hydrophobic Mica Surface

We used two silanes for hydrophobization of mica. One of them, methyltrichlorosilane (MTCS, Me-Si-Cl_3) belongs to a family of alkylchlorosilanes widely used for preparation of hydrophobic silicones [30, 41]. Compounds of this family were successfully used for activation of porous glass for binding mitochondria and whole blood cells [34]. Another compound, hexamethyldisilazane (HMDS, (CH_3)3Si-NH-Si-$(CH_3)_3$) is also used for hydrophobization of silicones [30, 40]. Schematically the reactions of both compounds with mica surface are depicted in figure 3A and B respectively.

Both chemicals are commercially available compounds and have a rather low boiling temperature (slightly above 100°C), making it possible to perform silylation of mica in gas phase at ambient conditions [21, 25]. We have found the following conditions for mica modification with both silanes: 5μl of chemical were placed into a 50 ml plastic vial and several pieces of mica were mounted at the top of the vial. Reaction was allowed to proceed at ambient conditions for 10 min. Then freshly cleaved mica strips were removed and investigated. A comparative study of contact angles for water for the freshly cleaved mica and methylated mica (Me-mica) indicated that surface of treated substrates became hydrophobic.

We used chlorosomes as a test sample for adhesion to Me-mica surface. Chlorosomes are intracellular structures of green photosynthetic bacteria containing

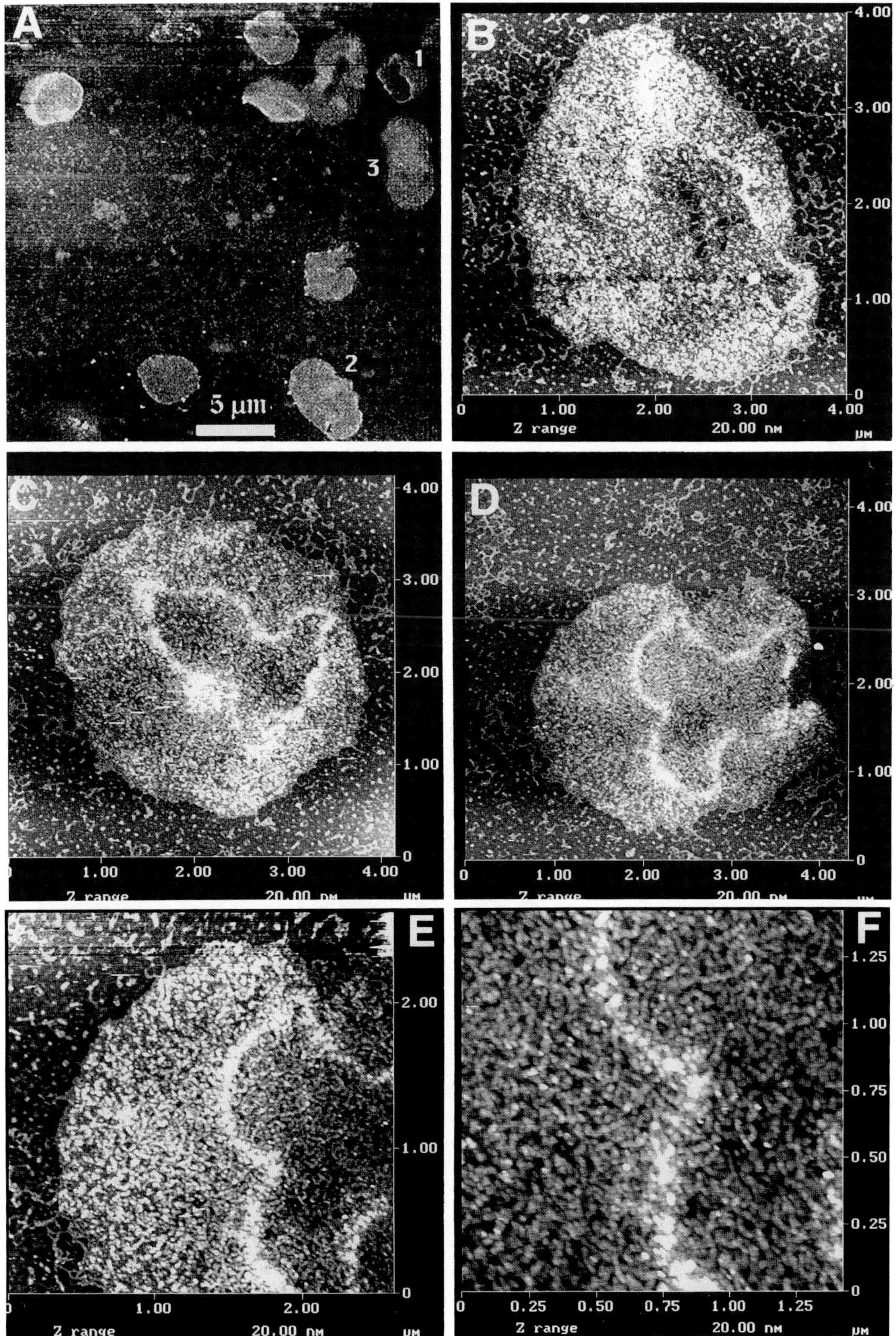

Figure 2: AFM images of kDNA deposited onto AP-mica. from TE buffer Image (**A**) was obtained by use of J-scanner. Other images were obtained using D-scanner. Scanning was done in air (**A**) and in propanol (**B-F**).

(A)

$$\text{mica}(-OH)_3 + Cl_3Si-CH_3 \rightarrow \text{mica}(-O-)_3Si-CH_3$$

(B)

$$\text{mica}(-OH)_3 + (CH_3)_3\text{-Si-NH-Si-}(CH_3)_3 \rightarrow \text{mica}(-O\text{-}Si\text{-}(CH_3)_3)_3$$

Figure 3: Schemes of mica hydrophobization with methyltrichlorosilane (**A**) and hexamethyldisilazane (**B**).

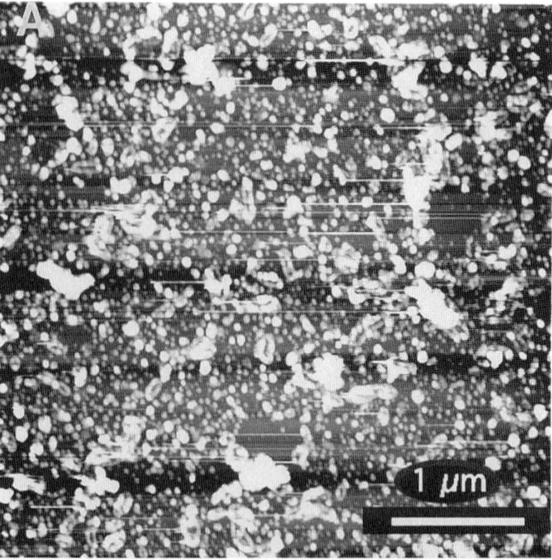

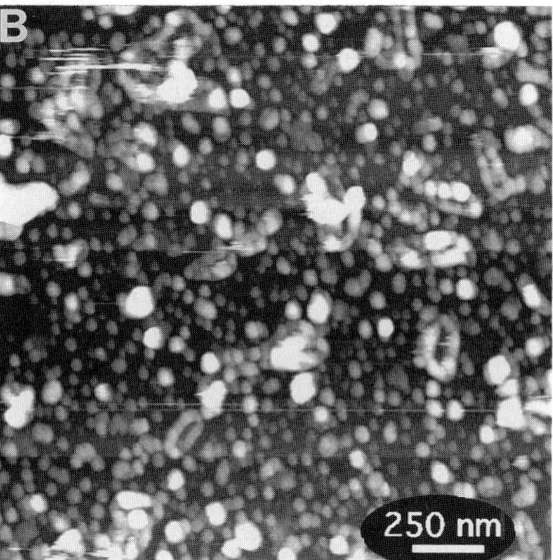

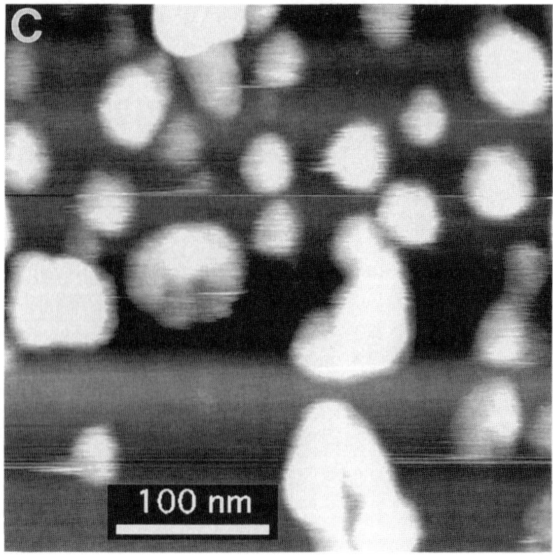

Figure 4 (*right column*): AFM images of the chlorosome sample. The samples were prepared by self deposition of chlorosome particles onto AP-mica in Tris-potassium thiocyanide buffer. Scanning was performed in propanol in contact mode.

----

bacteriochlorophylls [8]. These bag-like bodies are attached to the cytoplasmic side of the inner membrane. The chlorosome is surrounded by an envelope that is a lipid monolayer, which contains 18 and 11 kD polypeptides. Me-mica strips were immersed into a solution of chlorosomes prepared by diluting the stock solution (A_{240}=21) by 100 times in Tris-potassium thiocyanate buffer (10 mM Tris, 2M KSCN, pH 8.0). Chlorosome were allowed to bind to the surface for 30 min., then rinsed with water and nitrogen dried.

The AFM images of chlorosomes at different magnification are shown in Figure 4. Image A is a large scan showing the uniform coverage of Me-mica. Scanning over smaller area (image B) allows one to see individual spherical particles which often form rather large aggregates. Image C is a 350x350 nm scan showing several individual particles of 32-36 nm in diameter alongside bright particles of different shape that are presumed to be multimers of individual chlorosomes. According to electron microscopy data, chlorosomes *in situ* are ellipsoidal in shape, with dimensions 100x30x12 nm [8], and these particles may adopt a spherical shape when they are isolated from the cell. Structural studies of chlorosomes isolated from different types of cells are in progress. We estimated the surface roughness using data for small scan areas like image B and the mean roughness was around 0.06 nm.

A hydrophobic surface can be transformed into a surface with different macroscopic characteristics by treatment with an appropriate compound. To transform Me-mica surface into a positively charged one, we used tetraalkylammonium salts containing aliphatic chains which anchor the whole molecule to a hydrophobic surface. We used two salts with different lengths of a

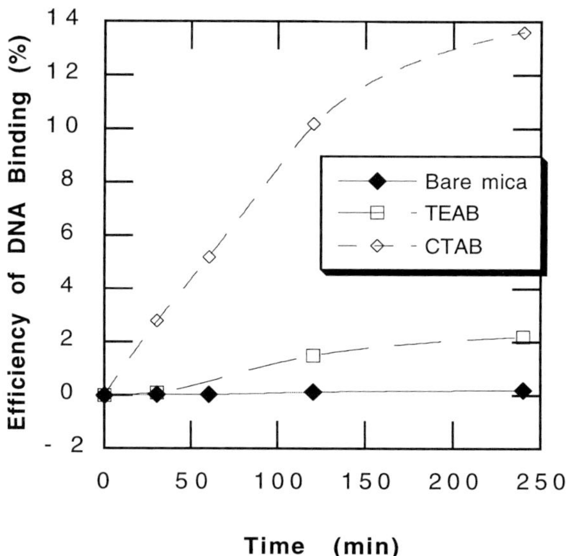

Figure 5: Kinetics of binding of ^{32}P-labeled DNA to Me-mica treated with CTAB and TEAB. The radioactivity of samples was normalized to a total radioactivity of DNA solution before immersing the mica substrates.

----

linker: tetraethylammonium bromide (TEAB, $(C_2H_5)_4N^+$ Br^-) and cetyltrimethylammonium bromide (CTAB, $[CH_3(CH_2)_{15}N^+(CH_3)_3]Br^-$). The latter contains a rather long aliphatic chain, and hence CTAB should exhibit better affinity for hydrophobic surfaces. After binding these salts, Me-mica is transformed into a positively charged surface, which should absorb DNA molecules strongly.

Me-mica strips were immersed into saturated water solutions of CTAB or TEAB for 30 min. Then wet CTAB or TEAB modified Me-mica strips were incubated in solutions of radiolabeled DNA for a specific period of time and radioactivity of rinsed and dried samples was measured. The result of DNA-binding experiments for both types of Me-mica modifications are shown in Figure 5. Freshly cleaved mica was used as a control in these experiments.

These data show that the treatment of a hydrophobic mica with TEAB or CTAB allowed us to obtain positively charged substrates which binds DNA molecules efficiently. Quantitatively the effect of DNA binding is higher for CTAB, supposedly due to its better binding to the Me-mica surface. The DNA binding efficiency of CTAB-Me-mica at least 10 times more higher than that of TEAB-Me-mica. Note that there is practically no binding of DNA to the Me-mica. Incubation of DNA bound with modified positively charged mica in high concentrated solution of NaCl results in removal of 90% of bound DNA. These results suggest that a non-specific electrostatic interaction is the predominant force holding DNA molecules on the surface.

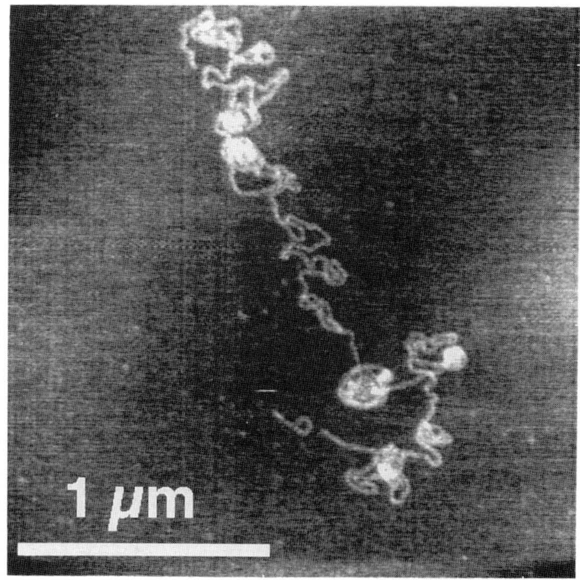

Figure 6: AFM image of lambda DNA deposited onto CTAB-Me-mica. Pieces of CTAB-Me-mica were immersed into the DNA solution in concentration 0.1 μg/ml dissolved in TE buffer (10 mM Tris-HCl, pH 7.6, 1mM EDTA) and incubated for 30 min. Rinsed with water and argon dried samples were imaged in air.

An FM image of lambda DNA deposited onto CTAB-Me-mica is shown in Figure 6. The molecule is rather convoluted (cf. Fig .1). This happens during deposition of lambda DNA molecules onto AP-mica as well [21], although convoluted molecules are present on CTAB-Me-mica surface in larger amount in comparison with AP-mica. This may be explained by the complex character of the surface which is hydrophobic macroscopically with attached positively charged groups. Systematic comparative AFM studies of the DNA deposition onto different types of positively charged surfaces may help understand the mechanism of their adsorption onto different types of modified surfaces.

Chemically Reactive Mica Surface

Keeping in mind the positive results obtained with AP-mica and Me-mica [21-26], we have worked out an alternative procedure for mica modification. The key step of this procedure is the mica activation with iodopropyltrimethoxy silane, Me_3-$Si(CH_2)_3$-I (IPTMS). The reaction of IPTMS with mica surface is shown schematically in Fig.7A. The chemical was synthesized according to a protocol described in [9].

IPTMS binds to SiOH groups of mica covalently similarly to APTES, but after reaction with mica leaves

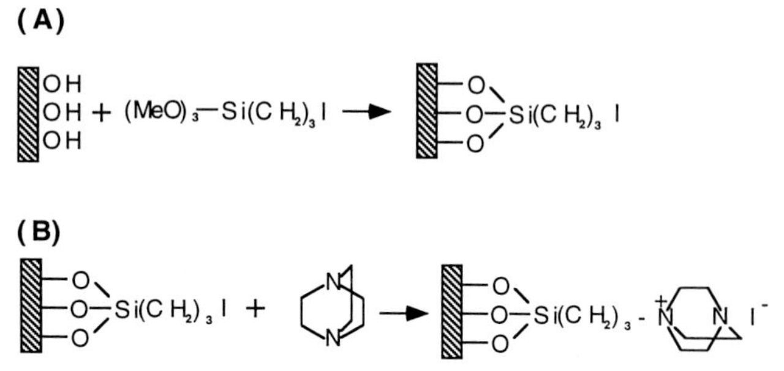

(A)

OH
OH + $(MeO)_3-Si(CH_2)_3I$ → O, O, O $-Si(CH_2)_3I$
OH

(B)

O, O, O $-Si(CH_2)_3I$ + DABCO → O, O, O $-Si(CH_2)_3-N^+$ N I^-

Figure 7: Schemes of mica modification with iodopropyltrimethoxy silane (**A**) followed by transformation of this IP-surface into a positively charged one by treatment with DABCO (**B**).

----

a chemically reactive iodopropyl group on the surface. This is a crucial point for preparation of the AFM substrates with any desirable characteristics, because the presence of a reactive group at the IP-mica surface enables one to perform chemical reactions on the surface directly. For example, 1,4-diazabicyclo [2,2,2] octane (DABCO) can be reacted with activated IP-mica in the way schematically shown in Fig. 7B.

As a result of this modification quaternary amine is attached to the mica surface covalently. 4N-mica surface should have characteristics similar to those of AP-mica, but 4N-mica should be positively charged even at extremely alkalic pH values. The 4N-mica allowed us to get high quality images of DNA, dsRNA, nucleoprotein complexes and proteins. Several examples are shown in Figure 8. Plate A shows images of dsRNA molecules extracted from retroviruses [cf. 21, 23]. Similarly to molecules deposited onto AP-mica samples, RNA molecules are visualized quite easily on 4N-mica substrate as well. Another example, Fig.8B, shows images of RNA polymerase *E. coli* (the sample was a gift of Dr. K. Severinov, Health Public Research Institute, New York). Deposition was done from solution containing 50 mM Tris-HCl, pH7.6, 40 mM KCl, 10 mM EDTA and 1mM mercaptoethanol at room temperature. As it is demonstrated by this picture, the surface is uniformly covered with sphere-like structures. Images are stable during repetitive scanning; no movement of the sample was noticed. The size of individual particles is 28 ± 3 nm. RNA polymerase is a rather large complex of 5 subunits, its molecular weight is 550 kD approximately and the exact quaternary structure of RNA polymerase holoenzyme is still unknown. Our data on polymerase globule sizes are consistent with the AFM results of Bustamante *et al.* [11]. Images of CAP protein (the sample of Dr. M. Fried, Penn State University) are shown in plate C on the scale close to that for images of RNA polymerase. Images are also very stable during repetitive scanning. In some cases individual particles can be measured. Their size is 6.1 ± 0.8 nm and these data are close to crystallographic data [34], about 5 nm for a dimer, keeping in mind the convolution between scanning probe and the sample [11, 16, 21]. The images shown in this Figure were obtained in propanol with regular Si_3N_4 tips. Note that similarly to AP-mica, 4N-mica retains its binding activity for several weeks.

Discussion

We suggested earlier to use aminopropyltrietoxy silane (APTES) to functionalize the mica surface with amine groups which protonate at neutral pH [21-26]. The major advantage of this procedure of sample preparation is that it works in a *wide variety of ionic conditions, pH and over a wide range of temperatures,* so AP-mica may be a suitable surface for imaging DNA complexed with different types of ligands. Here we also demonstrate that such large DNA complexes as kDNA can be deposited onto AP-mica for AFM imaging. Attachment of kDNA particles is very strong. First, we did not see any movement of the sample during scanning. Moreover there is no intra-particle movement revealed by scanning over part of the kDNA particle. Secondly, concentration of kDNA particles on the surface correlates with DNA concentration used for deposition. Third, images shown in Figure 2 were obtained by drying the water-rinsed samples in a rather strong argon flow, and we did not see any orientation of kDNA particles or its fragments that should happen for weakly bound molecules. We therefore conclude that AP-mica is a good substrate for structural studies of the kDNA network. The first AFM images of kDNA were reported by Thundat *et al.* [37], who developed a special procedure, critical point mounting, for kDNA samples, because they found shrinkage of samples and orientation of DNA molecules if a regular cation-assisted preparation method developed for imaging individual DNA molecules was applied for preparation of kDNA samples. Our AFM results suggest that there is no need to modify the DNA deposition procedure as long as AP-mica is used as a substrate. Similarly to other samples prepared on AP-mica (DNA, RNA, RecA-DNA and other nucleoprotein complexes [24-26]), kDNA samples on AP-mica are stable and do not absorb any contaminants for months without special precautions for storage.

In addition to positively charged AP-mica, we have developed procedures for preparing hydrophobic mica suitable for AFM studies. We checked their ability to hold samples for AFM imaging with chlorosomes. Coverage is rather uniform, indicating uniform modifica

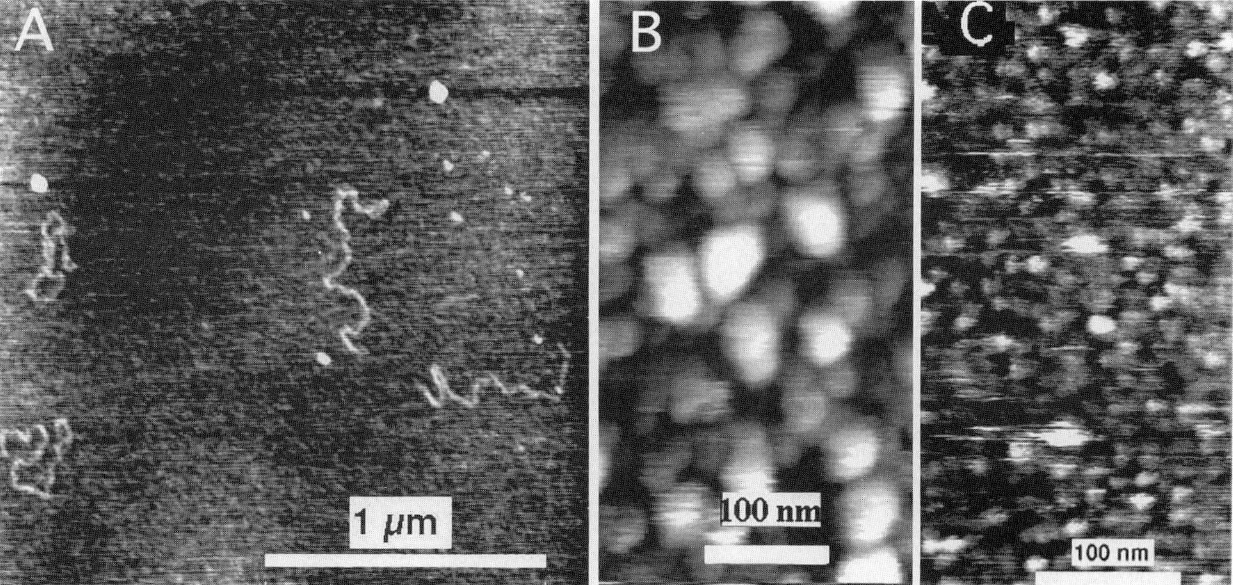

Figure 8: AFM images of samples deposited onto N4-mica: (**A**) dsRNA from reovirus; (**B**) RNA polymerase of *E. coli* and (**C**) CAP protein of *E. coli*. Images were obtained in propanol in contact mode.

tion of the surface. The Me-mica surface is also relatively smooth (roughness of non-covered spots is 0.06 ± 0.01 nm) so that individual hydrophobic molecules may be imaged on this substrate. Silylated porous silica and glass beads were used for binding whole cells and organelles (mitochondria, chloroplasts and microsomes) in a physiologically active state [4]. We therefore guess that hydrophobic surfaces (similar to Me-mica used in this work)may be used for AFM studies of whole cells and cell organelles. These samples are bound to hydrophobic substrates, and it is important for these studies that the binding process does not deteriorate the functional activity of the organelles: immobilized chloroplasts produced oxygen when exposed to light [4].

We also demonstrated that there is a simple means for transformation of a hydrophobic surface into a charged. We used organic salts containing alkyl groups for preparing a positively charged surface. An alkyl chain plays the role of an anchor holding the molecules at the surface while the charged group is exposed to the water solution. This type of positively charged surface allows the imaging of DNA, although large aggregates are formed at this surface in higher amount in comparison with AP-mica. The use of salts with different length of the carbohydrate chain enables one to change easily the length of the spacer between the surface and the charged group. A flexible spacer should help hold larger samples at the surface. It was shown in [4] that whole erythrocytes can be immobilized on glass beads modified with silanes with long alkyl spacer. Note that similar procedure of surface modification is usually applied for preparation of self-assembled monolayers (SAM, [38]).

The approach discussed above can be used for transformation of a hydrophobic surface into a negatively charged one. Dodecyl sulfate (SDS) may be an appropriate candidate for that. The dodecyl chain provides strong adhesion of SDS molecules to the Me-mica surface leaving a negatively charged sulfate group exposed to the water solution. Similarly, use of non-ionic detergents allows one to obtain surfaces with hydrophilic groups attached to a hydrophobic surface through an appropriate linker.

We also developed a procedure of mica modification allowing attachment of a chemically reactive group to the surface. We used IPTMS which leaves a reactive iodopropyl group on the surface. We demonstrated the reactivity of this immobilized group by treating the IP-mica with DABCO to make the surface strongly positively charged. This capability of IP-mica is important for development of procedures of covalent attachment of samples to the surface. First, by appropriate choice of reactive group bound to the surface one can attach the molecules to the surface with a high selectivity, so that the reaction may be limited to specific sites in the molecule (for example targeting of SH groups in a protein molecule or even specific aminoacid moieties in a protein or specific groups in nucleic acids bases). Secondly, in addition to topographic studies, the AFM is becoming a very promising tool for direct measurements of intermolecular interactions [12, 13, 19, 28]. This type of scanning probe microscopy, also termed chemical force microscopy (CFM), may find a very broad area of applications in surface science, physics, molecular and cell biology, immunology and eventually in pharmacology and medicine. An attachment of interacting components to the surface and a probe respectively is a technical, but a fundamental problem for measurements of intermolecular forces. Further studies of chemical modification of AFM probes is extremely important for routine use of AFM in this area.

Acknowledgements

We would like to thank A. Stasiak for preparation and sending us the samples of RecA-DNA complexes, M. Fried for providing us the sample of CAP protein and K. Severinov for sending us the sample of RNA polymerase; B. Jacobs for providing us the sample of dsRNA.

This paper was supported in part by NIH grants 1R21HG-0081801A1 and 1R43GM54991-01 (SBIR, Phase I), ONR grant N00014-90-J-1455 and grant F-017-95 from the Office of the Vice President for Research at ASU.

References

1. Allen MJ, Dong XF, O'Neil TE, Yau P, Kowalczykowski SC, Gatewood J, Balhorn R, Bradbury EM (1993) Atomic force microscope measurements of nucleosome cores assembled along defined DNA sequences. Biochemistry **32**: 8390-8396.

2. Allison DP, Bottomley LA, Thundat T, Brown GM, Woychick RP, Schrick JJ, Jacobson KB, Warmack RJ (1992) Immobilization of DNA for scanning probe microscopy. Proc Natl Acad Sci USA **89**: 10129-10133.

3. Allison DP, Thundat T, Jacobson KB, Bottomley LA, Warmack RJ (1993) Imaging entire genetically functional DNA molecules with the scanning tunneling microscopy. J Vac Sci Technol **A 11**: 818-819.

4. Arkles BC, Miller AS, Brininger WS (1978) In: Silylated Surfaces. Leiden DE, Collinns WT (eds). Gordon and Breach Science Publishers. New York, London, Paris. pp 363-375.

5. Bezanilla M, Manne S, Laney DE, Lyubchenko YL, Hansma HG (1995) Adsorption of DNA to mica, silylated mica, and minerals: Characterization by atomic force microscopy. Langmuir **11**: 655-659.

6. Binnig G, Rohrer H, Gerber CH, Weibel E (1982) Surface studies by scanning tunneling microscopy. Phys Rev Lett **49**: 57-60.

7. Binnig G, Quate CF, Gerber CH (1986) Atomic force microscope. Phys Rev Lett **56**: 930-933.

8. Blankenship RE, Brune DC, Wittmerhaus BP (1988) Chlorosome antennas in green photosynthetic bacteria. In: Light-Energy Transduction in Photosynthetic Higher Plants and Bacterial Models. Stevens SE Jr, Bryant DA (eds). Am Soc Plant Physiol, Rockville, MD. pp 32-46.

9. Booth BL, Ofunne GC, Stacey C, Tait PJT (1986) Silica-supported cyclopentadinyl-rhodium (I), -cobalt (I) and -titanium (IV) complexes. J Organometallic Chem **315**: 143-156.

10. Borst P, Hoeijmakes JHJ (1979) Kinetoplast DNA. Plasmid **2**: 20-40.

11. Bustamante C, Keller D, Yang G (1993) Scanning force microscopy of nucleic acids and nucleoprotein assemblies. Curr Opinion Struct Biol **3**: 363-372.

12. Bustamante C, Erie DA, Keller D (1994) Biochemical and structural applications of scanning force microscopy. Curr Opinion Struct Biol **4**: 750-760.

13. Frisbie CD, Rozsnyai LF, Noy A, Wrighton MS, Lieber CN (1994) Functional group imaging by chemical force microscopy. Science **265**: 2071-2074.

14. Gerola PD, Olson JM (1986) A new bacteriochlorophyll a protein complex associated with chlorosomes of greensulfur bacteria. Biochim Biophys Acta **848**: 69-76.

15. Hansma HG, Laney DE, Bezanilla M, Sinsheimer RL, Hansma PK (1995) Applications for atomic force microscopy of DNA. Biophys J **68**: 1672-1677.

16. Hansma HG, Hoh J (1994) Biomolecular imaging with the atomic force microscopy. Ann Rev Biophys Biochem Struct **23**: 115-139.

17. Hegner M. Wagner P, Semenza G (1993) Immobilizing DNA on gold via thiol modification for atomic force microscopy imaging in buffer solution. FEBS Letters **336**: 452-456.

18. Lal R, John SA (1994) Biological applications of atomic force microscopy. Am J Physiol **266**: C1-C21.

19. Lee GU, Chrisey LA, Colton RJ (1994) Direct measurements of the forces between complementary strands of DNA. Science **266**: 771-773.

20. Leuba SH, Yang G, Robert C, Samori B, van Holde K, Zlatanova J, Bustamante C (1994) Three-dimensional structure of extended chromatin fibers as revealed by tapping-mode scanning force microscopy. Proc Natl Acad Sci USA **91**: 11621-11625.

21. Lyubchenko YL, Gall AA, Shlyakhtenko L, Harrington RE, Jacobs BL, Oden PI, Lindsay SM (1992) Atomic force microscopy imaging of dsDNA and RNA. J Biomolec Struct Dyn **10**: 589-606.

22. Lyubchenko YL, Shlyakhtenko L, Harrington RE, Oden PI, Lindsay SM (1993) Atomic force microscopy of long DNA: Imaging in air and under water. Proc Natl Acad Sci USA **90**: 2137-2140.

23. Lyubchenko YL, Jacobs BL, Lindsay SM (1992) Atomic force microscopy of reovirus dsRNA: a

routine technique for length measurements. Nucleic Acids Res **20**: 3983-3986.

24. Lyubchenko YL, Oden PI, Lampner D, Lindsay SM, Dunker KA. (1993) Atomic force microscopy of DNA and bacteriophage in air, water and propanol: the role of adhesion forces. Nucleic Acids Res **21**: 1117-1123.

25. Lyubchenko YL, Lindsay SM (1995) DNA, RNA and nucleoprotein complexes immobilized on AP-mica and imaged with AFM. In: Procedures in Scanning Probe Microscopies. Engel A, Gaub H. (eds) John Wiley & Sons, Chichester, in press.

26. Lyubchenko YL, Jacobs BL, Lindsay SM, Stasiak A (1995) Atomic force microscopy of nucleoprotein complexes. Scanning Microsc **9**: 705-727.

27. Marchese-Ragona SP, Bucher R, Christie B (1994) Biological sample preparation. Application News Letters, TopoMetrix (Santa Clara, CA) **94-1**: 4-5.

28. Moy VT, Florin EL, Gaub HE (1994) Intermolecular forces and energies between ligands and receptors. Science **266**: 257-259.

29. Pierson BK, Castenholz RW (1992) In: The Prokaryotes. Balows A, Trüper HG, Dworkin M, Schleifer KH, Harder W (eds). Springer-Verlag, Berlin, pp. 3574-3774.

30. Plueddemann EP (1991) Silane Coupling Agents. Plenum Press, New York, London. pp. 1-53.

31. Rauch CA, Perz-Morga D, Cozzarelli NR, Englund PT (1993) The absence of supercoiling in kinetoplast DNA minicircles. EMBO J **12**: 403-411.

32. Simpson L (1979) Isolation of maxicircle component of kinetoplast DNA from hemoflagellate protozoa. Proc Natl Acad Sci USA **76**: 1585-1588.

33. Simpson L, Simpson AM, Blum B (1993) RNA editing in *Leishmania* mitochondria. In: RNA Processing - A Practical Approach. Hames D, Higgins S (eds) IRL Press, Oxford/New York. pp 69-105.

34. Steiz TA (1990) Structural studies of protein-nucleic acid interaction: the sources of sequence-specific binding. Quart Rev Biophys **23**: 205-280.

35. Thiemann OH, Maslov DA, Simpson L (1994) Disruption of RNA editing on *Leshmania tarentolae* by loss the minicircle-encoded guide RNA genes. EMBO J **13**: 5689-5700.

36. Thundat T, Allison DP, Warmack RJ, Brown GM, Jacobson KB, Schrick JJ, Ferrel TL (1992) Atomic force microscopy of DNA on mica and chemically modified mica. Scanning Microsc **6**: 911-918.

37. Thundat T, Warmack RJ, Allison DP, Jacobson KB (1994) Critical point mounting of kinetoplast DNA for atomic force microscopy. Scanning Microsc **8**: 23-30.

38. Ulman A (1991) Introduction to Ultra-Thin Films. Academic Press, New York. Chapter 5.

39. Vesenka J, Guthold M, Tang CL, Keller D, Bustamante C (1992) Substrate preparation for reliable imaging of DNA molecules with the scanning force microscopy. Ultramicrosc **42-44**: 1243-1249.

40. Weisenhorn, AL, Römer DU Lorenzi GP (1992) An atomic force microscope study of Langmuir-Blodgett films of a b-^{4,4} -helical pentadecavaline. Langmuir **8**: 3145-3149.

41. Wingren R, Elwing H, Erlandsson R, Welin S, Lundström I (1991) Structure of adsorbed fibrinogen obtained by scanning force microscopy. FEBS Letters **280**: 225-228.

42. Yang J, Takeyasu K, Shao Z (1992) Atomic force microscopy of DNA molecules. FEBS Letters **301**: 173-176.

43. Zlatanova J, Leuba SH, Yang G, Bustamante C, van Holde K (1994) Linker accessibility in chromatin fibers of different conformations: a reevaluation. Proc Natl Acad Sci USA **91**: 5277-5280.

Discussion with Reviewers

R. Lal: The cells, both in vivo and in culture, are surrounded by hydrophilic solutions. I am not sure why they adsorb to a hydrophobic surface.
Authors: The structure of the shell of isolated cells is very complex. It contains both hydrophobic and hydrophilic groups. The experiments with isolated cells and organelles immobilized on the surface, which we discussed, were performed with hydrophobic glass substrates [4]. The samples adsorbed on these silylated substrates remain active.

R. Lal: Does the functionalized substrate change the apparent height of the isolated macromolecules?
Authors: Height measurements of DNA under different scanning conditions (air, propanol and water) were performed in our early work [24]. Similar to the data obtained by others (e.g., [47]) the height of DNA is substantially less than expected. The mechanism of the apparent height reduction is not understood so far, but may be a combination of both adhesion of the tip to the sample and tip-induced compression of DNA.

J.A.R. Zasadzinski: How is the secondary and tertiary structure of a biological macromolecule altered by tight binding to a surface? Can surfaces be systematically modified to check just how much binding is required for the various AFM modes? Clearly, the softer the hold, the less denaturation might be expected. Can tapping or other non-contact modes be brought to bear here? What is the remaining limit of resolution? Can we ever expect superior resolution to TEM?
Authors: Generally speaking, any deposition process

should result in a deformation of a macromolecule. Even such a stable biomolecule as DNA can be stretched out during sample preparation [53]. In the case of AP-mica, DNA molecules retain their B-conformation [21]. An important observation was made in [4]. They showed that cells and cellular organelles immobilized on glass surfaces treated with silanes remain active.

The extent of surface modification with silanes can be controlled [30]. Clearly, the regime should be different for different samples and types of imaging. For example, AP-mica prepared by modification in APTES for 2 hours is sufficient for holding filamentous bacteriophage at the surface, so that imaging in situ can be performed [24]. However these images are quite streaky if contact mode in buffer solution is used. Our recent experiments have shown that stable images of fd phage in situ in buffer solution can be obtained in tapping mode AFM (Shlyakhtenko, unpublished data).

Resolution is typically higher for tapping mode AFM. For instance, double helical periodicity of DNA can be resolved [15]. There is no TEM data of DNA where this level of resolution was achieved. See below for more about other high resolution AFM data.

J.A.R Zasadzinski: What limits the resolution with contact mode AFM? Will non-contact modes - tapping, adhesion, etc. or lateral force provide higher resolution or other important data?

Authors: There are several factors, in addition to the tip geometry, that limit the resolution AFM. They are tip-sample/substrate interaction, the sample mobility and intramolecular thermal motion. Strong adhesion effects typically reduce the resolution. One type of tip-substrate adhesion is formation of a water layer between the tip and the sample while scanning is done air. For example, the capillary effect results in considerable broadening of the width DNA [20, 22]. Imaging in water in contact mode [22] or the use of tapping mode decrease the width of DNA by a factor of two. High resolution data for DNA molecules was achieved if imaging is done in propanol. Adhesion forces in propanol are much lower than in water; as a result, AFM images of DNA as thin as 2-4 nm can be obtained [15, 16, 22]. The use of tapping mode made it possible to resolve helical periodicity of DNA [15]. In cases when adhesion forces are small very high resolution can be achieved by the use of contact mode AFM. One of the recent examples is imaging of purple membrane with AFM by the group of A. Engel [48, 50]. The attachment of antibodies to C-terminal part of bacteriorhodopsin allowed the authors to obtain structural details of intracellular and extracellular sides of the membrane with subnanometer resolution. Similar high resolution AFM data for individual proteins are hard to obtain because of large thermal intramolecular motion. The development of cryo AFM in the group of Shao permitted them to get AFM images of individual IgG molecules with very high resolution [46].

J.A.R Zasadzinski: What are the major benefits to AFM studies of membrane proteins over cryo-electron microscopy? Are there certain systems that just cannot be done with TEM that can be examined by AFM? Are there systems in which the AFM resolution is superior to TEM?

Authors: Currently the resolution of AFM for membrane proteins is close to that for cryo-electron microscopy. However there is a number of advantages of AFM in comparison with cryo-EM. First, AFM topographical data can be obtained under physiological conditions. Examples include purple membrane [45], gap junctions [16], OmpF porin *E. coli* [52]. Very recently Müller *et al.* [48, 50] obtained AFM images of purple membranes in solution at subnanometer resolution. In particular they imaged individual polypeptide loops connecting transmembrane a-helices of bacteriorhodopsin. Moreover the use of AFM made it possible the same group to observe reversible force-dependent structural changes of bacteriorhodopsin molecules [49]. These changes were interpreted as the bending of individual loops of the protein. Such studies cannot be done with EM. There are several examples when both AFM and TEM were applied to similar systems. In the Engel group membrane protein OmpF was imaged with AFM with subnanometer resolution which is substantially higher than early EM data obtained in the same group [52]. Additional structural information for this system were obtained from the height AFM data; such information is unavailable for EM. This advantage of AFM was critical for EM and AFM studies of complexes of DNA with the heat-shock transcription factor 2 [54]. The height AFM data led the authors to conclude that DNA looping requires trimerization of the protein.

J.A.R Zasadzinski: How important is the environment or the substrate in the protein conformation? Is the problem with resolution more that the proteins moving over the time scales of the images, thereby blurring the images? What are the benefits to such kinetic information that obviously cannot be determined from fixed or frozen materials?

Authors: Conformation of protein is quite sensitive to the environmental conditions, therefore AFM studies in solution are so attractive. Binding a protein to the surface may influence its biochemical activity, but this undesirable effect may be decreased if the protein is tethered to the surface gently enough, but sufficiently strongly to withstand the tip-sweeping effect. The use of tapping or non-contact AFM is essential in this case. For example, lysozyme was imaged with AFM in solution.

The protein molecules remain active being bound to the mica surface, and this activity of the protein was visualized by the AFM [51]. Loose tethering of a macromolecule clearly results in lower resolution but is beneficial for imaging biochemical processes. This was shown in recent publications from the Hansma group, who imaged the process of DNA depolymerization induced by nuclease [44].

W. Fritzsche: Your SFM dimensions of CAP protein (ca. 6 nm) exceed hardly the crystallographic data (ca. 5 nm), in contrast to many SFM studies of small biomolecules (e.g., DNA) with stronger broadening effects. Is this low exaggeration due to a very sharp tip and/or shrinkage of the protein in alcohol, or is there another explanation?

Authors: The CAP protein was imaged in propanol to minimize adhesion forces [16]. At these scanning conditions we were able to get DNA and RNA molecules as thin as 3-4 nm [26], which is very close to the crystallographic sizes of DNA and RNA. We assume that the AFM probes usually have very fragile and sharp asperities, so by gentle approaching the tip at conditions of low adhesion of the tip to the sample one be able to use a sharp tip for imaging. Poor adhesion of the tip to the sample is another advantage preventing the tip contamination during scanning. We cannot exclude some shrinkage of the protein molecules at these scanning conditions in addition to the same effect induced by drying the sample.

Additional References

44. Bezanilla M, Drake B, Nudler E, Kashlev M, Hansma PK, Hansma HG (1994) Motion and enzymatic degradation of DNA in the atomic force microscope. Biophys J **67**: 2454-2459.

45. Butt H-J, Downing KH, Hansma PK (1990) Imaging the membrane protein bacteriorhodopsin with the atomic force microscope. Biophys J **58**: 1473-1480.

46. Han WH, Mou JX, Sheng J, Yang J, Shao ZF (1995) Cryo atomic force microscopy: a new approach for biological imaging at high resolution. Biochem **34**: 8215-8220.

47. Marsh TC, Vesenka J, Henderson E (1995) A new DNA nanostructure, the G-wire, imaged by scanning probe microscopy. Nucleic Acids Res **23**: 696-700.

48. Müller DJ, Schaber FA, Büldt G, Engel A (1995) Imaging purple membranes at aqueous solutions at subnanometer resolution by atomic force microscopy. Biophys J **68**: 1681-1686.

49. Müller DJ, Büldt G, Engel A (1995) Force-induced conformational change of bacteriorodopsin. J Mol Biol **249**: 239-243.

50. Müller DJ, Schoenenberger C-A, Büldt G, Engel A (1996) Immuno-atomic force microscopy of purple membrane. Biophys J **70**: 1796-1802.

51. Radmacher M, Fritz M, Hansma HG, Hansma PK (1994) Direct observation of enzyme activity with the atomic force microscope. Science **265**: 1577-1579.

52. Schabert FA, Henn C, Engel A (1995) Native *Escherichia coli* OmpF porin surfaces probed by the atomic force microscopy. Science **268**: 92-94.

53. Thundat T, Allison DP, Warmack RJ (1994) Stretched DNA structures observed with atomic force microscopy. Nucleic Acids Res **22**: 4224-4228.

54. Wyman C, Grotkopp E, Bustamante C, Nelson HCM (1995) Determination of heat-shock transcription factor 2 stoichiometry at looped DNA complexes using scanning force microscopy. EMBO Journal **14**: 117-123.

Scanning Microscopy Supplement 10, 1996 (Pages 111-121)
Scanning Microscopy International, Chicago (AMF O'Hare), IL 60666 USA
0892-953X/96$5.00+.25

MICROSCOPIC ANALYSIS OF DNA AND DNA-PROTEIN ASSEMBLY BY TRANSMISSION ELECTRON MICROSCOPY, SCANNING TUNNELING MICROSCOPY AND SCANNING FORCE MICROSCOPY

T. Müller-Reichert[1*] and H. Gross[2]

[1]European Molecular Biology Laboratory, Heidelberg, Germany,
[2] Institut für Zellbiologie, Eidgenössische Technische Hochschule-Hönggerberg, Zürich, Switzerland

(Received for publication October 18, 1996 and in revised form December 20, 1996)

Abstract

To investigate DNA and DNA-protein assembly, nucleic acids were adsorbed to freshly cleaved mica in the presence of magnesium ions. The efficiency of DNA adhesion and the distribution of the molecules on the mica surface were checked by transmission electron microscopy. In addition, various kinds of DNA-protein interactions including DNA wrapping and DNA supercoiling were analyzed using electron microscopy. In parallel, this Mg^{2+}/mica method can be applied (1) to analyze embedded DNA by scanning tunneling microscopy, (2) to visualize freeze-dried, metal coated DNA-protein complexes by tunneling microscopy, and (3) to image DNA or DNA-protein interaction in air or in liquid by scanning force microscopy. An advantage of such a correlative approach is that parallel imaging can reveal complementary information. The benefit of such a combined approach in analysis of protein-induced DNA bending is discussed.

Key Words: Transmission electron microscopy, scanning electron microscopy, scanning tunneling microscopy, scanning force microscopy, DNA, DNA-protein interaction, Repressor Activator Protein (RAP) 1.

*Address for correspondence:
T. Müller-Reichert
European Molecular Biology Laboratory
Postfach 10.2209
D-69112 Heidelberg, Germany
Telephone Number: +49-6221-387360
FAX Number: +49-6221-387 512
E-mail: reichert@embl-heidelberg.de

Introduction

The various conformational states which can be adopted by DNA molecules have been intensively studied by X-ray diffraction [40]. Visualization of individual DNA molecules, however, was only possible after the development of the transmission electron microscope (TEM) (Fig. 1a). Techniques had to be developed to visualize immobilized, dehydrated molecules under high vacuum conditions. One of the most widely used techniques for the preparation of nucleic acids for TEM was reported by Kleinschmidt and Zahn [29]. Homogenous spreading of the molecules was achieved by binding them to a thin layer of an unspecific DNA-binding protein (i.e., cytochrome C). DNA was then adsorbed to carbon coated TEM-grids prior to contrast enhancement by rotary shadowing at low elevation angles. By application of this "Kleinschmidt" technique, the contour length of individual molecules could be measured. However, when the "Kleinschmidt" technique is applied, the DNA is covered by the cytochrome C and due to this concomitant broadening, details of DNA bound proteins underneath are obscured! A technique for protein free spreading of DNA was first proposed by Hall *et al.* [26]. DNA molecules were sprayed onto the surface of freshly cleaved mica, shadowed with heavy metal and replicated with C or SiO. The adsorption of DNA by this technique, however, was rather uncontrolled and occasional. The technique has been modified by the application of Al^{3+} [24] and Mg^{2+} [30] to make the molecules adhere to the solid support. With this Mg^{2+}/mica technique, proteins bound specifically to nucleic acids could be visualized [30].

In the beginning of the 1980s a new class of microscope was invented, the scanning probe microscope (SPM). The most prominent microscopes of this new imaging technology are the scanning tunneling microscope (STM) [9] and the scanning force microscope (SFM) [7]. Common to both STM and SFM is the potential to measure surface related properties at atomic resolution. In general, topographic information of a

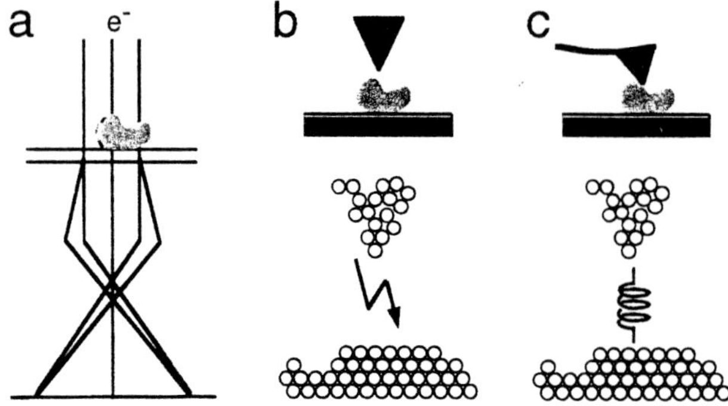

Figure 1. Imaging modes for analysis of DNA and DNA-protein complexes. **(a)** In TEM absorption and phase contrast are used to create a two-dimensional projection of a sample. **(b)** A tunneling current between metal tip and a conductive sample is recorded in STM to create a three-dimensional image of a scanned surface. **(c)** Three-dimensional image formation in SFM occurs by recording forces between a tip (cantilever) and a non-conducting surface.

---

given sample can be obtained in air and in liquid. This is in contrast to the TEM, where high vacuum conditions are necessary for imaging.

In STM a sharp, conductive probe (i.e., the tip) is attached to a piezoelectric XYZ-scanner and brought into close proximity to the surface of a conductive sample (Fig. 1b). During lateral scanning of biological objects, the tunneling tip is raised and lowered to keep the tunneling current constant (constant current mode). If the lateral position of the tip is plotted against the elevation of the probe, a three-dimensional non-contacting image of the surface features can be obtained. In contrast, SFM records interatomic forces between the atoms of the tip and the atoms of the sample during scanning of a cantilever over a given specimen (Fig. 1c). The tip, acting as a spring, is moved in raster fashion over the surface. Laser light is focussed on the back of the spring and deflections caused by the tip during scanning are measured by a segmented photodiode.

For both of the new imaging techniques, methods have to be available to immobilize biomolecules (i.e., make them adhere) onto solid supports. Moreover, when imaging under atmospheric conditions is intended, methods to properly dehydrate the specimens have to be applied. Techniques of preparation developed for TEM to immobilize and dehydrate biological specimens are of special importance in this respect. In this paper we discuss the application of mica as a support for TEM as well as for STM or SFM preparation. Particular attention will be paid to the fact that this technique offers the possibility to perform TEM and SPM experiments in parallel. Different kinds of DNA-protein interactions will be discussed. As an example the binding of the yeast Repressor Activator Protein 1 (RAP1) [22] to cloned DNA molecules containing telomeric, sequence-specific protein recognition sites will be described. In addition, advantages and disadvantages of using TEM, STM or SFM in the analysis of DNA-protein assemblies will be discussed.

Materials and Methods

Film thickness measurements

The metal coat thickness was measured with a quartz monitor (Balzers QSC 301; Balzers, Liechtenstein) which was oriented always perpendicular to the coating source. The film thickness was determined in the direction of the metal deposition and calculated according to the density of bulk platinum.

Preparation of DNA for TEM

For standard electron microscopy linear DNA fragments (409 bp; 10 ng/μl, diluted 7 times in 8 mM Mg-acetate) were adsorbed to freshly cleaved mica (Ruby B, BAL-TEC, Liechtenstein) and washed in double-distilled water for 3 hours [41]. Dehydration of the samples was carried out using ethanol. The specimens were rotary shadowed (BAL-TEC BAF 400T) with 5nm Pt/C at an elevation angle of about 3° and stabilized at 90° with a thin layer of carbon (6-7 nm). Replicas from the mica were floated onto the surface of bidistilled water, collected on 400 mesh copper grids, and analyzed in a Philips CM 12 (Philips Electron Optics, Eindhoven, The Netherlands) transmission electron microscope operated at an acceleration voltage of 100 kV.

Replica/anchoring technique for analysis of bare DNA

DNA fragments were deposited on freshly cleaved mica, washed, and dehydrated from ethanol in air as described above. Samples were rotary coated with 8 nm of Pt/C at room temperature at an elevation angle of 65° in the BAL-TEC BAF 400T. After evaporation of Pt/C a copper disk was glued with epoxy adhesive onto the Pt/C film. Finally, the copper disk with the DNA embedded in Pt/C was peeled off the mica. Imaging of the previously mica-exposed side of the Pt/C film allowed analysis of uncoated DNA molecules by STM. The mica exposed side has also been analyzed by SFM using a Nanoscope II [35].

Metal-coating of DNA for STM

With the Mg^{2+}/mica technique DNA was immobilized on freshly cleaved mica. Samples were washed (3 hours in distilled water) and dehydrated by ethanol/air or by freeze-drying. For freeze-drying samples were mounted on a specimen table under liquid nitrogen prior

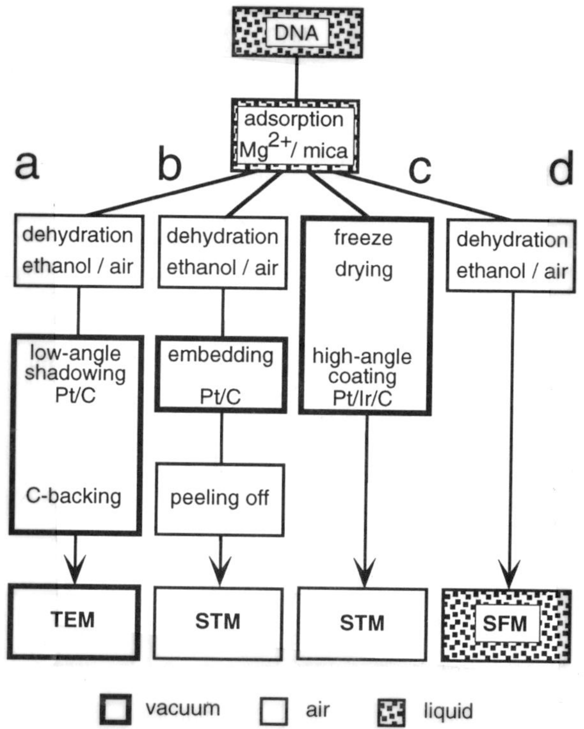

Figure 2. Flow diagram illustrating the main steps involved in the preparation of nucleic acids for TEM, STM and SFM. In the presence of magnesium ions DNA can routinely be adsorbed on freshly cleaved mica. **(a)** In standard TEM, DNA molecules then have to be dehydrated from ethanol in air, rotary shadowed at low elevation angles and coated with carbon (C backing). A replica of the preparation can be analyzed in the electron microscope. **(b)** For analysis by STM, DNA may be embedded in a thick layer of Pt/C after dehydration from ethanol in air. By peeling off the metal layer, the previously mica exposed side of the embedded DNA can be analyzed under atmospheric conditions. **(c)** Alternatively, freeze-drying can be applied to dehydrate the biomolecules. By high-angle metal coating reproducible imaging of DNA by STM in air can be achieved. **(d)** When analysis of DNA by SFM is intended imaging in buffer solution is the method of choice. To prevent lateral dislocation of the DNA on the support, dehydration of the DNA from ethanol in air prior to imaging might be necessary.

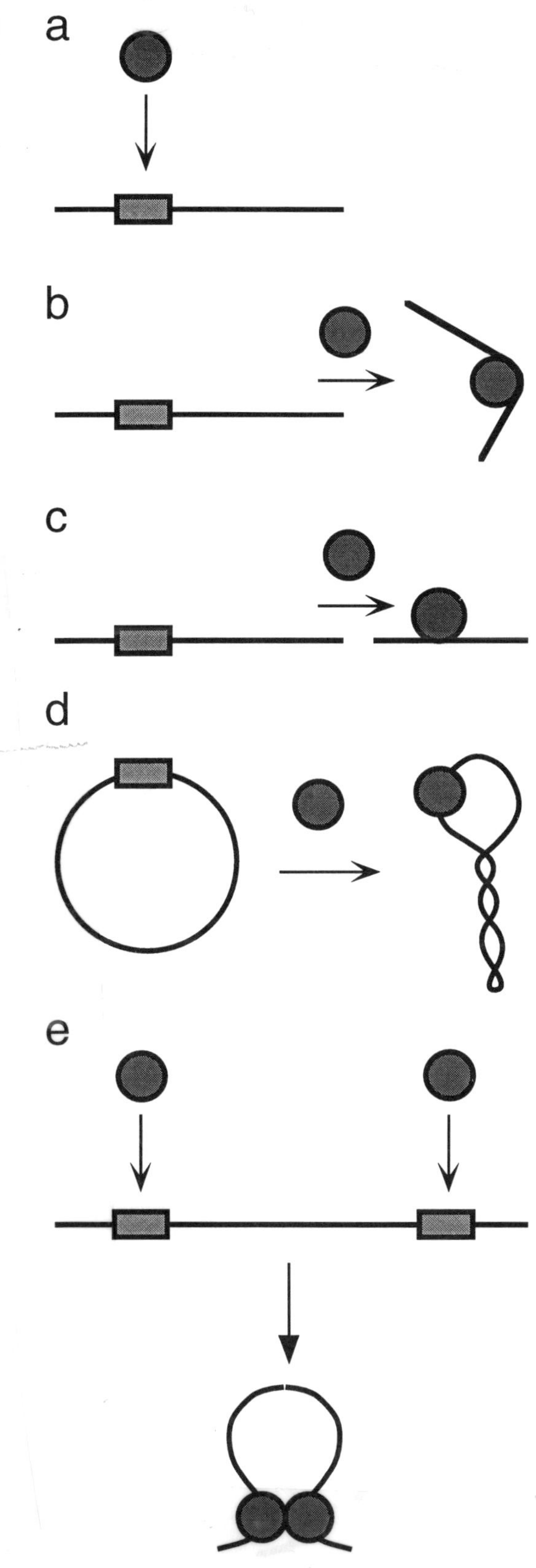

Figure 3. Types of DNA-protein interactions. Microscopic techniques may be applied to analyze the following features: **(a)** site specific occupation, **(b)** protein-induced DNA bending, **(c)** DNA wrapping, **(d)** DNA supercoiling, and **(e)** DNA higher order structure induced by protein-protein interactions. Black line: DNA molecule; grey rectangle: protein recognition site; grey circle: sequence specific DNA-binding protein.

to transferring the specimen onto the precooled cold stage (143K) of a freeze-etch unit (BAL-TEC BAF 400T). Freeze-drying was carried out for 3 hours at 193 K under high vacuum conditions (P < 10^{-6} mbar). Ethanol-dried samples were mounted on the specimen table at room temperature, transferred to the freeze-etch unit, and cooled down to 193 K prior to metal coating. Both ethanol-dried and freeze-dried samples were coated at 193 K by rotary shadowing them with 0.7-1 nm Pt/Ir/C at an elevation angle of 65° [3, 46]. The samples were warmed to room temperature after shadowing and removed from the vacuum.

STM imaging conditions

STM measurements were performed in the constant current mode under atmospheric conditions. A modified Park Scientific (Sunnyvale, CA) Tunneling Microscope, operated at a tunneling voltage in the range of 350 mV was used. The tunneling current was kept below 25 pA. The scan rate was usually 0.5-2 Hz. Tungsten tips (0.5 mm diameter) were electrochemically etched in a 1 M KOH solution by applying a DC voltage of about 15 V between the positively biased tungsten wire and a grounded Pt electrode.

Results and Discussion

Application of standard TEM to visualize DNA-protein assembly

The major steps involved in routine preparation of DNA for TEM are shown schematically in Fig. 2a. By addition of magnesium ions, DNA molecules are immobilized on a freshly cleaved mica surface. Dehydration of the sample is carried out from ethanol in air. To gain contrast, DNA has to be rotary shadowed with Pt/C at low elevation angles under high vacuum conditions. Carbon backing is applied for subsequent floating the carbon film on the surface of distilled water. Finally, the replica can be analyzed in the electron microscope [41]. When applying this routine preparation technique a homogenous spreading and a proper immobilization of DNA molecules on the solid support can be achieved.

Using low-angle shadowing by heavy metals to enhance the contrast of the small, fibrous DNA molecules on mica, a major disadvantage of this technique is the induction of self-shadowing. Owing to this effect, large metal clusters are growing not only in the background of the specimen, but also along the trace of the DNA molecules. As a consequence, the structure of the DNA itself (i.e., the sequence of bases) is obscured. Nevertheless, analysis of contour lengths can be carried out to determine the total length of linear DNA fragments and the correct positioning of bound protein complexes. Application of TEM imaging in analysis of different types of DNA-protein interactions is shown schematically in Fig. 3. The following types of interaction can be analyzed: (a) site specific occupation, (b) DNA bending, (c) DNA wrapping, (d) DNA supercoiling, and (e) DNA higher order structure. Site specific occupation of proteins can be analyzed by binding of a sequence specific DNA-binding protein to cloned DNA fragments containing a single, sequence specific binding site [34]. By this kind of analysis not only the accuracy, but also the efficiency of protein binding can be determined. Cloned DNA fragments containing a single, asymmetrically located binding site are also applied to analyze protein-induced DNA bending [34]. Interpretation of the images, however, is limited, because of the low-angle shadowing procedure (see below: STM of metal-coated DNA and DNA-protein complexes). DNA wrapping can be analyzed by binding of proteins to multiple sequence specific binding sites on relatively short DNA fragments [23]. Depending on the size of the DNA binding protein, wrapping of the substrate double helix should be accompanied by a considerable shortening in the length of a DNA fragment when analyzed by TEM. By this kind of analysis a possible wrapping of telomeric DNA by RAP1 could be excluded (Fig. 4). No shortening of the DNA was observed after multiple occupation of RAP1 binding sites. In contrast, Gyrase, a 200 kDa protein of *E. coli*, for example, is known to wrap a segment of 120 to 155 bp of DNA around itself [39]. TEM may also be applied to analyze protein-induced supercoiling [23]. For this purpose proteins have to be bound to sequence specific binding sites on circular DNA molecules. Alternatively, TEM may also be used to investigate the role of protein-protein interactions in establishing higher order DNA structures [42, 43].

The Mg^{2+}/mica technique allows a routine, protein-free (i.e., cytochrome C-free) spreading of single DNA molecules. Moreover, by the application of low-angle rotary shadowing for TEM imaging, a number of different DNA-protein interactions can be analyzed. The limitations of this kind of contrast enhancement, however, are twofold: First, analysis of DNA binding proteins is limited to polypeptides of a size ≥ 30 kDa [41]. Using this standard technique, smaller proteins can not be visualized on DNA fragments. Second, the trace of the DNA in close proximity of the protein complex might be obscured due to the low angle of shadowing. However, shadowing has to be performed at low angles to obtain sufficient contrast. To avoid the disadvantages of the shadowing procedure STM was applied soon after the introduction of the instrument to analyze uncoated samples. The advantages and major drawbacks of STM in analyzing DNA structure and DNA-protein assembly are discussed in the next paragraph.

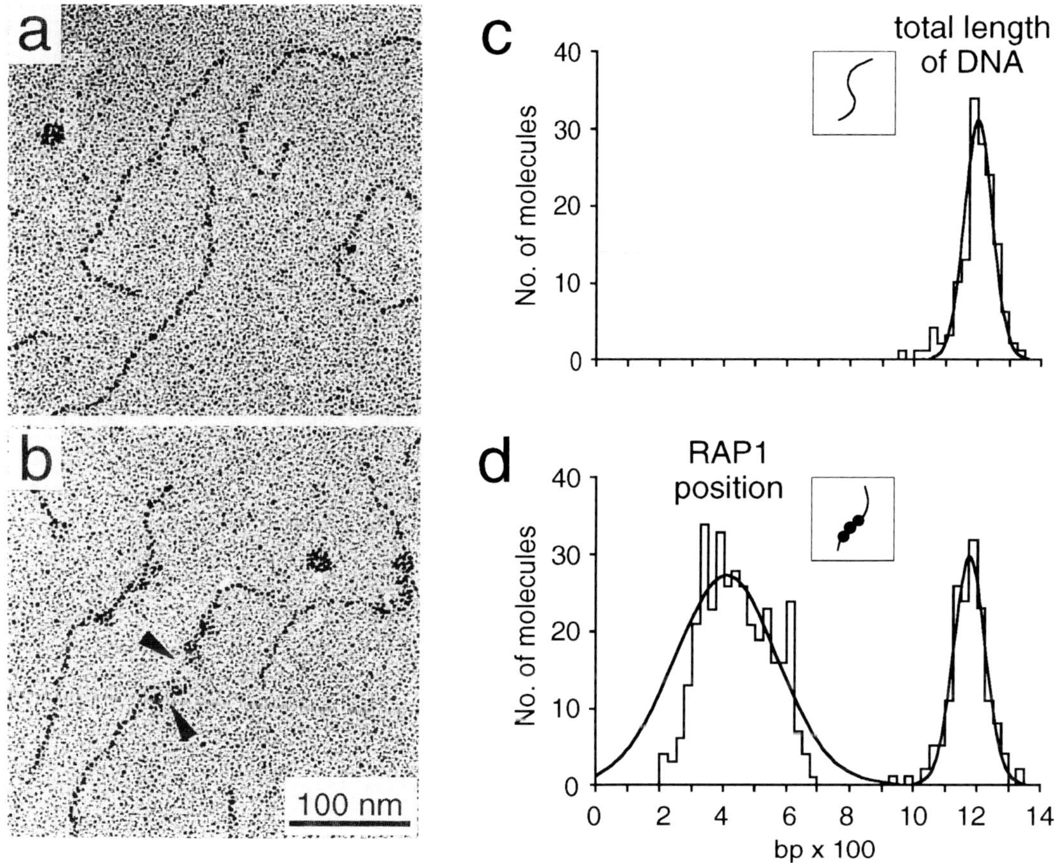

Figure 4. TEM analysis of DNA-protein assembly. (**a,b**) Electron micrographs illustrating the binding of the sequence specific, DNA-binding protein RAP1 to a cloned DNA fragment (1.2 kb) containing 14 binding sites for the yeast protein. (**c,d**) Statistical analysis of total DNA length and RAP1 positioning on the cloned DNA fragment. (**a**) Uncomplexed DNA fragment. (**b**) Multiple binding of RAP1 to the fragment. The trace of DNA between single complexes can not be followed (arrow heads). (**c**) Total length of uncomplexed DNA (1200 bp $\pm$ 35 bp; n = 45). (**d**) Total length of fragments with multiple bound RAP1 molecules was 1180 $\pm$ 39 bp (n = 155). The RAP1 complexes peak at position 408 $\pm$ 130 bp (n = 348).

Imaging of bare DNA by STM

Biologists were interested in applying STM to the study of macromolecular assemblies because of the potential of this novel imaging technique to provide atomic resolution on flat surfaces [10]. Due to this reason imaging of DNA by STM was started very early after the introduction of the microscope [8]. In most of the experiments where STM was applied to image DNA in air [1, 4, 5, 6, 17, 19, 28, 44], in vacuum [18] or in liquid [31, 32] highly oriented pyrolytic graphite (HOPG) was used as a support. Reproducibility in some of these experiments, however, was reported to be very poor. Most of the problems arose, because adsorption of nucleic acids to this conducting support was insufficient. In addition, chain-like structures mimicking nucleic acids were often observed on freshly cleaved graphite surfaces [16]. To avoid the HOPG-related problems, the following approach has been developed for analysis of bare DNA by STM (Fig. 2b). DNA fragments in the presence of Mg-acetate were deposited on freshly cleaved mica. Dehydration of the specimen was carried out by drying from ethanol in air. Subsequently, Pt/C was evaporated to embed the molecules in a layer of metal. Using this replica/anchoring technique [11, 15] the metal film was peeled off the mica and the previously mica exposed side of the embedded biomolecules was accessible for imaging by STM.

Under the given atmospheric conditions bare DNA could not be imaged by STM (Fig. 5). Only "hollow trenches" averaging 3.1$\pm$0.9 nm wide and 1$\pm$0.5 nm deep were visualized in a very smooth metal surface. By SFM measurements, however, it could be confirmed that DNA remained in the metal film during the peeling off procedure. Applying this alternative imaging technique, embedded DNA could be visualized using positive contrast [35]. Similar results were obtained by Dunlap *et al.* [20]. By partly masking the deposited DNA molecules before coating it was possible to produce adjacent segments of coated and uncoated regions. In uncoated regions DNA was observed as "empty furrows". In addition, DNA appeared as a "hollow trench" when adsorbed on gold surfaces [1]. In contrast, imaging of bare DNA molecules on mica with positive contrast has recently been reported by Guckenberger *et al.* [25]. By increasing the relative humidity of the ambient air to 70%, the authors demonstrated that a very thin film of water adsorbed to the surface was sufficiently conductive to allow STM imaging.

Application of mica for DNA immobilization turned out to be advantageous: (1) artefacts as described for HOPG have not been reported, (2) very clean, atomically flat areas can be obtained by cleaving the aluminum silicate structure along its crystalline planes [38], (3) the well established Mg^{2+}/mica technique can be applied for routine adsorption of the DNA [41], and (4) a control of the preparation (i.e., efficiency of DNA adhesion and distribution of molecules on the surface) can be obtained by standard TEM prior to imaging by STM or SFM [15, 34, 35]. Importantly, metal embedding by Pt/C appears to be sufficient to overcome the major disadvantage of mica (i.e., its low electrical conductivity). The surrounding Pt/C layer can provide sufficient conductivity for routine imaging. By applying this embedding technique, it becomes possible to use two major advantages of this method: First, the technique can be applied to establish the experimental conditions for imaging of bare DNA molecules. Because the intrinsic conductivity of DNA appears to be lower as expected, it is of importance to choose well defined, easy-to-use experimental conditions. Second, replica/anchoring can be used for studying artefacts accompanying adsorption of biomolecules. By application of freeze drying (see below) both the upper side and the previously mica exposed side of any biomolecule can be analyzed to investigate the structural implications of adhesion forces.

STM analysis of metal-coated DNA and DNA-protein complexes

Because the STM cannot visualize bare, embedded DNA molecules in air, metal coating was applied to achieve reproducible imaging conditions. In addition, this coating offered the possibility to investigate structural artefacts accompanying dehydration. This is of importance, because surface tension forces were shown to cause severe damaging of biological structures [36, 47]. Using the Mg^{2+}/mica technique, DNA molecules were immobilized on a freshly cleaved mica surface. Specimens were freeze-dried and rotary shadowed with Pt/Ir/C at a high angle of elevation and imaged by STM at atmospheric conditions (Fig. 2c). Alternatively, samples were dried from ethanol in air and coated with Pt/Ir/C under identical conditions. For ethanol/air-dried specimens measured values for DNA width and height were 5.1 ± 1.8 nm and 0.9 ± 0.2 nm, respectively. The width of freeze-dried DNA was 4.2 ± 1.3 nm and the height was 1.1 ± 0.1 nm [35]. Compared to freeze-drying, ethanol/air-drying appeared to broaden and flatten the DNA structure.

Metal coating for STM was first applied to image DNA-recA complexes [3]. The aim of this coating was to render the surface of both the sample and the substrate uniformly conductive [2]. The metal coat has to cover the specimen homogeneously without blurring fine structural details of the biological sample underneath. After freeze-drying of the specimen the coating film has to stabilize the preserved conformation of the dehydrated sample when transferring the sample from vacuum to atmospheric conditions. Pt/Ir/C films have a fine granularity and have proven to remain three-dimensionally stable during the transfer from vacuum to air [46]. A minimum thickness of 0.7-1 nm Pt/Ir/C was sufficient to achieve stable tunneling conditions [34]. By comparing ethanol/air-dried and freeze-dried DNA under stable tunneling conditions it turned out that even the rather rigid DNA molecules were structurally better preserved after freeze-drying. Freeze-drying was therefore applied in further STM imaging.

A combined approach using TEM and STM was applied to analyze protein-induced DNA bending (Fig. 6) [34]. In the presence of magnesium ions, DNA-RAP1 complexes were immobilized on a freshly cleaved mica surface (Fig. 6a). Different parts of a single piece of mica were used for parallel imaging by standard TEM and STM. For TEM the samples were ethanol/air-dried and rotary shadowed with Pt/C at a low elevation angle. Samples for STM were freeze-dried and coated with Pt/Ir/C at a high angle of elevation. The efficiency and accuracy of RAP1 binding was checked by TEM. Protein-induced DNA bending was then investigated by analysis of TEM (Fig. 6b) and STM (Fig. 6c) images. Investigations on DNA bending revealed a different distribution of bent angles for DNA-RAP1 complexes imaged by TEM or STM [34]. Analysis of TEM images revealed a Gaussian-shaped distribution of bent angles. In contrast, STM images of DNA-RAP1 complexes showed an increase in the frequency of higher angles. About 50% of the molecules showed an induced bend by RAP1 of 50° or greater.

How can this difference in bent angles be explained? In standard TEM, the complexes are contrast-enhanced by rotary shadowing at low elevation angles in order to visualize small filamentous structures like DNA. This

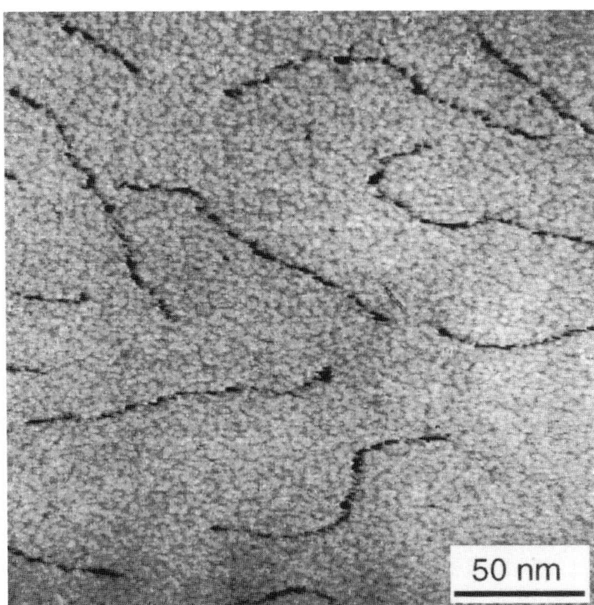

Figure 5. Use of the replica/anchoring technique to visualize bare, metal embedded DNA fragments. Owing to the low intrinsic electrical conductivity, DNA molecules are only visible as "hollow trenches".

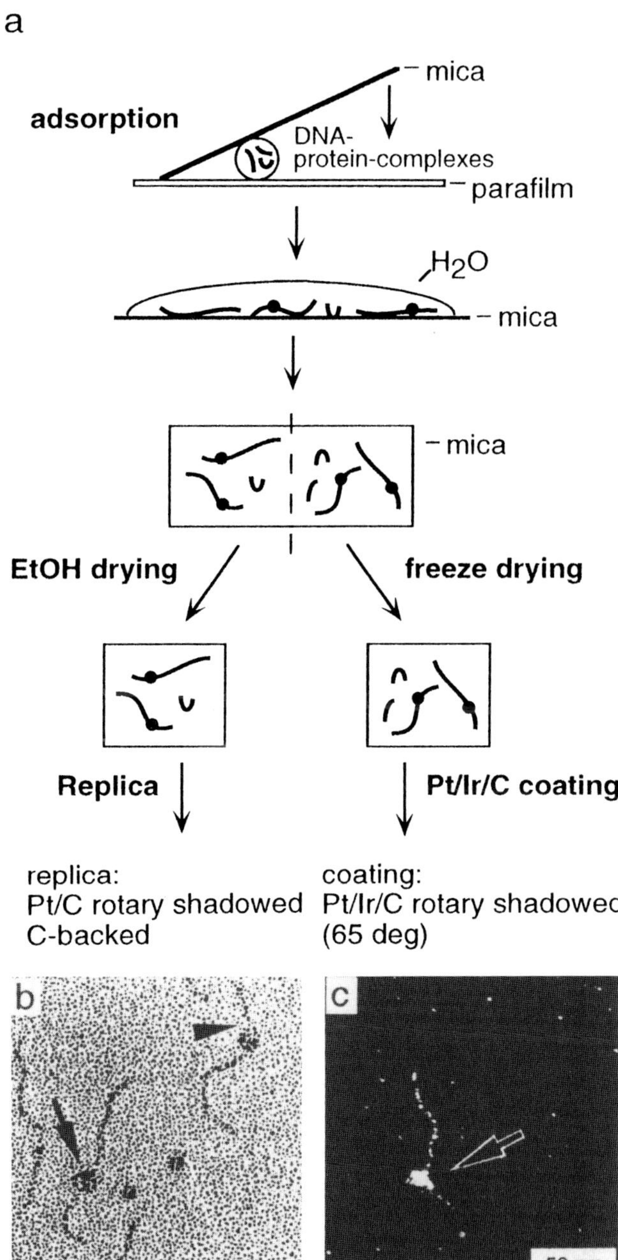

Figure 6. Combined approach for imaging of DNA-protein interaction by TEM and STM in parallel. (**a**) DNA-RAP1 complexes were immobilized on a freshly cleaved mica surface. Different parts of a single piece of mica were used for parallel imaging by TEM and STM. TEM samples were dried from ethanol in air and contrast enhanced with Pt/C at a low elevation angle. The carbon replica was analyzed in the TEM to check DNA adsorption and to determine correct positioning of the bound protein. STM analysis was carried out using freeze-dried specimens coated with Pt/Ir/C at a high elevation angle to investigate protein-induced DNA bending. (**b**) TEM micrograph of RAP1 molecules bound to a sequence specific binding site on the linear DNA fragment (arrow). Owing to the low-angle shadowing procedure the trace of the DNA close to the protein complex is obscured (arrowhead). (**c**) STM image of a DNA-RAP1 complex.

---

shadowing technique produced a relative large size of metal clusters and an indistinct protein structure due to "self-shadowing". The large metal grains can be followed along the DNA molecule, but are responsible for limiting the resolution of analysis near DNA-protein complexes. Owing to the protein size relative to the dimensions of the DNA, the trace of the fragment appears to be obscured in the vicinity of the polypeptide complex (Fig. 6b, arrowhead). Due to the high Z-resolution of the STM, a high elevation angle for coating of biological specimens can be applied. As a consequence, in STM images of DNA-protein complexes the trace of the DNA molecule can be followed in close proximity to the protein complex (Fig. 6c). This is of importance in view of the naturally occurring flexibility of DNA molecules as demonstrated in an STM image (Fig. 7a). The trace of the DNA close to the protein complex is shown schematically in Fig. 7b. In the vicinity of the RAP1 complex, a DNA bend of about 90° can be observed (arrows). Placing a "mask" at the position of the protein recognition site is considered to mimic conditions of low-angle shadowing (Fig. 7c). Owing to

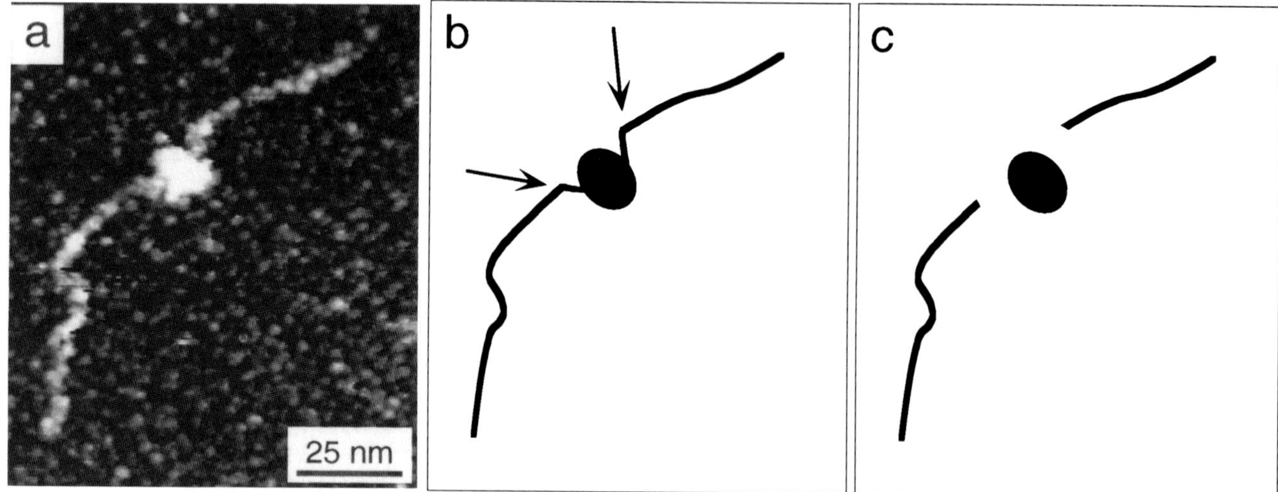

Figure 7. Analysis of protein-induced DNA bending. (**a**) STM image of a freeze-dried, Pt/Ir/C coated DNA-RAP1 complex. (**b**) Cartoon illustrating the trace of the DNA. In the vicinity of the RAP1 complex, a DNA bend of about 90° can be observed (arrows). (**c**) By placing a "mask" at the position of the RAP1 recognition site no DNA bend can be determined. This situation is thought to "mimic" low-angle rotary shadowing.

this contrast enhancement procedure, information close to the protein complex is lost and no DNA bend can be observed (Fig. 7c). As a result, high-angle coating in combination with STM imaging appears to be advantageous in analysis of protein-induced DNA bending.

Imaging nucleic acids by SFM

For biologists the SFM is attractive because it combines a potentially atomic resolution [37] with the opportunity to image non-conductive surfaces in aqueous media under native conditions (Fig. 2d). Despite these promising features, high resolution imaging of DNA by SFM is hampered by several problems such as: (1) the specimen-substrate attachment, (2) the choice of the specimen environment, and (3) the influence of tip curvature on resolution [12, 13, 14, 27]. By SFM different kinds of DNA-protein interactions were imaged: (a) protein-induced DNA bending, (b) DNA loop formation due to protein-protein interactions, (c) targeting of sequence specific DNA sites, and (d) assembly of non-sequence specific nucleoprotein complexes (for a review see: [33]).

For imaging of nucleic acids, the major advantage of the SFM is that imaging can be performed in electrolyte solutions. Beside the fact that electrolyte solutions are the natural environment of nucleic acids and proteins, the force between tip and sample can be reduced by a factor of 10 to 100 when imaging in liquids [45]. In most SFM experiments, however, it was reported that complexed or uncomplexed nucleic acids had to be completely dried and/or imaged in an alcohol solution to prevent translocation during scanning. Despite reducing forces between tip and sample, the major advantage of the SFM is lost, when dehydrated specimens are imaged in unphysiological solutions such as alcohol. In this respect the SFM offers no advantage compared to standard TEM or STM of metal-coated specimens, where freeze-drying can be applied to structurally preserve the DNA-protein complexes [3, 34]. On the other hand, SFM has been shown to be advantageous in imaging the interaction of nucleic acids with small proteins (≤ 30 kDa). Assembly of Cro protein from bacteriophage lambda with specific and non-specific DNA sequences has been imaged in air [21]. Cro, which has a molecular weight of 14.7 kDa, was visualized bound to the operator sites on a 1 kb DNA fragment. Owing to its size, Cro cannot be imaged applying standard, low-angle shadowing for TEM.

Conclusions

DNA can routinely be immobilized on a freshly cleaved mica surface in the presence of magnesium ions. This method opens up the possibility to perform STM or SFM experiments with TEM analysis in parallel. Applying the standard method of low-angle rotary shadowing various kinds of DNA-protein interactions can be analyzed. Using STM DNA embedded in a layer of metal can be visualized as a "hollow trench" only. However, this method is especially useful in establishing the experimental conditions for imaging of bare DNA in air. High-angle metal-coating of freeze-dried specimens appears to be useful in determination of protein-induced DNA bending. Both STM imaging of metal coated specimens as well as SFM analysis of DNA in solution appears to be advantageous in analyzing small proteins (≤ 30 kDa) bound to cloned DNA molecules.

References

[1] Allison DP, Bottomley LA, Thundat T, Brown GM, Woychik RP, Schrick JJ, Jacobson KB, Warmack RJ (1992) Immobilization of DNA for scanning probe microscopy. Proc Natl Acad Sci USA **89**: 10129-10133.

[2] Amrein M, Gross H, Guckenberger R (1993) STM of proteins and membranes. In: STM and SFM in Biology. Marti O, Amrein M (eds). Academic Press, San Diego, pp 127-175.

[3] Amrein M, Stasiak A, Gross H, Stoll E, Travaglini G (1988) Scanning tunneling microscopy of recA-DNA complexes coated with a conducting film. Science **240**: 514-516.

[4] Arscott PG, Lee G, Bloomfield VA, Evans DF (1989) Scanning tunneling microscopy of Z-DNA. Nature **339**: 484-486.

[5] Arscott PG, Lee G, Bloomfield VA, Evans DF (1990) Helical period of Z-DNA. Nature **346**: 706.

[6] Beebe TP, Wilson TE, Ogletree FD, Katz JE, Balhorn R, Salmeron MB, Siekhaus WJ (1989) Direct observation of native DNA structures with the scanning tunneling microscope. Science **243**: 370-372.

[7] Binnig G, Quate CF, Gerber C (1986) Atomic force microscope. Phys Rev Lett **56**: 930-933.

[8] Binnig G, Rohrer H (1984) Scanning tunneling microscopy. In: Trends in Physics. Janta J, Pantoflicek J (eds). European Physical Society, Petit-Lancy, Switzerland, pp 38-46.

[9] Binnig G, Rohrer H, Gerber C, Weibel E (1982) Surface studies by STM. Phys Rev Lett **49**: 57-61.

[10] Binnig G, Rohrer H, Gerber C, Weibel E (1983) 7 x 7 reconstruction on Si(111) resolved in real space. Phys Rev Lett **50**: 120-123.

[11] Blackford BL, Jericho MH (1991) A metallic replica/anchoring technique for scanning tunneling microscope and atomic force microscope imaging of large biological structures. J Vac Sci Technol B **9**: 1253-1258.

[12] Bustamante C, Erie DA, Keller D (1994) Biochemical and structural applications of scanning force microscopy. Curr Opin Struct Biol **4**: 750-760.

[13] Bustamante C, Keller D, Yang G (1993) Scanning force microscopy of nucleic acids and nucleoprotein assemblies. Curr Opin Struct Biol **3**: 363-372.

[14] Butt H-J, Guckenberger R, Rabe JP (1992) Quantitative scanning tunneling microscopy and scanning force microscopy of organic materials. Ultramicroscopy **46**: 375-393.

[15] Butt H-J, Müller T, Gross H (1993) Immobilizing biomolecules for scanning force microscopy by embedding in carbon. J Struct Biol **110**: 127-132.

[16] Clemmer CR, Beebe TP (1991) Graphite: A mimic for DNA and other biomolecules in scanning tunneling microscope studies. Science **251**: 640-642.

[17] Cricenti A, Selci S, Felici AC, Generosi R, Gori E, Djaczenko W, Chiarotti G (1989) Molecular structure of DNA by scanning tunneling microscopy. Science **245**: 1226-1227.

[18] Driscoll RJ, Youngquist MG, Baldeschwieler JD (1990) Atomic-scale imaging of DNA using scanning tunnelling microscopy. Nature **346**: 294-296.

[19] Dunlap DD, Bustamante C (1989) Images of single-stranded nucleic acids by scanning tunnelling microscopy. Nature **342**: 204-206.

[20] Dunlap DD, Garcia R, Schabtach E, Bustamante C (1993) Masking generates continuous segments of metal-coated and bare DNA for scanning tunneling microscope imaging. Proc Natl Acad Sci USA **90**: 7652-7655.

[21] Erie DA, Yang G, Schultz HC, Bustamante C (1994) DNA bending by Cro protein in specific and nonspecific complexes: Implications for protein site recognition and specificity. Science **266**: 1562-1565.

[22] Gilson E, Gasser SM (1995) Repressor Activator Protein 1 and its ligands: organising chromatin domains. In: Nucleic Acids and Molecular Biology. Lilley DMJ, Eckstein F (eds). Springer Verlag, Berlin/Heidelberg, vol 9, pp 308-327.

[23] Gilson E, Müller T, Sogo J, Laroche T, Gasser SM (1994) RAP1 stimulates single- to double-strand association of yeast telomeric DNA: implications for telomere-telomere interactions. Nucleic Acids Res **22**: 5310-5320.

[24] Gordon CN, Kleinschmidt AK (1969) Adsorption of DNA on aluminum-mica. Proc Twenty-Seventh Annual Meeting Electron Microscopy Society of America. Arceneaux C (ed). Claitor's Publishing Div, Baton Rouge, LA. pp 266-267.

[25] Guckenberger R, Heim M, Cevc G, Knapp HF, Wiegräbe W, Hillebrand A (1994) Scanning tunneling microscopy of insulators and biological specimens based on lateral conductivity of ultrathin water films. Science **266**: 1538-1540.

[26] Hall CE (1956) Visualization of individual macromolecules with the electron microscope. Proc Natl Acad Sci USA **42**: 801-805.

[27] Hansma HG, Hoh JH (1994) Biomolecular imaging with the atomic force microscope. Annu Rev Biophys Biomol Struct **23**: 115-139.

[28] Keller D, Bustamante C, Keller RW (1989) Imaging of single uncoated DNA molecules by scanning

tunneling microscopy. Proc Natl Acad Sci USA **86**: 5356-5360.

[29] Kleinschmidt AK, Zahn PK (1959) Über Desoxyribonukleinsäure-Molekeln in Protein-Mischfilmen (On Desoxyribo nucleic acid molecules in mixed protein films). Z Naturforsch **B14**: 770-775.

[30] Koller T, Sogo JM, Bujard (1974) An electron microscopic method for studying nucleic acid-protein complexes. Visualization of RNA polymerase bound to the DNA of bacteriophage T7 and T3. Biopolymers **13**: 995-1009.

[31] Lindsay SM, Tao NJ, DeRose JA, Oden PI, Lyubchenko YL, Harrington RE, Shlyakhtenko L (1992) Potentiostatic deposition of DNA for scanning probe microscopy. Biophys J **61**: 1570-1584.

[32] Lindsay SM, Thundat T, Nagahara L, Knipping U, Rill RL (1989) Images of the DNA double helix in water. Science **244**: 1063-1064.

[33] Lyubchenko YL, Jacobs BL, Lindsay SM, Stasiak A (1995) Atomic force microscopy of nucleoprotein complexes. Scanning Microsc **9**: 705-727.

[34] Müller T, Gilson E, Schmidt R, Giraldo R, Sogo J, Gross H, Gasser SM (1994) Imaging the asymmetric DNA bend induced by Repressor Activator Protein 1 with scanning tunneling microscopy. J Struct Biol **113**: 1-12.

[35] Müller-Reichert T, Butt H-J, Gross H (1996) STM of metal embedded and coated DNA and DNA-protein complexes. J Microsc **182**: 169-176.

[36] Nermut MV (1977) Freeze-drying for electron microscopy. In: Principles and Techniques of Electron Microscopy. Hayat MA (ed). Van Nostrand Reinhold, New York/London, pp 79-98.

[37] Ohnesorge F, Binnig G (1993) True atomic resolution by atomic force microscopy through repulsive and attractive forces. Science **260**: 1451-1455.

[38] Parker JL, Cho DL, Claesson PM (1989) Plasma modification of mica: Forces between fluorocarbon surfaces in water and a nonpolar liquid. J Phys Chem **93**: 6121-6125.

[39] Rau DC, Gellert M, Thoma F, Maxwell A (1987) Structure of the DNA gyrase-DNA complex as revealed by transient electric dichroism. J Mol Biol **193**: 555-559.

[40] Saenger W (1984) Principles of Nucleic Acid Structure. Springer, New York. pp 116-541.

[41] Sogo J, Stasiak A, De Bernardin W, Losa R, Koller T (1987) Binding of Protein to Nucleic Acids. In: Electron Microscopy in Molecular Biology. Sommerville J, Scheer U (eds). IRL Press, Oxford/Washington, pp 61-79.

[42] Thoma F, Koller T (1981) Unravelled nucleosomes, nucleosome beads and higher order structures of chromatin: Influence of non-histone components and histone H1. J Mol Biol **149**: 709-733.

[43] Thoma F, Koller T, Klug A (1979) Involvement of histone H1 in the organization of the nucleosome and of the salt-dependent superstructures of chromatin. J Cell Biol **83**: 403-427.

[44]. Travaglini G, Rohrer H, Amrein M, Gross H (1987) Scanning tunneling microscopy on biological matter. Surf Sci **181**: 380-390.

[45] Weisenhorn AL, Hansma PK, Albrecht TR, Quate CF (1989) Forces in atomic force microscopy in air and water. Appl Phys Lett **54**: 2651-2653.

[46] Wepf R, Amrein M, Bürkli U, Gross H (1991) Platinum/iridium/carbon: a high-resolution shadowing material for TEM, STM and SEM of biological macromolecular structures. J Microsc **163**: 51-64.

[47] Wildhaber I, Gross H (1985) The effects of air-drying and freeze-drying on the structure of a regular protein layer. Ultramicroscopy **16**: 411-422.

Discussion with Reviewers

J. Zasadzinski: What is the ultimate resolution you expect for STM of shadowed or replicated materials and what is the main factor limiting that resolution?

Authors: The expected resolution is highly dependent on the type of specimen imaged and the kind of image analysis applied. TEM experiments with 2-D protein crystals have shown that after coating with 0.3-0.5 nm Ta/W followed by image averaging a lateral resolution better than 1 nm can be obtained [51]. To my mind, the strength of STM imaging, however, is the analysis of those objects where averaging is not possible and thus a high signal-to-noise ratio is necessary (i.e., filamentous molecules). Unfortunately, due to the low electrical conductivity of biomolecules, reproducibility is hard to achieve in STM and interpretation of the data is not a trivial task. I speculate that the ultimate resolution on coated DNA samples is to resolve the pitch of the DNA helix. Nevertheless, structural details will be obscured by the metal coat. Despite of this fact metal coating can be applied to analyze protein-induced DNA bending by STM [34]. It has been shown recently that the STM information is in good agreement with data obtained from crystal structure [49]. In STM, metal shadowing is not the method of choice when high resolution imaging of biomolecules (e.g., DNA) is intended. Ultimate goal has to be the imaging of bare molecules. For this reason, an easy-to-use method was developed to establish the experimental conditions for analysis of uncoated samples [35]. This approach can be applied for analysis under both atmospheric and vacuum conditions. Broadening of specimens caused by the shape of the tip has also to be taken into account under both imaging conditions. Convolution is especially evident when imaging elevated or deepened structures [14].

J. Zasadzinski: How does STM resolution of shadowed materials compare to that of STM on hydrated materials as recently reported by Guckenberger *et al.* [48] or to the best new tapping mode SFM images? How does the resolution compare to standard TEM techniques?
Authors: The report on STM of biological specimens based on lateral conductivity of ultrathin water films [25] describes an interesting physical phenomenon. However, from the specimen preparation standpoint it is not clear to which extent the biological specimen is damaged during the preparation procedure. It seems likely that dehydration plays some role during imaging (i.e., analysis in a humid environment). So far no structural details of the DNA double helix could be detected. To my knowledge proper adhesion of the DNA to a solid support and translocation of the molecules during scanning is not fully solved even in tapping mode SFM. To my mind high resolution on DNA structure can not be achieved in SFM when disadvantageous air drying cannot be completely avoided. Structurally, SFM of air-dried DNA is of no advantage compared to standard TEM techniques.

J. Zasadzinski: Given that the STM or SFM resolution will likely never approach individual base pairs, what are the best questions to ask and expect to answer with probe microscopies?
Authors: In the beginning of STM imaging most of the researchers started out with the intention to sequence short DNA fragments. To date DNA sequencing can be performed fast and reliably using biochemical techniques. In my opinion sequencing of nucleic acids is not a question to be answered by scanning probe microscopy. However, questions can be answered concerning the conformational states of DNA molecules. Our knowledge on DNA structure is mainly based on analysis of DNA crystal structure. It will be interesting to analyse individual DNA molecules and the regularity of the helical arrangement under physiological conditions. Moreover, it would be of advantage to detect local regions in DNA fragments that flip from a right-handed B form to a left-handed Z form. To date Z-DNA regions can be detected using an anti-Z-DNA antibody [50]. Another interesting topic might be the structural analysis of local DNA unwinding due to binding of sequence specific transcription factors as reported for RAP1 [23]. Structural details of this kind of protein-DNA interaction cannot be expected by low-angle shadowing following TEM analysis.

J. Zasadzinski: Do you recommend that DNA researchers go out and invest in a STM, SFM or hybrid instrument, if they have never yet done so?
Authors: This strongly depends on the research project. Because of the low electrical conductivity of biological specimens most of the STM researchers continued to work with SFM [48]. However, it appears that the preparation of the biomolecules seems to limit also the analysis by SFM. Before high resolution data on DNA structure can be expected by SFM the proper attachment of the biomolecules without prior dehydration has to be solved. In parallel, imaging of bare DNA by STM has to be established for routine analysis.

Additional References

[48] Guckenberger R, Hartmann T, Knapp HF (1995) STM in biology. In: Scanning Tunneling Microscopy II. Wiesendanger R, Güntherodt H-J (eds). Springer Verlag, Berlin, Heidelberg. pp 312-318.

[49] König P, Giraldo R, Chapman L, Rhodes D (1996) The crystal structure of the DNA-binding domain of yeast RAP1 in complex with telomeric DNA. Cell **85**: 125-136.

[50] Pietrasanta LI, Schaper A, Jovin TM (1994) Probing specific molecular conformations with the scanning force microscope. Complexes of plasmid DNA and anti-Z-DNA antibodies. Nucleic Acids Res **22**: 3288-3292.

[51] Walz T, Tittmann P, Fuchs KH, Müller DJ, Smith BL, Agre P, Gross H, Engel A (1996) Surface topographies at subnanometer-resolution reveal asymmetry and sidedness of Aquaporin-1. J Mol Biol **264**: 907-918.

Scanning Microscopy Supplement 10, 1996 (pages 123-148)
Scanning Microscopy International, Chicago (AMF O'Hare), IL 60666 USA

0892-953X/96$5.00+.25

IMAGING SOFT MATERIALS WITH SCANNING TUNNELING MICROSCOPY

J.T. Woodward IV[1] and J.A. Zasadzinski[2*]

[1]Department of Physics and [2]Department of Chemical Engineering,
University of California, Santa Barbara, CA 93106-5080

(Received for publication August 6, 1995 and in revised form March 21, 1996)

Abstract

By modifying freeze-fracture replication, a standard electron microscopy fixation technique, for use with the scanning tunneling microscope (STM), a variety of soft, non-conductive biomaterials can be imaged at high resolution in three dimensions. Metal replicas make near ideal samples for STM in comparison to the original biological materials. Modifications include a 0.1 μm backing layer of silver and mounting the replicas on a fine-mesh silver filters to enhance the rigidity of the metal replica. This is required unless STM imaging is carried out in vacuum; otherwise, a liquid film of contamination physically connects the STM tip with the sample. This mechanical coupling leads to exaggerated height measurements; the enhanced rigidity of the thicker replica eliminates much of the height amplification. Further improvement was obtained by imaging in a dry nitrogen atmosphere. Calibration and reproducibility were tested with replicas of well characterized bilayers of cadmium arachidate on mica that provide regular 5.5 nm steps. We have used the STM/replica technique to examine the ripple shape and amplitude in the $P_{\beta'}$ phase of dimyristoylphosphatidyl-choline (DMPC) in water. STM images were analyzed using a cross-correlation averaging program to eliminate the effects of noise and the finite size and shapes of the metal grains that make up the replica. The correlation averaging allowed us to develop a composite ripple profile averaged over hundreds of individual ripples and different samples. The STM/replica technique is sufficiently general that it can be used to examine a variety of hydrated lipid and protein samples at a lateral resolution of about 1 nm and a vertical resolution of about 0.3 nm.

*Address for correspondence:
Joseph A. Zasadzinski
Department of Chemical Engineering,
University of California,
Santa Barbara, CA 93106

Telephone number: 805-893-4769
FAX number: 805-893-4731
gorilla@squid.ucsb.edu

Introduction

The scanning tunneling microscope (STM) (Binnig *et al.*, 1982; Quate, 1986) offers exciting new ways of imaging surfaces with resolution to the sub-molecular scale (Hansma and Tersoff, 1987). However, reproducible, three-dimensional STM images of biological or organic material are quite difficult to obtain as such materials are non-conductive and soft (Amrein *et al.*, 1988, 1989; Baró, *et al.*, 1985; Foster *et al.*, 1988; Foster and Frommer, 1988; Hansma *et al.*, 1988; Hörber *et al.*, 1988; Lang *et al.*, 1988; Travaglini *et al.*, 1987). Imaging with any probe microscope invariably involves mounting the sample on a substrate, which is often accompanied by its own set of artifacts (Hörber *et al.*, 1988; Lang *et al.*, 1988; Clemmer and Beebe, 1991; Patrick and Beebe, 1993). As a result, little new biological information has been obtained via STM in spite of a significant investment of research time and money. Much of the initial promise in this area has proven false; artifacts abound in the literature because direct STM images of insulating materials are difficult to understand based on any reasonable model of electron tunneling (Lang *et al.*, 1988; Lindsay and Barris, 1988; Spong *et al.*, 1989; Clemmer and Beebe, 1991). However, recent work suggests that many images of biological materials are actually maps of thin layers of loosely bound water with sufficient conductivity to provide a current for imaging (Sonnenfeld and Hansma, 1987; Yuan *et al.*, 1991; Guckenberger *et al.*, 1994). A better explanation might be that images of biological materials are primarily due to electrochemical reactions at the STM tip, as in the scanning electrochemical microscope (Arca *et al.*, 1994). However, such explanations are qualitative and it is difficult to directly relate image contrast to features in the image.

At present, most non-conductive and biological surfaces are imaged via atomic force microscopy (AFM) (Binnig *et al.*, 1986; Gould, *et al.*, 1988; Hansma *et al.*, 1988; Zasadzinski *et al.*, 1994a,b; Hui *et al.*, 1995) or one of its recent variants including non-contact mode (Manne *et al.*, 1994), lateral or friction force mode (Meyer *et al.*, 1992; Overney *et al.*, 1992), specific chemical or biochemical interactions (Florin *et al.*,

1994; Frisbie *et al.*, 1994), and tapping mode (Hansma *et al.*, 1994; Radmacher *et al.*, 1994). There are many benefits to these techniques, the most important being (1) non-conductors are readily imaged; (2) imaging can be done under solvents or in near *in vivo* conditions and (3) specific chemical information can be obtained via tip-sample interactions. However, there are several drawbacks to each of these techniques including the relative cost of the equipment, the time it takes to acquire images, and interpretation of specific interactions and relating these interactions to structural features. Moreover, all these force microscopy techniques are still limited to imaging materials bound to a substrate.

An alternative to these various AFM techniques that avoids much of the confusion of direct STM imaging of biomaterials is coating non-conductive surfaces with metal layers so as to make them conductive, then imaging with the STM (Travaglini *et al.*, 1987; Amrein *et al.*, 1988; Zasadzinski *et al.*, 1988; Obcemea and Vidic, 1992; Woodward and Zasadzinski, 1994). We have found that conventional freeze-fracture, which is extremely useful for imaging bulk organic materials with transmission electron microscopy (TEM), can be modified for reliable, high resolution, quantitative three dimensional imaging via the STM. In principle, a platinum replica of a fracture surface is an ideal sample for STM imaging - it is highly conductive, chemically homogeneous, inert, and easily manipulated. However, imaging with the STM is fundamentally different than with the TEM and has led us to modify the freeze-fracture technique to match the different requirements of the STM. The freeze-fracture STM techniques detailed in this paper are an important alternative to conventional AFM imaging of soft or insulating surfaces, and can be used to prepare substrate free samples for high resolution imaging. The technique works best for, and is illustrated by periodic membrane surfaces with various three dimensional contours. As we show here, on repetitive surfaces such as the ripples of the $P_{\beta'}$ phase of dimyristoylphosphatidylcholine (DMPC) in water, with the combination of freeze-fracture replication, careful STM imaging in controlled environments, and correlation averaging image analysis, we can achieve a 1 nm lateral and 0.3 nm vertical resolution. Such a high resolution, quantitative description of soft materials in solution is impossible by any other technique.

Sample Preparation for STM-Freeze-Fracture

Freeze-fracture replication is a four step thermal fixation technique originally developed by Steere (1957) to image thermally fixed as opposed to chemically fixed cells and suspensions. The basic fracture tables and evaporation equipment for the technique was first developed by Moor *et al.* (1961) (now commercially available from Balzers and others). Branton (1966) was the first to describe in detail the progression of the fracture through various materials including cell membranes. In the modern version of the technique, a thin layer of sample is trapped between two copper plates, quickly frozen, and then fractured under vacuum (Zasadzinski and Bailey, 1989). The fracture surface is replicated by evaporating a thin metal layer obliquely to the fracture surface; the metal film is followed by a thicker carbon layer for strength. Any remaining portions of the original sample is carefully cleaned away, and the replica is then examined under ambient conditions in the transmission electron microscope. In any individual replica of a non-repetitive surface examined in the TEM, the absolute resolution in a particular image is limited by the size of the metal film grains to about 2 nm (Chiruvolu *et al.*, 1994). Image analysis techniques can remove some of the influence of the granularity of the replica, especially for periodic surfaces.

Rapid freezing

The goal of rapid freezing is to remove heat at such a rate that (1) the details of the individual cell, liposome or vesicle "particles" are retained and (2) the distribution and orientation of the particles are not disturbed. For many systems, the second criterion is much more difficult to achieve, especially for particles less than 0.1 μm in size and usually requires that the solvent or continuous phase be vitrified. In freeze-fracture, vitrification generally is taken to mean that the solvent phase contains no recognizable crystals larger than the typical resolution of the images, although this is far from proving that the solvent phase is truly amorphous. Simple plunge freezing into liquid propane or ethane has been sufficient to vitrify water films up to about 50-100 μm in thickness for freeze fracture. Direct imaging of thin aqueous films at cryogenic temperatures show that the water can indeed be frozen as an amorphous solid without any crystallization and held in this phase indefinitely at low temperatures (Adrian *et al.*, 1984; Dubochet and McDowall, 1981; Dubochet *et al.*, 1982; Bellare *et al.*, 1988; Chiruvolu *et al.*, 1994).

However, if the solvent phase is crystallized, dispersed particles and solutes are swept to the crystalline grain boundaries and information on the original orientation and distribution of the particles in the solvent is lost. The distribution of the particles reflects the crystallization behavior of the solvent rather than any of the original properties of the system. Theoretical estimates of the cooling rates necessary to vitrify water range from 10^4 to 10^{10} K/sec (Bruegeller and Mayer, 1980), although experimentally, the minimum cooling rate to create vitrified water appears to be in the range of 10^5K/sec (Dubochet and McDowall, 1981; Dubochet *et al.*, 1982; Chiruvolu *et al.*, 1994). These cooling rates are readily accessible in the laboratory with a minimum investment via a variety of plunge and jet freezing equipment (Bellare *et al.*, 1988; Bailey *et al.*, 1991; Gilkey and Staehelin, 1986; Jahn and Strey, 1988; Müller *et al.*, 1980) .

As treatment with chemical cryoprotectants must be avoided, the physical parameters such as the heat capacity, thermal conductivity and crystallization behavior of the biomaterial system to be studied are fixed and usually cannot be modified without inducing chemical or ultrastructural changes in the sample. For freeze-fracturing a broad variety of liquid samples of any viscosity such as suspensions, solutions, and emulsions, the most practical method of rapid freezing is "sandwiching" the sample in a rigid container that is then contacted with a liquid cryogen. A wide variety of sample holders have appeared in the literature (Costello and Corless, 1978); however, the most commonly used are variations on the "copper sandwich" holders developed by Gulik-Krzywicki and Costello (1978) and commercialized by Balzers (BUO-12-056T and variations; Hudson, New Hampshire) to produce complementary replicas. Typically, from 0.1-0.5 ml of sample liquid is pipetted onto one of the planchettes, then a second planchette is used to spread the liquid to form a thin, 10-50 μm thick film. A variant of this sample holder with an annular opening in the top planchette is used by Jahn and Strey (1988). By far the most common method of rapid freezing is immersing the sample sandwich into a liquid cryogen, typically with a spring-loaded device to increase the relative specimen-cryogen velocity (Bailey *et al.*, 1991; Bellare *et al.*, 1988; Jahn and Strey, 1988).

Freezing a multicomponent, structured biomaterial sample sandwiched between metal sheets in a liquid cryogen is a complicated heat transfer process that involves convection at the cryogen-sandwich boundary, conduction through the sandwich, and conduction and possible phase changes, crystallization, etc. within the fluid specimen (Talmon *et al.*, 1981). A simplified model of cooling (Zasadzinski, 1988a) that has been experimentally validated (Bailey and Zasadzinski, 1991) shows that the cooling rate of samples is limited by convection from the cryogen to the specimen surface rather than by conduction through the specimen. The important criterion is that the Biot modulus of the specimen, hV/kA is $<<$ 1; (see Zasadzinski, 1988a, for discussion) h is the heat transfer coefficient from the cryogen to the sample, V is the sample volume, A is the surface area of the sample in contact with the cryogen, and k is the average sample thermal conductivity. For a typical freeze-fracture sample of thickness 100-200 μm, the Biot number, Bi, is 0.05 to 0.5. An important physical consequence of convection limited cooling (Bi $<$ 1) is that the temperature is *spatially* uniform within the sample during freezing (Bailey and Zasadzinski, 1991). This means that there are no spatial temperature gradients in the sample, and hence no driving force for spatial reorganization of the solutes or particles in the specimen, or localized phase separation. In this approximation, the cooling rate of a sample of heat capacity C_p, and density ρ, is:

$$\frac{dT}{dt} = \frac{-A}{V} h(T-T_C) \frac{1}{\rho C_p} \tag{1}$$

T is the sample temperature and T_C is the liquid cryogen temperature. Experimental measurements of the average cooling rate of a wide variety of specimens using various cryogens can be correlated using this simplified model (Zasadzinski, 1988a). The cooling rate is proportional to the ratio of specimen surface area to volume; this defines a characteristic sample dimension that governs the cooling rate. The only real dependence of the cooling rate on the sample composition is the inverse dependence on the thermal density, ρC_p, of the sample. The thermal density varies little between the typical materials encountered in freeze-fracture experiments. Surprisingly, the cooling rate is independent of the sample thermal conductivity, and hence, virtually all samples of the same characteristic dimensions freeze at the same rate (Bailey and Zasadzinski, 1991).

Optimization of the freezing process can be achieved by maximizing h, the heat transfer coefficient, while minimizing T_C, the cryogen melting point, by proper choice of cryogen and the velocity at which the cryogen contacts the sample. A limited amount of cryogen boiling enhances the heat transfer coefficient; too much boiling results in the formation of a vapor film around the sample that drastically reduces the rate of heat transfer (Bailey and Zasadzinski, 1991). Liquid nitrogen, and other cryogens at their boiling points should be avoided for this reason. The best practical cryogen appears to be ethane or propane cooled to near its freezing point by liquid nitrogen (Costello and Corless, 1978; Bailey and Zasadzinski, 1991). The distribution and orientation of dispersed tobacco mosaic virus in water has been preserved using both controlled plunge freezing in liquid propane and propane jet freezing (Zasadzinski and Meyer, 1986), although it is impossible to say if the water surrounding the colloidal particles is amorphous or microcrystalline.

Many interesting biological materials exist naturally either above or below room temperature. This requires that a temperature and environment controlled chamber be coupled to the freezing apparatus. Sealed, two stage ovens have been used to observe temperature dependent phase transitions in lyotropic nematic micellar phases (Sammon *et al.*, 1987) and to study phospholipid phase transitions (Zasadzinski and Schneider, 1987). Bellare *et al.* (1988) have constructed a temperature and humidity controlled cell in which samples can be equilibrated prior to plunge freezing and has used it to demonstrate a number of temperature dependent phases. Jahn and Strey (1988) have used a simple spring-loaded sample making device to look at composition and temperature sensitive materials. Thermotropic liquid crystalline phases stable over less than 1°C have been resolved using quick freezing techniques followed by freeze-fracture (Costello *et al.*, 1984; Zasadzinski *et al.*, 1986), provided that the equilibrium phase transition occurs by nucleation and growth. Thermotropic smectic phases have been successfully quenched from above 100°C for freeze-fracture investigation (Ihn *et al.*, 1992).

Because the chemical and physical properties of a biomaterial cannot be optimized for rapid freezing by chemical or physical cryoprotectants without changing the structure, a judicious choice of systems to investigate usually is the difference between success and failure. As avoiding solvent crystallization is of primary importance to successful images, it is useful to understand the solvent properties that affect crystallization. Crystallization consists of two steps, nucleation of small crystals of a critical size, then the growth of these crystals. Except for extremely pure liquids, nucleation occurs heterogeneously at insoluble impurities; such impurities are often the colloidal particles we wish to study, and hence are unavoidably present. The crystal growth velocity, u, is therefore the more important quantity to minimize by appropriate choice of sample properties. The crystal growth velocity is proportional to the degree of supercooling, ΔT, the entropy of fusion, α, and the fraction, $f\beta$, of acceptable sites on the interface (which reflects the steric constraints involved in packing solvent molecules into a different configuration), and inversely proportional to the solvent viscosity, μ:

$$u \propto \frac{f\beta}{\mu}\frac{\Delta T}{T_m} \qquad (2)$$

T_m is the equilibrium melting point.

Steric effects, are generally more pronounced than are viscous effects in Eqn. 2. Branched hydrocarbons such as isohexane ($(CH_3)_2CH(CH_2)_2CH3$) freeze at a much lower temperature (-153.7°C) than does straight chain n-hexane ($CH_3(CH_2)_4CH_3$)(-94°C), indicating that the steric restrictions imposed by the branching make crystallization much more difficult. Cyclohexane, on the other hand, adopts a fairly rigid conformation in the liquid phase that is easy to pack into a crystalline lattice and freezes at about 6°C (Roberts and Caserio, 1977). Clearly, as many of the physical properties of these solvents are similar, to optimize the system for rapid freezing, isohexane, which is sterically hindered, is a much better choice than cyclohexane, which crystallizes readily. In aqueous solutions, salts, solutes and macromolecules that tend to disrupt water structure by hydrogen bonding can hinder crystallization. For instance, ice formation in polyacrylamide gels is suppressed to below -17°C (Tanaka *et al.*, 1977).

Fracture and replication

Once frozen successfully, the sample must be replaced by a metal "replica" that is an accurate map of the biomaterial microstructure and is compatible with the requirements for STM imaging. The fracture, etching, and replication steps of the freeze-fracture technique are carried out at low temperature and high to ultrahigh vacuum. Typically, the "copper sandwich" samples are loaded under liquid nitrogen into a hinged brass block fracture stage. The fracture stage has sufficient thermal mass that the specimens do not heat up significantly during the brief time they are exposed to air during transfer into a vacuum chamber. The fracture stage is clamped to a temperature controlled coldfinger within the vacuum chamber. The specimen stage fractures the sandwiches on opening; a sharp, quick break is preferable to a long, steady pull for reasons discussed below. The stress on the specimen is primarily tensile.

The general behavior of any material under load can be classified as ductile or brittle depending on whether or not the material exhibits plastic deformation. A completely brittle material will fracture at the elastic limit. (The recovery of the original dimensions of a deformed body when the load is removed is known as elastic behavior. The limiting load beyond which a material no longer can recover its shape is known as the elastic limit). If a material is deformed beyond its elastic limit without fracturing, it is said to have undergone plastic deformation. For freeze-fracture replication, the goal is an ideal brittle fracture in which the sample fractures without deformation. An ideal brittle fracture limits deformation to only the molecules being pulled apart because a brittle material cannot redistribute local stresses to regions outside the fracture zone. The stress builds up in a very localized region, a crack forms at one or more points where the stress concentration overcomes the cohesive strength of the materials, then rapidly spreads, fracture the entire sample. It is important to point out that brittleness is not an absolute property; decreased temperature, increased rates of stress and the presence of cracks or notches increase the brittleness of a material (cf., Dieter, 1976). This is why it is necessary to fracture at the lowest practical temperature and to induce fracture by a quickly imposed, large load.

Griffith (cf., Dieter, 1976) proposed that, even prior to fracture, a brittle material contains a population of small cracks at impurities, boundaries, and other imperfections. When placed under sufficient stress, one or more of these cracks spreads into a brittle fracture, thereby decreasing the elastic energy at the expense of increasing the surface area, and hence the surface energy of the material. A crack will spread when the decrease in elastic energy is at least equal to the energy required to create the new crack surface. The surface energy can be thought of as the product of the new surface area created and the specific energy per unit area of the fracture surface. The Griffith theory states that the fracture will follow the path of least resistance (smallest specific surface energy or smallest molecular cohesion) provided that the fracture area created is not too large. In most two or more phase colloidal dispersions, the fracture surface propagates along the interface between the two phases, usually at particle-solvent boundaries. Apparently, solvent-solvent cohesion and particle-particle cohesion are larger than solvent-particle cohesion. Alternately, small differences in the volume contraction on cooling between the solvent phase and the particles could lead to debonding prior to fracture, or to the formation of cracks at the particle-solvent interface. In either case, the weak zone appears to be at the interface and interpretation of freeze-fracture images is greatly simplified. Branton (1966) has shown that the weak zone in bilayer membranes is along the hydrocarbon interior of the membrane; this also appears to be true for bilayer phases in general.

Etching

The controlled sublimation of the solvent, known as etching, can be used to enhance the topographic variations in a fractured, microstructured fluid. However, removing too much of the solvent can alter the apparent location and distribution of dispersed particles, hide evidence of crystallization induced reorganization, and make the replicas difficult to pick up and clean. If the dispersed phase is entangled and self-supporting, as are polymer solutions and gels, a limited amount of etching can bring out the network structure (Zasadzinski *et al.* 1987a). The important parameter in etching a sample is temperature. The sublimation pressure, hence the sublimation rate, is set once the sublimation temperature is fixed. The sublimation rate, S, in nanometers per second from a surface under vacuum can be obtained from gas kinetic theory:

$$S = \frac{P_s - P_v}{\rho_c} \left[\frac{M_c}{2\pi RT} \right]^{1/2} \times 10^7 \qquad (3)$$

in which P_s is the saturation vapor pressure and P_v the background pressure of the sublimating phase in dynes/cm^2, M_c is the molecular weight (g/mole) and ρ_c (g/cm^3) is the density of the sublimating phase. R is the gas constant, and T is the sublimation temperature (K). The saturation vapor pressure for most solvents can be found in general engineering handbooks. If the background pressure of the solvent is greater than the vapor pressure at the temperature chosen, material will condense from the vacuum onto the sample, obscuring surface details. Hence, care is necessary to understand the relative composition of the residual gases in the vacuum chamber.

Replication

The goal in the replication process is to reproduce the fracture surface as accurately as possible with an electron opaque shadowing layer backed by a continuous, electron transparent, backing layer. The resolution in freeze-fracture electron microscopy is limited by imperfections in replication (Akahori *et al.*, 1986; Gross *et al.*, 1985). Ideally, the evaporated metal atoms, which are usually a mixture of platinum and carbon, stick exactly where they land and form a structureless layer. However, the surface energy of the metal layer is much higher than that of the original fracture surface of water or hydrocarbons; hence, the metal film does not spread or "wet" the surface, but aggregates into small droplets (Adamson, 1990. Woodward, 1994). The aggregates grow in size, eventually merging with neighboring aggregates to form a continuous film. For most electron microscopy applications, the evaporation is stopped prior to the formation of a continuous metal layer. In our laboratory, about 1.5 nm of platinum carbon followed by about 15 nm of carbon backing gives optimum results. For proper interpretation of the replica, the sample material must be completely removed from the replica before viewing. In our laboratory, the cleaning method of Fetter and Costello (1985) has always given the best results.

Capillary condensation and feature height amplification in STM imaging in air

Up to this point, the preparation for TEM and STM examination are interchangeable; in fact, it was our original intention to be able to image the same sample with both TEM and STM to compare features and eliminate artifacts. However, we have found that there is a fundamental difference in the sample requirements for the two imaging techniques. In our initial studies of freeze-fracture replicas with the STM, we found that the height information was very erratic. Images of similar objects on the same replica could differ by more than a factor of ten in apparent height (Woodward *et al.*, 1991). Other times, the height of an object would grow or shrink with continued scanning (Hansma *et al.*, 1988). The literature also includes a number of accounts of unusual height measurements on a variety of samples (Coleman *et al.*, 1985; Hallmark *et al.*, 1987; Hansma and Tersoff, 1987) A number of theoretical models have proposed that specific quantum and atomic scale interaction scan lead to amplifications in specific tip-sample systems (Tersoff and Lang, 1990, Ciraci *et al.*, 1990; Yuan and Shao, 1990; Chen, 1992). Other theories propose that bulk compression by the scanning tip at regions of low conductivity or solid surface contamination could lead to amplifications of surface features (Soler *et al.*, 1986; Mamin *et al.*, 1986).

In most descriptions of scanning tunneling microscopy, there is no physical coupling between the STM tip and sample, even though there are at least three significant interactions to consider when imaging is done in air: (1) electrostatic interaction due to the potential difference between the tip and sample, (2) van der Waals attraction, and (3) capillary attraction due to the Laplace pressure generated by the formation of a highly curved fluid meniscus connecting the tip and sample (Woodward *et al.*, 1991; Woodward and Zasadzinski, 1994). These forces can cause the STM tip, while traversing a surface feature with an actual height, Z, to distort the surface being measured, resulting in an amplified height, Z^*. The role of capillary forces in scanning probe microscopy has been recognized in the AFM literature (Erlandsson, *et al.*, 1988; Weisenhorn *et al.*, 1989; Blackman *et al.*, 1990) while an appreciation of their importance during STM imaging has lagged (Yuan *et al.*, 1991; Anselmetti *et al.*, 1993, Guckenberger *et al.*, 1994).

The real forces that couple the tip to the sample can be modeled by a spring of spring constant, k_1. A second spring of spring constant, k_2, connects the sample to the STM base. Regardless of the functional form of k_1 and k_2, this combination of springs will amplify a surface feature of actual height Z by:

$$Z^* = Z(k_1 + k_2)/k_2 \tag{4}$$

where Z^* is the height measured by the STM. In most applications where the sample is a uniformly rigid solid firmly mounted to the base, k_2 is very large compared to k_1 and no amplification is evident (k_2 would be related to the bulk compressibility or elastic modulus of the sample being imaged (Mamin *et al.*, 1986)). However, if the sample is soft, non-uniform, or weakly connected to the base, as is often the case for organic or biological films, or even layered solids, k_2 can be much less than k_1, leading to large amplifications.

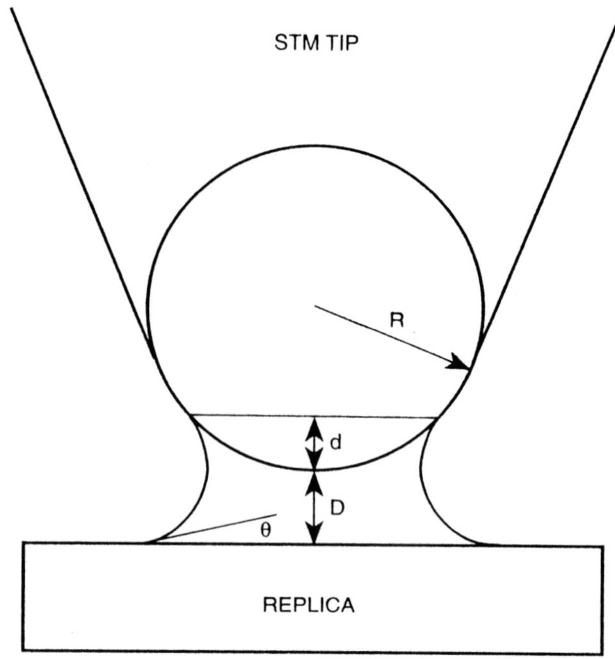

Figure 1: Schematic showing a fluid meniscus connecting the STM tip to the replica. The figure is not to scale.

To estimate these two important spring constants, we first assume that the force and spring constants between the tip and sample can be approximated by the linear combination of electrostatic, van der Waals and capillary forces. For a spherical tip of radius R and a tip-sample distance D, (Fig. 1) the van der Waals force is:

$$F_{vdW} = \frac{AR}{6D^2} \tag{5}$$

where A is the Hamaker constant (Israelachvili, 1991). The Hamaker constant for two platinum wires in air or an electrolyte is 20×10^{-20} J (Derjaguin *et al.*, 1978). The electrostatic force between a sphere held at a potential, V_o, and a grounded plane is given by

$$F_e = \frac{\pi \epsilon V_0^2 R}{D} \tag{6}$$

in the limit of R> >D.

The capillary force due to the meniscus between the

Table 1: Sample-Tip Spring Contants as a Function of Relative Humidity

Interaction	$P/P_s = 0.0$		$P/P_s = 0.10$		$P/P_s = 0.50$		$P/P_s = 0.75$	
	k_1(N/m)	F(nN)	k_1(N/m)	F(nN)	k_1(N/m)	F(nN)	k_1(N/m)	F(nN)
Electrostatic	0.01	0.01	1.1	1.1	1.1	1.1	1.1	1.1
van der Waals	3.3	1.7	3.3	1.7	3.3	1.7	3.3	1.7
Laplace	0.0	0.0	10.0	14.6	11.0	28.0	7.6	36.2
Total	3.31	1.71	14.4	17.4	15.4	30.8	12.0	39.0

Calculated values for the spring constants and forces of the three interactions we have considered both in a dry environment ($P/P_s = 0.0$) and for a water meniscus at various relative humidities. Note that k_1 is nearly constant over a wide range of relative humidities, and the dominant interaction is the Laplace pressure due to the meniscus.

----

sample and tip is:

$$F_L = \frac{4\pi R\gamma\cos\theta}{(1+D/d)} \qquad (7)$$

where γ is the surface tension of the condensate and θ is the liquid-solid contact angle, d is the distance the tip extends into the meniscus and is given by (V/RT) $\cos\theta/\ln(P/P_s)$ where V is the condensate molar volume, R the gas constant, T the temperature, and P/P_s is the relative humidity (Israelachvili, 1991). For water, γ = .073 J/m^2 and d = -(1.08 nm)/ln(P/P_s). Using these approximations, the spring constant $k_1 = -\delta F/\delta D$ is:

$$k_1 = \frac{\pi\epsilon V_0^2 R}{D^2} + \frac{AR}{3D^3} + \frac{4\pi R\gamma\cos\theta}{d(1+D/d)^2} \qquad (8)$$

and is governed by the Laplace pressure for relative humidities over 10% as shown in Table 1. From the exact meniscus profiles and interaction forces (Orr *et al.*, 1975; Zasadzinski *et al.*, 1987b), we calculate that the pressure within the meniscus can exceed 100 atm, which is sufficient to damage soft materials, although it would only lead to small deformations of crystalline solids such as graphite (Mamin *et al.*, 1986). The wide range of local pressures experienced by samples during scanning may explain the irreproducibility typically found in imaging biological materials.

We evaluate k_2, the spring constant between the sample and the STM base, for thin, metal surface replicas for TEM imaging as described above. However, the general principles used should be valid to estimate k_2 for many of the weakly supported samples common to STM imaging. Replicas for TEM are thin metal films mounted on porous metal mesh grids; where the replica is in direct contact with the mesh, k_2 is large. However, for the part of the replica that is loosely suspended over a pore of area, a:

$$k_2 = \frac{ET^3}{0.14a(1-\nu^2)} \qquad (9)$$

where E is the Young's modulus of the replica, ν is Poisson's ratio, and T is the thickness of the replica (Roark, 1965). Metal surface replicas prepared for transmission electron microscopy are about 25 nm thick with E = 2 x 10^{10} N/m^2 and ν = 0.3, and are mounted on a mesh with 30 μm spacings. Hence, k_2 = 0.003 N/m in the center of the mesh. Taken together with the estimate of k_1, Eq. 4 suggests a possible 1000 fold amplification, which is clearly outside the limits of our approximations, but also indicative of potential artifacts due to tip-sample interactions. Most importantly to reliable imaging, we cannot determine whether the particular area being imaged is well supported or not during a particular STM scan. Clearly, k_2 will vary significantly depending on the relative location of the sample with respect to the support mesh. As a result, we expect, and have found, that feature heights often vary with location on the surface. This effect will not be appreciable for chemically and physically homogeneous "bulk" samples (>1 mm thick). However, even if the sample is "bulk", any weak point in the mounting or in the local adhesion of the sample to the substrate or the substrate to the STM base can lead to amplification of surface features. This also suggests that certain areas of a sample might be highly amplified, leading to permanent local distortion of the surface, where other areas might not be distorted at all. As controlling the properties of the sample surface are difficult, and we do not know the details of the interaction a priori, it is important to identify the impact of capillary forces on test images, then minimize these forces for reliable imaging of unknown surfaces.

Modifications to freeze-fracture replicas for STM imaging

From Eqn. 9 above, it is clear that making the sample thicker and reducing the relative humidity are the best routes to increase k_2 and minimize sample height amplification. To properly test this ideas, we made two types of specimens, one "bulk" and one "thin film", with identical chemical composition and surface features and with well-known feature heights. We chose metal shadowed Langmuir-Blodgett multilayer islands of cadmium arachidate deposited on 1 mm thick mica substrates. The original specimen surface consisted of bilayer islands of cadmium arachidate of 1-2 μm in extent that vary in height by a bilayer from adjacent areas (Schwartz *et al.*, 1992a,b). The thickness of the bilayer is known from X-ray diffraction (Tippmann-Krayer, 1992) and AFM images to be 5.5 nm (Schwartz *et al.*, 1992a,b). To make the surfaces conductive and chemically uniform, 1.5-2 nm of platinum was applied by electron beam evaporation at a 45° angle while the sample was rotated in a Balzers 400K freeze-etch device (Balzers, Hudson, NH). This sample could then be imaged directly with the STM as the "bulk" sample. For the thin film samples, we deposited an additional 30 nm of carbon as a backing film to a second set of platinum coated LB island surfaces on mica, followed by a 200 nm thick layer of silver. These multilayer metal films were then stripped from the mica substrate using hydrofluoric acid, washed repeatedly in Millipore filtered water, and mounted, platinum side up, on silver wire mesh filters (SPI Supplies, West Chester, PA.) with a nominal 0.2 μm pore size. STM (Nanoscope II, Digital Instruments; Santa Barbara, CA) images were obtained with cut platinum/iridium tips at 100 mV bias voltage, a 1.0 nA tunneling current and a scan rate of 5.8 Hz. Step heights were measured by taking bearing plots of roughly equal areas on both sides of a step and taking the difference between the peak values to be the step height.

To control the environment and eliminate condensation between the tip and sample, we enclosed the STM in a bell jar with a base plate that allowed us to evacuate the chamber with a rotary pump to <0.1 torr and backfill with dry nitrogen (Woodward, 1994). For the "bulk" samples on mica, 46 bilayer steps were measured on 18 different images taken with different tips in both ambient conditions and a dry nitrogen atmosphere. Seventy percent of the measurements fell within eight percent of the mean, and there was no difference between samples imaged in air or in dry nitrogen. This variance in the step height was similar to that achieved measuring the bilayer steps directly with the AFM (Garnaes *et al.*, 1993). We attribute most of the scatter to nonlinear effects in the piezoelectric crystal in the scanning head as well as some coupling between the z calibration and the x and y position. As the sample is a bulk solid and the adhesion of the surface film to the substrate was good, k_2 is large in Eqn. 9 above and we did not expect nor did we find any amplification of feature heights. These bilayer steps of 5.5 nm also make excellent vertical calibration samples for the STM in a range difficult to find elsewhere.

We then imaged the thin film replicas of the LB bilayer steps in a dry atmosphere. In the dry atmosphere, the images were stable for more than one hour and lateral and vertical feature dimensions were reproducible for hundreds of scans. Ten images taken with three different tips gave 28 bilayer height measurements. The mean height was 91% of the mean height from the "bulk" sample. Seventy percent of the measurements fell within eight percent of the mean as with the coated sample. Again, the variance was similar to that achieved measuring the bilayer steps directly with the AFM (Garnaes *et al.*, 1993). There was no significant difference in heights measured with different tips.

However, when we changed the environment of the sample during imaging, we could record significant changes in feature heights due to contamination from the lab air. Fig. 2 shows one such series of images (Woodward and Zasadzinski, 1994). Imaging the replica gives an inverted image of the surface so the large feature resembling Australia is a bilayer higher than most of the replica surface and corresponds to a bilayer deep hole in the original LB film. When imaged in the dry nitrogen atmosphere (Figs. 2a-b), the image is stable. However, when lab air of 65% relative humidity is allowed into the bell jar, the height difference between the background and the bilayer 'hole' increased by from about 5.5 nm to more than 7 nm, (see Fig. 3) and several of the small features on the background showed a significant increase in height. The difference between the highest and lowest points (z-range) in the image increased from 18.6 nm in 1b to 40.0 nm in 2d. New "features" or high spots appear to grow on other areas of the replica as the replica is exposed to humid air for longer periods of time (Fig. 2d). The feature heights would vary from scan to scan when imaging in humid air, and would often appear and disappear. Upon flushing with dry nitrogen the effects of the condensation were quickly reversed. Figs. 2e, 2f show that the high spots vanish within minutes and the bilayer step returns to nearly the original height. However, although the "pseudo-features" that appeared in Figs. 2c and 2d have disappeared from the image and the z-range of the image was restored, some of the smaller bilayer islands seen in Fig. 1a have also shrunk or disappeared. Hence, the amplification of the sample during imaging in humid air

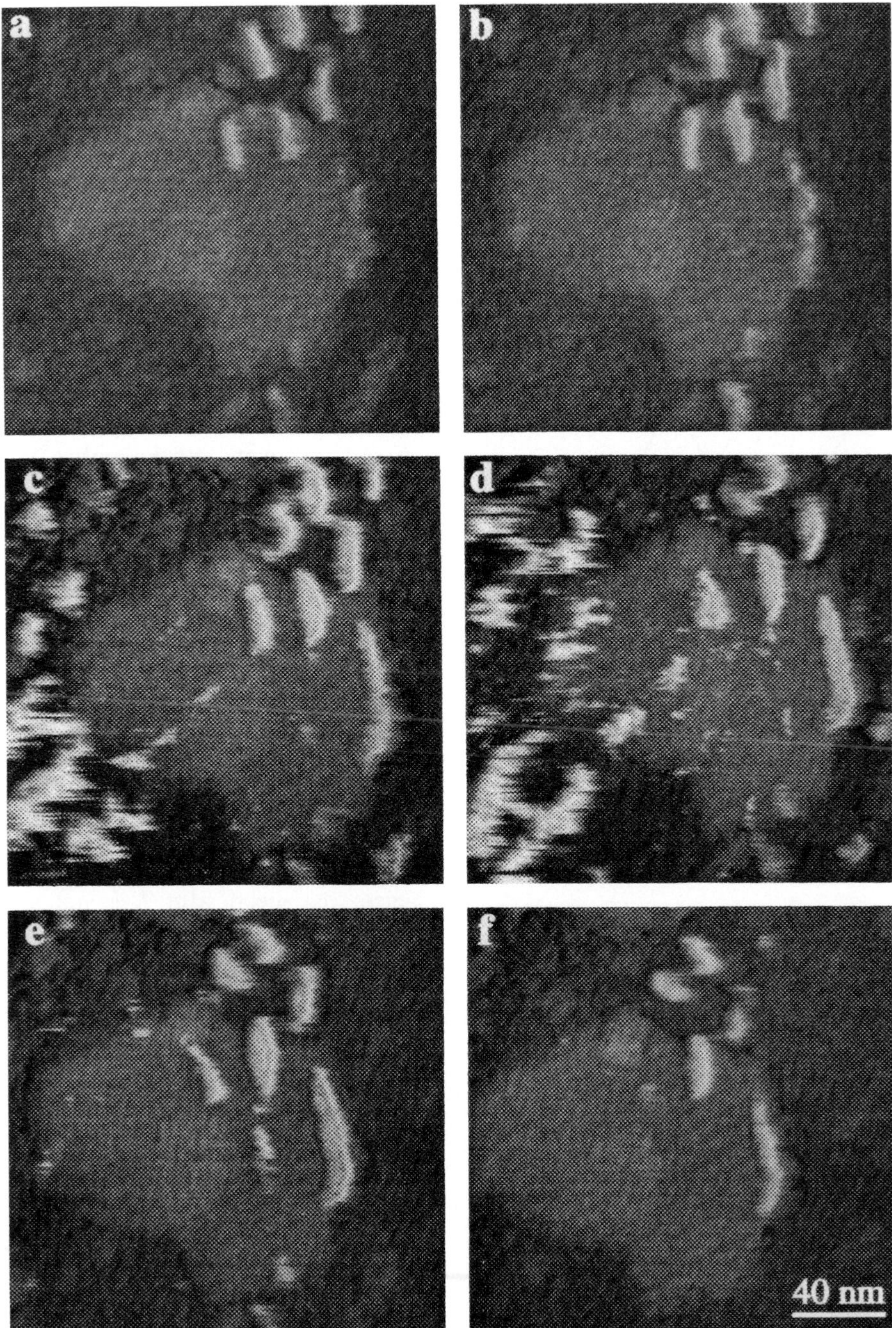

Figure 2: Series of six images of a platinum replica of a cadmium arachidate multilayer Langmuir-Blodgett film. Figs. **a-b** were taken under a dry nitrogen environment at time 15 and 31 minutes respectively and are nearly identical. The sharp differences in gray scale correspond to bilayer steps of 5.5 nm in height (See Fig. 3). Figs. **c-d** were taken after exposure to humid air at times 38 and 41 minutes and show significant variations in surface topology, most likely due to transient sample deformation due to capillary coupling of the STM tip to the sample (full arrows). Figs. **e-f** were taken with the sample under nitrogen again at times 50 and 57 minutes. Compare e-f with a-b; most of the surface features have returned to their original height and shape. All images are 200 nm x 200 nm and the height is linearly gray scaled over 44 nm from black (low) to white (high). Although most of the "pseudo-features" that appeared in c-d disappeared from the image and the z-range of the image was restored, some of the smaller bilayer islands seen in Fig. 2a have also shrunk or disappeared between 2a and 2f. Hence, the amplification of the sample during imaging in humid air can lead to permanent surface deformation, and hence should be avoided.

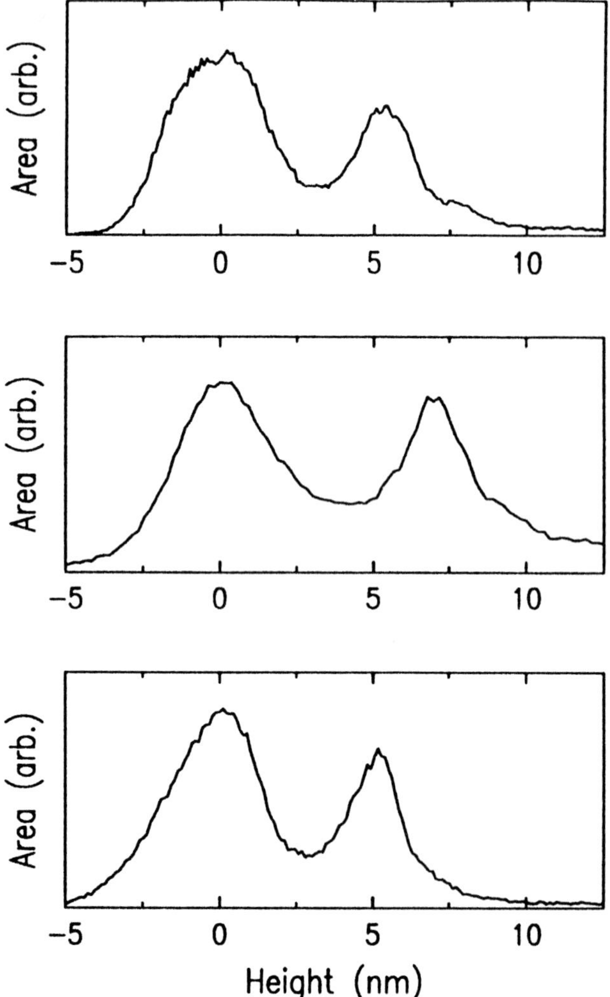

Figure 3: Height histograms of Figs. 2b (top), 2d (middle) and 2f (bottom) taken from areas on both sides of the large bilayer feature. The left peak in each plot corresponds to the background height and has been set to zero. The right peak corresponds to the height of the bilayer island. The plots are scaled so that the height difference between the peaks in the top figure is calibrated to the known bilayer height of 5.5 nm. The bilayer height increases to 7.0 nm upon exposure to air (middle) and returns to 5.4 nm (bottom) when returned to a dry nitrogen environment.

---

can lead to both transient and permanent surface deformation, and hence should be avoided. Fig. 3 shows histograms of the height difference between areas on either side of the large feature resembling Australia from the images in Figs. 2b, 2d and 2f. Fig. 3a shows that the step height (measured from the peak of the distributions) is 5.5 nm. (The width of the distribution is related to the finite size of the metal grains that make up the replica which are about 2 nm in diameter.) Fig. 3b shows that in humid air, the step height grows to more than 7 nm, and Fig. 3c shows that the step height returns to 5.4 nm on return to the dry atmosphere. Clearly, as these plots were from identical areas, capillary condensation from humid air is responsible for this amplification.

Quantitative Height Measurements by Correlation Averaging

From Figs. 2 and 3, it is clear that the resolution in any individual STM image of a freeze-fracture replica is limited by the granularity of the replicating film. In general, the quality of the replica is determined by how well the metal film replicates the surface on which it was deposited. The ideal replicating film would conform at the molecular level to the sample surface, be highly conductive and continuous, and would not reorganize over times long compared to typical imaging times. However, the metal replicas are made up of small metal grains several nanometers in diameter (Zasadzinski and Bailey, 1989; Ruben, 1989; Wepf *et al.*, 1991). Ideally, the evaporated metal atoms, which are usually a mixture of platinum and carbon, stick exactly where they land and form a structureless layer. However, the surface energy of the metal layer is much higher than that of the original fracture surface of water or hydrocarbons; hence, the metal film does not spread or "wet" the surface, but aggregates into small droplets (Adamson, 1990. Woodward, 1994). The aggregates grow in size, eventually merging with neighboring aggregates to form a continuous film. In STM images, the grains contribute a "noisy" background that can be removed by correlation averaging techniques or other image analysis techniques. However, as the grey scale in an STM image contains quantitative three-dimensional information, the type of correlation averaging technique becomes important to preserve this information unaltered (Woodward, *et al.*, 1995).

Correlation averaging is used to enhance images that contain a repeating pattern. Computer algorithms based on cross-correlation principles have been developed by several researchers (Frank *et al.*, 1978; Crepeau and Fram, 1981; Saxton and Baumeister, 1982; Henderson *et al.*, 1986) to locate and average patterns found in digitized images of electron micrographs. Correlation averaging is widely used to analyze images of biological structure with poor signal to noise ratios or those that do not form large crystal arrays. Correlation averaging schemes are a standard part of many commercially available software packages for electron microscopy image analysis (Hegerl, 1992). The dramatic improvement in image quality and resolution often reveals structural information that is obscured in any individual image. More recently, correlation averaging has been applied to images taken with scanning tunneling microscopes (Soethout *et al.*, 1988; Amrein *et al.*, 1989; Stemmer *et al.*, 1989; Wang *et al.*, 1990) and atomic force microscopes (Weigrabe *et al.*, 1991).

While the details of correlation averaging routines in the literature vary, they all follow the same basic outline:

(1) an image of the structure of interest is digitized,

(2) a small area containing at least one "unit" is chosen as the test image,

(3) this image is cross correlated with the main image,

(4) peaks in the resulting correlation correspond to the location of images similar to the test image,

(5) a criteria for selecting the largest peaks is then applied to determine which areas of the image are used to make the composite image, and

(6) these areas are averaged to form a composite image.

Many of the current image analysis routines contain a variety of additional techniques for enhancing images including corrections for the microscope transfer function, 3D reconstructions from a series of different tilts, and multivariate statistical analysis (Henderson *et al.*, 1990; Hegerl, 1992). The review by Hegerl describes many of the techniques included in commercially available image processing software. The following discussion will focus solely on correlation averaging.

Vertical distortions in correlation averaging

Although the composite images created by correlation averaging provide resolution superior to that in the original image, they also introduce a bias as described below. For electron micrographs, this bias is not important as the quantitative values of the pixel are rarely significant. The variety of contrast mechanisms in electron microscopy make it difficult to relate the absolute value of a pixel element to the properties of the original sample. In most images, the contrast in the composite is used primarily to determine the position of features in the image plane. In other cases, the ratio of two pixels is used; the absolute value of the pixel value is irrelevant (Zasadzinski and Bailey, 1989). In scanned probe microscopies, however, the pixel values from calibrated images represent absolute height measurements. Thus, any bias introduced by the correlation function leads to quantitative errors in "height" measurements made on the composite and could lead to possible misinterpretations of the image.

For calculation by computer, the discrete form of the correlation function

$$G(x',y') = \sum_{x=0}^{m-1}\sum_{y=0}^{m-1} T(x,y)M(x+x',y+y') \tag{10}$$

is usually used in conjunction with the fast Fourier transform (FFT) algorithm. Here T(x,y) is a (m)x(m) pixel test image, M(x,y) is an (n)x(n) pixel main image, and x, x', y and y' are all integers corresponding to the pixel location. Direct calculation using Eqn. (10) requires $(n\text{-}m+1)^2(m)^2$ multiplications while the use of the FFT algorithm requires $5(2\log_2 n)(n)^2 + (2\log_2 n)mn + 4m^2$ multiplications (Niblack, 1986). The value of T(x,y) or M(x,y) is the digitized pixel intensity of the original image for electron micrographs or relative height values for scanning probe images. In the correlation averaging techniques, the coordinates (x',y') of the peaks of the correlation function G(x',y') are taken to correspond to the locations in the original image that most closely resemble the test image. By averaging over many areas, random noise is minimized, giving a composite image that resembles the actual structure more closely than do any of the individual images. However, a close look at the correlation function shows that it does not treat positive and negative deviations from the test image in an equitable way.

There are two potential problems with the correlation function: (1) positive deviations from the actual structure lead to an increase in the value of the correlation function while negative ones lead to a decrease in the value, and (2) the relative effect of deviations from the test image strongly depends on the pixel value of the test image. The first problem tends to skew the entire image toward higher values while the second problem tends to increase higher valued pixels in the test image more than lower valued ones. To illustrate these problems we can examine the simple test case of a true image with pixel values (2, 5, 10, 5). For simplicity we will assume that these are also the values in our test image. Let us then imagine that there are two different areas in the main image, the first with each pixel value one lower than the test image, (1, 4, 9, 4), and the second with each pixel value one higher, (3, 6, 11, 6). The correlation value of the first area is 132 while that of the second area is 176. If the cut off for including a peak in the average falls between 132 and 176, the high valued area will be included in the average, while the low valued area will not. The higher pixel values are clearly favored. We should note that for our simple example, biasing the composite toward higher pixel values is not a problem for either TEM or SPM images as neither has an absolute reference scale. In all three of the simple images, the peak value is eight units above the lowest value.

To illustrate the second problem, take four more areas: (4, 5, 10, 5), (0, 5, 10, 5), (2, 5, 12, 5) and (2, 5, 8, 5). The respective correlation values with the test image are 158, 150, 174 and 134. The noise located at the pixel in the test image with the highest value has a much larger effect on the correlation value. When a threshold is selected, areas with positive valued noise will be selected over those with negative valued noise, leading toward a bias of the whole composite toward higher values. Furthermore, negative valued noise from high spots in the test image is more likely to be rejected than the same noise at a lower valued pixel. This causes the correlation to enhance the peaks in the test image. In the resulting composite, the image appears stretched with high areas extended disproportionately. In our example, if the threshold is ≥ 150 only the first three areas will be averaged to form the composite (2, 5, 10.7, 5). Note that the peak value has grown while the others have remained the same. The corresponding slight change in the visual appearance of the image is probably unimportant to the TEM user, but the SPM user will now measure the difference between the high and low point to be 8.7. Hence, the correlation averaging algorithm has the potential to cause a quantitative errors in our results.

Following Schulz-DuBois and Rehberg (1981), we propose that the structure function is superior to the correlation function when absolute pixel values are important. In its discrete form

$$S(x',y') = \sum_{x=0}^{m-1} \sum_{y=0}^{m-1} [T(x,y) - M(x+x',y+y')]^2 \quad (11)$$

the structure function is easy to recognize as the sum of the squared deviations. The main benefits of the structure function are: (1) it treats positive and negative noise equally, and (2) it does not discriminate based on the pixel value of the test image. The minima of the structure function give the best areas to be included in the composite. Direct computation of the structure function requires the same number of multiplications as the correlation function, but requires an extra addition for every multiplication. Calculation using the FFT algorithm requires $9(2\log_2 n)(n)^2 + (2\log_2 n)mn + 12n^2$ multiplications. For a 512x512 image with a 32x32 test area this is only twice the number required for the same calculation using the correlation function.

To test the relative merits of the correlation and structure functions we wrote a simple averaging program. The program calculates the correlation or structure function between an image file and a test area taken either from a portion of the image itself or from another file. The calculation is done in real space using Eqn. (10) or (11). The calculation is only done for points where the test image fits entirely within the main image, so there is no padding or wrapping around at the edge of the image. This results in a (m-n)x(m-n) field for the

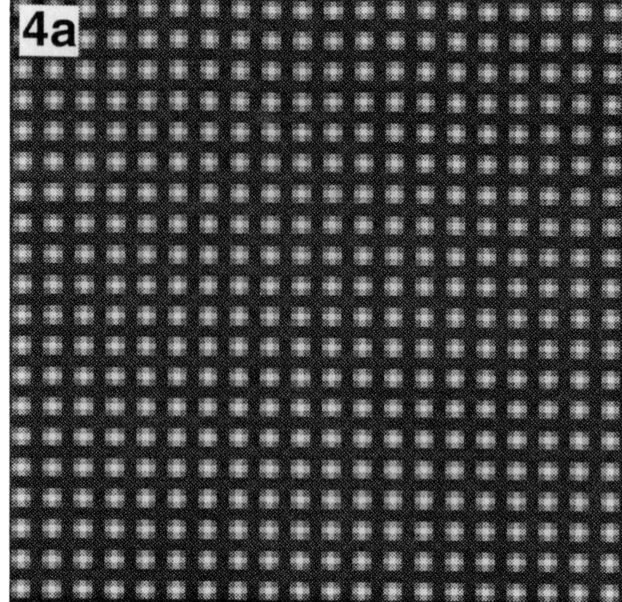

Figure 4: Computer generated images used to test the correlation averaging and structure averaging routines. (**a**) The image with no noise added. (**b**) The image with a noise level of 2.

Table 2: Errors in correlation and structure averaging of test sample

Correlation
Number of areas used in averaging

Noise Level	324	162	81	32
0	0.0±0.0 0.0±0.0	0.0±0.0 0.0±0.0	0.0±0.0 0.0±0.0	0.0±0.0 0.0±0.0
1	0.0±0.0 0.0±0.0	0.00±0.01 0.02±0.06	0.05±0.02 0.22±0.03	0.11±0.03 0.33±0.05
2	0.0±0.0 0.0±0.0	0.06±0.03 0.24±0.05	0.16±0.04 0.40±0.06	0.32±0.05 0.62±0.03
4	0.0±0.0 0.0±0.0	0.24±0.05 0.49±0.07	0.44±0.05 0.74±0.05	0.74±0.08 1.06±0.06
8	0.47±0.07 0.69±0.06	0.87±0.11 1.12±0.09	1.14±0.10 1.49±0.07	1.52±0.14 2.04±0.12

Structure

Noise Level	324	162	81	32
0	0.0±0.0 0.0±0.0	0.0±0.0 0.0±0.0	0.0±0.0 0.0±0.0	0.0±0.0 0.0±0.0
1	0.0±0.0 0.0±0.0	0.0±0.0 0.0±0.0	0.0±0.0 0.0±0.0	0.0±0.0 0.0±0.0
2	0.0±0.0 0.0±0.0	0.0±0.0 0.0±0.0	0.0±0.0 0.0±0.0	-0.01±0.02 0.06±0.09
4	0.00±0.01 0.02±0.06	0.00±0.01 0.02±0.06	-0.02±0.06 0.22±0.12	0.00±0.07 0.44±0.07
8	-0.06±0.10 0.50±0.03	-0.02±0.08 0.54±0.06	0.00±0.09 0.63±0.10	0.02±0.13 0.80±0.06

The results of the correlation and structure averaging routines. The programs were run on the same images and either 324, 162, 81 or 32 areas were used to form the composite images. The top value in each entry is the mean deviation in pixel value from the true image. The mean deviation is always near zero for the structure averaging routine, indicating that there is no bias induced by the averaging. The mean deviation is always positive for the correlation averaging routine, indicating that the correlation averaging tends to generally increase pixel values. The bottom value is the root mean square deviation per pixel which is also lower for the structure averaging routine. The test area used was the true image with no noise added.

---

correlation or structure function. The program then finds and sorts all the local maxima (correlation) or minima (structure) in the function. Extrema at the edge are ignored. The user pre-selects the number of areas to be averaged to form the composite image. The areas corresponding to highest local maxima in the correlation function, or lowest local minima in the structure function, are averaged and a file containing the output is created. Our program does not interpolate between data points. The program was written in C and all the results presented were obtained on a Silicon Graphics Indigo computer.

Test on a known image

To demonstrate the advantages of the structure function over the correlation function we made a test area and test image composed of a 20x20 array of test areas as can be seen in Fig. 4a. The test area is 5 pixels x 5 pixels with the following values:

$$\begin{matrix} 1 & 1 & 1 & 1 & 1 \\ 1 & 3 & 5 & 3 & 1 \\ 1 & 5 & 10 & 5 & 1 \\ 1 & 3 & 5 & 3 & 1 \\ 1 & 1 & 1 & 1 & 1 \end{matrix} \tag{12}$$

A pseudo-random number generator was then used to add varying degrees of noise to the image. A noise level of 1 corresponds to -1, 0 or 1 being added to each pixel, a noise level of 2 corresponds to -2, -1, 0, 1 or 2 being added, and so on. An image with a noise level of 2 is shown in Fig. 4b.

An averaging routine was used to make composite images using both the structure and correlation functions on images containing noise levels of 1, 2, 4 and 8. These correspond to images with a signal to noise ratio of 13.0, 4.3, 1.30 and 0.36 respectively. As our routine ignores the areas near the edge of the main image, both the correlation and the structure function find 324 peaks in the zero noise main image corresponding to the 18x18 matching areas on the interior. We ran the program using the best 324, 162, 81 and 32 maxima or minima to find the areas to be averaged for the composite. The average deviation per pixel between the composites and the true image and the root mean square (RMS) deviation per pixel were calculated. The average results of ten runs are shown in Table 2.

The results clearly show that for any appreciable noise level the correlation function skews the composite toward higher pixel values, while there is no significant skewing at any noise level for the correlation function. In addition the (RMS) deviation from the true image is significantly lower for the structure function generated composites than for the correlation function generated composites. The increase in the RMS deviation for both functions when fewer areas are included in the composite is to be expected, as it is less likely that the noise will average to zero when fewer data points are included in the average. Similarly, as the level of the noise increases, the typical deviation of the average from the true value grows.

In Fig. 5 we look at the extreme case of the results from taking the best 32 areas of the noise level 8 images. Fig. 5a shows the actual 5x5 unit cell. Figs. 5b and 5c are the composites from the correlation averaging and structure averaging programs respectively. The actual unit cell had pixels valued 1, 3, 5 and 10 as seen in (12). The average value of these pixels in the correlation composite is 1.8, 2.7, 7.8 and 14.5 respectively and in the structure composite they are 1.2, 3.0, 4.7 and 8.5. Although in practice it would be rare to restrict oneself to using so few of the available areas to form a composite, the example serves to illustrate the effect that the averaging routines can have on the composites. The

Figure 5. The results of the averaging routines on images with noise level 8 when only the top 32 areas are used in the average. **(a)** The true unit cell. **(b)** The composite unit cell formed by the structure averaging routine. Other than the central peak all pixels retained their true value. **(c)** The composite unit cell formed by the correlation averaging routine. The values of the pixels are generally greater than those in the true unit cell with the peak values especially enlarged.

error in the difference between the peak and low values is more than twice as large using the correlation function compared to the structure function.

The case we have presented assumes that the true image can be used as a test function, which is usually not the case in most applications. When we ran our averaging programs using a 10x10 pixel test area chosen from within the image we found that the biasing effect of the correlation function remains, although it is decreased by nearly half, while the RMS deviations are little changed. More surprisingly, the RMS deviations for the structure function increase to become comparable with, but still less than, those of the correlation function. There is still no bias with the structure function.

Measurements of ripple wavelength and amplitude in DMPC bilayers

One of the most difficult structures to examine by any technique are lyotropic liquid crystals - the general materials class into which bilayers, liposomes, and related lipid phases fall. Conventional x-ray diffraction methods usually cannot provide sufficient resolution as these materials are difficult to align, domain sizes and correlation lengths are small, and thermal fluctuations tend to obscure structural details (Sirota *et al.*, 1988). The freeze-fracture technique has been used primarily to investigate membrane structure and has led to a general acceptance of the fluid mosaic model of cell membranes, in which integral proteins are embedded in a lipid bilayer (Branton, 1966; Singer and Nicolson, 1972). Recent finding have shown that the lipid membrane is not only a passive matrix; protein function can be modified by the composition, phase and local structure of the lipids (Keller *et al.*, 1993).

Hence, it is important to study the phase behavior and structures of phospholipids, and saturated phosphatidylcholines (PC) in particular, as PC's are present in many cell membranes and are the major component of human lung surfactant (Longo *et al.*, 1993). Saturated PC's undergo three distinct structural transitions when dispersed in water: a subtransition, pretransition and main transition separating the phases L_c, $L_{\beta'}$, $P_{\beta'}$, and L_α (Tardieu *et al.*, 1972; Janiak *et al.*, 1979; Wack and Webb, 1989). In the high temperature L_α phase, the order within each bilayer is short-range and the trans-*gauche* intramolecular order is low (Tardieu *et al.*, 1972). The main transition is associated with lipid chain melting (Tardieu *et al.*, 1972). The bilayers are smooth and the molecules, on average, are normal to the bilayer (Sirota *et al.*, 1988). The $L_{\beta'}$ phase is characterized by flat bilayers with the lipid chains fully extended (all-*trans* configuration) and tilted with respect to the bilayer normal. The magnitude and direction of tilt depends on the water fraction (Sirota *et al.*, 1988). The low temperature $L_{\beta'} \rightarrow L_c$ transition involves a modification of the chain packing and dehydration of the head groups (Janiak *et al.*, 1979).

A satisfactory explanation of the $P_{\beta'}$ phase remains a theoretical and experimental challenge. In the $P_{\beta'}$ phase, the lipid chains retain much of their all-*trans* configuration, and the molecules are packed into a two-dimensional hexagonal lattice with long range correlations (Janiak *et al.*, 1979; Ruppel and Sackman, 1983; Zasadzinski and Schneider, 1987). The bilayers are characterized by regular three-dimensional corrugations, hence the common name of ripple phase. X-ray diffraction and freeze-fracture electron microscopy find the ripple wavelength in excess water to be about 10 - 15 nm (Tardieu *et al.*, 1972; Luna and McConnell, 1977; Janiak *et al.*, 1979; Ruppel and Sackman, 1983; Zasadzinski and Schneider, 1987; Zasadzinski *et al.* 1988; Wack and Webb, 1989). The ripples in the $P_{\beta'}$ phase are capable of aligning large molecules such as proteins along their length; diffusion within the plane of the bilayer appears to be anisotropic (Schneider *et al.* 1983). Replacing one or more methyl groups from the choline headgroup by hydrogen, thereby decreasing the size of the headgroup relative to the chains, eliminates the $P_{\beta'}$ phase (Zasadzinski and Schneider, 1987). Models of the $P_{\beta'}$ phase fall into two broad classes: phenomenological models based on modulations of membrane thickness (Goldstein and Leibler, 1988; Cevc, 1991) or curvature (Doniach, 1979; Lubensky and MacKintosh, 1993) and molecular models based on packing frustration between the lipid headgroup and chains (Carlson and Sethna, 1987; McCullough and Scott, 1990; Scott and McCullough, 1991; Schwartz *et al.*, 1994). The structural details of the ripples necessary to test these theories cannot be determined by X-ray diffraction or freeze-fracture electron microscopy. The particular features of this phase makes it an ideal system to examine by the freeze-fracture-STM technique (Woodward and Zasadzinski, 1995).

Metal replicas of DMPC bilayers in excess water, equilibrated at 16, 18, 20 and 23°C were prepared (as described above) for both TEM and STM examination to determine the temperature dependence of the ripple amplitude, waveform, and wavelength (the $P_{\beta'}$ phase of DMPC exists from 14-24°C). For each sample, 200 mg of DMPC (Avanti Polar Lipids, Atlanta, GA) was added to 0.2 ml Milli-Q water (Millipore, Bedford, MA); the resulting 50% DMPC/50% water mixture assured that the bilayers are fully hydrated (Wack and Webb, 1989). The samples were alternately centrifuged at low speed, vortexed, then heated at > 30°C for at least 24 hours to allow complete mixing. Thin films of the DMPC-water mixture were sandwiched between copper freeze-fracture planchettes, then equilibrated at 100% relative humidity

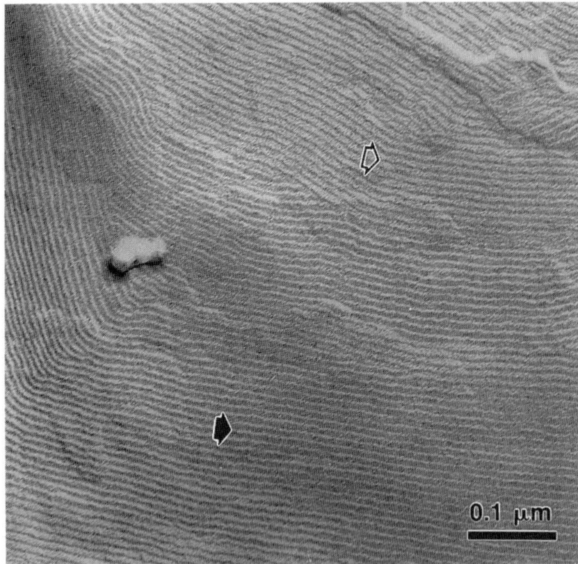

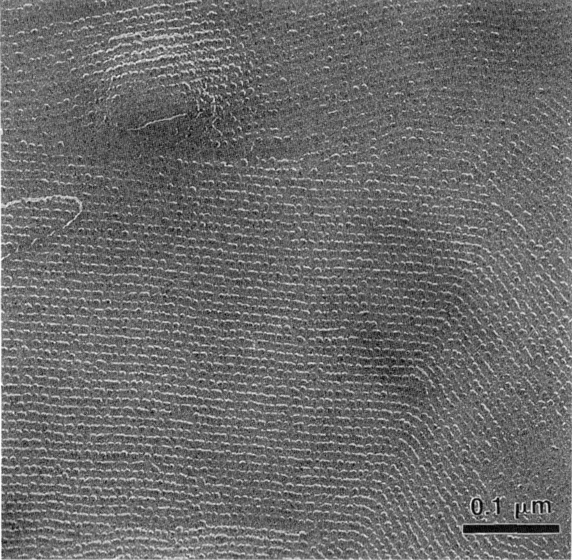

Figure 6: (**a**) TEM image of DMPC in excess water at 23°C. The ripples change orientation over μm length scales, typically at 120° angles, indicating an underlying hexagonal packing. The ripple asymmetry can be inferred from the shadow patterns - narrow, dark gray - wide, light gray lines in the upper left of the image (open arrow) adjacent to narrow, white - wide, dark gray lines in the lower right of the image (filled arrow). (**b**) TEM image of DMPC in excess water at 16°C (The sample at 18°C was similar). The ripples are much less well defined and appear like a string of beads. This beading is induced by a self-shadowing process that enhances small features. The ripple amplitude is much smaller and the asymmetry in the shadowing has disappeared.

Table 3: Amplitude of first two harmonics and wavelength of ripples from different images as determined by correlation averaging

Temperature	A_1(nm)	A_2(nm)	λ(nm)
20°C	0.62	0.12	10.2
	0.64	0.16	10.4
	0.44	0.07	10.2
	0.42	0.06	10.1
	0.44	0.06	10.3
	0.45	0.07	10.3
	0.64	0.10	12.9
23°C	1.20	0.25	10.9
	1.05	0.20	11.4
	1.03	0.20	10.8
	0.96	0.20	10.3
	1.02	0.20	11.2
	1.72	0.36	11.0
	1.82	0.38	11.0
	1.29	0.36	10.9
	0.98	0.26	10.9
	0.68	0.21	10.1
	0.54	0.16	9.8
	0.86	0.20	10.0
	1.16	0.28	10.5
	0.98	0.27	10.5
	1.03	0.30	10.3
	0.95	0.34	11.0
	1.08	0.40	11.0
	1.18	0.40	11.0
	0.74	0.14	10.4
	1.34	0.30	10.1
	1.19	0.25	10.1

at 23, 20, 18 or 16°C prior to rapid quenching in liquid propane cooled by liquid nitrogen. Freeze-fracture replication was done in a Balzers 400K freeze-etch machine as described above. The TEM samples were shadowed with 1.5 nm of Pt/C at a 45° angle relative to the surface, while the STM samples were coated normal to the surface while the sample table was rotated to insure a continuous coating. A 15 nm thick film of carbon was added to stabilize the shadowing film. Replicas for STM had an additional 0.5 μm thick layer of silver deposited by sputtering to increase film rigidity. Replicas of each sample were examined first with TEM (JEM 100CXII) to determine replica quality and identify the ripple phase. STM (Digital Instruments, Santa Barbara, CA) imaging was done with a 12 μm scanning head in the constant current mode under a dry nitrogen atmosphere. Each sample was examined with several

SYMMETRIC RIPPLES

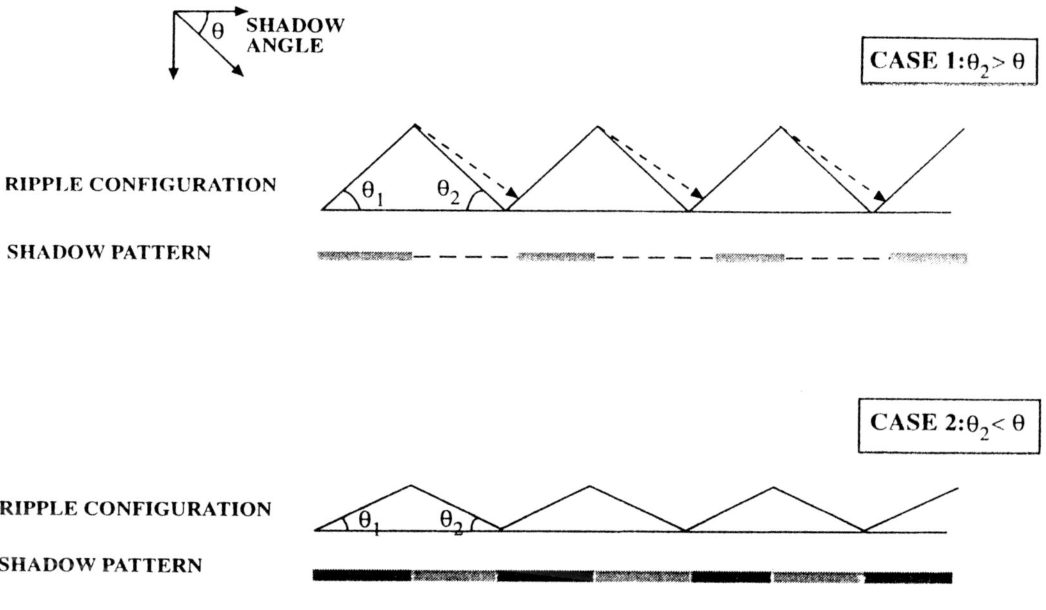

ASYMMETRIC RIPPLES

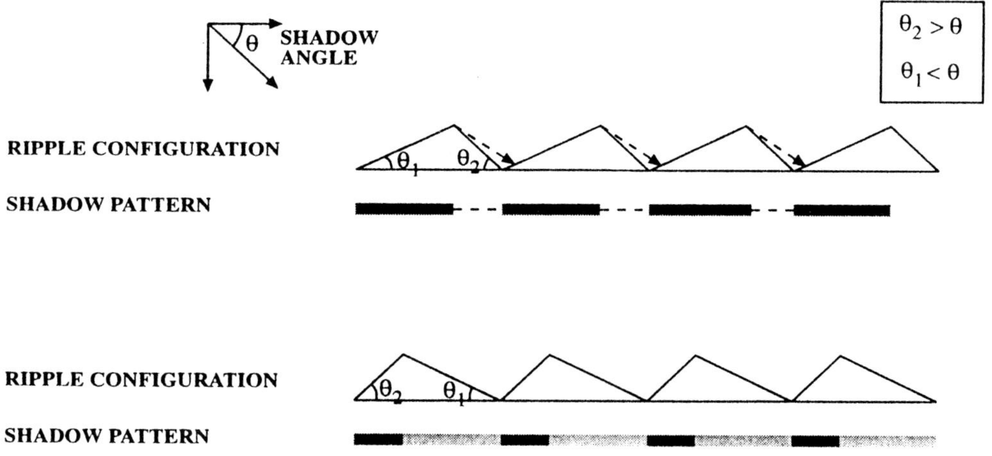

Figure 7: (**a**) Predicted ripple shadow pattern for symmetric ripples if $\theta_2 > \theta$, most of the ripple is uncoated by platinum and will appear gray-white in micrographs. If $\theta_2 \leq \theta$, all the ripple will be coated and will appear dark gray-light gray in micrographs. q is the macroscopic shadow angle, which is typically 45° in these experiments. (**b**) Asymmetric ripples will appear different depending on their relative orientation with respect to the shadow direction. Gray-white if the long side is oriented towards the shadow directions, dark gray-light gray if away. Compare to the patterns in Fig. 6a.

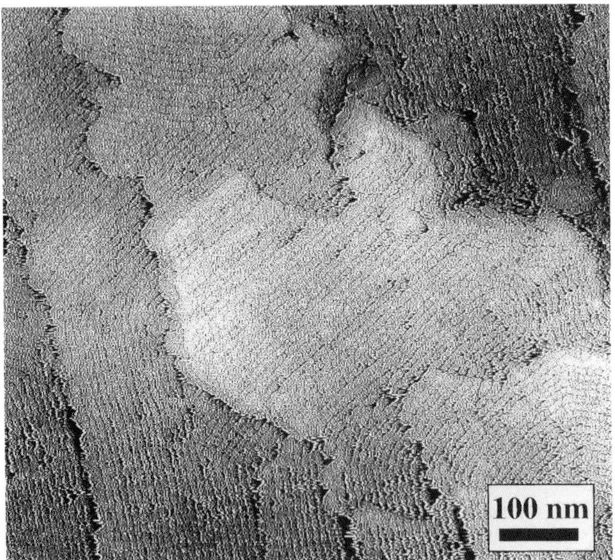

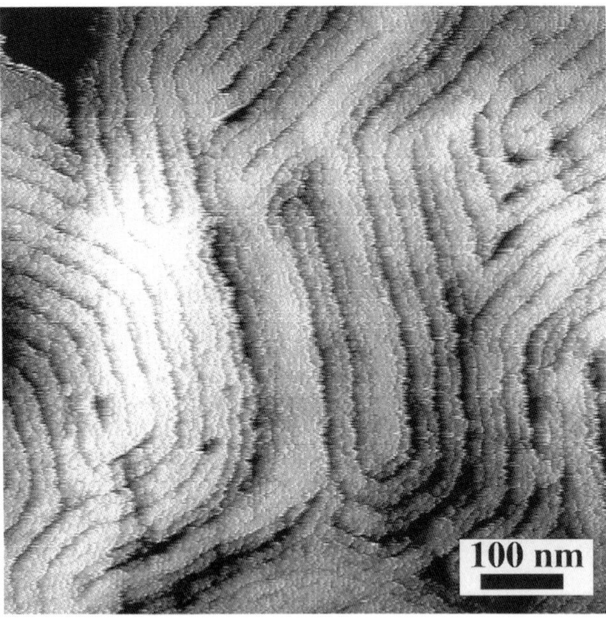

Figure 8: (**a**) Low resolution STM image of replica of DMPC ripples to compare to the TEM images in Fig. 6a. The general pattern of ripples is very similar to the TEM images. Note especially that the ripples appear to be very uniform in height over various patches on the surface, each of which is separated by a bilayer thickness. (**b**) Second type of ripple often observed in DMPC replicas. These ripples are significantly larger in amplitude and about twice the wavelength of the majority ripples shown in Fig. 8a.

----

differently cut Pt/Ir tips.

Representative TEM images of the 23°C and 16°C samples are shown in Fig. 6. In the 23 and 20°C samples, the ripples are continuous and the ripple asymmetry is apparent from the variation in shadowing across the image (Fig. 6a) (Zasadzinski and Schneider, 1987). Near the open arrow, the ripples appear as wider light gray lines with narrow, dark gray lines. At the filled arrow, the pattern is reversed - narrow white lines with wide, dark gray lines (Fig. 7). These patterns are only consistent with an asymmetric waveform for the ripples. In TEM images of the 18 and 16°C samples, the ripples are discontinuous and appear to be made up of a line of beads. This pattern of beads is consistent with a self-shadowing artifact; the oblique metal deposition enhances small features in the fracture surface (Ruben, 1989). It is clear that the ripple is of smaller amplitude, the shadowing pattern has lost its asymmetry and is less well defined, suggesting a temperature dependence of the ripple amplitude. When the replica is made by depositing the metal film normal to the surface (which eliminates self-shadowing), TEM images of the 18 and 16°C samples are featureless. The wavelength of the ripples is 11.0 ± 1.0 nm for all four temperatures. The measured values for the wavelength are in good agreement with the value of 11-12 nm found by x-ray diffraction (Janiak *et al.*, 1979; Wack and Webb, 1989) or freeze-fracture TEM of DMPC (Luna and McConnell, 1977; Ruppel and Sackmann, 1983; Zasadzinski and Schneider, 1987).

Survey STM scans of the 23°C and the 20°C samples show that the STM and TEM images have the same defect patterns and general features (Fig. 8a,b). To quantify the ripple amplitude and wavelength, roughly twenty 250 by 250 nm STM images with the ripples oriented in a single direction were analyzed for each sample to quantify the ripple features observed by TEM. At this image size, each ripple has about 20 pixels per wavelength (images are 512 by 512 pixels). Figs. 8a and b are representative images from the 23°C and the 20°C samples. Samples prepared at 18 and 16°C showed no modulated textures. Fourier transforms of both the smaller and larger scale images were used to evaluate the ripple wavelength, which was consistent with the TEM results (see Table 3). STM images suitable for analysis were averaged using a structure function program described earlier on a Silicon Graphics Indigo computer. A test area size of 32 by 32 pixels was used to insure that the test area contained at least one complete ripple. Areas that showed a minimum square deviation from the test image were then averaged to make a composite image (Woodward *et al.*, 1994). The composite image was then used as the test image to determine a new composite and eliminate any bias in the original choice of test image. Typically 1000-4000 areas were used to make a composite. For any given image,

Figure 9: (**a**) 250 nm x 250 nm STM image of a freeze-fracture replica of DMPC in excess water originally at 23°C. The ripple shape is obscured by the finite size of the metal grains that make up the replica film, imperfections in the ripples themselves, and noise from the STM. Correlation averaging helps eliminates these features to enhance the image quality. (**b**) 250 nm x 250 nm STM image of a freeze-fracture replica of DMPC in excess water originally at 20°C. The apparent amplitude of the ripples has decreased as compared with Fig. 9a.

---

we determined four different composites using 4 different test areas chosen arbitrarily from the original image. There was no significant difference in the composites that came from the different initial test areas.

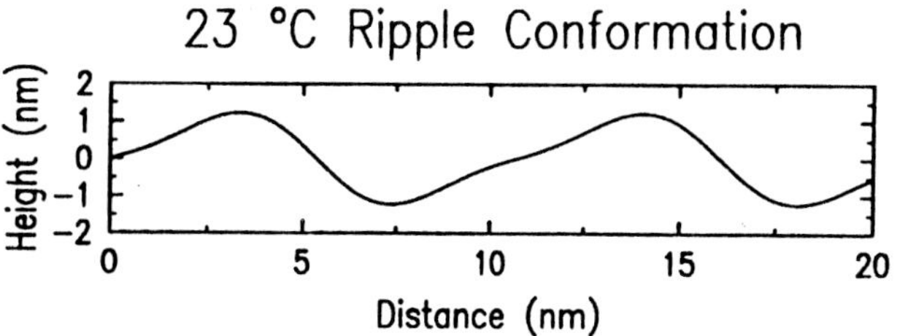

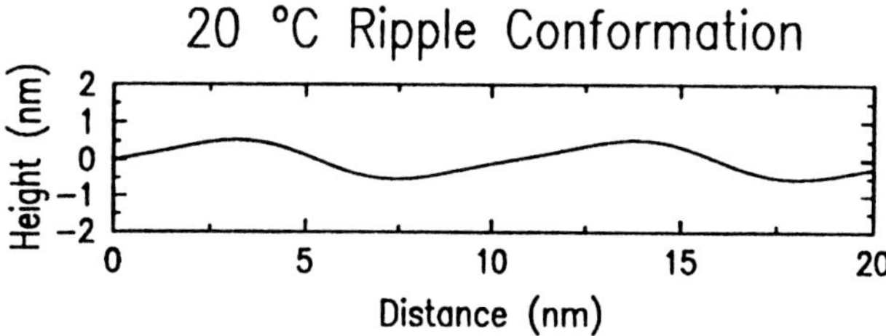

Figure 10: (**a**) Average ripple cross section at 23°C as determined from the average of the parameters in Table 3. The wavelength is 10.7±1 nm and the peak to valley amplitude is 2.4 nm. (**b**) Average ripple cross section at 20°C. The wavelength is 10.6 ±1 nm and the peak to valley height decreases to 1.1 nm. The asymmetry in waveform is still present.

The average ripple conformation determined from the composite image was then fit to a sum of harmonics using the non-linear least squares fitting routine found in C-PLOT (Cambridge, MA) to eliminate any systematic errors introduced by specific samples, images, STM tips, etc. The parameters fit were the amplitudes and phases of the first three harmonics of a sine wave, and the wavelength of the ripple. The data for both sets of samples are presented in Table 3. The average values of the first two harmonics for the 23°C samples are 1.1±0.3 nm and 0.3±0.1 nm, and the averaged wavelength is 10.7±1.0 nm. For the 20°C samples, the amplitude of the first harmonic is 0.5±0.2 nm, the second amplitude is 0.1±0.1 nm, and the wavelength is 10.6±1.0 nm. The amplitude of the third harmonic was negligible for both temperatures. The presence of a non-zero amplitude of the second harmonic shows that the ripples are asymmetric, as expected from the TEM micrographs. Fig. 10 shows a cross section of the ripple conformation calculated from the average values. The fits give a peak to valley height of 2.4 nm for the ripple at 23°C and 1.1 nm at 20°C. Combined with the zero amplitude we found at 18 and 16°C, this shows a distinct temperature dependence of the ripple amplitude in the $P_{\beta'}$ phase.

In our previous STM images of ripples, filtered images showed a significantly different amplitude of about 4.5 nm and what looked like a small secondary ripple perpendicular to the primary ripple (Zasadzinski

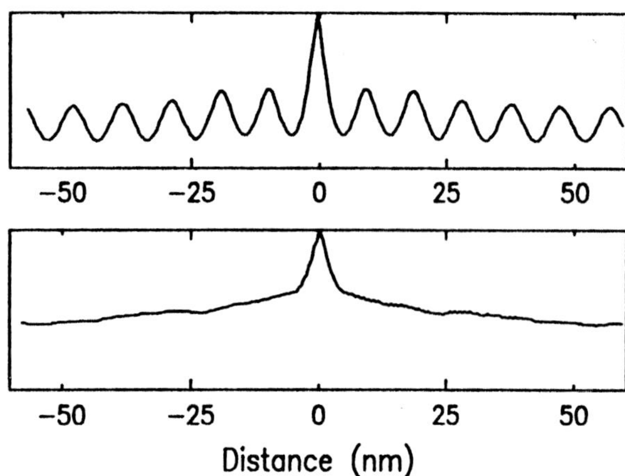

Figure 11: (**a**) Autocorrelation of the sample quenched from 23 °C shown in Fig. 9a. (**b**) Cross sections of (a) along (top) the ripple direction and (bottom) perpendicular to the ripple direction. The lack of modulation perpendicular to the ripple direction rules out any significant secondary modulation seen previously (Zasadzinski *et al.*, 1988).

---

et al., 1988). X-ray results have suggested that this secondary ripple may be the result of molecular tilt perpendicular to the ripple wavevector (Hentschel and Rustichelli, 1991). However, Fourier transforms of our STM and TEM images do not have any peaks other than those of the primary ripple. This secondary ripple, and the larger amplitudes are likely artifacts caused by the height amplification discussed earlier as those images were taken in air. As an additional check for the secondary ripple we looked at autocorrelations of the STM images and found no periodicity perpendicular to the primary ripple. Fig. 11 shows the autocorrelation and cross-sections from the image shown in Fig. 9a.

Our combined STM and TEM results show that in excess water, bilayers of DMPC have uniaxial, asymmetric ripples with a temperature dependent amplitude, varying from 2.4 nm peak to peak near the chain melting temperature to near zero near the chain crystallization temperature. However, the ripple wavelength of 11 nm does not change with temperature. The asymmetric ripple shape is consistent with a recent theory that suggests the ripple is the result of a coupling between molecular tilt and bilayer bending (Lubensky and MacKintosh, 1993). This theory depends, to some extent, on the chiral nature of DMPC although experimental results show that the ripple is unchanged when a racemic mixture of DMPC is used (Zasadzinski, 1988b; Katsaras and Raghunathan, 1995) However, the temperature dependence of the ripple amplitude has not previously been observed experimentally nor been predicted theoretically. However, it is not unexpected as the bend elasticity of the membrane is expected to decrease dramatically over the transition from crystalline to fluid bilayers as the temperature is raised through the $P_{\beta'}$ phase. Both STM and TEM show that the wavelength of the ripples is unchanged at all four temperatures. This indicates that models of the ripple phase based on a fixed offset of neighboring molecules due to packing frustration must be modified (Schwartz *et al.*, 1994). Such theories predict that the wavelength of the ripple would scale with the amplitude, in contradiction to our results. However, a change in molecular tilt with temperature might change the vertical component of a fixed molecular offset as the temperature increases from the tilted $L_{\beta'}$ phase to the untilted L_α phase. We have also carefully investigated the possibility of a second ripple normal to the first set of ripples by examining autocorrelations of the ripple images. No secondary ripple structure was found, indicating that previous indications of such ripples were likely artifacts (Zasadzinski *et al.*, 1988).

Summary

The freeze-fracture replica technique is uniquely valuable to extend the utility of STM to imaging bulk suspension, liquid crystals, and other biological materials in excess water. The technique is substrate-free and can examine interior interfaces impossible to see with other scanned probe microscopies at high resolution. The quantitative, three dimensional information available by this technique is a necessary extension of TEM images of replicas, especially for situations where a repetitive, three-dimensional structure is expected and quantitative information about the third, or vertical dimension is needed. By incorporating the necessary modifications described in this paper, replicas for STM can be made as easily as those for TEM with no significant additional investment. We have found that is it necessary to do our STM imaging of these replica surfaces in as dry an atmosphere as possible to inhibit amplification of surface features. However, images of the STM-replicas are reproducible and simple to interpret once these precautions are taken. On any particular replica, the granularity of the replica film limits resolution to about 2 nm laterally and about 1 nm vertically. However, the effects of the granularity can be removed by correlation averaging, especially for periodic surfaces. The ultimate resolution of the technique for the ripple phase of phospholipids is 1 nm lateral and 0.3 nm vertical resolution. The replica/STM technique is, therefore certainly as high resolution as direct STM or AFM imaging of supported biological materials, and can often be easier to interpret and reproduce. It can also be used as a check for tapping mode AFM images of soft surfaces.

References

Adamson AW (1990) The Physical Chemistry of Surfaces. John Wiley and Sons, New York, Ch. 13.

Adrian M, Dubochet J, Lepault, McDowell AW (1984) Cryo-electron microscopy of vitreous ice. Nature **308**: 32-36.

Akahori H, Nakajima Y, Terasawa K, Ishii H, Nonaka I (1986) High resolution freeze replica by means of high melting point metal shadowing. J Electron Microsc **35**: 202-207.

Amrein M, Stasiak A, Gross H, Stoll E, Travaglini G (1988) Scanning Tunneling Microscopy of recA-DNA complexes coated with a conducting film. Science **240**: 514-516.

Amrein M, Durr R, Winkler H, Travaglini G, Wepf R, Gross H (1989) STM of freeze dried and Pt-Ir-C-coated bacteriophage-T4 polyheads. J Ultrastructure Mol Struc Res **102**: 170-177.

Anselmetti D, Gerber Ch, Michel B, Wolf H, Guntherodt HJ, Rohrer J (1993) Deformation-free topography from combined scanning force and tunneling experiments. Europhys Lett **23**: 421-426.

Arca M, Bard AJ, Horrocks BR, Richards TC, Treichel DA (1994) Advances in Scanning Electrochemopy Microscopy. Analyst **119**: 719-726.

Bailey SM, Longo ML, Chiruvolu S, Zasadzinski JA (1991) A controlled temperature and environmental stage for rapid freezing fixation for freeze fracture. J Electron Microsc Tech **19**: 118-126.

Bailey SM, Zasadzinski JA (1991) Validation of convection limited cooling of samples for electron microscopy. J Microsc **163**: 307-320.

Baró AM, Miranda R, Alaman J, Garcia B, Binnig G, Rohrer H, Gerber Ch, Carrascosa JL (1985) Determination of surface topography of biological specimens at high resolution by scanning tunneling microscopy. Nature **315**: 253-254.

Bellare J, Davis HT, Scriven LE, Talmon Y (1988) The controlled environment vitrification system. J Electron Microsc Tech **10**: 87-104.

Binnig G, Rohrer H, Gerber Ch, Weibel E (1982) Surface studies by scanning tunneling microscopy. Phys Rev Lett **49**: 57-61.5.

Binnig G, Quate CF, Gerber Ch (1986) Atomic Force Microscope. Phys Rev Lett **56**: 930-933.

Blackman GS, Mate CM, Philpott MR (1990) Interaction forces of a sharp tungsten tip with molecular films of silicon surfaces. Phys Rev Lett **65**: 2270-2273.

Branton D (1966) Fracture faces of frozen membranes. Proc Nat Acad Sci USA **55**: 1048-1056.

Bruegeller P, Mayer E (1980) Complete vitrification in pure liquid water and dilute aqueous solutions. Nature **288**: 569-571.

Carlson JM, Sethna JP (1987) Theory of the ripple phase in hydrated phospholipid bilayers. Phys Rev A **36**: 3359-3374.

Cevc G (1991) Polymorphism of the bilayer membranes in the ordered phase and the molecular origin of the lipid pretransition and rippled lamellae. Biochim Biophys Acta **1062**: 59-69.

Chen CJ (1992) Effects of $m \neq 0$ tip states in STM: the explanation of corrugation reversal. Phys Rev Lett **69**: 1656-1659.

Chiruvolu S, Naranjo E, Zasadzinski JA (1994) Microstructure of complex fluids by electron microscopy. In: ACS Symposium Series Vol. 578, Structure and Flow in Surfactant Solutions (CA Herb, RK Prud'homme, eds) ACS, Washington, DC, Ch. 5.

Ciraci S, Baratoff A, Batra IP (1990) Tip-sample interaction effects in scanning tunneling and atomic force microscopy. Phys Rev B **41**: 2763-2775.

Clemmer CR, Beebe TP (1991) Graphite - A mimic for DNA and other biomolecules in scanning tunneling microscope studies. Science **251**: 640-642.

Coleman RV, Drake B, Hansma PK, Slough G (1985) Charge-density waves observed with a tunneling microscope. Phys Rev Lett **55**: 394-397.

Costello MJ, Corless JM (1978) The direct measurement of temperature changes within freeze-fracture specimens during rapid quenching in liquid coolants. J Microsc **112**: 17-37.

Costello MJ, Meiboom S, Sammon MJ (1984) Electron microscopy of a cholesteric liquid crystal and its blue phase. Phys Rev A **29**: 2957-2959.

Crepeau RH, Fram EK (1981) Reconstruction of imperfectly ordered zinc-induced tubulin sheets using cross-correlation and real space averaging. Ultramicroscopy **6**: 7-18.

Derjaguin BV, Rabinovich YI, Churaev NV (1978) Direct measurement of molecular forces. Nature **272**: 313-318.

Dieter GE (1976) Mechanical Metallurgy, 2nd Ed. McGraw-Hill, New York, Ch. 7.

Doniach S (1979) A thermodynamic model for the monoclinic (ripple) phase of hydrated phospholipid bilayers. J Chem Phys **70**: 4587-4596.

Dubochet J, McDowall AW (1981) Vitrification of pure liquid water for electron microscopy. J Microsc **124**: RP3.

Dubochet J, Lepault J, Freeman R, Berrimena JA, Homo J-C (1982) Electron microscopy of frozen water and aqueous solutions. J Microsc **128**: 219-237.

Erlandsson R, Chiang S, Mate CM, McClelland GM (1988) Atomic force microscopy using optical interferometry. J Vac Sci Technol A **6**: 266-270.

Fetter RD, Costello MJ (1985) A procedure for obtaining complementary replicas of ultra-rapidly frozen sandwich samples. J Microsc **141**: 277-290.

Florin E-L, Moy VT, Gaub HE, (1994) Adhesion forces between individual ligand-receptor pairs. Science, **264**: 415-418.

Foster J, Frommer JE, Arnett PC (1988) Molecular manipulation using a tunneling microscope. Nature **331**: 324-326.

Foster JS, Frommer JE (1988) Imaging liquid crystals using a tunneling microscope. Nature, **333**: 542-545.

Frank J, Goldfarb W, Eisenberg D, Baker TS (1978) Reconstruction of glutamine synthetase using computer averaging. Ultramicroscopy **3**: 283-290.

Frisbie CD, Rozsnyai LF, Noy A, Wrighton MS, Lieber CM (1994) Functional group imaging by chemical force microscopy, Science **265**: 2071-2074.

Garnaes, Schwartz JDK, Viswanathan R, Zasadzinski JA (1993) Nanoscale defects in Langmuir-Blodgett films observed by atomic force microscopy. J Synthetic Metals, **55**: 3975-3800.

Gilkey JC, Staehelin LA (1986) Advances in ultrarapid freezing for the preservation of cellular ultrastructure. J Electron Microsc Tech **3**: 177-210.

Goldstein RE, Leibler S (1988) Model for lamellar phases of interacting lipid membranes. Phys Rev Lett **61**: 2213-2216.

Gould S, Marti O, Drake B, Hellemans L, Bracker CE, Hansma PK, Keder NL, Eddy MM, Stuckey GD (1988) Molecular resolution images of amino acid crystals with the atomic force microscope. Nature **332**: 332-334.

Gross H, Müller T, Wildhaber I, Winkler H (1985) High resolution metal replication, quantified by image processing of periodic test specimens. Ultramicroscopy **16**: 287-304.

Guckenberger R, Heim M, Gevc G, Knapp HF, Wiegrabe W, Hillebrand A (1994) Scanning tunneling microscopy of insulators and biological specimens based on lateral conductivity of ultrathin water films. Science **266**: 1538-1540.

Gulik-Krzywicki T, Costello MJ (1978) Use of low temperature x-ray diffraction to evaluate freezing methods used in freeze-fracture electron microscopy. J Microsc **112**: 102-113.

Hallmark VM, Chiang S, Rabolt JF, Swalen JD, Wilson RD (1987) Observation of atomic organization on Au (111) by STM. Phys Rev Lett **59**: 2879-2882.

Hansma PK, Tersoff J (1987) Scanning tunneling microscopy. J Appl Phys **61**: R1-R23.

Hansma PK, Elings VB, Marti O, Bracker CE (1988) Scanning tunneling microscopy and atomic force microscopy: some applications to biology and technology. Science **242**: 209-216.

Hansma PK, Cleveland JB, Radmacher M, Walters DA, Hillner PE, Bezanilla M, Fritz M, Vie D, Hansma HG, Prater CB, Massie J, Fukunaga L, Gurley J, Elings V (1994) Tapping mode AFM in liquids. Appl Phys Lett **64**: 1730-1740.

Hegerl R (1992) A brief survey of software packages for image processing in biological electron microscopy. Ultramicroscopy **46**: 417-423.

Henderson R, Baldwin JM, Downing KH, Lepault J, Zemlin F (1986) Structure of purple membrane from halobacterium halobium: recording, measurement and evaluation of electron micrographs at 3.5 Å resolution. Ultramicroscopy **19**: 147-178.

Henderson R, Baldwin JM, Ceska TA, Zemlin F, Beckman E, Downing KH (1990) Model for the structure of bacteriarhodopsin based on high-resolution electron cryo-microscopy. J Mol Biol **213**: 899-929.

Hentschel MP, Rustichelli F (1991) Structure of the ripple phase in hydrated phosphatidylcholine multimembranes. Phys Rev Lett **66**: 903-906.

Hörber JKH, Lang CA, Hänsch TW, Heckl WM, Möhwald H (1988) Scanning tunneling microscopy of lipid films and embedded molecules. Chem Phys Lett **145**: 151-158.

Hui SW, Viswanathan R, Zasadzinski JA, Israelachvili J (1995) The structure and stability of phospholipid bilayers by atomic force microscopy. Biophys J **68**: 171-178.

Ihn KJ, Zasadzinski JA, Pindak R, Slaney AJ, Goodby J (1992) Freeze-fracture TEM observations of the liquid crystal analog of the Abrikosov phase. Science **258**: 275-278.

Israelachvili JN (1991) Intermolecular and Surface Forces. Academic Press, London, pp. 177, 330-332.

Jahn W, Strey R (1988) Microstructure of microemulsions by freeze-fracture electron microscopy. J Phys Chem **92**: 2294-2301.

Janiak MJ, Small DM, Shipley GG (1979) Temperature and compositional dependence of the structure of hydrated dimyristoyl lecithin. J Biol Chem **254**: 6068-6078.

Katsaras J, Raghunathan VA (1995) Molecular chirality and the "ripple" phase of phosphatidylcholine multibilayers. Phys Rev Lett **74**: 2022-2025.

Keller SL, Bezrukov SM, Gruner SM, Tate MW, Vodyanoy I, Parsegian VA (1993) Biophys J **65**: 23-27.

Lang CA, Hörber JKH, Hänsch TW, Heckl WM, Möhwald H (1988) Scanning tunneling microscopy of Langmuir-Blodgett films on graphite. J Vac Sci Technol A **6**: 368-370.

Lindsay SM, Barris B (1988) Imaging deoxyribose nucleic acid molecules on a metal surface under water by scanning tunneling microscopy. J Vac Sci Technol A **6**: 544-547.

Longo ML, Bisagno AM, Zasadzinski JA, Bruni R, Waring AJ (1993) A function of lung surfactant protein SP-B. Science **261**: 453-456.

Lubensky TC, MacKintosh FC (1993) Theory of 'ripple' phases of lipid bilayers. Phys Rev Lett **71**: 1565-1568.

Luna EJ, McConnell HM (1977) The intermediate monoclinic phase of phosphatidylcholines. Biochim Biophys Acta **466**: 381-392.

Mamin HGE, Ganz, Abraham DW, Thompson RE, Clarke J (1986) Contamination mediated deformation of graphite by the STM. Phys Rev B **34**: 9015-9018.

Manne S, Cleveland JP, Gaub HE, Stucky GD, Hansma PK (1994) Direct visualization of surfactant hemimicelles by force microscopy of the electrical double layer. Langmuir **10**: 4409-4413.

McCullough WS, Scott HL (1990) Statistical-mechanical theory of the ripple phase of lipid bilayers. Phys Rev Lett **65**: 931-934.

Meyer E, Overney R, Brodbeck D, Howald L, Frommer J (1992) Friction and wear of Langmuir-Blodgett films observed by friction force microscopy. Phys Rev Lett **69**: 1777-1780.

Moor H, Mühlethaler K, Waldner H, Frey-Wyssling A (1961) A new freezing ultramicrotome. J Biophys Biochem Cytology **10**: 1-13.

Müller M, Meister N, Moor H (1980) Freezing in a propane jet and its applications in freeze fracture. Mikroskopie **36**: 129-140.

Niblack W (1986) An Introduction to Digital Image Processing. Prentice/Hall International, Englewood Cliffs, New Jersey, pp. 106-110.

Obcemea CH, Vidic B (1992) Bilipid layer molecular organization in $P_{b'}$ phase studied by scanning tunneling microscopy. Ultramicroscopy **42-44**: 1019-1024.

Orr FM, Scriven LE, Rivas AP (1975) Pendular rings between solids: meniscus properties and capillary force. J Fluid Mech **67**: 723-742.

Overney RM, Meyer E, Frommer J, Brodbeck JD, Howald L (1992) Friction measurements on phase separated thin films with a modified atomic force microscope. Nature **359**: 133-135.

Patrick DL, Beebe TP (1993) On the origin of large scale periodicities observed during scanning tunneling microscopy studies of highly ordered pyrolytic graphite. Surface Sci **297**: L119-L121.

Quate CF (1986) Vacuum tunneling: a new technique for microscopy. Physics Today **39**: 26-33.

Radmacher M, Fritz M, Hansma HG, Hansma PK (1994) Direct observation of enzyme activity with the atomic force microscope. Science **265**: 1577-1579.

Roark RJ (1965) Formulas for Stress and Strain. McGraw-Hill, New York, p. 225.

Roberts JD, Caserio MC (1977) Basic Principles of Organic Chemistry. WA Benjamin Inc, Menlo Park, CA, Ch 12.

Ruben GC (1989) Ultrathin vertically shadowed platinum-carbon replicas for imaging individual molecules in freeze-etched biological DNA and material science metal and plastic specimens. J Electron Microsc Tech **13**: 335-354.

Ruppel D, Sackman E (1983) On the defects in different phases of two-dimensional lipid bilayers. J Phys **44**: 1025-1034.

Sammon MJ, Zasadzinski JA, Kuzma MR (1987) Electron microscope observation of the nematic-smectic transition in a lyotropic liquid crystal. Phys Rev Lett **57**: 2834-2837.

Saxton WO, Baumeister W (1982) The correlation averaging of a regularly arranged bacterial cell envelope protein. J Microsc **127**: 127-138.

Schulz-DuBois EO, Rehberg I (1981) Structure function in lieu of correlation function. Applied Phys **24**: 323-329.

Schneider MB, Chan WK, Webb WW (1983) Fast diffusion along defects and corrugations in phospholipid $P_{b'}$ liquid crystals. Biophys J **43**: 157-165.

Schwartz DK, Viswanathan R, Garnaes J, Zasadzinski JA (1992a) Surface order and stability of Langmuir-Blodgett films. Science **257**: 508-511.

Schwartz DK, Viswanathan R, Zasadzinski JA (1992b) Reorganization and crystallite formation in Langmuir-Blodgett films. J Phys Chem **96**: 10444-10447.

Schwartz DK, Viswanathan R, Zasadzinski JA (1994) Head-tail competition and modulated structures in planar surfactant (Langmuir-Blodgett) films. J Chem Phys **101**: 7161-7168.

Scott HL, McCullough (1991) Theories of the modulated 'ripple' phase of lipid bilayers. Int J Modern Physics **5**: 2479-2497.

Singer SJ, Nicolson GL (1972) The fluid mosaic model of the structure of cell membranes. Science **175**: 720-731.

Sirota EB, Smith GS, Safinya CR, Plano RJ, Clark NA (1988) X-ray scattering studies of aligned, stacked surfactant membranes. Science **242**: 1406-1409.

Soethout LL, Vankempen H, Nelissen BJ, Gerritsen JW, Groenevald PPMC (1988) STM measurements on graphite using correlation averaging of the data. J Microsc **152**: 251-258.

Soler JM, Baro AM, Garcia N, Rohrer H (1986) Interatomic forces in STM: giant corrugations of graphite. Phys Rev Lett **57**: 444-447.

Sonnenfeld R, Hansma PK (1987) Atomic resolution microscopy in water. Science **232**: 211-213.

Spong JK, Mizes HA, LaComb LJ, Dovek MM Jr, Frommer JE, Meyer E (1989) Contrast mechanism for resolving organic molecules with tunnelling microscopy. Nature **338**: 137-139.

Steere RL (1957) Electron microscopy of structural detail in frozen biological specimens. J Biophys Biochem Cytol **3**: 45-59.

Stemmer A, Engel A, Aebi U, Hefti A (1989) Scanning tunneling and transmission electron microscopy on identical areas of biological specimens. Ultramicroscopy **30**: 263-280.

Talmon Y, Davis HT, Scriven LE (1981) Progressive freezing of composities analyzed by isotherm migration methods. AIChE J **927**: 928-937.

Tanaka T, Ishiwata S, Ishimoto C (1977) Critical behavior of density fluctuations in gels. Phys Rev Lett **38**: 771-774.

Tardieu A, Luzzati V, Reman FC (1972) Structure and polymorphism of the hydrocarbon chains of lipids: a study of lecithin-water phases. J Mol Biol **75**: 711-733.

Tersoff J, Lang ND (1990) Tip dependent corrugation of graphite in scanning tunneling microscopy. Phys Rev Lett **65**: 1132-1135.

Tippmann-Krayer P, Kenn RM, Mohwald H (1992) Thickness and temperature dependent structure of Cd-arachidate Langmuir-Blodgett films. Thin Solid Films **210**: 577-582.

Travaglini G, Rohrer H, Amrein M, Gross H (1987) Scanning tunneling microscopy on biological matter. Surf Sci **181**: 380-390.

Wack DC, Webb WW (1989) Synchrotron x-ray study of the modulated lamellar phase in the lecithin water system. Phys Rev A **40**: 2712-2730.

Wang ZH, Hartmann T, Baumeister W, Guckenberger R (1990) Thickness determination of biological samples with a z-calibrated scanning tunneling microscope. Proc Nat Acad Sci **87**: 9343-9347.

Weigrabe W, Nonnenmacher M, Guckenberger R, Wolter O (1991) Atomic force microscopy of a hydrated bacterial surface protein. J Microsc **163**: 79-84.

Weisenhorn AL, Maivald P, Butt H-J, Hansma PK (1992) Measuring adhesion, attraction and repulsion between surfaces in liquids with an atomic-force microscope. Phys Rev B **45**: 11226-11232.

Wepf R, Amrien M, Burkli U, Gross H (1991) Platinum/iridium/carbon: a high-resolution shadowing material for TEM, STM and SEM of biological macromolecular structures. J Microsc **163**: 51-64.

Woodward JT (1994) Scanning tunneling microscopy on freeze-fracture replicas of ripple phase of dimyristoylphosphatidylcholine. Doctoral Thesis, University of California, Santa Barbara, CA.

Woodward JT, Zasadzinski JA, Hansma PK (1991) Precision height measurements of freeze-fracture replicas using the scanning tunneling microscope. J Vac Sci Technol B **9**: 1231-1235.

Woodward JT, Zasadzinski JA (1994) Height amplifications of scanning tunneling microscopy images in air. Langmuir **10**: 1340-1344.

Woodward JT, Kono C, Madsen LL, Zasadzinski JA (1995) Inherent bias in correlation averaged images. J Microsc **178**: 86-92.

Woodward JT, Zasadzinski JA (1996) Amplitude, waveform and temperature dependence of bilayer ripples in the $P_{\beta'}$ phase. Phys Rev E **53**: R3044-R3047.

Yuan J-Y, Shao Z (1990) Simple model of image-formation by scanning tunneling microscopy of nonconducting materials Ultramicroscopy **34**: 223-236.

Yuan J-Y, Shao Z, Gao C (1991) Alternative method of imaging surface topologies of nonconducting bulk specimens by scanning tunneling microscopy. Phys Rev Lett. **67**: 863-866.

Zasadzinski JA (1988a) A new heat transfer model to predict cooling rates for rapid freezing fixation. J Microsc **150**: 137-149.

Zasadzinski JA (1988b) Effect of stereoconfiguration on ripple phases of dipalmitoylphosphatidylcholine. Biochim Biophys Acta **946**: 235-243.

Zasadzinski JA, Meyer RB (1986) Molecular imaging of tobacco mosaic virus lyotropic nematic phases. Phys Rev Lett **56**: 636-639.

Zasadzinski JA, Schneider MB (1987) Ripple wavelength, amplitude, and configuration in lyotropic liquid crystals as a function of effective head group size. J Phys **48**: 2001-2011.

Zasadzinski JA, Meiboom S, Sammon MJ, Berreman DW (1986) Freeze-fracture electron microscope observations of the blue phase III. Phys Rev Lett **57**: 364-367.

Zasadzinski JA, Chu A, Prud'homme RK (1987a) Transmission electron microscopy of gel network morphology: Relating network microstructure to mechanical properties. Macromolecules **19**: 2960-2964

Zasadzinski JA, Sweeney JB, Davis HT, Scriven LE (1987b) Finite element calculations of fluid menisci and thin-films in porous media. J Coll Int Sci **119**: 108-116.

Zasadzinski JA, Schneir J, Gurley J, Elings V, Hansma PK (1988) Scanning tunneling microscopy of replicas of biomembranes. Science **239**: 1014-1016.

Zasadzinski JAN, Bailey S (1989) Applications of freeze-fracture replication to problems in materials and colloid science. J Electron Microsc Tech **13**:309-334.

Zasadzinski JA, Viswanathan R, Madsen LL, Schwartz DK (1994a) Langmuir-Blodgett Films. Science **263**: 1726-1733.

Zasadzinski JA, Viswanathan R, Schwartz DK, Garnaes J, Madsen LL, Chiruvolu S, Woodward JT, Longo ML (1994b) Applications of AFM to structural characterization of organic thin films. Colloids and Surfaces A **93**: 305-333.

Discussion with Reviewers

H.J.K. Hörber: It is stated that Chiruvolu *et al.* (1994) determined the resolution of freeze-fracture electron microscopy, due to the film grains, was limited to about 2 nm. In this paper, the authors claim 1 nm lateral and 0.3 nm vertical resolution. How does this fit together?

P.M. Frederik: In the abstract the resolution of the STM replica technique is stated to be better than 1 nm, whereas in the summary a more precise statement is given; 1 nm lateral resolution and better than 0.5 nm vertical resolution. In the Introduction this is further refined and better than 0.5 nm turns out to be 0.3 nm. Please clarify.

Authors: On any individual image, the apparent resolution is limited by the granularity of the replicating film to about 2 nm laterally, and somewhat less than this vertically. However, for periodic surfaces, such as the ripples examined here, image analysis techniques such as correlation averaging, can be used to remove much of the effect of the replica grain. By averaging over many images in a consistent way, we have found that the average shape of the ripple can be determined much more precisely - a statistical analysis of the data shows that after fitting the ripple shape to a series of harmonics, the standard deviation of the amplitude and wavelength of the ripple reduce to $\pm$ 1 nm laterally and $\pm$ 0.3 nm vertically. Similar enhancements in lateral resolution are possible in TEM imaging as well by correlation averaging.

H.J.K. Hörber: If one tries to clean a metal surface that once was in contact with organic material, one notices that the only way to clean it completely is to remove some layers of the metal. Is this done in the cleaning procedure for replicas? If yes, how does this affect the resolution? If not, is it really possible to be sure that contaminations will not affect height measurements, which only on a homogeneous material will give exact results for STM height measurements.

Authors: Our cleaning procedures are designed to be quite gentle, and we have no indication that we are removing any of the metal of the replica; there is only about 1-2 nm of platinum to start with, and even removing a few layers would destroy the replicas. However, we do not need the surfaces extremely clean; by extremely we mean that all organic materials are removed from the metal. The STM treats adsorbed organic layers as just part of the tunneling gap between the metal tip and the metal sample; insulating contaminants do not appear to have any effect on the topography of the surface, as long as they are sufficiently thin that tunneling can occur. Chemically and physically, the metal replica itself is homogeneous at the resolution we require. We further insure that our height measurements on replicas are accurate by careful calibration to other replicas with known features - such as our Langmuir-Blodgett films (See Fig. 2). The precision and reproducibility of our measurements suggest that our height measurements are accurate.

H.J.K. Hörber: Is there a way to detect crystallization artifacts of the water, which really might occur, as the sample provides in most cases crystallization seeds?

Authors: Crystallization artifacts can occur in poorly prepared samples, and typically manifest themselves by segregation of liposomes to ice crystal grain boundaries. We checked for this with the TEM replicas that were prepared simultaneously with the STM replicas. This was one of the main benefits of the combined STM/-TEM study - the TEM quickly showed which samples were frozen properly, fractured properly, and replicated properly. The question of whether or not ice crystals exist at the nanometer scale is much harder to address. It is clear from direct TEM images of water frozen in a similar way that the water is amorphous (Dubochet and McDowall, 1981). The lipid bilayers have a significantly higher viscosity and lower diffusivity that water, hence, we do not expect any significant rearrangement of the lipid structure.

P.M. Frederik: A roof-tile shape of ripples in DMPC is described and analyzed. How to the authors accommodate areas with line defects in their averaging routine.

Authors: The first step of the correlation averaging

routine is to identify areas that have a strong correlation to the "average" feature. Areas of imperfection, either due to defects in the ripples or to defects in replication, have low correlation to the average feature and are eliminated from further averaging by using a cutoff value for the correlation.

Scanning Microscopy Supplement 10, 1996 (pages 149-164) 0892-953X/96$5.00+.25
Scanning Microscopy International, Chicago (AMF O'Hare), IL 60666 USA

ACCESSING NUCLEAR STRUCTURE FOR FIELD EMISSION, IN LENS, SCANNING ELECTRON MICROSCOPY (FEISEM)

T.D. Allen*, G.R. Bennion, S.A. Rutherford, S. Reipert, A. Ramalho, E. Kiseleva and M.W. Goldberg

CRC Department of Structural Cell Biology, Paterson Institute for Cancer Research, Christie Hospital NHS Trust, Wilmslow Road, Manchester M20 9BX, UK.

(Received for publication September 25, 1995 and in revised form April 18, 1996)

Abstract

Scanning electron microscopy (SEM) has had a shorter time course in biology than conventional transmission electron microscopy (TEM) but has nevertheless produced a wealth of images that have significantly complemented our perception of biological structure and function from TEM information. By its nature, SEM is a surface imaging technology, and its impact at the subcellular level has been restricted by the considerably reduced resolution in conventional SEM in comparison to TEM. This restriction has been removed by the recent advent of high-brightness sources used in lensfield emission instruments (FEISEM) which have produced resolution of around 1 nanometre, which is not usually a limiting figure for biological material.

This communication reviews our findings in the use of FEISEM in the imaging of nuclear surfaces, then associated structures, such as nuclear pore complexes, and the relationships of these structures with cytoplasmic and nucleoplasmic elements. High resolution SEM allows the structurally orientated cell biologist to visualise, directly and in three dimensions, subcellular structure and its modulation with a view to understanding its functional significance. Clearly, intracellular surfaces require separation from surrounding structural elements *in vivo* to allow surface imaging, and we review a combination of biochemical and mechanical isolation methods for nuclear surfaces.

Key Words: nucleus, nuclear envelope, nuclear pore-complex, nucleocytoplasmic transport, nucleoplasmin, field emission in lens scanning electron microscopy, chromium coating, colloidal gold, backscatter electron imaging.

*Addres for correspondence:
TD Allen,
CRC Department of Structural Cell Biology, Paterson Institute for Cancer Research, Christie Hospital NHS Trust, Wilmslow Road, Manchester M20 9BX, UK.
Phone: 0161 446 3116
Fax: 0161 446 3109
e-mail ulttda @ picr.cr.man.ac.uk.

Introduction

Current studies on the nucleus by a variety of novel microscopic techniques are producing something of a renaissance in our understanding of nuclear structure and function. The organisation of peripheral structure, namely the nuclear envelope and nuclear pore complexes [3, 7, 15-19, 14, 25, 29-31, 36, 41, 42]; the internal organisation with respect to subjects such as laminas [1, 14, 18, 19, 20, 21, 24], coiled bodies [11], transcriptional activity [38, 39] and chromosomes [32,38] have all shown striking progress over the last few years. Nuclear transport [2, 10, 11, 12, 13, 22, 23, 26, 29] is under active investigation, both biochemically and structurally: *in vitro* systems have produced significant progress in accessing the dynamics of mitosis [18, 40], and the control, mechanics and role of the nucleus in programmed cell death (apoptosis) is also being addressed [35]. This communication to reviews our own recent attempts to visualise, at macromolecular resolution, directly and in 3D, the organisation of the nuclear envelope and associated structures [4-8, 17, 18, 34] using field emission, in lens scanning electron microscopy (FEISEM) supported by conventional transmission electron microscopy (CTEM).

FEISEM, well established with electron microscopists, has only relatively recently begun to realise its potential in cell biology, perhaps as a result of relatively limited accessibility to investigators in this field. Field emission instruments with specification resolutions of around 1 nanometre, provide essentially the same effective resolution in biological material which has been available in transmission electron microscopy (TEM) for the past 25 years. Whereas conventional TEM (CTEM) tends to be restricted to relatively thin (30-60 nm) sections, or isolated molecules spread thinly on a support film, (with the exception of high voltage electron microscopy (HVEM)) the advantage of field emission in scanning instruments is that specimens of infinite depth can be surface imaged to much the same effective resolution as TEM. Two criteria must be fulfilled to achieve TEM resolution in a scanning electron microscope (SEM) - firstly a high brightness source, allowing sufficient signal to be collected by rastering the specimen with a very small spot size (0.7 to 1.0 nm diameter), secondly efficient collection of that signal to provide relatively noise free images at 200 to 300 thousand times magnification. The high brightness criterion has been met by the production of field emission (FE) sources, which fall into 2 categories, namely cold FE, or thermally assisted ('Schottky') sources, which vary in their properties and requirements. Cold field emission, as its name suggests, works at low temperatures, but to overcome the 'work function' required to produce field emission in these conditions from a pointed tungsten tip, requires 10^{-8} pascals operating pressure in the gun region, and regular 'flashing' of the tip to disperse contamination on the tip surface, which significantly reduces emission. Cold field emission is inherently unstable, but this has been overcome as an operating difficulty by a sophisticated feedback loop to the photomultiplier.

Thermally assisted, or 'Schottky' field emission maintains the tip at around 1800°C, repelling any potential contamination, and obviating the requirement for flashing. In combination with doping the tip with Zirconium oxide, the higher temperature allows the work function for field emission to be overcome at 2 orders of magnitude lower vacuum (10^{-6} pascal) in the gun area. Schottky emission is stable, and is consequently the FE tip of choice for beam writing and microanalysis. In terms of absolute resolution, the virtual source size at the point of emission from cold FE sources is smaller than the Schottky, but this seems to produce very little difference in performance by the time the beam hits the specimen. In general, costs of production of Schottky systems are cheaper than cold FE systems, and Schottky tips may well exceed 10,000 hrs of consistent emission.

The second requirement for achievement of 1 nm resolution has been the positioning of the specimen in the electron optical column in the field of the final lens (like conventional TEM), where spherical and chromatic aberrations are at their minimum, and also the secondary electron emission from the specimen surface is effectively 'trapped' in the magnetic field of the lens, allowing a very high efficiency of collection from the 'in lens' detector. More recent FE instruments however have come close to the same resolution using close (3 mm) working distances in a conventional 'pinhole' final lens arrangement.

Both cold FE and Schottky sources allow good resolution (around 4 nm) at very low (1.0-1.5 kV) accelerating voltages, allowing surface imaging with minimal penetration and beam damage. However, this is still short by a considerable amount of the 1 nm resolution achievable at around 30 kV, which has been aided by the current developments in thin coatings (1.5-3 nm) with refractory metals (eg. chromium, tungsten and tantalum). Thus several types of FE instruments have a performance which in general cannot be considered limiting for biological specimens, leading to the continual quest for optimal preservation procedures, but with enormous potential for direct, macromolecular three-dimensional (3D) imaging of the surface of bulk specimens.

Observation of the surface of an internal membrane bound structure such as the nucleus, clearly requires direct access by the electron beam. Once this has been achieved, the ability to view structure on that surface is not limited by section thickness, nor uncertainty of plane of section and limited contrast dependent on electron dense staining (as in TEM), and a 'panorama' of fresh surface detail becomes apparent. This was graphically demonstrated by Hans Ris [36, 37] in his description of the asymmetry of the nuclear pore complex on either face of the nuclear envelope, for 30 yrs assumed identical on the available TEM evidence. This evidence also included the excellent freeze fracture studies which established much of what we know of nuclear pore complex numbers, distribution and dynamics, but because of the nature of a frozen fracture to split pore complexes at the planes of the nuclear membrane, could not observe peripheral structures. As Ris [36] pointed out in his original article, the nuclear pore complex is particularly unsuited to either thin section or whole mount observation. Due to its size (110 nm in diameter, 80-100 nm high) bulk (estimated mol.wt. 120 million Da) any thin section (30-60 nm) will only present a portion of the structure, and the contrast imparted by negative staining of whole mounts is also limited, as 80 nm depth of negative stain would be impenetrable by TEM. More recently however, we must allow for the 'thin ice' approach coupled to powerful computerised image reconstruction from minimal contrast, but this approach still produces an 'averaged' structure, which in the case of nuclear pore complexes (NPCs) involved in dynamic transport processes, may 'average out' crucial structural variations related to different functional states.

Accessing the Nucleus for Surface Imaging

In choosing to partition its genetic material (unlike bacteria) the eukaryotic cell (presumably unwittingly) created many difficulties by the separation of the nucleus and cytoplasm, mainly in the provision of efficient communication and transport between them. This giant evolutionary leap also led to the frustration of future investigators finding the surface of the nucleus in which they were interested to be firmly embedded in cytoplasm. The approaches taken to get round this problem fall into two categories, nuclear isolation and nuclear exposure.

Nuclear isolation of large nuclei

This involves wholesale removal of the nucleus from the rest of the cell, a process clearly involving disruption of the cell membrane, dispersion of the cytoplasm, and the exposure of a 'clean' nuclear surface. This is all clearly less than physiologically optimal, and as the nucleus may well have peripheral cytoplasmic attachments, will involve a separation of any links close to its surface. Care should also be taken with the milieu into which the nucleus is isolated. Using amphibian material, isolation buffers (based on unbuffered 5:1 mixtures of 0.1 M KCl and 0.1 M NaCl, pH 6.0-7.5) will be varied with different and amphibian species. With insect polytene nuclei, isolation is performed in insect Ringers solution (7.5 g NaCl, 0.35 g KCl and 0.21 g $CaCl_2$ [27]. Clearly, once elements of the cell are isolated, they should be maintained in as near physiological conditions as possible. [5, 27]. Manual removal of part of a cell is made considerably easier the bigger the cell is. For this reason, the amphibian oocyte (usually *Xenopus* which is easily isolated from 'parental' tissue) has become the cell of choice for much nuclear investigation. Viewed with a bench stereomicroscope and maintained in isolation buffer [5, 27] the oocyte membrane is slit with a fine needle, the yolk squeezed through the slit, and the nucleus (0.2 mm in diameter - within the resolution of the unaided eye) is recognised as a 'pearly' sphere during its exit, and can then be washed using pasteur pipettes and its contents (lampbrush chromosomes) dispersed, and the 'empty' nuclear envelope utilised for observation. In our case, the envelope is deliberately ripped as it is spread onto the surface of a 5 mm square silicon chip, to which it adheres naturally in much the same way as it would to a glass cover slip. The advantage of a silicon substratum (although opaque to transmitted light) is that because it is a good electrical conductor (unlike glass) it helps to facilitate imaging at 30 kV acceleration voltage in the FEISEM (Fig. 1). Once attached and spread - allowing observation of both nucleoplasmic and cytoplasmic sides of the nuclear envelope (NE), the preparation can be extensively washed (20-60 mins in isolation buffer) then fixed, or alternatively structurally modulated by proteolytic or detergent extraction, or labelled by immunocytochemistry prior to fixation. Fixation of a relatively fragmentary piece of tissue is usually 10 mins in 2.0% glutaraldehyde (possibly with added tannic acid (0.2%), 10 mins 0.1% OsO_4 (both in isolation buffer [4, 15]) followed by 1% aqueous uranyl acetate for 10 mins. This is followed by dehydration through an ethanol series, and critical point drying (CPD) from high purity CO_2 using 'Arklone' (Arcton 112, ICI Chemicals, Runcorn, UK) as transitional fluid. 100% ethanol and arklone are stored over molecular sieve to facilitate absolute dehydration, and the 99.999% pure CO_2 ($H_2O < 5$ ppm) passed through a molecular filter (Tousimis, Rockville, MD, USA). Critical point drying, was performed in a Balzer's (Liechtenstein) CPD unit, and the chips were transferred directly from the CPD chamber to the coating unit (Edwards Auto 306, Crawley, U.K.) where they were pumped to 5x 10^{-8} using a cryopumped system. Once this vacuum was achieved (40-50 mins) the system was backfilled with high purity argon to a pressure of 75 x 10^{-3} mbar, and sputter coated with either chromium or tantalum to a thickness of 1.5 to 2.0 nm as measured by thin film monitoring at a deposition rate of around 0.1 nm/sec. Oxidation of the target was removed by 30 secs pre-sputtering with a shutter covering the specimens. During coating the specimens were rotated at 150 rpm at an angle of 45° to the target. Wherever possible, specimens were examined as soon as possible after coating. Despite the fact that chromium and tantalum coatings can provide good signal for a few days to a few weeks, there is no doubt that optimum imaging occurs soon after coating. The post-fixation stages of this protocol can be considered constant for all subsequent preparations, unless stated.

Manual isolation of nuclei from other oocytes has also been applied to a variety of other species. Avian oocytes from both Pigeon and Chick have generated successful NE preps, as have polytene nuclei from insect salivary glands such as *Chironomus* [7], establishing the nuclear pore basket as an evolutionarily conserved structure. As nuclei get progressively smaller (eg. *Chironomus* polytene nuclei - approx. 50 μm in diameter) the manual isolation of nuclear envelopes demands a large increase in skill (and patience!)

Isolation of small nuclei

Mammalian nuclei from tissue culture cells such as CHO (Chinese Hamster Ovary) rarely exceed 10 μm in diameter, and although they are routinely microinjected and can be considered manipulable to some degree, they really fall below the lower limit for manual isolation. Because of their small size, working 'one at a time' is clearly precluded, and consequently it is best to start with large numbers (millions) to achieve reasonable numbers on each preparation, ideally around 10,000 per chip. Thus the approach has a biochemical, as well as a mechanical element. A pellet of suspension grown tissue culture cells (eg. K562) is resuspended in a hypotonic buffer designed to permeabilise the cell membrane and swell the cytoplasm, with the actual 'release' of nuclei achieved by 2-3 strokes of the suspension between the sides of a test tube and plunger (Dounce Homogeniser)

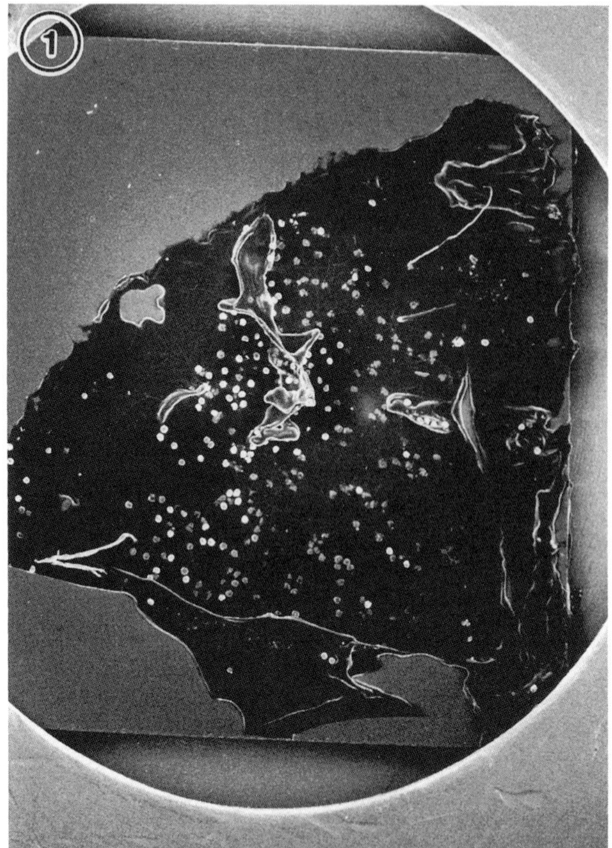

Figure 1: Low magnification image (x10 in instrument) (30 kV accelerating voltage) of area of 5 mm^2 silicon chip used for mounting of biological specimens. This particular chip has been used in dry fracture and shows adhesive and fractured nuclei. (Bar=2 nm).

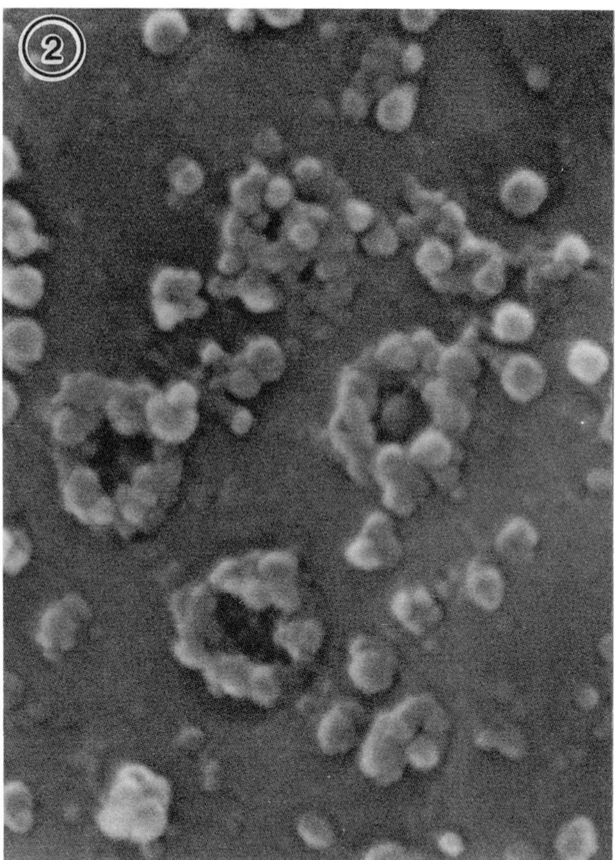

Figure 2: Nucleus from human cell line (K562) after detergent free isolation and purification on an isotonically balanced Percoll gradient showing the surface of the NE and NPC structure. (Bar=100 nm).

----

creating a shearing force which disrupts the cell membrane and cytoplasm, releasing the nucleus into suspension [34]. At this stage the nuclear surface retains an extensive coating of cytoplasm and endoplasmic reticulum which is then removed by spinning through a sucrose gradient, 'cleaning' the nuclear surface. In our hands the osmotic consequences of exposure to sucrose for unfixed nuclei were not acceptable, and consequently percoll was found to be a considerable improvement [30]. Nuclei were isolated from the correct part of the gradient, resuspended in a physiological buffer, [34] and spun onto poly-l-lysine coated silicon chips, fixed and dehydrated, critical point dried, and coated as above, but in this case with 5-6 nm Cr, as charging was more apparent at 30 kV on spherical nuclei than on spread NEs on silicon chips. In these preparations good structural preservation of NE and NPCs was achieved (Fig. 2), and surface visualisation of apoptotic clustering of NPCs was observed [34, 35].

Exposure of the nuclear surface *in situ*

This methodology was evolved to retain the structural polarity of nucleus and cytoplasm, for tissue culture cells grown attached to the substratum rather than in suspension culture (above). As with suspension cells however, the membrane must be permeabilised and cytoplasm 'diluted' before the nucleus can be exposed, in this situation after fixation, in contrast to the previous method. This process involves a brief 'stabilisation' fixation of short duration and low strength fixative, followed by extraction with detergent. These processes are related, the shorter the fixation consistent with NE stabilisation, the shorter the subsequent detergent extraction is required. We have found this to be variable for different cell lines, but in the case of *Xenopus* fibroblasts grown in tissue culture, prefixation with 1.0% paraformaldehyde and 0.01% glutaraldehyde in phosphate buffered saline (PBS) for 1 minute required 1 - 3 hrs extraction with 0.5% Triton X-100 (also in PBS). For mammalian tissue culture cells extraction needed to

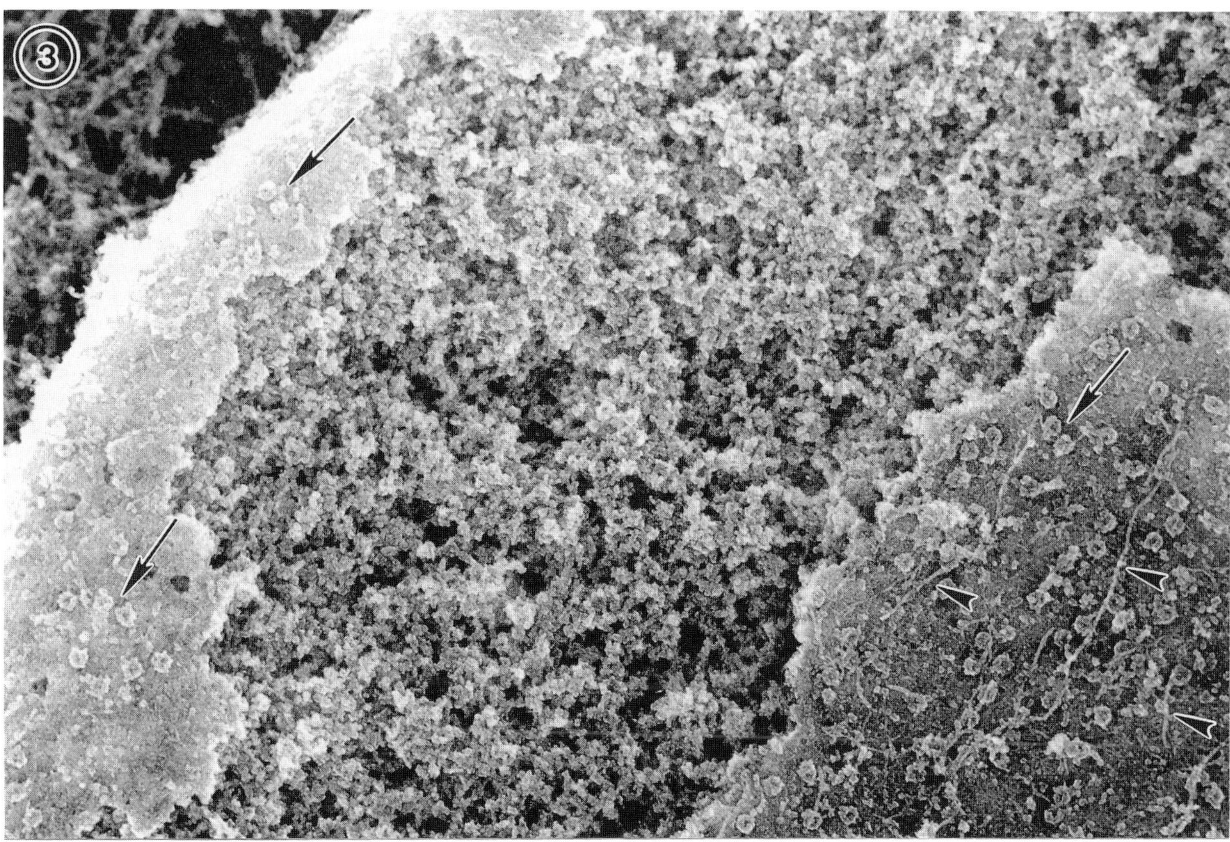

Figure 3: Surface view of *Xenopus* tissue culture cell nucleus after dry fracture subsequent to the protocol described in text. Two areas of nuclear envelope, either side of exposed chromatin, (centre) display nuclear pore complexes, (arrowed) and remnant cytoskeletal filaments (arrowheads). Differential packing of the chromatin filaments is also apparent. (Bar=500 nm).

---

be longer, 4 - 6 hrs (but this can be checked by observing decreasing birefringence using phase contrast microscopy) the danger being that an extended extraction will cause the cells to detach from the substratum.

After prefixation and extraction, the extracted cells are re-fixed normally in 2% glutaraldehyde followed by OsO_4 (sometimes with one OsO_4-thiocarbahydrazide-OsO_4 rotation) for 20 - 30 minutes each. This is followed by conventional dehydration and CPD. At this stage the cover slip (or Si chip) on which the cells were grown is touched to double-sided adhesive (Agar 'Sticky Tabs', Agar, Stansted, UK) and separated without 'sliding' the surfaces against each other. Both the original cells and the chip with adhesive should be retained as both carry fractured surfaces. These are coated normally (1-2 nm chromium) and observed by FEISEM. In most cases, despite 'firm' contact with the adhesive, there are only scattered patches of cells which have fractured, but these are easily found. Separation through the cell can occur at all planes, but nuclear surfaces, displaying a similar quality of preservation to isolated unfixed nuclei [14, 19] are regularly observed, often with adjacent adherent cytoskeletal elements, possibly indicative of NPC-cytoskeleton attachment. Although regular exposure of the nuclear envelope surface occurs with this type of fracture, the planes of fracture are not limited to following membranous 'planes of weakness' as occurs in frozen fracture [28]. In the same cell, a whole series of step-like fractures across upper nuclear surface, through several levels of nuclear contents, through the underside nuclear membrane and into the cytoskeletal remnants of the cytoplasm is a common event (Fig. 3). This allows 3D imaging of chromatin packaging '*in situ*' and does present the relatively 'loose' and tight packaging in surface imaging of fibre organisation that would be envisaged from the electron lucent and electron dense regions of euchromatin and heterochromatin respectively in TEM thin sections. Nucleoli appear to be too dense for fractures to occur through their dense cored regions, so that the fractures in nucleolar areas tend to leave them as a spherical 'outcrop' also indicating some sort of interfa-

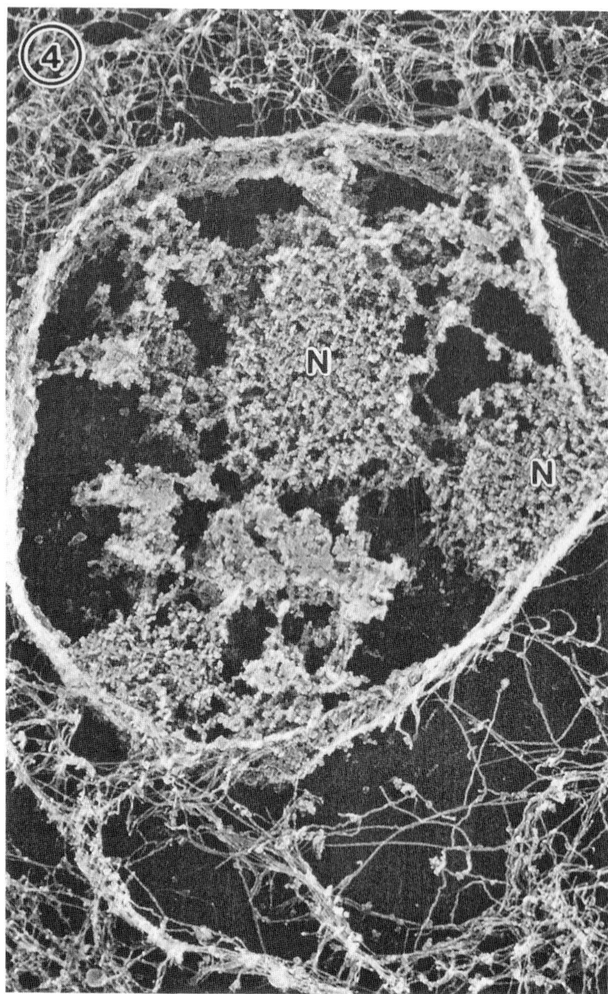

Figure 4: Deresined section of HeLa cell after detergent extraction, and extraction of nuclear DNA showing the remaining nucleoskeletal structure, which also has prominent nucleoli (N). (Bar=2.0 μm).

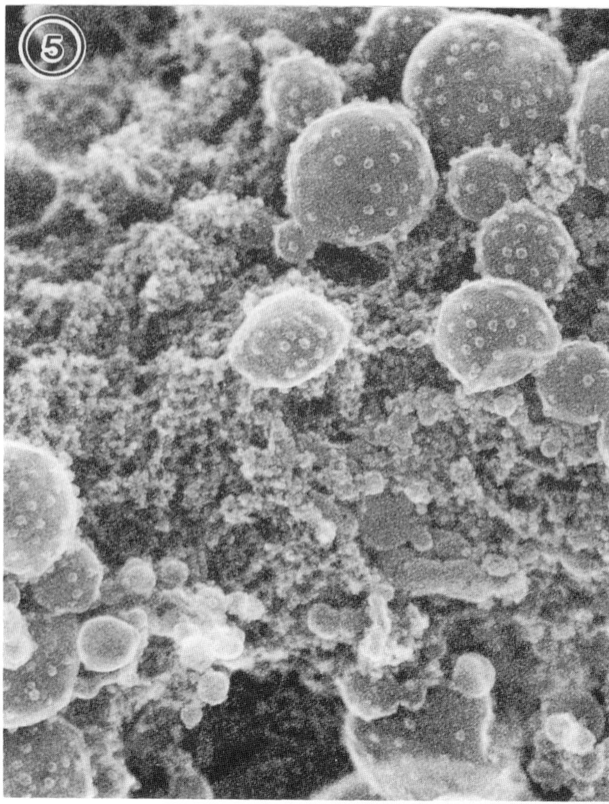

Figure 5: Early stages of *in vitro* nuclear formation, showing *Xenopus* demembranated sperm head surface, with binding of numerous vesicles from the cytoplasmic extract to the surface of the sperm chromatin. (Bar=100 nm).

cial separation between the nucleolar surface and surrounding nucleoplasm.

Exposure of 3D Nuclear Structure in Deresined Sections

This technology allows the use of resin embedded material to generate specimens for 3D imaging, by accessing the area of interest by exposure at the face of a cut section, and removal of resin embedding exposing 3D detail within the depth of the specimens, which are usually cut in the region of 200-500 μm thick. To observe 'whole' fixed and embedded material in this way is usually limited, as resin removal produces little surface topography, on account of the overall fixation of the cell contents. If however, the specimens have been extracted, removing or retaining specific structures, then they may well be imaged to advantage to complement the transmitted image produced in the TEM [9]. In the case of nucleoskeletal preparations for example, the DNA and associated proteins are removed from the nucleus using endonucleases to cleave the DNA, followed by removal of the fragments by electroelution, starting from small numbers of cells encapsulated in agar beads [21] which protect their biochemical functions during cytoplasmic and nuclear extraction. Structures remaining in these preparations are the cytoplasmic cytoskeletal filaments, nuclear envelope/pore complex residues, and the nucleoskeleton itself, organised as a network of filaments of different sizes [23], clearly attached to the peripheral remnants of the nuclear envelope at one end, and often attached to retained nuclear structures such as the nucleolus, at the other (Fig. 4). FEISEM imaging of these preps gives some idea of the texture of the nucleolar e-dense regions too electron opaque for TEM, and allows individual fibres to be followed over relatively long distances within the thickness of the deresined section. Backscattered electron imaging of colloidal gold labelling to areas of DNA replication also has advantages in visualising gold colloid in sites of high electron density that would not be visualised in transmitted electron images.

In Vitro Nuclear Formation

This cell free system obviates the need for separation of the nucleus from its surrounding cytoplasm, as it relies on the properties of amphibian or mammalian cytoplasmic extracts to reform functional nuclear structure *in vitro*. Briefly, demembranated sperm heads from *Xenopus* are incubated in cytoplasmic extracts from *Xenopus* eggs, in the presence of an adenosine triphosphate (ATP) regenerating system. Nuclear formation takes place at room temperature over 1 - 2 hrs [40]. Initially the sperm chromatin is decondensed at the surface, and binds numerous vesicles present in the extract, which flatten, fuse and form a new, complete nuclear envelope over a period of around 1 hr [18] (Fig. 5). At the same time nuclear pore complexes are formed in the NE, and once NE formation is complete, exact DNA replication takes place. This is dependent upon, and controlled by the nuclear envelope, which ends replication at the exact duplication of the DNA. Experimental permeabilisation of the nuclear envelope at this stage will result in a further blip of DNA replication. In some extracts, replication is followed by chromatin condensation akin to the early stages of mitosis. The breakdown of completed nuclei can however be routinely stimulated by addition of recombinant cyclin B protein to interphase extracts to activate the universal mitotic kinase p34cdc2 which generates an *in vitro* mitotic breakdown of pre-formed nuclei. Clearly, this is an extremely attractive system for surface imaging of the interactions between forming NE and chromatin stages in NPC formation, and the processes occurring at mitotic breakdown. All that is required prior to fixation is a gentle resuspension of the extract, washing away excess material at the nuclear surface, spinning down onto Si chips, and fixation *in situ*, allowing comparative observations of hundreds of nuclei in each of a series of time course fixations separated if necessary by no more than 60 secs. One further advantage of the system is the scope for fractionation of the components of the extracts, and reformation in the absence of specific components by depleting the extract using antibodies linked to magnetic beads [20]. FEISEM allows 3D imaging of the interactions between all components and sequential monitoring of assembly, and its intermediate stages, as well as the stages of *in vitro* mitotic breakdown [18]. Addition of specific inhibitors such as GTPγS (inhibiting fusion of membranes) or Brefeldin A (prevents coating of membranes) are other easily performed experimental modulations. Once attached and fixed to the Si chip, the nuclei are accessible to post CPD dry fracturing protocols to expose internal structural organisation.

Further Access to Nuclear Envelope Structure

Once isolated and attached to an adhesive substratum, the amphibian (or insect) nuclear envelope is available for further molecular dissection, which we have pursued in 3 ways, proteolytic or detergent extraction (or a combination of both) and various mechanical manipulations which also expose fresh structure [17, 19].

Proteolysis

Isolated NEs from *Xenopus* or *Triturus* were incubated for 5 mins in concentrations of 1, 10 or 100 μg/ml trypsin (Worthington Biochem Corp, Freehold NJ, USA) in 5-1 buffer at room temperature. Membrane structure of the NE was unaffected, but the proteins of the NPC were dismantled in a sequential and progressive manner. With 50 million pores on each NE, it is a relatively simple matter to achieve suitably representative morphology at different concentrations. The cytoplasmic face of the NPC showed sequential removal of cytoplasmic granules, leading to exposure of the surface of the cytoplasmic coaxial ring, followed by breakdown of the ring itself, starting from a single point and proceeding circumferentially from that point, leading to 'horseshoe' and then C shaped remnants (Fig. 6). On no occasion have we observed more than a single break in the cytoplasmic coaxial ring, indicating the existence of an initial trypsin sensitive site followed by successive removal of building blocks from that point circumferentially, with the whole disassembly possibly indicative of the 'physiological' breakdown stages in normal mitotic disassembly of NPCs. At the same time, internal elements of the NPC become successively exposed, and the overall breakdown continues until only NPC sized holes are visible in a largely 'depopulated' cytoplasmic face of the NE. On the nucleoplasmic side, similar changes are mirrored in the breakdown of NPC baskets and the nucleoplasmic coaxial ring, although where remnants of the nuclear envelope lattice are retained, it appears to be more trypsin resistant than the NPCs [19].

Detergent extraction

Incubation of unfixed isolated NEs for 5 minutes in 0.5% Triton X 100 (Sigma Chemicals, St. Louis, USA) in 5 : 1 buffer produces a 'converse' of proteolytic incubation, in that the NPCs are unaffected, but the membranes of the NE are removed, exposing the underlying filamentous network of the nuclear laminas.

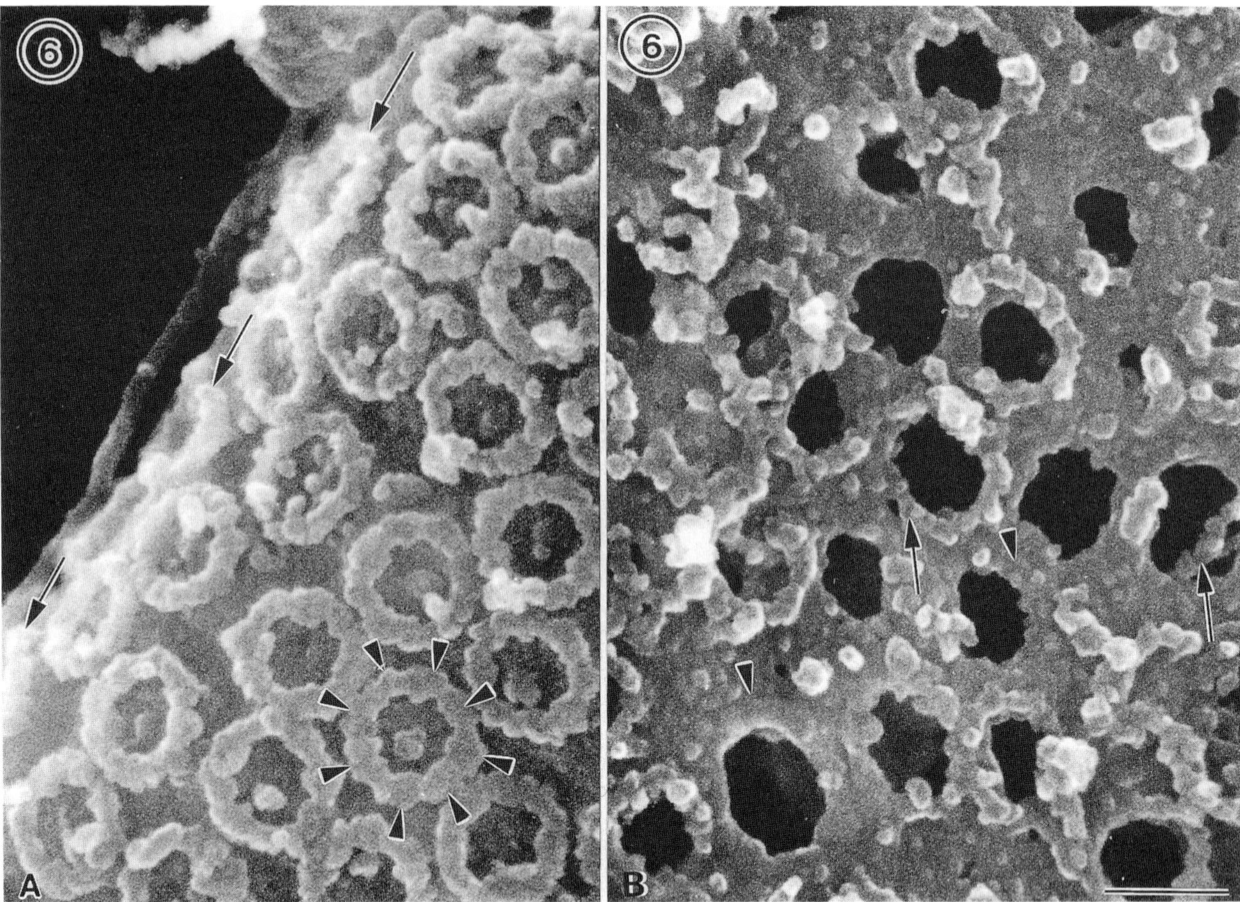

Figure 6: Cytoplasmic view of individual nuclear pore complexes after incubation with increasing concentrations of trypsin (a=10 μg/ml, b=100 μg/ml) for 5 mins. The cytoplasmic ring clearly exposed by removal of surface granules is a, shows a regular 8 sub-unit structure (arrowheads) and the side elevation of the surface profile is also apparent (arrows). In B, the coaxial ring is largely removed, with large holes appearing in the centre of the disassembling pore complexes. (Bar=100 nm).

This is apparent from both cytoplasmic and nucleoplasmic sides of the NE, although as around two-thirds of NPC structure sits 'above' the lamin network, less details of lamin organisation and interaction with NPC structure is apparent from the cytoplasmic side. Occasionally, probably due to mechanical disruption, NPCs become partially detached and are seen in 'side elevation' lying on the surface of the lamin network. Viewed from the nucleoplasmic side, the interaction of the NPCs with the lamin network is clearly visualised. Although many individual lamin filaments have become detached and disrupted, the remainder appear to insert into the base of the nucleoplasmic coaxial ring, often in intermediate positions between the attachments of the fibres of the NPC basket structure (Fig. 7).

Combined proteolysis and detergent extraction

Use of trypsin and detergent in combination allows further access to the 'core' structures of the NPC, as they may be 'hidden' by a combination of both surface proteinaceous material and NE membranes. Visualisation of NPCs after both brief detergent and trypsinisation has clearly indicated the radial arm structure at the periphery of the NPC core, exposed by removal of both peripheral proteins (the coaxial rings) and the NE membranes, particularly the pore membrane, [19].

Fracturing techniques for nuclear envelopes

'Non-biochemical' access to internal NPC structure has been achieved by two simple, apparently crude, but undoubtedly effective ways of separating subunit NPC structure. If freshly isolated unfixed nuclear envelopes are 'rolled' over the surface of the Si chip during isolation, the NPCs separate in the plane of the membrane, leaving their cytoplasmic coaxial rings attached to the surface of the chip, and exposing underlying struc-

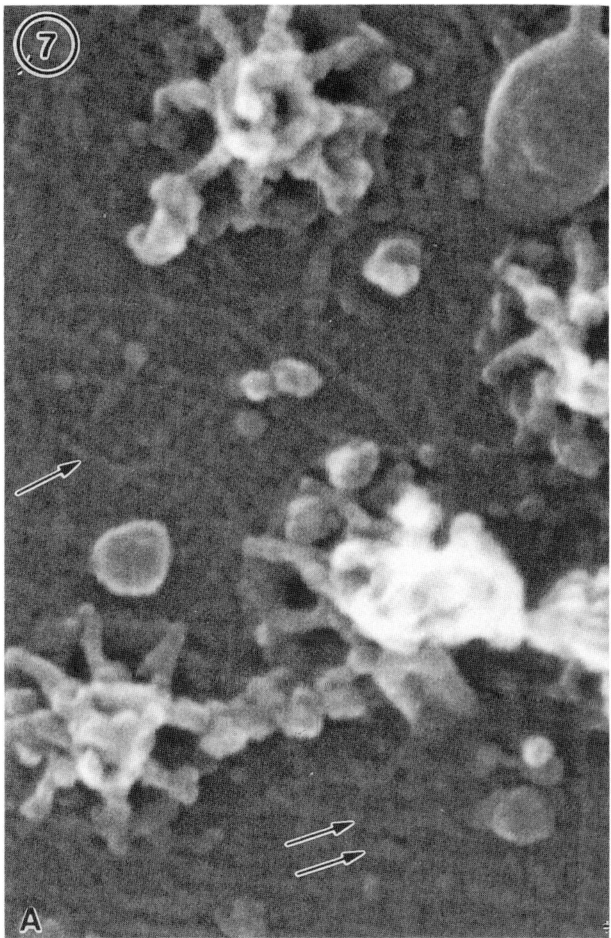
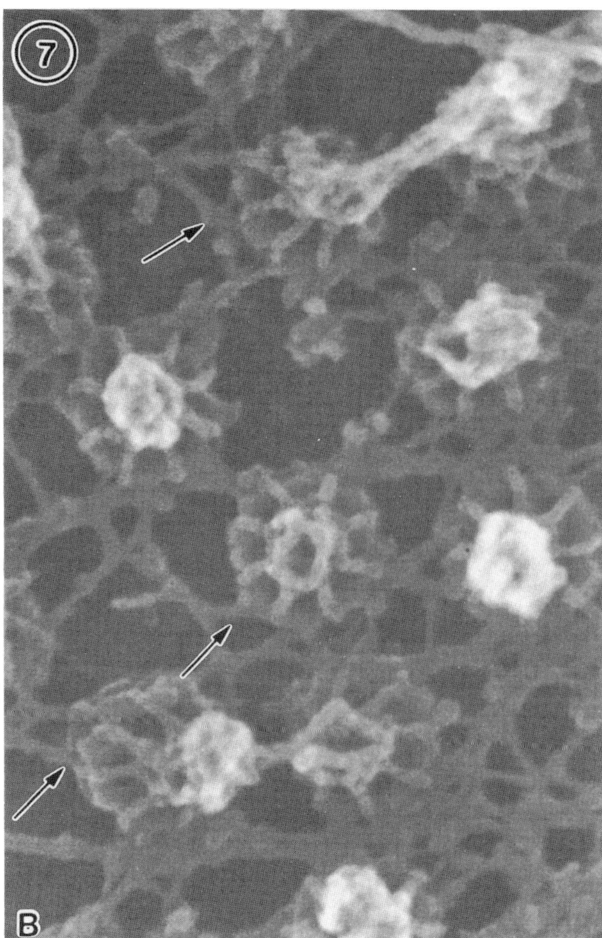

Figure 7: Unextracted (7A) and Extracted (7B) views of the nucleoplasmic face of *Xenopus* oocyte nuclear envelope. In 7A, lamin filaments can be seen in relief (arrowed) at the surface of the inner nuclear envelope membrane. In 7B, after Triton 100 X removal of this membrane, the lamin filaments are apparent, attached to the nucleoplasmic coaxial ring of the NPCs. Mechanical stress has resulted in some disorganisation of the lamin organisation overall, which appears to have an orthogonal arrangement in undisturbed NE membranes. (Bar = 100 nm).

ture on the intact NE (Fig. 8). Scraping of the fixed, dehydrated, critical point dried NE with a fine needle, or touching the NE to an adhesive surface such as double-sided sellotape will cause a similar separation, exposing the same underlying structure, namely the star ring, which is remarkably consistent in appearance whether exposed pre- or post-preservation (Fig. 9). Although the 'incurved' nature of isolated envelopes makes rolling of the nucleoplasmic side of the NE difficult (not achieved so far), scraping or adhesive has successfully removed NPC baskets, exposing underlying structure. In this case the NPC baskets have been distorted or flattened, allowing internal visualisation of internal structure without removal of the baskets themselves, the advantage here being that the internal structures exposed cannot be confused with remnants of disrupted baskets.

The Visualisation of Labelled Molecules by FEISEM

As well as exposing structural details of the NE and NPC, we have also attempted to characterise these observations by using antibodies or lectins to identify specific nucleoporins (nuclear pore proteins). Although there are probably 100 different proteins contributing to the NPC structure, only a few have been characterised *in situ* using immuno-TEM, and this has often been in cryosections, which further inhibit exact localisation to specific pore substructures due to their relatively limited contrast [30]. Whereas vertical sections through the NPC show profiles, and FEISEM could be considered limited to surface imaging, FEISEM preparations will routinely expose enormous numbers of both cytoplasmic and nucleoplasmic faces of the NPC, as well as indicating

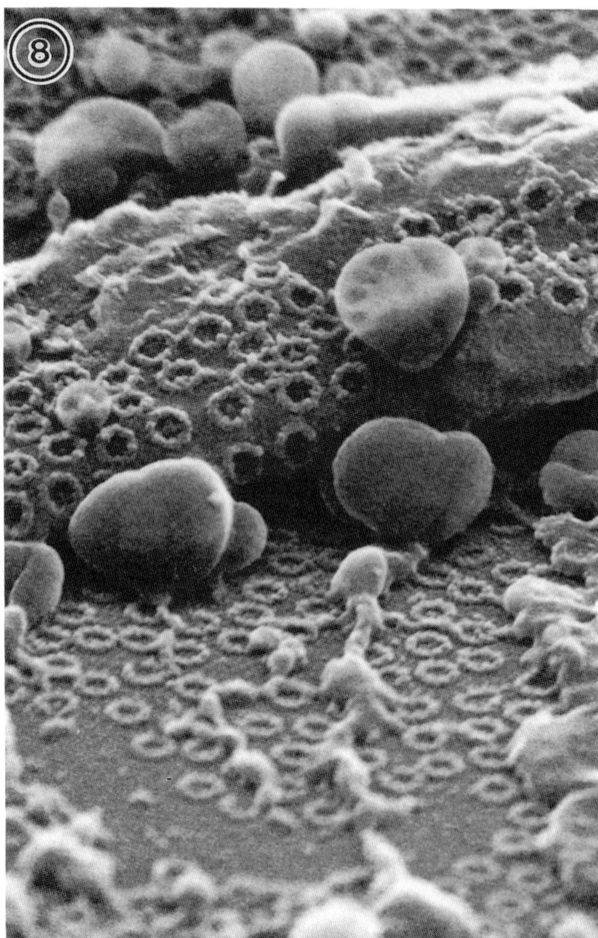

Figure 8: View of isolated NE from *Xenopus* oocyte that has been 'rolled' over the surface of the Si chip during isolation, leaving cytoplasmic components of the NPCs attached to the surface of the chip (foreground). (Bar=200 nm).

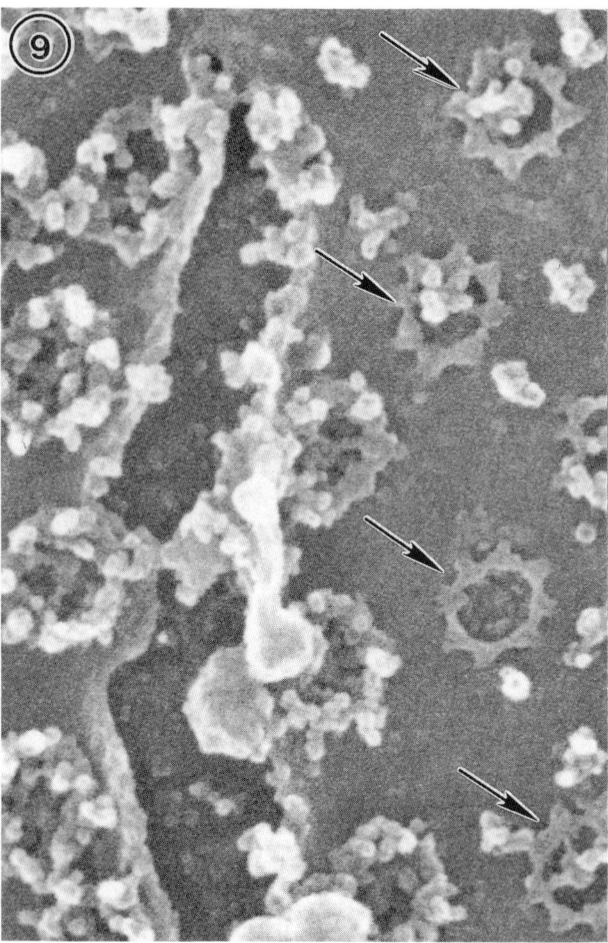

Figure 9: 'Dry fracture' by touching post critical point dried NE to adhesive to produce a surface separation of pore components from whole NPCs (left) exposing the underlying 'star' ring (arrowed). (Bar=100 nm).

the sites of direct binding to the individual structures of the NPC, and adjacent structures. As many of the nucleoporins are characterised by the presence of n-acetyl-glucosamine, and active transport through the pore is blocked by wheat germ agglutinin (WGA) [42], (which binds to n-acetyl-glucosamine), we have also been interested to pinpoint as precisely as possible the sites of binding of WGA.

Clearly, whole primary antibody molecules which are 15 nm in length, with secondary antibody molecules bound to them will potentially have a "radius of uncertainty" of several nanometres, with respect to the precise site of binding of the primary antibody to the antigen. In this situation, spatial resolution of around 2 nm may not be the limiting factor in pinpointing the molecular site of antibody binding. Our own approach has been initially to use conventional primary antibody binding, followed by secondary antibodies bound to colloidal gold, usually of 5 or 10 nm diameter, as well as WGA directly complexed to 5 or 10 nm gold. We have visualised the colloidal gold using a retractable solid state backscatter detector (KE Developments, Cambridge, UK). In order to retain the maximum amount of structural detail in the same image as the gold localisation, we have coated the gold labelled specimens with approximately 2 nm of Chromium, so that the optimal secondary electron signal from the specimen is retained. Because of the relative atomic Z numbers (Cr, 24; Au, 79) there is little interference from the backscatter electron (BSE) signal generated by the Cr coating compared with BSE signal from the colloidal gold. In fact, there is more BSE signal from areas of increased osmiophilic fixation (Os, Z=76), but as this is diffuse rather than particulate, it is not usually a problem with respect to gold localisation. Indeed, in NE preparations, the dense concentrations of

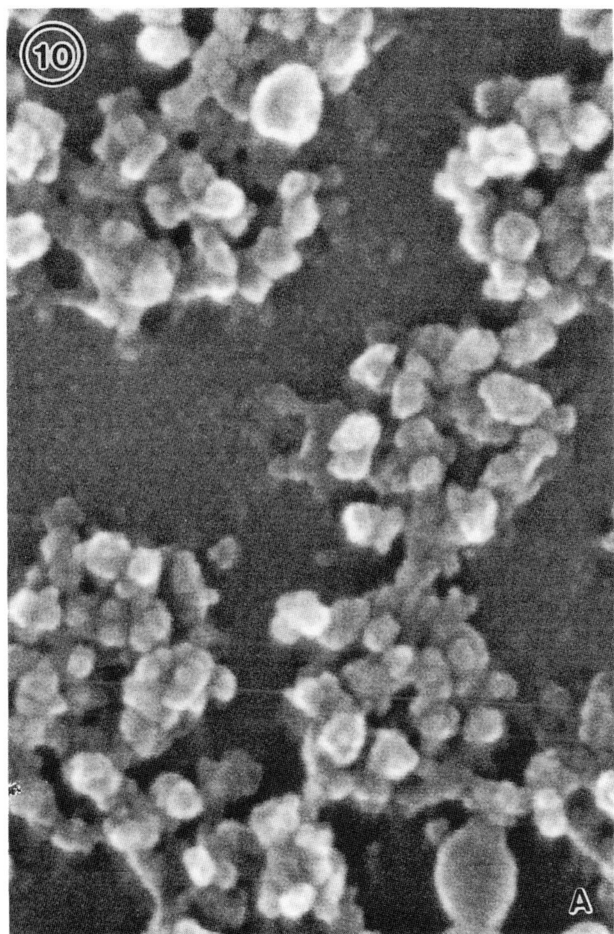

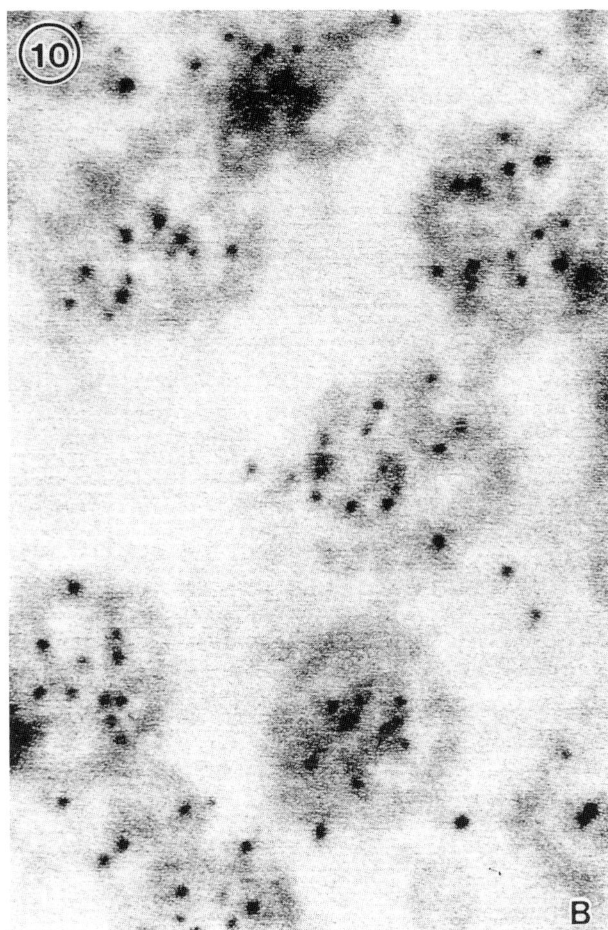

Figure 10: **(A)** SE image of cytoplasmic face of *Xenopus* labelled NE with antibody to P180 (a gift of Dr U Aebi). **(B)** BSE image of same area. The colloidal gold is apparent as a clearly discrete signal, whereas the increased BSE signal from the osmiophilia of the pore complex is more diffuse. (Bar=100 nm).

osmiophilic material in the NPC such as the coaxial rings, are clearly observed by BSE imaging, although with significantly less topographical detail than secondary electron (SE) imaging (Figs. 10, 12abc). In order to optimise the visualisation of the gold colloid relative to the SE image of the structure to which it is bound, we have reversed the BSE signal, and used a standard mixing box to (ETPRA, Australia) superimpose it in the SE signal. In this way the gold colloidal appears as typical 'black dots' familiar from TEM gold labelling, superimposed on the typical SE visualisation of 3D surface topography (Figs. 11, 12). This type of visualisation will also often demonstrate the protein 'shell' which coats the gold colloid, not usually observed in TEM (Figs. 12de). To date, we have only been successful with gold down to 5 nm in diameter with this approach, due to limitations in our system for optimising and mixing BSE and SE signal levels, but we feel that this could be improved to smaller gold colloid, if not down to 1.4 nm nanogold (Nanoprobes, Stony Brook, NY, USA), at least to nanogold which has been silver enhanced to around 3 nm in diameter. These smaller gold probes have the advantage of direct chemical bonding to the molecule of interest, rather than depending on a less stable electrostatic interaction between colloid and protein, but are best used with Fab fragments rather than whole antibodies, again reducing 'the radius of uncertainty'.

The Visualisation of NPC Transport by FEISEM Nuclear Import

For many years, nuclear import has been visualised by microinjection of colloidal gold carrying a necessary nuclear localisation sequence (NLS), mainly in the excellent studies of Feldherr *et al.* [12, 13]. Thin section TEM studies show binding of gold colloid, up to a maximum diameter of 25 nm, first to the periphery of

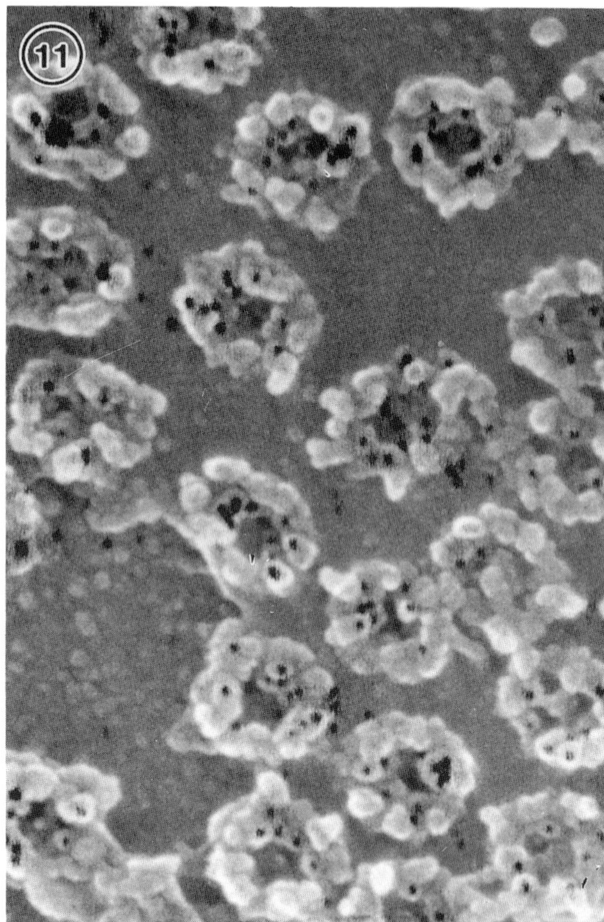

Figure 11: Mixed image of reversed contrast BSE signal mixed (about 60:40) with SE signal during signal detection, guaranteeing that registration between the two images is retained. WGA - gold labelling of cytoplasmic face of NPCs on *Xenopus* NE, 10 nm diameter gold. (Bar=100 nm).

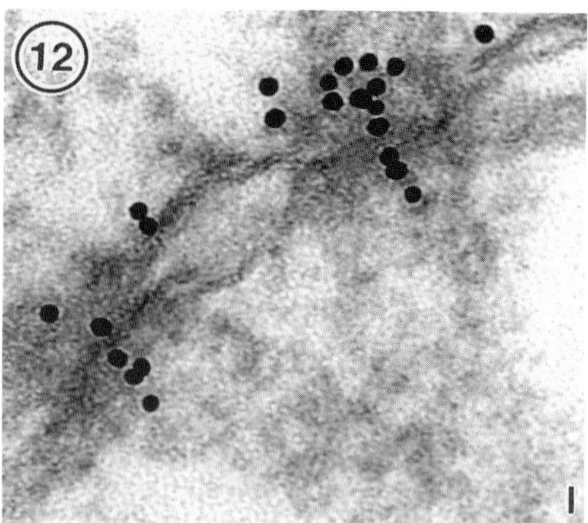

the NPC at the tips of the cytoplasmic filaments, followed by movement down these filaments, across the surface of the NPC, and movement through the central core (Fig. 12i). As TEM does not readily visualise NPC basket structure, the route through this inner aspect of the pore complex remains uncharted. In our own studies, we have microinjected 10-15 nm colloidal gold, complexed with nucleoplasmin, which has 4 NLS (nuclear location sequences) per molecule. We have cut sections at a thickness which is equivalent to the diameter of the pore (100-120 nm) in an attempt to generate whole profiles of NPCs somewhere in the section. Making stereopairs in the TEM has helped with 3D reconstruction of the passage of gold through the pore (Allen, unpublished). We have also isolated NEs from microinjected nuclei, and observed gold binding to the surface of the cytoplasmic filaments and migration to the centre of the pore. Although passage through the NPC core is 'single file' 15 or more nucleoplasmin gold particles can be bound at once to the cytoplasmic filaments. Passage through to the nucleoplasmic side, again using SEM stereopairs appears to result in emergence through the centre of the NPC basket ring (Fig. 12fgh).

Nuclear pore export

The molecular requirements for nuclear export are less well characterised than those for import, although mRNA complexed gold will be exported after microinjection [10, 29]. We have investigated a system which allows direct visualisation of nuclear export in physiologically normal circumstances. Polytene nuclei in insect salivary glands contain giant (polytene) chromosomes which generate enormous numbers of mRNA transcripts for the production of salivary proteins. These transcripts are liberated from permanently amplified regions of the polytene chromosome (Balbiani Rings (BR)) in the form of RNA and ribonucleic protein (RNP) granules 50 nm in diameter. These Balbiani ring granules attach to the NPC baskets, where they appear to 'dock' in the correct orientation, before becoming 'unravelled' to allow passage through the NPC with the RNP fibre exiting the pore rather like toothpaste out of a tube. We have visualised both the binding to the NPC basket, and export directly from surface imaging, and using deresin methods (see above) have observed BR granule binding to the basket in 'side elevation'. Our findings are in agreement with those of Mehlin *et al.* [25] who showed unravelling of the BR granule by TEM tomography, but did not demonstrate the involvement of the NPC basket, as this is not normally apparent in TEM thin sections.

Conclusions

Field emission in lens scanning EM (FEISEM) offers the possibility of equivalent effective resolution to

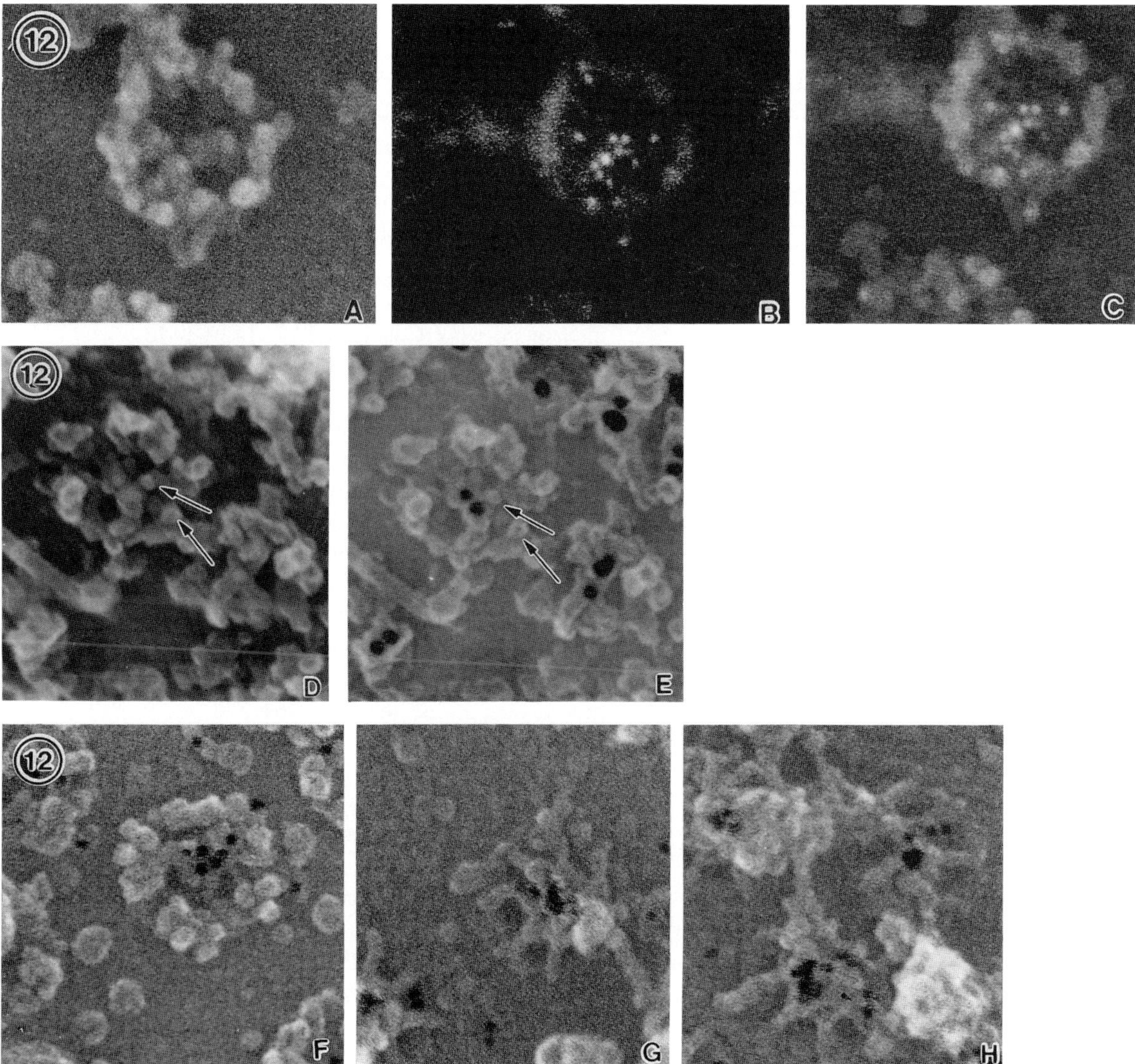

Figure 12: FEISEM of microinjected 5 nm nucleoplasmic gold, visualised by **a**) secondary electron (SE) imaging; **b**) Backscatter electron (BSE) imaging; and **c**) electronically mixed imaging of a and b. Mixing of unreversed contrast BSE imaging with SE imaging is not optimal with respect to exact localisation of the colloidal gold. **de:** SE and BSE/SE mixed images of microinjected 10 nm nucleoplasmin gold in a central position on the cytoplasmic face of the NPC. Topography is insufficient for gold colloid characterisation, as similar sized spheres, (arrowed) are clearly not gold in the mixed signal, in which the contrast reversed BSE image from the colloidal gold is more readily apparent than A-C. **fgh:** Microinjected nucleoplasmin 10 nm gold colloid in passage through the NPC. 12**f** shows the cytoplasmic face of the NE, 12**g** and **h**, the nucleoplasmic face, with gold 'exit' through the inner basket ring of the NPC basket (arrowed). **i** (*on page 160, top right column*): Conventional thin section TEM showing NPC transport of Nuceloplasmin gold during passage through the central region of the NPC. (Bar (all cases) = 100 nm).

---

conventional TEM for biological specimens, with the advantages of bulk specimens and surface imaging, which allows direct 3D visualisation of interfaces within the cell such as the nuclear envelope. Without the constraints of thin sections or single layers of molecules, surface imaging can be applied to any aspect of the cytoplasm or nucleus that can be accessed, either as a result of isolation, proteolytic or detergent exposure, de-resining of sectioned material, or simple fracturing techniques not requiring the high tech approach associated with cryofracturing. It also offers the possibility of antibody characterisation using colloidal gold labelling, and also direct observation of nucleo-cytoplasmic transport *in situ* within the cell, viewed from either the nuclear interior or the cytoplasmic face of the NE. Thus, FEISEM can provide useful complementary information to extend our understanding of the 3D organisation of nuclear surfaces and interfaces.

References

1. Aebi U, Cohn J, Buhle L, Gerace L (1986) The nuclear lamina is a meshwork of intermediate-type filaments. Nature **323**: 560-564.

2. Akey CW, Radermacher M (1993) Architecture of the *Xenopus* nuclear pore complex revealed by three-dimensional cryo-electron microscopy. J Cell Biol **122**: 1-19.

3. Allen TD, Jack EM, Harrison CJ (1988) The three dimensional structure of Human Metaphase chromosomes determined by Scanning Electron Microscopy. In: Chromosomes and Chromatin, Vol 2. Adolph KW (ed). CRC Press, Boca Raton. pp. 51-72.

4. Allen TD, Goldberg MW (1993) High resolution SEM in cell biology. Trends in Cell Biol **3**: 205-208.

5. Allen TD, Goldberg MW (1995) High resolution scanning electron microscopy in cell biology. In: Cell Biology: A Laboratory Handbook, Vol 2. Celis J (ed). Academic Press, San Diego, CA. pp 193-204.

6. Allen TD, Goldberg MW, Reipert S (1994) FEISEM of the cell nucleus, a consideration of parameters. In: Microbeam Analysis. Proc. 28th Annual MAS Meeting. Friel J (ed.), VCH Press, New York. pp 371-372.

7. Allen TD, Goldberg MW, Pelling C, Solovei I (1994) Evolutionary conservation of basket structure in nuclear pore complexes, and their possible role in mRNA transport. In: Proc. 2nd Meeting Microscopy Society of America. Bailey G, Garrat-Reed A (eds). San Francisco Press, San Fransisco. pp 12-13.

8. Allen TD, Goldberg MW (1995) Field emission scanning electron microscopy in studies of nuclear pore complex structure and function. Scanning **17**: 38-49.

9. Capco DG, Krochmalnic G, Penmans S (1984) A new method of preparing embeddment-free sections for transmission electron microscopy. J Cell Biol **98**: 1878-1885.

10. Fabre E, Hurt EC (1994) Nuclear transport. Curr Op Cell Biol **6**: 335-342.

11. Ferreira JA, Carmo-Fonseca M, Lamond AL (1994) Differential interaction of splicing snRNPs with coiled bodies and interchromatin granules during mitosis and assembly of daughter cell nuclei. J Cell Biol **126**: 11-23.

12. Feldherr CM, Akin D (1993) Regulation of nuclear transport in proliferating and quiescent cells. Exp Cell Res **205**: 179-186.

13. Feldherr CM, Akin D (1994) Variations in signal mediated nuclear transport during the cell cycle in BALB/C 3T3 cells. Exp Cell Res **215**: 206-210.

14. Georgatos SD, Meier J, Simos G (1994) Laminas and lamin-associated proteins. Curr Op Cell Biol **6**: 347-353.

15. Goldberg MW, Allen TD (1992) High resolution scanning electron microscopy of the nuclear envelope: demonstration of a new, regular, fibrous lattice attached to the baskets of the nucleoplasmic face of the nuclear pores. J Cell Biol **119**: 1429-1440.

16. Goldberg MW, Allen TD (1992) High Resolution Scanning Electron Microscopy (HRSEM) of the nuclear envelope, nuclear pore substructure, baskets and fibrous components of the inner nuclear envelope. In: Proc. 50th Annual Meeting of Electron Microscopy Society of America. Bailey GW, Bentley JA, Small A (eds). San Francisco Press, San Fransisco. pp 492-494.

17. Goldberg MW, Allen TD (1993) The nuclear pore complex: three-dimensional surface structure revealed by field emission, in-lens scanning electron microscopy with underlying structure uncovered by proteolysis. J Cell Sci **106**: 261-274.

18. Goldberg MW, Blow JJ, Allen TD (1992) The use of field emission in-lens scanning electron microscopy to study the steps of assembly of the nuclear envelope *in vitro*. J Struct Biol **108**: 257-268.

19. Goldberg MW, Allen TD (1995) Structural and functional organisation of the nuclear envelope. Curr Op Cell Biol **7**: 301-310.

20. Goldberg MW, Jenkins H, Allen TD, Hutchison CJ (1995) *Xenopus* lamin B3 has a direct role in the assembly of a replication competent nucleus: evidence from cell free egg extracts. J Cell Sci **108**: 3451-3461..

21. Hozak P, Hassan AB, Jackson DA, Cook PR (1993). Visualisation of replication factories attached to a nucleoskeleton. Cell **73**: 361-373.

22. Hutchison CJ, Bridger JM, Cox LS, Kill, IR (1994) Weaving a pattern from disparate threads: lamin function in nuclear assembly and DNA replication. J Cell Sci **107**: 3259-3269.

23. Jackson DA, Cook PR (1988) Visualisation of filamentous nucleoskeleton with a 23 nm axil repeat. EMBO J **7**: 3667-3677.

24. Jarnik M, Aebi U (1991) Towards a more complete 3D-structure of the nuclear pore complex. J Struct Biol **107**: 291-308.

25. Jenkins H, Whitfield WGF, Goldberg MW, Allen TD, Hutchison CJ (1994) Evidence for direct involvement of laminas in the assembly of replication competent nuclei. Acta Biochemica Polonia **42**: 133-144, 195.

26. Mehlin H, Daneholt B, Skoglund V (1992) Translocation of a specific premessenger ribonucleoprotein particle through the nuclear pore studied with electron microscope tomography. Cell **69**: 605-613.

27. Macgregor H, Varley J (1983) Working with Animal Chromosomes. J Wiley & Sons, New York. pp 115-144.

28. Malecki M, Ris H (1992) Surface topography and intercellular organisation of human cells in suspension as revealed by SEM. Scanning **14**: 76-85.

29. Melchior F, Gerace L (1995) Mechanisms of nuclear protein import. Curr Op Cell Biol **7**: 3, 310-318.

30. Pante N, Baston R, McMorrow I, Burke B, Aebi U (1994) Interactions and three-dimensional localization of a group of nuclear pore complex proteins. J Cell Biol **126**: 603-617.

31. Pante N, Aebi U (1994) Towards understanding the 3D structure of the nuclear pore complex at the molecular level. Curr Op Struct Biol **4**: 187-196.

32. Pelling C, Allen TD (1993) Scanning electron microscopy of polytene chromosomes 1. Chromosome Res **1**: 221-237.

33. Reichelt R, Holzenburg A, Buhle EL, Jarnik M, Engel A, Aebi U (1990) Correlation between structure and mass distribution of the nuclear pore complex, and of distinct pore complex proteins. J Cell Biol **110**: 883-894.

34. Reipert S, Reipert BM, Allen TD (1994) Preparation of isolated nuclei from K562 haemopoietic cell line for high resolution scanning electron microscopy. Microsc Res Techn **29**: 54-61.

35. Reipert S, Berry J, Hughes ME, Hickman JA, Allen TD (1995) Changes in mitochondrial mass in the haemopoietic stem cell line FDCP-Mix after treatment with etoposide. A comparative study by multiparameter flow cytometry, confocal and electron microscopy. Exp Cell Res **221**: 281-288.

36. Ris H (1989) Ultrastructure with high resolution low voltage SEM. EMSA Bull **21**: 54-56.

37. Ris H (1991) The 3D structure of the Nuclear Pore Complex as seen by high voltage electron microscopy and high resolution low voltage scanning electron microscopy. Inst Phys Conf Ser No **98**: 657-662.

38. Solovei I, Gaginskaya E, Allen TD, Macgregor H (1992) A novel structure associated with a lampbrush chromosome in the chicken *Gallus domesticus*. J Cell Sci **107**: 759-772.

39. Spector DL (1993) Nuclear organisation of the pre-MRNA processing. Curr Op Cell Biol **5**: 442-447.

40. Wiese C, Wilson KL (1993) Nuclear membrane dynamics. Curr Op Cell Biol **5**: 387-394.

41. Wilken N, Kossner U, Senecal J-L, Scheer U, Dabauvalle M-C (1993) Nup180, a novel nuclear pore complex protein localizing to the cytoplasmic ring and associated fibrils. J Cell Biol **123**: 1345-1354.

42. Yoneda Y, Imamoto-Sonobe N, Yamaizummi M, Uchida T (1989) Reversible inhibition of protein transport into the nucleus of wheat germ agglutinin injected into cultured cells. Exp Cell Res **173**: 586-595.

Discussion with Reviewers

M. Malecki: How would you modify protocols for isolation of nuclei from cells cultured in suspensions versus on substrates?
Authors: For nuclear isolation, the cells could be removed from the substratum (either scraping or trypsin) and then homogenised after swelling in much the same way. Because of the high probability however of the attachment to the substratum generating polarising influences on the cell and its nucleus, I think the approach illustrated in Fig. 3, where the nucleus and its contents are exposed *in situ* is potentially a valuable approach. Cells in suspension may well be polarised in some way, but obviously there is no method of either ascertaining, or maintaining such polarity. Attached tissue culture cells are known to require a certain amount of spreading over the substratum before crucial processes such as DNA replication can be achieved.

M. Malecki: How does spinning down of isolated nuclei on the poly-l-lysine (PLL) coated silicon chips affect nuclear architecture? What is the adhesion rate?
Authors: As the nuclei are usually fixed in suspension prior to attachment to PLL coated silicon chips, we do not envisage any alteration as a result of nuclear attachment. The nuclei show no distortion after attachment, retaining their spherical morphology. Usually a brief spin (5 min, 1000 g) is adequate for a high yield of nuclear attachment.

M. Malecki: Did you attempt to reduce the radius of uncertainty by conjugating gold beads to primary FABs?
Authors: We have not attempted this yet, but fully agree that small gold (1-3 nm) directly conjugated to FAB fragments of primary antibodies should considerably improve the resolution of labelling. The resolving power of FEISEM is capable of this, although the immunology could be demanding.

M. Malecki: Would you be willing to share your observations on nuclear pore architecture prepared by: 1) isolation in buffers followed by fixation: 2) fixed in situ followed by their exposure: 3) rapid freezing followed by their exposure?
Authors: We can respond to parts 1 and 2, we have not tried 3 yet. Isolation in buffers followed by fixation is

our standard approach to NPC structure by FEISEM (see refs. 13-18 and 43). 'Fixed *in situ* followed by exposure'- this is a difficult approach, as the fixation will preclude the separation of structures such as the nuclear envelope from their surrounding material, namely cytoplasm and nucleoplasm. We have however isolated nuclei in the presence of glutaraldehyde, and noticed a significant increase in the length of cytoplasmic filaments retained on the cytoplasmic face of the nuclear pore complex (NPC). It is important to bear of this type of approach in mind when considering methods which will allow visualisation of organelles such as NPCs in interaction with elements of both the nucleus and cytoplasm.

Additional Reference

43. Goldberg MW, Allen TD (1996) The nuclear pore complex and lamina: three dimensional structures and interactions determined by field emission in-lens scanning electron microscopy. J Mol Biol **257**: 848-865.

Scanning Microscopy Supplement 10, 1996 (pages 165-176)
Scanning Microscopy International, Chicago (AMF O'Hare), IL 60666 USA

0892-953X/96$5.00+.25

PROBLEMS IN PREPARATION OF CHROMOSOMES FOR SCANNING ELECTRON MICROSCOPY TO REVEAL MORPHOLOGY AND TO PERMIT IMMUNOCYTOCHEMISTRY OF SENSITIVE ANTIGENS

A.T. Sumner*

MRC Human Genetics Unit, Western General Hospital, Edinburgh EH4 2XU, U.K.

(Received for publication August 6, 1995 and in revised form June 24, 1996)

Abstract

Although much information about chromosome structure and behaviour has been obtained using light microscopy, greater resolution is needed for a thorough understanding of chromosome organisation. Scanning electron microscopy (SEM) can provide valuable data about these three-dimensional organelles. The introduction of methods using osmium impregnation of methanol-acetic acid-fixed chromosome spreads revolutionised matters, producing life-like images of chromosomes. Nevertheless, it became clear that osmium impregnation introduced various artefacts, although the resulting images were still useful. Methanol-acetic acid-fixed chromosomes are, in fact, flattened on the glass substratum, and the 3-dimensional appearance obtained after osmium impregnation is the result of swelling during this process. At the same time, the fibrous substructure of the chromosomes becomes much coarser. More recently a number of alternative methods have become available for studying chromosomes by SEM. Isolated chromosomes, that have not been allowed to dry during preparation, retain a 3-dimensional appearance without osmium impregnation, and the same is true of methanol-acetic acid-fixed chromosomes that have been treated with 45% acetic acid and processed without drying; however, these methods do not permit the routine production of intact metaphase spreads. Use of cytocentrifuge preparations obviates the use of acetic acid fixation and osmium impregnation, produces intact metaphase spreads, and permits the immunocytochemical detection of antigens that are easily destroyed by routine fixation procedures.

Key Words: Chromosomes, scanning electron microscopy, immunocytochemistry, methanol-acetic acid fixation, osmium impregnation.

*Address for correspondence:
A.T. Sumner
35 West Street, Penicuik, Midlothian EH26 9DG, U.K.
Telephone/FAX number: (+44)-1968-672265

Introduction

Because of their 3-dimensional structure, disposition in the cell, and behaviour, chromosomes should be highly appropriate objects for study by scanning electron microscopy (SEM). However, it has not proved to be a simple matter to prepare chromosomes for SEM, as this necessarily involves freeing them from the surrounding cytoplasm, with the probability of altering their structure. In fact, early attempts to examine chromosomes by SEM provided little useful information (Christenhuss *et al.*, 1967; Neurath *et al.*, 1967; Smith 1970; Pawlowitzki and Blaschke, 1971). The introduction of osmium impregnation techniques, which provide apparently lifelike images of chromosomes, represented a great advance (Harrison *et al.*, 1981; Maruyama 1983; Mullinger and Johnson 1983; Takayama *et al.*, 1985), and a substantial amount of work has been done using these procedures. There are, nevertheless, grounds for supposing that the osmium impregnation methods may introduce a number of artefacts. For a start, the chromosomes have to be fixed in methanol-acetic acid, which is known to extract histones and other proteins from chromosomes (Dick and Johns, 1968; Sivak and Wolman, 1974; Retief and Rüchel, 1977; Hancock and Sumner, 1982), and in general renders them unsuitable for immunocytochemical procedures. Secondly, the results of impregnating chromosomes with osmium can be quite variable, and there is evidence that a significant part of this variability could be due to the osmium impregnation itself (Sumner and Ross, 1989; Sanchez-Sweatman *et al.*, 1993). It is therefore desirable to consider carefully what artefacts might be produced during preparation of chromosomes for SEM.

In the work to be described in this paper, changes that occur in chromosome morphology during osmium impregnation of methanol-acetic acid-fixed chromosomes for SEM are described, and alternative methods of chromosome preparation, that may preserve morphology or immunogenicity better, are investigated. It has been a particular concern to preserve the immunogenicity of certain antigens (the kinetochore antigens that react with

CREST serum - Tan, 1989, and the antigen recognised by AC1 - Holland *et al.*, 1995) that are easily destroyed by fixation. Preliminary results show that with suitable methods of preparation, immunolabelling of sensitive antigens on chromosomes prepared for SEM can be carried out, thus adding compositional information to purely morphological observations.

Material and Methods

Chromosome preparations

Conventional methanol-acetic acid fixed chromosome spreads were made from human lymphocyte cultures according to standard procedures (e.g., Watt and Stephen, 1986; Macgregor and Varley, 1988), or from CHO (Chinese hamster ovary) cells, cultured in RPMI 1640 medium until nearly confluent, and accumulated in metaphase using Colcemid. After making the spreads on 22 mm square coverslips, they were allowed to dry, usually overnight, before processing further.

For treatment with 45% acetic acid, the above procedure was modified using a method derived from that described by Martin *et al.* (1994). Chromosome preparations from CHO cells, fixed in methanol-acetic acid, were spread on coverslips in the usual way, but instead of letting the cell suspension dry, the coverslips were flooded with 45% acetic acid immediately before the methanol-acetic acid finally dried out. After a few seconds they were plunged into glutaraldehyde (2.5% in cacodylate buffer, pH 7.4, containing 0.1M sucrose), left overnight, and either dehydrated and critical point dried from carbon dioxide, or impregnated with osmium as described below.

Cytocentrifuge preparations were also made from cultures of human lymphocytes or CHO cells, grown in the same way as for methanol-acetic acid fixation. However, treatment at the end of culture was different. After pelleting the cells and decanting off the supernatant, the cells were resuspended in the hypotonic solution described by Stenman *et al.* (1975) for 10 min in the refrigerator. This hypotonic solution consists of 10 mM HEPES[4-(2-hydroxyethyl)-1-piperazineethanesulphonic acid, sodium salt], 30 mM glycerol, 1 mM calcium chloride, and 0.8 mM magnesium chloride. After this treatment, 0.3 ml of the cell suspension was added to each chamber of a Shandon Cytospin centrifuge, and the cells were spun down on to slides for 15 min at 1500 rpm. After centrifuging, the slides were allowed to dry, and either used immediately, or left overnight before further processing.

Isolated human chromosomes were prepared using the polyamine method of Sillar and Young (1981). The resulting chromosome suspension was diluted approximately 5-fold with PBS (phosphate-buffered saline, Oxoid). Coverslips of 13 mm diameter were loaded into the wells of a multiwell plate (Falcon 24-well plate, cat. no. 3047, Becton Dickinson), and approximately 0.5 ml of the chromosome suspension added. The multiwell plates were then centrifuged for 10 min at 1000 rpm in a Sorvall ST6000 refrigerated centrifuge at 0-4°C. Subsequent processing was carried out without letting the specimens dry, by pipetting off the supernatant and adding the next solution to the well.

Further treatments

Methanol-acetic acid fixed chromosome spreads were treated with trypsin (Difco Bacto trypsin, reconstituted according to the manufacturer's instructions, and then diluted further 100-fold with distilled water). This solution was always used fresh, and digestion of chromosome preparations was for 5-30 s. After digestion, the chromosome preparations were washed thoroughly with distilled water. Some preparations were left undigested with trypsin. The slides were then transferred to glutaraldehyde for further processing (see below).

Cytospin preparations were treated with Triton X-100 (0.1% in PBS) for 5, 15 or 30 min, to remove cytoplasm, and then washed in PBS (3 lots, each for 5 min), before fixation with glutaraldehyde or immunocytochemical labelling (see below).

Isolated chromosomes were transferred to glutaraldehyde before osmium impregnation (see below) or critical point drying.

Osmium impregnation

Chromosome preparations were left in glutaraldehyde (2.5% in cacodylate buffer, pH 7.4, containing 0.1 M sucrose), either for 30 min or overnight, whichever was more convenient for that particular experiment. After washing thoroughly in tap water, the preparations were transferred to freshly prepared osmium tetroxide (1% in distilled water, 5 min), and then again washed very thoroughly with running tap water, before transfer to a freshly prepared solution of thiocarbohydrazide (TCH, 0.5% in distilled water, 5 min), followed by another thorough wash in running tap water. Thiocarbohydrazide acts as a bifunctional ligand, binding to osmium already in the tissue, and in turn binding further osmium at the next stage of osmication (Murphy, 1978). This cycle of osmium tetroxide and TCH was repeated several times (3 to 11 times, according to the requirements of the experiment), always finishing with an osmium tetroxide treatment. In experiments in which the number of cycles of treatment was not varied, it was standardised at nine.

After the last wash, the specimens were dehydrated through graded acetone solutions (25%, 50%, 75% and 100%), and critical point dried from liquid carbon dioxide. The slides or coverslips were broken into small enough pieces, and the pieces bearing chromosomes attached to stubs with double-sided adhesive tape. They were then coated lightly with platinum in a Polaron E5100 sputter coater, and examined in a Hitachi S-800 field emission scanning electron microscope, at accelerating voltages between 1 and 25 kV.

Immunocytochemistry

CREST serum was obtained from Professor G. Nuki (Department of Rheumatology, University of Edinburgh) and before use was diluted at between 1:25 and 1:50 with PBS containing 1% bovine serum albumin (BSA). Monoclonal antibody AC1 (Holland *et al.*, 1995) was a gift from Dr. G. Hadlaczky (Institute of Genetics, Biological Research Centre, Szeged, Hungary), and was used without dilution. Chromosome preparations were incubated with the antibody solution overnight (up to 19 h), and then washed with PBS containing 1% BSA (3 × 5 minutes). Preparations treated with CREST serum were then incubated for approximately 2 h with horseradish peroxidase (HRP)-labelled anti-human IgG (Sigma) diluted at between 1:25 and 1:50 in PBS containing 1% BSA, while chromosomes that had been incubated with monoclonal antibody AC1 were incubated with HRP-labelled anti-mouse IgM (Sigma) diluted as above. After this the slides were washed again in PBS containing 1% BSA (3 × 5 minutes). Peroxidase activity was detected using diaminobenzidine (DAB) solution (Sigma: 0.5 mg/ml in PBS), to which 50 μl of 1 volume hydrogen peroxide was added immediately before use. Incubation was for 1 h, after which the reaction product was intensified with silver, as described by Burns *et al.* (1985).

For colloidal gold labelling, chromosome preparations which had been incubated with monoclonal antibody AC1 were transferred to Tris-HCl buffer, pH 8.2, containing 0.9% sodium chloride and 1% BSA, and then incubated with anti-mouse IgM labelled with 10 nm colloidal gold (British Biocell International, Cardiff, UK) diluted 1:25, for 2 h. After incubation the chromosomes were washed again in the Tris buffer, then in distilled water, and the colloidal gold particles enhanced using a Silver Enhancement Kit (British Biocell International), according to the manufacturers instructions.

Immunolabelled chromosome preparations were dehydrated and critical point dried as described above, without osmication.

Results

Morphology of methanol-acetic acid fixed chromosomes

Chromosomes fixed in methanol-acetic acid, spread on glass, and prepared for SEM without further treatment are only slightly raised and show no fine structure (Fig. 1 a, b), in agreement with early scanning electron microscope observations of chromosomes, in which there was no osmication (Christenhuss *et al.*, 1967; Neurath *et al.*, 1967; Smith, 1970; Pawlowitzki and Blaschke, 1971). The chromosomes appear similar if they are fixed in glutaraldehyde before critical point drying (Fig. 1 c, d); note that interphase nuclei are only very slightly raised and do not appear as the expected nearly spherical objects. If, however, the chromosomes are treated briefly with trypsin, or indeed merely washed with PBS, before glutaraldehyde fixation and critical point drying, the chromosomes are still relatively flattened, but are seen to consist of a network of fine fibres (Fig. 1 e, f), as described by Squarzoni *et al.* (1994) and by Rizzoli *et al.* (1994). Subsequent impregnation with osmium results in the disappearance of the fine fibrillar structure, to be replaced with a more granular appearance; at the same time the profile of the chromosomes becomes raised, with an approximately semi-circular cross-section (fig. 1 g-l). Note that with the maximum degree of osmium impregnation, the nuclei remain relatively flattened (Fig. 1k), while chromosomes cross each other in a way that appears most unnatural (Fig. 1l).

Size of fibres in methanol-acetic acid fixed chromosomes

Sizes of chromosome fibres were measured on high resolution micrographs taken at ×120 000. No significant differences were found between the sizes of fibres imaged at 5 kV or 25 kV, or between coated and uncoated fibres. Chromosomes subjected to increasing numbers of stages of osmication showed a steady increase in the size of their fibres, or, for the more heavily osmicated chromosomes, their surface granularity (Fig. 2). On the other hand, trypsin treatment before osmium impregnation produces a progressive decrease in the size of the chromosome substructures (Fig. 3). Note that in all cases there is substantial variability in the size of the objects being measured, as well as differences between separate experiments. Nevertheless, the general trends are repeatable.

Isolated chromosomes

Chromosomes isolated using the polyamine method appear to be well raised above the glass substratum, with

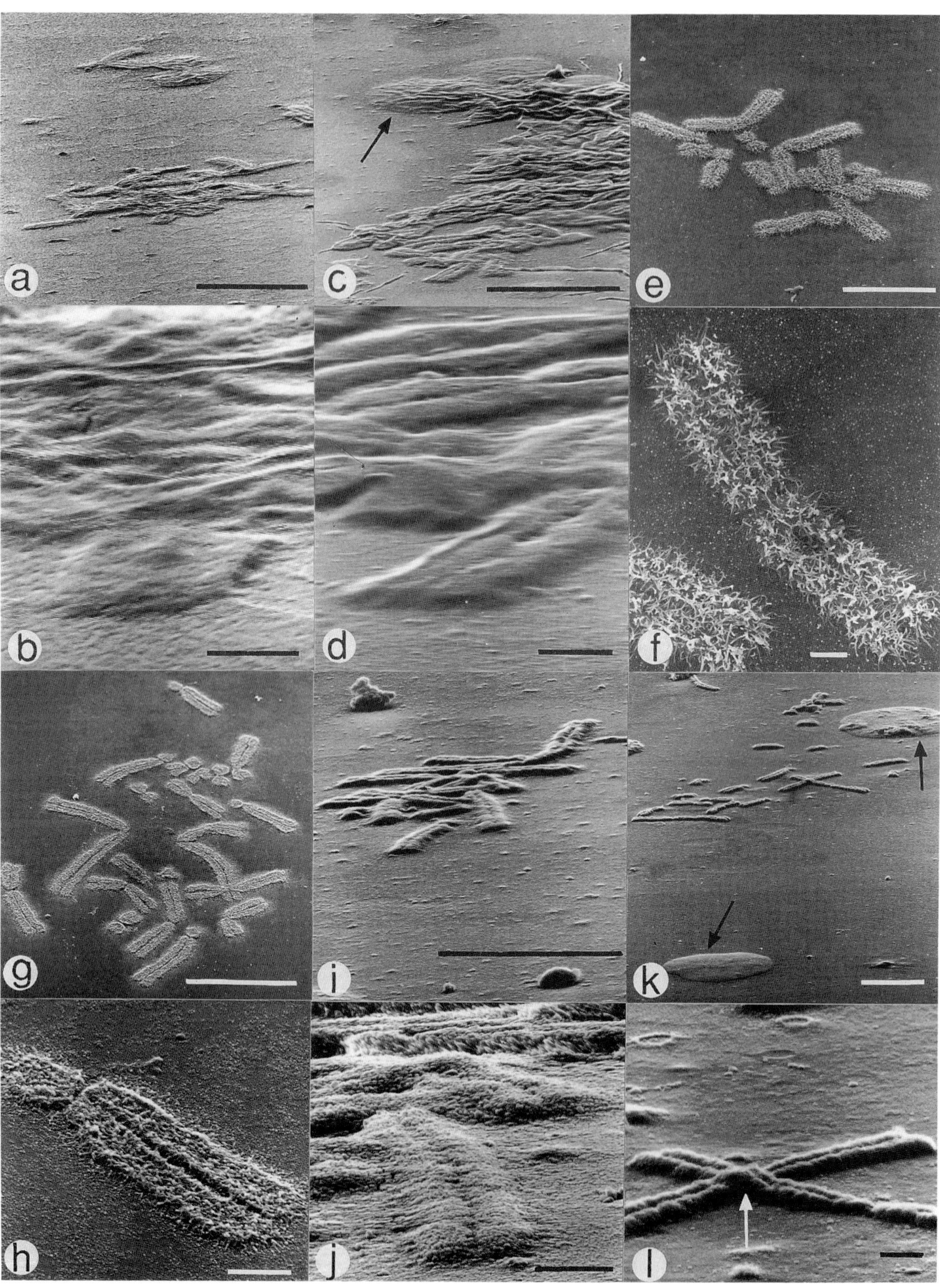

a circular cross-section, but tend not to show a clear split into sister chromatids (Fig. 4). Chromosomes prepared by critical point drying without osmium impregnation (Fig. 4a) appear generally similar to those that have received osmication (Fig. 4 b, c), although the unosmicated chromosomes appear slightly smoother.

Figure 1 on facing page

Figure 1. Human lymphocytes chromosomes, fixed in methanol-acetic acid, and prepared for scanning electron microscopy. **a,b**: chromosomes spread on glass and examined without further treatment; the chromosomes are flattened and featureless. **c,d**: chromosomes fixed in glutaraldehyde and critical point dried; the appearance is quite similar to that of untreated chromosomes. Note the flattened appearance of the interphase nuclei (arrow). **e,f**: chromosomes digested with trypsin, 5 sec, fixed in glutaraldehyde, and critical point dried; these chromosomes show a network of fine fibres. **g,h**: chromosomes digested with trypsin, fixed in glutaraldehyde, and impregnated with 3 cycles of osmium tetroxide/-thiocarbohydrazide (OTOTO) treatment. Chromosomes similar to those in e and f, without osmium impregnation. **i,j**: as g and h, but with 5 cycles of OTOTO; the chromosomes are distinctly raised with a semi-circular profile. **k,l**: as g-j, but with 7 cycles of OTOTO; the chromosomes are raised still higher than in i and j, but nuclei are still flattened (arrows in k), and chromosomes cross each other in an unnatural-looking way (arrow in l). Scale bars equal 10 μm in a, c, e, g, i, and k and equal 1 μm in b, d, f, h, j and l.

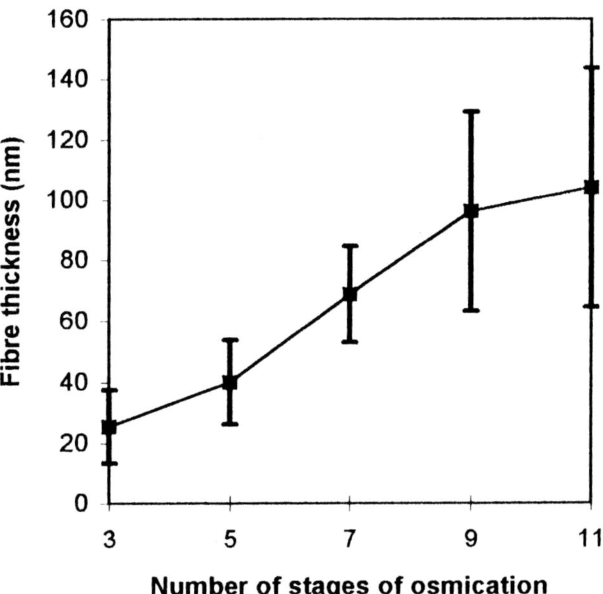

Figure 2. Graph showing the increase in thickness of chromosome fibres with increasing number of cycles of osmication. Methanol-acetic acid-fixed chromosomes, no trypsin treatment. Error bars represent ± 1 standard deviation.

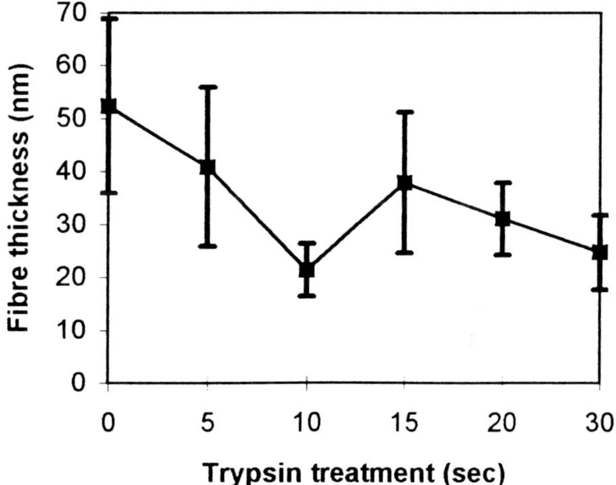

Figure 3. Graph showing the decrease in thickness of chromosome fibres with increasing length of trypsin treatment. Methanol-acetic acid-fixed chromosomes, subjected to 9 cycles of osmication after the trypsin treatment and glutaraldehyde fixation. Error bars represent ± 1 standard deviation.

Chromosomes prepared with 45% acetic acid

Methanol-acetic acid fixed chromosomes treated with 45% acetic acid immediately before drying, and then plunged into glutaraldehyde, show good morphology, with an approximately circular cross-section (Fig. 5). Chromosomes overlap each other in a natural way (Fig. 5a), in contrast to the appearance given by the standard methanol-acetic acid/osmication procedure (Fig. 1l). Unosmicated chromosomes appear relatively smooth (Fig. 5a), while osmicated chromosomes have a more fluffy appearance. Complete metaphases are rarely seen. Interphase nuclei are well raised, with an approximately spherical shape (Fig. 5c), in complete contrast to nuclei prepared by the standard method (e.g. Fig. 1k).

Cytocentrifuged chromosomes

Metaphase cells prepared by cytocentrifugation do not reveal chromosomes unless the surrounding cytoplasm is removed. Treatment with Triton X-100, followed by glutaraldehyde fixation and critical point drying, shows chromosomes that are only slightly raised above the surrounding material, and that are largely featureless (Fig. 6a). Subsequent osmication, however, produces metaphases in which the chromosomes have a good 3-dimensional structure and show a fibrous substructure (Fig. 6 b, c).

Immunocytochemistry

CREST serum labels the kinetochores of mammalian chromosomes (see Tan, 1989, for a review). When applied to unfixed cytocentrifuge preparations of human

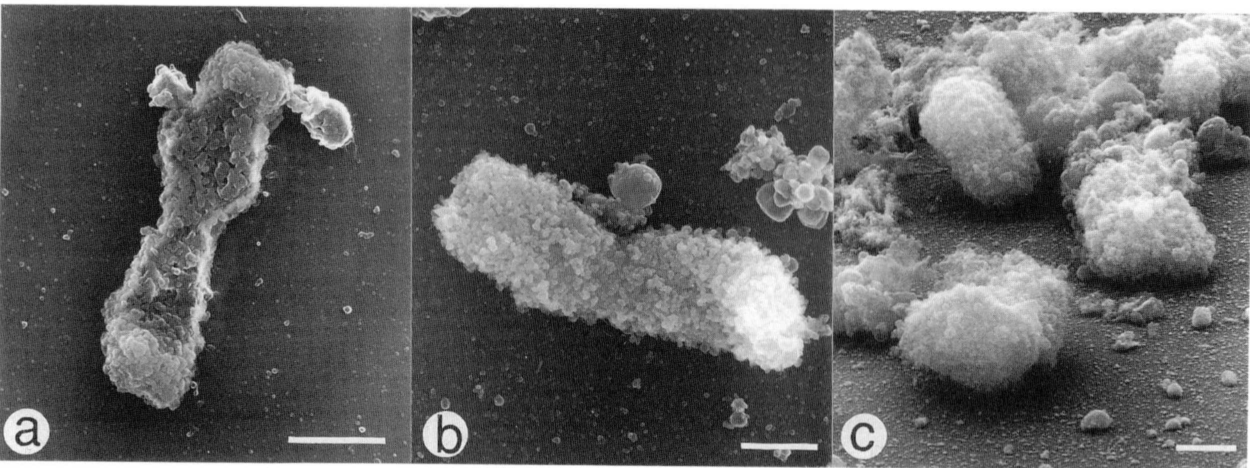

Figure 4. Scanning electron micrographs of isolated chromosomes. **a**: critical point dried without any prior treatment. **b**: osmicated before critical point drying. **c**: as b, showing that the chromosomes are approximately circular in cross-section. Scale bars = 1 μm.

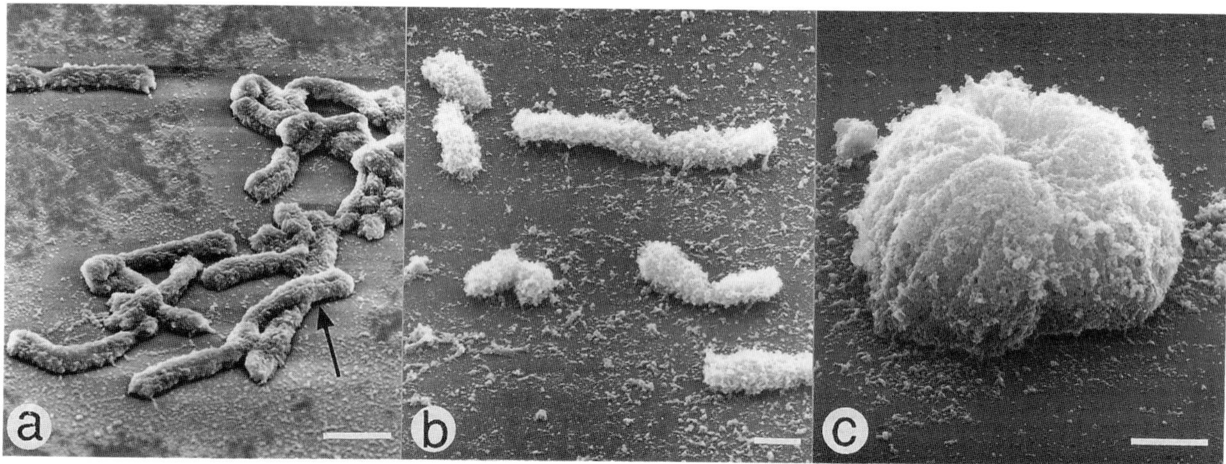

Figure 5. Methanol-acetic acid-fixed chromosomes, spread on glass, and treated with 45% acetic acid, and not allowed to dry. **a**: non-osmicated chromosomes; note the approximately circular cross-section, and the natural way in which one chromosome lies across another (arrow). **b**: osmicated chromosomes; similar to a, but with a fluffier appearance. **c**; a nucleus from an osmicated preparation; note that it is well raised and approximately spherical, not flattened like those shown in Figure 1. Scale bars = 2 μm.

chromosomes, and detected using a horseradish peroxidase-labelled secondary antibody, silver intensification of the DAB reaction product revealed paired dots at the centromeres (Fig. 7a), just as can be demonstrated by, for example, immunofluorescence. These dots are clearly visible using back-scattered electrons (Fig. 7 a, c), but are virtually invisible using the secondary electron signal (Fig. 7b). Chromosome structure is reasonable even though the chromosomes were not fixed before labelling, and no osmication was carried out afterwards.

Monoclonal antibody AC1 labels another centromeric antigen, which in some cases appears, by fluorescence microscopy, to form a ring round the centromeric constriction (Holland *et al.*, 1995), although the resolution is scarcely good enough to be confident of this. Such a centromeric ring can be seen clearly on the chromosome shown in Fig. 7d, and is especially clear when a series of micrographs at different angles of tilt is examined. In Fig. 7e, labelling is demonstrated using back-scattered electrons: on some chromosomes it extends right across the centromeres, while on others it forms two discrete blocks, one on each side. In both Fig. 7d and 7e, the area occupied by the DAB reaction

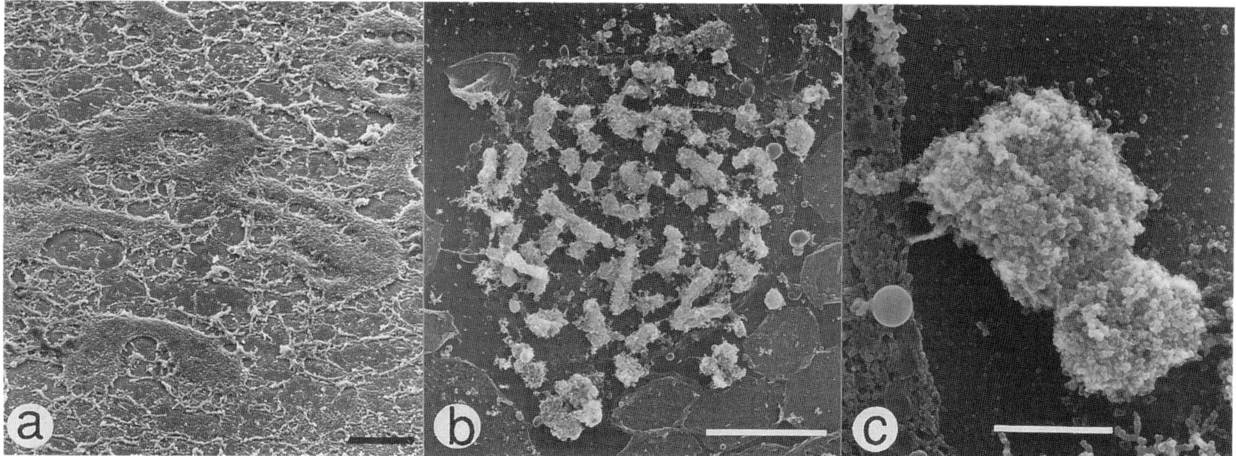

Figure 6. Cytocentrifuged preparations of chromosomes. **a**; CHO chromosomes treated with Triton X-100, fixed in glutaraldehyde and critical point dried. The chromosomes are only slightly raised above the substratum. Scale bar = 2 μm. **b**: human chromosomes treated with Triton X-100, fixed in glutaraldehyde, and subjected to 5 cycles of osmication. The chromosomes are now well raised. Scale bar = 10 μm. **c**: a single chromosome treated as in b, showing the fibrous substructure. Scale bar = 1 μm.

---

product is quite extensive, and in Fig. 7e also rather diffuse, as if the DAB reaction product has spread from the actual site of labelling. Results with an alternative method of labelling, silver-enhanced colloidal gold, are shown in Fig. 7f. Although there is considerable, presumably non-specific, scatter of colloidal gold particles, at least two regions are visible where there is a concentration of gold particles extending across the width of the chromosomes.

Discussion

Chromosomes fixed in methanol-acetic acid, and spread on glass, appear featureless and only very slightly raised above the substratum, when examined by SEM. Similar results have previously been reported by others who have examined fixed, but otherwise untreated chromosomes by SEM (Christenhuss *et al.*, 1967; Neurath *et al.*, 1967; Smith, 1970; Pawlowitzki and Blaschke, 1971) and observations using Atomic Force Microscopy show that such chromosomes are between 50 nm and 350 nm high (de Groot and Putman, 1992; Fritzsche *et al.* 1994). Exposure of the fixed chromosomes to buffer (with or without trypsin) reveals a network of fibres throughout the chromosomes (as described by Squarzoni *et al.* (1994) and by Rizzoli *et al.* (1994), but this tends to be obscured by subsequent osmium impregnation, which also swells the chromosomes. In fact, exposure of chromosomes to phosphate-buffered saline (PBS) also results in their height increasing, to between 300 nm and 900 nm (de Groot and Putman, 1992; Fritzsche *et al.*, 1994). Osmium impregnation has also been shown to deposit enough material on subcellular structures to produce a significant increase in size (Kelley *et al.*, 1973; Ip and Fischman 1979), which could well account for the increase in size of chromosome substructure reported here, as well as the qualitative change in surface structure observed. Although the continuous, granular appearance of heavily osmicated chromosomes might simply be the result of gross swelling of individual fibres, which as a result have become fused to form a continuous but rough surface, other explanations are possible. Rizzoli *et al.* (1994) regard the surface as being formed from precipitates of osmium with the thiocarbohydrazide used for impregnation. On the other hand, it is now well established that chromosomes do have a surface coating of ribonucleoprotein (Hernandez-Verdun and Gautier, 1994), and it has been proposed that this forms the surface coat seen after osmium impregnation (Sumner and Ross, 1989). Thus there remain many problems in interpreting the images obtained by SEM from methanol-acetic acid-fixed chromosomes. Although they appear much as one might expect, they are clearly the result of a series of artefactual changes, and cannot be a precise representation of the morphology and fine structure of the chromosomes as they were in life. Although much valuable work has been done with such preparations, results must be interpreted with caution, and probably it is only at the grossest level of chromosome structure that reliable conclusions can be drawn. In particular, the dimensions of the fibrous or granular substructure seen after osmium impregnation are very variable, and are dependent on both the degree of osmium impregnation

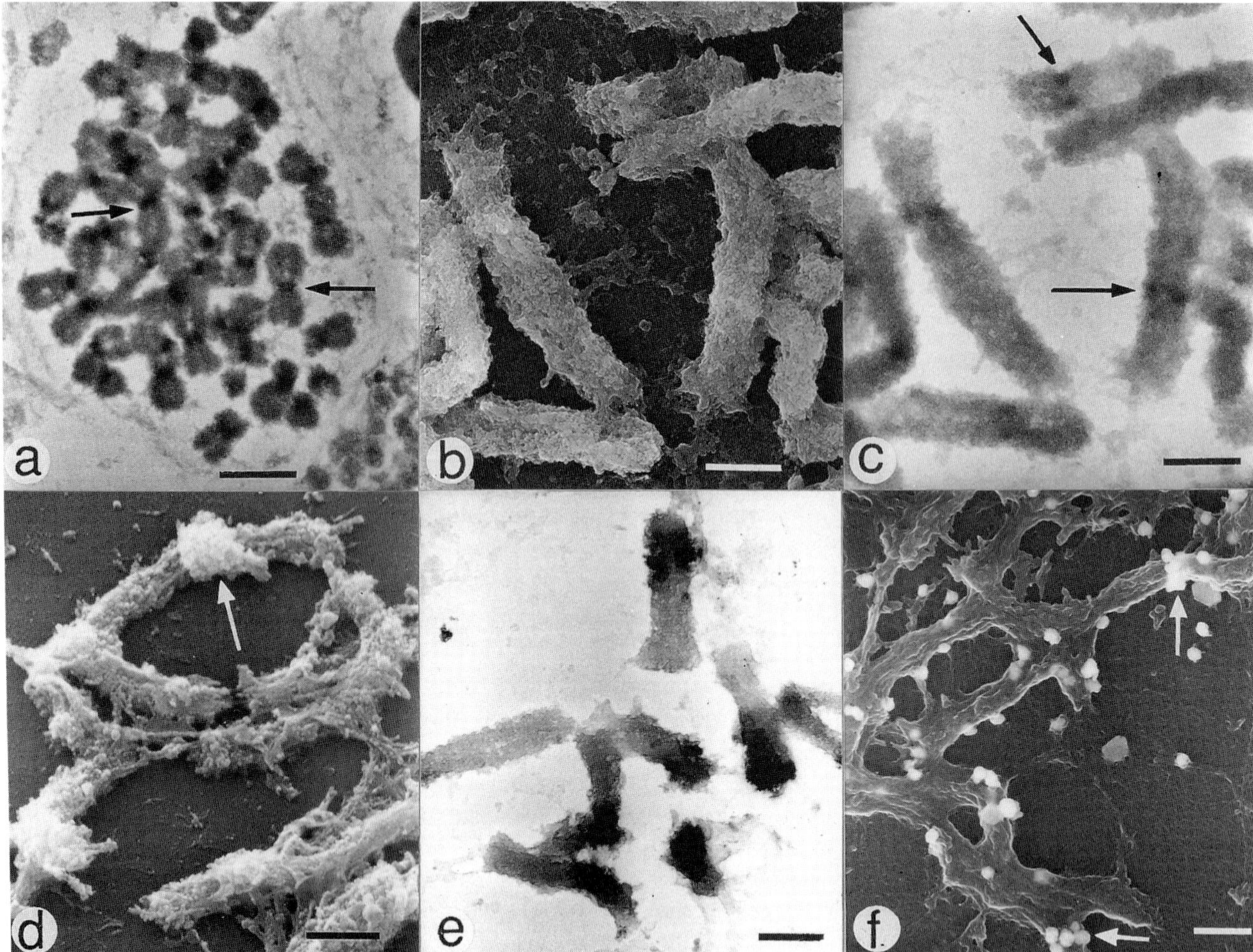

Figure 7. Scanning electron micrographs of immunolabelled human chromosomes, prepared by cytocentrifugation. **a-c**: CREST labelling, detected using horseradish peroxidase (HRP)-labelled second antibody, diaminobenzidine (DAB), and silver intensification. a: metaphase spread, imaged using back-scattered electrons (BSE) in reverse contrast (i.e. strong signals appear dark). Note the reaction at the centromeres of the chromosomes (arrows). Scale bar = 5 μm. b and c: the same chromosomes imaged using secondary electrons (b) and BSE (c); the reaction product is clearly visible as two spots at the centromeres using BSE (arrows), but is scarcely visible using secondary electrons. Scale bars = 2 μm. **d-f**: chromosomes labelled with monoclonal antibody AC1. d and e: antibody detected using HRP, DAB and silver intensification. In d, the reaction product is visible using secondary electrons as a substantial ring round the centromere (arrow), while in e, using BSE, the reaction appears as a diffuse mass. Scale bars = 1 μm. f: sites of antibody binding detected using colloidal gold, followed by silver intensification, and imaged using BSE. Sites of apparently specific labelling arrowed. Scale bar = 1 μm.

---

and the amount of trypsin treatment.

The flattening of chromosomes on to the substratum can be avoided if the chromosome spreads are never allowed to dry until preparation is finished (i.e. when critical point drying is completed). This can be achieved either by using isolated chromosomes, in which case it is impossible to obtain metaphase spreads, or by using treatment with 45% acetic acid. The latter procedure, pioneered by Martin *et al.* (1994), not only involves the use of quite concentrated acetic acid, but also causes disruption of many metaphases, although the chromosomes do appear to have a lifelike configuration. With both the isolated chromosomes and those prepared using 45% acetic acid, osmium impregnation seems to produce little change to the general morphology of the chromosomes.

Of the methods discussed so far, only the use of isolated chromosomes is likely to be compatible with immunocytochemical labelling, since methods involving acetic acid will destroy or extract most chromosomal antigens (Jeppesen, 1994). However, the use of isolated chromosomes for immunocytochemistry in the SEM has not been pursued, since the impossibility of obtaining metaphase spreads makes identification of specific chromosomes more difficult. Cytocentrifuge preparations are usually used for light microscopical immunocytochemistry of chromosomes (Jeppesen, 1994), and it has proved practicable to use similar preparations for SEM. The main difference between making cytocentrifuge preparations of chromosomes for light microscopy and for SEM is that in the latter case it is necessary to remove the surrounding cytoplasm to visualise the chromosomes; use of a mild detergent such as Triton X-100 is adequate to achieve this, and does not appear to cause any morphological degradation of the chromosomes, nor does it extract the antigens that have been tested. On the other hand, it is still not entirely clear how well preserved the chromosome morphology is. There is some evidence that the centrifugation process may leave the chromosomes somewhat flattened, even though they are surrounded by a supporting layer of cytoplasm. Nevertheless, well raised chromosomes with good morphology can be obtained without osmication. Although such preparations clearly make possible the immunolabelling of specimens to be viewed in the SEM, with the advantage of being able to study the 3-dimensional arrangement of antigens on chromosomes, further refinement of the chromosome preparation, resulting in greater consistency, and optimisation of the immunolabelling process, are clearly desirable. While horseradish peroxidase in combination with DAB can produce a strong, clear reaction, with apparently minimal non-specific background, it appears to be essentially a low resolution technique, owing to spreading of the DAB reaction product during development. On the other hand, colloidal gold, although it should theoretically produce much higher resolution, limited only by the size of the gold particles, seems to be prone to producing excessive background labelling on this type of specimen, and further work is required to optimise labelling.

From the point of view of producing chromosome preparations for SEM with the minimum of treatment that might disrupt structure or extract chromosomal components, cytocentrifuge preparations are probably best, but the metaphases so produced tend to lack the clarity of conventional metaphases fixed in methanol-acetic acid and impregnated with osmium. It is, however, clear that the latter type of preparations involve so many artefacts that they cannot be acceptable for studying anything more than the grosser features of chromosome organisation. In fact, osmium impregnation does not seem to be necessary for obtaining good chromosome preparations for the SEM or even for producing adequate contrast. Preparations of chromosomes made by treatment with 45% acetic acid, without drying, seem to be superior morphologically to standard air-dried, methanol-acetic acid-fixed spreads, but so far it has proved to be difficult to retain intact metaphases from mammalian cells, presumably because drying is important for the adherence of the chromosomes to the substratum. Perhaps a universal method of preparing chromosomes for SEM that retains them in the configuration that they had in life, while permitting immunolabelling of even the most delicate of antigens, is not possible, and it may, for the foreseeable future, be necessary to use specific procedures depending on the application required.

Acknowledgements

I should like to thank Sheila McBeath for human lymphocyte cultures used for preparing chromosome spreads, Pat Malloy for the isolated chromosomes, and William Christie for cultures of CHO cells; Andrew Ross and Elizabeth Graham for preparing the chromosomes for scanning electron microscopy; and Sandy Bruce for preparing the illustrations.

References

Burns J, Chan VTW, Jonasson JA, Fleming KA, Taylor S, McGee JOD (1985) Sensitive system for visualising biotinylated DNA probes hybridised *in situ*: a rapid sex determination of intact cells. J Clin Pathol **38**: 1085-1092.

Christenhuss R, Büchner T, Pfeiffer RA (1967) Visualization of human somatic chromosomes by scanning electron microscopy. Nature **216**: 379-380.

de Groot B, Putman CAJ (1992) High-resolution imaging of chromosome related structures by atomic force microscopy. J Microsc **168**: 239-247.

Dick C, Johns EW (1968) The effect of two acetic acid containing fixatives on the histone content of calf thymus deoxyribonucleoprotein and calf thymus tissue. Exp Cell Res **51**: 626-632.

Fritzsche W, Schaper A, Jovin TM (1994) Probing chromatin with the scanning force microscope. Chromosoma **103**: 231-236.

Hancock JM, Sumner AT (1982) The role of proteins in the production of different types of chromosome bands. Cytobios **35**: 37-46.

Harrison CJ, Britch M, Allen TD, Harris R (1981) Scanning electron microscopy of the G-banded human karyotype. Exp Cell Res **134**: 141-153.

Hernandez-Verdun D, Gautier T (1994) The chromosomes periphery during mitosis. Bioessays **16**: 179-185.

Holland KA, Keresõ J, Zákány J, Pravnovskzy T,

Monostori E, Belyaer N, Hadlaczky GY (1995) A tightly bound chromosome antigen is detected by monoclonal antibodies in a ring-like structure on human centromeres. Chromosoma **103**: 559-566.

Ip W, Fischman DA (1979) High resolution scanning electron microscopy of isolated and *in situ* cytoskeletal elements. J Cell Biol **83**: 249-254.

Jeppesen P (1994) Immunofluorescence techniques applied to mitotic chromosome preparations. In: Methods in Molecular Biology, vol. **29**: Chromosome Analysis Protocols. Gosden JR (ed) Humana Press, Totowa, NJ, pp. 253-285.

Kelley RO, Dekker RAF, Bluemink JG (1973) Ligand-mediated osmium binding: its application in coating biological specimens for scanning electron microscopy. J Ultrastruct Res **45**: 254-258.

Macgregor H, Varley J (1988) Working with Animal Chromosomes. John Wiley & Sons, Chichester. pp. 13-24.

Martin R, Busch W, Herrmann RG, Wanner G (1994) Efficient preparation of plant chromosomes for high-resolution scanning electron microscopy. Chromosome Res **2**: 411-415.

Maruyama K (1983) Stereoscopic scanning electron microscopy of the chromosomes in *Vicia faba* (broad beans). J Ultrastruct Res **82**: 322-326.

Mullinger AM, Johnson RT (1983) Units of chromosome replication and packing. J Cell Sci **64**: 179-193.

Murphy JA (1978) Non-coating techniques to render biological specimens conductive. Scanning Electron Microsc **1978**; II : 175-193.

Neurath PW, Ampola MG, Vetter HG (1967) Scanning electron microscopy of chromosomes. Lancet **2**: 1366-1367.

Pawlowitzki IH, Blaschke R (1971) Preparing chromosomes for scanning electron microscopy without enzymic digestion. J Microsc **93**: 119-122.

Retief AE, Rüchel R (1977) Histones removed by fixation. Their role in the mechanism of chromosomal banding. Exp Cell Res **106**: 233-237.

Rizzoli R, Rizzi E, Falconi M, Galanzi A, Baratta B, Lattanzi G, Vitale M, Manzoli L, Mazzotti G (1994) High resolution detection of uncoated metaphase chromosomes by means of field emission scanning electron microscopy. Chromosoma **103**: 393-400.

Sanchez-Sweatman OH, De Harven EP, Dubé ID (1993) Human chromosomes: evaluation of processing techniques for scanning electron microscopy. Scanning Microsc **7**: 97-106.

Sillar R, Young BD (1981) A new method for the preparation of metaphase chromosomes for flow analysis. J Histochem Cytochem **29**: 74-78.

Sivak A, Wolman SR (1974) Chromosomal proteins in fixed metaphase cells. Histochem **42**: 345-349.

Smith BVK (1970) The application of scanning and transmission electron microscopy to a study of whole chromosomes. Micron **2**: 39-57.

Squarzoni S, Cinti C, Santi S, Valmori A, Maraldi NM (1994) Preparation of chromosome spreads for electron (TEM, SEM, STEM), light and confocal microscopy. Chromosoma **103**: 381-392.

Stenman S, Rosenqvist M, Ringertz NR (1975) Preparation and spread of unfixed metaphase chromosomes for immunofluorescence staining of nuclear antigens. Exp Cell Res **90**: 87-94.

Sumner AT, Ross A (1989) Factors affecting preparation of chromosomes for scanning electron microscopy using osmium impregnation. Scanning Microsc Suppl. **3**: 87-99.

Takayama S, Taniguchi T, Kitasumi H (1985) Scanning electron microscopic examination of the secondary constriction of Vero cells. Exptl Cell Res **157**: 556-560.

Tan EM (1989) Antinuclear antibodies: diagnostic markers for autoimmune diseases and probes for cell biology. Adv Immunol **44**: 93-151.

Watt JL, Stephen, GS (1986) Lymphocyte culture for chromosome analysis. In: Human Cytogenetics: a Practical Approach. Rooney DE, Czepulkowski BH, (eds) IRL Press, Oxford, pp 39-55.

Discussion with Reviewers

B. Hamkalo: Why is the chromosome overlap shown in Fig. 11 considered "unnatural"?
Author: The chromosome indicated appears to be floppy, that is, it has no natural stiffness. Configurations seen in living and in fixed cells suggest that bends of the kind illustrated do not occur naturally, and the chromosome in Fig. 5a also appears to be stiffer. The implication is that the chromosome in Fig. 11 has either been distorted during the drying process after spreading, or has been rebuilt in an artificial configuration during osmication.

B. Hamkalo: Does the author have an explanation for why PBS treatment results in so much better definition in the methanol-acetic acid-fixed preparations?
Author: It is assumed that washing the fixed chromosomes in PBS (or indeed probably almost any buffer) washes away a layer of material, perhaps cytoplasmic protein, but possibly a superficial layer of the chromosome itself (see Hernandez-Verdun and Gautier, 1994), thereby exposing the chromosomal fibres. So far as I know, this hypothesis has not been formally tested.

E. De Harven: With reference to Figure 2, do you imply that the average diameter of chromosomal fibres

after three stages of osmication is only 30 nm? If this is the case, what do you regard as the diameter of the un-osmicated fibre?
Author: In the particular experiment the results of which are illustrated in Figure 2, un-osmicated specimens were not prepared. In specimens that have not been treated in any way after spreading on glass, no fibres are visible. Treatment with trypsin or PBS, followed by glutaraldehyde fixation, give fibre diameters in the region of 20-30 nm. This would suggest that (i) methanol-acetic acid-fixed chromosomes show much the same fibre diameter that would be expected in a native chromosome, and (ii) that **light** osmication does not cause any significant increase in fibre diameter. There is, however, some variation between and within individual experiments, and it would not be justifiable to combine results from separate experiments, although trends found in repeated experiments of the same type are reproducible.

B. Hamkalo: The morphology of chromosomes in Fig. 7f is considerably poorer than others shown in this figure. Is this due to the additional steps required for immunogold labelling and intensification? If so, has the author tried to introduce a brief fixation step prior to adding immunogold or to modify chemically the primary antibody (with, for example, biotin) so that after this reaction chromosomes could be fixed with glutaraldehyde and detected with streptavidin gold?
Author: The relative degradation of morphology seen in Fig. 7f could well be due the additional treatments that this specimen was subjected to, although there is quite a bit of variation in this type of preparation regardless of the immunolabelling treatment used. It would certainly be valuable to try the modifications suggested, and it might well be expected that an intermediate fixation stage would result in improved morphology at the end of the procedure. Obviously, in any procedure as complicated as this, the number of possible variations is rather large, and it has not been possible to try all the useful ones yet.

B. Hamkalo: Given all the morphological changes described, what does the author think about the chances of defining the *bona fide* structure of the chromosome?
Author: This is a very speculative matter! It is relatively easy to suggest what the best procedures to maintain a lifelike image of chromosomes would involve. Firstly, it is clearly necessary to avoid flattening on to the substrate, and this almost certainly means that drying must be avoided. Secondly, the surrounding cytoplasm must be removed, without extracting chromosomal components or altering chromosome structure. Thirdly, it will be necessary to find some reliable way of attaching to the substrate chromosomes prepared according to these principles. If a number of independent procedures, based on such principles, produce similar images of chromosomes, then it is probably reasonably safe to say that a good representation of chromosome organisation has been attained.

T.D. Allen: You rightly state that 3:1 methanol-acetic fixation and spreading results in chromosome flattening, and that cytocentrifugation obviates this step. However, in your materials and methods, you state that after centrifuging, the slides were left to dry - surely air-drying in this situation is going to flatten the chromosomes via the surface tension as the air-water interface passes through the chromosomes to a far greater effect than the evaporation of the more volatile methanol-acetic mixture (75% methanol). While avoiding the problems of methanol-acetic fixation, you have replaced them with air-drying from an aqueous medium. In our own experiments, we have deliberately 'overfilled' the cytocentrifuge well, and carried out the preparation direct from moist without air-drying. This would also perhaps remove the need for detergent extraction that you state for your cytocentrifuge preparations.
Author: Your point is well taken - drying of any sort would be best avoided, although it clearly helps to attach cells and chromosomes to the substrate. With cytocentrifuge preparations, it was suspected that cytoplasmic proteins might nevertheless help to support the chromosomes, although the image in Fig. 6a casts doubt on this. I suspect that, however the chromosomes might be prepared by cytocentrifuging, detergent extraction of cytoplasm would still be necessary. On an historical note, the method described for preparing chromosomes for scanning electron microscope immunocytochemistry was derived directly from methods used for light microscope immunocytochemistry; in view of the sensitive nature of the antigens being investigated, it was necessary to proceed cautiously, and adapt existing procedures for light microscopy to the scanning electron microscope, without attempting any radical new approaches.

T.D. Allen: Does it really need nine repetitions to be consistent in osmium impregnation? During the later stages of our own studies, we reduced the protocol to OsO_4 fix, TCH impregnation, OsO_4 fix, i.e. one cycle only.
Author: In practice we determine the number of cycles of osmium impregnation empirically, and the optimal degree of impregnation varies across the specimen, and from one specimen to another. Precise counting of the number of cycles of osmication is obviously necessary for the sort of studies described in this paper, but for routine use, the osmication is repeated until the techni-

cian thinks it 'looks right'. I suspect the number of cycles of osmication required may also be influenced strongly by the precise details of the methanol-acetic acid-fixation, which is something that appears to vary considerably from one laboratory to another.

T.D. Allen: 'Lightly coated' with platinum. How thick was this coating - platinum is known to be grainy and not to form a continuous coat with maybe 3-4 nm thickness at least.
Author: Partly because of the problems mentioned by the questioner, we did not attempt to measure the thickness of the platinum coating. We suspect that it would in fact be very variable on the rough and fibrous surfaces that we were dealing with. In practice, we found no differences in detailed appearance or in dimensions of the surface structure of the chromosomes whether they were coated or not, suggesting that the coating was very thin. On a technical point, it appeared that the coating was only necessary to provide conductivity across the glass substrate; uncoated chromosomes that had been spread on pre-coated glass coverslips did not charge up.

T.D. Allen: DAB as the end point is not a 'point' localisation, as colloidal gold is - why was it chosen in preference to colloidal gold? Why was it felt necessary to enhance with silver the 10 nm colloidal gold, which is well within the resolution of the field emission instrument without silver intensification?
E. de Harven: Why was it necessary to silver enhance these 10 nm gold markers so much, apparently about 30 times? Using a Hitachi S-800 FE instrument, 10 nm gold particles can be resolved almost without any enhancement.
Author: Cytocentrifuge preparations, as used here, were originally developed for light microscopy, and the DAB end point could be monitored conveniently by light microscopy, thus provided a quick check on whether any particular procedure had worked. The same principle was also applied when using colloidal gold; in fact, although the gold particles should indeed have been visible in the scanning electron microscope without intensification, the actual number of particles was so low (see Fig. 7f) that it would have been impossibly tedious to locate them without some degree of intensification. It should be emphasised that, at this stage, it was not the intention to demonstrate high resolution, but merely to show that the sensitive antigens being detected could be preserved in chromosomes prepared for scanning electron microscope. High resolution would, of course, be a long term goal.

M. Malecki: Considering the limits in preparation of chromosomes by immunolabelling for SEM which you demonstrated in your paper, and taking into account progress in specimen preparation, including multiple labelling and energy transfers, for laser scanning fluorescence microscopy, would you consider it possible to have some of your questions concerning distribution of the AC1 antibody on chromosomes to be answered by this latter technique?
Author: Confocal laser scanning microscopy would indeed be a potential alternative method for studying the three-dimensional distribution of antibody AC1 (and indeed any other type of label) on chromosomes, albeit with a reduced ultimate resolution compared with SEM. In practice there might be two possible problems. The first concerns the resolution of the microscope in the Z direction, which according to some reports would be scarcely adequate (e.g. Brakenhoff *et al.*, 1989). However, the fact that the distribution of AC1 sometimes appears as a ring round the centromeric constriction by focusing the fluorescence microscope up and down suggests that axial resolution should not, in fact, be a problem. The second point concerns the three-dimensional preservation of the specimen. This is easy to assess using SEM, but not so easy using confocal microscopy, although the latter has the advantage that the specimen can be examined wet.

Additional Reference

Brakenhoff GJ, van der Voort HTM, van Spronsen EA, Nanninga N (1989) Three-dimensional imaging in fluorescence by confocal scanning microscopy. J Microsc **153**: 151-159.

Scanning Microscopy Supplement 10, 1996 (pages 177-187) 0892-953X/96$5.00+.25
Scanning Microscopy International, Chicago (AMF O'Hare), IL 60666 USA

ELECTRO-OPTICAL IMAGING OF F-ACTIN AND ENDOPLASMIC RETICULUM IN LIVING AND FIXED PLANT CELLS

Nina Stromgren Allen[1,2*] and Marty N. Bennett[2]

[1]Department of Botany, North Carolina State University, Raleigh, NC, and
[2]Department of Biology, Wake Forest University, Winston-Salem, NC

(Received for publication August 28, 1995 and in revised form December 9, 1996)

Abstract

Confocal and video micrographs of living and fixed alfalfa roots, onion epithelial and pear pollen cells illustrate the architecture of the cytoskeleton and endoplasmic reticulum in plant cells. Fixation of plant tissues to preserve cytoplasmic structure poses special problems. When possible, emphasis should be placed on the imaging of structures in stained living cells over time. The early events that occur when Nod factors or bacteria elicit nodule formation in alfalfa roots will illustrate several approaches to plant cell fixation, staining and imaging. The first observable events after Nod factor stimulation occur in root hairs and are changes in rates of cytoplasmic streaming, nuclear movements, and changes in the shape of the vacuole. Within ten minutes, the endoplasmic reticulum shifts position towards the tip of the root hair. For comparison, the endoplasmic reticulum localization in pollen tubes and onion epithelial cells will be illustrated. The actin cytoskeleton undergoes a series of changes over a twelve hour period. These changes in the cytoskeleton are spatially and temporally correlated with the observed growth changes of the root hairs. This dynamic change of the actin filament and endoplasmic reticulum and associated secretory vesicles in these root hairs suggests a mechanism for the observed root hair growth changes.

Key Words: actin, endoplasmic reticulum, alfalfa, video microscopy, onion epithelial cells, pollen tubes, cytoskeleton, Nod factors.

*Address for correspondence:
Nina Stromgren Allen
Department of Botany, Box 7612
North Carolina State University
Raleigh, NC 27695-7612
Telephone Number: 919-515-8382
FAX Number: 919-515-3436
E-mail: nina_allen@ncsu.edu

Introduction

Video microscopy and computer manipulations have permitted better visualization of dynamic changes in the internal structures of plant cells [2] in both normal cells and in cells stimulated by internal or external agents. Differential interference contrast (DIC) light microscopy methods give an overall impression of the activities and structures in a living cell, but to insure the best identification of specific structures, fluorescently labeled antibodies or other specific tags are used [8]. Some labels easily enter living plant cells, but for other probes and antibodies the cells must be fixed. It is generally agreed that visualization and fixation can be more problematic in plant cells than in animal cells. Contributing to these problems are the presence of cell walls with strongly autofluorescent lignin, large often highly acidic vacuoles as well as autofluorescent chloroplasts. Dyes, such as the Ca^{2+} indicator fura-2, will quickly enter the vacuole of the plant cell, a structure with a high Ca^{2+} content. Hence the desired Ca^{2+} measurements in the cytoplasm will be obscured by vacuolar fluorescence. Vacuolar membrane disruption can occur during fixation resulting in the release of compounds disruptive to the cytoplasm and the cytoskeleton. It is important to select plant tissues with low autofluorescence as has been done here.

Plants are stationary and they respond to the environment by changing their growth. Early responses by plant cells leading to growth changes would appear to be linked to changes in the localization of actin filaments, microtubules, and the endoplasmic reticulum (ER) [9]. In order to detect such changes in the cell architecture we and others have tried to improve methods of staining and fixation of plant cells. Some of these will be described and their usefulness will be briefly discussed.

The plant cytoskeleton consists of actin, microtubules, and intermediate filaments and the actin filaments often colocalize with the endoplasmic reticulum [2, 12]. In animal cells the distribution of actin filaments is well known for fixed and living cells through the use of electron and fluorescence microscopy. Actin filaments have been imaged in a number of plant cells with particular success in cryofixed specimens [11, 12, 18]. The method of tip growth and its relation to the actin cytoskeleton is of interest to plant biologists and a controversy exists regarding the presence or absence of actin at the tip of growing cells of germinating pollen and root hairs. Few actin-associated proteins (myosins, profilin), which might effect the actin arrays, have been characterized in plant cells. So far this field of study lags far behind what can be described for animal cells [6].

We have been particularly interested in the imaging over time of growing alfalfa root hairs after they have been challenged by the alfalfa specific factor NodRm-IV(S) [3, 5 and M. Bennett, D. Erhardt, S. Long and N. Allen, in preparation] in order to understand parts of the signal transduction pathway from the reception of Nod factors by the plant ending in the observed growth change. Nod factors produced by rhizobia are lipo-chitooligosaccharides with different long-chain fatty acids at the non-reducing glucosamine moiety [10, 14, 26]. Nod factors, such as NodRm-IV(s), specific to *Medicago sativa* (alfalfa) are produced by the soil bacteria *Rhizobium meliloti* in response to plant flavonoids [13]. These factors can cause many developmental changes in the plant eventually leading to the formation of pseudonodules on the roots [7]. Nod factors have recently been postulated to be plant growth regulators in that they can redirect plant growth in non-leguminous plants [19]. One of the first observed morphological changes is the deformation of growing root hairs and previous work implicated a role for the cytoskeleton in this change in growth [7, 18]. Hairs arise by tip growth from epidermal cells on the root (Fig. 1) and are long tubular projections often containing the cell nucleus. Within 3-4 hours after roots are exposed to *Rhizobia* or Nod factors, they will form a characteristic curl or show other deformities. Our aim was to image changes in the location of actin filaments and the endoplasmic reticulum in Nod factor challenged root hairs over time to further understand the basis for the growth form change.

The location of the ER in growing pollen tube tips was imaged in order to compare the location of the ER in tip growing root hairs and pollen tubes. The ER location obtained in living onion cells using a cooled charged coupling device (CCD) camera coupled with image enhancement illustrates the increase in resolution that can be obtained with this imaging mode in optically favorable specimens.

Materials and Methods

Plant material

Alfalfa seeds, *Medicago sativa* L., (AS13, Ferry-Morse Seed, Modesto, CA) were surface sterilized by soaking for 30 min in 70% ethanol followed by 100% bleach for 30 min. Seeds were rinsed thoroughly in distilled water and germinated on Whatman No. 4 filter paper moistened with distilled water in an inverted Petri plate at room temperature. Seedlings (24 hours old) were transferred to the surface of square Petri plates containing buffered nodulation medium (BNM) and 13% Bacto agar (Difco Laboratories, Detroit, MI). The BNM was composed of 2 mM $CaSO_4.2H_2O$, 2 mM 2-(N-morpholino)ethane sulfonic acid (MES), 0.5 mM $MgSO_4.7H_2O$, and 0.5 mM KH_2PO_4, with the minor salts 50 nM $Na_2Mo_4.2H_2O$, 50 nM H_3BO_3, 50 nM $FeSO_4.7H_2O$, 50 nM $MnSO_4.H_2O$, 16 nM $ZnSO_4.7H_2O$, 1 nM $Na_2MoO_4.2H_2O$, 0.1 nM $CoCl_26H_2O$, and 0.1 nM $CuSO_4.5H_2O$. The pH was adjusted to 6.0 using 2N KOH. The plates were sealed with Parafilm, and the seedlings were grown in the dark at 16°C. The inner onion epidermal peels were prepared as described in [2]. Pear pollen tubes were germinated on the microscope slide in pollen germination medium (PGM) [25].

Nod factor treatment

Tissues were treated with or without 10^{-8} M NodRm-IV(S) (obtained from Drs S.R. Long and D.W. Ehrhardt, Department of Biological Sciences, Stanford University, Stanford, CA) for time periods of 5, 10, and 15 min, and between 1 and 12 hours. Agar plates, containing the growing seedlings were placed horizontally in the growth chamber and 10^{-8} M Nod factor or buffer was applied along the root using a transfer pipette. The plated seedlings were kept horizontal, which ensured that the Nod factor remained in contact with the tissue for the duration of the treatment.

Fixation and staining

Actin. The lower 3 cm of treated and untreated alfalfa roots, which include both the young root hairs (48-72 hours old) and the zone of emerging root hairs (Fig. 1) were excised and placed individually on microscope slides containing 0.1 mM m-Maleimidobenzoyl-N-hydroxysuccinimide ester (MBS) (Pierce, Rockford IL) [22] in actin stabilizing buffer (ASB) for 30 min. ASB was a modified version of the cytoskeletal stabilization buffer described in Abe and Davies [1] consisting of 5 mM 4 (-2- Hydroxyethyl)-1-Piperazineethanesulfonic Acid (HEPES), 10 mM Mg acetate, 2 mM Ethylene Glycol Bis-(b-Aminoethyl Ether) N,N,N',N'-Tetraacetic Acid (EGTA), and 1 mM Phenylmethyl-Sulfonyl Fluoride (PMSF) (Sigma, St. Louis, MO), adjusted to a pH of 7.5 with 1N KOH.

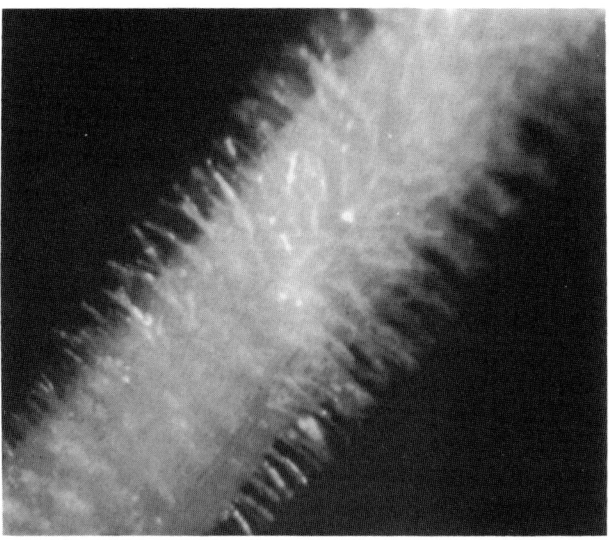

Figure 1. Micrograph of a root of *Medicago sativa* (alfalfa) showing root hairs emerging from the epidermal cells. The nod receptive hairs used in this study varied in length from 120-150 μm long and 10-22 μm wide. Field width = 550 μm.

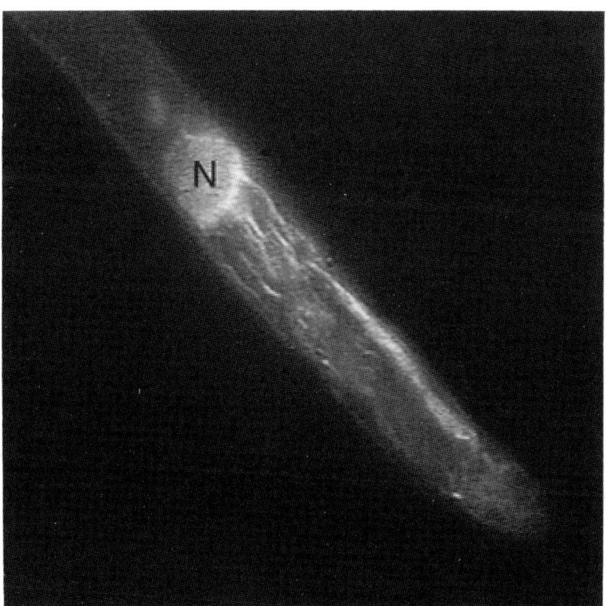

Figure 2. Alfalfa root hair (18 μm in diameter) stained for actin with rhodamine phalloidin. The nucleus is surrounded by actin filaments with one filament bundle extending from the nucleus. Figure 3F is an optical section taken a few microns below the section depicted in this micrograph.

After primary protein crosslinking with MBS, No. 1 coverslips were carefully placed over the root tips. By using filter paper as wicks, the tissue was perfused with freshly prepared 2.5% paraformaldehyde (Trioxymethylene) (Purified) (Fisher, Fair Lawn, NJ) and 0.05% glutaraldehyde (Ted Pella, Inc., Redding, CA) in ASB. The roots were fixed for 45 min in the dark. After the root tips were washed with ASB, they were digested for 30 min with 1.0% Cellulase (Sigma) followed by a 30 min extraction in 0.5% Triton X-100 in ASB. The root tips were briefly rinsed with ASB prior to staining.

The stabilized actin cytoskeleton was labeled by perfusing 100 μl of 0.17 μM rhodamine phalloidin (Molecular Probes, Eugene, OR) in ASB. Cells were placed in a moistened chamber for 4-6 hours in the dark and then rinsed. The root tips were mounted in Vectashield mounting medium for fluorescence (Vector Laboratories, Inc., Burlingame, CA) and gently crushed between microscope slide and coverslip and sealed with fingernail polish.

Endoplasmic reticulum. A stock solution containing 1 mg 3,3'-dihexyloxacarbocyanine iodide [$DiOC_6(3)$] in 1 ml dimethylsulfoxide (DMSO) (Molecular Probes) was diluted 1000:1 in BNM or PGM. Onion or alfalfa specimens were incubated in BNM stain solution for 10 min, then rinsed thoroughly three times in BNM and mounted for observation [20, 21]. Pollen grains were placed on microscope slides in PGM solution containing the $DiOC_6(3)$ stock solution in a 1000:1 dilution and germinated and recorded over time on the confocal microscope.

Microscopy

The alfalfa root hairs stained with rhodamine phalloidin were viewed on an Axiophot microscope (Carl Zeiss, Inc., Thornwood, NY) equipped with epifluorescence and AVEC-DIC optics [4]. Most observations were made with a 63 x, 1.4 N.A. planapochromatic objective and further magnified with a 1.25 x zoom lens. The root hairs were found to have very little autofluorescence. Fluorescent images were visualized with a Hamamatsu Silicon Intensified Target camera (Hamamatsu Photonic Systems, Division of Hamamatsu Corp., Bridgewater, NJ) and then computer enhanced using the Image-1 system (Universal Imaging, West Chester, PA). The images were then recorded on a Panasonic TQ2028F optical memory disk recorder (OMDR), digitized, stored as TIFF files and later printed on Ep-2 photographic paper at 312 pixels per inch with a PowerMacintosh 8100 with Adobe Photoshop and Fuji or Codonics (Middleburg Heights, OH) Color printer.

The alfalfa root hairs and pear pollen grains stained with $DiOC_6(3)$ were imaged with a Biorad (Hercules, CA) 600 confocal laser scanning on a Zeiss IM35

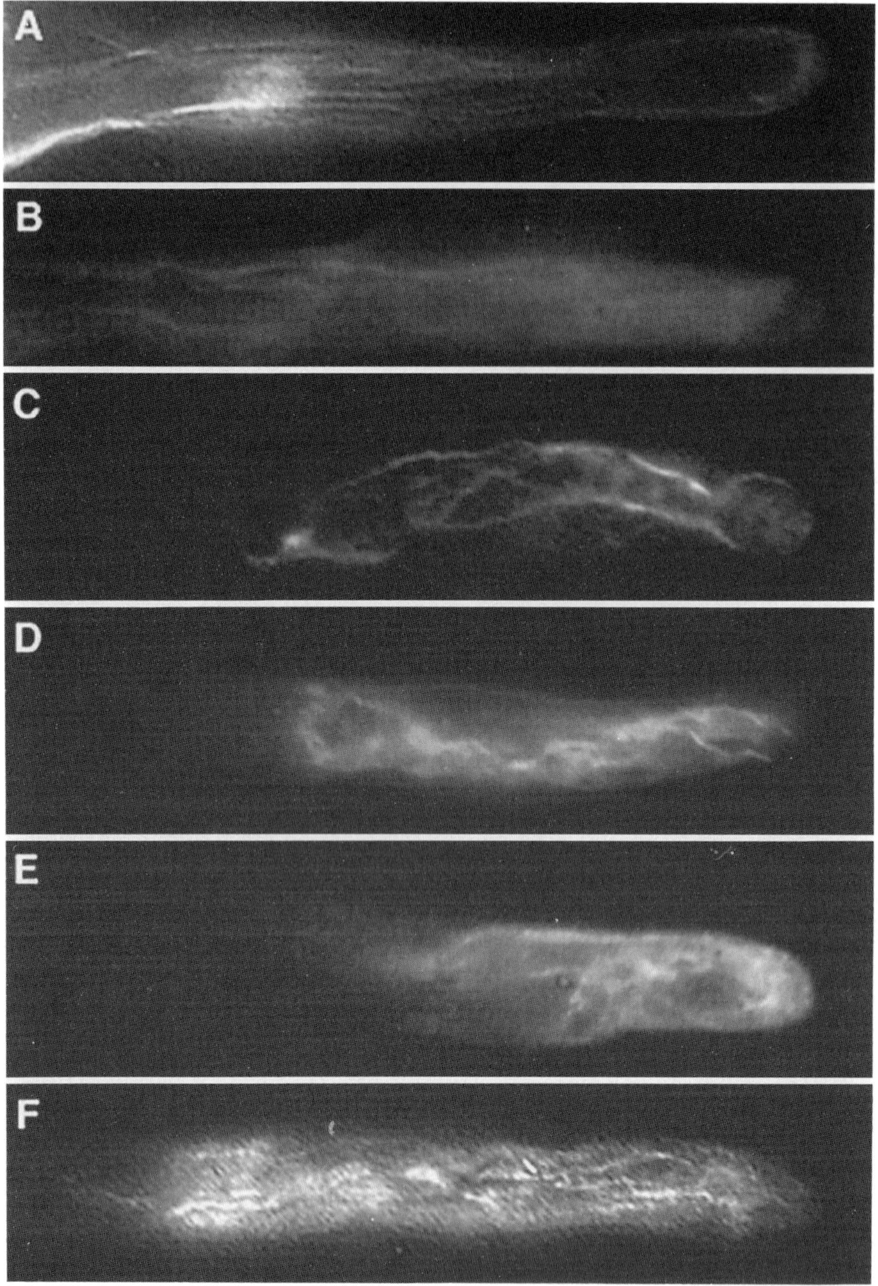

Figure 3. A to F. A montage of representative actin root hairs with rhodamine phalloidin stained actin filaments. Note that some hairs contained actin staining in the very tip. Long filaments of actin ran the length of the hairs and often connected into a more diffuse network of actin at the tip. When focusing through the tip, the network appeared to be connected to the long bundles entering the dome area.

microscope. The fluorescent images are confocal and the DIC images were recorded simultaneously, but they are not confocally derived. These images were time-lapsed on the OMDR, and some were digitized and stored on Image I as TIFF files. The onion epithelial cell was imaged on the Zeiss Axiophot equipped with a 63x, 1.4 N.A. objective and a matching oiled 1.4 N.A. condenser and a Hamamatsu Cooled CCD camera. A 4x zoom lens was present between the microscope and camera. The image was further enhanced and digitized using the Metamorph Image Analysis System (Universal Imaging). The TIFF files were then processed with the Adobe Photoshop and printed using a Codonics Color printer as above.

Results

Actin localization in alfalfa root hairs

25 seedlings with many root hairs were fixed, stained with rhodamine phalloidin, observed and recorded as described in Materials and Methods. The transparent hairs ranged in diameter from 10-22 μm and extended out in all directions from the thick, opaque root making imaging difficult. Representative images of such

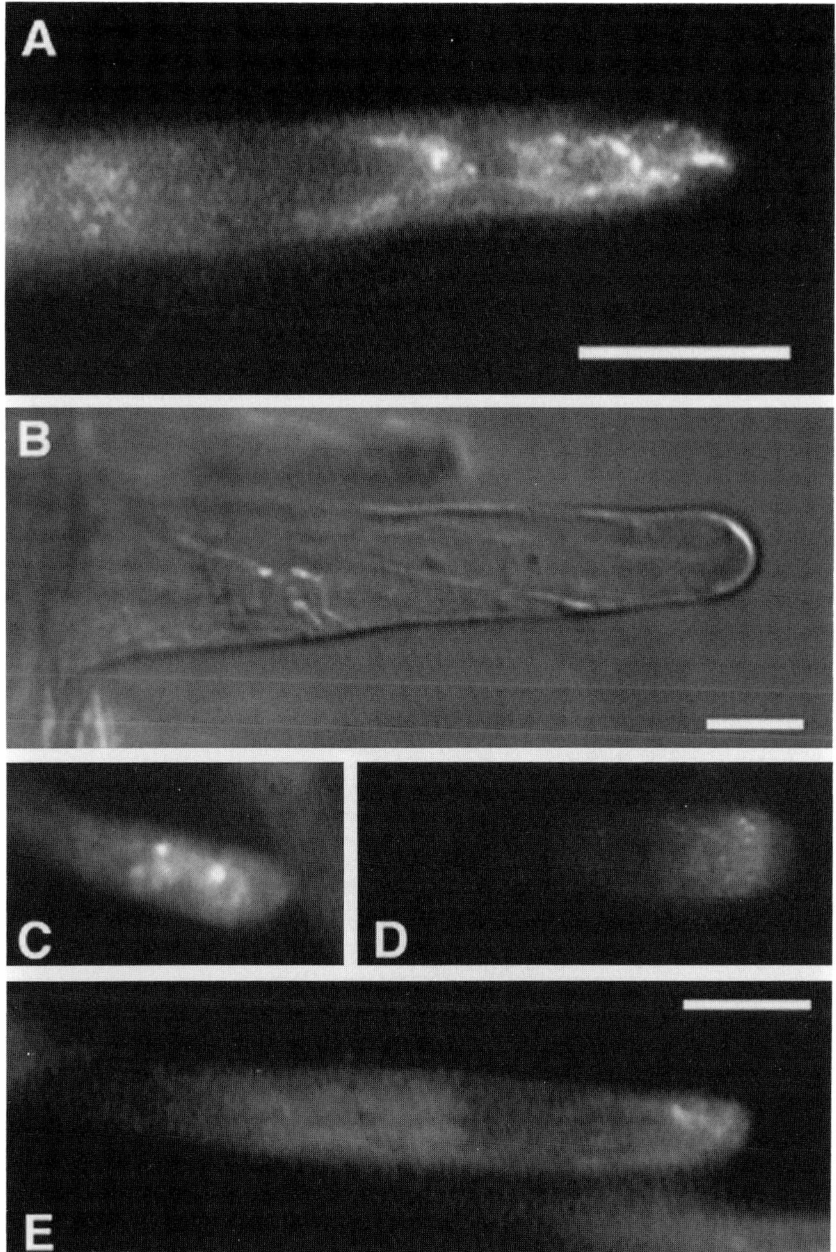

Figure 4. Changes in the actin cytoskeleton 10 min after exposure to 10^{-8} M NodRm-IV(S). The effectiveness of the fixation process was demonstrated by the fact that the appearance of the fixed hair imaged in DIC microscopy (B) did not change from that of living cells. In micrographs A, C, D, and E discrete foci of actin staining occurred in the tips of the root hair. The staining pattern indicated actin filament fragmentation. This striking pattern was observed only in cells exposed to Nod factor and was not present in older root hairs (those greater than 200 μm in length) or epidermal cells. The observed cytoskeletal alterations occurred prior to any observable changes in cell morphology. Bar = 20 μm.

stained and fixed root hairs can be seen in Fig. 2 and 3. Fig. 2 is a representative sample in a root hair of the intense actin staining that always surrounded the nucleus, whether it was present in the root hair or in the cell body (not shown). Actin filaments also course in linear arrays parallel to the long axis of the hair. Fig. 3 depicts 6 different root hairs in which the parallel actin bundles are best observed in A, B, C and E in the area of the cytoplasm where the microscope was in focus just below the plasma membrane and above the vacuole. No actin was seen in the vacuole. In some hairs a distinct cap of actin staining is clearly visible in the apical dome of the root hair as in A, C, E, and Fig. 2. Similar patterns of actin distribution have recently been described for *Hydrocharis* root hairs [21].

Actin distribution in NodRm-IV(S) stimulated root hairs

25 seedlings were exposed to 10^{-8} M NodRm-IV(S) for 10 min and Fig. 4 illustrates the dramatic shift in the rhodamine phalloidin staining pattern when compared to untreated root hairs. As seen in A, C, D, and E, a punctate staining pattern appeared at and near the tip of the root hairs. This pattern could be observed in 25% of

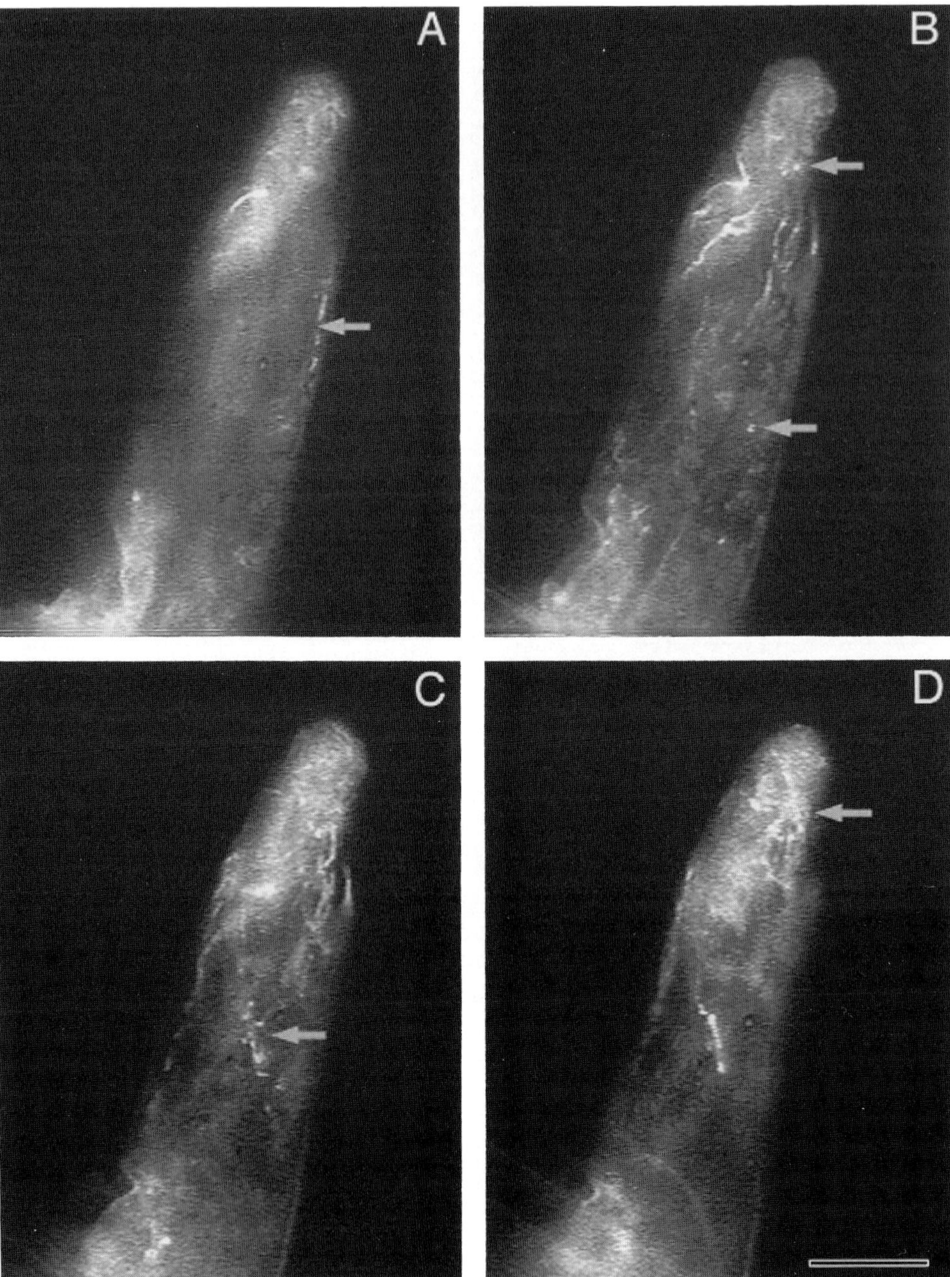

Figure 5. The effects of Nod factor on the cytoskeleton 15 min post-perfusion. This series of optical sections demonstrated the initiation of actin cable fragmentation marked by discontinuities in the longitudinal cables (arrow A, C) and the formation of discrete foci staining throughout the cytoplasm (arrow B, C). Although some actin filaments appear to remain, these micrographs show a widely occurring breakdown of the actin cytoskeleton. Bar = 20 μm.

all root hairs demonstrating actin labeling and was indicative of actin fragmentation. A qualitative assessment would indicate a somewhat higher diffuse actin staining than what was seen in unstimulated root hairs (Fig. 3). Fig. 4B is a DIC image of a root hair demonstrating that the cytoplasm of fixed and stained root hairs appeared very much like that of living hairs, thus indicating that very little apparent damage occurred during the preparation process. Fig. 5 is a series of four optical sections taken through a root hair 15 min post Nod factor perfusion. Fig. 5A presents the view just beneath the plasma membrane with the arrow indicating a fragmenting actin filament. Fig. 5B is an image taken with the microscope focused further into the cell demonstrating that actin foci were present (arrows). "Curled" actin filaments were seen near the left tip of the root hair. Fig. 5C again shows actin foci at the arrow. These images would indicate a widespread breakdown of the actin filaments into shorter fragments that may not be tightly stretched.

Although no images are shown for root hairs one hour after Nod factor exposure, we did find that the actin staining appeared normal and in DIC we could observe no differences between treated and untreated

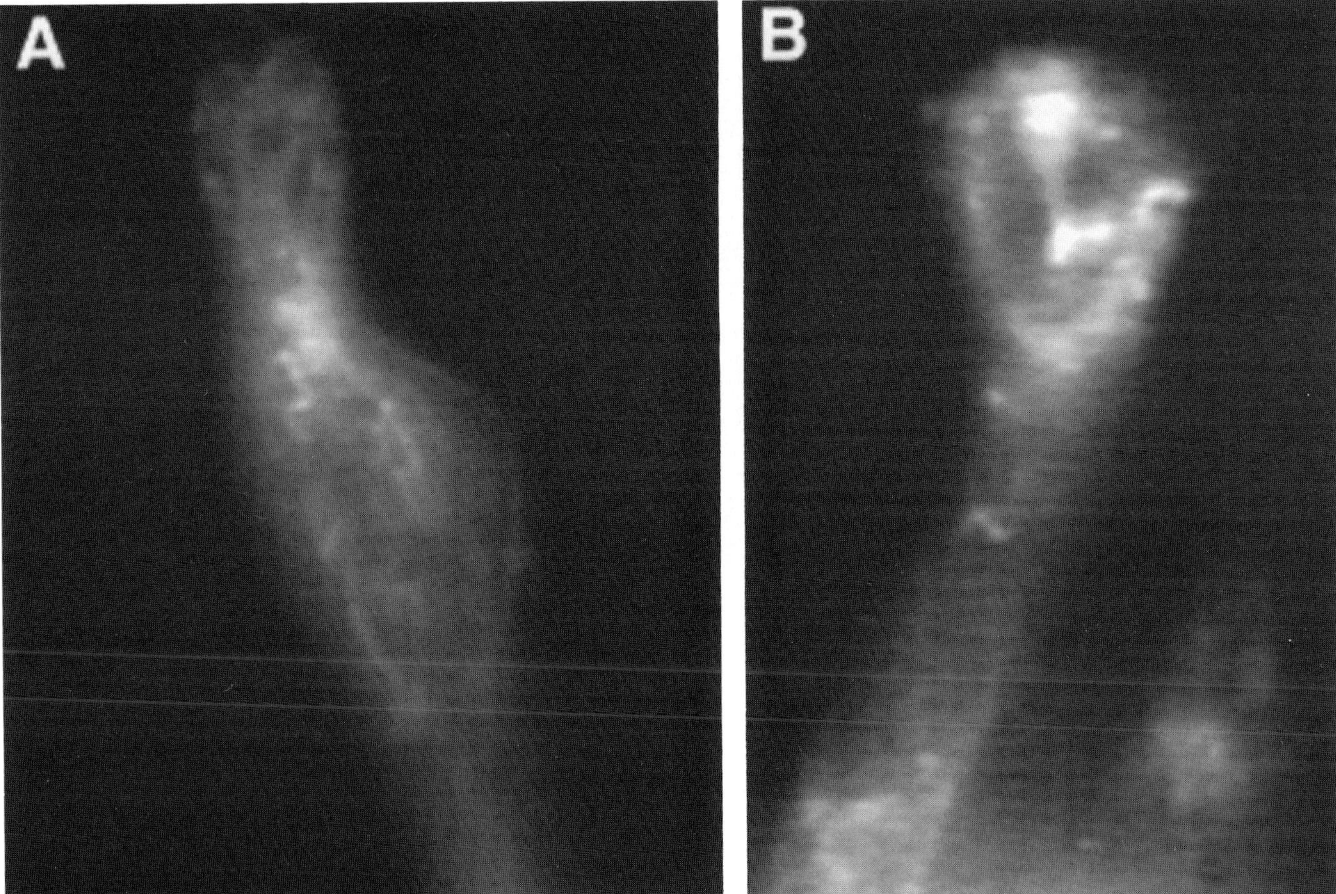

Figure 6. Rhodamine phalloidin labeled root hairs 3 hours post-perfusion with Nod factor with clearly visible root hair deformations. The root hairs were no longer cylindrical and became crooked or bulbous with obvious distortion. Within the tips of these structures, intense foci of actin staining were present very close to the plasma membrane. Within the same optical plane of these foci, a definite anastomosing pattern of filaments was often detected (A).

---

hairs. Both return to a normal streaming pattern. However, 3 hours after Nod factor exposure and just when root hair deformations were visible, the actin localized into foci and very few actin cables were present (Fig. 6 A,B). Fig. 6A is a root hair of the type that shows a bulge below the root tip with very distinct actin foci. Fig. 6B represents a common deformation in which the root hair will branch. The actin foci seem to occur in an upside down horseshoe shape and no filaments were observed. The staining was only seen when the microscope was focused near the plasma membrane (the magnification is high and the optical sections were estimated to be ca. 350 nm) and suggests that in areas where the root hairs deform, there could be focal aggregations of actin filaments. The exact time course and length of time during which the actin filaments break down is not clear. Observations (not shown here) of root hairs exposed to Nod factors ten hours earlier revealed the normal patterns of long parallel filaments, nuclear staining, and apical dome staining. It has been possible to develop a method that has allowed the imaging of actin staining in Nod factor perfused root hairs over time and to gain some insight into the changing actin distribution in such cells.

Detection of the endoplasmic reticulum in normal and nod factor stimulated alfalfa root hairs

$DiOC_6(3)$ is a lipid stain that fairly selectively visualizes the endoplasmic reticulum [16, 17, 20, 23] in both living animal and plant cells. Alfalfa root hairs were allowed to take up the dye for 10 min, placed in a perfusion chamber in BNM, and recorded at a rate of 1 frame every 4 seconds on the OMDR. Fig. 7a is a dual image video micrograph of the same root hair imaged in DIC and fluorescence microscopy modes. A large vacuole can be seen with no dye content. The unstimulated cells were observed and recorded over 30 min and

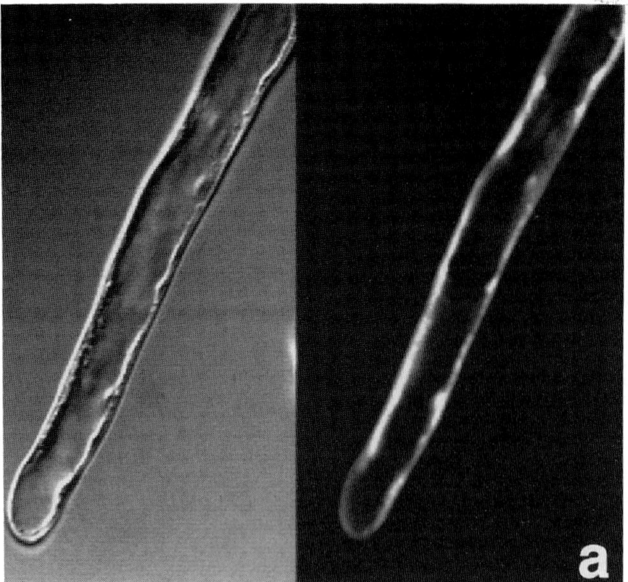

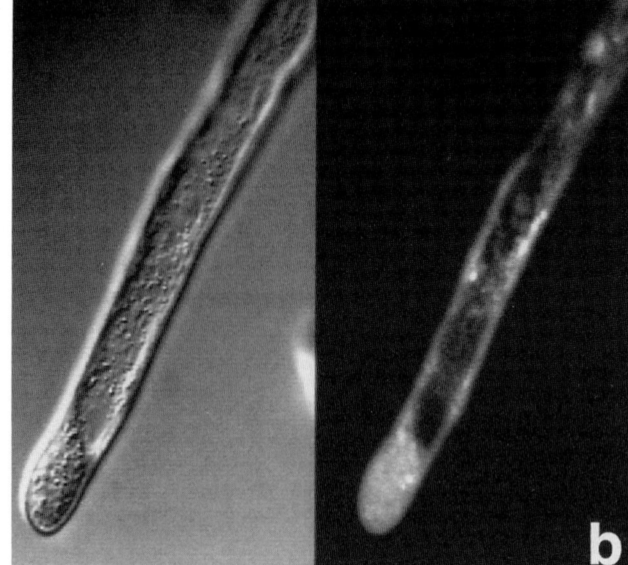

Figure 7. Confocal laser scanning micrograph of an alfalfa root hair stained with $DiOC_6(3)$ specific for the ER. Figure 7a depicts the unstimulated hair in DIC and fluorescence mode. The vacuole fills most of the root hair tip and the cytoplasm show general staining for ER. After NodRm-IV(S) stimulation, the cytoplasm rich in ER staining migrates to the tip of the root hair displacing the vacuole. The diameter of the root hair was 18 μm.

---

at this point the Nod factor was perfused across the root. Fig. 7b was recorded 12 min after Nod factor perfusion. The vacuole had shifted position as compared to Fig. 7a and many vesicles (DIC) as well as the stained ER had shifted towards the tip of the root hair. The time frame of this shift correlated well with the observed changes in the actin architecture after Nod factor perfusion.

Imaging of the endoplasmic reticulum in pear pollen tubes

Pollen tube and root hair tip growth may be based on the same phenomena [9] and it is felt that both actin and the ER are involved in this growth extension. The same staining methods were employed to image the ER in pollen tubes as was described for the alfalfa root hairs. Dual, simultaneous images (DIC and confocal fluorescence) of the tip and lower end of a germinating pollen tube from an hour-long OMDR recording of the germination of the pollen grain and growth of the pollen tube are seen in Figure 8. In that video recording, the cytoplasm in DIC was seen to stream vigorously carrying many small vesicles and particles to the tip of the pollen tube and the ER also was shifting position rapidly over time. The ER was present throughout the cell. The DIC image demonstrated that the very tip of the pollen tube was filled with cytoplasm, but the fluorescently labeled ER was excluded from that very tip. This differed from the alfalfa root, in which the ER occupied the very tip of the root hair.

Imaging of the endoplasmic reticulum in onion epithelial cells

The first and most commonly used plant cell for visualization of the ER has been the onion epithelial cell [2, 16] . These cells are optically favorable since they are present in very thin, transparent sheets and their cytoplasm is less dense than that found in most plant cells and in alfalfa root hairs or pollen tubes. It is possible to distinguish two different types of ER in onion cells [2]. Fig. 9 depicts a small area of an onion cell in focus just below the cell wall and plasma membrane. The cells were stained with $DiOC_6(3)$ and imaged using the Hamamatsu Cooled CCD followed by computer-enhancement with the Metamorph image analysis system. Staining made it possible to see the typical network of ER tubules, the saccules and a clump of putative secretory vesicles. This image would indicate that the dye is quite specific for the ER. We have not yet obtained such images from alfalfa root cells.

Conclusions

It is now possible to image with greater clarity some of the cytoskeletal elements present in plant cells [22] including now alfalfa and *Hydrocharis* [21]. This will lead to further understanding of many processes involving the cytoskeletal architecture. The shape of a plant cell is in part due to the underlying cytoskeleton [15, 27] and it is known that actin plays a significant role in tip

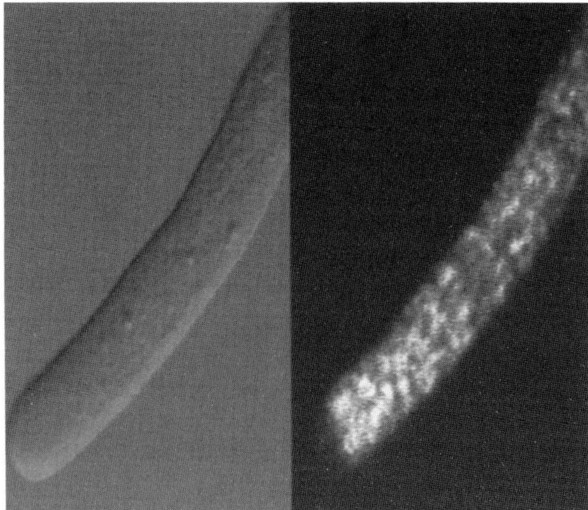

Figure 8. Confocal laser scanning micrograph of a pollen tube germinating from a pear pollen grain and vitally stained with $DiOC_6(3)$. Note that the ER staining does not extend to the very tip of the pear pollen tube. This is in contrast to the staining pattern observed for ER in alfalfa root tips.

----

growth processes in certain cells as well as in cellular elongation in general [9]. Hence if there are changes in the cytoskeleton during the normal growth process, one might expect to see changes in form. We have demonstrated that there is a breakdown of the actin cytoskeleton after exposure to NodRm-IV(S), which is correlated in time with the observed growth change (root curling) at three to four hours. We also show that the changes in distribution of the actin and ER in response to Nod factors occurred within ten minutes after exposure and resulted in a fragmentation of actin filaments and a movement of the ER and/or ER derived vesicles towards the root hair tip. It is not clear if the actin filaments are acting as tracks for vesicle movements to discrete locations or if the cytoskeleton itself serves as a template for morphogenesis. In either case, it is reasonable to predict that a change in the cytoskeleton would result in a change in the shape of the root hair.

The dome of the alfalfa root hair tip contains a meshwork of actin filaments, which has also been seen in *Hydrocharis* root hairs [21]. Ridge [18] using a different preparation method does not see actin at the tip of alfalfa root hairs. It will be important in the future to label the actin filaments in living alfalfa roots and to observe the precise actin localization at the tip of the root hairs and the general change in actin localization over time.

There is a difference in staining of the ER of pollen tube tips and alfalfa root hairs, which may indicate that

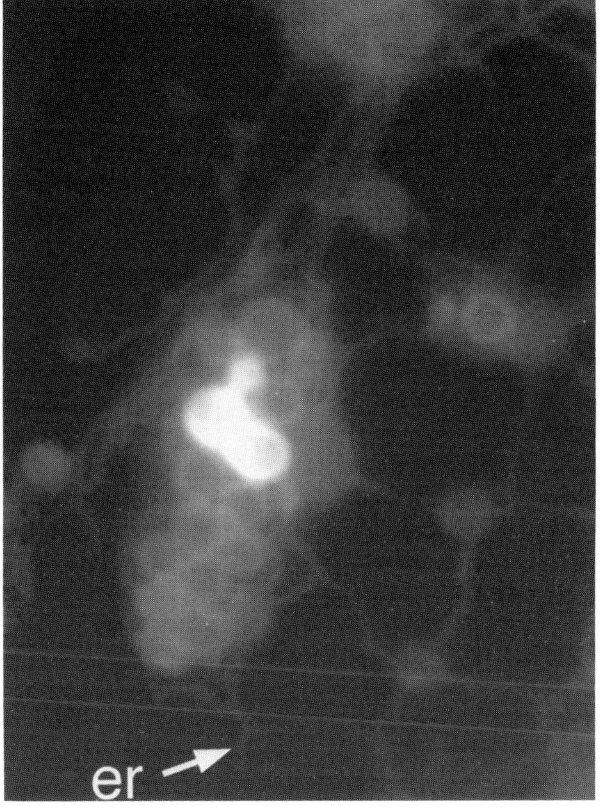

Figure 9. Onion epithelial cells from the adaxial side of an onion bulb scale stained with $DiOC_6(3)$. The ER tubules form a polygonal network (arrow) tightly appressed to the plasma membrane. A cluster of putative secretory vesicles are seen near the center of the image. Saccules are present where the tubules join each other. Field width = 24 μm.

----

the deposition of ER derived vesicles may differ in these two types of cells and hence tip growth itself may differ in some ways in these two cell types.

Acknowledgements

I would like to thank Sharon Long and David Ehrhardt for the NodRm-IV(S) and for introducing me to the alfalfa/*Rhizobium* system. I would like to thank Stephen Smith for the use of his confocal microscope. Thanks to Universal Imaging and Hamamatsu for the loan of the Metamorph and Cooled CCD camera as well as to Mark Hobson (Universal Imaging) for technical assistance. Also thanks to Linda Ross, Dennis Kubinski and Dana Moxley for technical assistance and to Suresh Tiwari for teaching me the pear pollen system.

References

1. Abe S, Davies E (1991) Isolation of F-actin from pea stems. Evidence from fluorescence microscopy. Protoplasma **163**: 51-61.

2. Allen NS, Brown DT (1988) Dynamics of the endoplasmic reticulum in living onion epidermal cells in relation to microtubules, microfilaments and intracellular particle movement. Cell Motil Cytoskeleton **10**: 153-163.

3. Allen NS, Bennett MN, Cox DN, Shipley A, Ehrhardt DW, Long SR (1994) Effects of nod factors on Alfalfa root hair Ca^{++} and H^{+} currents and on cytoskeletal behavior. In: Advances in Molecular Genetics of Plant-Microbe Interactions, Vol 3. Daniels MJ, Downie JA, Osbourn AE (eds). Kluwer Academic, Dordrecht. pp 107-113.

4. Allen NS, Jones N (1992) National resources. Video microscopy and motion analysis at Wake Forest University. Electron Microsc Soc America Bull **22**: 82-84.

5. Bennett MN (1995) Nodulation Factor Induced Early Responses of the Actin Cytoskeleton in Alfalfa Root Hairs. Master's Thesis. Wake Forest University, Winston-Salem, NC, USA.

6. Bohn W, Etzrodt D, Foisner R, Wiche G, Traub P (1995) Cytoskeleton architecture of C6 rat glioma cell subclones. Scanning Microsc Suppl. **10**: 285-294.

7. Brewin NJ (1993) The *Rhizobium*-legume symbiosis: plant morphogenesis in a nodule. Semin Cell Biol **4**: 149-156.

8. Haugland RP (1996) Handbook of Fluorescent Probes and Research Chemicals. Molecular Probes, Eugene, OR, USA.

9. Heath IB (1990) Tip Growth in Plant and Fungal Cells. Academic Press. San Diego. pp 91-210.

10. Lerouge P, Roche P, Faucher C, Maillet F, Truchet G, Promé JC, Dénairié J (1990) Symbiotic host-specificity of *Rhizobium meliloti* is determined by sulphated and acylated glucosamine oligosccharide signal. Nature **344**: 781-784.

11. Lancelle SA, Cresti M, Hepler PK (1987) Ultrastructure of the cytoskeleton in freeze-substituted pollen tubes of *Nicotiana alata*. Protoplasma **140**: 141-150.

12. Lichtscheidl IK, Lancelle SA, Hepler PK (1990) Actin endoplasmic reticulum complexes in *Drosera*. Their structural relationship with the plasmalemma, nucleus, and organelles in cells prepared by high pressure freezing. Protoplasma **155**: 116-126.

13. Long SR (1989) *Rhizobium*-legume nodulation: Life together in the underground. Cell **56**: 203-214.

14. Long SR (1990) *Rhizobium* sweet talking. Nature **344**: 712-713.

15. Parthasarathy MV, Perdue TD, Witztum A, Alvernaz J (1985) Actin network as a normal component of the cytoskeleton in many vascular plant cells. Amer J Bot **72**: 1318-1323.

16. Quader H, Schnepf E (1986) Endoplasmic reticulum and cytoplasmic streaming: Fluorescence microscopical observations in adaxial epidermis cells of onion bulb scales. Protoplasma **131**: 250-252.

17. Quader H (1990) Formation and disintegration of cisternae of the endoplasmic reticulum visualized in live cells by conventional fluoresence and confocal scanning microscopy: evidence for the involvement of calcium and the cytoskeleton. Protoplasma **155**: 166-175.

18. Ridge RW (1992) A model of legume root hair growth and *Rhizobium* infection. Symbiosis **14**: 359-373.

19. Rohrig H, Schmidt J, Walden R, Czaja I, Miklasevics E, Wieneke U, Schell J, John M (1995) Growth of tobacco protoplasts stimulated by synthetic lipo-chitooligosaccharides. Science **269**: 841-843.

20. Schumm J, Allen N (1990) Endoplasmic reticulum, calciosomes and their possible roles in signal transduction. Protoplasma **154**: 172-178.

21. Shimmen T (1995) Roles of actin filaments in cytoplasmic streaming and organization of transvacuolar strands in root hair cells of *Hydrocharis*. Protoplasma **185**: 188-193.

22. Sonobe S, Shibaoka H (1989) Cortical fine actin filaments in higher plant cells visualized by rhodamine-phalloidin after treatment with m-maleimidobenzoyl N-hydroxysuccinimide ester. Protoplasma **148**: 80-86.

23. Teresaki M, Chen LV, Fujiwara K (1986) Microtubules and the endoplasmic reticulum are highly interdependent structures. J Cell Biol **103**: 1557-1568.

24. Thimann KV, Reese K, Nachmias VT (1992) Actin and the elongation of cells. Protoplasma **171**: 153-166.

25. Tiwari SC, Polito VS (1988) Organization of the cytoskeleton in pollen tubes of *Pyrus communis*: a study employing conventional and freeze-substitution electron microscopy, immunofluorescence, and rhodamine-phalloidin. Protoplasma **147**: 100-112.

26. Truchet G, Roche P, Lerouge P, Vasse J, Camut S, deBilly F, Dénarié J (1991) Sulphated lipo-oligosaccharide signals of *Rhizobium meliloti* elicit root nodule organogenesis in alfalfa. Nature **351**: 125-130.

27. Wernicke W, Jung G (1992) Role of cytoskeleton in cell shaping of developing mesophyll of wheat (*Triticum aestivum* L.). Eur J Cell Biol **57**: 88-94.

Discussion with Reviewers

E. Ralston: From the experiments described in the paper it is clear that actin filaments and ER distribution

are modified after stimulation by the Nod factor, but it is not possible to access how critical these changes are. Have you tried to depolymerize the actin filaments before applying the Nod factor? What happens to the microtubules?
M. Terasaki: Does cytochalasin block the accumulation of the endoplasmic reticulum at root tips after application of the NOD factor?
Authors: If cytochalasin is added, tip growth stops, so no curl would be observable in response to nod factors. Furthermore, the roots respond to DMSO by growing no root hairs. It is not a good experiment, but we tried it. Experiments on the microtubules are being carried out by Dr Sharon Long.

E. Ralston: Fig. 4 shows the effect of 10 min exposure to Nod factor on actin distribution. But the pattern described is observed in only 25% of the root hairs showing actin labeling. Does this mean that only a small fraction of the root hairs respond to the stimulation, or that the pattern is transient and that different root hairs are not in phase? Would a higher fraction be observed to respond after longer exposure?
Authors: Many root hairs break during fixation and we cannot be sure that they are properly stained. We do not know the effect of longer exposure.

E. Ralston: In Fig. 3, the different panels show patterns of a wide range of brightness. Were all recorded with the same camera settings and were they all computer enhanced to the same degree?
Authors: Similar, but not exactly identical settings were used.

M. Terasaki: Is it possible that there is a Ca^{2+} gradient in the root tip, such as seen in growing pollen tubes and could that be the function of the accumulation of endoplasmic reticulum at the root tip?
Authors: There is no Ca^{2+} gradient [28]; the Ca^{2+} is high over the nuclear area and there are oscillations. We have seen a gradient occasionally from the tip with calcium green, again with most of the staining over the nucleus. So roots and pollen tubes differ.

Additional Reference

28. Ehrhardt DW, Wais R, Long SR (1996) Calcium spiking in plant root hairs responding to *Rhizobium* modulation signals. Cell **85**: 673-681.

Scanning Microscopy Supplement 10, 1996 (pages 189-199)
Scanning Microscopy International, Chicago (AMF O'Hare), IL 60666 USA
0892-953X/96$5.00+.25

NEURAL TRANSPLANT STAINING WITH DiI AND VITAL IMAGING BY 2-PHOTON LASER-SCANNING MICROSCOPY

Steve M. Potter[1*], Jerome Pine[2], and Scott E. Fraser[1]

[1]Division of Biology and [2]Department of Physics, California Institute of Technology, Pasadena, CA

(Received for publication August 7, 1995, and in revised form January 7, 1996)

Abstract

We are developing a multielectrode silicon "neuroprobe" for maintaining a long-term, specific, two-way electrical interface with nervous tissue. Our approach involves trapping a neuron (from an embryonic rat hippocampus) in a small well with a stimulation/ recording electrode at its base. The well is covered with a grillwork through which the neuron's processes are allowed to grow, making synaptic contact with the host tissue, in our case a cultured slice from a rat hippocampus. Each neuroprobe can accommodate 15 neurons, one per well. As a first step in studying neurite outgrowth from the neuroprobe, it was necessary to develop new staining techniques so that neurites from the probe neurons can be distinguished from those belonging to the host, without interference from non-specific background staining. We virtually eliminated background staining through a number of innovations involving dye solubility, cell washing, and debris removal. We also reduced photobleaching and phototoxicity, and enhanced imaging depth by using a 2-photon laser-scanning microscope. We focused on using the popular membrane dye, DiI, however a number of other membrane dyes were shown to provide clear images of neural processes using pulsed illumination at 900 nm. These techniques will be useful to others wishing to follow over time the growth of neurons in culture or after transplantation in vivo, in a non-destructive way.

Key Words: Neural prosthesis, cell suspension staining, cellular debris, phototoxicity, non-destructive imaging, rat hippocampus, carbocyanine dye, grafting, two-photon fluorescence, neuroprobe.

*Address for correspondence:
Steve Potter
Division of Biology 156-29
California Institute of Technology
Pasadena, CA 91125
Telephone number: (818) 395-6787
Fax number: (818) 564-8709
e-mail: spotter@gg.caltech.edu
http://www.caltech.edu/~pinelab/pinelab.html

Introduction

One of the enduring problems associated with the study of neural transplants is discerning the transplanted cells from the host cells, to assess their survival and integration with the host tissue. In cases where the transplanted cells are immunologically distinguishable from the host, many immunohistochemical methods exist for visualizing them post-mortem. The increasing popularity of organotypic culture of neural tissue (Gahwiler, 1988; Stoppini *et al.*, 1991) and the use of embryos capable of growing on a microscope stage (O'Rourke and Fraser, 1990) have created the possibility of studying transplant integration while the tissue is still living, in order to understand the dynamic changes transplanted cells undergo. Attempts at vital labeling and imaging of neural transplants have met with limited success (e.g., see monumental efforts by Pyapali *et al.*, 1992 and Onifer *et al.*, 1993) due to the difficulty of finding a labeling method that (1) is persistent for days or weeks, (2) is specific to the transplanted cells, (3) labels fine neuronal processes, and (4) is compatible with non-destructive imaging.

As a tool for basic research into the relationship between synaptic plasticity and neuronal morphology, and eventually as an aid to handicapped people, we are developing a long-term, specific, two-way electrical connection to neural tissue called the neuroprobe (Fig. 1). Fabricated from silicon using technology developed for microelectronics (Tatic-Lucic, 1994; Tatic-Lucic and Tai, 1994), the neuroprobe has cell-sized wells with electrodes in them, and grillwork on top designed to keep the neuron's cell body in the well, next to the electrode, while allowing the neuron's processes to grow out and make synaptic contact with the host tissue. To develop and test this technology, we chose the embryonic rat hippocampus as a source of donor cells, and have been implanting neuroprobes into cultured slices of hippocampus from newborn rats. The well-studied physiology of cultured rat hippocampal slices (Stoppini *et al.*, 1991; Buchs *et al.*, 1993; Muller *et al.*, 1993; Stoppini *et al.*, 1993), the role of the hippocampus in learning and memory, and its deterioration in pathological states such as Alzheimer's disease (Etcheberrigaray *et al.*, 1994), all make this system a fertile and relevant testing ground.

Our collaborators Gyuri Buzsaki and Anatol Bragin at Rutgers University in New Jersey make use of the present labeling techniques with their studies of neuroprobes implanted into living rats. Although our neuroprobe neurons are not conventional transplants as such, we share the same concerns as most of the transplant community regarding the clear observation of the integration of donor cells with the host tissue. At present, membrane dyes, such as the carbocyanine dye DiI (1,1'-dioctadecyl-3-3-3'-3' tetramethylindocarbocyanine, reviewed in Honig and Hume (1989)), give the most hope for satisfying the above constraints. DiI brightly labels fine neuronal processes, can persist in neurons for weeks, and is relatively harmless to the labeled cells (Honig and Hume, 1986; Vidal-Sanz *et al.*, 1988). The greatest difficulty with the use of membrane dyes is to ensure that the label does not spread to host neurons. We refined techniques for the staining of donor cells in suspension in a way that minimizes artifactual staining of host cells and maximizes cell viability and brightness.

With confocal and wide-field fluorescence microscopy, the non-destructive observation of stained, living transplants is extremely difficult, if not impossible, due to the problems of photobleaching of the dye and dye-induced phototoxicity. To greatly reduce these difficulties, we constructed a 2-photon laser-scanning microscope (TPLSM). Two-photon microscopy, pioneered by Webb and co-workers (Denk *et al.*, 1990; reviewed in Williams *et al.*, 1994) takes advantage of the 2-photon absorption effect. High-energy pulses of infrared (IR) light are used, at approximately twice the wavelength usually used for a given fluorophore. Two IR photons excite the fluorophore, which then emits a single visible photon. Because of the squared dependence of 2-photon emission on excitation intensity, only at the focus of the laser beam within the specimen is the photon flux great enough to induce fluorescence. Thus, the tissue above and below the plane of focus is spared from phototoxicity and photobleaching. This is especially helpful if numerous images are taken, e.g., for time-series analysis and three-dimensional reconstructions. Because there is no source of out-of-focus fluorescence, no confocal aperture is necessary in the detector light path. This allows more sensitive detection of weak labeling than with confocal microscopy. An additional advantage is provided by the greater penetration of IR light into the specimen, which allows imaging of deeper structures than hitherto possible, including DiI-stained neurons transplanted to cultured hippocampal slices.

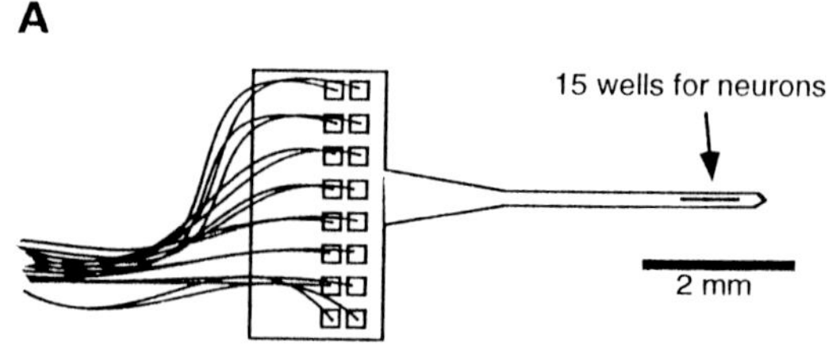

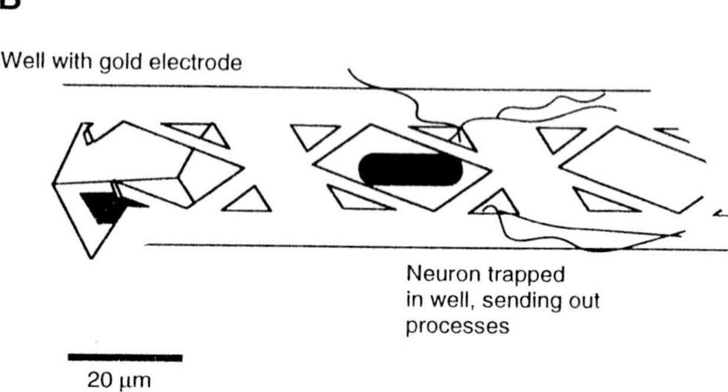

Figure 1: The neuroprobe. **A.** A schematic drawing of the neuroprobe is shown, with 16 leads (15 well electrodes plus ground) connected to bonding pads on the handle. **B.** Close-up schematic of the neuroprobe wells, showing grillwork that allows an immature neuron to be dropped through the center hole with a microsyringe, and smaller corner holes that permit the outgrowth of neurites.

Materials and Methods

Preparation of cell suspensions

Unless otherwise indicated, all donor cells were obtained from E18 Wistar rat embryo hippocampi. Pregnant rats were killed by inhalation of carbon dioxide (CO_2), and embryos were immediately removed by cesarean section. Hippocampi were rapidly removed using a stereomicroscope under sterile conditions, cut into 1 mm pieces, and digested with 0.25% trypsin, 0.02 mg/mL DNase (Sigma, St. Louis, MO), in Hanks balanced salt solution, no calcium or magnesium (Gibco, Grand Island, NY), at 37°C for 15 min. Pieces were gently washed twice in plating medium (Neurobasal with B27 supplement (Gibco), with 500 μM glutamine, 25 μM glutamate), and gently triturated in 1 mL of plating medium with 5 passes through the 0.78 mm opening of a blue tip of a P-1000 Pipetman. Suspended cells were decanted and the remaining pieces were triturated once more. Cell suspensions were gravity-filtered through a 70-μm nylon mesh (Falcon, Becton-Dickinson, Franklin

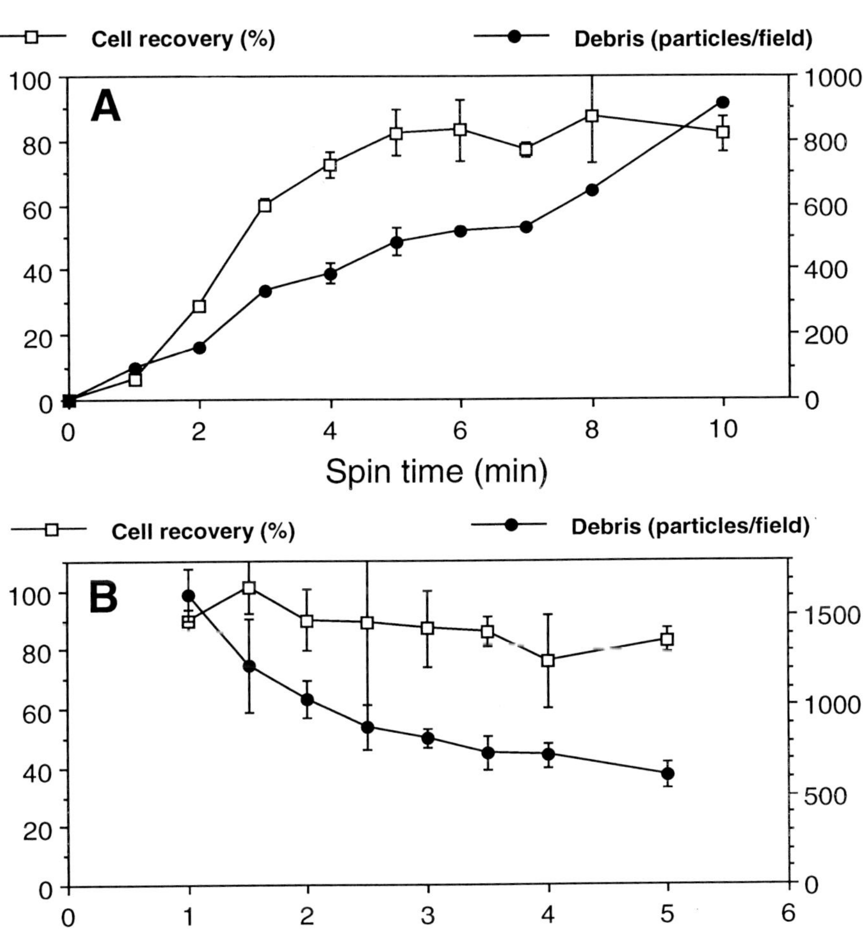

Figure 2: Centrifugation of cell suspensions. **A.** Spin-time vs. cell recovery and debris counts. Cells and debris from whole cerebral hemispheres of newborn rats, in 2 ml of suspension, were centrifuged through 0.5 ml of a 5% BSA solution, and cells and debris were quantified as described in Materials and Methods. Although recovery of cells at 2 min was only about 30%, we found unacceptable amounts of debris in samples spun for longer. The majority of the debris was still in the supernatant fraction after 10 min, hence the approximately linear increase in debris with spin time. Means of duplicate spins are plotted, with error bars indicating the range. **B.** Volume of suspension vs. cell recovery and debris counts. All samples were spun for 6 min. Means±SEM of triplicate spins are plotted.

Lakes, NJ) to remove large debris, and centrifuged for 2 min at 160 x g through a layer of 5% bovine serum albumin in phosphate-buffered saline (BSA/PBS), to remove small debris. Pellets were resuspended by gentle trituration (3 passes) as above.

For debris removal studies, we used whole cerebral hemispheres (including the hippocampus) of newborn rats (postnatal day 1) with meninges and choroid plexus removed. 2-4 hemispheres were minced, digested, triturated, and filtered through a 70 μm mesh as above, producing 20-50 million cells. These were diluted to approximately one million cells/mL in plating medium, and kept on ice in a 50-mL centrifuge tube that was stirred by inverting every 15 min, to prevent cells from settling during the experiments. Cell suspension (1 ml) was added to the indicated volume of plating medium in a 15-ml centrifuge tube and mixed by inverting. This was underlaid with 0.5 ml of BSA/PBS using a 1 ml syringe with a 16-gauge needle. All spins were conducted in triplicate (volume study) or duplicate (time study) at 160 x g (calculated at 1 cm from the end of the tube) in a swinging-bucket rotor at room temperature. Spin-up time (30 sec) was included in spin times, and spin-down (10 sec) was not. After spinning, all but 30 μl of the supernatant was aspirated, transferred to a separate tube and mixed by inverting a few times. Approximately 0.5 ml of this was plated under a 25 mm diam. coverslip raised 1.0 mm off the base of a Petri dish using pieces of glass capillary tubing. Pellets were immediately resuspended in 1 ml of plating medium with 3 passes of a P-1000 Pipetman, and 0.5 ml of this was plated into the coverslip dishes, as above. Cells and debris were allowed to settle for 2 h in the incubator, and then video images were taken, one from a random spot on each dish, with phase-contrast optics using a 20x/0.4 NA (numerical aperture) objective on a Nikon Diaphot inverted microscope. The coverslips allowed excellent homogeneity of plating density across the dish (less than 7% variance from field to field), since there was no meniscus effect. The video images were analyzed using NIH Image software (National Institutes of Health, available free by anonymous FTP at zippy.nih.gov) on a Macintosh Power PC computer. Cells in each 492 x 375 μm field were counted manually, and debris and

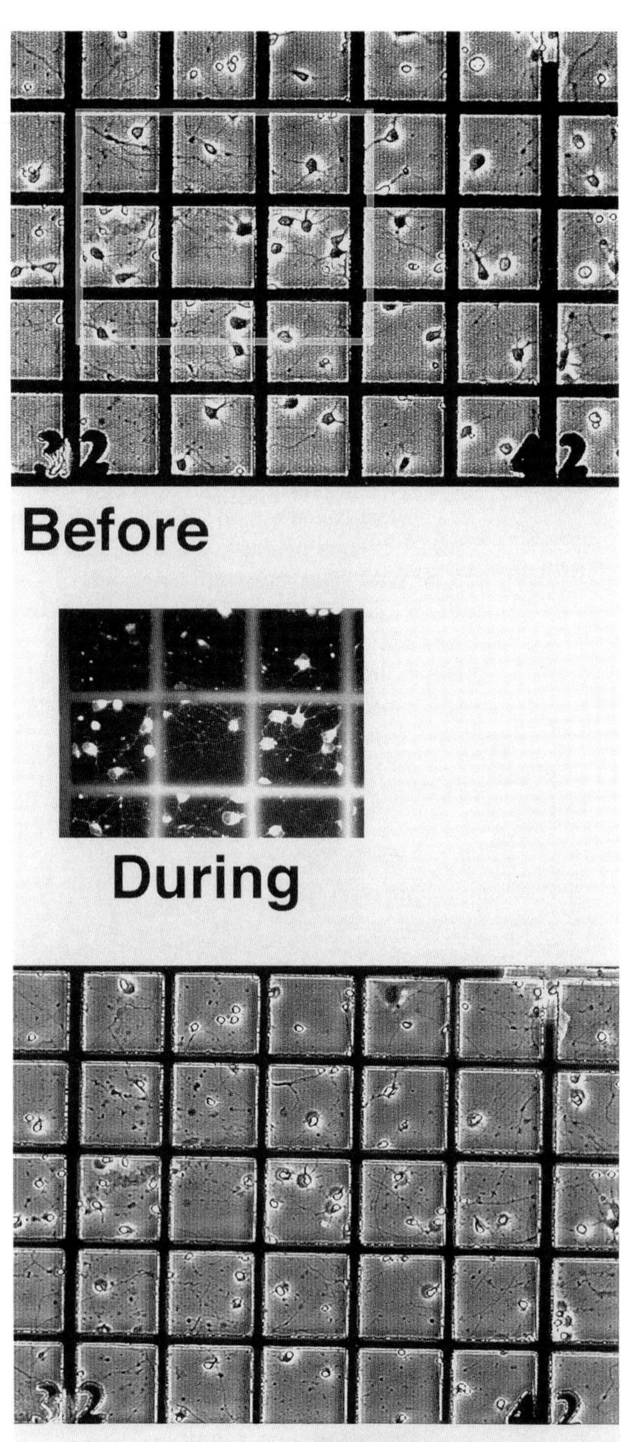

Figure 3 (*at left*): Phototoxicity of DiI. Stained embryonic rat hippocampal cells were cultured for 3 days on numbered-grid dishes, and exposed to green light from the mercury arc lamp of the Nikon Diaphot microscope for varying amounts of time from 10 sec to 6 min. Shown here are video images of a culture before, during, and 1 day after a 30 sec exposure. This was lethal to all the cells in the illuminated area, indicated by condensation of cell nuclei, swelling of the somata, and disintegration of the processes. Cells in unexposed areas remained healthy, as did unstained cultures exposed for as long as 6 min. Each square on the grid is 100 μm.

---

cells were quantified together using the "analyze particles" function. Cell counts were subtracted to produce final debris counts. Platings of the uncentrifuged cell suspension were used to set recovery maxima.

Staining of neurons with DiI

A 40 mg/ml stock solution of DiI-C12 (Molecular Probes, Eugene, Oregon) was prepared by dissolving the tar-like dye in dimethylformamide containing 2.5% (w/v) of Pluronic F127 (BASF, Parsippany, New Jersey; also available from Molecular Probes, Eugene, Oregon). Pluronic F127 catalyzes dye transfer to the cell membranes. Dye stock solution was stored at -20°C and dissolved easily upon warming to room temperature. Staining solutions were prepared by adding 2 μl of dye stock to 2 ml of plating medium, producing a final concentration of 40 μg/ml DiI, 0.0025% F127, and 0.1% dimethylformamide. Approximately 0.5 ml of cell suspension was added with gentle mixing to the staining solution after it was warmed to 37°C. Cells were stained for 15 min at 37°C. Staining at lower temperatures, with reduced membrane turnover, results in incomplete staining of intracellular membranes. BSA/PBS (0.5 ml) was layered under the cells in staining solution, using a long Pasteur pipet, and the cells were pelleted with a 6 min spin at 160 x g. The staining solution and most of the BSA/PBS was carefully decanted, and the stained cells were resuspended by trituration (3 passes) in 1 ml of plating medium. Another centrifugation through BSA/PBS and resuspension, as above, was performed to remove any remaining dissolved dye.

Phototoxicity testing

DiI-stained cells, or control cells "stained" in vehicle lacking the dye, were plated at a density of approximately 300 cells/mm^2 onto custom-made numbered-grid Petri dishes coated with polylysine and laminin (Sigma) (Banker and Goslin, 1991). After 3 days in culture, the stained cultures and unstained control cultures (viabilities of 69$\pm$10% and 73$\pm$10%, respectively) were exposed for varying periods (10 sec to 6 min) to the green line of a 200 W mercury arc lamp on the Nikon Diaphot epifluorescence microscope with a 20x/0.75 NA objective. Illuminating with the 50 W tungsten lamp, video images of the exposed regions, and

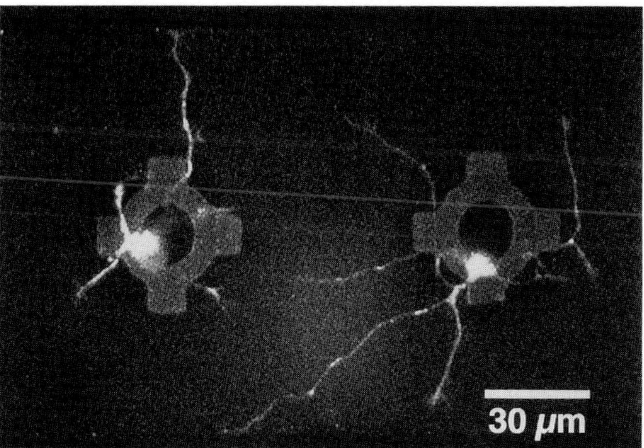

Figure 4: Outgrowth of neurites from neuroprobe. Stained embryonic rat hippocampal cells were implanted into neuroprobes and grown in culture for one day, as described in Materials and Methods. The top panel shows a Nomarski optics photograph of a pair of wells from which living neurons are sending their processes. The same neuroprobe was imaged using the TPLSM, shown in the bottom panel. This image is a lookthrough projection of unfiltered data, along the z-axis of 10 slices taken at 1 μm intervals, clearly showing the cell bodies trapped under the transparent grillwork, as well as the extending processes and their growth cones. The neuroprobe was returned to the incubator, and the neurons were still healthy and growing the next day.

----

unexposed control regions in the same dish, were recorded using phase contrast optics (10x/0.25 NA objective) just before and 1 day after arc lamp exposure.

Preparation of neuroprobes

Neuroprobes were fabricated at Caltech in the laboratory of Yu-Chong Tai by Svetlana Tatic-Lucic and John Wright (Tatic-Lucic, 1994). They were glued to plastic Petri dishes with a tiny drop of silicone rubber, with the shank resting on a piece of silicon wafer for support. They were sterilized by soaking in 95% ethanol overnight, washed with sterile water, and coated with polylysine and laminin. Neuroprobes were submerged in a drop of plating medium after rinsing off the laminin. Stained cells were plated in a 50 μl drop onto a silicone rubber-coated 10 mm diameter coverslip which was placed in the Petri dish next to the neuroprobe. The rubber discouraged the cells from adhering to the coverslip before they could be transferred to the neuroprobe. Stained cells were transferred from the coverslip to the neuroprobe wells using a microsyringe consisting of a 1 mm capillary pulled (using a Brown-Flaming pipet puller; Sutter Instruments, Novato, CA), cut (using a diamond knife), and polished (using a hot tungsten filament) to have a smooth opening of approximately 100 μm diam. The microsyringe was connected with Intramedic tubing to a 250 μl micrometer syringe (Gilmont, VWR Scientific, San Francisco, CA), used to pick up and aspirate cells by varying the air pressure in the tubing. The tubing was also connected to a 60 cc syringe, to adjust the magnitude of the pressure difference created by the micrometer syringe that picked up and expelled the cells. The microsyringe was moved using a Leitz micromanipulator, while observing using Nomarski optics on an Olympus upright BHMJ metallurgical microscope with a 20x/0.4 NA ultra-long working distance objective. Tungsten lamp epi-illumination was filtered to remove green light using a red photographic filter. Routine inspections and Nomarski photography were conducted with unfiltered tungsten light. After filling the neuroprobe with cells (which took 20-40 min/neuroprobe), the coverslip was removed from the dish, and the dish was flooded with plating medium. Cells in control dishes or neuroprobes were incubated at 37°C in a 5% CO_2 atmosphere. Great care was taken to assure that the air in the incubator remained moist and of the proper CO_2 level, by limiting the frequency and duration of door-openings.

Preparation of cultured slices

Hippocampal slices (400 μm) were prepared from postnatal day 7-11 Wistar rat pups, according to the method developed by Stoppini and co-workers (Stoppini *et al.*, 1991). Membranes inserts upon which the slices were grown (Millipore CM, Bedford, MA) were cut down to allow them to fit into 35 mm Petri dishes, and to allow imaging from above with a large objective. Slice culture medium was Gibco Minimum Essential Medium (Hank's salts, no glutamine) with 25% horse serum (Hyclone, Logan, UT), supplemented with 5 mM sodium bicarbonate, 30 mM HEPES (hydroxyethylpiperazine ethanesulfonic acid), 30 mM glucose, 3 mM glutamine, 2.5 mM magnesium sulfate, 2 mM calcium chloride, and 1 mg/l insulin (Sigma), 10 ml/l penicil-

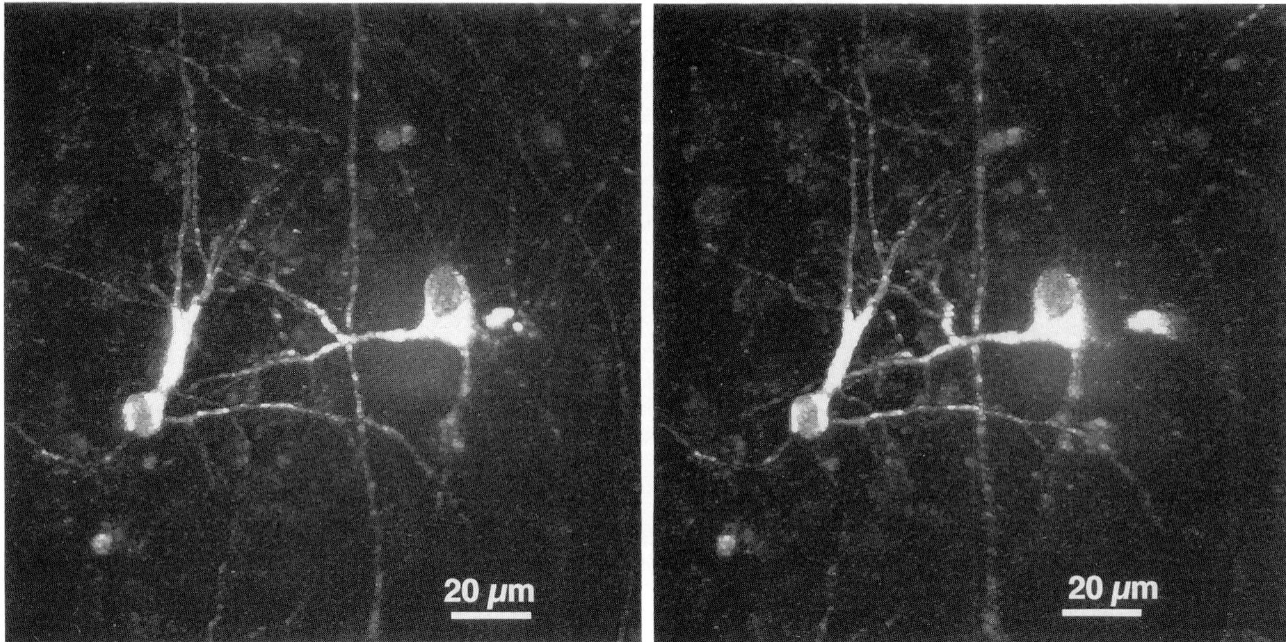

Figure 5: Stereo 3D image of DiI-stained neurons transplanted to a cultured slice. Stained embryonic hippocampal neurons were seeded onto cultured hippocampal slices as described in Materials and Methods. This image was made 4 days after seeding, using the TPLSM at 930 nm with a 63x/1.2 NA water-immersion objective, from a z-series of 31 slices, 2 μm apart, averaging two 5-second scans per slice. The depth is from 10 to 70 μm below the slice surface. No photobleaching or phototoxicity was detectable during or after scanning.

---

lin/streptomycin (Sigma P0781), pH 7.2 at 37°C. Slices were maintained at 37°C in a 5% CO_2 atmosphere and fed twice weekly by replacing half of the medium. They remained healthy for at least 6 weeks. DiI-stained cells were diluted in slice medium to a density of 20,000-50,000 cells/ml, and a 5 μl drop of them was placed on each slice (at least one day after preparing the slice culture) which gave approximately 20-50 neurons/slice, depending on how far the drop spread before soaking through the membrane.

2-photon imaging

The TPLSM was built from a Molecular Dynamics (Sunnyvale, CA) Sarastro 2000 upright confocal laser-scanning microscope, using a Titanium:sapphire mode-locked IR laser (Coherent Mira 900, Palo Alto, CA) as the excitation source. With the use of three different mirror sets, the laser is continuously tunable across a range from 710 to 970 nm (10 nm bandwidth), emitting 200 fs pulses with a peak amplitude of 50 kW, a repetition rate of 76 MHz, and a mean power of a few tens of milliwatts at the sample, which did not cause detectable sample heating. The Mira 900 is pumped by an 8 W argon-ion laser (Coherent Innova 310) operating in multiline visible mode. Visible and IR laser beams were covered with metal tubing to allow safe operation by biologists. One of the mirrors in the Sarastro 2000 was replaced with a 680 nm long-pass dichroic mirror, which allowed us to use either the Ti:sapphire or the Sarastro's on-board 25 mW argon-ion laser (if standard confocal operation is desired). The scanning mirrors and one stationary mirror in the Sarastro were replaced with broad-band mirrors (Newport, Irvine, CA) to enhance the IR throughput. We used no confocal aperture, a 680 nm short-pass primary beamsplitter and a 680 nm short-pass IR-blocking filter (Chroma, Brattleboro, VT) when operating in TPLSM mode. The Sarastro 2000 incorporates a Nikon Optiphot-2 upright microscope with an epifluorescence attachment. The stage height is adjusted by a computer-controlled stepper motor to 0.1 μm accuracy. Microscope control, data collection and image analysis were carried out on an Indigo workstation (Silicon Graphics, Mountain View, CA) using Image-Space software (Molecular Dynamics).

Neuroprobes were transferred to Hibernate (air-buffered Neurobasal) medium (Gibco) before imaging, since the pH drift of Neurobasal in air harmed cultures left out for more than 30 min. Because the focus of the IR beam is different than that of the visible light used for positioning the specimen (10-40 μm), and because the optical sectioning of the 2-photon effect is so extreme, it was easiest to find the neural processes by first executing a vertical (y-z) scan through the neuroprobe surface, and setting the stage height to bring the IR focus to the level of the neuroprobe grillwork. For neuroprobe imaging, we used pulsed excitation at 900 nm and a 40x/0.65 numerical aperture (NA) water-immersion objective. 512 x 512 pixel images were acquired, with a pixel size of 0.5 μm. For z-series scans, 10 z-steps of 1-2 μm were used. Two scans were averaged per frame, and raw data were projected along the z-axis using the "lookthrough" function.

For slice imaging, a 20x/0.75 NA air objective was used. When we tuned the laser from 900 to 960 nm, we reduced autofluorescence from slice cells (presumably due to endogenous flavins), increasing our signal-to-noise ratio substantially. For z-series scans, 30 z-steps of 2 μm were used, with 5 scans averaged/frame. Three-dimensional data were processed with a 3x3x3-pixel median filter and projected along the z-axis using the "maximum intensity" function. Microscope images were printed using the Kodak XLS 8300 digital printer.

Results

Preparation of stained cells

Any method for producing cell suspensions by mechanically dissociating neural tissue is bound to produce debris, consisting of pieces of broken cells, neurites from the surviving dissociated cells, and clumps of undissociated cells. If this debris is stained and transplanted along with the cells of interest, unacceptable background staining of host tissue is inevitable (Onifer *et al.*, 1993). DiI has generally *not* been observed to be transferred from labeled to unlabeled neurons in culture (St. John, 1991), however, using the TPLSM, we observed non-specific labeling of host neurons in rats injected with DiI-labeled cells and debris. It is important, therefore, to prepare a cell suspension that is as free of debris as possible. We found that suspensions prepared from embryonic day 18 rats (E18, 21 day gestation) had far less debris than suspensions prepared from newborn (postnatal day 1-3) rats. The embryonic hippocampal preparation also has the advantage of fewer non-neuronal cell types. We routinely observed that 85-90% of viable cells in dissociated E18 cultures had neuronlike morphology.

Preparing a clean suspension of cells is unfortunately more of an art than a science, requiring a customized approach for each new tissue, animal age, etc. We found trituration of brain tissue to be especially variable in cell yield and viability. To reduce some of this variability, we triturated trypsin-digested hippocampal pieces using a P-1000 Pipetman, so that the tip opening (0.78 mm) was well-defined, unlike that of the ubiquitous fire-polished Pasteur pipet. This also made it easy to control the volume of each trituration pass. After trituration, debris larger than the cells (10-20 μm diam) was removed by gravity filtration of the suspension through a 70 μm nylon mesh. The small debris was more troublesome to eliminate. It is traditional to centrifuge cell suspensions to remove small debris, but a plethora of different protocols exist, so we quantified debris removal and cell recovery under a variety of different conditions. At 160 x g, we observed that essentially all of the cells were pelleted from a suspension of 2.5 ml in 6 min (Fig. 2A), however, a substantial proportion of the debris was also pelleted. Thus, one must sacrifice some of the healthy cells by spinning for a shorter time, in order to leave most of the debris behind in the supernatant fraction. Diluting the suspension beforehand reduced the debris in the pellet, while having little effect on cell recovery (Fig. 2B). We routinely centrifuged our suspensions before staining for 2 min at 160 x g in a volume of 2-3 ml. Repeated centrifuging adversely affected cell recovery and viability, although this effect was not quantified.

DiI is most commonly used as a neural tracer (in living and fixed tissue) by placing a small crystal or suspension of crystals in contact with the cells to be labeled. It is also dissolved in oil which is then injected in small droplets into the tissue. This has allowed numerous excellent studies in which living processes of endogenous neurons were traced (for example, see: Hosokawa *et al.*, 1992; Dailey *et al.*, 1994; Hosokawa *et al.*, 1994; O'Rourke *et al.*, 1994). Crystals or oil solutions of DiI are used because the most common form of the dye (DiI-C18) is only slightly soluble in aqueous media. This presents some problems for staining of cells in suspension. The worst is that DiI crystals are difficult if not impossible to wash from the cells by centrifugation. If any crystals of DiI are transferred to the host along with the stained cells, then background staining of host cells will result. Furthermore, we and others (Paramore *et al.*, 1992) observed the lethal clumping of cells around DiI crystals. We deemed unacceptable the filtration of DiI-C18 to remove crystals, due to clogging of filters, uncertainty of the concentration of dye in the final "solution," and subsequent crystal formation during staining.

We completely eliminated the problem of DiI insolubility by (1) using the more soluble form of DiI with a shorter alkyl tail, DiI-C12 instead of DiI-C18, and (2) using the surfactant Pluronic F127 in the staining solutions. We observed substantially brighter staining with DiI-C12, compared to an equal concentration of DiI-C18, presumably because all of the dye was dissolved and available to stain the cells. Pluronic F127, a non-ionic macromolecular polyol, serves to "catalyze" the transfer of dye molecules to the cell membrane

(Lojewska and Loew, 1986, 1987), and has been used by researchers to harmlessly stain neurons with voltage-sensitive membrane dyes at ten times the concentration we use (Davila *et al.*, 1973). The DiI-C12/F127 solution passed freely through a 0.2 μm filter, with no diminution in its light absorption, indicating it was completely dissolved in cell culture medium at the highest concentration used, 40 μg/ml.

Unlike crystals, dissolved DiI can be removed from the suspension of stained cells by centrifugation, although this is not trivial. Even with three washings with 1.5 ml of medium, we observed unacceptable background staining of cell culture dishes, and of cultured slices onto which stained cells were plated. In order to almost completely remove dissolved DiI from the cell suspension in two washings, we centrifuged (in a swinging-bucket rotor) the stained cells through a 0.5 ml layer of 5% bovine serum albumin (BSA) in phosphate-buffered saline. The dye solution remained above the dense protein solution, and was easily decanted after spinning without disturbing the pellet of stained cells. An additional, unexpected advantage of using a BSA layer was that the cells were easier to resuspend. Cells pelleted in culture medium often clump, and one must re-filter the suspension through a nylon mesh or triturate more vigorously, killing cells and generating more small debris. Cells pelleted in BSA were easily dispersed with one or two passes of the P-1000 Pipetman, with few if any clumps remaining. The two centrifugations to remove dissolved DiI also serve to further reduce the level of small debris in the final suspension.

In a study of DiI-stained rat motor and sensory neurons (St. John, 1991), the longer chain C18 form of DiI was found to be toxic, killing all labeled neurons by 10 days in culture, while the DiI-C12-stained neurons remained healthy. Unlike St. John (who did not use F127), we did not find DiI-C12 to be completely harmless. We observed toxicity of DiI-C12 in cultured hippocampal neurons if they were stained for longer than 30 min in a 40 μg/ml solution. No adverse effects from a 15 min stain were observed, compared to unstained control cultures. Although much of the dye was concentrated in intracellular vesicles after one week in culture, all of the neurons' fine processes were still clearly visible, including growth cones. Thus, it is crucial to adjust staining conditions such that control cultures of stained donor cells show good viability and staining.

Cells brightly stained with DiI by the above procedures remained healthy for at least two weeks in culture, unless we attempted to photograph them using the 545 nm line of a 200 W mercury arc lamp as the excitation source (a common arrangement for wide-field fluorescence microscopy of rhodamine-like dyes). Significant fading of the dye was observed during 30-sec exposures, and the photographed cells usually died by the next day. This was surprising, since DiI has a reputation for being stable and non-phototoxic (Honig and Hume, 1989). To quantify the effect, we grew stained hippocampal neurons on numbered-grid Petri dishes and exposed a 500 μm-diameter region to green light from the arc lamp for varying amounts of time. Upon returning to the same regions the next day, we found that exposures as short as 10 sec (using a 20x/0.75 NA objective) adversely affected cell morphology, and 30 sec exposures were lethal to all exposed cells (Fig. 3). Exposures as long as 6 min had no effect on unstained control cultures, and unexposed regions in the stained cultures remained healthy.

TPLSM imaging

To avoid the above problems with photobleaching and phototoxicity, we used a custom-made TPLSM to non-destructively image stained cells (see Materials and Methods). Stained cells were placed into the wells of neuroprobes and checked for outgrowth after one day in culture, using Nomarski optics on an upright microscope with a tungsten lamp. Even the light from the tungsten lamp caused phototoxicity in cells that were exposed for over 30 min (during the filling of the wells), so we used a green-blocking filter (Hoya 25A) whenever possible. A representative Nomarski photograph of outgrowth one day after filling the wells is shown in Fig. 4. The same neuroprobe was subsequently imaged using the TPLSM, with 900 nm excitation, shown in the lower panel of Fig. 4. The grillwork, made from silicon nitride, is transparent, but slightly autofluorescent. Repeated scanning (at least one hour) had no adverse effect on the neurons, which were still healthy and growing the next day. Little, if any photobleaching was observed, except when vertical scans were made in which several lines were scanned in rapid succession (to enhance the signal-to-noise ratio) before advancing the stage. This produced a thin line of bleaching visible for several minutes in subsequent horizontal scans. When all of the debris removal and dye rinsing steps were rigorously followed, there was virtually no background staining on the surface of the probes or control dishes.

Eventually, we intend to use the TPLSM to image the outgrowth of neurites from neuroprobes after they have been placed on cultured hippocampal slices. In order to optimize imaging parameters, we plated stained cells directly onto the slices, at low density (20-50 cells/slice). These grew and extended processes, some reaching across the entire slice (>2 mm) by one week after plating. The TPLSM allowed us to obtain clear images of stained processes deeper than 100 μm below the surface of the slice, more than twice as deep as possible using visible excitation in confocal mode. Many of the stained cells attained pyramidal cell morphology, with apical and basal dendrites, by 6 days post-transplant (Fig. 5). Fine processes were still easily visible at 17 days, the longest time after plating that 2-photon imaging was carried out.

Discussion

We developed a procedure for producing hippocampal cells stained brightly with DiI, with a minimum of stained debris, that will be useful to anyone wishing to study prelabeled neural transplants. Although most of the methods described have been used by others in the past, such as the use of DiI-C12, Pluronic F127, centrifugation through BSA solutions, and imaging with pulsed IR light, they seem to be greatly underappreciated. We combined them for the first time and applied them to the study of neural cells transplanted to cultured brain slices while they were still alive and growing.

We showed that DiI can have substantial phototoxicity and photobleaching in cultured hippocampal neurons, using light doses commonly used for film recording of membrane fluorescence. It was demonstrated for the first time that 2-photon laser-scanning microscopy of DiI-stained neurons is possible using pulsed IR light at 900 nm. Repeated scanning of stained neurons with the TPLSM resulted in no significant photobleaching or phototoxicity. The TPLSM allowed the visualization of processes of labeled neurons transplanted to cultured hippocampal slices to a depth of over 100 μm.

With the combination of careful DiI staining methods and 2-photon imaging, we accomplished persistent labeling and non-destructive imaging of fine processes of transplanted living neurons without significant background staining of the host tissue. We are now facing the serious task of increasing the viability and outgrowth of neurons from our neuroprobe. Since each neuroprobe holds only 15 cells, we want to maximize the chance that the implanted cells are neurons (not glia, which are not electrically excitable) and that they are healthy. We consider ourselves lucky when half of the wells show neurite outgrowth after a couple of days in culture, and have not yet seen outgrowth after transfer of probes to cultured slices. Others routinely observe 5% survival of transplanted neurons (Demierre *et al.*, 1990; Ruiz-Flandes *et al.*, 1993); the remaining 95% break up and may cause troublesome background staining. To address the issue of cell type and health, we are making time-lapse movies of cells immediately following plating in Petri dishes, in order to observe the fate of each cell, and follow it back to time-zero to see if the healthy neurons have any distinguishing features immediately after dissociation and staining. This may help us to select only those healthy neurons for implantation into the neuroprobes.

We focused our efforts on DiI because it is the most popular membrane dye for tracing neuronal processes. DiI has a one-photon absorption maximum of 550 nm (and a one- or two-photon emission maximum of 570 nm). Thus, we were hitting the short-wavelength tail of its excitation curve with 2-photon excitation at 900 nm. Although the Titanium:sapphire laser we used is continuously tunable from 710 to 970 nm, the decrease in output of the laser as it was tuned to wavelengths longer than 900 nm outweighed the increase in absorption by DiI (however, see note in last paragraph of Materials and Methods). Thus, we chose 900 nm in most cases because it gave us the maximum signal, given our instrumentation. The 2-photon absorption spectrum of DiI is unknown. In preliminary studies using dissociated cultures of rat hippocampal neurons, we observed that other membrane dyes normally excited by blue light provided an even brighter signal than DiI when excited by pulsed IR light at 900 nm. The dyes tested include BODIPY®-ceramide (503 nm excitation maximum (ex. max.)), DiO (3-3' dioctadecyloxacarbocyanine perchlorate, 484 nm ex. max.), and DiA (4-(4-(dihexadecylamino)styryl)-N-methylpyridinium iodide, 491 nm ex. max.) (all available from Molecular Probes). With 2-photon excitation, there is no difficulty separating excitation light from emitted light, since they are hundreds of nm apart. By using a 680 nm short-pass beamsplitter, we can detect virtually all of the emitted light from any visible-emitting fluorophore. In contrast, with normal fluorescence microscopy, one must select filters carefully suited to each fluorophore used, accepting the loss of some emitted light due to the overlap between one-photon excitation and emission spectra.

2-photon microscopy has also been proven successful for the imaging of fruit fly neurons expressing Green Fluorescent Protein (GFP; Potter *et al.*, 1996), a naturally fluorescent protein from jellyfish (Chalfie *et al.*, 1994) normally excited by blue light (480 nm ex. max.). We are currently working on expressing GFP in rat hippocampal neurons. Onifer and co-workers (Onifer *et al.*, 1993) used a retrovirus to transfer the bacterial LacZ gene to a dividing neuronally-derived cell line (RN33B cells). Transplanted cells expressing b-galactosidase, the LacZ gene product, were observed post-mortem, and no b-galactosidase reaction was observed in tissue from animals that received injections of transfected but lysed cells. This gives hope that GFP may provide a background-free, long-lasting way to label transplanted neurons.

Acknowledgments

We thank Sheri McKinney for her expert technical

assistance; Pete Vanderklish for teaching us hippocampal slice preparation; Yu-Chong Tai, Svetlana Tatic-Lucic and John Wright for neuroprobe design and fabrication; Gyuri Buzsaki and Anatol Bragin for their continued collaboration on the Neural Prosthesis project; The NINDS at the NIH for funding the Neural Prosthesis project; and Molecular Dynamics, a Silvio Conte Center award from the NIMH, and the Beckman Institute for support of the 2-photon imaging facility.

References

Banker G, Goslin K (1991) Culturing Nerve Cells. MIT Press, Cambridge, Mass, pp 251-282.

Buchs PA, Stoppini L, Muller D (1993) Structural modifications associated with synaptic development in area CA1 of rat hippocampal organotypic cultures. Dev Brain Res **71**: 81-91.

Chalfie M, Tu Y, Euskirchen G, Ward WW, Prasher DC (1994) Green fluorescent protein as a marker for gene-expression. Science **263**: 802-805.

Dailey ME, Buchanan J, Bergles DE, Smith SJ (1994) Mossy fiber growth and synaptogenesis in rat hippocampal slices *in vitro*. J Neurosci **14**: 1060-1078.

Davila HV, Salzberg BM, Cohen LB, Waggoner AS (1973) A large change in axon fluorescence that provides a promising method for measuring membrane potential. Nature **241**: 159-160.

Demierre B, Martinou JC, Kato AC (1990) Embryonic motoneurons grafted into the adult CNS can differentiate and migrate. Brain Res **510**: 355-359.

Denk W, Strickler JH, Webb WW (1990) 2-photon laser scanning fluorescence microscopy. Science **248**: 73-76.

Etcheberrigaray R, Gibson GE, Alkon DL (1994) Molecular mechanisms of memory and the pathophysiology of Alzheimers-disease. Ann NY Acad Sci **747**: 245-255.

Gahwiler BH (1988) Organotypic cultures of neural tissue. Trends Neurosci **11**: 484-489.

Honig MG, Hume RI (1986) Fluorescent carbocyanine dyes allow living neurons of identified origin to be studied in long-term cultures. J Cell Biol **103**: 171-187.

Honig MG, Hume RI (1989) DiI and DiO - versatile fluorescent dyes for neuronal labeling and pathway tracing. Trends Neurosci **12**: 333-341.

Hosokawa T, Bliss TVP, Fine A (1992) Persistence of individual dendritic spines in living brain slices. Neuroreport **3**: 477-480.

Hosokawa T, Bliss TVP, Fine A (1994) Quantitative 3-dimensional confocal microscopy of synaptic structures in living brain-tissue. Microsc Res Techn **29**: 290-296.

Lojewska Z, Loew LM (1986) Pluronic F127: An effective and benign vehicle for the insertion of hydrophobic molecules into membranes. Biophys J **49**: 521a.

Lojewska Z, Loew LM (1987) Insertion of amphophilic molecules into membranes is catalyzed by a high molecular weight non-ionic surfactant. Biochim Biophys Acta **899**: 104-112.

Muller D, Buchs PA, Stoppini L (1993) Time course of synaptic development in hippocampal organotypic cultures. Dev Brain Res **71**: 93-100.

O'Rourke NA, Fraser SE (1990) Dynamic changes in optic fiber terminal arbors lead to retinotopic map formation - an in vivo confocal microscopic study. Neuron **5**: 159-171.

O'Rourke NA, Cline HT, Fraser SE (1994) Rapid remodeling of retinal arbors in the tectum with and without blockade of synaptic transmission. Neuron **12**: 921-934.

Onifer SM, White LA, Whittemore SR, Holets VR (1993) In vitro labeling strategies for identifying primary neural tissue and a neuronal cell-line after transplantation in the CNS. Cell Transpl **2**: 131-149.

Paramore CG, Turner DA, Madison RD (1992) Fluorescent labeling of dissociated fetal cells for tissue culture. J Neurosci Meth **44**: 7-17.

Potter SM, Wang C-M, Garrity PA, Fraser SE (1996) Intravital imaging of green fluorescent protein using two-photon laser-scanning microscopy. Gene **173**: 25-31.

Pyapali GK, Turner DA, Madison RD (1992) Anatomical and physiological localization of prelabeled grafts in rat hippocampus. Exp Neurol **116**: 133-144.

Ruiz-Flandes P, Demierre B, Mattenberger L, Kato AC (1993) Migration of purified embryonic motoneurons grafted into adult-mouse CNS. Int J Dev Neurosci **11**: 525-533.

St. John PA (1991) Toxicity of "DiI" for embryonic rat motoneurons and sensory neurons *in vitro*. Life Sci **49**: 2013-2021.

Stoppini L, Buchs PA, Muller D (1991) A simple method for organotypic cultures of nervous tissue. J Neurosci Meth **37**: 173-182.

Stoppini L, Buchs PA, Muller D (1993) Lesion-induced neurite sprouting and synapse formation in hippocampal organotypic cultures. Neuroscience **57**: 985-994.

Tatic-Lucic S (1994) Silicon Micromachined Devices for *In Vitro* and *In Vivo* Studies of Neural Networks. Doctoral Thesis, California Institute of Technology, Pasadena, CA.

Tatic-Lucic S, Tai YC (1994) Novel extra-accurate method for 2-sided alignment on silicon-wafers. Sensors and Actuators: A Physical **42**: 573-577.

Vidal-Sanz M, Villegas-Perez MP, Bray GM, Aguayo AJ (1988) Persistent retrograde labeling of adult-rat retinal ganglion-cells with the carbocyanine dye DiI. Exp Neurol **102**: 92-101.

Williams RM, Piston DW, Webb WW (1994) 2-photon molecular-excitation provides intrinsic 3-dimensional resolution for laser-based microscopy and microphotochemistry. FASEB J **8**: 804-813.

Discussion with Reviewer

M. Malecki: What is your experience in labeling of neurons grown on substrates as compared to those treated in suspensions?
Authors: Labeling neurons after they are growing on a substrate would not be applicable to experiments involving transplants of cell suspensions to slices or living animals. In trials where we stained probe neurons after placing them in the wells, we observed unacceptable levels of background staining, and dye transfer to slice neurons. Staining and washing the the cells in suspension completely eliminated this problem.

M. Malecki: What are the DiI's spectral shifts upon binding to cell membranes?
Authors: DiI (550 nm excitation/570 nm emission) does not change its spectra appreciably upon membrane binding. The two-photon excitation maximum is not known, since our laser only excites up to 970 nm.

M. Malecki: Can you share your observations on phototoxicity induced by imaging with confocal laser scanning microscopy as compared with two-photon-excitation fluorescence microscopy?
Authors: We did not perform phototoxicity tests using regular CLSM, although recently we have made 2-photon time-lapse movies in 3D for over 7 hours without phototoxicity, which has never been accomplished using CLSM. Attempts to produce a phototoxic effect using TPLSM have failed. Continuous scanning for over an hour did not adversely affect the short- or long-term health of cultured, stained neurons.

M. Malecki: The final intensity of the neurons' fluorescence may be the result of many factors including: dispersion of fluorochrome within membranes of outgrowing axons, intracellular turnover, fading of fluorochromes due to illumination, etc. Can you discuss mechanisms of fluorescence fading which might occur in your experiments?
Authors: Dye fading in long-term experiments is most likely due to dilution by newly synthesized membranes during neuronal outgrowth. Cultures left in the incubator also showed fading over 2-3 weeks, ruling out photobleaching as a primary cause. We have not observed loss of dye to adjacent unstained cells, except in the case of macrophages engulfing the remains of dead stained cells.

M. Malecki: Would you consider application of Green Fluorescence Proteins (GFP) in your studies?
Authors: Exposure to dye and the GFP vector (adenovirus) could take place simultaneously, with excess dye and viral particles being washed out together. We have not tried this, but it may be useful for imaging labeled neurons during the 2-3 day lag time it takes for maximal GFP production to occur.

Scanning Microscopy Supplement 10, 1996 (pages 201-211)
Scanning Microscopy International, Chicago (AMF O'Hare), IL 60666 USA
0892-953X/96$5.00+.25

VIDEO RATE CONFOCAL LASER SCANNING REFLECTION MICROSCOPY IN THE INVESTIGATION OF NORMAL AND NEOPLASTIC LIVING CELL DYNAMICS

Pavel Vesely[1]* and Alan Boyde[2]

[1]Institute of Molecular Genetics, Academy of Sciences of the Czech Republic, Prague, Czech Republic,
[2]Department of Anatomy and Developmental Biology, University College London, London, UK

(Received for publication September 25, 1995 and in revised form October 1, 1996)

Abstract

The introduction of video rate confocal laser scanning microscopes (VRCLSM) used in reflection mode with high magnification, high aperture objective lenses and with further magnification by a zoom facility allowed the first detailed observations of the activity of living cytoplasm and offered a new tool for investigation of the structural transition from the living state to the specimen fixed for electron microscopy (EM). We used a Noran Odyssey VRCLSM in reflection (backscattered) mode. A greater degree of oversampling and more comfortable viewing of the live or taped video image was achieved at zoom factor 5, giving a display monitor field width of 10 μm. A series of mesenchyme derived cell lines - from normal cells to sarcoma cells of different malignancy - was used to compare behaviour of the observed intracellular structures and results of fixation. We contrasted the dynamic behaviour of fine features in the cytoplasm of normal and neoplastic living cells and changes induced by various treatments. The tubulo-membraneous 3D structure of cytoplasm in living cells is dynamic with motion observable at the new limits of resolution provided by VRCLSM. All organelles appear integrated into one functional compartment supporting the continuous 3D trafficking of small particles (vesicles). This integrated dynamic spatial network (IDSN) was found to be largest in neoplastic cells.

Key Words: reflection confocal microscopy, backscattered light imaging, high spatial and temporal resolution, functional integrity of cytoplasm, speed of cytoplasmic motility, fixation, neoplastic cell compartmentalization

*Address for correspondence:
Pavel Vesely
Institute of Molecular Genetics,
Academy of Sciences of the Czech Republic
Flemingovo 2, 166 37 Prague 6, Czech Republic
Telephone Number: +42(2)3312 288
FAX number: +42(2)2431 0955
E-mail: pvy@img.cas.cz

Introduction

A fully confocal, unitary (laser origin) beam confocal laser scanning microscope (CSLM) for the reflection mode was developed by Draaijer and Houpt (1988). The later commercial development of the design principles involved - AOD (acousto-optic deflector) scanning on the line and galvanometer mirror scanning on the frames axes - led to the release of the Noran Odyssey instrument (with which we have worked). Very much higher scan rates are achieved in the multiple beam scanning confocal microscopes designed by Petran *et al.* (1968). These "tandem scanning" (the priority term for, and synonym for, confocal) microscopes gave excellent reflection images. However, their advantage in high speed scanning was lost in attempts at recording at high speed. Furthermore, although there are far fewer pixels in a TV scanning arrangement than in a mechanical scanning, Nipkow disc, confocal microscope, the temporal resolution at each pixel is maximal in the direct video scanning arrangement. Information is only written into a pixel at the time that the focussed laser beam is illuminating the corresponding point in the specimen. Because of these considerations, because video-tape recording is cheap and simple and reliable (if not free of problems!) and because it is simple to achieve a comfortable degree of empty magnification, the Odyssey type of video rate CLSM (VRCLSM) represented a major revolution in the evolution of light microscopical techniques for live cell biology.

VRCLSM with high magnification, high aperture objective lenses and with further magnification given by a zoom facility allowed the first detailed observations of activity of living cytoplasm. Rapid motion in fine granular 3D structure of the interior of a living cell, some of which apparently exceeded the frequency which had been previously possible to register at the time, was revealed in a variety of cell types: these included rat and chick bone derived cells (Boyde and Jones, 1992; Boyde *et al.*, 1994) and rat tumour cell lines (Vesely *et al.*, 1993; Vesely and Boyde (1994).

In spite of the potential of the VRCLSM in the investigation of intracellular processes - and in correlative microscopy as a tool for investigations of the structural transition from the living state to the fixed specimen for electron microscopy (EM) - many questions about the nature of the observed intracellular dynamic structures remained to be answered. The conditions for reliable imaging had therefore to be defined, a technique (or better an art) for performing real time experiments to be developed and managed, and comparative biological examination of various cells to be performed. The purpose of the present paper is to review our findings with this technique (to the date of the 14th Pfefferkorn Conference, August 1995), and to discuss the necessary future developments before this technique can become an ordinary tool of cell biology.

Materials and Methods

Microscopy

We have used a Noran (Middleton, WI) Odyssey video rate laser scanning microscope (VRCLSM) with an argon ion laser 488nm line in the reflection mode, mainly with Nikon NA 1.4, 160 mm tubelength, oil-immersion objective lenses with nominal magnifications of 100X and a 60X, NA 1.2 water immersion lens. The first has been strongly preferred for the type of study considered here. To minimize spherical aberration, living cells were grown on coverslips so that no additional water layer might distort the image (Inoué, 1990).

Image recording and analysis

Video tape was used throughout the experiments to record all images at real time as well as the synchronous voice commentary on the actions and effects. Tapes were evaluated by thorough visual inspection and were also used for measuring the timing of activity and for the estimation of the frequency of dynamic phenomena.

Further magnification given by scanning a reduced area was regularly used in the majority of the present studies, usually to give a zoom factor of 5 corresponding to a display monitor field width of 10 μm.

We used two image analysing computers with the Odyssey: first, the Noran TN8502, a Unix based machine which permitted sequential frame capture and averaging, and secondly, the FENESTRA system of Kinetic Imaging Ltd (Liverpool, UK), which allowed sequential single field or frame capture, but could not Kalman average images in a sequence. Because rapid motion is blurred if the odd and even fields of one TV scan are interlaced, sets of 8 individual half frame (single "field", half the interlaced TV frame) images were grabbed at video rate giving half the number of lines and a time interval between subsequent images of 1/25s ("FENESTRA", Kinetic Imaging Ltd, Liverpool, UK).

For printing, the FENESTRA images were transformed to TIFF format. The image.tif files were processed in Adobe Photoshop by stretching the histogram and using sharpening before Typesetter printing at resolution of 3000 dpi.

(Figures 1-4 on facing page)

Figure 1: Rat sarcoma K4 cell edge focussed 0.5 μm deep in the cell. A "hairy" surface and peripheral cytoplasm are shown. Picture width 10 μm, zoom factor 5, pinhole size 5.66 mm.

Figure 2: Cell process of normal (rat bone derived) cell shows close contact (dark areas) with the coverslip and the structure of the peripheral cytoplasm. Picture width 10 μm, zoom factor 5, pinhole size 2.98 mm.

Figure 3 A,B: Rat sarcoma K4 cell focussed 3 μm deep into the cell. In the upper right corner, the granular structure of the spatial network is seen: this changes to interference fringes produced by interference between light reflected from the coverslip and the cell surface at the steep side slope of the cell. Relative lateral displacement of small particles within 1/25s can be seen as 3D effect if images A and B are viewed as stereopairs. Picture width 10 μm, zoom factor 5, pinhole size 2.62 mm.

Figure 4 A,B: 3-4 μm deep in a rat sarcoma K4 cell the spatial network between nuclei is seen and motility after 1/25s is visualized on pseudo-3D inspection. Picture width 30 μm, zoom factor 2, pinhole size 4.36 mm.

Cells

Mesenchyme derived cell lines from normal fibroblasts (WEF) and rat bone derived cells (kindly prepared by Dr. C.M. Gray, Department of Anatomy and Developmental Biology, University College London) to sarcoma cells of different malignancy (K2, T15, A870N, A72N) and Rous sarcoma transformants K4 cells were used (Vesely *et al.*, 1993). The latter cells were grown in Prague and transported to London in tissue culture flasks as monolayer cultures, and, after recovery at 37°C, resuspended by trypsinisation, and seeded on to 22 mm square borosilicate glass coverslips (BDH Lab Supplies, Poole, UK; Cat. No. 406/0187/33, thickness No. 1). The cleaning procedure for the coverslips was double washing in distilled water, dipping twice in absolute ethanol, and flaming. The medium used was Eagle's Minimal Essential Medium (MEM) with Hank's balanced salt solution (BSS) and all non- essential amino acids (Institute of Molecular Genetics, AS CR, Prague) supplemented with 10% calf serum (Bioveta, Ivanovice, Czech Republic). Cells were cultured in 3% CO_2 incubator for approximately 48 hours before examination. The coverslips were mounted inverted, using silicone grease, as a bridge supported by two piers made

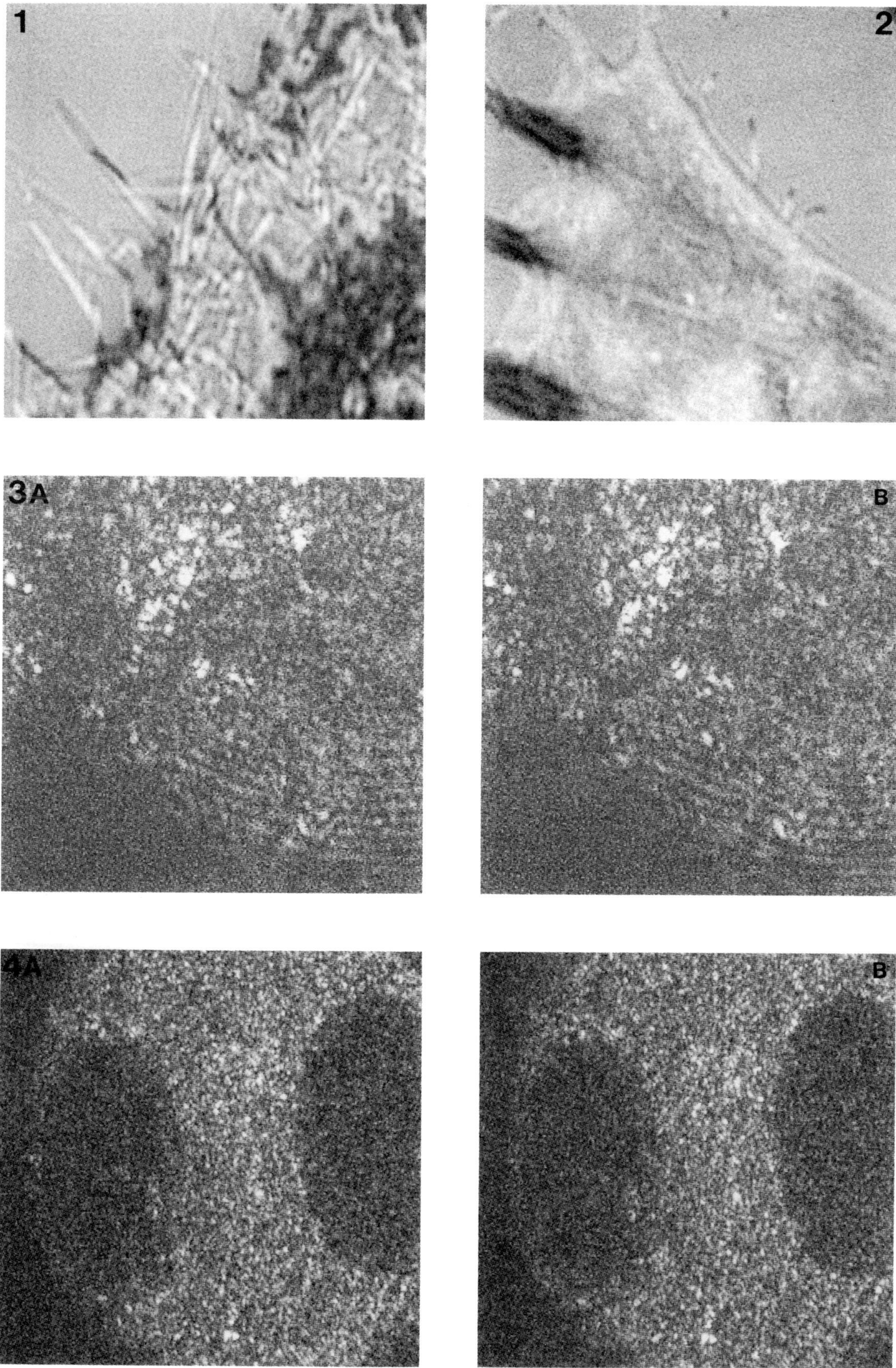
1
2
3A
B
4A
B

of sections of glass slides glued parallel to the long axis of standard 25 mm wide slides. This arrangement proved to be surprisingly stable, with very little focus drift over periods of several minutes. It permitted the rapid exchange of fluid media by pipetting at one end and blotting at the other.

Before examining a few cells in each sample at this, the highest useful magnification, we scanned each sample with the focus adjusted to the plane of contact of the cells with the substrate in order to examine whether the cells were in standard condition.

Results

Both temporal and spatial components of the final resolution have to be assessed against the level of noise and approached in different ways. Standard interlaced video leads to an unfortunate downgrading of temporal resolution in that adjacent lines in the image are not nearest neighbours in time. This deficiency can be partly compensated for by throwing away half the lines in the image so that adjacent data points in the frame direction may be spaced physically (by an empty line), but are as close as possible in time: they are only separated by 1/256 of 1/50th second in our case. Pixels in the line direction are only separated by 1/(512*512)sec. In either case, this excellent time resolution adds to the well known enhancement of XY, and particularly, Z, resolution, greatly to extend the possibilities of visualizing activity in living cytoplasm (whilst still respecting the Shannon-Kotelnikov theorem about the need for a 2 to 3 times higher sampling frequency than the frequency observed). Real temporal resolution currently can also be estimated visually from observing biological activity and its changes during experimental manipulation.

Spatial resolution can be measured using established procedures in light microscopy. To assess the zoom factor necessary to view the phenomena of rapid motion in comfort, we used mounted specimens of the diatom *Amphipleura pellucida* (purchased from Northern Biological Supplies, Ipswich, UK). Using the 100/1.4 lens, we found that the 0.25 μm transverse periodicity could be seen at zoom 2, and the 0.2 μm longitudinal striping at zoom 3. However, more comfortable viewing of the live or taped video image was achieved with a greater degree of oversampling at zoom factor 5, and this was the standard we adopted for examining intracytoplasmic motion: this gave a display monitor field width of 10 μm.

In the types of biological experiments which were undertaken, there are intrinsic and extrinsic sources of noise (Pawley, 1995). Intrinsic noise is exemplified by the mainly thermal, Brownian motion seen as shimmering mainly over the cellular compartments: this can be dealt with by appropriate setup of the experiments and examination of the effect of blocking particular types of motility and evaluating the residual activity. Extrinsic noise is due mainly to the electronic instability of the whole imaging system: it can be explored by using solid, motionless specimens (grids, fixed cells). Thus true cellular activity can be recognized in spite of the presence of background noise of thermal origin which constantly slightly blurs the entire image, and of an irregular flickering of the image caused by the imaging system. True motility is revealed by movements which (a) extend over distances larger than just a few pixels, and which (b) also proceed in oblique directions (and are therefore not related to TV scan problems) and/or (c) contrast with the relative stability of other parts of the cell, e.g., the cell edge. The cell edge can be used with advantage as a point of reference not only for relative motions in X,Y axes but also for controlling the stability of the focus in Z-axis. This very important aspect as the dynamic behaviour of intracellular structure could be to some extent simulated by slight focus wobbling.

(*Figures 5-8 on facing page*)

Figure 5 A,B: Interior of K4 cell, showing part of the nucleus in the lower left and IDSN to the right, focussed 7 μm deep in the cell from the surface of the coverslip. Two images 1/25s apart. To detect the relative motion of features (only easily achieved by viewing the live dynamic image!), the images may be viewed as a stereo-pair, when features which have not moved will appear to lie in one plane: this will demonstrate the overall translocation of particles of 0.2 μm size of the IDSN. Picture width 10 μm, zoom factor 5, pinhole size 2.94 mm.

Figure 6 A,B: Interior of nucleus of K4 cells 3 μm deep in the cell from the surface of the coverslip showing differences in the image of the cell interior depending on the confocal detector pinhole size. The same field taken with pinhole 5.66 mm for (A) and 2.62 mm for (B). Picture width 10 μm, zoom factor 5.

Figure 7: Interior of a normal (rat bone derived) cell 1 μm deep to the coverslip. Nucleus to top right, to the left of which lies IDSN: 'communication' layer seen in bottom left corner. Picture width 10 μm, zoom factor 5, pinhole size 4.08 mm. (Cells kindly provided by Dr. Colin M. Gray)

Figure 8: Edge (top) and interior of a normal (rat bone derived) cell 0.5-1 μm deep to the coverslip. Thin microspikes along the edge of the cell (which could be used as an indicator of focus stability) and structures in the communication layer are seen. Picture width 10 μm, zoom factor 5, pinhole size 2.98 mm.

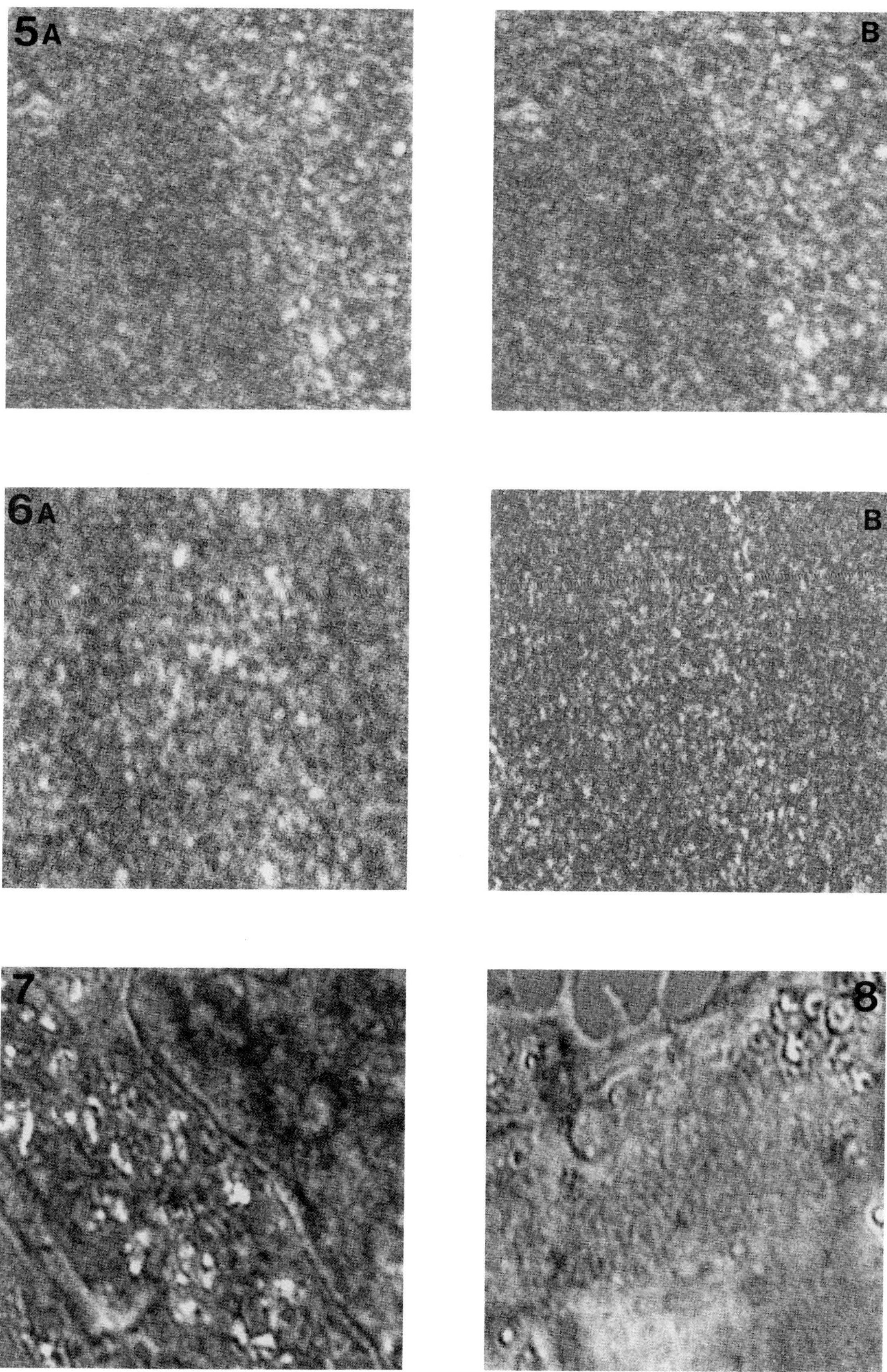
5A
B
6A
B
7
8

The way in which the image is created in the reflection mode of VRCLSM strongly depends on the level of focus inside the cell and particularly on the distance in the Z-axis from the surface of coverslip on which the observed cell is attached:

(a) At a level of up to 0.5 μm from the coverslip, the image is dominated by interference reflection contrast. It essentially shows the structure of the cell edge (Fig. 1) and maps the focal contacts and other regions (Fig. 2) in which the ventral plasma membrane is fairly close to the coverslip. For inspection of cell-to-substratum contacts, we used zoom 1 with the 100X objective, a physical pinhole size of about 3 mm and photomultiplier gain of 1000 arbitrary units.

(b) When the focus is moved 1-2 μm deep into the cell in thin (2-3 μm) normal cells, the image is composed from overlying processes: namely, simple reflection-cum-scattering contrast from the intracellular cytoplasmic structures (Fig. 9) and interference between reflection from the surface of the coverslip and from the dorsal surface of the cell (the side away from the coverslip and facing the culture medium). The dynamics of this type of image indicate rather fast fluctuation of the cell dorsal surface seen as rapid changes in darker sinuous stripes in the center of the image (Fig. 10).

(c) If focussed between 2 and 6 μm into thicker, in our case usually neoplastic cells, and on increasing the PM gain to about 1400, the signal from the central parts of the cell is dominated by simple reflection and backscattering contrasts (backscattered light imaging) (Fig. 4). If the steep side slope of a thick neoplastic cell becomes a part of the image, then parallel contour lines produced by interference between strong reflection from the coverslip and from the side slope of the cell are seen and can be used as a measure of the height of the cell ($\lambda/2$ = 244 nm gives the difference in height between neighbouring black stripes) (Fig. 3).

Using the highest possible resolution (spatial and temporal) of light microscopy achieved so far in the VRCLSM, with the Nikon (Kingston, Surrey, UK) 100/1.4 lens and in the reflection mode, the known structures of the living cell are regularly seen, including all types of contacts between the cell and substratum, surface microvilli and filopodia and ruffles, and, inside the cell periphery, elongated mitochondria, stress fibres and various granules, and, in the main body of the cell, the nucleus with nucleoli. In addition to these structures, a much sharper image of the cell edge is obtained which clearly shows the difference in "hairiness" and cell-to-substrate contacts between neoplastic (Fig. 1) and normal (Fig. 2) cells.

Deep inside the cell, the spatial network of the cytoplasm (Fig. 4) is seen, the dynamics of which can be revealed if A and B images in this text are looked at as stereopairs, so that temporal motion is appreciated as parallax. The perinuclear cytoplasm appears to be organized into a 3D network of tubulo-membraneous character with repeating details at well below 0.5 μm intervals with rapidly moving small features (Fig. 5). When such an image is obtained inside the cell, then the image character depends on the physical size of the confocal detector pinhole and thus on the actual thickness of the optical section visualized (as seen in images of the nucleus of a K4 cell in Fig. 6). The contrast derives from variations in the backscattering of light (Cheng and Kriete, 1995). The darker areas may correspond to cytosol. In comparison, large very dark areas inside some nuclei contain white spots which move at rather high speed and with less apparent restriction.

Fig. 3 shows the internal network described as an integrated dynamic spatial network (IDSN) and the side slope of the cell as presented in an interference reflection image: 13-14 periods between pairs of black stripes give a depth of 3 μm (14*244nm), which also corresponds with the reading of the microscope micrometer.

The periphery of the lamellar cytoplasm of normal cells appears to be "empty" (Fig. 9) when compared with neoplastic cells (Fig. 3), having only a few particles moving in various directions. The normal cell (Fig. 10) also registers wobbling of the dorsal cell surface, which is a rather unexpected phenomenon. The image of

(*Figures 9-11 on facing page*)

Figure 9 A,B: Cell periphery of normal (rat bone derived) cell 1 μm deep in the cell where reflection interference image from cell-to-substrate contacts is mixed with reflection scattering image of the intracellular cytoplasmic features. Viewing the image pair as if they were a stereoscopic pair reveals motile changes in 1/25s. Picture width 10 μm, zoom factor 5, pinhole size 2.98 mm.

Figure 10 A,B: Cell periphery of normal (rat bone derived) cell 2 μm deep in the cells shows complicated image which arises from superimposition of the reflection cum backscattering image of the internal cytoplasm over the reflection interference image: the latter is produced by interference of the light reflected from the coverslip and the dorsal surface of the cells (facing the medium side). Rapid changes in 1/25s of the interference pattern seen as darker sinuous stripes in the center of the image indicate rather fast fluctuation of the cell dorsal surface. Picture width 10 μm, zoom factor 5, pinhole size 2.98 mm.

Figure 11 A,B: Nucleus of normal (rat bone derived) cell shows very fine structures. 3D viewing reveals motile changes accomplished in 1/25s. Picture width 13 μm, zoom factor 4, pinhole size 2.98 mm.

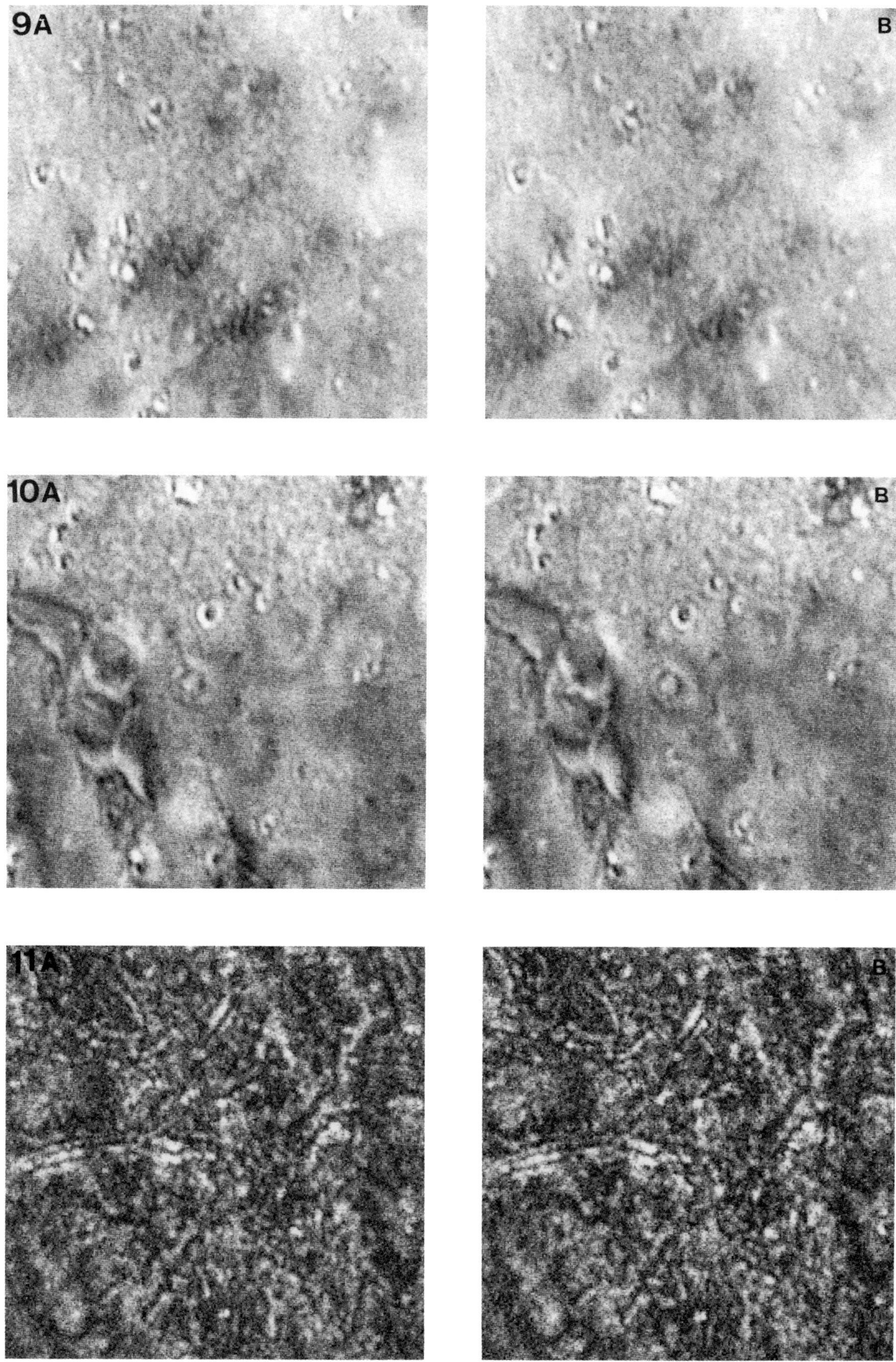
9A
B
10A
B
11A
B

the internal structure of the normal cell nucleus (Fig. 11) is even more surprising: it shows structures for which identification will depend on dynamic analysis.

The rate of motile exchanges in the cell internal network is rather stable and robust, but there appear to be some differences between the types of cell examined. Its frequency can be estimated to be within a range which extends from a distinguishable 5-10 Hz to over 25 Hz. The measurements are difficult as the background noise slightly blurs the entire image. 3D trafficking of particles means that particles move in all directions. A proportion of them therefore pass the plane of focus and their coming and going out of this plane thereby contributes falsely to the noise level.

The speed of these intracellular motions does not change quickly after (a) instantaneous exchanges of full medium; (b) change to serum-free medium; or (c) phosphate buffered saline (PBS); or (d) a slight drop of pH to 6.5; or (e) moderate temperature changes; or even (f) during the first 10-20 seconds after treatment with distilled water (hypotonic shock). This behaviour indicates a functional stability of this type of motility which appears not to be easily influenced by the presence, or the short absence, of growth factors or by rapid changes of microenvironmetal conditions. Only changing to ice-cold full medium slowed down the overall rate of motion after some minutes. This remarkable stability, together with the fact that the range of speeds seen in the IDSN fills the gap in the spectrum of observable speeds in the cell, raises the question as to what extent the regulation of speeds of exchanges and trafficking within the IDSN can be of importance in the epigenetic regulation of cellular processes.

There is also some indication of the mechanical robustness of the IDSN which may be related to internal tensions: it takes more than 10 seconds for hypotonic treatment to start to show the swelling and bursting in the IDSN which leads to its disruption.

Several modes of 3D motile activity in the IDSN were observed. Overall motility spans the entire cell interior, and this may make it difficult to detect the nucleus. In primary rat fibroblasts incubated for 5 hours in serum free Eagle's MEM pH 6.5 prior to examination, the motile activities in the nucleus and cytoplasm were found to be asynchronous. In some cases, "particles" appeared to move across the nuclear membrane in one or more regions, whilst they were apparently blocked in others. Within the cytoplasm, some areas showed higher rates of motility compared to their surroundings, reminiscent of the motility domains (of "nucleating" type) visualized in motile cells by the differencing mode in scanning acoustic microscopy (Vesely *et al.*, 1994). Other than the variations in motile activity noted within the nucleus of some cells, images of rather stable nucleoli surrounded by very active motion in the peri-nucleolar area were seen.

The course of fixation with 2.5% glutaraldehyde in PBS and/or in 96% ethanol was studied in order to differentiate between active intracellular motility and thermal motion and also to examine the usefulness of this approach for correlative microscopy in the control of specimen preparation for electron microscopy. There are differences in the effects of these two fixatives. Glutaraldehyde stabilises structures within approximately 5 seconds and they appear unchanged as the consequence. There is a slight variation in the time (4-6 seconds) needed for full immobilization of the cell: it depends on cell thickness even in a monolayer culture. Ethanol fixation is even more rapid: it fixed within 3-5 seconds, but induced changes seen as released tension and partial restructuring. The granularity of structure was preserved, but differed slightly from the living state.

Glutaraldehyde did not stop all motility, but the remaining motility (bound Brownian motion) was stopped by additional ethanol treatment. It may be that glutaraldehyde crosslinking is less potent in fixing cell fluid compartments, whereas ethanol may possess greater potential to coagulate proteins from solution and, in this way, change everything and even embed loose structures left oscillating on internal surfaces after glutaraldehyde fixation. When hypotonic treatment with water had preceded ethanol fixation by several seconds, then the expansion of the internal network was slightly delayed behind the overall swelling of the cell, which again indicates how resistant the internal network is. After fixation with both fixatives, the optical contrast of the fine 'granular' structure characteristic of cytoplasm is immediately increased, possibly by rapid sequestration of water from proteinaceous structures.

The IDSN varies in size relative to the cell body, being largest in the epithelial-like, fully transformed, rat RSV (Rous Sarcoma Virus) sarcoma K4 cells, where it almost fills the cell body (Fig. 5). It is smallest, confined to the center of the cell, in primary fibroblasts, and almost missing in some rat calvarium bone-derived cells (Fig. 7). The IDSN in K4 cells attached to a solid substratum *in vitro*, where it is most developed, appears to extend up to the "dorsal" side (i.e., that side remote from the objective lens, facing the medium) of the cell membrane. It does not extend into either the cell periphery or into cellular processes. Its extent varies with cell type and with the time in culture after trypsinization, which may in turn be related to the position in the cell cycle.

From observations by optical sectioning and from the effects of fixation with glutaraldehyde and ethanol, it can be deduced that there is a very narrow fluid compartment between the "ventral" side (facing the substratum) of the cell membrane and the core of the IDSN. This represents a communicating layer (Fig. 8), approximately 1 μm thick, in which both centripetal and centrifugal trafficking and clustering of small particles can be observed, sometimes along, or in, what appear to be preformed channels. Moving particles were sometimes seen to arise from above or below the current focus level, apparently through small openings, but to continue to move horizontally.

The overall image of the internal structural and dynamic organization of the cell creates a novel view of compartmentalization along the height (Z) axis. The (< 1 μm) contacting layer of cell membrane and rather stable structures close to the underlying substratum give rise to the known reflection interference image of cell-to-substratum contacts. Above the contacting layer (further from the objective lens) lies an approximately 1 μm thick layer (Fig. 8), apparently engaged in every kind of particulate transport, which spans almost from one edge of the cell to the other and also occupies the cell center underneath the IDSN. This communication layer does not extend into the peripheral rim of expanding lamellar cytoplasm. Next, in normal cells, only the nucleus with almost negligible IDSN in its vicinity can be found (Fig. 7). Lastly, parts of the top ("dorsal") surface of the cell are seen at the same level as prominent nucleoli. In neoplastic cells, an IDSN (Fig. 5) of up to several micrometers in thickness is found. The IDSN appears to integrate the cytoskeleton and all perinuclear cellular organelles from the nuclear membrane, through endoplasmic reticulum, cis Golgi network, Golgi stacks, trans Golgi network, through to the "dorsal" cell membrane into one dynamic transport compartment.

Discussion

The intracellular spatial network apparently supporting 3D movement of small particles may be described as an integrated or integrating dynamic spatial network (IDSN) because this is how it is observed in video rate reflected/backscattered imaging mode. From a functional viewpoint, it seems interesting that the IDSN is much more developed in sarcoma cells than in normal mesenchymal cells. Spatial periodicity in the IDSN was estimated to be well below 0.5 μm: measurement with two dimensional Fast Fourier Transform (using Noran TN8502 software) in a T15 cell returned major periodicities between 0.1337 and 0.168 μm^{-1}.

The static morphological image of the IDSN is reminiscent of the networks in extracted cells shown by Bell and Safiejko-Mroczka (personal communication), of those of cytoplasmic filaments seen in stereoscopic high voltage electron micrographs of whole cell mounts (Ris, 1985), and/or of actin meshworks of the types demonstrated in cultured cells (Small, 1981) and in the terminal web of intestinal epithelial cells (Hirokawa and Heuser, 1981). A perinuclear structure of a similar type was also imaged in high pressure frozen and freeze-fractured leukemia cells by Malecki (1991). Finally, a new video-rate confocal scanning laser microscope (built for imaging human skin) using 100x objective lens showed a perinuclear granularity of a type similar to the IDSN in still images of epidermal cells (Rajadhyaksha *et al.*, 1995). It seems that the IDSN can also support endocytotic function as well as exocytotic export of proteolytic enzymes (Krepela *et al.*, 1989): the dorsal surface of the K4 cell is very actively engaged in micropinocytosis (Vesely, 1972). The IDSN is rather stable in respect of mild temperature changes or a short absence of growth factors contained in full tissue culture medium, and apparently enables 3D trafficking of small "particles" (≤0.2 μm) over distances that cannot currently be assessed.

IDSN also seems to confer mechanical stability, which in turn may have functional impact on important growth properties such as anchorage dependent growth (Stoker *et al.*, 1968). The network may provide the cell with a mechanical self-sustaining structure for intracellular 3D transport. Thus it may free the cell from the dependence on attachment and consequent stretching that is apparently needed for the communication layer to resume function, which in turn seems to be vital for the growth of normal cells. Elucidation of a possible biomechanical role for the IDSN may enhance understanding of the differences between normal and neoplastic cells. The methodology employed here may then contribute in the rapid evaluation of the malignant state of living neoplastic cells in biopsies.

Fixation with glutaraldehyde stabilizes the IDSN without changes and fast enough to allow experiments in correlative microscopy to develop further. Particularly it should be of interest to try to compare in the EM the visualization of the IDSN in various reactive stages of the cell or phases of the cell cycle as seen by the reflection mode of VRLSCM.

There is a kinetic difference between fixation with glutaraldehyde and ethanol. It is possible that ethanol disrupts some linkages in the trabecular structure of the IDSN causing its rearrangement by the residual forces in the net. Ethanol apparently possesses a greater potential to coagulate proteins from solution, stabilizing pendulous structures still moving on internal surfaces after glutaraldehyde fixation.

Exploratory experiments comparing video rate single frame presentation (and grabbing) and averaging of 2, 4, 8, and 16 frames indicated that averaging of 2 frames decreases both noise and the time resolution. Averaging 4 and more frames, the loss of time resolution sharply increases.

Spherical aberration due to the use of the oil immersion objective in water is hardly a problem at the very short ranges, past glass (up to 10 μm deep), which we employed. To obtain images less influenced by this possibility, we also examined some samples using water immersion objectives, but found no difference in the way the IDSN was imaged and in the intracellular dynamics.

The requirements for correlative examination of the living cell by the reflection mode of VRCLSM followed by fixation and analysis of the observed structure in the EM can be summarized into several steps of accomplishment. The basic setup is composed of a VRCLSM based on an upright or an inverted microscope (less suitable for water immersion lenses) with reflection mode and a suitable lens, video recorder (date and time, including seconds), an open through-flush chamber of the blotting type and 37°C heating of a plastic "tent" incorporating the microscope. Image acquisition should consist of non-interlaced image grabbing and/or digital videodisc. Accommodation of cells and experimental manipulations can be further improved by introducing a closed through-flush chamber, temperature control of the microscope stage and the lens and finally by computer controlled operations. A substantial drawback in the design of contemporary light microscope stands is in the rather sloppy control of specimen positioning in X, Y and especially Z. Anything cheaper than feedback control (Lani, 1993) is not really suitable for the job. Another necessity is presentation of the contextual information about the present state of experimental manipulations with the cell under observation, or in another part of the same cell. An ordinary inverted microscope arranged to allow observation of the specimen at lower magnification from the opposite direction would be the easiest solution. Employing a system for bidirectional viewing (Maly and Vesely, 1979) should allow for much more sophisticated approach than that described here.

Acknowledgements

Odyssey was purchased with financial support of The Wellcome Trust. We thank S.J. Jones for the use of a 60/1.2 Nikon WI lens purchased with her Royal Society Grant. The experimental work was supported by grants from the Grant Agency of the Czech Republic Reg. No. 312/94/1481 and a Royal Society exchange visit to Pavel Vesely. We thank Dr. Colin M. Gray for rat calvaria cells. The technical assistance of R. Radcliffe was supported by Veterinary Advisory Committee of the The Horserace Betting Levy Board.

References

Boyde A, Jones SJ (1992) Real time confocal microscopy. Binary Comput Microbiol **4**: 119-123.

Boyde A, Vesely P, Gray C, Jones SJ (1994) High temporal and spatial resolution studies of bone cells using real-time confocal reflection microscopy. Scanning **16**: 285-294.

Cheng PC, Kriete A (1995) Image contrast in confocal light microscopy. In: Handbook of Biological Confocal Microscopy, 2nd edition. Pawley JB (ed). Plenum Press, New York. pp 281-310.

Draaijer A, Houpt PM (1988) A standard video-rate confocal laser-scanning reflection adn fluorescence microscope. Scanning **10**: 139-145.

Hirokawa N, Heuser JE (1981) Quick-freeze, deep-etch visualization of the cytoskeleton beneath surface differentiations of intestinal epithelial cells. J Cell Biol **91**: 399-409.

Inoué S (1990) Foundations of confocal scanned imaging in light microscopy. In: Handbook of Biological Confocal Microscopy, 2nd edition. Pawley JB (ed). Plenum Press, New York. pp 12-25.

Krepela E, Vesely P, Chaloupkova A, Zicha D, Urbanec P, Rasnick D, Vicar D (1989) Cathepsin B in cells of two rat sarcomas with different rates of spontaneous metastasis. Neoplasma **36**: 529-540.

Lani F (1993) Feed-back stabilized focal plane control for light microscopes. Rev Sci Instrum **64**: 1474-1477.

Malecki M (1991) High voltage electron microscopy and low voltage scanning electron microscopy of human neoplastic cell culture. Scanning Microsc Suppl **5**, S53-S73.

Maly M, Vesely P (1979) A new light microscopic method for the synchronous bidirectional illumination and viewing of living cells in different contrast modes, and/or at different focal levels or magnifications. J Microsc **117**, 411-416.

Pawley JB (1995) Fundamental limits in confocal microscopy. In: Handbook of Biological Confocal Microscopy, 2nd edition. Pawley JB (ed). Plenum Press, New York. pp 19-37.

Petran M, Hadravsky M, Egger MD, Galambos R (1968) Tandem scanning reflected light microscope. J Opt Soc Am **58**: 661-664.

Rajadhyaksha M, Grossmann M, Esterowitz D, Webb RH, Anderson RR (1995) In vivo confocal scanning laser microscopy of human skin: melanin provides strong contrast. J Invest Dermatol **104**: 946-952.

Ris H (1985) The cytoplasmic filament system in critical point-dried whole mounts and plastic-embedded sections. J Cell Biol 100, 1474-1487.

Small JV (1981) Organization of actin in the leading edge of cultured cells: influence of osmium tetroxide and dehydration on the ultrastructure of actin meshworks. J Cell Biol **91**: 695-705.

Stoker MPG, O'Neil C, Berryman S (1968) Anchorage and growth regulation in normal and virus-transformed cells. Int J Cancer **3**: 683-693.

Vesely P (1972) Tumour cell surface specialization in the uptake of nutrients evidenced by cinemicrography as a phenotypic condition for density independent growth. Folia Biologica (Praha) **18**, 395-401.

Vesely P, Boyde A (1994) Dynamics of intracellular structures revealed in living neoplastic cells by reflection on interference imaging with three types of video rate laser scanning confocal microscopes. Scanning **16** (Suppl IV), p IV-52 (abstract).

Vesely P, Jones SJ, Boyde A (1993) Video-rate confocal reflection microscopy of neoplastic cells: Rate of intracellular movement and peripheral motility characteristic of neoplastic cell line (RSK4) with high degree of growth independence in vitro. Scanning **15**: 43-47.

Vesely P, Luers H, Riehle M, Bereiter-Hahn J (1994) Subtraction scanning acoustic microscopy reveals motility domains in cells *in vitro*. Cell Motil Cytoskel **29**: 231-240.

Discussion with Reviewers

M. Malecki: Phototoxicity is a major plague in light microscopy of living cells. What was the energy delivered to the cells in your experiment? Did you evaluate the viability of the cells?

Authors: It was not our aim to preserve cells, except in the very short term. We did not evaluate the viability of the cells. The patterns of microscopic movement which we observed could be seen when we first found a field - and immediately, as could be checked from the videotape record - from which we may conclude that they were not induced by the process of observation. There was no change on changing the zoom factor, whilst any radiation damage would be expected to square with doubling the magnification. We are not surprised by this evident lack of a phototoxic effect, because only a tiny fraction of the incident radiation is absorbed in the specimen: that part which is reflected and backscattered (1 in 10^6 according to M. Petran, personal communication, 1970) possibly being of no consequence because it is reflected.

From measurements made with a laser power meter, the maximum power given in the 488 nm line with the system we used was 570 W at the objective lens.

Scanning Microscopy Supplement 10, 1996 (pages 213-224) 0892-953X/96$5.00+.25
Scanning Microscopy International, Chicago (AMF O'Hare), IL 60666 USA

EMERGING APPLICATIONS OF FLUORESCENCE SPECTROSCOPY TO CELLULAR IMAGING: LIFETIME IMAGING, METAL-LIGAND PROBES, MULTI-PHOTON EXCITATION AND LIGHT QUENCHING

Joseph R. Lakowicz*

Center for Fluorescence Spectroscopy, Department of Biochemistry and Molecular Biology, University of Maryland School of Medicine, Baltimore, MD

(Received for publication August 22, 1995 and in revised form September 2, 1996)

Abstract

Advances in time-resolved fluorescence spectroscopy can be applied to cellular imaging. Fluorescence lifetime imaging microscopy (FLIM) creates image contrast based on the decay time of sensing probes at each point in a two-dimensional image. FLIM allows imaging of Ca^{2+} and other ions without the need for wavelength-ratiometric probes. Ca^{2+} imaging can be performed by FLIM with visible wavelength excitation. Instrumentation for FLIM is potentially simple enough to be present in most research laboratories. Applications of fluorescence are often limited by the lack of suitable fluorophores. New, highly photostable probes allow off-gating of the prompt autofluorescence, and measurement of rotational motion of large macromolecules. These luminescent metal-ligand complexes will become widely utilized. Modern pulse lasers allow new experiments based on non-linear phenomena. With picosecond and femtosecond lasers fluorophores can be excited by simultaneous absorption of two or three photons. Hence, Ca^{2+} probes, membrane probes, and even intrinsic protein fluorescence can be excited with red or near infrared wavelengths, without ultraviolet lasers or optics. Finally, light itself can be used to control the excited state population. By using light pulses whose wavelength overlaps the emission spectrum of a fluorophore one can modify the excited state population and orientation. This use of non-absorbed light to modify emission can have wide reaching applications in cellular imaging.

Key Words: Time-resolved fluorescence spectroscopy, fluorescence lifetime imaging microscopy (FLIM), calcium imaging, fluorescent probes, metal-ligand probes, pulse lasers, two-photon-induced fluorescence.

*Address for correspondence:
J.R. Lakowicz
Center for Fluorescence Spectroscopy
Dept. Biochemistry Mol. Biology, Univ. Maryland Sch. Medicine, 725 W. Lombard S., Baltimore, MD 21201
Telephone Number:410 706 8409
FAX Number:410 706 8408
E-mail: jf@cfs.umbi.umd.edu

Recent Advances in Fluorescence Microscopy and Spectroscopy

Fluorescence Lifetime Imaging Microscopy

Fluorescence microscopy is widely used in cell biology to study the spatial distribution of ions and macromolecules [1, 10, 20, 32]. At present almost all microscopic imaging is accomplished by intensity imaging. However, it is well known that quantification of intensities is difficult in a microscope due to the unknown probe concentration at each point in the image, unknown quantum yield, photobleaching and other confounding factors. To circumvent these difficulties we developed a method for imaging based on the decay time or lifetime at each point in the image. We chose the lifetime as the contrast mechanism because we knew that decay times were mostly independent of the probe concentration and photobleaching, and that lifetimes could be measured under optically compromised conditions [18, 27]. Lifetime imaging can be advantageous compared to intensity imaging if one wishes to localize intracellular ions. Many ion indicators are now known to display ion-sensitive lifetimes. If one images the fluorescent intensity one observes where the indicator is localized in the cell. If one measures the lifetime at each point in the cell, that is creates a lifetime image, one learns the local ion concentrations independent of indicator localization.

Assume that the lifetime of the probe is different in the two regions of the cell (Fig. 1, top). If one could create a contrast based on the lifetime at each point in the image, one would resolve two regions of the cell, each with an analyte (Ca^{2+}) concentration which was revealed by the lifetime image. The creation of such fluorescence lifetime images, in which the contrast is based on lifetimes, appeared to be a daunting challenge. Consider the difficulties of performing 2.62 x 10^5 lifetime measurements for a typical 512 x 512 image. Given the difficulties of measuring even a single lifetime in a cuvette, such a task seems nearly impossible. However, image intensifiers and CCD (charge-coupled device) camera technology now makes this possible [15, 16]. Fig. 2, (right) shows the Ca^{2+} lifetime image of

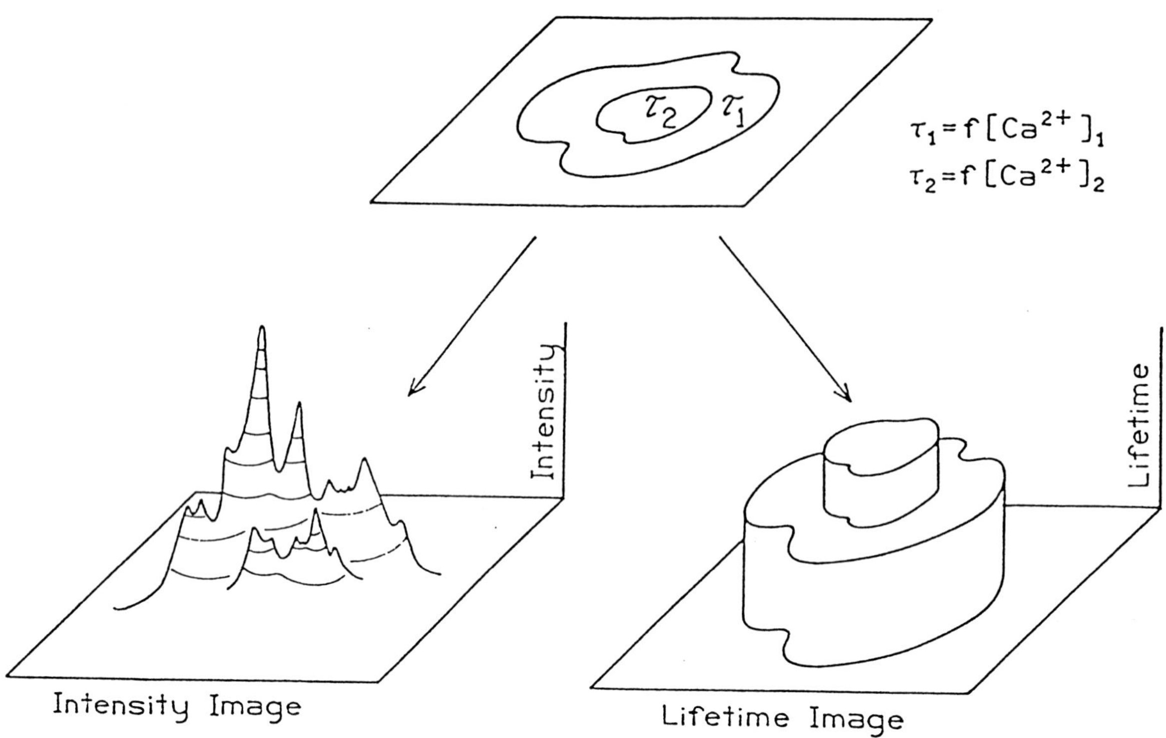

Figure 1. Fluorescence lifetime imaging microscopy (FLIM).

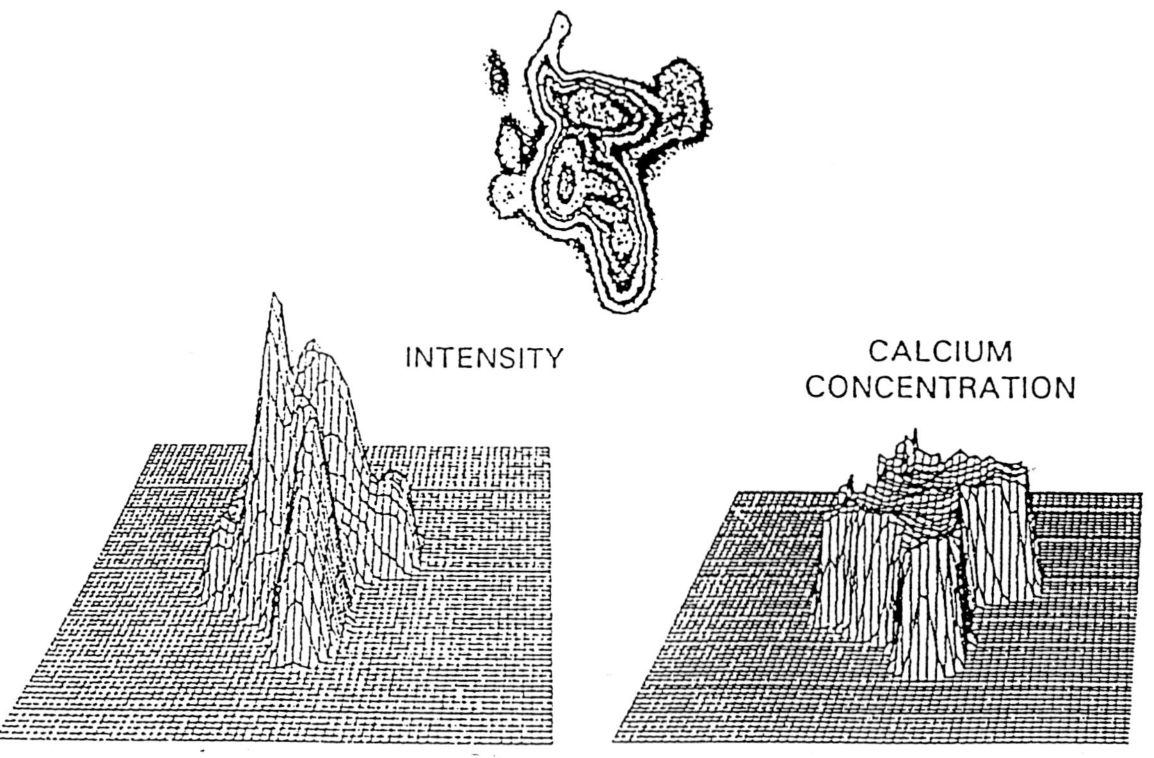

Figure 2. Intensity (top and left) and calcium images (right) of Quin-2 fluorescence in COS cells.

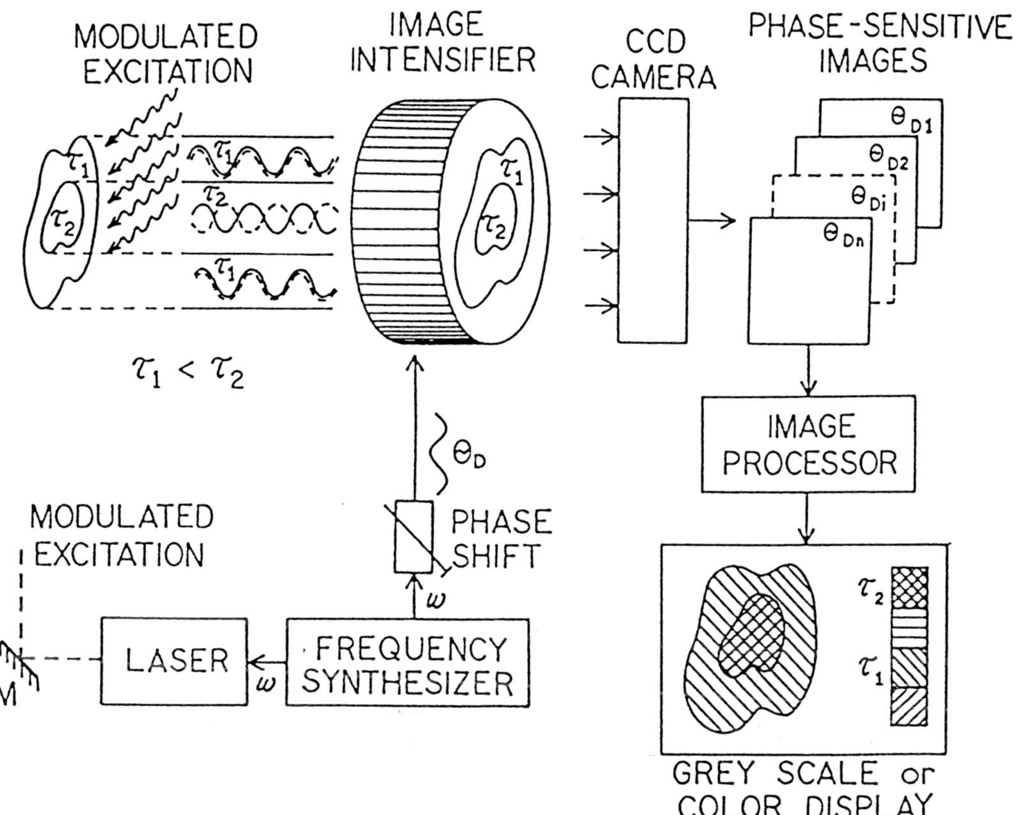

Figure 3. Apparatus for fluorescence lifetime imaging microscopy (FLIM).

COS cells based on the probe Quin-2 [19], along with the intensity image (left). The intensity images show the expected spatial variations due to probe localization, and the Ca^{2+} (phase angle) image shows the expected uniform concentration of intracellular calcium. As predicted, the lifetime imaging provides chemical imaging, which within limits is insensitive to the local probe concentration.

The cellular FLIM images in Fig. 2 were obtained using moderately complex instrumentation, which consists of a ps dye laser, a gain-modulated image intensifier, and a slow-scan scientific grade CCD camera (Fig. 3). However, the FLIM instruments in the future can be compact, even mostly a solid state device. This possibility is shown in Fig. 4, where we show that the light source can be a laser diode, assuming the fluorescent probes are available. An image intensifier is a moderately simple device, but is delicate and requires high voltages. Reports have appeared on gatable CCD detectors [24]. Present gatable CCDs are too slow (50 ns gating time). This time response is likely to improve, and probes can be developed with longer decay times, which is described in the next section. Then the FLIM apparatus will consist of only modest additions to a standard fluorescence microscope.

What type of chemical imaging will be possible using FLIM technology? Based on our current understanding of FLIM, and factors which affect fluorescence lifetimes, we can predict that lifetime imaging will allow imaging of a variety of cellular properties such as proximity, binding and microviscosity (Table 1). Proximity imaging can be possible based on the phenomena of fluorescence resonance energy transfer (FRET). Energy transfer occurs when a fluorescent donor is within a given distance of an acceptor. The distances for FRET are typically 30-60Å, which is typical of the size of biological macromolecules. FRET decreases the lifetime of the donor; thus allowing the proximity of a donor and acceptor to be determined from the donor decay time. Lifetime-based fluorescence probes are also known for a wide variety of ions [26], including Ca^{2+}, Mg^{2+}, K^{+}, pH, Cl^{-} and O_2. Hence, FLIM can have wide-ranging applications in cell physiology.

Probe Chemistry - Development of Long-Lived Metal-Ligand Probes

The application of fluorescence to analytical chemistry, clinical chemistry, flow cytometry, and imaging are

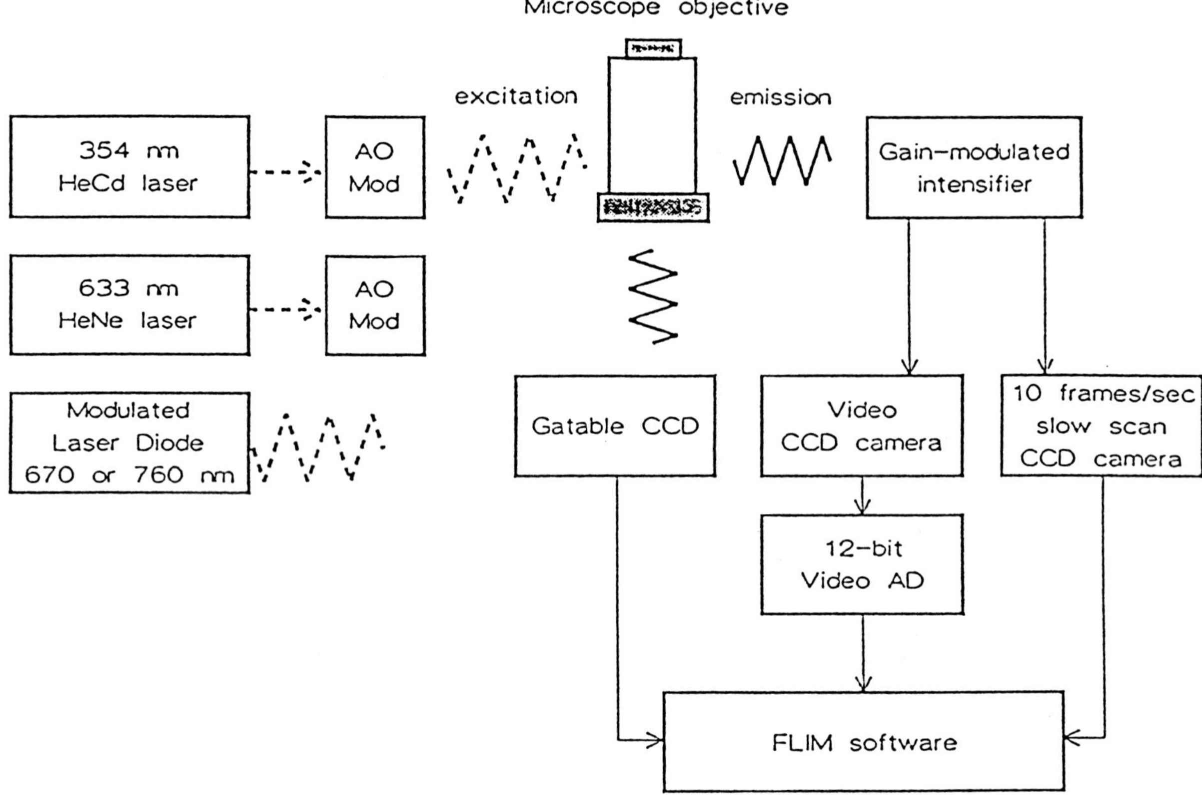

Figure 4. Future apparatus for FLIM.

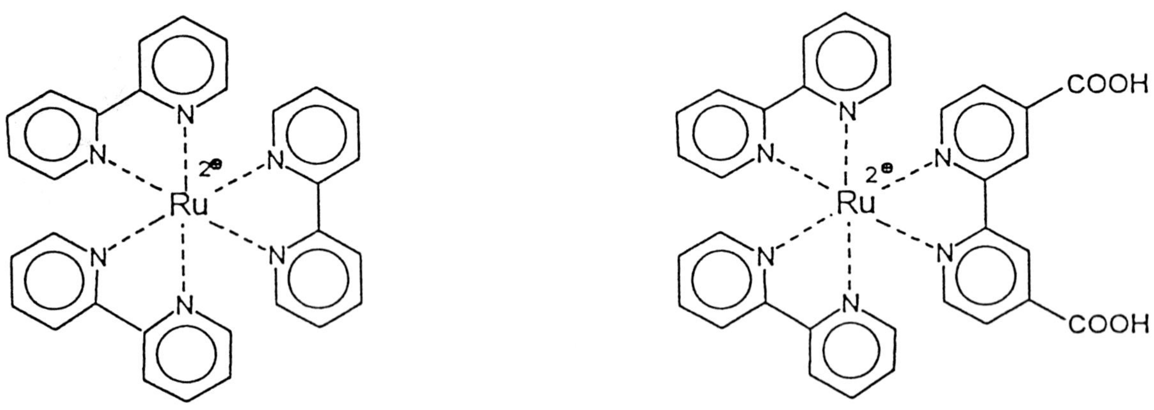

Figure 5. Chemical structure of $[Ru(bpy)_3]^{2+}$ and of $[Ru(bpy)_2(dcpby)]$.

limited not by the instrument technology, but by the available probes. There are only a limited number of conjugatable long wavelength probes, and none which display specific analyte sensitivity. What is needed is an arsenal of probes, all of which can be excited with laser diodes, and which are specifically sensitive to cations, anions, and other analytes. While several laboratories are working in this topic, the total effort is minor in comparison to the number of scientists engaged in instrument development, technology development, theory or applications.

In an attempt to circumvent the limitations of available probes we have been developing long-lived metal ligand complexes for measurement of the hydrodynamics of high molecular weight biomolecules [30, 31].

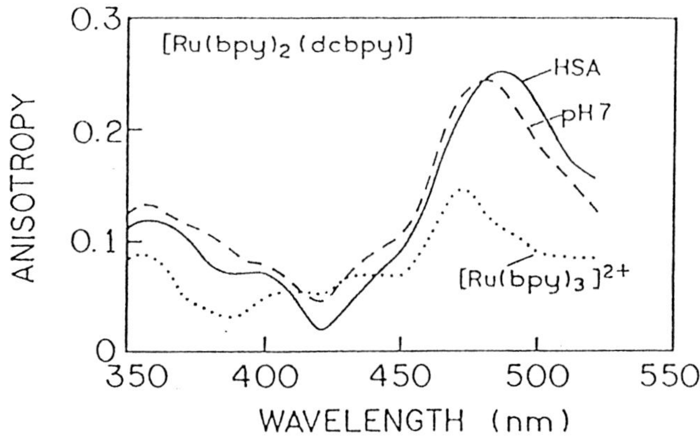

Figure 6. Excitation anisotropy spectra (bottom) of $[Ru(bpy)_2(dcpby)]$, free (— — —) and when conjugated to HSA (————). The anisotropy spectra are in glycerol:buffer (9:1, v/v) at -55°C. This viscous solvent is used to prevent rotational diffusion during the excited state lifetime. The dotted lines (·····) represent the symmetrical $[Ru(bpy)_3]Cl_2$.

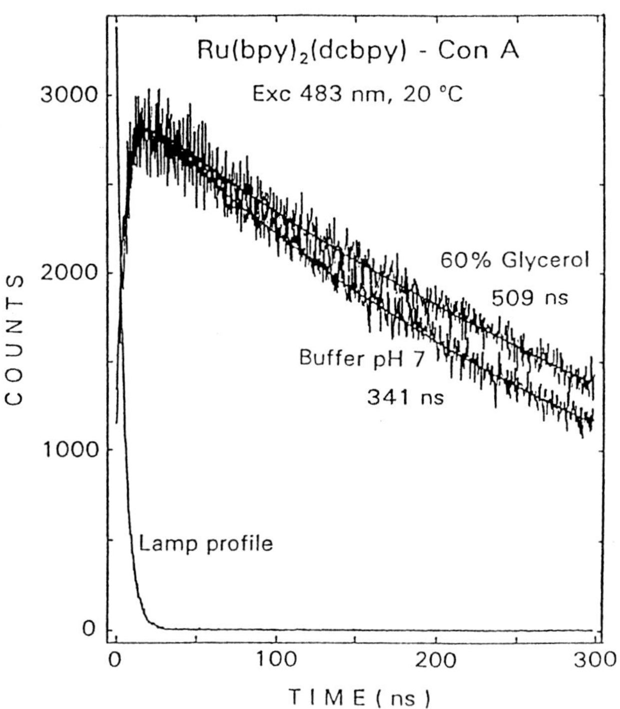

Figure 7. Intensity decays of $[Ru(bpy)_2(dcbpy)]$ conjugated to ConA.

For instance, almost all known fluorophores display lifetimes ranging from 1 to 10 ns. This decay time limits the timescale of the events which alter the emission. Metal-Ligand complexes (MLC) of the type $[Ru(bpy)_3^{2+}]$

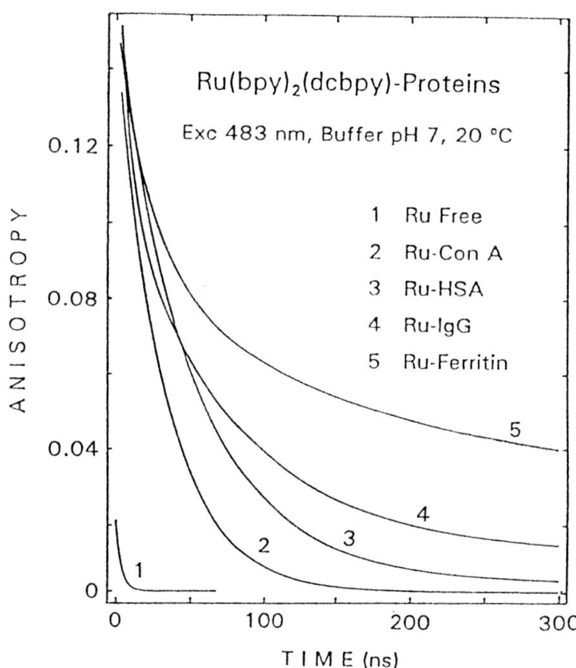

Figure 8. Anisotropy decays of $[Ru(bpy)_2(dcbpy)]$ in buffer and conjugated to proteins.

(Fig. 5) display lifetimes ranging from 100 ns to microseconds, and thus allow the detection of slower processes.

The usefulness of such metal-ligand complexes (MLCs) is derived from their favorable anisotropy properties. Suitable non-symmetrical complexes, such as $[Ru(bpy)_3(dcbpy)]^{2+}$ (Fig. 5, right) display high anisotropy in the absence of rotational diffusion (Fig. 6). The high anisotropy makes this probe useful for measuring rotational motion of proteins. A valuable feature of the MLCs is their long decay times, over 300 ns when conjugated to a protein (Fig. 7). The long decay times in turn allow measurement of long rotational correlation times, as shown for MLC-labeled proteins in Fig. 8.

The polarized emission from metal-ligand complexes offers numerous experimental opportunities in biophysics and cellular imaging. A wide range of lifetimes, absorption, and emission maxima can be obtained by careful selection of the metal and the ligand. For instance, long wavelengths are desirable for clinical applications, such as fluorescence polarization immunoassays. Absorption wavelengths as long as 700 nm can be obtained using osmium [11], and lifetimes as long as 100 μs can be obtained using rhenium as the metal in such complexes [25]. The rhenium complexes also display good quantum yields and high initial anisotropies in aqueous solution. At present it is difficult to obtain long lifetimes, long wavelengths and high quantum yields all in a single

Table 1. Cell biology applications of FLIM

Type of Imaging	**Analyte or Property**
Chemical imaging	Ca^{2+}, Mg^{2+}, Cl^-, pH, O_2, Na^+ and K^+
Ligand binding to proteins	NADH, TNS
Chromosome imaging	Acridine lifetimes depend on DNA base composition
Microviscosity imaging	Identify viscosity-lifetime probes
Proximity imaging by energy transfer	Protein-protein binding Protein-membrane association

Table 2. Enabling technologies for the biomedical applications of time-resolved fluorescence spectroscopy

Lasers	Laser diodes Two-photon excitation with picosecond-femtosecond lasers
Image intensifiers	Gatable on nanosecond timescale Red-sensitive for optical tomography
CCDs	Fast frame rates Gatable on nanosecond timescale
Computers	Allow processing of numerous images from CCD images
Probe chemistry	Need long-wavelength probes to take advantage of laser diodes and low autofluorescence

metal-ligand complex. Additional research is needed to identify which of these metal-ligand complexes displays the most favorable spectral properties for a particular application, and to synthesize conjugatable forms of the desired probes.

Advanced Applications of Fluorescence Spectroscopy

Two Photon-Induced Fluorescence (TPIF)

We are all familiar with one-photon induced fluorescence (OPIF), which is the common occurrence in our experiments. With intense laser sources, it is possible to observe the emission resulting from the simultaneous absorption of two long wavelength photons. For instance, tryptophan in proteins, which normally absorbs light at 290 nm, can be excited by the simultaneous absorption of two 580 nm photons (Fig. 9), which shows the emission spectra of human serum albumin (HSA) excited at 295 or 590 nm. This remarkable phenomenon occurs only with high light intensity because the two photons must be in the same place at the same time to allow simultaneous absorption. It should be emphasized that there is no direct relationship between the absorption spectra and absorption cross sections for one- and two-photon excitation. In principle, one- and two-photon absorption occurs to excited states of different symmetry due to the selection rules for optical absorption.

Because two-photon excitation requires two-photons, the fluorescence intensity is proportional to the square of the light intensity. This dependence on the square of the intensity is shown in Fig. 10. The emission intensity of HSA, is linearly dependent on the incident light intensity at 295 nm, and quadratically dependent on the intensity at 590 nm [13]. Two-photon induced fluorescence has also been observed for fluorophores bound to membranes [14] and nucleic acids [12].

The fact that TPIF depends on the intensity squared provides an important opportunity for fluorescence microscopy. In fluorescence microscopy, confocal optics are often used to eliminate fluorescence from outside the focal plane of the lenses [22, 34]. Removal of this out-of-focus light provides remarkable improve-

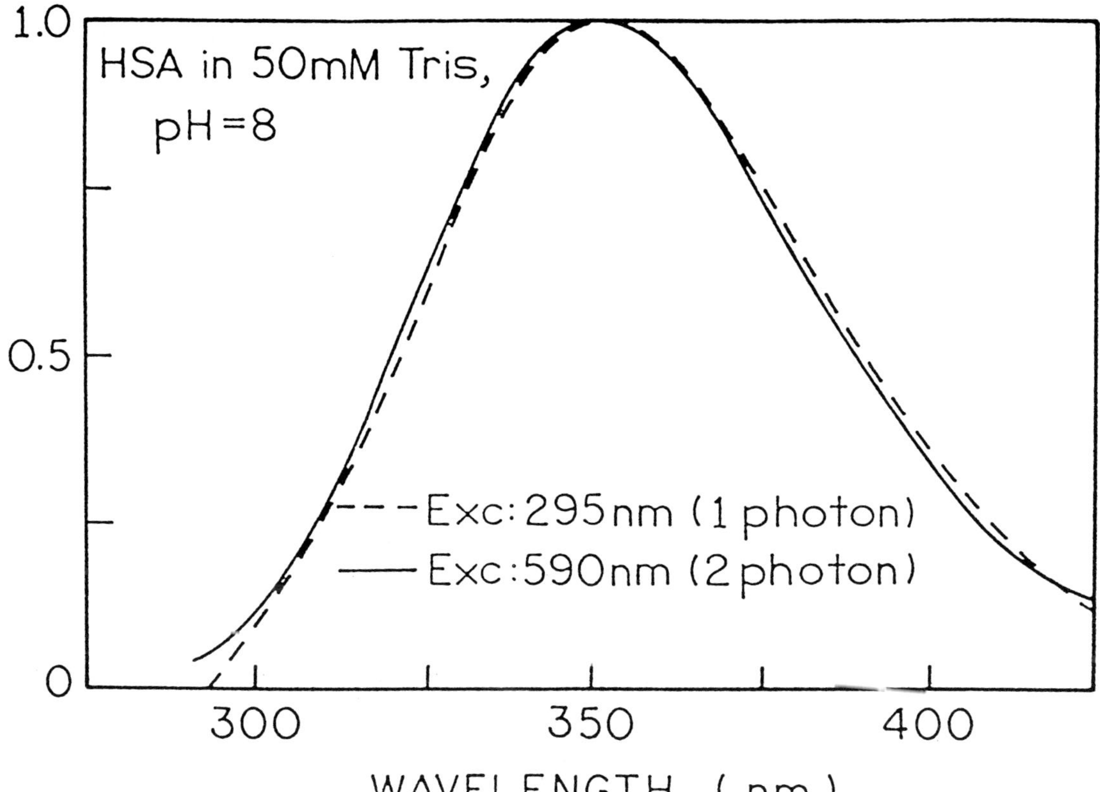

Figure 9. Emission spectra of HSA for one- and two-photon excitation.

ment in image quality because for one-photon excitation the fluorescence occurs from the entire thickness of the sample. Much of this emission is devoid of spatial information and only serves to degrade image contrast and resolution.

Professor Webb and colleagues at Cornell University, recognized that the intensity-squared dependent of TPIF provided the opportunity for intrinsic "confocal" excitation [2, 33]. The sample can be excited only at the desired depth based on the position of the focal point (Fig. 11), and the signal comes dominantly from this region. Perhaps more importantly, the fluorophores which are not in the focal plane are not excited, consequently, are not photobleached and are thus available for imaging when the focal plane is moved (Fig. 11). There are additional advantages of two-photon microscopy, such as the greater availability of optical components and increased transmission of the optics for the longer wavelengths. It is possible that the sample auto-fluorescence will be lower with two-photon excitation, but at present, we do not know if the endogenous fluorophores in cells will display high or low cross-sections for two-photon excitation. It is already known that some Ca^{2+} probes display good two-photon absorption [23], and Dr. Webb has reported lifetime images with two-photon excitation [23]. Hence, we can now imagine the creation of three-dimensional (3D) chemical images of cells (Fig. 10), which could display the local Ca^{2+} concentration as seen for the two-dimensional (2D) FLIM imaging in Fig. 2.

Three-Photon Excitation

We have recently extended the use of high intensity laser pulses to allow three-photon excitation. Three-photon excitation has been observed for the scintillator 2,5-diphenyloxazol (PPO) [7], for a tryptophan derivative [8] and for the calcium probe Indo-1 [9, 29]. Three-photon excitation was accomplished with femtosecond pulses of a mode-locked Ti:Sapphire laser at wavelengths above 800 nm. The emission spectra of Indo-1 were the same for excitation at 351 and 860 nm (Fig. 12). At 885 nm the intensity of Indo-1 depended on the cube of the laser power, but at a slightly shorter wavelength of 820 nm the intensity depended on the square of the laser power (Fig. 13). This indicates that the mode of excitation (two- or three-photon) can change with small changes in wavelengths. Remarkably, the intensity of Indo-1 with three-photon excitation is within a factor of 10 of that observed with two-photon excitation (Fig. 13), suggesting that three-photon excitation can be practical in fluorescence microscopy.

An important aspect of three-photon excitation is that it can be accomplished with wavelengths near 800 nm, which is the peak of the Ti:Sapphire laser tuning

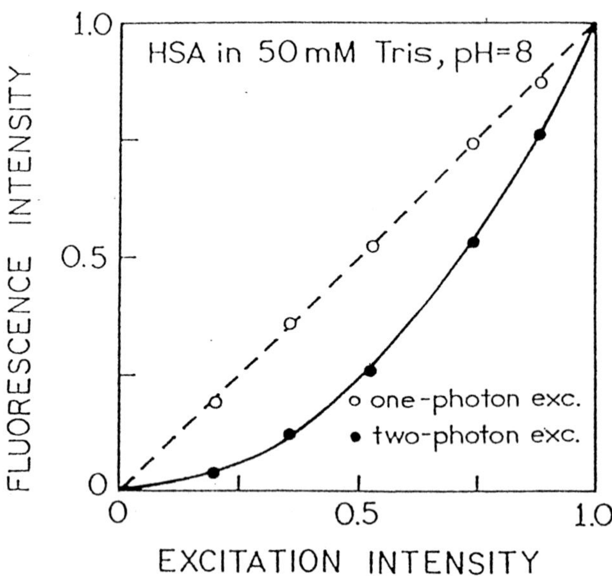

Figure 10. Dependence of the emission intensity of HSA on the incident light intensity (right). The excitation intensities are normalized to one at the highest values.

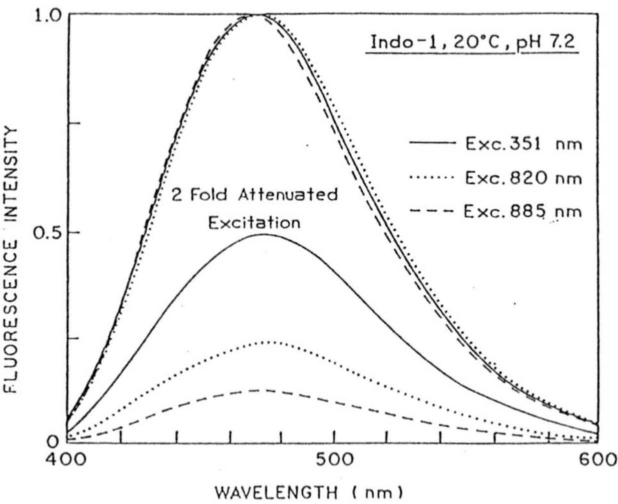

Figure 12. Emission spectra of Indo-1 with excitation at 351, 820 and 885 nm.

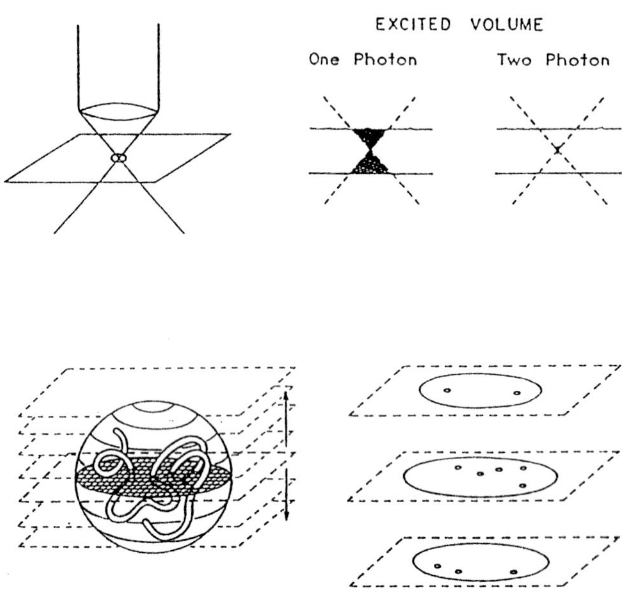

Figure 11. Intrinsic "confocal" excitation using TPIF in microscopy (top) and three-dimensional optical imaging (bottom).

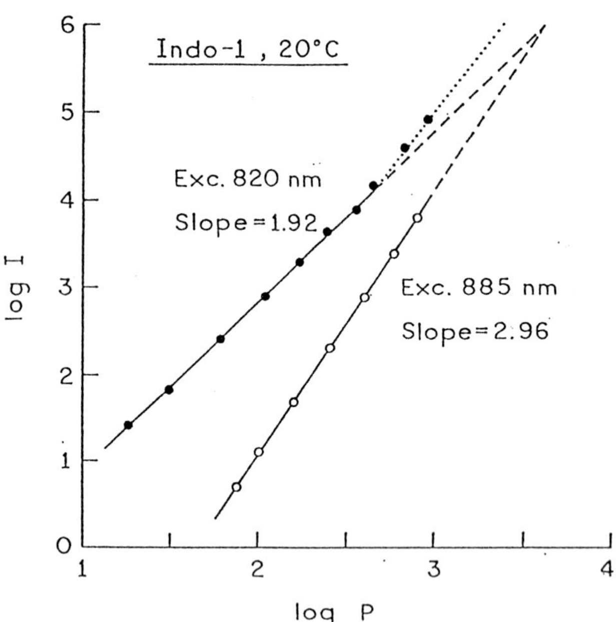

Figure 13. Dependence of the emission intensity of Indo-1 on laser power at 820 and 885 nm.

curve. Also, the spatial profile of the excited fluorophores can be smaller than available with two-photon excitation. This is shown experimentally in Fig. 14, which shows the width of the excited volume of Indo-1 with one-, two- and three-photon excitation. It seems probable that simple Ti:Sapphire lasers will be available in the near future, making two- and three-photon excitation possible in many microscopy laboratories.

Light Quenching of Fluorescence

Measurements of time-resolved fluorescence, and particularly the recent interest in two- and three-photon

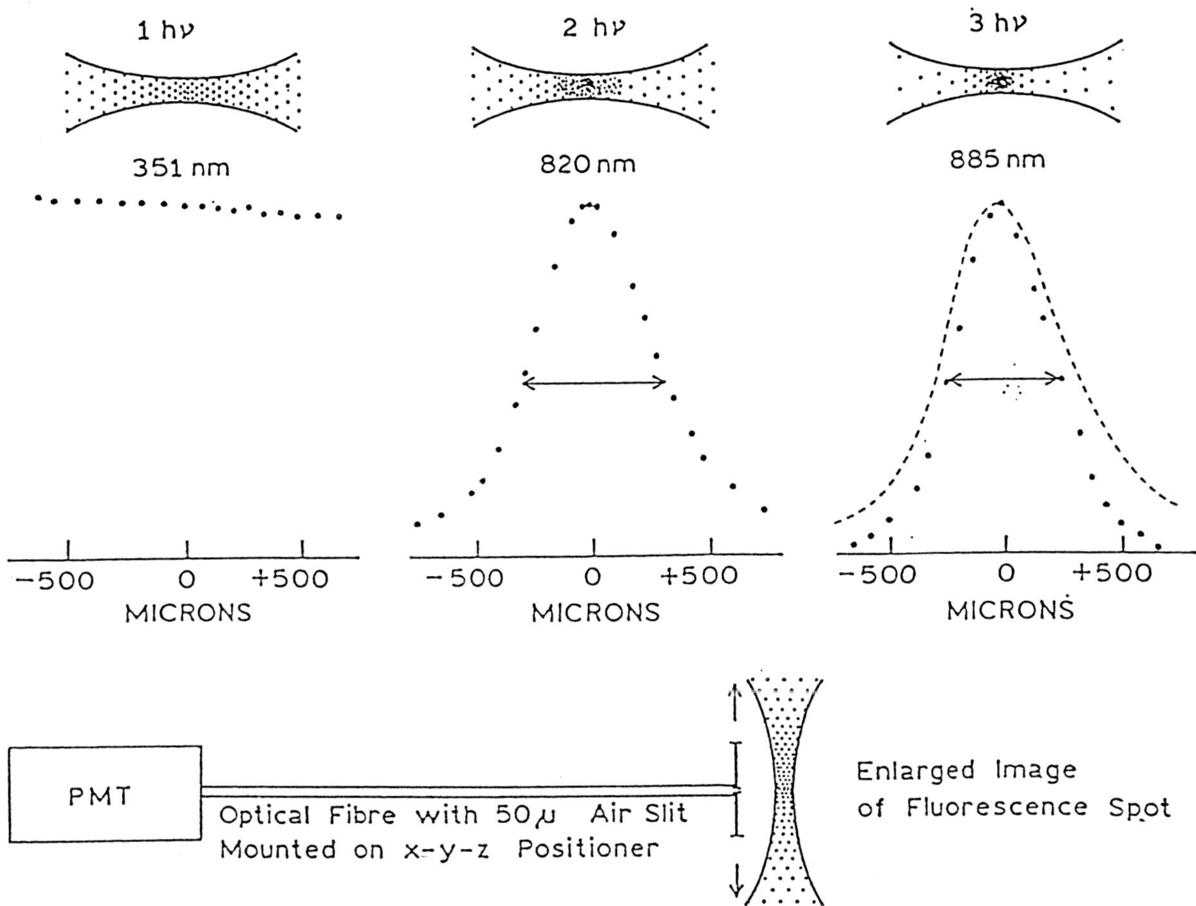

Figure 14. Spectral profile of the excited volumes of Indo-1 with one-, two- and three-photon excitation.

---

excitation, require the use of intense laser sources. The use of these intense laser sources allows observation of the phenomena of stimulated emission. If a fluorophore is illuminated at a wavelength which overlaps its emission spectra, the fluorophore can be stimulated to return to the ground state (Fig. 15). Since the stimulated photon travels parallel to the "quenching beam," and since the emission is generally observed at right angles to the illumination, the emission appears to be quenched.

Of course, light quenching or stimulated emission was predicted by Einstein in 1917 for atoms in the gas phase [3]. Historically, light quenching has only been observed using the very intense pulses from Q-switched Ruby lasers [4, 21]. The fact that we now know that light quenching can be observed with modern ps lasers results in numerous opportunities for novel fluorescence experiments [5, 6, 17].

Consider that the sample is excited with one pulse, followed by a second longer wavelength quenching pulse (Fig. 15, left). The quenching pulse can result in an instantaneous change in the excited state population. It is important to recognize that this change in population should be non-destructive since we are not depleting the ground state or bleaching the sample. Hence, the experiment may be repeated numerous times for improved signal-to-noise, if needed to measure small effects.

One remarkable opportunity of light quenching, and there are other opportunities which are not described in this article, is that light quenching displays the same $\cos^2 \theta$ dependence as does light absorption [6, 21]. This means that not only is the total excited state population altered by the quenching pulse, but that selectively oriented parts of the excited state population are quenched. Consequently, depending on the polarization of the quenching light, the polarization of the emission can be altered from 1.0 to -1.0 (Fig. 16), resulting in a high degree of orientation of the excited state population. In contrast, one-photon excitation of randomly oriented fluorophores can only result in polarization values for 0.5 to -0.33. It may even be possible to break the Z-axis symmetry, which heretofore has been pervasive in

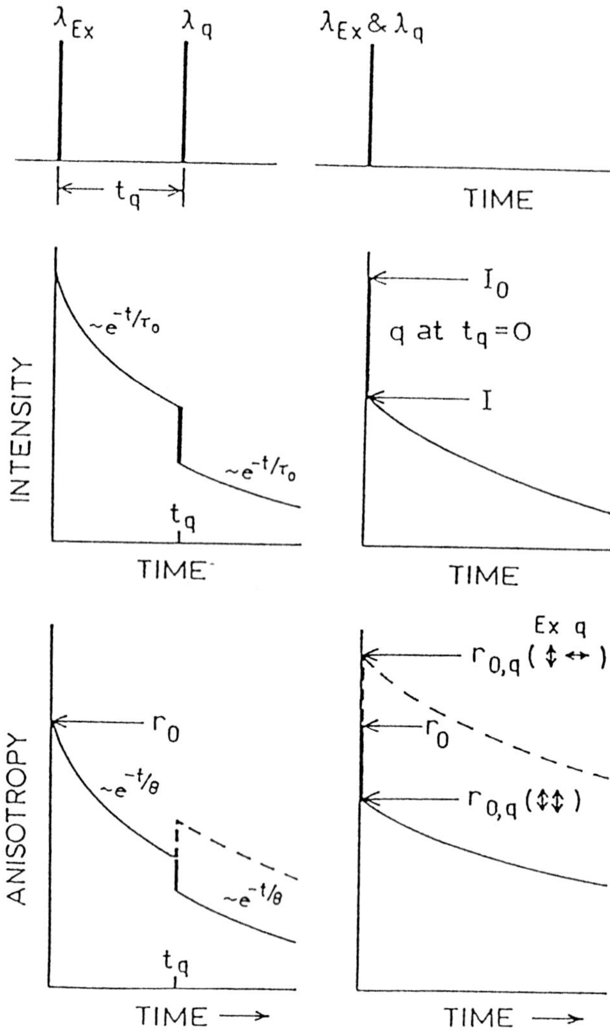

Figure 15. Schematic of light quenching by a time-delayed quenching pulse. The dotted lines refer to light quenching with perpendicular polarized light.

---

the optical spectroscopy of randomly oriented solutions. In our opinion, the phenomena of light quenching can result in a new class of fluorescence experiments in which the sample is excited with one pulse, and the excited state population is modified by the quenching pulse(s) prior to measurement.

In closing, we wish to reiterate that time-resolved fluorescence is now moving out of the research laboratory and into the world of cellular imaging and numerous other sensing applications. Advances in laser sources, CCD detection, and other technologies is resulting in the possibility of simple instrumentation for previously complex measurements. These enabling technologies are summarized in Table 2. The increasing availability of intense picosecond and femtosecond lasers, and laser systems which provide multiple time-delayed pulses, will result in the increased use of two-photon excitation and stimulated emission to control and/or modify the excited state population.

Acknowledgements

The concepts and results described in this article were developed over a number of years. Dr. Joseph R. Lakowicz wishes to thank the National Institutes of Health (RR-08119, RR-10416, GM-39617, GM-35154) and the National Science Foundation (MCB-8804931, BIR-9319032, DIR-8710401) for their continued support.

References

[1] Bach PH, Reynolds CH, Clark JM, Mottley J, Poole PL (eds.) (1993) Biotechnology Applications of Microinjection, Microscopic Imaging, and Fluorescence. Plenum Press, New York, pp. 1-255.

[2] Denk W, Strickler JH, Webb WW (1990) Two-photon laser scanning fluorescence microscopy. Science **248**: 73-76.

[3] Einstein A (1917) On the quantum theory of radiation. Phys Z, **18**: 121.

[4] Galanin MD, Kirsanov BP, Chizhkova ZA (1969) Luminescence quenching of complex molecules in a strong laser field. Soviet Phys JETP Lett **9**: 502-507.

[5] Gryczynski I, Bogdanov V, Lakowicz JR (1993) Light quenching of tetraphenylbutadiene fluorescence observed during two-photon excitation. J Fluoresc **3**: 85-82.

[6] Gryczynski I, Bogdanov V, Lakowicz JR (1994) Light quenching and depolarization of fluorescence observed with laser pulses. A new experimental opportunity in time-resolved fluorescence spectroscopy. Biophys Chem **49**: 223-232.

[7] Gryczynski I, Malak H, Lakowicz JR (1995) Three-photon induced fluorescence of 2,5-diphenyloxazole with a femtosecond Ti:sapphire laser. Chem Phys Letts **245**: 30-35.

[8] Gryczynski I, Malak H, Lakowicz JR (1995). Three-photon excitation of a tryptophan derivative using a fs Ti:sapphire Laser. Biospectroscopy **2**: 9-15.

[9] Gryczynski I, Szmacinski H, Lakowicz JR (1995) On the possibility of calcium imaging using Indo-1 with three-photon excitation. Photochem Photobiol **62**: 804-808.

[10] Herman B, Jacobson K (eds.) (1989) Optical Microscopy for Biology. Proc Intl Conf Video Microscopy. Wiley-Liss, New York, pp. 1-658.

[11] Kober EM, Marshall JL, Dressick WJ, Sullivan

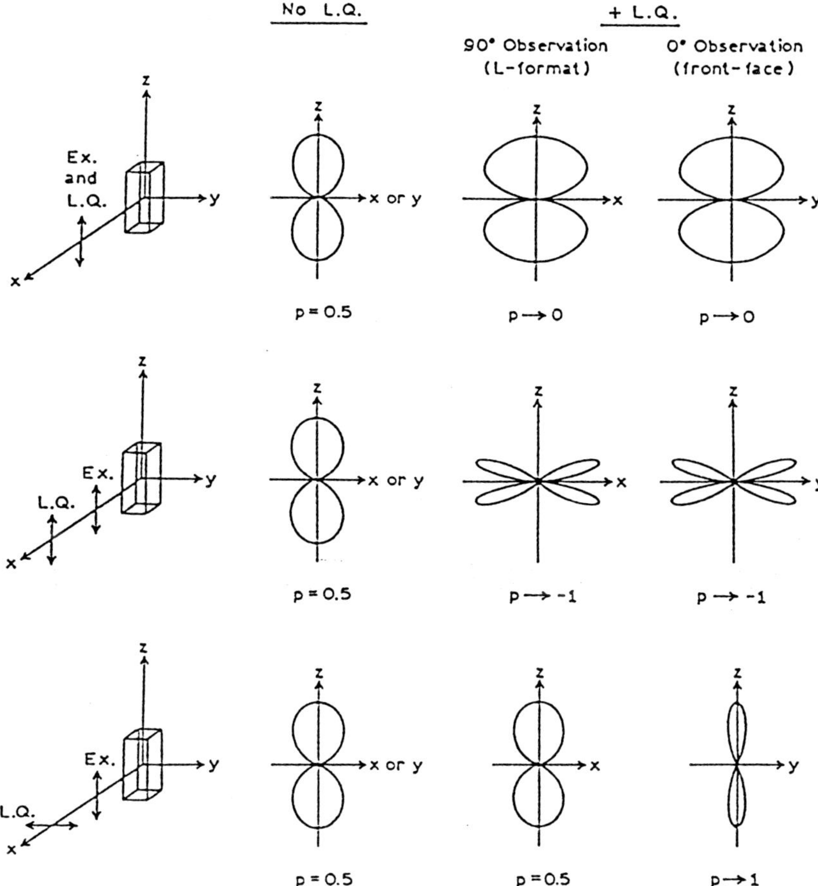

Figure 16. Effects of polarized light quenching on the orientation of the excited state population.

BP, Caspar JV, Meyer TJ (1984). Synthetic control of excited states. Nonchromophoric ligand variations in polypyridyl complexes of Osmium(II). Inorg Chem **24**: 2755-2763.

[12] Lakowicz JR, Gryczynski I (1992). Fluorescence intensity and anisotropy decay of the 4',6-diamidino-2-phenylindole-DNA complex resulting from one-photon and two-photon excitation. J Fluoresc **2**: 117-122.

[13] Lakowicz JR, Gryczynski I (1993) Tryptophan fluorescence intensity and anisotropy decays of human serum albumin resulting from one-photon and two-photon excitation. Biophys Chem **45**: 1-6.

[14] Lakowicz JR, Kuśba J, Danielsen E (1992). Two-photon induced fluorescence intensity and anisotropy decays of diphenylexatriene in solvents and lipid bilayers. J Fluoresc **2**: 247-258.

[15] Lakowicz JR, Szmacinski H, Nowaczyk K, Berndt K, Johnson ML (1992) Fluorescence lifetime imaging. Anal Biochem **202**: 316-330.

[16] Lakowicz JR, Szmacinski H, Nowaczyk K, Johnson ML (1992) Fluorescence lifetime imaging of calcium using Quin-2. Cell Calcium **13**: 131-147.

[17] Lakowicz JR, Gryczynski I, Kuśba J, Wiczk W, Szmacinski H, Johnson ML (1994) Site-to-site diffusion in proteins as observed by energy transfer and frequency-domain fluorometry. Photochem Photobiol **59**: 16-29.

[18] Lakowicz JR, Koen PA, Szmacinski H, Gryczynski I, Kuśba J (1994) Emerging biomedical and advanced applications of time-resolved fluorescence spectroscopy. J Fluoresc **4**: 117-136.

[19] Lakowicz JR, Szmacinski H, Nowaczyk K, Lederer WJ, Kirby MS, Johnson ML (1994) Fluorescence lifetime imaging of intracellular calcium in COS cells using Quin-2. Cell Calcium **15**: 7-27.

[20] Matsumoto B (ed.) (1993) Methods in Cell Biology. Vol. 38: Cell Biological Applications of Confocal Microscopy. (1993) Academic Press, New York, pp. 1-380.

[21] Mazurenko YT, Danilov VV, Vorontsova SI (1973). Depolarization of photoluminescence by powerful excitation. Opt Spectrosc **35**: 107-108.

[22] Pawley JB (ed.) (1990) Handbook of Biological Confocal Microscopy. Plenum Press, New York.

[23] Piston DW, Sandison DR, Webb WW (1992)

Time-resolved fluorescence imaging and background rejection by two-photon excitation in laser scanning microscopy. Proc SPIE **1640**: 379-389.

[24] Ricch RK, Mountain RW, McGonagle WH, Huning CM, Twichell JC, Kosicki BB, Savoye ED (1991) An integrated electronic shutter for back-illuminated charge-coupled devices. IEEE **91**: 171-174.

[25] Sacksteder LA, Lee M, Demas JN, DeGraff BA (1993) Long-lived, highly luminescent Rhenium(I) complexes as molecular probes: intra- and intermolecular excited-state interactions. J Am Chem Soc **115**: 8230-8238.

[26] Szmacinski H, Lakowicz JR (1994) Lifetime-based sensing. In: Topics in Fluorescence Spectroscopy, Volume 4: Probe Design and Chemical Sensing. Lakowicz JR (ed). Plenum Press, New York, pp. 295-334.

[27] Szmacinski H, Lakowicz JR (1994) Frequency-domain lifetime measurements in highly scattering media. Sensors and Actuators B **30**: 207-215.

[28] Szmacinski H, Gryczynski I, Lakowicz JR (1993) Calcium-dependent fluorescence lifetimes of Indo-1 for one- and two-photon excitation of fluorescence. Photochem Photobiol **58**: 341-345.

[29] Szmacinski H, Gryczynski I, Lakowicz JR (1996) Three-photon induced fluorescence of the calcium probe Indo-1. Biophys J **70**: 547-555.

[30] Terpetschnig E, Szmacinski H, Lakowicz JR (1995) Fluorescence polarization immunoassay of a high-molecular-weight antigen based on a long-lifetime Ru-ligand complex. Anal Biochem **227**: 140-147.

[31] Terpetschnig E, Szmacinski H, Malak H, Lakowicz JR (1995) Metal-ligand complexes as a new class of long-lived fluorophores for protein hydrodynamics. Biophys J **68**: 342-350.

[32] Wang Y-L, Taylor DL (eds.) (1989) Fluorescence Microscopy of Living Cells in Culture, Part A: Fluorescent Analogs, Labeling Cells, and Basic Microscopy. Academic Press, New York.

[33] Webb WW (1990) Two photon excitation in laser scanning fluorescence microscopy. Micro '90, vol. 13 (London). pp 445-450.

[34] Wilson T (ed.) (1990) Confocal Microscopy. Academic Press, London.

Scanning Microscopy Supplement 10, 1996 (pages 225-236)
Scanning Microscopy International, Chicago (AMF O'Hare), IL 60666 USA
0892-953X/96$5.00+.25

COMPARATIVE SCANNING, TRANSMISSION AND ATOMIC FORCE MICROSCOPY OF THE MICROTUBULAR CYTOSKELETON IN FENESTRATED LIVER ENDOTHELIAL CELLS

Filip Braet[1*], Ronald De Zanger[1], Wouter Kalle[2], Anton Raap[2], Hans Tanke[2] and Eddie Wisse[1]

[1]Laboratory for Cell Biology and Histology, Vrije Universiteit Brussel, Brussels-Jette, Belgium, [2]Department of Cell Biology - Laboratory for Cytochemistry and Cytometry, University of Leiden, The Netherlands

(Received for publication August 6, 1995, and in revised form December 27, 1995)

Abstract

Endothelial fenestrae control the exchange of fluids, solutes and particles between the sinusoidal lumen and the microvillous surface of the parenchymal cells. Fenestrae have a critical dimension in the order of 150-200 nm, making it necessary to use microscopes with a resolution better than the light microscope. Comparative whole-mount preparations of isolated, purified and cultured rat liver sinusoidal endothelial cells (LEC) were studied by scanning electron microscopy (SEM), transmission electron microscopy (TEM) and atomic force microscopy (AFM). Examination of detergent-extracted LEC by SEM and TEM shows an integral cytoskeleton: sieve plates are delineated by a sieve plate-associated cytoskeleton ring and fenestrae by a fenestrae-associated cytoskeleton ring. By using microtubule altering agents we could demonstrate: (1) the architectural role of microtubules in arranging fenestrae, (2) the existence of a population of microtubules resistant against low temperature and colchicine, (3) the ability of LEC to shift the microtubule assembly-disassembly steady state under various conditions, (4) and the necessity of an intact microtubular cytoskeleton to support the increase in the number of fenestrae after cytochalasin B. Topographical examinations of AFM images revealed that sieve plates are delineated by elevated borders, probably projections of the underlying tubular cytoskeleton.

Key Words: liver, endothelium, fenestrae, sieve plates, cytoskeleton, cytochalasin B, taxol, colchicine, microtubules, fenestrae-associated cytoskeleton, electron microscopy, atomic force.

*Address for correspondence:
Filip Braet,
Laboratory for Cell Biology and Histology,
Free University of Brussels (VUB), Laarbeeklaan 103,
1090 Brussels-Jette, Belgium.
Tel number:32-2-477.44.18
Fax number:32-2-477.40.00
e-mail:filipbra@cyto.vub.ac.be

Introduction

In 1970, the first description and electron microscopic observation of sinusoidal liver endothelial cell (LEC) fenestrae was made by Wisse [40]. The application of perfusion fixation of the rat liver revealed the presence of groups of fenestrae arranged in sieve plates. In subsequent reports, Widmann [38] and Ogawa [31] verified the existence of fenestrae in LEC by using transmission electron microscopy (TEM). The liver sinusoids can be regarded as unique capillaries which differ from other capillaries, because of the presence of fenestrae lacking a diaphragm and a basal lamina underneath the endothelium [40]. Knook *et al.* [11, 24, 25] pioneered the isolation and cultivation of LEC and they described the presence of fenestrae in freshly isolated and cultured LEC by TEM and scanning electron microscopy (SEM). In the following years, the number of reports about the ultrastructure of LEC fenestrae increased enormously [for an extended review, see reference 12].

On the basis of morphological and physiological evidence, several authors reported that the grouped fenestrae act as a dynamic filter [13, 15, 29]. Fenestrae filter fluids, solutes and particles that are exchanged between the sinusoidal lumen and the space of Disse, allowing only particles smaller than the fenestrae to reach the parenchymal cells or to leave the space of Disse [for an extended review, see reference 42]. Interestingly, the diameter and the number of fenestrae can be influenced by different agents, such as ethanol [10], serotonin [41], nicotine [16], endotoxin [14] and cytochalasin B [34]. Cytochalasin B is the most frequently described agent regulating LEC fenestrae [4, 28, 30, 34, 36]. Cytochalasin B which acts on the actin cytoskeleton increases the frequency of fenestrae enormously, as was first reported by Steffan *et al.* [34]. They found that the cytoskeleton is involved in the fenestrae dynamics [for a review, see 1]. In the last decennium, several investigators have postulated hypotheses about the dilatation and contraction of fenestrae. Briefly, the presence of an calcium-calmodulin-actomyosin system is

Abbreviations used: EGTA; ethylene glycol bis [2-aminoethylether]-N,N,N',N' tetra-acetic acid, LEC; rat liver sinusoidal endothelial cells, PIPES; piperazine-N,N'-bis [2-ethanesulfonic acid], SEM; scanning electron microscopy, TEM; transmission electron microscopy, AFM; atomic force microscopy.

---

postulated to be responsible for the dynamics of LEC fenestrae [17, 19]. However, it remains to be elucidated whether the LEC cytoskeleton is involved in the regulation of fenestrae diameter and how the cytoskeleton is involved in the fenestrae formation. By using whole mount TEM we described that fenestrae are surrounded by a fenestrae-associated cytoskeleton ring which alters in diameter after different treatments also known to change fenestrae diameter [6]. We also postulated that microtubules stabilize sieve plates and reported that cytochalasin B treatment changes the organization of microtubules.

It was the purpose of this study to investigate the cytoskeleton of LEC, especially with regard to microtubules and their relation to fenestrae, by using detergent-extracted whole mounts [6] of cultured LEC [5]. Atomic force microscopy (AFM) was used to obtain topographical data of the fenestrated areas and surrounding cytoplasm.

Materials and Methods

Isolation, purification and culture of LEC

Male Wistar rats (weighing approximately 250 g) were fed a standard diet *ad libitum*. The method for the isolation of LEC has been described earlier [5], and was based on the method by Smedsrød *et al.* [33]. Briefly, the liver was perfused with collagenase A (Boehringer Mannheim Biochemica, Belgium). After incubation of the fragmented tissue in the same solution, the resulting cell suspension was centrifuged at 100 x g for 5 minutes to remove the parenchymal cells. The supernatant, containing a mixture of sinusoidal liver cells, was then layered on top of a two step Percoll® gradient (25-50%) and centrifuged for 20 minutes at 900xg. The intermediate zone, located between the two density layers was enriched in LEC. LEC purity was further enhanced by selective removal through adherence of the Kupffer cells and spreading on collagen of the LEC. LEC were further cultivated in 24-multiwell plates on collagen-coated nickel grids (300 mesh) for TEM. Formvar (1%) coated nickel grids (300 mesh) were used, later coated with diluted collagen. 10 μl of Collagen-S stock solution (Boehringer Mannheim, Belgium), was diluted with 900 μl sterile water. For SEM, LEC were cultivated on collagen-coated thermanox coverslips instead of nickel grids. Serum free LEC culture medium consisted of RPMI-1640 with 2 mM L-glutamine, 100 U/ml penicillin, 100 mg/ml streptomycin and 10 ng/ml endothelial cell growth factor. The cultures were estimated to have > 95% purity, since less then 5% of the cells examined by SEM and TEM were devoid of fenestrae.

Scanning electron microscopy

LEC were rinsed twice with phosphate-buffered saline (PBS). To visualize the sieve plate-associated cytoskeleton [6], LEC were extracted for 1 min at 21°C in cytoskeleton buffer, consisting of 1 mM EGTA, 100 mM PIPES, 4% polyethylene glycol 6000 and 0.1% Triton X-100 in PBS at pH 6.9. After extraction, cells were fixed with 2% glutaraldehyde in 0.1M Na-cacodylate buffer with 0.1 M sucrose for 1 hour. Postfixation was done with 1% osmium tetroxide in 0.1 M Na-cacodylate for 1 hour. SEM samples were dehydrated in graded ethanol solutions, critical point dried and sputter coated with 10 nm gold. The samples were examined with a Philips SEM 505 (Philips Eindhoven, The Netherlands) at an accelerating voltage of 30 kV.

To study the effect of colchicine or taxol, we treated the cells with 200 μM colchicine (Sigma Chemicals, C3915, Germany) [39] for 2 hours and counted the number of microtubules in relation to the number of fenestrae and sieve plates or with 10 μM taxol (Sigma Chemicals, T7402, Germany) [22] for 2 hours. In order to determine whether the microtubular cytoskeleton is involved in the increase of fenestrae caused by cytochalasin B [6], we treated LEC with colchicine or taxol as described above, followed by cytochalasin B treatment at a concentration of 20 μM for 2 hours (Sigma Chemicals, C6762, Germany) [34]. Taxol and cytochalasin B were dissolved with dimethyl sulfoxide (DMSO) and then diluted with serum free LEC culture medium. The DMSO concentration in the assay was ≤0.4% and these levels had no effect on the ultrastructure or viability of LEC as determined by EM, trypan blue and propidium iodide test. Control media also contained DMSO in the same concentration as the treated LEC and were incubated in serum free culture medium without the cytoskeletal agents.

After incubation, LEC were prepared for routine SEM investigation [5] or were used to visualize their cytoskeleton in SEM as described above.

Transmission electron microscopy

LEC were rinsed twice with PBS. In order to preserve the fine structure of the cytoskeleton, samples were slightly fixed with freshly prepared 4% formaldehyde in PBS for 1 min at 21°C. Extraction with cytoskeleton buffer for 1 min at 21°C was performed as

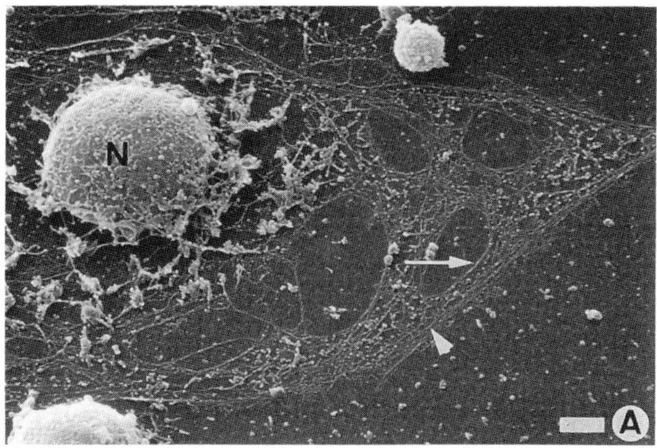

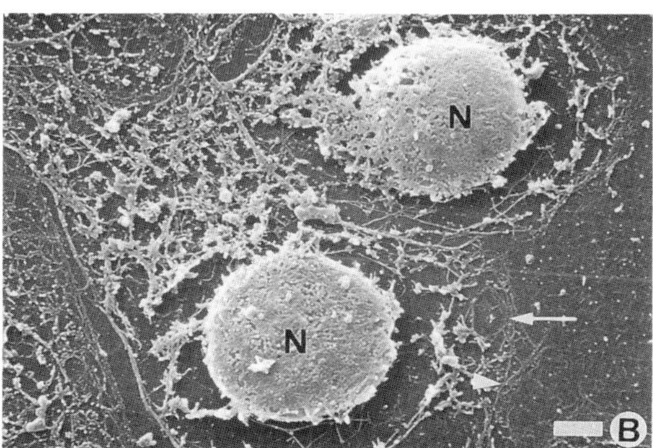

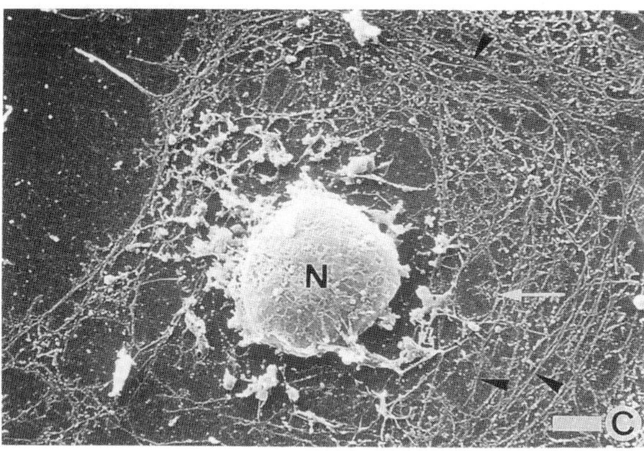

Figure 1: Scanning electron micrographs of cytoskeleton buffer extracted LEC. These cells were not prefixed before extraction. Notice the centrally located, bulging nucleus (N) and the sieve plate-associated cytoskeleton (->) composed of a circle of tubular structures. **(A)** Control cells reveal a normal distribution of microtubules (>), that can be recognized by their straight outline. They are scattered throughout the cytoplasm in variable directions. Notice the close relation between microtubules and sieve plates. **(B)** Treatment of LEC with colchicine results in the disappearance of many microtubules. However, microtubules (>) were still scattered around sieve plates. Taxol treatment of LEC **(C)** results in abundant microtubules (>), bars 1μm.

described above. After extraction, cytoskeletons were fixed, dehydrated and critical point dried as for SEM. The specimens were examined in a Philips EM 400 (Philips Eindhoven, The Netherlands) at an accelerating voltage of 120 kV.

To investigate the effect of different agents and low temperature on the number of microtubules, LEC were treated with colchicine or taxol, as described in the previous paragraph, followed by a 2 hours incubation at 37°C or 4°C. After incubation, LEC were prepared to visualize their cytoskeleton in TEM as described above.

Atomic force microscopy

The AFM we used, is the Explorer™ (Topometrix TMX 2000, Darmstadt, Germany). The cantilever of the AFM was positioned in the optical axis of a Leica inverted microscope. Standard silicon nitride tips (Topometrix SFM-Probes, Ref 1520-00, USA) with a spring constant of 0.032 N/m and a 4 μm on 4 μm pyramidal base were used. LEC cultured on cover slips were scanned in contact mode, either in air (first set of preparations) or in 0.1 M Na-cacodylate buffer supplemented with 0.1 M sucrose (second set of preparations) [9]. The first set of preparations for AFM were identical to the samples for SEM, except that they were not extracted with cytoskeleton buffer. A second set of cells, was fixed with glutaraldehyde as described above and examined with the AFM. Images, taken in sensor current mode or topographic mode, were analyzed by the Topometrix Image Analysis Software. Images of preparations, scanned in sensor current mode were directly recorded from the computer screen. The x-y-z calibration was regularly checked with a calibration grid.

Morphometric analysis

Analysis of the cytoskeleton was performed on randomly taken electron micrographs with magnifications of 5000 x and 20.000 x for SEM, calibrated with a 28.800 lines/inch grating stub. For TEM, micrographs were taken at the calibrated magnifications of 12.000 x and 44.000 x using a 54.800 lines/inch grating replica. An Ibas I computer (Kontron, Munich, Germany), with a digitizer tablet was used to count and measure the number of sieve plates, fenestrae-associated cytoskeleton rings and microtubules. For each experiment, about 15 micrographs were taken in randomly selected fields for each experimental variable and magnification. All experiments were repeated three times. Statistical analysis was performed with the Mann-Whitney U test.

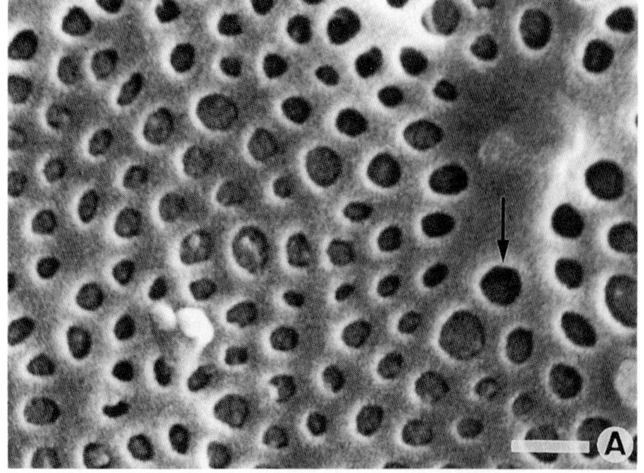

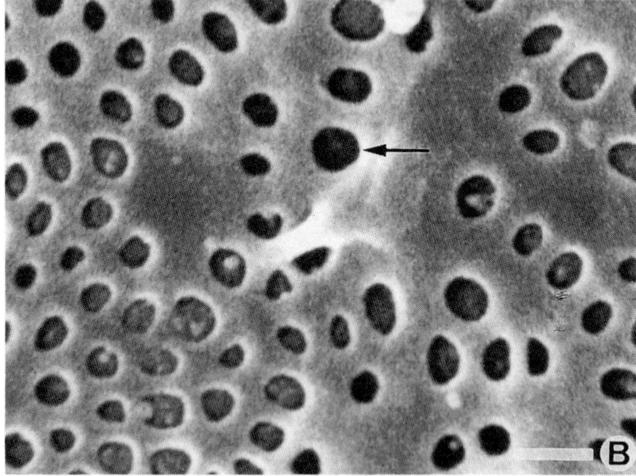

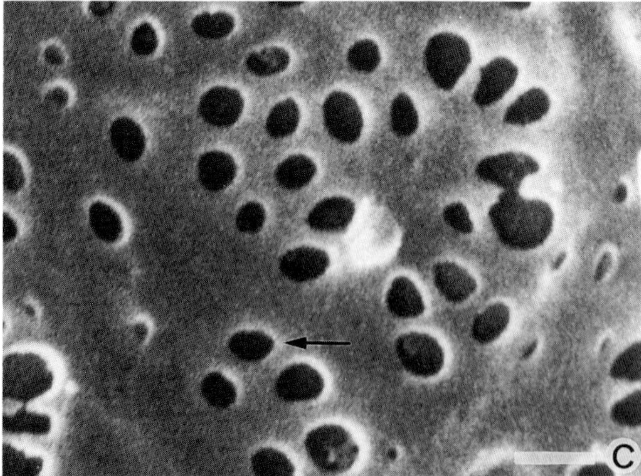

Figure 2: High-power scanning electron micrographs of intact fenestrae of non-extracted LEC. Note that the number of fenestrae (->) is higher in the cytochalasin B treated LEC **(A)** when compared with figures **(B)** (colchicine/cytochalasin B) and **(C)** (taxol/cytochalasin B). However, the number of fenestrae in the colchicine/cytochalasin B treated cells **(B)** is higher than in the taxol/ cytochalasin B treated cells **(C)**, bars 200 nm.

Results

Scanning electron microscopy

The extraction of LEC without formaldehyde treatment, showed a well preserved cytoskeleton (Fig. 1A-1C). In control cells (Fig. 1A), microtubules could be recognized by their diameter and straight outline. Typical sieve plates were encircled by tubular structures; apparently sieve plate-associated cytoskeleton [6]. When LEC were treated with colchicine, the number of microtubules per area decreased significantly. However, microtubules were still present, scattered alongside the sieve plates (Fig. 1B). In the non-fenestrated areas of LEC most of the microtubules were broken or disappeared. When cells were treated with taxol (Fig. 1C) an abundance of microtubules were revealed, scattered throughout the cytoplasm and alongside the sieve plates.

As shown in Fig. 2A and Table 1, cytochalasin B treatment resulted in an increase in the number of fenestrae. Treatment of LEC with colchicine followed by cytochalasin B administration resulted also in a moderate increase of the number of fenestrae (Fig. 2B and Table 1). Taxol treatment followed by cytochalasin B administration resulted in an inhibition of the cytochalasin B effect (Fig. 2C and Table 1). The number of sieve plates was unchanged, except for the colchicine / cytochalasin B-treated LEC, were a slight difference was measured (Table 1). A redistribution of the microtubules was observed in the cytochalasin B treated cells. Most of the microtubules were lying parallel to the cell periphery [6]. In the case of the colchicine / cytochalasin B-treated LEC, we could not observe this redistribution. Most of the microtubules disappeared or were lying alongside fenestrae or sieve plates (data not shown). In the taxol / cytochalasin B-treated cells, a high number of microtubules was present (Table 1). These microtubules had the same distribution as described and illustrated in Fig. 4C-4D.

Transmission electron microscopy

Examination of extracted control LEC (Fig. 3A) showed results comparable to SEM (Fig. 1A). Sieve plates were delineated by dense cytoplasmic arms (Fig. 3A and 3B). Microtubules could be recognized at higher magnification within this border (Fig. 3B). Three different spatial organizations of fenestrae were found. Some were arranged in sieve plates, others were oriented linearly and a third group was formed by fenestrae lying single in the cytoplasm (Fig. 3A). It seems that microtubules determine these different patterns, because they closely delineate the fenestrae (Fig. 3A-3B). Incorporation of a fixative in the extraction method

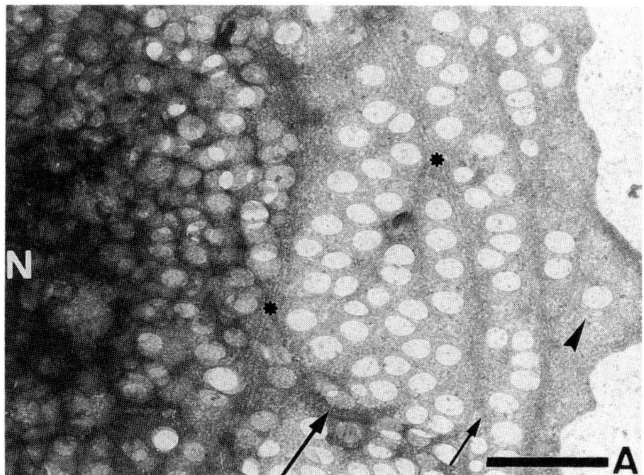

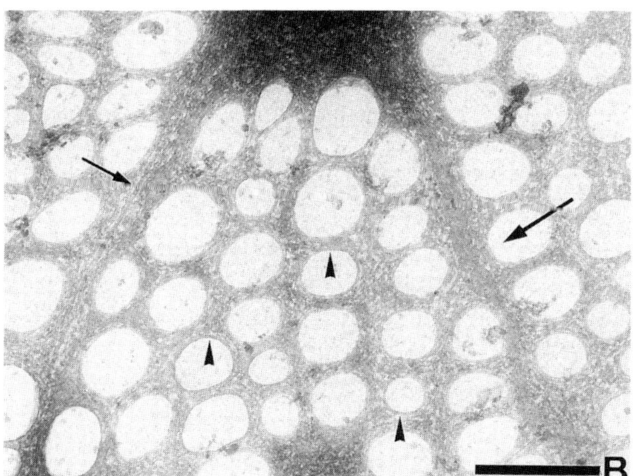

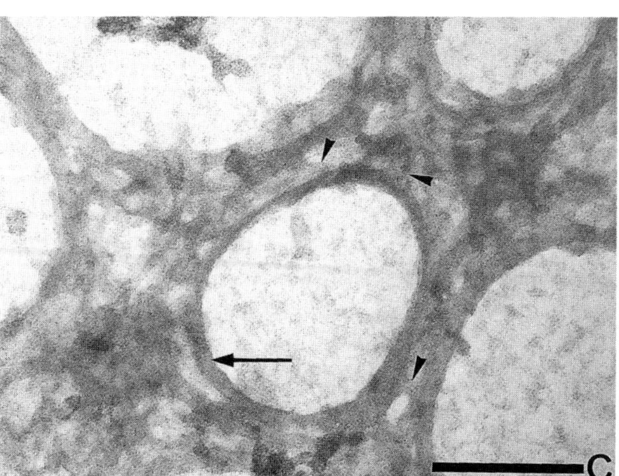

Figure 3: Transmission electron micrographs of whole mount formaldehyde prefixed, cytoskeleton- buffer extracted control LEC. (A) Low magnification showing the cell nucleus (N) and extracted cytoplasm. Three different organizations of fenestrae were found, some are arranged in sieve plates (large arrow), others linearly (small arrow) and some are lying single in the cytoplasm (>). Note also the microtubules which scatter throughout the cytoplasm (*), bar 1000 nm. (B) Higher magnification showing a part of a sieve plate indicated by large arrows. Sieve plates are defined by a darker border. Note the long tubular structures that are running close to the sieve plate (small arrow). Inside the sieve plate, numerous fenestrae-associated cytoskeleton rings (>) can be observed, bar 500 nm. (C) Higher magnification of a fenestra, showing that the fenestrae-associated cytoskeleton ring (->) is connected to the cytoskeleton elements (>) surrounding the fenestrae, bar 100 nm.

---

enabled the visualization of the fenestrae-associated cytoskeleton ring (Fig. 3C). This ring is approximately 30 nm thick and is clearly connected to the surrounding cytoskeleton by branching structures.

LEC treated with colchicine and incubated at 4°C, showed a decrease in the number of microtubules (Table 2). Only straight microtubules were lying in the neighborhood of sieve plates and linearly arranged fenestrae (Fig. 4A). In the non-fenestrated cytoplasm most of the microtubules disappeared, but when present, they now showed a zigzag pattern (Fig. 4B). When cells were treated with taxol and 4°C, the number of microtubules quadrupled as compared to the control (Table 2), whereas microtubules scattered throughout the cytoplasm in variable directions (Fig. 4D). Accumulation of microtubules could be observed along sieve plates (Fig. 4C). During experiments on the effect of temperature, we found that cold (4°C) incubation of LEC resulted in a lower number of microtubules per area as compared with 37°C (Table 2). However, taxol treated cells revealed an equal number of microtubules at all temperatures (Table 2). We also observed a difference in the number of fenestrae when incubated at different temperatures (Table 2). Measurements revealed that the number of fenestrae decreased significantly when incubated at 4°C in control, colchicine and taxol-treated cells. In contrast, the number of sieve plates was constant, except for the taxol treated cells.

Atomic force microscopy

LEC prepared for SEM without extraction were examined by AFM (Fig. 5A). The cell surface displayed fenestrae arranged in sieve plates at high magnification. These sieve plates are well delineated by an elevated border. This border probably corresponded to the underlying tubular structures as observed in SEM (see Fig. 1) and TEM (see Fig. 3A, 4A and 4C). Height measurements between the lowest point of the sieve plates and the elevated border, reveal that the sieve plates lie approximately 200 nm lower than the surface of the nearby cytoplasm. In addition, measurements on the elevated borders reveal a width of 176 $\pm$ 33 nm (n

Table 1. The effect of cytochalasin B on the number of microtubules, fenestrae and sieve plates, after colchicine and taxol treatment.

Parameter	n of *Microtubules* per 10 μm²	n of *Fenestrae* per 10 μm²	n of *Sieve plates* per 100 μm²
Control	2.3 ± 1.3	61.4 ± 12.1	8.1 ± 1.8
Colchicine	2.6 ± 1.1	44.4 ± 7.3***	8.7 ± 2.6
Taxol	16.9 ± 7.0***	26.4 ± 12.9***	7.6 ± 1.8

Note: Data on the number of microtubules, fenestrae and sieve plates obtained by SEM. LEC were first treated with colchicine or taxol for 2 hours, followed by cytochalasin B treatment for 2 hours. The control LEC were only treated with cytochalasin B. The number of fenestrae and sieve plates were counted on non-extracted LEC, while the number of microtubules were counted by using detergent-extracted LEC. Results are expressed as mean ± S.D.. Data are obtained from 3 experiments. Notice the significant difference in number of microtubules and fenestrae after the different treatment (***p ≤ 0.001 Mann-Whitney U test, two-sided).

Table 2. Effect of temperature on the number of microtubules, fenestrae and sieve plates, after colchicine and taxol treatment.

Parameter	n of *Microtubules* per 10 μm²		n of *Fenestrae* per 10 μm²		n of *Sieve plates* per 100 μm²	
at 37°C:						
Control	8.3 ± 1.9		34.5 ± 10.0		6.9 ± 1.9	
Colchicine	2.4 ± 1.2***		35.3 ± 8.8		7.1 ± 2.1	
Taxol	15.6 ± 2.6***		38.3 ± 15.1		9.1 ± 3.6	
at 4° C:		***		**		
Control	2.9 ± 0.9	**	22.3 ± 6.4	** *	7.5 ± 2.7	**
Colchicine	1.1 ± 0.8**		24.3 ± 6.4		6.9 ± 1.4	
Taxol	13.7 ± 4.2***		28.8 ± 6.4**		6.7 ± 1.9	

Note: Morphometric data on the number of microtubules in the fenestrated areas, studied by whole mount TEM. LEC were treated with colchicine or taxol for 2 hours, followed by a two hours incubation at 37°C (upper row) or 4°C (lower row). Results are expressed as mean ± S.D.. Data are obtained from 3 experiments. Notice the significant difference in number of microtubules, fenestrae and sieve plates after colchicine and taxol treatment (***p ≤ 0.001 or **p ≤ 0.01, *p ≤ 0.05 Mann-Whitney U test, two-sided).

= 100). Examination of wet-fixed LEC results in comparable observations with the ones obtained after critical point drying and gold sputtering. Sieve plates are well delineated by an elevated border, suggesting the underlying tubular cytoskeleton (Fig. 5B).

Discussion

The cytoskeleton, microtubules in particular, are thought to play an important role in cell architecture, as expressed in shape, organization of organelles and vesicle traffic [23]. As for LEC, it has been demonstrated that microtubules take part in intracellular trafficking of vacuolar transport [21] and general cyto-architecture [36]. In our previous reports [6, 8] we described that microtubules have a close topographical relation with sieve plates and the fenestrae-associated cytoskeleton. This observation made us anxious to study the alterations of LEC microtubules under different experimental conditions, using microtubule altering agents.

Colchicine prevents further polymerization of microtubules in cells [39], whereas taxol stabilizes or enhances the polymerization of microtubules [22]. Our results confirm the disruption of microtubules by colchicine (Fig. 1B) and the polymerizing ability of taxol (Fig. 1C). However, after treatment with colchicine, a low number of microtubules was still found scattered alongside sieve plates (Fig. 1B). The presence of microtubules after colchicine (Fig. 1B and Table 2) or cold treatment (Fig. 4A-4B and Table 2) demonstrates that a fraction of the microtubules is stable and insensitive against depolymerizing conditions, which has also been demonstrated

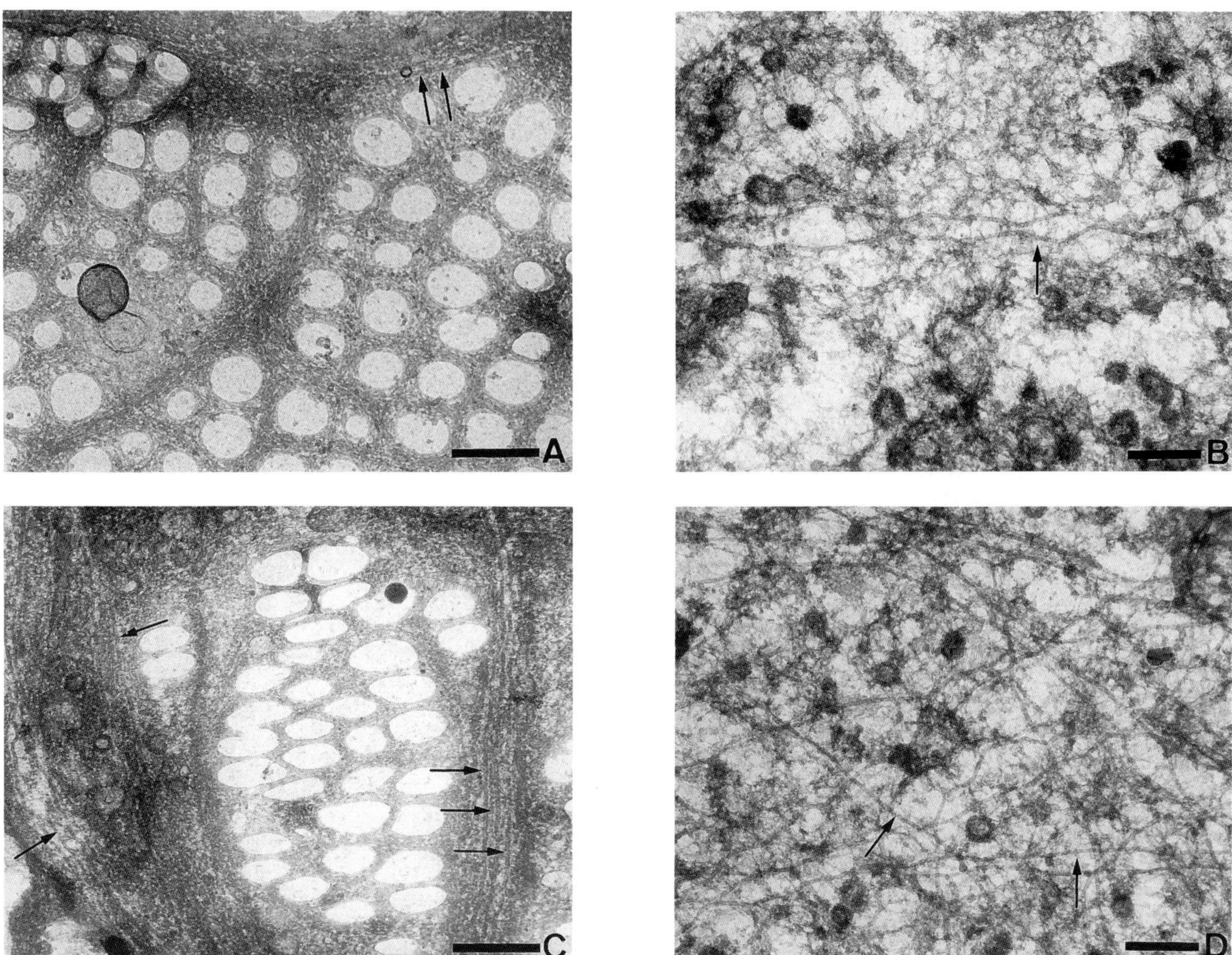

Figure 4: Transmission electron micrographs of whole mount formaldehyde prefixed, cytoskeleton buffer extracted LEC. Figures **A** (bar 500 nm) and **B** (bar 250 nm) represents LEC treated for 2 hours with colchicine followed by a 2 hours incubation at 4°C. Figures **C** (bar 500 nm) and **D** (bar 250 nm), are LEC treated with taxol for 2 hours followed by a 2 hours incubation at 4°C. Note that the occurrence of microtubules is low in the colchicine/4°C treated cells as compared to the taxol/4°C treated cells. **(A)** only straight (->) microtubules in the neighborhood of sieve plates and fenestrae were preserved, in the non-fenestrated areas **(B)** a low number of microtubules (->) could be observed which lost their straight outline. **(C)** Taxol/4°C treatment revealed an accumulation of parallel microtubules (->) around sieve plates. **(D)** However, in the non-fenestrated areas straight microtubules scatter throughout the cytoplasm in variable directions.

---

in other cell types [27, 32]. These data illustrate that stable microtubules lie around sieve plates or linearly arranged fenestrae, which demonstrates the intrinsic architectural role of these microtubules in maintaining LEC structure [6, 7, 8]. Taxol with or without cold treatment resulted in an unusual abundance of parallel arranged microtubules along sieve plates (Fig. 1C and 4C). Parallel arrangements of newly assembled microtubules were only found in the fenestrated areas (Fig. 4C). The occurrence of ordered microtubular arrays under influence of taxol was also described in other cells [22]. LEC treated with taxol show a doubling of the number of microtubules, whereas a four fold loss was found after colchicine treatment (Table 2). These measurements demonstrate the ability of LEC to shift the microtubule assembly-disassembly steady state under various conditions. The disassembly of the LEC-cytoskeleton under external stimuli, for instance after contact with invading cancer cells, probably leads to an increased permeability of these cells. The loss of cytoskeletal elements in endothelial cells [20] has been

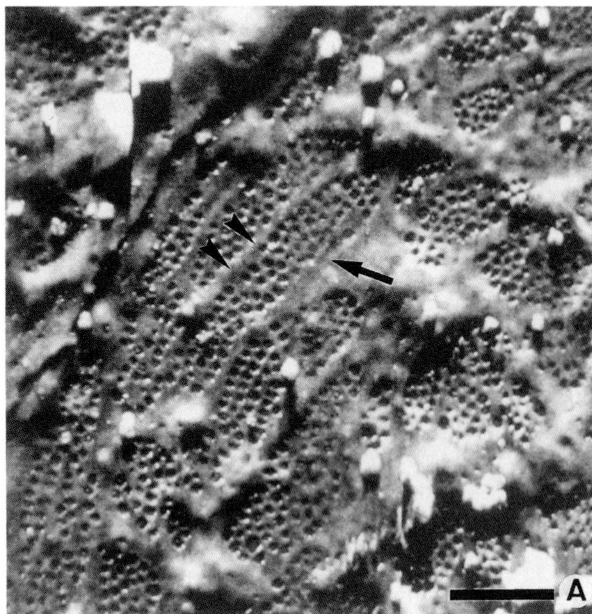

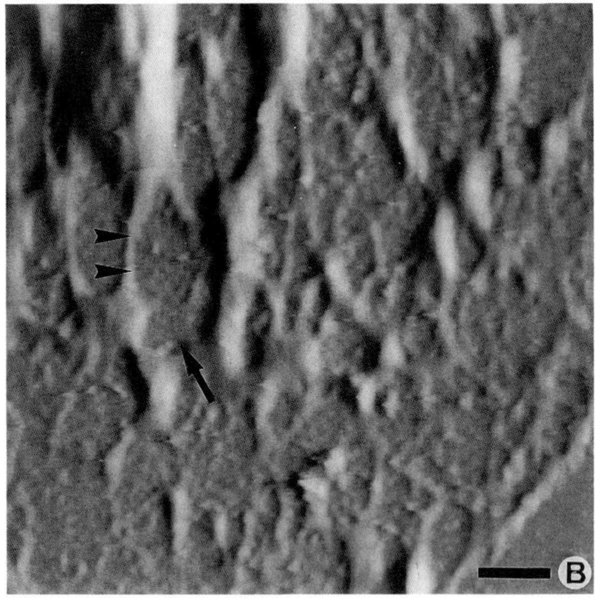

Figure 5: **(A)** Atomic force image of a critical point dried and gold sputtered LEC, bar 2 μm. This high resolution image of a part of the LEC cytoplasm clearly shows fenestrae in sieve plates (->). Note that these sieve plates are well delineated by elevated borders (>). This border probably corresponds to the underlying tubular structures as observed in SEM (see Fig. 1) and TEM (see Fig. 2A, 3B and 3C). **(B)** Atomic force image of a wet-fixed LEC, bar 2 μm. This image shows the same details as in figure A, indicating that the SEM preparation did not cause preparations artifacts. Note also that the sieve plates (->) are well delineated by an elevated border (>), suggesting the underlying tubular cytoskeleton.

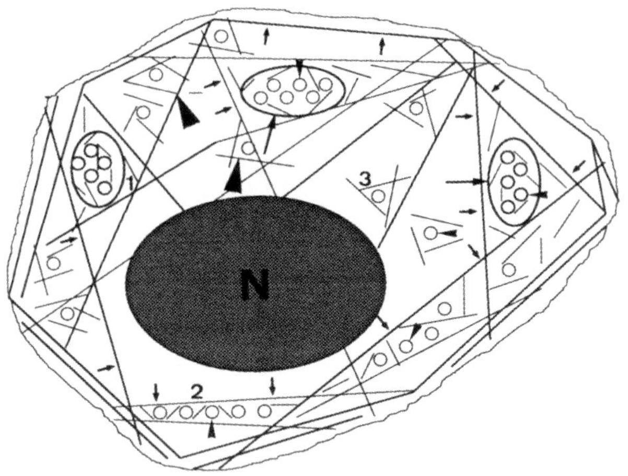

Figure 6: Proposed model of LEC fenestrae and its associated cytoskeleton, based on whole mount cytoskeletal preparations. The drawing shows the following cytoskeletal components: microtubules (small arrows), sieve plate-associated cytoskeleton (large arrows), fenestrae-associated cytoskeleton (small arrowheads), and a centrally located nucleus (N). Three different organizations of fenestrae were found, some were arranged in sieve plates (1), others linearly (2) and some were lying single in the cytoplasm (3). Additionally, in the neighborhood of sieve plates and fenestrae, microtubules always were running nearby the sieve plate- or fenestrae-associated cytoskeleton. From the microtubules small interconnecting structures (large arrowheads) seem to cross-link the surrounding cytoskeleton.

---

considered as a possible cause of the increased mobility of cancer cells [35].

Cytochalasin B is well known to increase the number of fenestrae [34]. Our results confirm this increase of fenestrae (Table 1 and Fig. 2A). When LEC were treated with colchicine, followed by cytochalasin B, we could also observe a moderate increase of the number of fenestrae. However, taxol treatment followed by cytochalasin B, resulted in an inhibition of the cytochalasin B effect (Fig. 2C and Table 1). We can conclude that the assembly of microtubules, resistant against depolymerization, blocks the effect of cytochalasin B. In addition, removal of microtubules leads to an inhibition of the de novo formation of fenestrae. It seems, therefore, that an intact and functional microtubular cytoskeleton is necessary for the increment of fenestrae by cytochalasin B.

When LEC were incubated at 4°C, the number of fenestrae decreased in normal, colchicine or taxol treated cells (Table 2). We do not have an explanation for this phenomenon. However, Fratté *et al.* [18] reported that the first preservation damage of livers stored for 2 hours at 4°C occurred in LEC. The reduction in fenestrae is probably one of the first signs of cold storage. It seems that this reduction in fenestrae is independent of microtubules. This was illustrated when LEC were treated with taxol and incubated at 4°C or 37°C. The number of microtubules was equal in both groups, but the number of fenestrae decreased when incubated at 4°C (Table 2).

It was reported that measurements on isolated microtubules by AFM showed a width of 80-110 nm [37]. Measurements on the elevated borders around sieve plates (Fig. 5A) reveal a width of approximately 180 nm. We postulate that these elevated borders are projections of the underlying tubular cytoskeleton. Further studies on LEC-cytoskeleton by AFM are necessary to prove this statement. However, at the moment no literature data are available about AFM and detergent-extracted cells. Probably, AFM studies on the naked cytoskeleton will suffer from technical problems; such as shadowing and lateral deformation of structural details [9]. Fenestrae visualization has hitherto been restricted to electron microscopy [40, 42], due to their limited size. We consider, therefore the possibility of visualizing fenestrae and the projections of the underlying cytoskeleton by AFM as an important achievement, particularly because preparation steps beyond fixation are not needed [9].

In this study, we used detergent-extracted whole mounts of LEC visualized by SEM (Fig. 1) and TEM (Fig. 3-4) based on a modification of the method of Bell *et al.* [2, 3, 26]. In addition, non-extracted LEC were visualized by SEM (Fig. 2) and AFM (Fig. 5). Based on the data obtained in this report and in our previous studies [6, 7, 8] we can conclude that whole mount TEM is a powerful method for studying the organization of the LEC-cytoskeleton. We demonstrated the possibility to visualize the fenestrae-associated cytoskeleton by SEM using prefixation before the extraction [6, 7, 8]. However, the best results obtained on fenestrae-associated cytoskeleton was obtained by TEM (Fig. 3) [6, 7, 8]. With this method it was possible to observe the closely spaced branching filaments which interconnect the fenestrae-associated cytoskeleton and microtubules (Fig. 3C) [6]. Whole mount SEM can be used to visualize the LEC-cytoskeleton. However, a lower resolution is obtained because of the gold-coating of the cytoskeleton and the lower resolving power of the SEM. For example, it was impossible to resolve closely spaced filaments which interconnect the surrounding cytoskeleton. In addition, it was impossible to count the number of microtubules in relation to the number of fenestrae-associated cytoskeleton rings in SEM-preparations when prefixation before the detergent-extraction was used. Prefixation enhances the retention of cytoskeletal proteins which mask other cytoskeletal structures, i.e. microtubules, which lie deep in the cell [3]. When SEM was used, we counted microtubules on detergent-extracted LEC without prefixation, and fenestrae on non-extracted LEC (Table 1). It is clear that whole mount TEM is a more convenient method to depict the entire cytoskeleton without restriction in resolving power (resolution) or penetration power (masking of cytoskeleton proteins).

Bell *et al.* [2] postulated the possibility of introducing artifacts by using aldehydes. Without prefixation prior the extraction, we could clearly observe precipitates as remnants of fenestrae in SEM [6]. In addition, fixation with 4% formaldehyde for 1 minute at 21°C resulted in well preserved cytoskeleton rings around fenestrae [6, 7, 8]. These results demonstrated that structures which are present, but not well preserved could be visualized when prefixation prior the crude detergent-extraction was used. We can conclude, that prefixation markedly improves the preservation of the cytoskeleton. However, we must be aware about the different results which can be obtained when different preparative methods are been used [26].

In general, a model for the microtubular cytoskeleton of LEC can be proposed as presented in figure 6: (1) Sieve plates and linearly arranged fenestrae are stabilized by microtubules, (2) LEC possess a stable population of microtubules resistant against cold and colchicine, (3) LEC have the ability to shift the microtubule steady state of assembly under various conditions, (4) And it seems that an intact microtubular cytoskeleton is necessary to promote the increase of fenestrae under influence of cytochalasin B.

Acknowledgements

We are grateful to Chris Derom for her photographic contribution. Marijke Baekeland and Carine Seynaeve are acknowledged for their technical assistance. This investigation was supported by the Belgian National Fund for Scientific Research, grant N°3.0053.92 and N°3.0050.95, and VUB/OZR grant N° 195.332.1310. Philip Braet was an aspirant of the National Fund for Scientific Research-Belgium.

References

1. Arias IM (1990) The biology of hepatic endothelial cell fenestrae. In: *Progress in Liver Disease IX* (Schaffer F, Popper H, eds) WB Saunders, Philadelphia, pp 11-26.

2. Bell PB (1981) The application of scanning electron microscopy to the study of the cytoskeleton of cells in culture. Scanning Electr Microsc **2**: 139-157.

3. Bell PB, Lindroth M, Fredricksson BA, Liu XD (1989) Problems associated with preparations of whole mounts of cytoskeletons for high resolution electron microscopy. Scanning Electr Microsc **3**: 117-135.

4. Bingen A, Gendrault JL, Kirn A (1989) Cryofracture study of fenestrae formation in mouse liver endothelial cells treated with cytochalasin B. In: *Cells of the Hepatic Sinusoid 2* (Wisse E, Knook DL, Decker K, eds) Kupffer Cell Foundation, Leiden, pp 466-470.

5. Braet F, De Zanger R, Sasaoki T, Baekeland M, Janssens P, Smedsrød B, Wisse E (1994) Assessment of a method of isolation, purification and cultivation of rat liver sinusoidal endothelial cells. Lab Invest **70**: 944-952.

6. Braet F, De Zanger R, Baekeland M, Crabbé E, Van Der Smissen P, Wisse E (1995) Structure and dynamics of the fenestrae-associated cytoskeleton of rat liver sinuoisdal endothelial cells. Hepatology **21**: 180-189.

7. Braet F, De Zanger R, Van Der Smissen P, Smedsrød B, Wisse E (1995) Fenestrae-associated cytoskeleton in isolated rat liver sinusoidal endothelial cells. In: *Cells of the Hepatic Sinusoid 5* (Wisse E, Knook DL, Wake K, eds) Kupffer Cell Foundation, Leiden, pp 256-259.

8. Braet F, De Zanger R, Crabbé E, Wisse E (1995) New observations on cytoskeleton and fenestrae in isolated rat liver sinusoidal endothelial cells. J Gastroen Hepatol **10**: S3-S7.

9. Braet F, Kalle WHJ, De Zanger RB, de Grooth BG, Raap AK, Tanke HJ, Wisse E (1996) Comparative atomic force and scanning electron microscopy; an investigation on fenestrated endothelial cells in vitro. J Microsc **181**: 10-17.

10. Charels K, De Zanger R, Van Bossuyt H, Van Der Smissen P, Wisse E (1986) Influence of acute alcohol administration on endothelial fenestrae of rat livers: An in vivo and in vitro scanning electron microscopic study. In: *Cells of the Hepatic Sinusoid 1* (Kirn A, Knook DL, Wisse E, eds) Kupffer Cell Foundation, Rijswijk, pp 497-502.

11. De Leeuw AM, Barelds RJ, De Zanger R, Knook DL (1982) Primary cultures of endothelial cells of the rat liver. Cell Tissue Res **223**: 201-215.

12. De Leeuw AM, Brouwer A, Knook DL (1990) Sinusoidal endothelial cells of the liver: fine structure and function in relation to age. J Electr Micr Tech **14**: 218-236.

13. De Zanger R, Wisse E (1982) The filtration effect of rat liver fenestrated sinusoidal endothelium on the passage of (remnant) chylomicrons to the space of Disse. In: *Sinusoidal Liver Cells* (Knook DL, Wisse E, eds) Elsevier, Amsterdam, pp 69-76.

14. Dobbs BR, Rogers GWT, Xing H-Y, Fraser R (1994) Endotoxin-induced defenestration of the hepatic sinusoidal endothelium: a factor in the pathogenesis of cirrhosis? Liver **14**: 230-233.

15. Fraser R, Bosanquet AG, Day WA (1978) Filtration of chylomicrons by the liver may influence cholesterol metabolism and atherosclerosis. Atherosclerosis **29**: 113-123.

16. Fraser R, Clark SA, Day WA, Murray FEM (1988) Nicotine decreases the porosity of the rat liver sieve: a possible mechanism for hypercholesterolemia. Br J Exp Pathol **69**: 345-350.

17. Fraser R, Dobbs BR, Rogers GWT (1995) Lipoproteins and the liver sieve: the role of the fenestrated sinusoidal endothelium in lipoprotein metabolism, atherosclerosis, and cirrhosis. Hepatology **21**: 863-874.

18. Fratté S, Gendrault JL, Steffan AM, Kirn A (1991) Comparative ultrastructural study of rat livers preserved in Euro-Collins or University of Wisconsin solution. Hepatology **13**: 1173-1180.

19. Gatmaitan Z, Arias IM. Hepatic endothelial cell fenestrae (1993) In: *Cells of the Hepatic Sinusoid 4* (Knook DL, Wisse E, eds) Kupffer Cell Foundation, Leiden, pp 3-7.

20. Gottlieb AI, Langeville BL, Wong MKK, Kim DW (1991) Biology of disease: structure and function of the endothelial cytoskeleton. Lab Invest **65**: 123-137.

21. Hellevik T, Bondevik A, Smedsrød B (1993) Intracellular trafficking of endocytosed collagen in rat liver endothelial cells. In: *Cells of the Hepatic Sinusoid 4* (Knook DL, Wisse E, eds) Kupffer Cell Foundation, Leiden; pp 431-433.

22. Horwitz SB, Schiff PB, Parness J, Manfredi J, Mellado W, Roy SN (1986) Taxol: a probe for studying the structure and function of microtubules. In: *The cytoskeleton, a target for toxic agents* (Clarkson TW, Sager PR, Syversen TLM, eds) Plenum Press, New York, pp 53-65.

23. Kelly RB (1990) Microtubules, membrane traffic and cell organization. Cell **61**: 5-7.

24. Knook DL, Sleyster EC (1976) Separation of Kupffer and endothelial cells of the rat liver by centrifugal elutriation. Exp Cell Res **99**: 444-449.

25. Knook DL, Blansjaar N, Sleyster EC (1977) Isolation and characterization of Kupffer and endothelial cells from the rat liver. Exp Cell Res **109**: 317-329.

26. Lindroth M, Bell PB, Fredriksson BA, Xiao-Dong L (1992) Preservation and visualization of molecular structure in detergent-extracted whole mounts

of cultured cells. Microscopy Res Techn **22**: 130-150.

27. Mareel MM, De Bruyne GK, Segers JL, Rabaey ML, De Neve WJ, Storme GA (1985) Cold sensitivity of cytoplasmic microtubules in malignant and non-malignant mouse cells. In: *Microtubules and microtubule inhibitors* (De Brabander M, De Mey J, eds) Elsevier Science Publishers BV, Amsterdam, pp 327-333.

28. McGuire RF, Bissell DM, Boyles J, Roll FJ (1992) Role of extracellular matrix in regulating fenestrations of sinusoidal endothelial cells isolated from normal rat liver. Hepatology **15**: 989-997.

29. Naito M, Wisse E (1978) Filtration effect of endothelial fenestrations on chylomicron transport in neonatal rat liver sinusoids. Cell Tissue Res **190**: 371-382.

30. Oda M, Kazemoto S, Kaneko H, Yokomori H, Ishii K, Tsukada N, Watanabe N, Suematsu M, Tsuchiya M (1993) Involvement of Ca^{2+}-calmodulin-actomyosin system in contractility of hepatic sinusoidal endothelial fenestrae. In: *Cells of the Hepatic Sinusoid 4* (Knook DL, Wisse E, eds) Kupffer Cell Foundation, Leiden, pp 174-178.

31. Ogawa K, Minase T, Enomoto K, Onoé T (1973) Ultrastructure of fenestrated cells in the sinusoidal wall of rat liver after perfusion fixation. Tohoku J Exp Med **110**: 89-101.

32. Schulze E, Kirschner M (1987) Dynamic and stable populations of microtubules in cells. J Cell Biol **104**: 277-288.

33. Smedsrød B, Pertoft H, Eggertsen G, Sundström C (1985) Functional and morphological characterization of cultures of Kupffer cells and liver endothelial cells prepared by means of density separation in Percoll, and selective substrate adherence. Cell Tissue Res **241**: 639-649.

34. Steffan AM, Gendrault JL, Kirn A (1987) Increase in the number of fenestrae in mouse endothelial liver cells by altering the cytoskeleton with cytochalasin B. Hepatology **7**: 1230-1238.

35. Stossel TP (1994) The machinery of cell crawling. Sci Am **271**: 54-63.

36. Van Der Smissen P, Van Bossuyt H, Charels K, Wisse E (1986) The structure and function of the cytoskeleton in sinusoidal endothelial cells in the rat liver. In: *Cells of the Hepatic Sinusoid 1* (Kirn A, Knook DL, Wisse E, eds) Kupffer Cell Foundation, Rijswijk, pp 517-522.

37. Vinckier A, Heyvaert I, D'Hoore A, McKittrick T, Van Haesendonck C, Engelborghs Y, Hellemans Y (1995) Immobilizing and imaging microtubules by atomic force microscopy. Ultramicrosc **57**: 337-343.

38. Widmann JJ, Cotran RS, Fahimi HD (1972) Mononuclear phagocytes (Kupffer cells) and endothelial cells in regenerating rat liver. J Cell Biol **52**: 159-170.

39. Wilson L (1986) Microtubules as targets for drug and toxic chemical action: the mechanisms of action of colchicine and vinblastine. In: *The cytoskeleton, a target for toxic agents* (Clarkson TW, Sager PR, Syversen TLM, eds) Plenum Press, New York, pp 53-65.

40. Wisse E (1970) An electron microscopic study of the fenestrated endothelial lining of rat liver sinusoids. J Ultrastruct Res **31**: 125-150.

41. Wisse E, Van Dierendonck JH, De Zanger RB, Fraser R, McCuskey RS (1980) On the role of the liver endothelial filter in the transport of particulate fat (chylomicrons and their remnants) to parenchymal cells and the influence of certain hormones on the endothelial fenestrae. In: *Communications of Liver Cells* (Popper H, Gudat F, Bianchi L, Reutter W, eds) MTP Press, Basel, pp 195-200.

42. Wisse E, De Zanger RB, Charels K, Van Der Smissen P, McCuskey RS (1985) The liver sieve: Considerations concerning the structure and function of endothelial fenestrae, the sinusoidal wall and the space of Disse. Hepatology **5**: 683-692.

Discussion with Reviewers

K.R. Robinson: What is your hypothesis regarding the role of F-actin in fenestrae and sieve plate dynamics, since inhibition of G-actin polymerization shows dramatic effects on fenestrae number? Could F-actin be the "small interconnecting structures" mentioned in the legend for Fig. 6, or "branching structures" seen in the whole-mount preparations?

P.B. Bell Jr: Do you have any ideas as to the nature of the short filaments that you observe connecting microtubules to the fenestrae-associated cytoskeleton and interconnecting the cytoskeletal filaments?

Authors: At the moment we have no idea how filamentous actin is involved in the fenestrae dynamics. However, it has been postulated by others [4, 34] that the inhibition of actin polymerization leads to the fusion of lipid bilayers, resulting in the formation of fenestrae. The small branching structures are fine and short interconnecting filaments, which have a thickness of approximately 8 nm [6]. In detergent-extracted cells, the reported thickness of microfilaments is 5 to 10 nm [26]. These branching structures fulfil therefore the morphologic characteristic of filamentous actin [6]. However, at the moment we have no data available about the immunological identification of these structures.

K.R. Robinson: Does the size of the fenestrae (mean size or distribution) change with the experimental treatments? The appearance of the fenestrae in the SEMs of Fig. 2 suggests this.

Authors: Different reports show conflicting data about the alterations in fenestrae diameter after treatment with cytochalasin B, i.e. decreased [4], increased [30] or unchanged [34]. However, all reports confirm the increase in the number of fenestrae when LEC were treated with cytochalasin B [4, 6, 30, 34]. In our hands, fenestrae diameter decreases 3-5% when LEC were treated with cytochalasin B (unpublished results). Unfortunately, we have no data available about the alterations in fenestrae diameter after taxol / cytochalasin B and colchicine / cytochalasin B-treatment.

K.R. Robinson: Would an alternative, ultrathin metal coating strategy such as 1nm chromium deposition as employed by Apkarian and others, enhance the visualization of fine detail of the cell surface as well as extracted preparations? This might be useful for instance, in delineating the fenestrae-associated cytoskeleton ring or obtaining more accurate measurements of fenestral dimensions.

Authors: Of course, structural surface information is limited by the quality of the metal film deposited onto it. Coating of the sample with a thin layer of chromium or tungsten enhances the visualization of fine structural details of the cytoskeleton when observed by SEM [3]. Our experience is that gold-coating with a Balzers sputtering device (Type 07120/160) induces melting artifacts caused during sputtering. However, we have no experience with LEC-cytoskeletons coated with a thin layer of chromium or tungsten using advanced sputtering methods.

K.R. Robinson: What is the potential for aldehyde fixation to alter the cell surface morphology? Have you performed any living-cell studies using the AFM?

Authors: AFM allows high-resolution imaging of biological samples under wet conditions. When living cells were scanned, images with artifacts were obtained, mainly caused by the softness of the samples. When cells underwent a short aldehyde fixation, the image quality improved. It seems that aldehyde fixation increases the rigidity of cells, resulting in a better image acquisition (Hoh JH and Schoenenberger CA (1994) Surface morphology and mechanical properties of MDCK monolayers by atomic force microscopy. J Cell Sci **107**: 1105-1114). This phenomenon was also discussed in our previous study [9]. Further studies will be conducted to find the optimal conditions for visualizing living LEC by AFM.

R.M. Albrecht: Our experience has been that extraction of cellular and cytoskeletal associated material in order to clean the cytoskeletal elements sufficiently to permit good observation by SEM can cause partial disruption of the cytoskeleton. (ie. the cleaner you get it the less there is of it). The extraction required for whole mount TEM is less rigorous and affords better retention of the overall structure since all the cytoskeletal associated material need not be completely removed. Did the authors carry out any comparisons along these lines? Have they tried observing the cytoskeletal organization of unextracted, stained cells by HVEM or IVEM and compared this to TEM and SEM observations of extracted cells?

Authors: We agree with these statements, which are consistent with our experience. When LEC were slightly fixed, followed by detergent-extraction, a better retention of the overall cytoskeleton was observed in TEM [6]. When these samples were observed by SEM, cytoskeletal structures which lie deep in the cell couldn't be observed (see also discussion). Previously we discussed the visualization of the cytoskeleton and the different results when preparation procedures for whole mount-SEM and TEM are used [6]. Additionally, in an attempt to elucidate the possible effects of detergent-extraction on the LEC-cytoskeleton, we tried to visualize the cytoskeleton of fixed-LEC in a cryo-EM study (see also this volume, F. Braet & P. Frederik). We found evidence that the fenestrae are again delineated by cytoskeleton rings in cryo-preparations, in accordance with results obtained with detergent-extraction.

Scanning Microscopy Supplement 10, 1996 (pages 237-247)
Scanning Microscopy International, Chicago (AMF O'Hare), IL 60666 USA
0892-953X/96$5.00+.25

CORRELATED CONFOCAL AND INTERMEDIATE VOLTAGE ELECTRON MICROSCOPY IMAGING OF THE SAME CELLS USING SEQUENTIAL FLUORESCENCE LABELING, FIXATION, AND CRITICAL POINT DEHYDRATION

Lee D. Peachey[1*], Harunori Ishikawa[2] and Tohru Murakami[1,2]

[1]Department of Biology, University of Pennsylvania, Philadelphia, PA,
[2]Department of Anatomy, Gunma University School of Medicine, Maebashi, Gunma 371, Japan

(Received for publication October 15, 1995 and in revised form December 31, 1996)

Abstract

Confocal laser scanning microscopy (CLSM) and intermediate voltage transmission electron microscopy (IVEM) each has its own particular advantages. CLSM can examine living cells, but is particularly useful when applied to cells that have been lightly fixed, permeabilized, and stained with fluorescent-labeled antibodies for localization of specific molecular species at the resolution of the light microscope while still in the hydrated state. IVEM provides much higher resolution images, but requires more drastic preparation procedures, including dehydration. This paper presents methods for combining these complementary approaches to examine exactly the same cells sequentially by CLSM and IVEM. Cells are grown in culture on sterile formvar films spread over gold index grids on cover glasses, which are mounted on larger cover glasses or microscope slides with spacers to prevent compression of the cells. Light and epifluorescence microscopy, and CLSM are performed concentrating on cells in grid openings. Then the grids are fixed with aldehydes followed by OsO_4, dehydrated and critical point dried (CPD) from liquid CO_2. Immediately following CPD, the grids are ready for examination in the IVEM. Low magnification (300-600x) survey images allow correlation of the IVEM images with the light microscopic images. In higher power images, structures that are fluorescent labeled can be related to corresponding regions in the IVEM images.

Key Words: Confocal microscopy, high voltage electron microscopy, correlated microscopy, specimen preparation, cell cultures, fluorescence immunocytochemistry, fluorescence labeling.

*Address for correspondence:
Lee D. Peachey
Department of Biology, University of Pennsylvania
Philadelphia PA 19104-6018
Telephone number: (215) 898-5788
FAX number: (215) 898-8780
E-mail: ldp@ivem.bio.upenn.edu

Introduction

The light microscope has a singularly important advantage over most other kinds of microscopes in that it can be used to visualize dynamic processes in living cells. Such live preparations produce images that can be both beautiful and informative, within the limits of resolution of the light microscope and the investigator's abilities to stain cellular components while still maintaining the cells in the living state with normal physiological functions. Movies of cells in action are possible and often necessary for adequate description of dynamic cellular activities.

At the other end of the spectrum, with respect to specimen preparation, is the sequence of preparative steps that a cell usually is subjected to prior to imaging with transmission electron microscopy (TEM) as a thin section. For TEM, biological specimens may be subjected to fixation with aldehydes and osmium tetroxide, alcohol dehydration, embedding in epoxy resins, and microtomy. Exquisite in their detail, the resulting images have much higher resolution of fine structure than is possible in light microscopy. Such images also potentially are very revealing about the mechanisms underlying physiological activities in these cells, provided that the effects of specimen preparation procedures can be evaluated, and also provided that the detailed information present in the TEM images can be understood in relation to the living cell images. One major difficulty in understanding the relationship between light and electron images is to account for possible effects on cellular structure of the preparation procedures used. Another difficulty arises because one is required to relate images of one cell observed in the light microscope to images of a different cell seen in the TEM. While this may be straightforward for larger, well-formed structures with easily recognized forms (mitochondria or striated myofibrils, for example), the correlation can be difficult at best for smaller structures, structures that are found only infrequently, and structures with more subtle forms. Thus electron microscopists and cell biologists are presented with a real challenge when they wish to bridge the gap between light and electron microscopy in a precise way, in order to combine the special information each type of microscope provides. This is the bridge this paper will focus on.

One important plank in the bridge between light and electron microscopy of biological cells has been available for 20 years or more: the high (or intermediate) voltage TEM (see Peachey *et al.*, 1974 for several early examples). These instruments, few in number but powerful in their penetration of thick specimens, are readily available thanks to the wisdom and funding of the Biomedical Research Technology Program of the National Institutes of Health (Bethesda, MD). Whole cells, grown on or otherwise mounted on electron microscope specimen grids, can be imaged intact in such higher voltage electron microscopes, avoiding some of the more drastic preparation steps required in thin-section electron microscopy, in particular the embedding and sectioning steps. Thus one can examine whole, intact cells with all the resolution available in the TEM with minimal mechanical disturbances, and with reduced, if not exactly minimal, chemical treatment.

This paper describes some methods and results from an effort to provide a direct correlation between light and electron microscopy, by imaging the same cells sequentially in microscopes of both types.

Materials and Methods

All of the examples shown are from cells grown in culture. The procedure followed was similar to that in use in several laboratories for preparing cells grown in culture for high voltage (HVEM) or intermediate voltage electron microscopy (IVEM) imaging, in which cells are released from their host tissues using chemical or enzymatic treatment and then plated on formvar-coated electron microscope (EM) specimen grids for culture (Wolosewick and Porter, 1976; Murakami *et al.*, 1993).

The overall goal was to examine the same cells, first without any form of treatment, by light microscopy, and finally, after fixation and dehydration, by TEM. Phase microscopy was often used first, and was useful for judging the state of the cultures and the maturity of the cells. Fluorescence microscopy then followed, using fluorescent probes or fluorescent labels conjugated to specific antibodies to identify and locate specific proteins in cells, and brief fixation and permeabilization of the cells to allow penetration of the labels. Confocal fluorescence scanning laser microscopy provided greater 3D resolution of the distribution of labels and provides images free of background, out of focus fluorescence. Then the cells were ready for the more drastic specimen preparation procedures, including dehydration, required for transmission electron microscopy.

In order to facilitate examination of cells using various forms of microscopy at various stages in a sequence of specimen preparation steps, and especially in order to provide for localization and imaging of the exact same cells through all these steps, several modifications were introduced into the usual procedures for electron microscopy of whole mounts of cultured cells. These procedures, including the modifications, will now be described in some detail. This description will include some procedures not used for the illustrations presented here, in order to provide a broader picture of the specimen preparation methods available for sequential confocal and electron microscopy.

Cell culture

Cells were grown on glass coverslips prepared with gold EM grids, as follows. Clean glass slides were dipped in an approximately 1% solution of formvar in ethylene dichloride, and dried in air. The plastic film was floated off on a clean water surface, and the slide was discarded. Three or four gold EM specimen grids, previously cleaned with dilute acid and rinsed, were placed on top of the floating plastic film, in an asymmetric pattern so that individual grids could be identified uniquely later, even when the film and grids had been inverted. Gold grids do not poison or otherwise contaminate the cultures, and the use of "finder grids" with index marks to identify the grid openings is useful in tracking individual cells through the various microscopic examinations. Next, a square cover glass was pressed down on the plastic film carrying the grids, pushed through the water surface, inverted under water, and brought back into air, with care not to wrinkle the film or lose the grids. Air drying and sterilization overnight under UV light completed the preparation of the culture substrates.

All of the cells illustrated here are from cultures of primary heart explants from 8-day chick embryos. Cells were dispersed by treatment with 0.05-0.1% trypsin in calcium- and magnesium-free Hank's balanced salt solution at 37°C for 10-20 min followed by filtration to remove undissociated tissue. The resulting mixed cardiac myocytes and fibroblasts were cultured for 3-7 days in Eagle's minimum essential medium (Flow Laboratories, McLean, VA) plus 10% fetal bovine serum (Sigma, St. Louis, MO) at a density of about 2.5×10^4 cells per dish containing one cover glass. Cells begin to attach and spread within one day, and begin to take on characteristic shapes: flat and multipolar for fibroblasts and more rectangular and usually bipolar for myocytes. By day three, some myocytes form cell attachments, or fuse to form myotubes and start to contract.

Phase microscopy

Phase microscopy was done directly on cells growing over holes in the EM grids, while the cells continued to grow in the culture medium. Cells obscured by grid bars or not growing over grids can be examined at this stage, but these cells cannot be examined later by electron microscopy. When the desired stage of growth had been achieved, cover glasses were removed from the culture dishes and the preparation for sequential microscopy started.

Epifluorescence microscopy

Typically the cells, still on the grids and cover glasses, were fixed briefly (e.g., 10 min at room temperature) in 4% paraformaldehyde in PBS (phosphate buffered saline). The cells were permeabilized by immersion in 0.1% Triton-X-100 in PBS at room temperature for about 10 min, followed by a brief rinse in PBS. Labeling with fluorescent probes followed. This could be a direct fluorescent label, such as fluorescein-phalloidin, or a sequence of primary and fluorescent-labeled secondary antibodies, such as rabbit anti-myosin followed by goat anti-rabbit antibody conjugated to fluorescein. The incubation times and concentrations of probes or antibodies used depended on the particular probes and antibodies. Fluorescein isothiocyanate (FITC)-phalloidin (Sigma), used for the cells illustrated in this paper, was utilized by adding 3.3 μl of a 100 μg/ml stock in ethanol to 400 μl of PBS, with incubation at room temperature for 20 min. After a 10 minute rinse in PBS, the cover glasses were mounted on larger glass coverslips, with the grids and cells facing the larger coverslip and the two coverslips held apart with thin plastic spacers or otherwise held in place so the grids and the cells would not be crushed. Various other configurations of slides and cover glasses are possible, with the important points being not to crush the preparations, to have the cells as close as possible to the cover glass through which images are made, and to be able to disassemble the mount in order to recover the grids for further processing. The mounting medium used at this stage was a glycerol-water mixture, with the addition of anti fading agents. A standard fluorescence microscope, either inverted or upright, is suitable for visual examination of the preparations and photography. Images were recorded on black and white or color film, with records being kept of the locations of the cells photographed in relation to the index marks on the grids and the grid location on the cover glass. Epifluorescence images shown here were obtained using a 35 mm camera on the microscope contained in the confocal microscopy setup, using the standard epifluorescence filter set and a Zeiss (Oberkochen, Germany) PlanApo oil immersion objective, numerical aperture (n.a.) 1.3. The instrument was then switched to the confocal scanner for confocal microscopy, without moving the specimen.

Confocal laser scanning microscopy

We used both a Biorad MRC-600 confocal scanner with a Zeiss Axiophot upright microscope and an argon-ion laser and a Leica upright confocal microscope with an argon-krypton laser. The mounting of the specimens was not changed from that used for epifluorescence microscopy. The same cells were located and, typically, one or more Z-series scans were made and stored for future analysis and comparison to the IVEM images. Details are provided in the figure legends for the images shown here.

Intermediate voltage electron microscopy

Following all steps of light microscopy, the slides and/or cover glass sandwiches containing the specimen grids and cells were opened so that the grids could be processed further for electron microscopy. One method used involved immersing the assembly under water, and gently lifting the cover glass carrying the grids away from the opposing slide or cover glass. In an alternate method, the assembly was set, with the cover glass carrying the grids downward, on pedestals in a glass container, such that the pedestals contacted only the upper, larger glass slide or cover glass. Water was then added to the container until the bottom surface of the assembly was immersed. After some time, the lower cover glass floated away from the upper one, and the cover glass with the grids attached floated to the bottom of the container. These methods eliminated surface tension effects, and the separation of the assembly without damage to the preparations was facilitated. Gentleness is paramount at this stage, as the thin support films on the grids can easily broken be or dislodged.

The cover glasses and attached grids with cells were fixed with 1% glutaraldehyde in 0.1 M cacodylate buffer, pH 7.2 at room temperature for 20 minutes, rinsed with buffer alone, and post fixed in 1% osmium tetroxide in the same buffer at 4°C. The preparations are then rinsed briefly with distilled water, and optionally stained with 2% uranyl acetate in water for 10 minutes followed by another rinse in distilled water. Next, the grids were gently detached from the cover glass by cutting around them, through the plastic film, while under water. The grids were then placed in a grid holder designed to fit into the critical point dryer (CPD) for the remainder of the processing.

Dehydration was done in a graded series of ethanol-water mixtures, about 3 minutes in each concentration, using a series of 15%, 30%, 50%, 75%, 95%, and 100% ethanol (3 times in 100%). CPD was carried out starting with the chamber filled with just enough 100% ethanol to cover the grids in the holder, and with approximately 5 cycles of introduction of liquid carbon dioxide and bleeding, with 5 minute soaks between. The chamber was slowly bled to atmospheric pressure after warming to about 60°C. Finally, the grids were lightly coated with carbon in a vacuum evaporator to stabilize the films.

Intermediate voltage electron microscopy was done on JEOL (Tokyo, Japan) 4000-EX instruments either at Chiba University, Japan or at the University of Pennsylvania in Philadelphia, at 300-400 kV.

Results

Figs. 1-4 show portions of the same two early myoblasts in a 4-day cultured heart explant from an 8-day old chick embryo. The preparation was stained with FITC-labeled phalloidin to stain f-actin. Fig. 1 is the epifluorescence image, and shows arrays of actin-containing stress fibers. The arrowhead points toward a group of such actin-containing fibers, just inside the cell membrane of one of the cell processes. Some of these actin-containing fibers are seen to lie close to the lateral cell membranes of the cell processes, while some extend across the cell processes, in a pattern that is typical for these cells. Fig. 2 is a confocal projection along the Z-axis of a series of 16 images of the same specimen, taken at a Z-spacing of 0.48 μm between scans. The total vertical thickness included in this projection is approximately 8 μm, and covers the entire thickness of the distal portions of these cell processes. Thus all the same structures that appear in the non-confocal, epifluorescence image also appear in the confocal image. However, the pattern of stress fibers is shown more clearly in the extended focus confocal projection than in the epifluorescence image because of the removal of out of focus light by the confocal aperture. For the same reason, evidence of very early myofibril formation, not seen in the epifluorescence image, is visible in the confocal projection as linear arrays of fluorescent dots in the region indicated by the asterisk. The arrowhead outside the cell in Fig. 2 has the same placement with respect to the cell processes as in Fig. 1. In addition, the arrow in Fig. 2 points to a small, lateral extension from an internal stress fiber, which is not as clearly seen in Fig. 1. Such fine details often become apparent in the epifluorescence image only after they have been seen in the confocal images.

Fig. 3 is an IVEM image from the tip region of the upper-left cell process in Figs. 1 and 2, at higher magnification that the earlier images, and rotated slightly in the counter-clockwise direction. For orientation, an arrowhead points in the same direction and to the same region of the cell as the arrowhead in the earlier figures. It should be noted that, unlike in the fluorescence images, all structures are visible in the IVEM image, since specificity of staining with the phalloidin stain is lost in these images. The resolution of structure is, of course, much better in the IVEM. The arrow in Fig. 3 points to the same extension of a stress fiber as in Fig. 2. One can clearly see this extension in the IVEM image. One can also conclude that nearby structures, similar in appearance in the IVEM to this extension, do not contain f-actin, since they are not stained in the confocal fluorescence image. Examples of such structures are the two dense objects indicated by the arrowheads in Fig. 4, which do not appear at all in the fluorescence images, and which are tentatively identified by size and shape as mitochondria. Fig. 4 is at higher magnification, and shows more clearly the structures

(Figures 1-4 on facing page)

Figure 1. Epifluorescence image of portions of two attached myoblastic cells in a 4-day culture of an heart explant from an 8-day old chick embryo. This preparation was stained with FITC-phalloidin. Each cell has a cytoplasmic process extending upward from the bottom of the figure. The arrowhead points toward a group of actin-containing fibers near the end of one of these processes. Bar indicates 10 μm.

Figure 2. Confocal projection, shown as a negative, along the optic axis of a series of 16 fluorescence confocal images of the same cell as seen in Fig. 1, taken with a focal change of 0.48 micrometers between scans. The arrowhead is positioned in the same location as in Fig. 1. The arrow points to a fine lateral projection from one of the stress fibers. This projection is only barely visible in the epifluorescence image, but very clear in the confocal projection. Pixel size 0.17 micrometers. Bar indicates 10 μm.

Figure 3. IVEM image of the same cell at low magnification. Again, the arrowhead points at the equivalent position marked in the previous figures. The arrow points to the same lateral projection as indicated in Fig. 2. Bar indicates 5 μm.

Figure 4. Higher magnification IVEM image of the same cell. The arrow points to the same lateral projection indicated earlier. The arrowheads point to other structures, presumed to be mitochondria from their morphology, that are similar in size and density to the lateral projection indicated. These clearly are not actin-containing structures, since they are not visible in the epifluorescence and confocal fluorescence images, demonstrating the importance of having the specificity of the fluorescence labels and the ability to identify the same structures in the two kinds of images. Line indicates 1 μm.

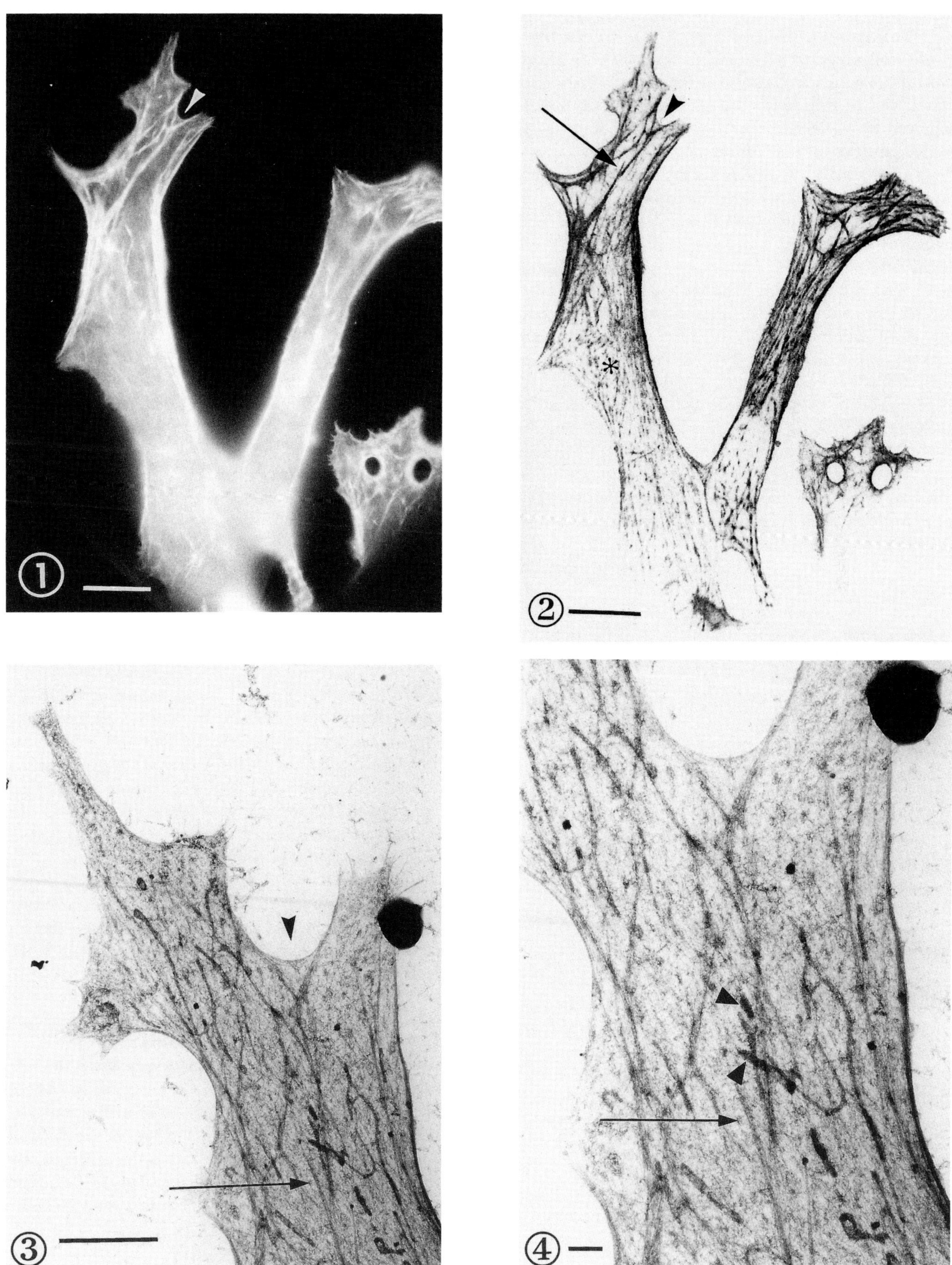
①
②
*
③
④

indicated.

The next set of images, Figs. 5-8, is from a fibroblastic cell prepared in the same way. These images illustrate that, under certain conditions, identification and correlation of even fairly fine structures easily can be achieved by visual inspection of corresponding images from epifluorescence, confocal, and IVEM microscopes. Four arrows mark the tips of actin stress fibers projecting from the main body of the cell toward a tapering cell process, which points toward the left edge of each figure. From the epifluorescence image (Fig. 5), one could describe the four processes, from top to bottom, as #1 long and narrow, #2 shorter and broader, #3 long and narrow, with a forked tip, and #4 broad and possibly representing a bundle of several individual and separate stress fibers. In Fig. 6, the confocal image, these fibers are easily identified and their descriptions, as given above, are confirmed. Additionally, variations in fluorescence along the length of #1, and the composition of #4 as consisting of multiple fibers, are considerably more apparent because of the greater clarity of the confocal image compared to the epifluorescence image. The IVEM images (Figs. 7 and 8) provide additional resolution of these features, e.g., the branching of fiber #2 near its end, and details of the subfiber composition of #4.

Part of the reason we have been able unambiguously to identify the corresponding structures in the light and electron micrographs in these figures is that the selected structures have characteristic shapes and are relatively free from closeby and superimposed other structures. The next example will show how information from the fluorescence images can provide information and structural identification not readily available in the IVEM images because of additional material obscuring this information.

Fig. 9 is a confocal projection of a cell in the same type of culture that yielded the cells shown in Figs. 1-8. In this cell, the fluorescence image reveals circular, banded structures with a periodicity of approximately two micrometers (arrow). We identify these periodic structures as myofibrils early in development. In the corresponding locations of the IVEM image (Fig. 10, arrow), dense, curved structures are found around the nucleus, but these do not appear obviously banded, and would not have been thought to be myofibrils on the basis of the IVEM images alone. Without the correlated image from the confocal microscope, with its fluorescent antibody identification of actin-containing striated fibrils, this cell could have been misidentified in the IVEM as a fibroblast rather that a developing myoblast. The confocal fluorescence image, in this case, provides a useful map to lead the investigator to a specific subset of structures of interest in the often more cluttered and

(*Figures 5-8 on facing page*)

Figure 5. Epifluorescence image of portion of a fibroblastic cell in a 4-day culture of an heart explant from an 8-day old chick embryo. A tapered cell process extends downward at the bottom of the figure. Arrows indicate the distal tips of 4 stress fibers extending toward and into this cell process. Bar indicates 10 μm.

Figure 6. Confocal projection along the optic axis of a series of 11 fluorescence confocal images of the same cell as seen in Fig. 5, taken with a focal change of 0.48 micrometers between scans. The same four stress fibers are indicated by arrows. Pixel size 0.14 micrometers. Bar indicates 10 μm.

Figure 7. IVEM image of the same cell at low magnification, with arrows indicating the same four stress fibers. Greater detail of fine structure of the fibers is available in this image, including branching near the tip of the third fiber from the left and the subfibrillar nature of the fiber on the right. Bar indicates 10 μm.

Figure 8. Higher magnification IVEM image of the same cell as in Figures 5-7. The fiber on the right now is clearly resolved into several finer fibers. Bar indicates 5 μm.

----

confusing image from the electron microscope.

Discussion

We have demonstrated that a single cell can be imaged through several forms of microscopy, including fluorescence confocal microscopy and 400 kV IVEM, and that intracellular structures can be identified in the corresponding images. One potential value of this kind of correlated microscopy on the same cells is the possibility of identifying structures early in their formation in the cell, as during cell differentiation. For example, striated myofibrils are easy to identify in either light or electron microscope images when they are fully formed (Ishikawa *et al.*, 1990) by virtue of their characteristic and stereotyped pattern of striations. Prior to the emergence of this pattern, however, it can be very difficult to identify forming structures in the absence of information on their protein content, which information may be easily available using fluorescent antibodies. Correlation, in a spatially precise way, between fluorescent images using labeled antibodies and IVEM images of the same cells allows this chemical identification to be carried over to the high resolution images of the electron microscope, where very small and perhaps irregularly shaped and incompletely formed structures can be sorted out from a maze of similar structures and then be studied at high resolution.

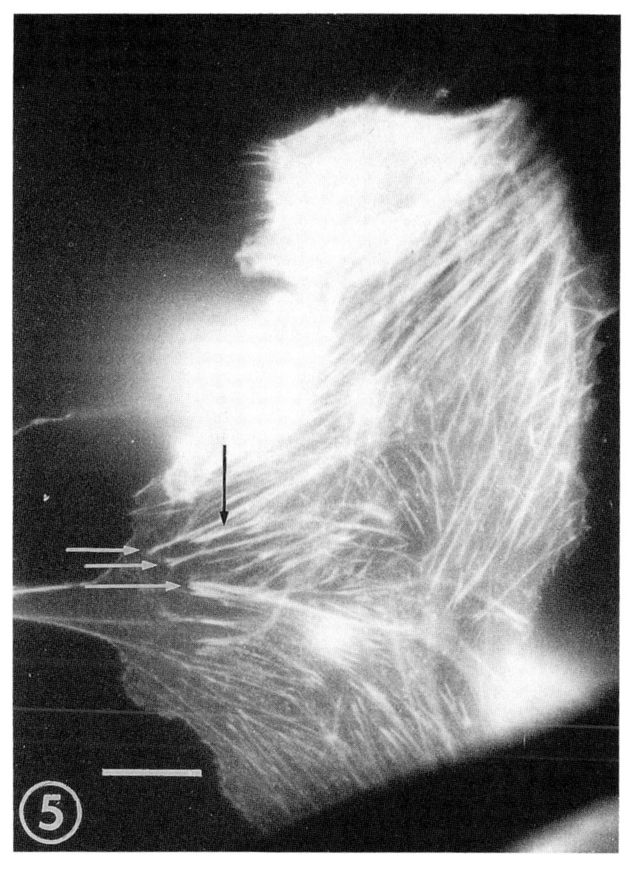
5

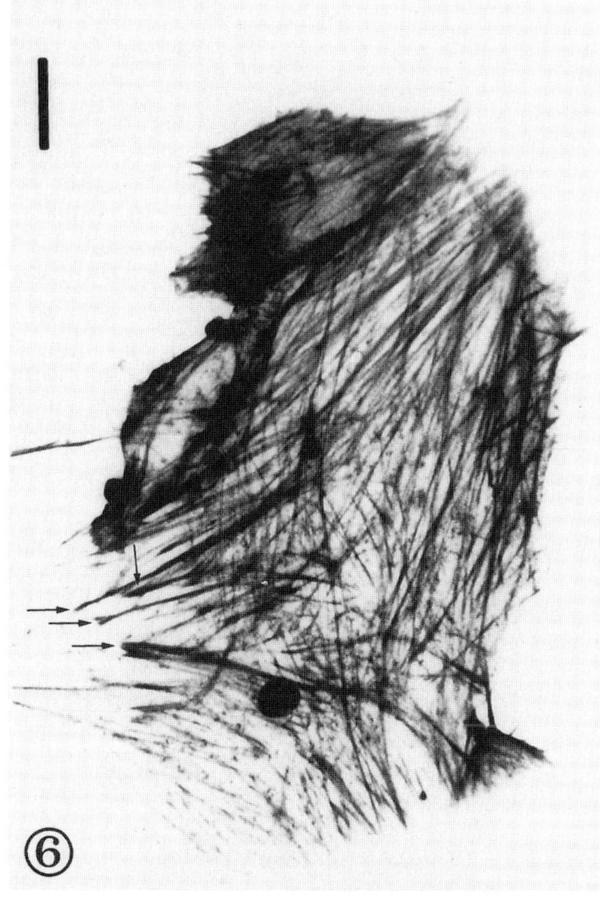
6

7

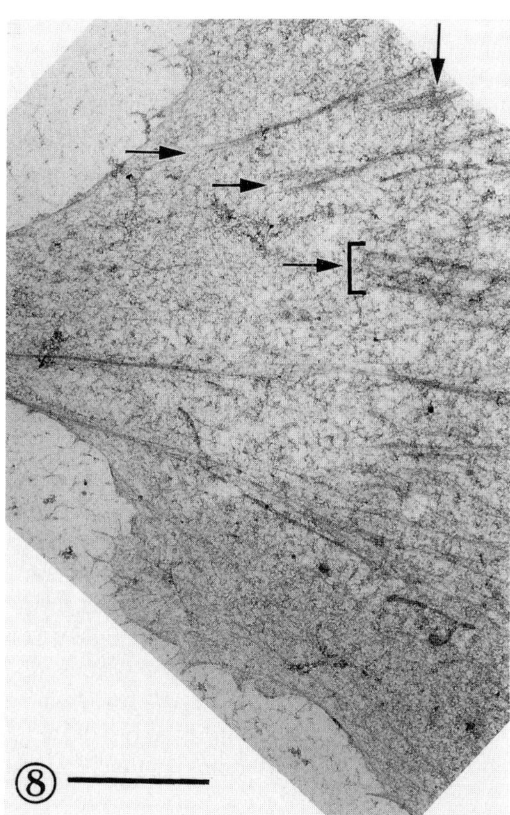
8

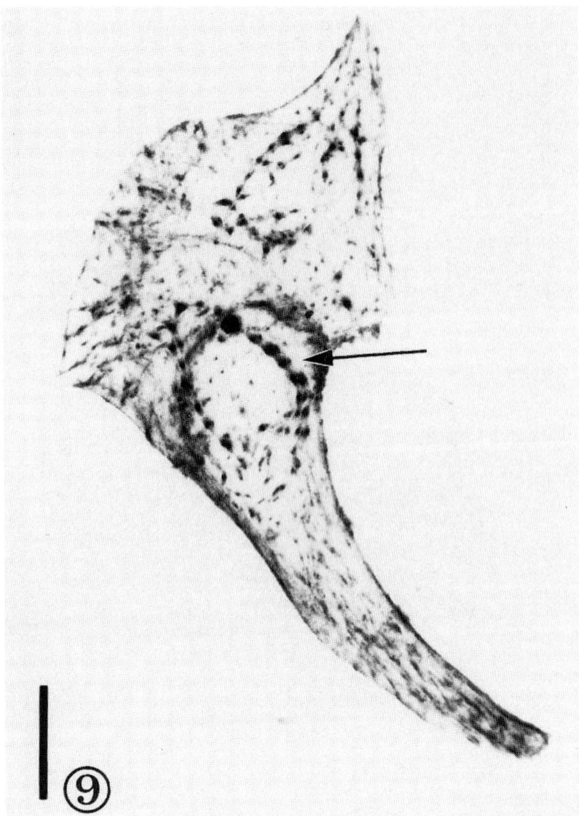

Figure 9. Confocal projection along the optic axis of a series of 13 fluorescence confocal images of a cell in a similar preparation, taken with a focal change of 0.48 μm between scans. Several curved structures are seen, with periodic fluorescence at a spacing of about 2 micrometers (arrow). These are identified as developing myofibrils. Pixel size 0.10 μm. Bar indicates 10 μm.

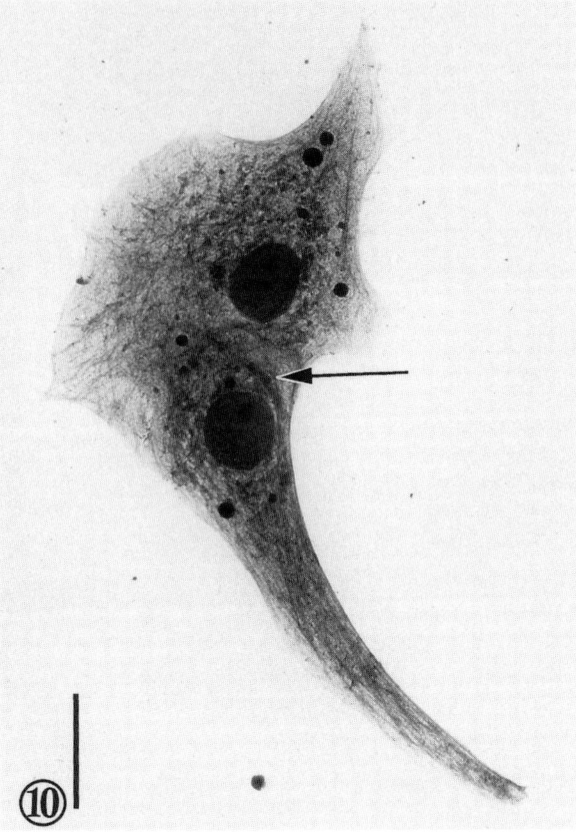

Figure 10. IVEM image of the same cell as in Fig. 9 at the same magnification. While the curved structure indicated with the arrow in Fig. 9 can be found, and while it can be seen to curve around a nucleus in the cell, this structure has no apparent periodicity in this electron image. Bar indicates 10 μm.

A significant factor in our ability to relate images of the same cells taken using different microscopes to each other, and visually to identify with certainty the same structures in the different images, is the high contrast of the structures with respect to their immediate background as well as the relative absence of superimposed structure in the IVEM images, where all material shows up, not only that which has been stained with the fluorescent tag. We have shown one example (Figs. 9 and 10) where details of the correspondence was not clear, based on appearance in the two images, but the structures still could be identified by overall shape and by location in the cell. We have found many examples, in our studies, where the structural field was to complex and too cluttered with different classes of structures, especially in the IVEM images, to be able to identify with certainty which structures in the IVEM corresponded to fluorescent structures in the light micrographs. Clearly it would be valuable, in such cases, to exactly superimpose the two images, so that structures in the fluorescence image would lie directly over the structures in the same regions of the IVEM image, allowing for their identification. We have started to do this, but have not advanced far enough at this time to present the results. We need to precisely match images with respect to magnification, position, and orientation, and to compensate for shrinkage or other spatial distortions in one or another image, before this approach will bear fruit.

We are not unaware, of course, of the possibility of using antibodies tagged with, for example, gold particles, which can be imaged directly in the electron microscope and which can potentially provide the sort of chemical identification we need without the necessity of combining light and electron images. We consider our approach and the immuno-EM approach as complementary, and we plan to make comparisons between the two approaches in the future. As is usually the case, we expect that having two solutions to the same problem will prove valuable, and that there will be specific instances in which each method proves to be superior to the other.

Acknowledgments

This work was supported by the Biomedical Research Technology Program, National Center for Research Resources, National Institutes of Health, U.S.A. (RR-2483) and by grants from the Ministry of Education, Science, and Culture (International Scientific Research Program: Joint Research) of Japan. We also thank Dr. Prof. T. Nagano of Chiba University, Japan for use of his IVEM for part of this work.

References

Ishikawa H, Peachey LD, Schultheiss T, Holtzer H (1990) Correlated confocal and intermediate voltage electron microscopy of myofibrillogenesis in muscle cells: methods and preliminary results. Proc. 12th Int. Congr. Electron Microscopy, San Francisco Press, San Francisco. pp 178-179.

Murakami T, Ono M, Ishikawa H (1993) Confocal interference reflection microscopy of cultured cells. Bioimages **1**: 1-8.

Peachey LD, Fotino M, Porter KR (1974) Biological applications of high voltage electron microscopy. In: High Voltage Microscopy. Swann PR, Humphries CJ, Goringe MJ (eds). Academic Press, New York. pp 405-413.

Wolosewick JJ, Porter KR (1976) Stereo high-voltage electron microscopy of whole cells of the human diploid line, WI-38. Am J Anat **147**: 303-324.

Discussion with Reviewers

M. Malecki: Did you estimate the degree of shrinkage of cells exposed to dehydration and critical point drying? How is this going to affect attempts to superimpose images from LSCM and IVEM described in your future plans?

Authors: We have started to address the question of spatial distortions between confocal and EM images by comparing test specimens with easily identified and located fiducial objects, such as fluorescent-stained fluorescent beads, either alone on a film or combined with cell preparations. Shrinkage during dehydration is one possible source of such distortion, and would take place after confocal and before IVEM imaging. Optical distortion in one or the other microscope is the another. So far we have only evaluated the magnitude and nature of the distortion, without trying to distinguish between distortion during preparative procedures and optical distortion, though we expect to do that in the more distant future. While the results are still preliminary, they are somewhat encouraging in that it appears that the residual differences between images after adjusting for magnification and orientation are small, and they may be correctable with a simple, affine transformation. However, we wish to leave the possibility of needing higher-order transformations as an open possibility for the future.

M. Malecki: In this project you were using the transmission electron microscope operating at 300-400 kV. Can you share your experience on using lower accelerating voltages to study structure of cell whole-mounts? Particularly interesting would be your comments concerning the beam damage of cell whole-mounts observed at various accelerating voltages.

Authors: We, and others, have in fact studied whole-cell mounts at lower voltages, as low as 100 kV. In general terms, the results in terms of image quality are good in the thinner parts of well-spread cells, and not as good where the cells are thicker, as one would expect. We have not systematically studied beam damage at different voltages, though the expectation would be that there would be less damage at higher voltages because of reduced inelastic scattering.

M. Malecki: In your other paper (Heath and Peachey, 1987) you were successful in imaging of muscles within 0.75 μm thick sections using an energy filtering transmission electron microscope (EFTEM) operating at 80-100 kV. What is your experience in studying cell whole-mounts with EFTEM?

Authors: Energy filtering electron microscopy at lower accelerating voltages could also be used in studies of whole cells such as we have reported here, though we have not tried it. As you point out, we have used an 80 kV energy filtering microscope to visualize sections of embedded cells up to 0.75 μm thick with good success (Peachey *et al.*, 1987), and this should work on at least the thinner parts of whole cell mounts. The limit on thickness that we encountered with sections in the EFTEM was that beyond about half micrometer in thickness, there was so much inelastic scattering at 80 kV that there were not enough electrons within a reasonable width of energy window to be able to form an image in an acceptable exposure time. Whole mounts of cells do not have embedding material, so thicker specimens should be acceptable, but probably not the region of a tissue culture cell close to the nucleus, which can be 10 micrometers or more in thickness.

M. Malecki: You provide an example of studies on myofibril assembly as the justification for this project. In such a case preservation of filaments is of utmost importance. Can you include images demonstrating preservation of filaments e.g. actin and myosin prepared according to the protocols you describe or provide comments concerning that preservation?
Authors: We have published some images, both from confocal microscopy and IVEM, of more mature cardiac myocytes prepared using the same protocol (Ishikawa *et al.*, 1990). These images show well-preserved myofibrils with essentially the adult form of banding and the usual arrangement of thick and thin filaments, indicating that the preservation of these filaments is satisfactory.

J. Turner: The critical point drying method has been largely abandoned, I believe for the wrong reasons, and it is nice to see it revived here. However, to ignore the fact that it is a controversial method does not serve the readers. If nothing else, more detail concerning the CPD method is essential and comments related to these details would be helpful.
Authors: We took the following precautions to keep the residual water in the specimens low during and after the CPD procedure. The absolute alcohol used was stored over molecular sieve, to remove water. The minimum amount of ethanol needed to cover the specimens was used in the CPD chamber at the start of the drying procedure, making its removal easier. The liquid carbon dioxide was of the "bone dry" classification from the supplier, but was not further dried by us. When carbon dioxide was introduced into the chamber, we watched (through a window in the top of the chamber) the swirling of the liquid and ran as much carbon dioxide through the chamber on each cycle as was consistent with not damaging the grids. The grids were held in a special holder that allowed good circulation of liquid around the grids, while still preventing them from becoming lost. This active, rinsing and stirring procedure lasted at least 5 minutes on each cycle, and was followed by a minimum of a 5 minute soak before the next cycle was initiated. In our experience, this protocol results in cells with good three dimensionality, which we check by stereo imaging of small cell processes such as ruffles or microspikes from the cell surface. Occasionally we do not see such good preservation of three dimensionality, and we suspect that this most often is due to a poor batch of liquid carbon dioxide, since a new tank usually solves the problem. It also is important to note that even though we immediately transfer the dried grids to a vacuum desiccator with phosphorous pentoxide for storage, and are careful not to expose the grids to the atmosphere any longer than necessary when loading them into the IVEM, within a couple of days the cells begin to have an altered appearance in which very fine structures are not crisply delineated. It is difficult to describe the appearance in a very distinct way, but is easily recognized with experience. We say that the cells have a "mushy" look, and we believe that this may be due to a small amount of water picked up by the specimens, since the problem seems to be worse in the summer than in the winter.

J. Turner: Do the authors have any experience or comments about a comparison of CPD with freezing methods?
Authors: No, I am afraid we don't.

Reviewer III: Does using a mounting medium of glycerol-water, and then removing the specimens in water, then fixing in 0.1 M cacodylate buffer have an effect on the structural integrity of the sample? In the removal process how do you avoid getting objective oil in the water mixture?
Authors: We do not see any obvious swelling of these cells, which previously have been permeabilized with detergent and fixed with glutaraldehyde, when we remove the cover glass using water after light microscopy and further fix in buffer. We conclude that these cells are no longer osmotically active. Still, this removal procedure must be carried out very slowly and carefully, or there can be mechanical damage to the cells and/or the support films. We clean the cover glasses with ether-ethanol or another solvent mixture to remove all the immersion oil before carrying out the cover glass removal procedure.

Reviewer III: The formvar film is highly hydrophobic and deteriorates with prolonged exposure to UV. Could you discuss alternative methods available for sterilization of the formvar film and how to make the film a better substrate for cell adhesion?
Authors: Our formvar films are thicker than what one usually uses for electron microscopy at 100 kV. The higher accelerating voltages that we use allow us to use thicker support films, and this makes the preparations more likely to survive the multiple handling steps we subject them to. This also may help to reduce the effects of UV sterilization, though we have not specifically investigated this point. Presumably one could sterilize these grids prior to plating the cells using chemicals, but we have not found it necessary to use any method other than UV irradiation. Contamination of cultures has not been a problem in these studies. We do modify the procedure for making the films when we have trouble with cell spreading or adhesion. If the cells do not spread well or do not adhere to the film, then we try to make the films more hydrophilic. First, films cast

on a water surface, rather than on glass, and then picked up so that the cells will be plated on the side of the film that was facing the water when the film was cast seem to aid cell spreading and adhesion. Second, glow discharge treatment of the grids also makes them more hydrophilic and thus improves cell adhesion and spreading. Occasionally, for some types of cells, the problem seems to the opposite, that the films are too hydrophilic, and in this case we plate the cells on the air-contact surface of water surface-spread films, which we believe to be more hydrophobic.

Reviewer III: Correlative microscopy is extremely powerful by the ability of gaining direct ultrastructural information on previously dynamic processes, however to date the limitations are the labor intensive manual process and the ability to unequivocally superimpose two images. Elaborate on the obstacles that must be overcome for computerized image analysis of correlative microscopy and the future of correlating 3-D data from CSLM and whole mount EM.

Authors: A complete discussion of these important points would more than double the length of this paper, and will have to wait for another occasion. We will try to respond briefly, however, and provide some references to other publications from our laboratory. Our initial approach to 3-D analysis of electron micrographs of thick specimens has been to accept that considerable human labor will be necessary, at least at first. We like to put a more positive spin on this by saying that the human eye-brain system is very powerful at 3-D recognition of classes of objects when seen in 3-D, as in the case of stereo pairs of EM images made by specimen tilting between exposures. Therefore we use this powerful recognition capability rather that to try to develop entirely computer-based recognition algorithms and programs (in fact, however, we also are doing the latter in parallel). In our initial approach to this, we use the computer to display multiple digitized IVEM stereo images and to generate 3-D cursors, which the operator can use to point to or trace structures that he/she has identified visually (Ishikawa *et al.*, 1990; Peachey *et al.*, 1995). The computer then uses information obtained from the placement of these cursors by the operator to calculate 3-D positions, sizes, and shapes of the selected objects, and further can generate graphic depictions of the objects so selected, located, and traced. The reviewer's request emphasizes the correlation, in 3-D, of this information from IVEM with parallel information obtained by confocal microscopy, and this is exactly the emphasis of our work at present. The confocal microscope gets 3-D information not as stereo pairs of images, but as serial optical slices through the volume of the specimen. 2-D projections can be calculated through these 3-D data sets, and software is provided with confocal microscopes to do just that. Pairs of 2-D confocal projections can be viewed stereoscopically, and then treated the same way we treat stereo pairs from the IVEM of the same specimens. Additionally, there are programs available for extracting 3-D models from serial section images (e.g., Young *et al.*, 1987). The future of correlating these two kinds of 3-D data sets from the two kinds of microscopes will depend on finding ways to merge these 3-D representations of the same structures extracted from the two kinds of images. It is relatively easy to think of ways of doing this by returning to 2-D representations, e.g. by making projections in the equivalent directions through the two extracted data sets (models, reconstructions, etc.), and then superimposing these projections for comparison. The two projections can be color-coded, and the merging, or lack of merging, of the colors will reveal what matches and what does not. Ideally, however, one would prefer to stay in the 3-D world of the reconstructions, and to merge these in some 3-D equivalent of 2-D superposition. Whether this can be done in a useful way, using color or some other form of coding to separate the two data sets in the mind of the observer, remains to be seen in the future.

Additional References

Peachey LD, Heath JP (1987) Electron microscopy of thick sections of embedded cells using the Zeiss EM 902 energy filtering electron microscope. Mag Electr Microsc **5**: 15-21.

Peachey LD, Heath JP, Lamprecht GG, Bauer R (1987) Energy filtering electron microscopy (EFEM) of thick sections of embedded biological tissues at 80 kV. J Electr Microsc Techn **6**: 219-230.

Peachey LD, Fodor L, Haselgrove JC, Dunn SM, Huang J (1995) Proc Microsc Microanal (Bailey GW, Ellisman MH, Hennigar RA, Zaluzec NJ, eds.). Jones & Begell Publ, New York, NY. pp 630-631.

Young SJ, Royer SM, Groves PM, Kinnamon J C (1987) Three-dimensional reconstructions from serial micrographs using the IBM PC. J Electr Microsc Techn **6**: 207-217.

Scanning Microscopy Supplement 10, 1996 (pages 249-260) 0892-953X/96$5.00+.25
Scanning Microscopy International, Chicago (AMF O'Hare), IL 60666 USA

PRE-EMBEDDING STAINING OF SINGLE MUSCLE FIBERS FOR LIGHT AND ELECTRON MICROSCOPY STUDIES OF SUBCELLULAR ORGANIZATION

Evelyn Ralston[1*] and Thorkil Ploug[2,3]

[1]Laboratory of Neurobiology, NINDS, [2]Experimental Diabetes, Metabolism and Nutrition Section, Diabetes Branch, NIDDK, National Institutes of Health, Bethesda and
[3]Copenhagen Muscle Research Centre, Rigshospitalet, Copenhagen, Denmark

(Received for publication June 6, 1996 and in revised form July 29, 1996)

Abstract

Skeletal muscle fibers are large, multinucleated cells which pose a challenge to the morphologist. In the course of studies of the distribution of the glucose transporter GLUT4, in muscle, we have compared different preparative procedures, for both light (LM) and electron microscopy (EM) immunocytochemistry. Here we show that pre-embedding staining of single teased fibers, or of single enzymatically dissociated fibers, has several advantages over the use of sections for observing discrete patterns that extend over long distances in the cells. We report on an optimization study carried out to establish fixation and permeabilization conditions for EM immunogold labeling of the fibers. We find that a simple fixation with depolymerized paraformaldehyde alone, followed by permeabilization with 0.01% saponin, offers the best compromise between the conflicting demands of unhindered tissue penetration and morphology preservation.

Key Words: Skeletal muscle, soleus, immunocytochemistry, immunogold, nanogold, permeabilization, pre-embedding, glucose transport, GLUT4, trafficking.

*Address for correspondence:
Evelyn Ralston
Laboratory of Neurobiology, Bldg 36, Rm 2A-21,
NINDS, NIH, Bethesda MD 20892-4062
Telephone number: (301) 496-6164
FAX number: (301) 480-1485
E-mail: esr@codon.nih.gov

Introduction

Studies of protein localization and trafficking by immunocytochemistry require optimal preservation of cell morphology and access to all intracellular membrane systems by ligands such as antibodies. The two requirements are often in conflict, since fixatives optimal for morphological preservation may impede penetration of reagents, while permeabilizing agents may compromise the morphology. They are even more difficult to satisfy for large cells, such as skeletal muscle fibers.

Tissue sections are frequently used as a solution to the morphology vs. penetration dilemma. Cryostat sections of muscle, for light microscopy (LM), are simple to prepare and to stain. Although sections are extremely useful, for example for screening antibodies, it may be difficult to convert the information they provide into a three-dimensional pattern, especially for markers that are very unevenly distributed. This is especially critical for muscle fibers which are organized in several domains, both across and along the fiber axis. Just below the plasma membrane lies a thin rim of cytoplasm containing the myonuclei and enriched in endoplasmic reticulum, Golgi complexes, and vesicles of the endosomal-lysosomal and other endocytotic and exocytotic pathways. Beneath this superficial layer of cytoplasm is the dense core of highly organized myofibrils which give striated muscle its characteristic appearance and name. A network of internal tubules, the T-tubules, cruise through both domains, penetrating the depth of the cell, while being continuous with the plasma membrane, and in communication with the extracellular space. Along the fiber, the neuromuscular junction (NMJ) and the myotendinous junctions constitute small but biologically very important differentiated domains. Because it occupies a large fraction of the volume of the fibers, the myofibrillar core of the muscle fibers is evident in both transverse and longitudinal muscle sections. In contrast, the layer of subsarcolemmal cytoplasm, thin and mostly perpendicular to the direction of sectioning, is under-represented.

Another problem with the use of muscle sections for localization work in light microscopy, is that the cellular origin of the staining is not always clear. Skeletal muscle includes fibroblasts, endothelial and smooth muscle cells, neurons and Schwann cells at the NMJ and in intramuscular nerve branches, as well as satellite cells within the basement membrane of muscle fibers themselves. In sections, thin processes of these cells may be so close to the muscle fibers that it is difficult to resolve them.

Better resolution can of course be achieved by electron microscopy (EM). With more awareness of the limitations of light microscopy for studying colocalization of proteins in subcellular compartments (Griffiths *et al.*, 1993), the practice of immunoelectron microscopy in trafficking and localization studies is increasing. Post-embedding staining of ultrathin cryosections (Tokayasu, 1986) is frequently used in such work (see, for example, Slot *et al.*, 1991). A large number of sections can be obtained from a small volume of tissue, the staining procedure is simple and rapid, there is no penetration problem, and double staining with secondary antibodies conjugated to colloidal gold particles of various sizes is possible. A major drawback is that large sections are technically difficult to obtain. In the case of skeletal muscle, the ultrathin cryosections give a very limited view of the cell and the sampling problem can become overwhelming. In addition, only the antigens exposed on the surface of sections are accessible to antibodies, thus limiting the sensitivity of the technique. This becomes a problem with less abundant proteins, such as receptors and transport proteins.

Recently, we have been confronted with these problems while studying the distribution of the glucose transporter GLUT4 (Birnbaum, 1992; Holman and Cushman, 1994) in skeletal muscle, and its redistribution in response to stimulation of the muscle by insulin and contractions (Ploug *et al.*, in preparation). Several membrane systems are involved: the plasma membrane, the T-tubules, the Golgi complexes, the endosomal pathway, and specialized intracellular vesicles subjected to regulated exocytosis. It is important to achieve unrestricted access to these membrane systems and the ideal preparation would allow to visualize them all simultaneously.

After comparing several preparative procedures for LM and EM immunocytochemistry, we conclude that single fibers, separated by mechanical teasing, in many respects offer the system best suited to such studies. We show that careful optimization of fixation and permeabilization conditions allow labeling of fibers in a pre-embedding mode that is satisfactory for both light and electron microscopy, and we give examples of applications of the technique.

Materials and Methods

Antibodies and reagents

P-1 is a polyclonal rabbit antibody raised against a synthetic peptide corresponding to the last 13 amino-acids of the C-terminal part of GLUT4 (Ploug *et al.*, 1990). P-1 was affinity-purified on immobilized peptide and used at a concentration corresponding to an approximately 1:1000 dilution of crude serum. The anti-dystrophin antibody (NCL-DYS2) was obtained from Novocastra (Newcastle, UK) and used at a 1:50 dilution. Anti-tubulin DM1α was given by Dr. Steve Doxsey (University of Massachusetts). Rhodamine-α-bungarotoxin was provided by Dr. Herman Gordon (University of Arizona). All fluorescent or biotinylated secondary antibodies and fluorescent streptavidin were from Vector Laboratories (Burlingame, CA). Nanogold-conjugated Fab fragments and silver enhancement kits were purchased from Nanoprobes Inc (Stony Brook, NY)

Cryostat sections

Male Wistar rats (300-350 g) were anesthetized by intraperitoneal injection of sodium pentobarbital (5 mg/100 g body weight). The soleus muscle was dissected out, cut into small pieces, each of which was mounted in a small drop of Tissue-Tek (Miles, Elkhardt, IN) and frozen by rapid immersion in a beaker with isopentane, cooled to near-freezing point in liquid nitrogen. 10 μm cryostat sections were placed on 8-well Teflon coated glass slides and fixed for 10 min in 2% formaldehyde in 0.1 M Sorensens phosphate buffer. Incubation with primary antibody was for 2 hours, with secondary biotinylated antibody for 1 hour, with fluorescein-conjugated streptavidin for 20 min and washes were for 3x5 min each. Sections were mounted in a drop of Vectashield (Vector Labs. Inc., Burlingame, CA).

Preparation of single FDB fibers by enzymatic dissociation

Fibers from the flexor digitorum brevis of rats were prepared according to the basic procedure of Bekoff and Betz (1977), and stained as described in Ralston (1993). Briefly, the muscle was incubated with collagenase A immediately after dissection. The fibers were released by trituration into Pasteur pipettes and plated onto glass coverslips coated with Matrigel, then fixed for staining. The staining procedure was similar to that described below for soleus fibers, except that shorter incubations and washes were used.

Teased fibers preparation and staining for fluorescence and electron microscopy

Male Wistar rats (300-350 g) were anesthetized by an intraperitoneal injection of sodium pentobarbital (5 mg/100 g body weight). A canula was then placed in the

Table 1: Evaluation of various fixation and permeabilization conditions for GLUT4 staining of teased, single rat muscle fibers

A detailed description of the procedure is in the Methods section. Briefly, the muscle was first fixed by perfusion for 10 min, then fixed by immersion for a additional 1 to 3 hours, with the fixative and at the temperature indicated in the first column (rt = room temperature). The detergent used for permeabilization was present throughout the staining. The quality of the morphology was judged on the basis of the following criteria: mitochondria with well-delineated cristae, sharp membranes, clear vesicles, Golgi complexes with stacks of cisternae, absence of holes in the cytoplasm.

Fixation	Permeabilization	LM intensity[1]	EM intensity	EM intensity
2% formaldehyde at rt	0.25-0.5% TX100	+	-	-
	0.25-0.5% saponin	++	-	-
(2% formaldehyde + 0.1-0.5% glutaraldehyde) at rt	0.1% saponin	0	0	+(+)
	50% ethanol in H_2O for 1h	+(+)	+	+(+)
	50% acetone in H_2O for 1h	+(+)	+	+(+)
PLP[2] at rt	0.1% saponin	+	0	+
2% formaldehyde + 0.2% picric acid at rt	0.1% saponin	++	+	+
2% formaldehyde at 4°C	0.1% saponin	+++	++	++
	0.01% saponin	+++	+++	+++

[1] intensity of GLUT4 staining; 0: no staining; + to +++ indicate increasing level of staining/quality; -: not done.
[2] paraformaldehyde-lysine-periodate (McLean and Nakane, 1974)

---

descending aorta and the inferior vena cava was cut open. The hindquarter was perfused with 150 ml of room temperature Krebs-Henseleit bicarbonate buffer containing procaine hydrochloride (1 g/l) for 1.5 min and then with 900 ml 2% freshly depolymerized paraformaldehyde in 0.1 M Sorensens phosphate buffer at 2-6°C for 9 min. The soleus muscle was excised and fixed at 4°C for an additional 12 hours in the perfusion fixative and then transferred to PBS (phosphate buffered saline; 10 mM sodium phosphate (pH 7.3), 150 mM NaCl). Under a stereomicroscope, bundles of fragments of 1-3 individual muscle fibers were teased with fine forceps, and transferred to 50 mM glycine in PBS in a 24-well tissue culture plate. Further blocking of non-specific binding sites was done by incubating with IMB (Immuno buffer: 50 mM glycine, 0.2% bovine serum albumin, 1.5 mM AEBSF (4-(2-aminoethyl)-benzene-sulfonyl fluoride, hydrochloride; Boehringer Mannheim, Indianapolis, IN), and 0.05% sodium azide in PBS) containing 0.01% saponin for 30 min. Primary antibodies were diluted in the same buffer supplemented with 200 μg/ml of goat IgG and fibers were incubated on an orbital shaker for 12-16 hours. Fibers were washed for 3x30 min with blocking buffer followed by a 2 hour incubation with appropriate biotinylated secondary antibody (Vector Laboratories, Burlingame, CA) diluted 1:1000 in blocking buffer. After 3x30 min washes in blocking buffer and 2x10 min washes with PBS, fibers were incubated for 20 min with FITC (fluorescein isothiocyanate) labeled streptavidin (2μg/ml, Pierce, Rockford, IL) and Hoechst 33242 (1 μg/ml) in PBS and finally mounted in Vectashield (Vector Laboratories) on glass slides. For mounting, bundles of fibers were placed in the center of a glass slide in a drop of Vectashield. Under a stereomicroscope, individual fibers were then grasped with fine forceps and aligned parallel to the longitudinal axis of the slide with 10-20 fibers in each of two columns. Fibers were examined and photographed with a Leitz DMRD microscope.

Usually, fibers were processed in parallel for EM. In this case, the secondary antibody incubation was for 2 hours with goat anti-rabbit Fab fragments conjugated to 1.4 nm gold (Nanoprobes Inc., Stony Brook, NY) diluted 1:200 in blocking buffer. Fibers were washed

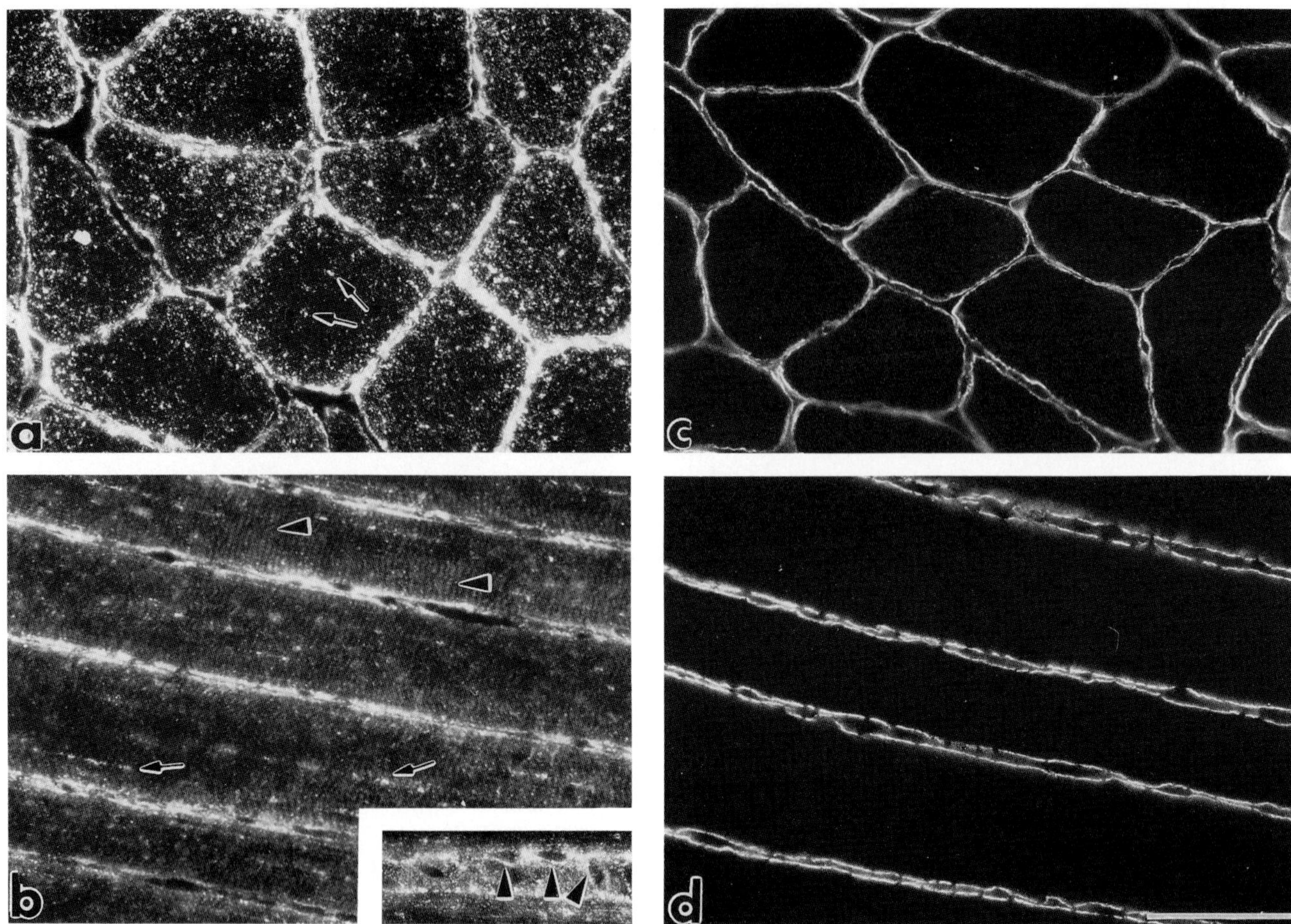

Figure 1: Immunofluorescent staining of transverse (a, c) and longitudinal (b, d) cryostat sections of rat soleus muscle with antibodies to GLUT4 (a-b) and dystrophin (c-d), followed by biotinylated secondary antibodies and fluorescein-conjugated streptavidin. GLUT4 staining is concentrated along the periphery of the fibers, as well as in rows of punctate staining in the core of the fibers (small arrows) and in a faint striated pattern (arrowheads, in b). The insert, in b, shows a fragment of a fiber in another area of the same section, where the nuclei (arrowheads) have been exposed en face by the cut. A perinuclear staining can be seen. Dystrophin only stains the plasma membrane. Size bar: 50 μm.

2x20 min in blocking buffer, 2x15 min in PBS, and then fixed for 1 hour at room temperature in 2.2% glutaraldehyde in 0.1 M phosphate buffer, pH 7.3. After 3x10 minutes washing in water, silver enhancement (HQ Silver, Nanoprobes Inc.) was done for 4-5 minutes (or 7-8 minutes for light microscopy), followed by 3x10 minutes washing in water and storage overnight in 0.1 M phosphate buffer at 4°C. Fibers were treated with 1% osmium tetroxide in 0.1 M phosphate buffer for 20 min, rinsed once in buffer and twice in 50% acetone, en block mordanted for 15-20 min in 2% uranyl acetate in 50% acetone, dehydrated in a graded acetone series, infiltrated with epoxy resin Polybed 812 (Polysciences Inc., Warrington, PA), cut down to a length of about 3 mm, horizontally embedded in flat molds and cured for 2-3 days at 50°C. Silver-gold interference colored sections were cut en face, stained for 5 min with aqueous lead citrate and examined in a JEOL 1200 microscope at 60 kV.

Confocal Microscopy

Confocal images were collected on a laser scanning confocal microscope equipped with a krypton/argon mixed gas laser (model MRC-600; Bio-Rad Laboratories, Richmond, CA), using a 63x Zeiss lens. Enhancement of the images for presentation was done in Photoshop 3.0 (Adobe Systems, Mountain View, CA).

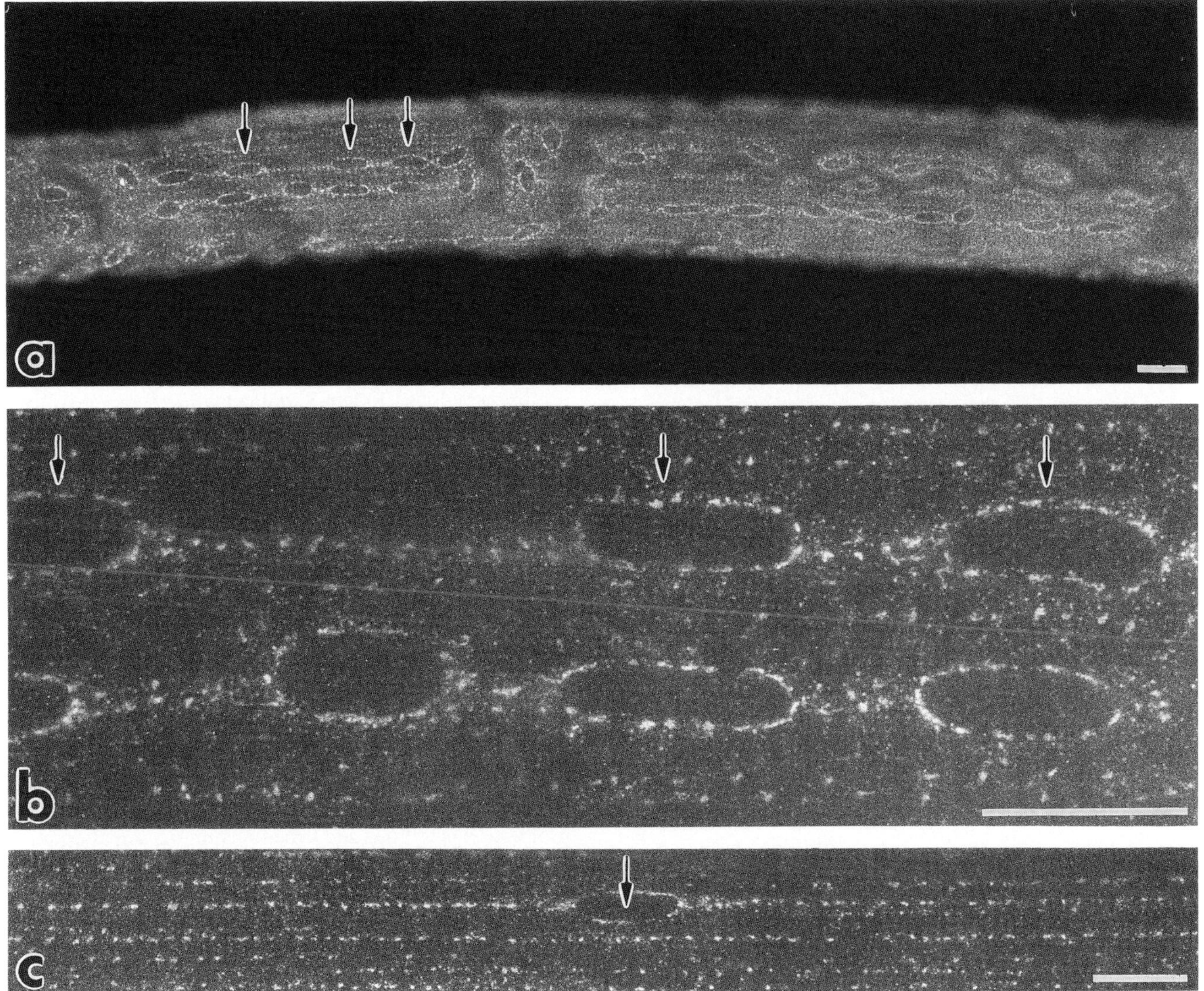

Figure 2: Staining of single teased rat soleus fibers with anti-GLUT4 antibodies. At low magnification (a), all nuclei, some of which are marked by a small arrow, are highlighted. Dark channels in transverse or longitudinal position reflect the presence of blood vessels on the surface of the muscle. At higher magnification (b), details of the staining show up. Punctate staining surrounds the nuclei and extends between them in rows. The nuclei shown by the small arrows are those shown in a. The rows of punctate staining can be very long as shown in c, where only one nucleus is present (small arrow). Size bars: 20 μm.

Results

Light microscopy observation of immunostained muscle sections and muscle fibers

Fig. 1 shows a transverse (a) and longitudinal (b) section of soleus muscle, stained with the anti-GLUT4 antibody, P-1. We expect GLUT4 to be nearly entirely sequestered in intracellular vesicles in these basal fibers from muscles neither stimulated with insulin nor with contractions (Bornemann *et al.*, 1992; Marette *et al.*, 1992). The transverse section shows that staining for GLUT4 appears to be concentrated at the edge of the fibers, on or just beneath the plasma membrane. There is also a punctate staining inside the fibers (small arrows). In longitudinal sections, GLUT4 shows staining on or beneath the plasma membrane, and, occasionally, in long rows inside the fibers (small arrows), that probably correspond to the punctate pattern seen in transverse sections. There is also a fine punctate staining

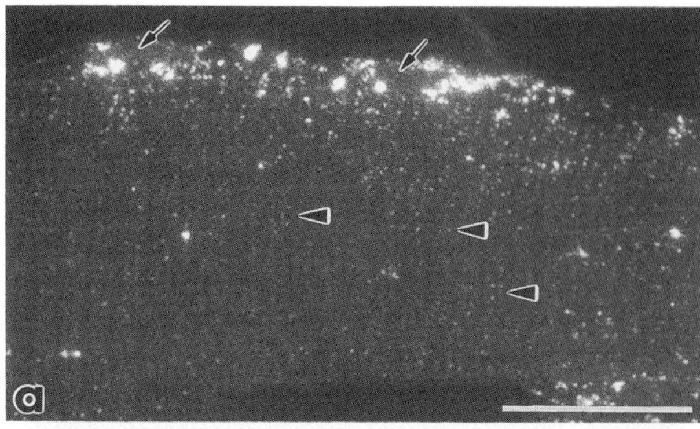

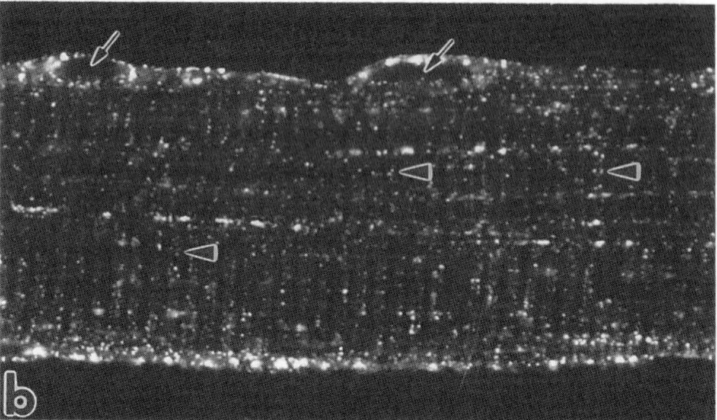

Figure 3: Comparison between semi-thin cryosection of soleus (a) and confocal optical section of a soleus fiber (b), both stained for GLUT4. A strong punctate staining surrounds the nuclei which are indicated by small arrows. There is also a weaker punctate staining throughout the fiber thickness (arrowheads). In (b), the confocal image is taken in a plane about 10-15 μm inside the fiber. Size bar: 20 μm.

----

that forms a striated pattern (arrowheads). Occasionally, the section plane coincides with the surface of the fiber and a perinuclear staining can be observed (arrowheads in the insert). In parallel, fibers were stained with an antibody to the protein dystrophin (Fig. 1 c and d), which is known to be localized just beneath the sarcolemma (Squarzoni *et al.*, 1992). The staining for dystrophin appears much sharper than that for GLUT4, suggesting that GLUT4, in contrast to dystrophin, is spread out in a relatively wide band of cytoplasm below the plasma membrane. We can thus conclude that GLUT4 is concentrated near the plasma membrane, but the resolution of the staining is very limited in these sections at the light microscopy level.

In contrast, when single teased fibers are stained and observed, the appearance of the staining is very different (Fig. 2). At low magnification, each nucleus of the fiber stands out, due to a perinuclear staining (Fig. 2a). At higher magnification, this perinuclear staining appears punctate (Fig. 2b). We also observe rows of dots that often course very long distances between the nuclei (Fig. 2c). The cross-striated staining is barely visible in these views that are focused on the surface of the fiber. Inside the fibers (not shown), the striated pattern can be seen, together with occasional longitudinal rows of stronger punctate staining that correspond to the rows observed on longitudinal cryostat sections (Fig. 1b).

Differences between staining patterns of whole-mount fibers and sections might be due to insufficient permeabilization of the former. To verify that antibodies are in fact able to penetrate fibers that have been permeabilized with a low concentration of saponin (0.01%), we compare the staining of a semi-thin cryosection of soleus (500 nm), stained without permeabilization, to that of an optical section of a fiber, obtained with a confocal microscope (Fig. 3). Both images show some perinuclear staining (small arrows point to the nuclei) and finer punctate staining throughout the fiber (arrowheads). It is clear that the penetration through the whole fiber is very good. This optical section is typical of a depth of 10-15 μm below the surface of the fiber. Closer to the belly of the largest fibers (up to 50 μm in diameter), staining decreases, though it remains observable throughout the fiber. These results show that the thin superficial layer of cytoplasm (1 to 2 μm deep) just beneath the plasma membrane, and several microns of the myofibril core, can be stained without restrictions.

One disadvantage of the teased fibers is that their production is tedious and time-consuming. An alternative is to prepare single fibers by enzymatic dissociation (Bekoff and Betz, 1977). In this preparation, the muscle is not fixed by perfusion but the single fibers are fixed after plating onto coverslips. Rat flexor digitorum brevis (FDB) fibers are prepared and stained for various markers. FDB fibers are very short and whole fibers can be observed without changing field in the microscope. Fig. 4 shows a rat FDB fiber stained for tubulin and observed with a confocal microscope at two different depths levels, at the surface (a) and deeper inside the fiber (b). The staining shows that microtubules surround nuclei (small arrows) and form a longitudinal and transverse (arrowheads) network in the fibers. The continuity of the staining over long distances can be observed easily. This pattern suggests a long-range cytoskeletal organization of the superficial layer of cytoplasm. A comparison with GLUT4 staining suggests that GLUT4 vesicles may be aligned along microtubules.

Pre-embedding EM immunogold staining of muscle fibers

Because we are interested in a detailed quantitative and qualitative analysis of GLUT4 distribution, the staining also needs to be observed and interpreted at the

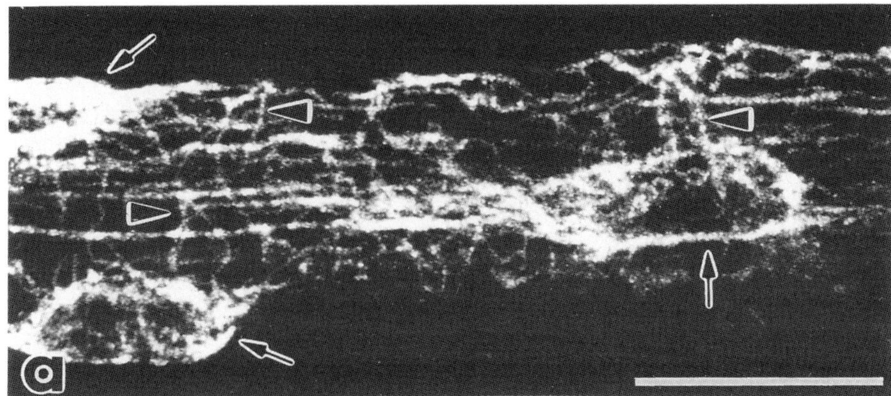

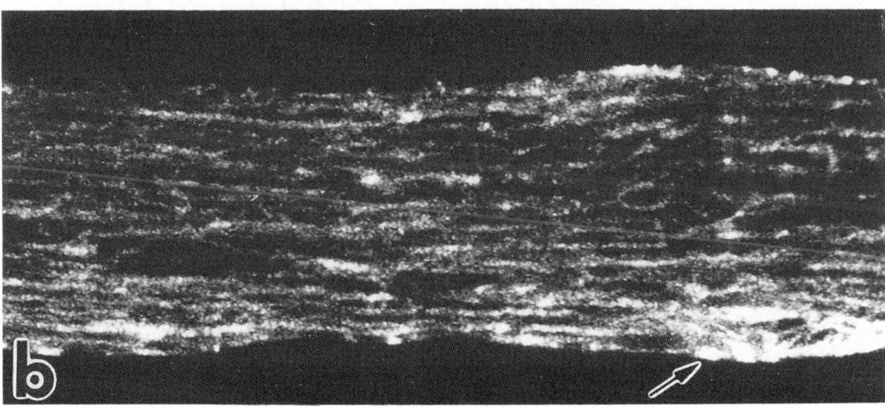

Figure 4: Rat FDB fiber, prepared by enzymatic dissociation, stained with anti-tubulin, and observed with a confocal microscope: (a) image recorded at the surface; (b) image recorded within the fiber. Tubulin surrounds the nuclei (small arrows) and forms a longitudinal network (a and b). At the surface, it also forms a transverse network (arrowheads). Size bar: 20 μm.

electron microscopy level. We reasoned that the staining of the whole fibers in pre-embedding mode could be extended to electron microscopy. Soleus fibers are stained with anti-GLUT4 and subsequently by a Fab fragment covalently coupled to a 1.4 nm gold cluster, followed by silver enhancement. With silver enhancement for 7-8 min, the staining can be observed in bright -field light microscopy (Fig. 5). The pattern is entirely comparable to that observed in immunofluorescence (Fig. 2).

To decide on the best possible conditions of fixation and permeabilization of the fibers, an optimization study was carried out, that took advantage of parallel processing for LM and EM. Combinations of fixatives and permeabilization methods (see Table 1) were first tested in LM to observe the overall degree of staining. For conditions that appeared satisfying, the fibers were then embedded, sectioned, and viewed in the electron microscope. As a standard to assess the general morphology of the samples, we examined sections from muscle fixed by perfusion with 2.5% glutaraldehyde, and processed without teasing the fibers apart. Among the various combinations tested, we found that fixation with cold formaldehyde alone and permeabilization with 0.01% saponin give the best results as judged by the degree of staining and the quality of the morphology. Fig. 6 shows electron micrographs of GLUT4-stained soleus fibers. The first panel (a) demonstrates that the plasma membrane is devoid of staining. A high density of gold/silver grains is found in vesicles close to the nucleus and in the trans-Golgi area. Inside the fiber (b), mitochondria and elements of sarcoplasmic reticulum, some of which are labeled, are aligned between bundles of myofibrils. T-tubules are especially noticeable at triad junctions, shown in (b) and (c), and are unlabeled in this unstimulated muscle.

The level of staining with the pre-embedding technique is much higher than that obtained with ultrathin cryosections (Ploug *et al.*, in preparation). Another attractive feature of the technique is that very long stretches of fibers can be followed in each section.

Junctional vs. extra-junctional staining in muscle fibers

There is great interest in proteins concentrated at the NMJ, since they may be involved in the formation and maintenance of the nerve-muscle synapse (Hall and Sanes, 1993). Therefore, we were intrigued to see GLUT4 concentrated at the NMJ, in cryostat sections double-stained with the anti-GLUT4 antibody and with rhodamine-α-bungarotoxin, a ligand of the acetylcholine receptor clustered at the NMJ (not shown). However, since GLUT4 is concentrated around myonuclei, increased staining at the NMJ may just reflect the cluster-

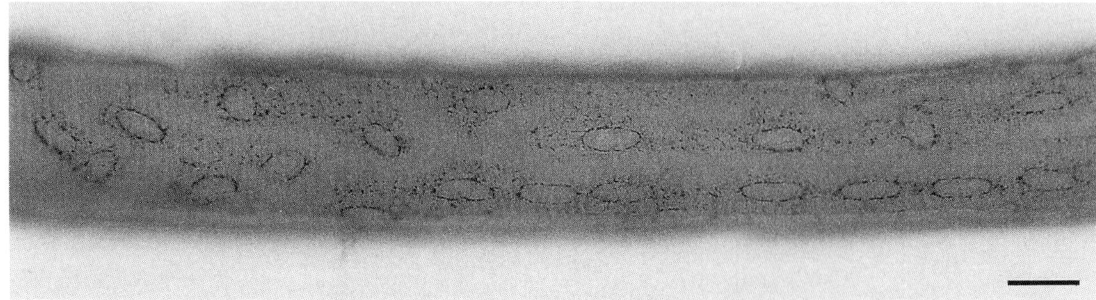

Figure 5: Single fiber stained with anti-GLUT4, followed by a nanogold-conjugated Fab fragment and extended silver enhancement, and photographed in bright-field optics. The pattern compares well with that observed in Fig. 2 by immunofluorescence. Size bar: 50 μm.

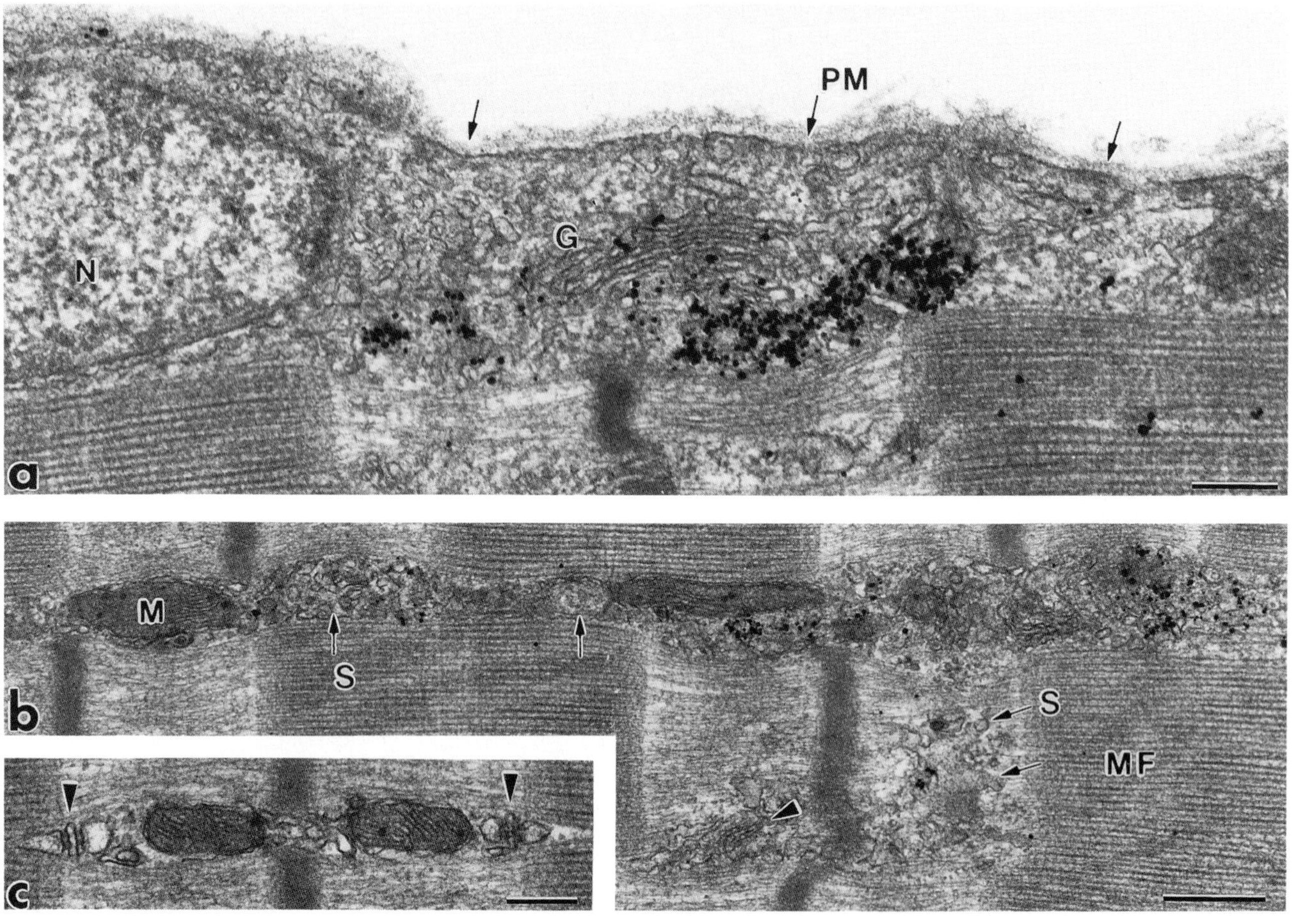

Figure 6: Electron micrograph of a single fiber stained with anti-GLUT4, followed by a nanogold-conjugated Fab and short silver enhancement: (a): close to the plasma membrane (PM), which is unlabeled, heavy gold labeling of the trans region of a Golgi complex (G) located near a nucleus (N); (b): deeper in the fiber, sarcoplasmic reticulum (S and small arrows) is specifically labeled but junctional T-tubules (arrowhead) are not. Some staining over the myofibrils (MF) may result from staining of underlying SR or from low level of background staining, though non-specific staining of nuclei and mitochondria is very low; (c): detail of (b) showing unstained triad junctions (arrowheads), at the border between the A and I bands; M: mitochondria. Size bars: 200 nm (a); 500 nm (b); 200 nm (c).

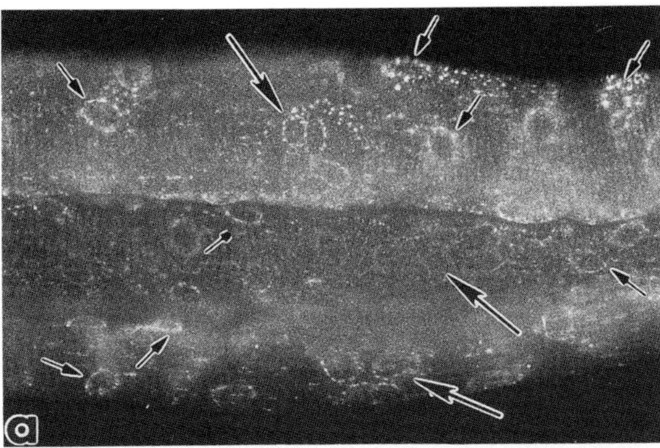

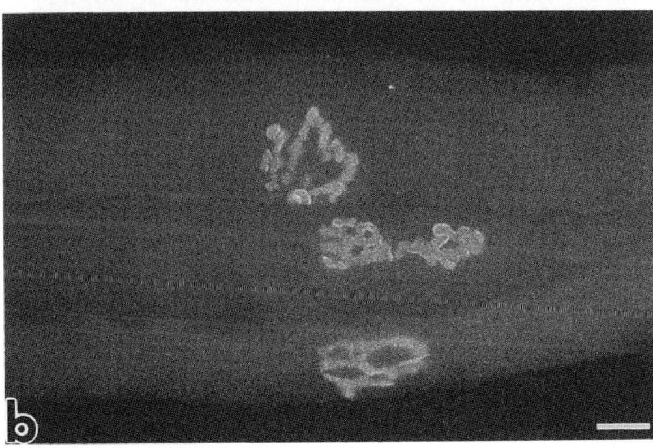

Figure 7: A bundle of three soleus fibers double-stained with anti-GLUT4 (a) and rhodamine-α-bungarotoxin (b). Junctional nuclei (large arrows) are clustered but do not show more GLUT4 staining than extra-junctional nuclei (small arrows). Size bar: 20 μm.

----

ing of the several endplate nuclei at the NMJ. In contrast, extra-junctional nuclei are generally single and more or less evenly spaced along the fibers.

When whole fibers are stained with anti-GLUT4 and rhodamine-α-bungarotoxin, the NMJ nuclei are easily detectable (Fig. 7). The level of staining around these nuclei appears no different from that around the extra-junctional nuclei in the same fiber. In muscle fibers processed for EM, the NMJ can be recognized by the pre-synaptic elements which often remain attached to the muscle fibers during teasing, and by the characteristic folds of the post-synaptic muscle membrane (Fig. 8a). GLUT4 staining at the NMJ is mostly seen in the trans Golgi area (Fig. 8b) and does not appear more intense than elsewhere. These data suggest that GLUT4 concentration at the NMJ does not result from increased local transcription or translation of the GLUT4 gene, in contrast to the acetylcholine receptor and a number of other genes (Hall and Sanes, 1993).

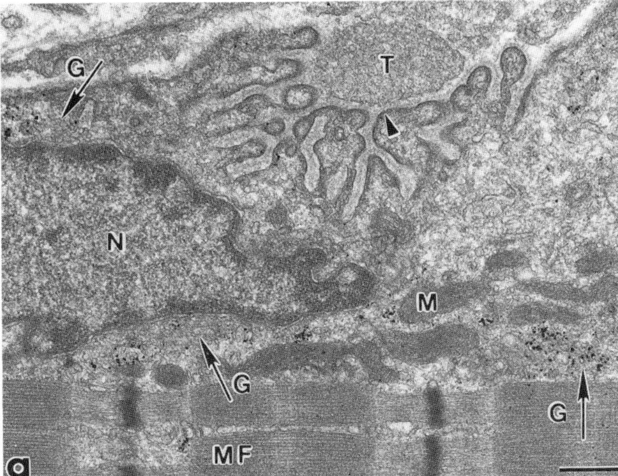

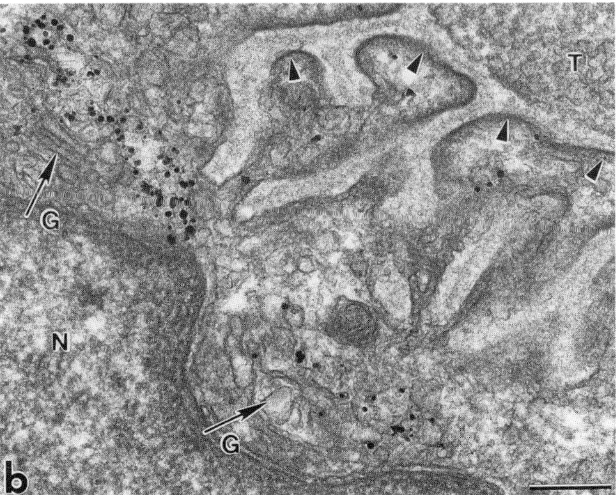

Figure 8: EM view of the NMJ in a soleus fiber immunolabeled for GLUT4. (a): the characteristic shape of the NMJ is easily recognized at low magnification, with folds of the muscle membrane (arrowhead) close to a junctional nucleus (N), surrounding a nerve terminal (T). GLUT4 is mostly in the trans region of the Golgi complexes (G). M: mitochondria; MF: myofibrils.(b): enlarged fragment of a neighboring section of (a) showing GLUT4 labeling in the Golgi complex (G) between the nucleus (N) and the muscle membrane folds (arrowheads) near the nerve terminal (T). The layer of extracellular matrix present between pre- and post-synaptic membranes and in the post-synaptic membrane folds can be seen. Size bars: 500 nm (a); 200 nm (b).

Discussion

This paper demonstrates that whole muscle fibers are preferable in many respects to sections, whether thin or thick, for studies of subcellular organization in muscle; and that fixation and permeabilization conditions can be optimized to allow penetration of antibodies throughout the muscle fibers, while maintaining a satisfactory morphology.

We believe the first point is well illustrated by a comparison of the information provided by the sections shown in Fig. 1 a-b and the whole fibers shown in Fig. 2. The latter demonstrate much more strikingly the perinuclear pattern of GLUT4, which corresponds to staining in the trans part of the Golgi complex, or in the trans Golgi network (Ploug *et al.*, in preparation). The elusiveness of this pattern in sections results from its spatial geometry. From many observations on both conventional and confocal microscopes, we know that the perinuclear staining and the long rows observed in Fig. 2, b-c, are in the thin rim of cytoplasm, just below the plasma membrane. If we assume, for simplicity, that the fibers are cylindrical, then this cytoplasmic staining covers the surface of a cylinder. No single planar section through this cylinder will encounter more than a few elements of the staining. Since the fibers, in reality, are more polygonal than cylindrical (see Fig. 1), occasional sections will expose the staining pattern that is routinely seen in whole mounts (see insert in Fig. 1b). From our experience, however, this is rare.

Similarly, the simultaneous observation of junctional and extra-junctional areas in whole fibers (Figs. 7 and 8), explains the apparent enrichment of GLUT4 at the NMJ observed in transverse muscle sections. In this as in the previous example, results from sections appear misleading.

Although teased fiber preparations still contain non-muscle cells, for example in fragments of blood vessels (not shown) or at the NMJ (Fig. 8), non-muscle staining can be resolved from staining of the muscle fibers, because the spatial organization of the cellular structures can be observed in these whole-mounts.

At the EM level as well, at least in our hands, pre-embedding labeling, which allows penetration of the antibodies throughout the fiber, gives a much stronger labeling than that obtained with ultrathin cryosections, even if we use the same combination of secondary antibody and of silver enhancement (Ploug *et al.*, in preparation). This is not totally unexpected since labeling of sections is limited to epitopes exposed on the surface. The size of the secondary antibody and of the gold is critical : we did not observe any staining in muscle fibers when we used a 5 nm gold-conjugated secondary antibody. By using a Fab fragment with a small gold cluster (Hainfeld, 1987; Hainfeld and Furuya, 1992), instead of a complete IgG adsorbed to colloidal gold, we obtain much better penetration.

The choice and concentration of detergent used for permeabilization is important. We find a better retention of labeling and of cytoplasmic vesicles with saponin than with Triton X-100 and with lower detergent concentrations (0.01-0.02% compared to 0.1-0.5%; see Table). The low degree of staining when Triton or higher concentrations of saponin are used, is probably due to extraction, which is not prevented by the relatively light fixation (Goldenthal *et al.*, 1985). There may be some interaction between saponin and serum since we find that saponin concentration has to be increased to 0.02% if the blocking/incubation buffer contains normal goat serum (3%). Generally we use purified goat IgG instead. Other techniques used for permeabilization, for example treatment with 50% ethanol or acetone in water, also work, at least to some extent. Ethanol permeabilization has been successfully employed in studies of CNS vibratome sections (Llewellyn-Smith and Minson, 1992). In fact, ethanol and acetone allow penetration of antibodies even after fixatives containing up to 0.5% glutaraldehyde. In contrast, even 0.1% saponin does not allow penetration in these conditions. However, the combination of saponin and cold formaldehyde gives better overall results than combinations that include some glutaraldehyde.

It is worth mentioning that some muscle studies bypass the problem of penetration by using skinned muscle fibers, generally from the rabbit psoas, prepared by mechanical (Wang and Wright, 1988) or chemical (Brenner, 1993) skinning of the sarcolemma. These preparations, useful as they may be, cannot be used for studies of trafficking from and to the plasma membrane itself.

Enzymatically dissociated muscle fibers represent an interesting alternative to teased fibers. The rapidity and ease of obtaining many fibers is remarkable and the fibers can be maintained alive and cultured for several days. However, in its original formulation (Bekoff and Betz, 1977), this technique appears solely suited to the preparation of fibers from the flexor digitorum brevis muscle of rodents , a multi-pennate muscle with very short fibers, predominantly of type IIa. Recently, however, a modified protocol has been proposed that extends enzymatic dissociation to other muscles as well (Rosenblatt *et al.*, 1995). It is possible, then, that the preparation of whole fibers will become easier, making it definitely the tool of choice in many cell biological studies.

Acknowledgements

We thank the colleagues who generously provided us with antibodies or reagents. The EM work greatly benefited from the expert assistance and support of the staff of the NINDS EM Facility, Ms. Virginia Tanner and Dr. Jung-Hwa Tao-Cheng, who also helped with constructive criticism of the manuscript. We thank Drs. T. S. Reese and S. W. Cushman for support through this work. T. Ploug was supported by the Danish National Research Foundation and by a fellowship from the Weimann Foundation, Denmark. We acknowledge a NATO Collaborative Research Grant.

References

Bekoff A, Betz WJ (1977) Physiological properties of dissociated muscle fibres obtained from innervated and denervated adult rat muscle. J Physiol **271**: 25-40.

Birnbaum MJ (1992) The insulin-sensitive glucose transporter. Int Rev Cyto **137A**: 239-297.

Bornemann A, Ploug T, Schmalbruch H (1992) Subcellular localization of GLUT4 in nonstimulated and insulin-stimulated soleus muscle of rat. Diabetes **41**: 215-221.

Brenner B (1993) Technique for stabilizing the striation pattern in maximally calcium-activated skinned rabbit psoas fibers. Biophys J **41**: 99-102.

Goldenthal KL, Hedman K, Chen JW, August JT, Willingham MC (1985) Postfixation detergent treatment for immunofluorescence suppresses localization of some integral membrane proteins. J Histochem Cytochem **33**: 813-820.

Griffiths G, Parton RG, Lucocq J, van Deurs B, Brown D, Slot JW, Geuze HJ (1993) The immunofluorescent era of membrane traffic. Trends Cell Biol **3**: 214-219.

Hainfeld JF (1987) A small gold-conjugated antibody label: improved resolution for electron microscopy. Science **236**: 450-453.

Hainfeld JF, Furuya FR (1992) A 1.4-nm gold cluster covalently attached to antibodies improves immunolabeling. J Histochem Cytochem **40**: 177-184.

Hall ZW, Sanes JR (1993) Synaptic structure and development: the neuromuscular junction. Cell suppl. **72**: 99-121.

Holman GD, Cushman SW (1994) Subcellular localization and trafficking of the GLUT4 glucose transporter isoform in insulin-responsive cells. BioEssays **16**: 753-759.

Llewellyn-Smith IJ, Minson JB (1992) Complete penetration of antibodies into vibratome sections after glutaraldehyde fixation and ethanol treatment: light and electron microscopy for neuropeptides. J Histochem Cytochem **40**: 1741-1749.

Marette A, Richardson JM, Ramlal T, Balon TW, Vranic M, Pessin JE, Klip A (1992) Abundance, localization, and insulin-induced translocation of glucose transporters in red and white muscle. Am J Physiol **263**: C443-C452.

McLean IW, Nakane PK (1974) Periodate-lysine-paraformaldehyde fixative, a new fixative for immunoelectron microscopy. J Histochem Cytochem **22**: 1077-1083.

Ploug T, Stallknecht BM, Pedersen O, Kahn BB, Ohkuwa T, Vinten J, Galbo H (1990) Effect of endurance training on glucose transport capacity and glucose transporter expression in rat skeletal muscle. Am J Physiol **259**: E778-E786.

Ralston E (1993) Changes in architecture of the Golgi complex and other subcellular organelles during myogenesis. J Cell Biol **120**: 399-409.

Rosenblatt JD, Lunt AI, Parry DJ, Partridge TA (1995) Culturing satellite cells from living single muscle fiber explants. In Vitro Cell Dev Biol Anim **31**: 773-779.

Slot JW, Geuze HJ, Gigengack S, Lienhard GE, James DE (1991) Immuno-localization of the insulin regulatable glucose transporter in brown adipose tissue of the rat. J Cell Biol **113**: 123-135.

Squarzoni S, Sabatelli P, Maltarello MC, Cataldi A, Di Primio A, Maraldi NM (1992) Localization of Dystrophin COOH-terminal domain by the fracture-label technique. J Cell Biol **118**: 1401-1409.

Tokuyasu KT (1986) Application of cryoultramicrotomy to immunocytochemistry. J Microsc **143**: 139-149.

Wang K, Wright J (1988) Architecture of the sarcomere matrix of skeletal muscle: immunoelectron microscopic evidence that suggests a set of parallel inextensible nebulin filaments anchored at the Z line. J Cell Biol **107**: 2199-2212.

Discussion with Reviewers

N. Jones: How adaptable are these techniques with other tissue types?

Authors: These techniques have also been used with cells in culture: differentiated PC12 cells (Tanner *et al.*, 1996) and myotubes of the skeletal mouse muscle cell line C2 (Ralston and Ploug, 1996). For each, some optimization of the fixation/ incubation conditions was necessary.

The limitations are that for each tissue there is a minimum degree of fixation below which the morphology is not acceptable for electron microscopy, and there is a maximum degree of fixation beyond which penetration of antibodies and/or antigenicity are lost. For some tissues and antibodies it may be difficult to optimize both at the same time. There are, however, many different fixatives and ways of permeabilizing tissues, and where the conventional ones fail, it may be well worth trying new ones. Then the limit shifts to one of time and effort required.

N. Jones: Why does the fixation step at 4°C versus

room temperature retain better antigenicity?

Authors: Fixation at lower temperatures prevents extraction of the lipids, which are not fixed by formaldehyde, and autolysis (see, for example, the discussion by M. A. Hayat in Principles and Techniques of Electron Microscopy, CRC Press, 1989). Again, the fixation temperature may be more or less critical depending on tissue and antigen.

We have not tried lower percentages of formaldehyde. Recently, we have used shorter antibody incubation times for muscle fibers, with success. For the thinner cells in cultures, we always use shorter incubation times (2 hours for primary antibody, one hour for secondary antibody).

N. Jones: You indicated that the superior results with the detergent saponin versus Triton X-100 may be due to greater extraction with Triton X-100. Have you biochemically determined that this is the case?

Authors: No, we have not done so. Indirect arguments supporting extraction rather than, for example, change in antigenicity of the epitopes, can be found in the paper by Goldenthal *et al.* (1985).

C. Franzini-Armstrong: I am concerned about an apparent discrepancy in the pattern of antibody labeling that is obtained in the whole fiber approach versus the cryosections. There are two clear differences between the images presented here for the two approaches: the images from cryosections totally fail to show the ring of antibody positive dots that so clearly surround the nuclei in the whole fibers. Similarly, the punctate nature of the "staining" is not visible in the cryosections. I do not believe that the geometrical argument used in the manuscript is correct. For example, the subsarcolemmal pattern of small antibody positive dots should be very clearly visible in cross sections such as those shown in Fig. 1a, provided that the sections are observed in a confocal microscope. Have the authors observed the cryosections under the same optical conditions as the whole fibers?

Authors: We do not believe there is any major qualitative discrepancy in the labeling obtained in the whole fiber approach compared to semithin cryosections. The advantage of the former approach is complete viewing access to staining patterns of structures localized in the thin subsarcolemmal cytoplasmic rim. In contrast, cryosections only render this view in rare areas where the section plane grazes the surface of a fiber. This is the essence of our "geometrical argument". The thin sheet-like character of the subsarcolemmal GLUT4 pattern can best be appreciated by projection and 3D reconstruction of serial confocal images, which are not shown here. An insert in Fig. 1 now shows a fragment from another area in the same section, in which the perinuclear staining appears. However, the staining is not very crisp, in part because these images are not confocal, and in part because the morphology of these 10 μm-thick cryostat sections of unfixed and non-cryoprotected tissue is generally less well preserved than that of semithin (500nm) cryosections of fixed and cryoprotected tissue. For semithin cryosections (Fig. 3a), the staining pattern appears similar to what is seen in optical sections of whole fibers (compare Fig. 3a and 3b), with the note of caution that, at least in our hands, the morphology of sections, even semithin and ultrathin, is still not as good as that of the whole fibers. Furthermore, it should be kept in mind that in well preserved semithin cryosections, only antigens close to the upper section surface will be labeled (for a discussion of labeling efficiency of sections see "Fine Structure Immuno-cytochemistry" by G. Griffiths (Springer-Verlag, 1993)). In contrast, the whole thickness of an optical section from a permeabilized fiber is stained and the confocal section shown in Fig. 3b may be thick compared to the stained depth of the cryosection.

Finally, it is worth pointing out that the observation of whole fibers, confocal or not, offers an enormous sampling advantage over the observation of sections. In the former case, the whole surface and interior of the fiber can be scanned, laterally and in-depth, in order to find areas that have the best orientation and the best examples of typical staining elements. Cryosections, in contrast, are small to start with, and, as mentioned earlier, their overall labeling efficiency is not as high as in the whole fiber approach.

Additional References

Ralston E, Ploug T (1996) GLUT4 in cultured skeletal myotubes is segregated from the transferrin receptor and stored in vesicles associated with the Golgi complex. J Cell Sci **109**: 2967-2978.

Tanner VA, Ploug T, Tao-Cheng J-H (1996) Subcellular localization of SV2 and other secretory vesicle components in PC12 cells by an efficient method of pre-embedding EM immunocytochemistry for cell cultures. J Histochem Cytochem **44**: 1481-1488.

Scanning Microscopy Supplement 10, 1996 (pages 261-271)
Scanning Microscopy International, Chicago (AMF O'Hare), IL 60666 USA
0892-953X/96$5.00+.25

MULTIPLE LABELING IN ELECTRON MICROSCOPY: ITS APPLICATION IN CARDIOVASCULAR RESEARCH

M.M.H. Marijianowski, P. Teeling, K.P. Dingemans, and A.E. Becker*

Department of Cardiovascular Pathology, Academic Medical Center, University of Amsterdam, Meibergdreef 9, 1105 AZ Amsterdam, The Netherlands

(Received for publication October 11, 1995, and in revised form January 19, 1996)

Abstract

The heart is a muscular pump kept together by a network of extracellular matrix components. An increase in collagens, as in chronic congestive heart failure (CHF), is thought to have a negative effect on cardiac compliance and, thus, on the clinical condition. Conventional electron microscopy allows for the study of cellular and extracellular components and scanning electron microscopy (SEM) can put these structures in three-dimensional perspective. However, in order to study extracellular matrix components in relation to cells, immunoelectron microscopy is superior. We have used this technique in our studies on heart failure. Heart specimens were fixed in 4% paraformaldehyde and 0.1% glutaraldehyde in sodium cacodylate buffer, dehydrated by the method of progressive lowering of temperature and embedded in LR Gold plastic. Immunolabeling could be achieved with different sized gold-conjugated secondary antibodies or protein-A gold conjugates. Depending on the objective, ultra small gold (USG) conjugates or a regular probe size can be used. Labeling efficiency could be increased by bridging antibodies. The double and triple staining procedures were based on single staining methods using one- and two-face labeling. The choice of antibodies and gold conjugates depended on the objectives. Immunoelectron microscopy, using multiple labeling, allowed a detailed study of the organization of the extracellular matrix and its relationship with cardiac myocytes. This may prove to be a useful tool for the study of chronic heart failure.

Key Words: Immunoelectron microscopy, double-labeling, triple-labeling, collagen type I, collagen type III.

*Address for correspondence:
Anton E. Becker
Department of Cardiovascular Pathology,
Academic Medical Center, University of Amsterdam,
Meibergdreef 9, 1105 AZ Amsterdam, The Netherlands
Telephone number: (+31) 20 5665646
FAX number: (+31) 20 6914738

Introduction

The extracellular matrix plays an important role in the regulation of tissue function and a close relation exists between the various proteins and cells. Collagens are a source of strength to the tissue, elastin and proteoglycans are essential to matrix resiliency, and the structural glycoproteins help to create tissue cohesiveness. The extracellular matrix of the heart is comprised of various components which are organized into a three-dimensional network (Caulfield and Borg, 1979; Borg and Caulfield, 1981; Borg *et al.*, 1982; Robinson *et al.*, 1983). The epimysium is defined as the sheath of connective tissue that surrounds the entire muscle; endomysium is the connective tissue that surrounds and interconnects cells; perimysium connects epimysium to endomysium and surrounds groups of myocytes. The form and distribution of the connective tissue of the heart is such that it may play an important role in the elastic properties of the ventricles (Borg *et al.*, 1981). The collagen network in particular is thought to play an important role in maintaining the integrity of the heart muscle. The collagen fibers provide a scaffolding that supports muscle cells and blood vessels. They act as lateral connections between cells and muscle bundles and by virtue of this architecture govern the delivery of force generated by the myocytes. In this setting the collagens determine myocardial stiffness (Weber, 1989). The major collagens are collagen type I and type III, which possess different physical properties: collagen type I provides rigidity whereas collagen type III contributes to elasticity. Hence, the ratio of these two types of collagen is important. In hearts of patients with chronic congestive heart failure (CHF) there is an increase in extracellular matrix components (Heneghan *et al.*, 1991; Schaper and Speiser, 1992; Yoshikane *et al.*, 1992; Marijianowski *et al.*, 1995). Moreover, the ratio between collagen type I and type III is changed in favor of collagen type I.

Light and conventional electron microscopy have established these observations. Scanning electron microscopy (SEM) has revealed the three-dimensional relation between matrix components, cardiac myocytes and connective tissue cells. However, the exact identification of components and cells is possible only by immunolabeling. Because of the functional implications a detailed study of the changes in extracellular matrix components in hearts of patients with chronic CHF may contribute to an understanding of the progressive and often irreversible nature of the disease.

This article discusses the technical aspects of multiple labeling techniques, with an emphasis on the techniques specifically applied in our studies on chronic heart failure.

Multiple Labeling Techniques in Immunoelectron Microscopy

Many monoclonal antibodies will not recognize an epitope in plastic-embedded sections because fixation and embedding may alter the tertiary structure of the epitope. In this case immunolabeling on cryosections is an option, although the morphology is less well preserved and it takes more effort to obtain good quality contrast. The present review will focus on plastic post-embedding procedures, instead of pre-embedding procedures, because the antigen detection using gold conjugates and gold-conjugated antibodies generally gives better results compared to detection with enzymes. The gold particle is smaller and more discrete than the enzyme precipitate in the tissue and, therefore, allows for a more precise localization. Furthermore, once the tissue is embedded in plastic the antigenicity is preserved well enough to use the tissue specimen for detection of several antigens. This contrasts with pre-embedding procedures that are done for the detection of one antigen only and do not allow further researches for other epitopes in the same tissue block.

Fixation, dehydration and embedding in plastic

The choice of fixation is dependent on the balance between optimal antigen detection and acceptable preservation of morphology (Craig and Goodchild, 1982). A fixation with 4% paraformaldehyde and 0.1% glutaraldehyde in sodium cacodylate buffer (pH 7.4) for about 1 hour at 4°C often meets both demands. The most propitious procedure to overcome loss of antigenicity is either embedding in plastic or processing for cryoultramicrotomy immediately after fixation. However, it is possible to store tissue specimens in 0.5% buffered paraformaldehyde at 4°C for several weeks without complete loss of antigenicity; a feature which depends on the nature of the antigen. Another way of storing the specimens is to immerse the tissue blocks in 2.3 M sucrose after washing them in 50 mM glycine in phosphate buffered saline (PBS) at 4°C. These tissue specimens are then used for cryoultramicrotomy or for freeze substitution.

Dehydration of the tissue samples is done either by using progressive lowering of temperature (PLT) or by freeze substitution. Ethanol, methanol, acetone, dimethylformamide or ethylene-glycol can be used as the dehydrating agents (Quintana, 1994). The PLT method is based on a lowering of the temperature during dehydration and a subsequent impregnation with the appropriate plastic (see below). Dehydration by freeze substitution requires cryoprotection of the tissue specimen by immersion in 2.3 M sucrose at 4°C after which the tissue specimens are frozen in liquid nitrogen and dehydrated in methanol containing 0.5% uranyl acetate at -80°C. The temperature is slowly raised to -45°C, after which the tissue is impregnated with the plastic.

Several plastics are available for immuno-electron microscopy, such as Lowicryl K4M, HM20, LR White, and LR Gold. Both Lowicryl K4M and HM20 are made of acrylate and methacrylate and are still liquid at very low temperatures. K4M is a polar (hydrophylic) plastic and can be used at -30°C. HM20 is apolar (hydrophobic) and can be used at -50°C. Polymerization of both plastics is carried out under ultraviolet light at -30°C and -50°C, respectively. LR Gold is a plastic based on acrylate with hydrophylic properties and stays liquid at -20°C. Polymerization is also done under ultraviolet light but at -20°C. The polymerization at low temperatures under ultraviolet light is followed by further polymerization at room temperature for 48 to 72 hours.

The plastics differ in cutting qualities, preservation of morphology and antigenicity (Herken *et al.*, 1988a; Herken *et al.*, 1988b; Kann and Fouquet, 1989; Benichou *et al.*, 1990; de Mesy Jensen and di Sant'Agnese, 1992; Hamilton *et al.*, 1992; Migheli *et al.*, 1992; Robertson *et al.*, 1992; Kasper and Migheli, 1993). The published results regarding LR White and LR Gold are contradictory. Most authors favor LR Gold, although LR White is usually preferred when membrane contrast is important (Taatjes *et al.*, 1994).

Sectioning

Tissue sections of 50 to 70 nm thickness are cut and collected on Formvar-coated 50 mesh copper grids. When the detection with the primary antibody is performed with ultra-small gold (USG) conjugates and visualization using silver enhancement, the tissue sections must be collected on Formvar-coated 50 mesh nickel grids. In case of multiple immunolabeling on both sides of the grid, uncoated (200-700 mesh hexagonal) grids are used.

Immunolabeling and detection

The following paragraphs are based on the procedures that we use. Only a general outline of the methods can be provided, since detailed instructions are highly

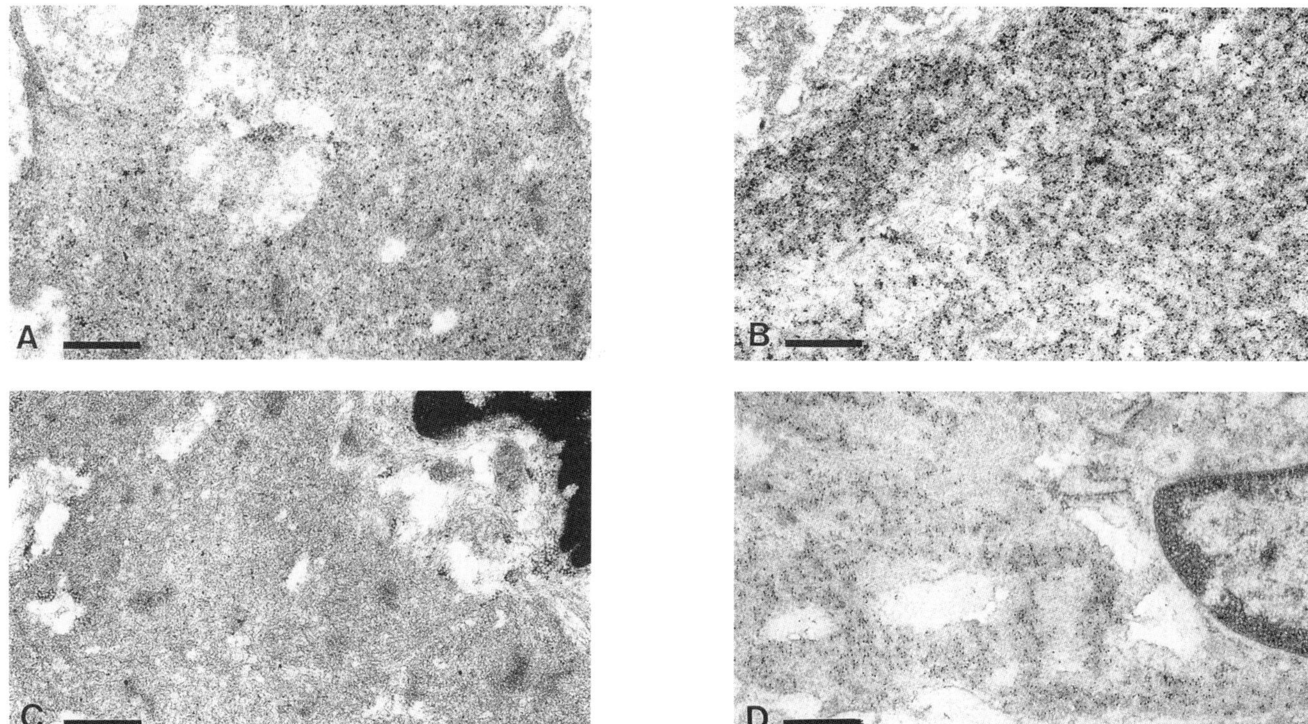

Figure 1: Immunolabeling for actin in smooth muscle cells from aortic tissue. **A.** Labeling with anti-actin antibody (clone 1A4) and protein-A/10 nm. **B.** Labeling with the same primary anti-actin antibody, a rabbit anti-mouse secondary antibody and finally protein-A/10 nm. **C** and **D.** The primary antibody is anti-actin (clone HHF35). **C** shows direct labeling of the primary antibody with protein-A/10 nm and **D** using a bridging antibody. Bar=0.5 μm.

dependent on variables like tissue and antigen characteristics, tissue fixation and the plastic used for embedding. We usually start by using single-immunostaining to identify the staining characteristics of the antigens to be studied. This allows for the proper sequence in the multiple staining procedures. This procedure also appears to be in line with that promoted by Bendayan *et al.* (1987), who stated that no single procedure can be recommended as *the* best approach in cytochemistry.

The detection of the primary antibody can be done with either gold-conjugated secondary antibodies, USG conjugates or protein-A or protein-G gold conjugates.

For the gold-conjugated secondary antibodies, several sizes of gold particles are available. USG conjugates have a probe size smaller than 1 nm, which comes so close to the resolution of the electron microscope that it starts to interfere with adequate imaging (Shimizu *et al.*, 1992). Enlargement of the probe (sizes larger than 40 nm are possible) with silver enhancement is therefore necessary. The use of USG conjugates is functional because of the high labeling efficiency of the small probe and a good overview at low magnification because of the large probe size after silver enhancement. The silver enhancement procedure according to Danscher (1981) gives reproducible results and the amplification time corresponds well with the total enlargement of the probe. Protein-A or -G gold conjugates react with the immunoglobulin (Ig)G portion of the primary antibody and have different affinities for several species (Bendayan, 1982; Roth, 1982; Silver and Hearn, 1987; Holm *et al.*, 1988; Roth, 1989; Roth *et al.*, 1989). Both have a high affinity for rabbit and guinea pig IgG. However, with antibodies raised in goats or mice and particularly with mouse monoclonal antibodies, protein-G gold conjugates yield intense and specific labeling, whereas protein-A gold conjugates yield more variable results (Bendayan and Garzon, 1988). In the case of multiple immunolabeling procedures, a brief treatment with 1% glutaraldehyde in PBS after the incubation with protein-A or protein-G is necessary to inhibit the antibody complex.

For all three options, i.e., protein-A, protein-G gold conjugates and gold-conjugated secondary antibodies, the size of the gold conjugate influences the labeling efficiency, because steric hindrance increases with probe size which in turn results in a decreased labeling (Yokota, 1988).

Table 1. Progressive lowering of temperature dehydration and embedding protocol.

		A*	B*	C*
1.	Cut tissue in blocks not larger than 2 mm^3			
2.	Dehydrate for 10 min. in ethanol 30%	4°C	0°C	0°C
3	Dehydrate for 10 min. in ethanol 50%	0°C	-10°C	-10°C
4.	Dehydrate for 30 min. in ethanol 70%	-10°C	-20°C	-20°C
5.	Dehydrate for 30 min. in ethanol 90%	-20°C	-30°C	-30°C
6.	Dehydrate for 30 min. in ethanol 100%	-20°C	-30°C	-50°C
7.	Dehydrate for 30 min. in ethanol 100%	-20°C	-30°C	-50°C
8.	Impregnate for 1 h. in ethanol:plastic 1:1	-20°C	-30°C	-50°C
9.	Impregnate for 1 h. i n ethanol:plastic 1:2	-20°C	-30°C	-50°C
10.	Impregnate in pure plastic overnight at appropriate temperature.			
11.	Replace plastic twice.			
12	Place tissue blocks in gelatine capsules, fill with appropriate plastic, seal and polymerize at appropriate temperature under UV light.**			

*A: LR Gold plastic
*B: Lowicryl K4M plastic
*C: Lowicryl HM20 plastic

**After polymerization under UV light at low temperatures the blocks are cured for 48 to 72 hours under UV light at room temperature.

Due to the (possible) toxicicity of the fumes and plastics themselves, one should take precautions while working with these plastics. Always wear gloves and work in a well ventilated room. If one performs the PLT in a Balzers TTP 010 cabinet, make sure it stands under a fume hood; when using a Balzers LTE 020 embedding unit make sure that the exhaust does not end in the room.

Detection of a single epitope

A high labeling efficiency and an easy detection is usually achieved using a small gold probe (for instance 10 or 15 nm).

Table 2. Single staining with gold conjugated secondary antibody

1.	Collect sections on Formvar coated grids.		
2.	Block free aldehyde groups with PBS/glycine 50 mM	15	min
3.	Wash grids with PBS/BSA-C (0.2%)	3×2	min
4.	Incubate with primary antibody in PBS/BSA-C (0.2%)	60	min
5.	Wash with PBS/BSA-C (0.2%)	5×2	min
6.	Incubate with gold conjugated secondary antibody*	30	min
7.	Wash with PBS	3×2	min
8.	Jetwash with distilled water and dry at room temperature.		
9.	Contrast lightly with uranyl acetate and lead citrate.		

*Gold-conjugated antibodies:

Goat anti-rabbit gold conjugate with sizes ranging from 5 to 20 nm.
Goat anti-mouse gold conjugate with sizes ranging from 5 to 20 nm.
Goat anti-rabbit or goat anti-mouse ultra-small (< 1 nm) gold conjugate.

or: Protein-A gold conjugate with sizes ranging from 5 to 20 nm.

All steps are performed on a clean sheath of Parafilm with the grids, section face downwards, on a 30-50 μl droplet. During the above labeling protocol the sections should not be allowed to dry since this may give rise to non-specific staining and contamination (dirt) of the contrasted section.

Protein-A binds strongly to rabbit, guinea pig and human IgG, with mouse IgGs it binds with different affinity.
If specific antibodies with weak binding capacity for protein-A are used (sheep, goat and some mouse IgGs) or when multiplication of the gold signal is desired, a bridging antibody (rabbit anti-goat or mouse or swine anti-rabbit) should be included. The incubation time is 30 minutes.

Detection of two epitopes

The mode of detection is dependent on the primary antibodies used. The detection of two epitopes with primary antibodies from different species is performed by diluting both antibodies in a mixture. The same is valid for the corresponding secondary antibodies (using different probe sizes). In this case the standard protocol for single-staining can be used.

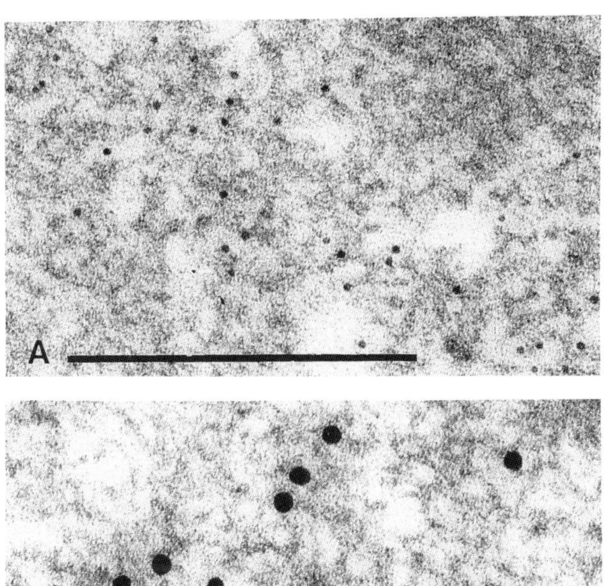

Figure 2: Immunolabeling of actin in aortic tissue using the same primary anti-elastin antibody (clone 1A4). The secondary labeling is a rabbit anti-goat antibody conjugated with 6 nm **(A)** and 12 nm **(B)** gold. Bar=0.5 μm.

The protocol for double staining is used when one of the gold conjugates is protein-A. Protein-A will react with the second primary antibody and therefore an inhibition step is necessary. The protocol for double staining is also used when the primary antibodies are from the same species. A one-face method can be performed using a combination of protein-A or protein-G and a gold-conjugated secondary antibody. The use of different sized gold conjugates of the same species is possible when a two-face method is used.

For the two-face immunolabeling method, sections are collected on uncoated 200-700 mesh hexagonal copper grids. A single staining is performed on either side of the grid, using two different-sized gold conjugates to detect the primary antibody. After labeling the grid is coated with Formvar on the side where the sections have been collected to prevent damage by the electron beam.

Another possibility is detection of the first epitope with USG and enhancement of the probe. The enlargement of the USG will mask the epitope-antibody complex. This method cannot be used when both epitopes are situated at a short distance from each other. The enlargement of the USG can cover both epitopes, or the detection of the second epitope is not possible because of steric hindrance. For this case, the two-face labeling method has to be used.

Table 3. Double staining with gold conjugated secondary antibody and primary antibodies from the same species

Step		Time
1.	Collect sections on Formvar-coated grids	
2.	Block free aldehyde groups with PBS/glycine 50 mM	15 min
3.	Wash with PBS/BSA-C (0.2%)	3×2 min
4.	Incubate with primary antibody diluted in PBS/BSA-C (0.2%)	60 min
5.	Wash with PBS/BSA-C (0.2%)	5×2 min
6.	Incubate with gold conjugated secondary antibody*	30 min
7.	Wash with PBS	3×2 min
8.	Fix/inhibit the antibody complex with 1% GA in PBS	10 min
9.	Wash with PBS	5 min
10.	Block free aldehyde groups with PBS/Glycine 50 mM	3×5 min
11.	Wash with PBS/BSA-C (0.2%)	3×2 min
12.	Incubate with second primary antibody diluted in PBS/BSA-C 0.2%	60 min
13.	Wash with PBS/BSA-C (0.2%)	5×2 min
14.	Incubate with different sized gold conjugated secondary antibody*	30 min
15.	Wash in PBS	3×2 min
16.	Jetwash with distilled water and dry at room temperature	
17.	Contrast lightly with uranyl acetate and lead citrate	

*Gold-conjugated antibodies (see Table 2).

Detection of three epitopes

In general, the protocol for the double staining can be used. The choice of either a one-face or a two-face detection method depends on the primary antibodies used.

Background reduction

The use of BSA-C (Aurion, Wageningen, The Netherlands) will prevent background. BSA-C is an acetylated and linearized form of bovine serum albumin and competes with the negative charge of the gold particles, thus preventing charge-determined background. Furthermore, the hydrophobic binding sites which are introduced in many plastics are blocked.

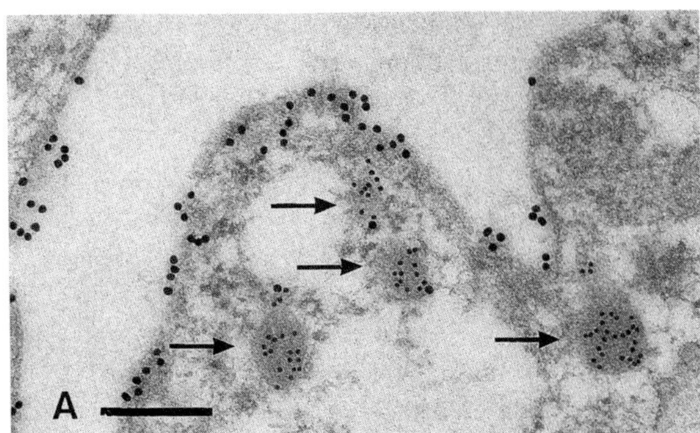

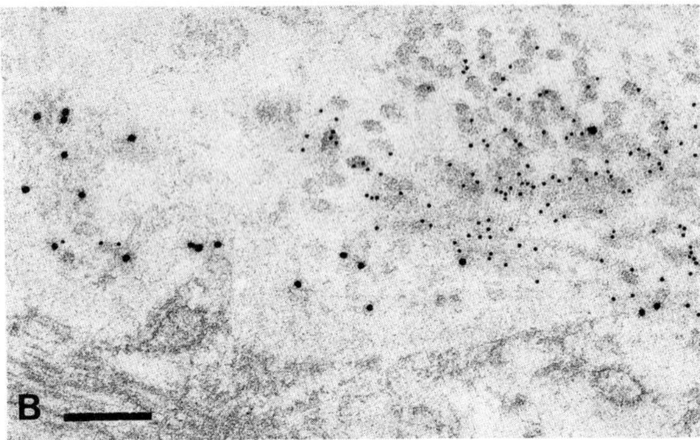

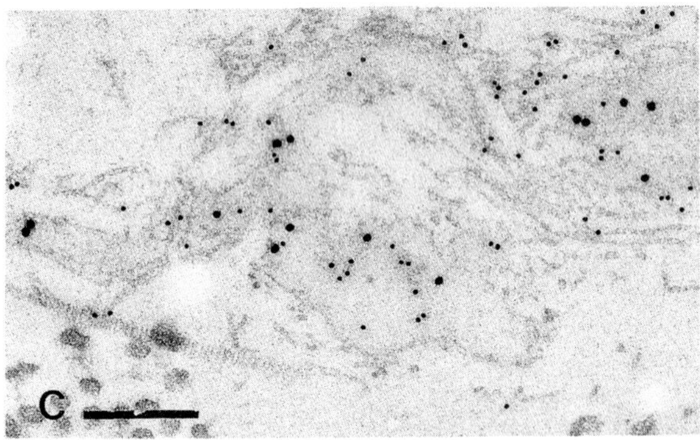

Figure 3: Immuno-double labeling on human heart tissue. **A.** Immunolabeling of a vessel wall. The Weibel-Palade bodies (indicated with an arrow) are stained with anti-vWf antibody using protein-A/10 nm. The lectins are stained with UEA-1 and a protein-A gold/15 nm. Bar represents 0.5 μm. **B.** Staining of the extracellular matrix of the heart. Collagen type III is stained with a secondary rabbit anti-goat antibody detected with 10 nm gold and collagen type VI with 15 nm gold. A two-face method is used. Bar represents 0.5 μm. **C.** Staining of collagen type VI and elastin in the extracellular matrix of a heart. Collagen type VI and elastin antibodies are labeled with protein-A/10 and 15 nm, respectively. Bar=0.5 μm.

Materials and Methods

In this section, we will outline the specific procedures used in our studies in cardiovascular research.

Tissue sampling

All labeling experiments shown in this paper are performed on aortic and heart tissue specimens.

The hearts used were obtained from patients with chronic CHF on the basis of dilated cardiomyopathy (DCM, n=4) and ischemic heart disease (IHD, n=4), clinically all in the New York Heart Association classes IV and V, and collected as cardiac explants or at autopsy. The autopsy hearts all became available within 10 hours after death. Heart tissue from adult patients (n=4) who died of noncardiovascular-related diseases, served as reference.

The hearts were cut perpendicular to the long axis of the left ventricle. Tissue samples were obtained from the left ventricular wall at the level of the base of the papillary muscles. Aortic tissue was obtained from the same hearts.

The tissue specimens were cut into small pieces (1-2 mm^3), fixed in 4% paraformaldehyde and 0.1% glutaraldehyde in sodium cacodylate buffer (pH 7.4), embedded in LR Gold plastic after dehydration in ethanol using the PLT method. The protocol for the dehydration and embedding that we have used is shown in Table 1.

We prefer the use of LR Gold plastic because we have most experience with this particular plastic and, furthermore, it fitted our research objectives well.

Immunolabeling procedures

Our protocol for the detection of a single epitope is shown in Table 2 and that for the detection of two epitopes in Table 3.

All primary antibodies used in this study are listed in Table 4. Protein-A conjugated with 10 nm gold (protein-A/10) and 15 nm (protein-A/15) were obtained from the Department of Cell Biology, University of Utrecht (The Netherlands), rabbit anti-goat gold-conjugated secondary antibodies from Jackson Immunoresearch laboratories (West Grove, Pennsylvania, USA), rabbit anti-goat and rabbit anti-mouse from Dako (Glostrup, Denmark).

Results

Detection of a single epitope

The combination of the primary and secondary antibodies will determine the result. Not only the primary antibody, which reacts with the antigen, but also the secondary gold conjugate determines labeling

Table 4. Antibodies used

Antibody	Host	Clone	Source
Smooth muscle actin	mouse	1A4	Dako, Glostrup, Denmark
Smooth muscle actin	mouse	HHF35	Dako, Glostrup, Denmark
Ulex europaeus agglutinin 1			Dako, Glostrup, Denmark
von Willebrand factor	rabbit		Dako, Glostrup, Denmark
Collagen type I	goat		SBA, Birmingham, AL, USA
Collagen type III	goat		SBA, Birmingham, AL, USA
Collagen type IV	goat		SBA, Birmingham, AL, USA
Elastin	rabbit		Ron Wanders, Specieel Lab. Procreatie, AMC, Amsterdam, The Netherlands
Fibronectin	rabbit		Dako, Glostrup, Denmark

efficiency.

As an example we will discuss labeling of actin, present in smooth muscle cells in aortic tissue, which can be achieved using a variety of techniques. When a monoclonal actin antibody (clone 1A4) in combination with protein-A/10 is used, the affinity of protein-A for the primary antibody is a limiting factor. The labeling efficiency for monoclonal mouse antibodies can be increased by introducing a "bridging antibody" (in this case a rabbit-anti-mouse antibody), since protein-A reacts strongly to rabbit antibodies. In this way, the labeling of smooth muscle cells in the aorta is increased. This is shown in Fig. 1A and 1B.

The same labeling efficiency can be achieved using a different primary antibody (clone HHF35) in combination with a bridging antibody (Fig. 1D). However, the combination of direct labeling of the anti-actin antibody (clone HHF35) and a secondary protein-A/10 conjugate shows almost no labeling activity (Fig. 1C). The choice of gold label is a second limiting factor. Adjacent tissue sections are incubated with the same primary antibody against actin and a secondary rabbit anti-goat gold-conjugated secondary antibody using different gold sizes of 6 nm and 12 nm. The labeling efficiency is increased using the smaller gold-sized secondary antibody. The results are shown in Fig. 2.

Single staining procedures are not only important to localize one specific antigen, but also to determine the sequence in the multiple labeling procedures. It is preferable to use the antibody which shows the least labeling as a first step in a double or triple labeling and to use a small gold conjugate to reduce steric hindrance.

Detection of two epitopes

Fig. 3 shows the results of a double-labeling experiment. Fig. 3A shows the labeling of a blood vessel in human heart tissue using two antibodies raised against vascular endothelial cells. *Ulex europaeus* agglutinin 1 (UEA-1), a membrane marker, and an antibody against the von Willebrand factor (vWf), staining the Weibel Palade bodies, were both used to stain endothelial cells. The staining protocol is as follows: UEA-1; rabbit anti-UEA-1 antibody; Protein-A/15; fixation with 1% glutaraldehyde in PBS; vWf primary antibody; protein-A/10, according to the protocol for double labeling (Table 3).

Fig. 3B shows a staining with primary antibodies against collagen type III and type VI in a human heart. The method used is a two-face method: both antibodies are goat anti-human. The second antibody used is a rabbit anti-goat, followed by protein-A.

Fig. 3C shows a staining with anti-collagen type VI antibody and anti-elastin antibody in a human heart. The staining protocol is as follows: collagen type VI (rabbit anti-human); protein-A/10; fixation; anti-elastin antibody (rabbit anti-human).

Fig. 4 shows a double-staining of collagen type I and type III. Two-face immunolabeling was done to immunolocalize both types of collagen. The secondary antibodies are rabbit anti-goat. The 15 nm gold particles

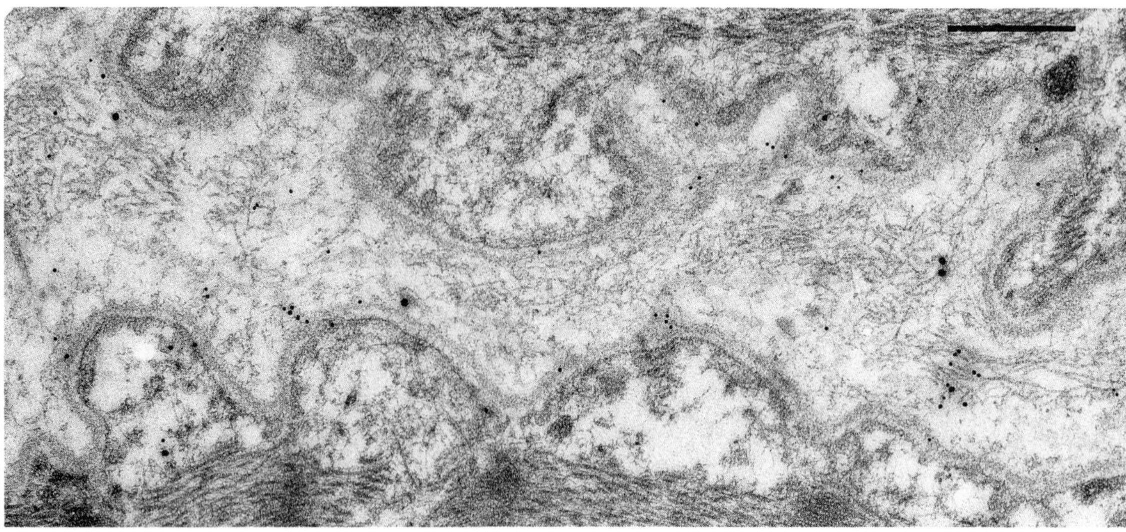

Figure 4: Immuno-double labeling of collagen type I and type III in the extracellular matrix of a human heart with chronic congestive heart failure. The 15 nm gold particles represent collagen type I; the 30 nm gold particles represent collagen type III. The method is a two-face immunolabeling. Bar=0.5 μm.

---

represent collagen type I; the 30 nm gold particles represent collagen type III.

Detection of three epitopes

Fig. 5A shows a triple labeling using antibodies against elastin (rabbit anti-human), fibronectin and actin. The staining protocol is: anti- fibronectin antibody labeled with protein-A/10 nm and after fixation a mixture of the two remaining antibodies, using different-sized gold-conjugated antibodies. The fixation step between the first and second primary antibody is necessary to prevent cross reaction. The 5 nm gold particles represent actin; the 15 nm gold particles represent elastin.

Fig. 5B shows a staining with anti-collagen type VI antibody, anti-collagen type III antibody and anti-elastin antibody. A two-face method has to be used since all antibodies are polyclonal, with two from the same species (goat). The protocol is as follows: anti-elastin antibody; protein-A/5; fixation; anti-collagen type VI antibody; rabbit anti-goat; protein-A/15. The remaining anti- collagen type III antibody is applied at the other side of the grid and labeled with protein-A/10.

Discussion

The extracellular matrix is a relatively stable structure that surround cells and can act as a support. It is considered of great importance for the regulation of cell shape, cell migration, control of cell growth and differentiation (Hay, 1991). A diversity of extracellular matrix components exists, each possessing different functional characteristics: collagens are a source of strength to the tissue, elastin and proteoglycans are essential to matrix resiliency, and the structural glycoproteins help to create tissue cohesiveness. Together these components constitute an intricate network that is difficult to unravel by light microscopy. Electron microscopy can therefore greatly contribute to the study of the fine organisation of this matrix.

Unfortunately, extracellular matrix components are often poorly visible in routinely stained electron microscopic specimens. One method to improve the situation is a brief prestaining of sections with diluted tannic acid general to increase the overall contrast of extracellular matrix components (Dingemans and van den Bergh Weerman, 1990). However, this method is not effective in the differentiation of elements that are morphologically identical (e.g., collagen types I and III), or nearly identical (e.g., fibronectin and elastin-associated microfilaments). The use of immunolabeling techniques, therefore, can be of importance for the study of extracellular matrix components. This is especially so when multiple labeling is applied for the simultaneous identification of several components in one specimen. As we have discussed, these methods can be of importance to reveal the subtle extracellular matrix changes that occur in some pathological processes.

Chronic CHF is characterized by an increase in extracellular matrix components (Heneghan *et al.*, 1991; Schaper and Speiser, 1992; Yoshikane *et al.*, 1992; Marijianowski *et al.*, 1995). Cyanogen bromide analysis

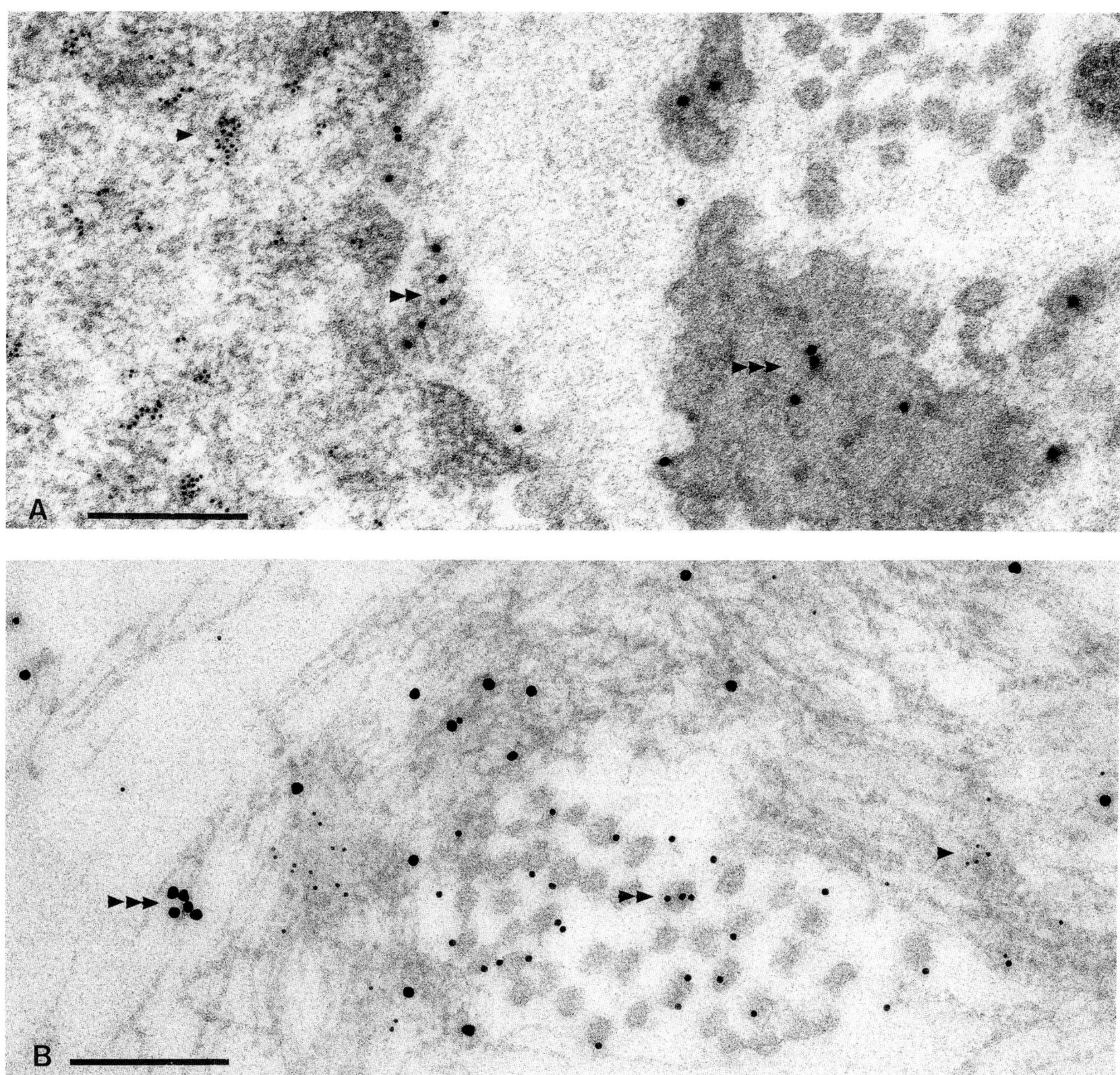

Figure 5: **A.** Immuno-triple labeling of elastin, fibronectin and actin in aortic tissue. The anti-fibronectin antibody is labeled with protein-A gold/10 nm (two arrowheads); anti-actin and anti-elastin antibodies are used in a mixture and labeled with 5 nm (one arrowhead) and 15 nm (three arrowheads) gold conjugated antibodies, respectively. **B.** Immuno-triple labeling of collagen type VI, collagen type III and elastin in the extracellular matrix of a human heart. A two-face method is used. The anti-elastin antibody is labeled with protein-A/5 nm (one arrowhead), anti-collagen type VI is labeled with a rabbit anti-goat antibody and subsequently protein-A/15 nm (three arrowheads). On the other side of the grid anti-collagen type III antibody labeled with protein-A/10 nm (two arrowheads) has been applied. Bar=0.5 μm.

---

and immunohistochemical analysis followed by microdensitophotometric quantification (Marijianowski *et al.*, 1995) showed that the interstitial collagens are increased and the ratio between collagen type I and type III is changed in favor of collagen type I. The localization of both type I and type III collagen is established at the electron microscopic level using immunolabeling techniques. This technique provides more detailed information on the exact changes in hearts of patients with chronic CHF, since the endomysium and perimysium can be identified more easily at the ultrastructural than at the light microscopic level, especially since the distribution into endomysium and perimysium is lost in hearts of patients with chronic heart failure.

Immuno-electron microscopy clearly showed that the increase in endomysial collagens is a diffuse phenomenon, whereas that in the perimysium was much more inhomogeneous. Our studies also show an increase in type I collagen over type III, although these observations were not quantified (Marijianowski *et al.*, 1995; see also Fig. 4).

We have shown that hearts of patients with chronic CHF not only display an increase in collagens but, in fact, present distinct alterations in the composition of the interstitial collagen network. It is likely that these structural changes provide the substrate for the progressive increase in myocardial stiffness. It thus appears that labeling of several antigens, other than types I and III collagen, at the ultrastructural level may contribute further to understand the changes in the extracellular matrix components in chronic CHF.

References

Bendayan M (1982) Double immunocytochemical labeling applying the protein A-gold technique. J Histochem Cytochem **30:** 81-85.

Bendayan M, Garzon S (1988) Protein G-gold complex: comparative evaluation with protein A-gold for high resolution immunocytochemistry. J Histochem Cytochem **36:** 597-607.

Bendayan M, Nanci A, Kan FWK (1987) Effect of tissue processing on colloidal gold cytochemistry. J Histochem Cytochem **35:** 983-986.

Benichou JC, Frehel C, Ryter A (1990) Improved sectioning and ultrastructure of bacteria and animal cells embedded in Lowicryl. J Electron Microsc Techn **14:** 289-297.

Borg TK, Caulfield JB (1981) The collagen matrix of the heart. Fed Proc **40:** 2037-2041.

Borg TK, Ranson WF, Moslehy FA, Caulfield JB (1981) Structural basis of ventricular stiffness. Lab Invest **44:** 49-54.

Borg TK, Sullivan T, Ivy J (1982) Functional arrangement of connective tissue in striated muscle with emphasis on cardiac muscle. Scanning Electron Microsc 1982; **IV**: 1775-1784.

Caulfield JB, Borg TK (1979) The collagen network of the heart. Lab Invest **40:** 364-372.

Craig S, Goodchild DJ (1982) Post-embedding immunolabeling. Eur J Cell Biol **28:** 251-256.

Danscher G (1981) Localization of gold in biological tissue. A photochemical method for light and electron microscopy. Histochemistry **71:** 81-88.

de Mesy Jensen KL, di Sant'Agnese PA (1992) Large block embedding and "pop-off" technique for immunoelectron microscopy. Ultrastruct Pathol **16:** 51-59.

Dingemans KP, van den Bergh Weerman MA (1990) Rapid contrasting of extracellular elements in thin sections. Ultrastruct Pathol **14:** 519-527.

Hamilton G, Hamilton B, Mallinger R (1992) Effects of monomeric acrylic embedding media on the antigenicity of two epitopes of the MIC2-encoded Ewing's sarcoma cell membrane antigen. Histochemistry **97:** 87-94.

Hay ED (1991) Cell Biology of Extracellular Matrix. Plenum Press, New York.

Heneghan MA, Malone D, Ervan PA (1991) Myocardial collagen network in dilated cardiomyopathy. Morphometry and scanning electron microscopy study. Int J Med **160:** 399-401.

Herken R, Fussek M, Barth S, Gotz W (1988a) LR-white and LR-gold resins for postembedding immunofluorescence staining of laminin in mouse kidney. Histochem J **20:** 427-432.

Herken R, Fussek M, Thies M (1988b) Light and electron microscopical postembedding lectin histochemistry for WGA-binding sites in the renal cortex of the mouse in polyhydroxy aromatic resins LR-white and LR-gold. Histochemistry **89:** 277-282.

Holm R, Nesland JM, Attramadal A, Johannessen JV (1988) Double-staining methods at the ultrastructural level applying colloidal gold conjugates. Ultrastruct Pathol **12:** 279-290.

Kann ML, Fouquet JP (1989) Comparison of LR white resin, Lowicryl K4M and epon postembedding procedures for immunogold staining of actin in the testis. Histochemistry **91:** 221-226.

Kasper M, Migheli A (1993) LR Gold and LR white embedding of lung tissue for immuno-electron microscopy. Acta Histochemica **95:** 221-227.

Marijianowski MMH, Teeling P, Mann J, Becker AE (1995) Dilated cardiomyopathy is associated with an increase in the type I to type III collagen ratio. A quantitative assessment. J Am Coll Cardiol **25:** 1263-1272.

Migheli A, Attanasio A, Schiffer O (1992) LR Gold embedding of nervous tissue for immunoelectron microscopy studies. Histochemistry **97:** 413-419.

Quintana C (1994) Cryofixation, cryosubstitution, cryoembedding for ultrastructural, immunocytochemical and microanalytic studies. Micron **25:** 63-99.

Robertson D, Monaghan P, Clarke C, Atherton AJ (1992) An appraisal of low-temperature embedding by progressive lowering of temperature into Lowicryl HM20 for immunocytochemical studies. J Microsc **168:** 85-100.

Robinson TF, Cohen-Gould L, Factor SM (1983) Skeletal framework of mammalian heart muscle: arrangement of inter- and pericellular connective tissue structures. Lab Invest **49:** 482-488.

Roth J (1982) The preparation of protein A-gold complexes with 3 nm and 15 nm gold particles and their use in labelling multiple antigens on ultra-thin sections. Histochem J **14:** 791-801.

Roth J (1989) Immunolabeling with the protein A-gold technique: an overview. Ultrastruct Pathol **13:** 467-484.

Roth J, Taatjes DJ, Warhol MJ (1989) Prevention of non-specific interactions of gold-labeled reagents on tissue sections. Histochemistry **92:** 47-56.

Schaper J, Speiser B (1992) The extracellular matrix in the failing human heart. Basic Res Cardiol **87(suppl 1):** 303-309.

Shimizu H, Ishida-Yamamoto A, Eady RA (1992) The use of silver-enhanced 1-nm gold probes for light and electron microscopic localization of intra- and extracellular antigens in skin. J Histochem Cytochem **40:** 883-888.

Silver MM, Hearn SA (1987) Postembedding immunoelectron microscopy using protein A-gold. Ultrastruct Pathol **11:** 693-703.

Taatjes DJ, Mount SL, Trainer TD, Tindle BH (1994) Post-embedding immunocytochemistry for adhesive proteins and clatrin in LR white- and LR Gold-embedded human platelets. Anat Anz **176:** 67-73.

Weber KT (1989) Cardiac interstitium in health and disease: the fibrillar collagen network. J Am Coll Cardiol **13:** 1637-1652.

Yokota S (1988) Effect of particle size on labeling density for catalase in protein A-gold immunocytochemistry. J Histochem Cytochem **36:** 107-109.

Yoshikane H, Honda M, Goto Y, Morioka S, Oshima A, Moriyama K (1992) Collagen in dilated cardiomyopathy. Scanning electron microscopic and immunohistochemical observations. Jpn Circ J **56:** 899-910.

Discussion with Reviewers

K. Robinson: In the "double-face" labelling experiments, can section thickness be a problem in terms of focusing on the images for obtaining micrographs? Is there a maximum section thickness for which this is possible?

Authors: No, the depth of field of a transmission electron microscope is considerable greater than the thickness of any specimens normally examined at the level of resolution used, and thus the specimen appears equally sharp throughout its thickness (Glauert, 1974). This means that, in practice, the section thickness never constitutes a limitation.

K. Robinson: You assert that it is preferable in multiple labelling procedures to first apply the antibody which gives the lowest labelling index, and to use small gold probes to minimize steric hindrance from subsequently applied antibodies. Can such effects either entirely mask antigens, or give misleading results with respect to labelling intensity? If so, what steps could be taken to correct or at least adjust for this?

Authors: Yes, the possibility that antigens are masked partially, which may give misleading results cannot be overlooked. This effect can be avoided by the use a two-face method.

G. Pasquinelli: The authors state that when a multiple labelling is going to be performed, one should employ the smallest gold probe in the first step of the labelling procedure to avoid any steric hindrance at the next step; however, in the shown example (Fig. 3) it seems that the reverse is true (15 nm gold particles followed by the 10 nm ones).

Authors: This comment is basically correct. In the present example, however, it is known that the two antigens have different localizations (cell surface and Weibel-Palade bodies, respectively). Steric hindrance, therefore, seems unlikely in this particular case. Moreover, the two antibodies used in this case show a high and comparable degree of labeling, so that no problems caused by partial masking are to be expected.

Additional Reference

Glauert AM (ed) Principles and practice of electron microscope operation: practical methods in electron microscopy. (1974) (Glauert AM, ed) North-Holland Publishing Company, Amsterdam, The Netherlands.

Scanning Microscopy Supplement 10, 1996 (pages 273-284) 0892-953X/96$5.00+.25
Scanning Microscopy International, Chicago (AMF O'Hare), IL 60666 USA

COVALENT LABELING OF PROTEINS WITH FLUORESCENT COMPOUNDS FOR IMAGING APPLICATIONS

Darl R. Swartz*

Department of Anatomy, Indiana University Medical Center, Indianapolis, IN 46202

(Received for publication October 10, 1995 and in revised form August 16, 1996)

Abstract

The labeling of proteins with fluorescent compounds for microscopy has allowed a greater understanding of biological processes. The preparation of fluorescent proteins is the first step in development of their use in microscopy. Methods are described to label and characterize a protein as an example of the general approach for other proteins. Skeletal muscle alpha-actinin was labeled with either fluorescein-5-maleimide or 5-iodoaceamidofluorescein and the reaction characterized. The maleimide reaction was much more rapid and efficient than the iodoacetamide reaction giving a coupling efficiency of 65% under the given ration conditions. The fluorescein-5-maleimide alpha-actinin was functionally characterized and there was essentially no influence on the fluorescein label on the F-actin binding properties of alpha-actinin. The fluorescein alpha-actinin was also shown to specifically bind to the Z-line of isolated myofibrils. A general outline and discussion are presented on how to label and characterize proteins for use in microscopy.

Key Words: fluorescence microscopy, covalent labeling, alpha-actinin, fluorescein alpha-actinin, myofibrils.

*Address for correspondence:
Darl R. Swartz
Department of Anatomy
Indiana University Medical Center,
635 Barnhill Dr., Indianapolis, IN 46202
Telephone Number: 317 274-8188
FAX Number: 317-278-2040
E-mail: dswartz@indyvax.iupui.edu

Introduction

Covalent labeling of ligands for imaging applications is essentially an extension of classical histochemical staining methods in that it is a method which enhances the contrast of the specimen. The important exception is that covalent labeling allows for much higher resolution images and much more refined interpretations of a given image. An example of this is the detection of specific proteins using immunological techniques. One of the first applications of a covalently-labeled protein was the immunofluorescent detection of pneumococcal antigen in tissues using fluorescein-labeled primary antibodies (Coons *et al.*, 1942). Since that time there have been many developments in the methodology for preparing covalent adducts of various ligands for use in imaging applications (Haugland, 1992; Wang and Taylor, 1989). By far the most common use of labeled proteins is in immunofluorescence microscopy using fluorescently-labeled secondary antibodies. There are many excellent commercial suppliers of labeled secondary antibodies so the researcher need not spend valuable laboratory time preparing their own labeled secondary antibodies.

Another application of covalent labeling of proteins is in labeling of cytoskeletal proteins for cell biology applications. One of the first uses of this approach was the labeling of heavy meromyosin for the demonstration of actin involvement in cytokinesis (Aronson, 1965). Developments in the reaction chemistries (Hartig *et al.*, 1977; Haugland, 1992) allowed for use of much more efficient and milder reaction conditions which widened the spectrum of proteins which could be labeled. This, coupled with enhancements in light microscopy imaging technology, afforded the opportunity of following cytoskeletal dynamics in living cells using covalently labeled proteins (Wang and Taylor, 1980, 1989). Covalent labeling of proteins with undecagold resulted in high resolution electron microscopy applications of covalently labeled proteins (Safer *et al.*, 1986; Milligan *et al.*, 1990) and developments in fluorescence photo-oxidation have allowed for dual use of fluorescently labeled proteins at both the light and electron microscope level (Deerinck *et al.*, 1994). The recent development of molecular biological means to prepare fusion proteins containing green fluorescent protein (Cubitt *et al.*, 1995) offers an alternative to covalent labeling of purified proteins. Thus, covalently labeling of proteins has wide usage in imaging applications both at the light and electron microscope level.

This paper is limited to discussion of methods for covalent labeling of proteins. This labeling employs a compound which is essentially bivalent in that one part of the molecule is reactive with the ligand of interest while the other part is a reporter molecule. The reactive moiety is a chemical reagent which reacts with the ligand to form a stable covalent linkage with the ligand. The reporter molecule is a chemical moiety which allows for detection of the ligand. Classical examples of the reporter molecules are fluorochromes, biotin, enzymes, and other selective haptens to which there are specific antibodies. The labeling of proteins takes advantage of reactive groups present on the proteins. These can be reactive amino acids present in the polypeptide such as primary amines (e.g., lysine; both intracellular and extracellular proteins) or sulfhydryls (e.g., cysteine; intracellular proteins). Most intracellular proteins have free sulfhydryl side chains (cysteines) while extracellular proteins have disulfide crosslinks (cystines) which are not readily reactive. Most all proteins have primary amines, either as the amino terminus or from lysine side chains. Although many other side chains or reactive groups on the proteins are potential sites for covalent modification, the reaction conditions are usually not desirable. A major concern for chemical modification of proteins is the maintenance of biological function. Thus, reactions which proceed at room temperature or lower and at neutral or slightly alkaline pH are desired and this in turn dictates to some extent what reaction chemistry one can utilize.

Below, I will describe methods for the labeling of skeletal muscle alpha-actinin with thiol-reactive fluorescein compounds as an example of how to covalently label a protein. This will be followed by assaying a biological function of the labeled protein and use of the protein in an imaging application.

Materials and Methods

Protein purification

Alpha-actinin was purified from rabbit skeletal muscle using modifications to the procedure described by Feramisco and Burridge (1980). The $MgCl_2$ precipitation step was omitted and the low ionic strength/37°C supernatant was cooled to 4°C and 21 g solid $(NH_4)_2SO_4$ was added per 100 ml supernatant and the solution stirred for 20 min. The $(NH_4)_2SO_4$ precipitate was collected by centrifugation at 5,000g for 30 min, resuspended in a minimal volume of dialysis buffer [1 M KCl, 10 mM KH_2PO_4 (pH 7.0), 1 mM EDTA, 0.1% 2-mercaptoethanol (2-ME), 0.1 mM phenylmethylsulfonyl fluoride (PMSF)] and dialyzed overnight against 10 vol of dialysis buffer relative to initial tissue mass. The dialyzed solution was clarified by centrifugation at 100,000g for 1.5 hour at 4°C and crude alpha-actinin in the supernatant was precipitated by adding 21 g solid $(NH_4)_2SO_4$ per 100 ml supernatant. The precipitate was collected by centrifugation and further purified by ion-exchange chromatography on diethylaminoethyl(DEAE)-cellulose (DE-52), hydroxylapatite, and size-exclusion chromatography on Sephacryl S-400. Alpha-actinin was stored as a suspension in 50% $(NH_4)_2SO_4$ at 4°C. Alpha-actinin retains it actin-binding properties for at least a year when stored under these conditions. Actin was purified from rabbit skeletal muscle as described by Pardee and Spudich (1982) and rabbit skeletal muscle myofibrils were purified as described by Swartz *et al.* (1990).

Protein labeling

Alpha-actinin was dialyzed at 4°C against 100 mM KCl, 10 mM HEPES (pH 8.0), 1 mM EGTA, 0.1% 2-ME to remove $(NH_4)_2SO_4$. The protein solution was made to 10 mM dithiothreitol (DTT) and incubated at room temperature for 30 min to reduce any oxidized thiols. The solution was cooled on ice and then desalted on a Sephadex G-25 column equilibrated with reaction buffer [100 mM KCl, 10 mM HEPES (pH 8.0), 1 mM EGTA] at 4°C. Elution of the protein fractions was monitored by ultra violet (UV) absorbance. To ensure separation of the reducing agents (2-ME and DTT) from the protein fractions, the level of thiol was monitored by the addition of 5,5'-dithiobis-(2-nitrobenzoic acid (DTNB) (Ellman, 1959) to fractions trailing the protein peak. The protein concentration was determined by absorbance at 280 nm using an extinction coefficient of 1.0 ml/mg.cm and a molecular weight of 100,000 per alpha-actinin monomer (Suzuki *et al.*, 1976). The dyes [fluorescein-5-maleimide (F-5M) and 5-iodoaceamido fluorescein (5-IAF), Molecular Probes, Eugene, OR] were prepared by dissolving about 1 mg dye/100 μl dimethylformamide, and the concentration of the dye was determined spectrophotometrically. A small aliquot of the stock solution was diluted in 10 mM Tris (pH 9.5), 1 mM NaN_3, 0.1% 2-ME and the absorbance read at 490 nM. The concentration was calculated using an extinction coefficient of 83,000 /M.cm (Haugland, 1992). For the reaction, the dye was first diluted in about 1/20 volume of reaction buffer then rapidly mixed with the protein solution and incubated at room temperature in the dark. The protein concentration was 1.5 mg/ml and 15 ml of protein was used for each dye giving 0.22 μmoles of alpha-actinin monomer per reaction. The conjugation reaction was done using 2 mole dye/mole alpha-actinin monomer thus, 0.44 μmole of each dye was added. After incubation at room temperature (22°C) for 16 hours, the reaction was quenched by addition of DTT to a final concentration of 1 mM. The protein was precipitated by mixing the solution with an equal volume of saturated $(NH_4)_2SO_4$, collected by centrifugation (a yellow-orange pellet was observed) and dissolved in a minimal volume of rigor buffer (RB: 75 mM KCl, 10 mM imidazole (pH 7.2), 2 mM MgCl2, 2 mM EGTA, 1 mM NaN_3) made to 1 mM DTT. Unconjugated dye and $(NH_4)_2SO_4$ were removed by desalting on a Sephadex G-25 column equilibrated with RB. The conjugated protein can be stored at this point for 1 week on ice or as an $(NH_4)_2SO_4$ pellet at 4°C in the dark for over a year and still maintain functionality.

The time course of the reaction was followed by removing aliquots at timed intervals and mixing with an equal volume reaction buffer containing 0.2% 2-ME to quench the reaction. Samples were then analyzed by SDS/PAGE (sodium dodecyl sulfate/polyacrylamide gel electrophoresis) followed by observation of the unstained gel using a UV light box. Images of the fluorescent protein in the gels were documented using Polaroid type 667 film with camera settings at f5.6 and exposure times of 5-15 sec.

Binding studies

The biological activity of F-5M alpha-actinin was compared to unlabeled alpha-actinin using an actin sedimentation assay similar to that described by Meyer and Aebi (1990). Different levels of alpha-actinin dimer (0.2-15 μM) were mixed with a fixed level of G-actin (18 μM final) in RB made to 1 mM adenosine triphosphate (ATP) and 1 mM DTT and incubated at room temperature for 1 hour. The mixtures were centrifuged at 13,000g for 20 min then the supernatant was carefully removed. The pellets were resuspended in SDS/PAGE sample buffer and aliquots of the supernatants were mixed with sample buffer. Both supernatants and pellets were analyzed for alpha-actinin content by densitometry of the alpha-actinin bands in SDS/PAGE. After electrophoresis, the gels were stained with Coomassie brilliant blue, and alpha-actinin content determined by densitometry of gels using unlabeled alpha-actinin as the protein standard. Binding isotherms were generated by plotting free [alpha-actinin] in the supernatant against bound [alpha-actinin] in the pellet.

Glycerinated rabbit skeletal muscle myofibrils were washed 3 times with 10 vol of RB containing 1 mg/ml bovine serum albumin (BSA) and 1 mM DTT (RB-BSA-DTT) then diluted to 1 mg/ml myofibrillar protein. Myofibrils (50 μl) were mixed with 150 μl RB-BSA-DTT containing 0.1 μM F-5M alpha-actinin (final) and incubated 1h at room temperature in the dark. The labeled myofibrils were collected by centrifugation for 15 sec in a microfuge at maximum speed (13,000g). The supernatant was removed and the myofibrils were resuspended in 200 μl RB-BSA-DTT; then 100 μl were plated on a coverslip and the myofbrils were fixed by pooling with 400 μl of 3% formaldehyde in RB for 15 min. The coverslips were drained, rinsed in RB, then mounted on 50 μl of mounting medium [75% glycerol (v/v), 75 mM KCl, 20 mM Tris (pH 8.5), 2 mM $MgCl_2$, 2 mM EGTA, 1 mM NaN_3, 1 mg phenylenediamine/ml] and sealed to slides with "Wet 'n' Wild" nail polish.

Microscopy

Myofibrils were imaged using a Zeiss axiovert microscope equipped with a 100X (NA 1.3) phase contrast objective, epifluorescence illumination with a 100W Hg lamp, a fluorescein narrow band pass filter and a CCD (Photometrics 200, KAF 1300 chip, Photometrics Inc., Tucson, AZ). The CCD was controlled from a matrix board in a Macintosh Quadra 840AV using IPLab Spectrum software (version 1.3, Signal Analytics, Vienna, VA). Images were obtained using 0.2-0.8 sec exposures giving a minimum dynamic range of 400 (max-min intensity). Images were processed using IPLab software by rotation to the horizontal using bilinear interpolation, cropping out a six sarcomere region of interest (ROI), converting the 16 bit image to 8 bit, and montaging the phase and fluorescence ROI's. Further text and graphic information was added to the images in Canvas (version 3.5, Deneba Software, Miami, FL).

Assays

Protein concentration was measured with the bicinchoninic acid (BCA) assay as modified by Hill and Straka (1988) for use with reducing agents. BSA was used as the standard for myofibrils and unlabeled alpha-actinin was used as the standard for F-5M and 5-IAF alpha-actinin. SDS/PAGE was done on 10% gels as described in Fritz *et al.* (1989).

Results

Coupling reaction

The coupling of a dye to a protein is highly dependent upon the chemical reaction employed and the buffer conditions. Fig. 1 shows the dye structures and the chemistry of the reactions. Fig. 2 shows the time course of alpha-actinin conjugation with either F-5M (panel A, A') or 5-IAF (panel B, B'). At different times, the reaction was quenched with excess 2-ME followed by separation of labeled protein from unlabeled using SDS/

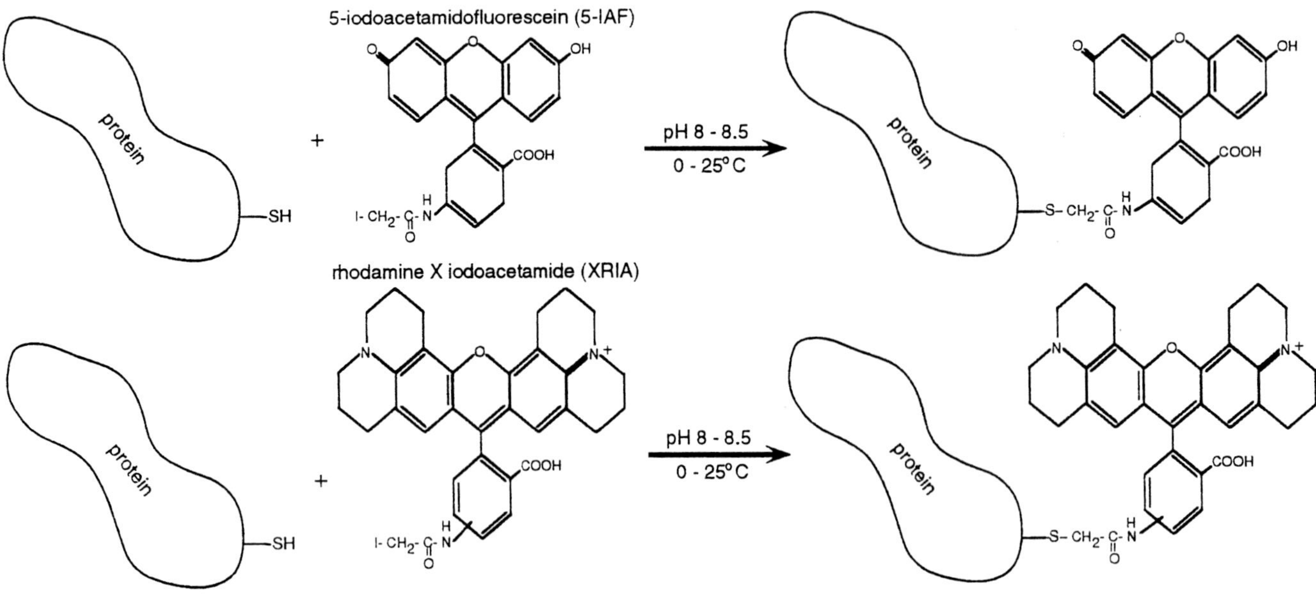

Figure 1. Diagram of covalent labeling proteins with thiol-specific dyes.

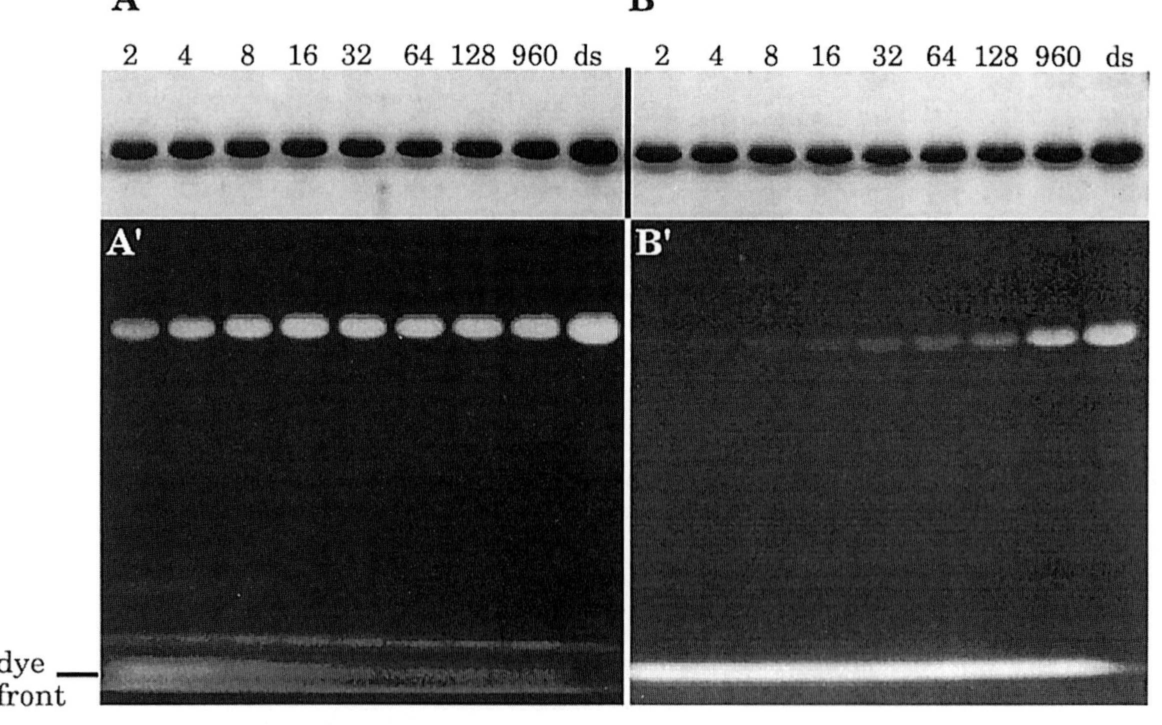

Figure 2. Time course of alpha-actinin labeling with thiol-specific dyes followed with SDS/PAGE. Time course of F-5M (panel A, A') and 5-IAF (panel B, B') labeling of alpha actinin. The dyes were mixed with the protein in 100 mM KCl, 10 mM HEPES (pH 8.0), 1 mM EGTA at 20°C. At specified times, aliquots were removed, and the reaction quenched with 2-ME. Samples were separated from unconjugated dye by SDS/PAGE followed by imaging under UV illumination (panels A' and B') and white light after Coomassie staining (panels A and B). Text above lanes designates incubation time in min and ds is de-salted protein after 960 min incubation. The fluorescence at the bottom of gels (A' and B') was the unconjugated dye.

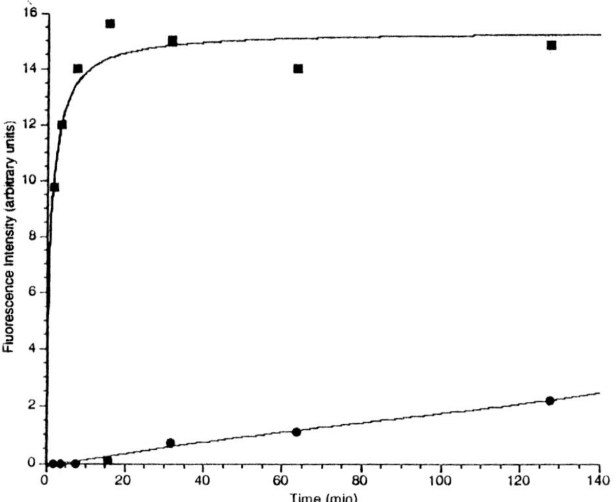

Figure 3. Kinetics of covalent labeling of alpha-actinin with thiol-specific dyes. Photographs of the UV gels were digitized using a calibrated flatbed scanner and light-intensity measured from the digital images. The integrated intensity was determined for each band after correction for background intensity and plotted as a function of time. Note that reaction of alpha-actinin with F-5M (filled squares) was much more rapid than with 5-IAF (filled circles).

PAGE. The progress of the conjugation reaction was determined by photographing the gels under UV illumination (panels A' and B') and analysis of the pictures for intensity using a flatbed scanner. After UV photography, the gels were stained with Coomassie, destained and imaged using a flatbed scanner (panels A and B). The UV gels show that the conjugation with F-5M was very rapid and that there was very little dye at the dye front after about 32 min incubation. Conversely, the reaction with 5-IAF proceeded very slowly with protein conjugate bands only being easily visible after 32-64 min of reaction while there was substantial fluorescence at the dye front even after 960 min. The continual slow reaction of 5-IAF with protein suggests that the reagent was not exhausted by hydrolysis. The last lane of each gel (ds) shows the labeled protein after 960 min of the reaction followed by desalting to remove unconjugated dye. Note that there was essentially no fluorescence at the dye front for either the F-5M or 5-IAF conjugate. The progress of the reaction was also monitored by measuring the integrated intensity of the bands from the photographs using a flatbed scanner and image analysis using IPLab Spectrum. This is shown graphically in Fig. 3. The reaction with F-5M is more than halfway done after 2 min incubation and is essentially complete at 32 min while the reaction with 5-IAF was much less complete even at 128 min.

Estimation of the dye to protein ratio by spectrophotometry showed that the F-5M conjugate had 1.3 mole fluorescein/mole alpha-actinin monomer while the 5-IAF conjugate had 0.34 mole fluorescein/mole protein. Thus, under the buffer and temperature conditions used for the conjugation reaction, the F-5M reaction was much more efficient giving 0.65 coupling efficiency (1.3 mole coupled/2 mole added) while the 5-IAF gave only a 0.17 coupling efficiency. Only the F-5M conjugate was used in further studies because of its higher labeling ratio.

Functionality test

Alpha-actinin is characterized as an actin bundling protein; thus, one can characterize the ability of alpha-actinin to induce bundling using a sedimentation assay which pellets the alpha-actinin/actin complex. Comparing the bundling ability of unlabeled alpha-actinin with labeled allows for determination of whether the labeling influences a biological property of the molecule. Fig. 4 shows gels which contain the supernatant and pellet fractions of samples with different levels of alpha-actinin but constant actin. The unlabeled alpha-actinin samples are in panel A while the F-5M labeled alpha-actinin samples are in panel B. The unlabeled panel shows that at 0.4 μM alpha-actinin, most of the actin and alpha-actinin was in the pellet. As alpha-actinin was increased, there was an increase in the amount of alpha-actinin and actin in the pellets suggesting that the alpha-actinin bound to and bundled the actin into aggregates. Neither actin or alpha-actinin was found in the pellet when incubated in buffer alone (data not shown). There was very little difference in the appearance of the gels between the unlabeled and labeled alpha-actinin.

Quantitative analysis of the bound and free alpha-actinin by densitometry of the gels is shown in Fig. 5. The curves were the best fit to the hyperbolic equation:

$$B=(B_{max}*free)/(K_d + free) \quad (1)$$

The unlabeled alpha-actinin gave a B_{max} of 2.7 μM and a K_d of 1.2 μM while the F-5M alpha-actinin gave a B_{max} of 3.7 μM and a K_d of 2.5 μM for the alpha-actinin dimer. The B_{max} value corresponds to saturation at 1 alpha-actinin dimer per 5-7 actin monomers (18 μM actin for the assay). These values were close considering the scatter of the data suggesting that the labeling did not dramatically alter a biological property of the alpha-actinin. A more heavily labeled alpha-actinin is likely to result in changes in biological function.

Myofibril binding

Previous studies have shown that labeled alpha-actinin (smooth or skeletal) will bind to the Z-line of skeletal myofibrils (Sanger *et al.*, 1984; Swartz *et al.*, 1993). Thus, rigor myofibrils were incubated with F-5M labeled alpha-actinin to demonstrate this property of

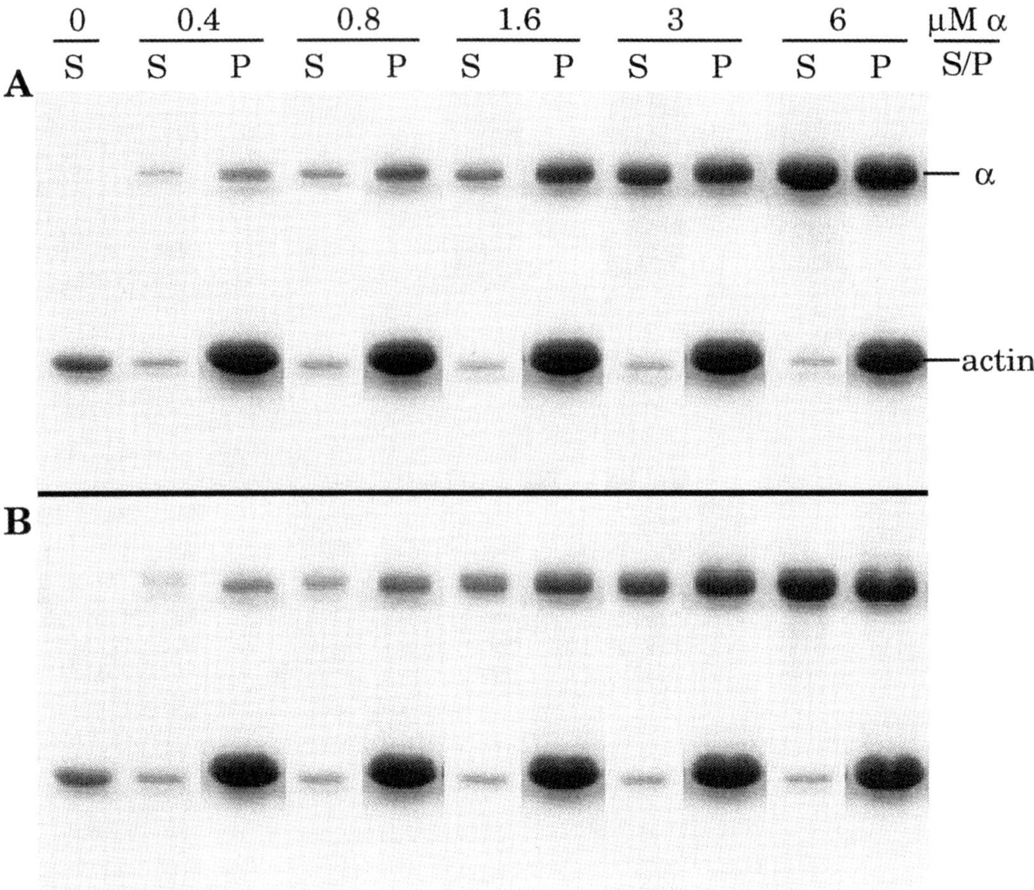

Figure 4. Functionality assay of F-5M alpha-actinin. Different concentrations of either unlabeled (panel A) or F-5M alpha-actinin (panel B) were mixed with a fixed concentration (18 μM) of G-actin under polymerization conditions and incubated for 1h at 20°C. Samples were centrifuged at 13,000g for 30 min at 20°C and aliquots of the supernatant (S) and pellet (P) were assayed by SDS/PAGE followed by Coomassie staining. Numbers above lanes designate alpha-actinin dimer concentration. Note that the supernatant and pellet samples were run on separate gels and the images were made by digitally montaging the appropriate lanes giving the odd appearance of the actin band in the pellet lanes.

labeled alpha-actinin. Fig. 6 shows phase-contrast (1P, 2P) and fluorescein-fluorescence (1F, 2F) images of two different rabbit psoas myofibrils at different sarcomere lengths. The F-5M alpha-actinin localized to the Z-line of the rigor myofibrils with some minor binding to regions just adjacent to the Z-line. The Z-lines within a myofibril did not seem to show the same level of incorporation of F-5M alpha-actinin for reasons unknown at present. Pretreatment of myofibrils with unlabeled alpha-actinin does not prevent incorporation of F-5M alpha-actinin into the Z-line suggesting that the mechanism of incorporation is that of exchange of unlabeled for labeled alpha-actinin and this is currently under study. These imaging studies demonstrate the that the labeled protein bound to the predicted in situ location with high specificity.

Discussion

In this paper, I have briefly demonstrated a method to covalently label a protein for use in microscopy. The specific example of covalent labeling of alpha-actinin was presented as well as a functionality test for the labeled protein and application of the label protein in light microscopy. The major steps and factors to be considered in labeling a protein to be used for imaging applications are discussed below.

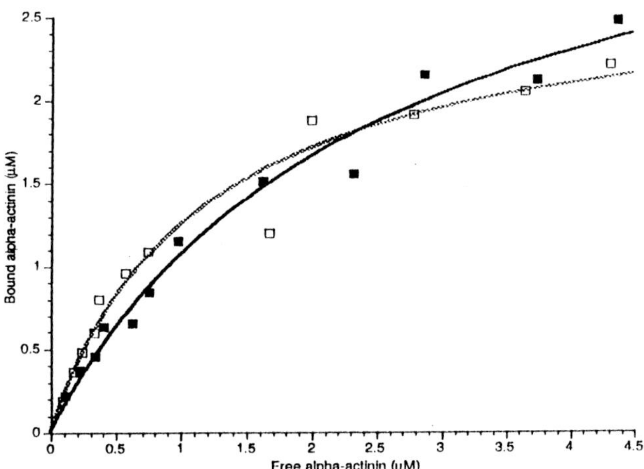

Figure 5. Binding isotherm of unlabeled and F-5M alpha-actinin. The concentration of alpha-actinin in the supernate and pellet were determined by densitometry of Coomassie-stained gels of samples separated by SDS/PAGE. Data were analyzed by non-linear least-square fitting to a hyperbolic function of the form $B = B_{max}$ * [alpha-actinin]/(K_d+[alpha-actinin]) where B = [bound], B_{max} = [bound] at saturation and K_d = dissociation constant. The B_{max} values were 2.7 and 3.7 μM for unlabeled (open squares) and F-5M alpha-actinin (filled squares) respectively while K_d values were 1.2 and 2.5 μM for unlabeled and F-5M alpha-actinin respectively.

Step 1: Select a reaction chemistry

A good starting point is a reference on chemical modification of proteins which can be found in books such as Means and Feeney (1971) or Hermanson (1996) and a brief review is presented in the handbook by Haugland (1992). When selecting a reaction chemistry, you need to decide whether the dye will be covalently attached to either a primary amine or a cysteine residue. As a general rule, cysteine-specific dyes are the preferred route, because the chemical reactions for coupling to cysteines are usually more efficient under reasonably mild conditions. Thus, the decision is based on whether the protein has reactive cysteines or not and whether cysteine modification alters biological activity. Most all intracellular proteins have available cysteines so this is the chemistry of choice. Some of the cysteine can be buried in the protein structure and not reactive with thiol reagents. In this work, alpha-actinin has a total of 9 cysteines/monomer and of these, 3 are available for reaction, thus, 6 of the thiols are buried and only available under denaturing conditions (Suzuki *et al.*, 1976). Extracellular proteins usually do not have available cysteines as they are oxidized to disulfide crosslinks to aid in protein stability. Thus, for extracellular proteins, reactions involving primary amines are usually employed. In some cases, primary amines may be readily reactive under mild pH conditions but this can be protein specific. Thus, although the reaction conditions are usually more harsh for conjugation at primary amines, the extracellular proteins are usually more stable than intracellular proteins.

There are some alternatives for labeling of extracellular proteins which do not directly utilize primary amine reactions. For example, intramolecular disulfides can be mildly reduced to give available cysteines followed by labeling with thiol-specific dyes. Alternatively, primary amines can be modified with 2-iminothiolane under mild conditions resulting in an thiolated protein (Jue *et al.*, 1978). This protein can then be modified with a thiol-specific dye. We have utilized this approach to label antibodies with thiol-specific dyes giving good results (Swartz, 1989). Although these alternative methods require some additional steps, they can be utilized to great advantage if one needs a particular labeled protein which is not commercially available.

Once you have decided whether you are going to use a amine-specific or thiol specific dye, your next decision is which reagent to utilize and what reaction conditions to try. For amine-reactive dyes, the general choices are isothiocyanate, sulfonylchloride, and succinimidyl ester derivatives. The general order of preference is succinimidyl ester > isothiocyanates > sulfonylchlorides in terms of reactivity under the same conditions. The pH for reactions with amine-reactive dyes should be in the 8.5-9.0 region such that the amine is unprotonated because this is the reactive species. The buffer for the reaction should be a non-amine containing buffer. Buffers of choice include borate, phosphate, or possibly high pKa Good's buffers. It must be realized that hydroxyl ions compete with the unprotonated amines for the reactive dye. Thus, at higher pH's, more dye will react with hydroxyl ions. As a general rule, the reaction can be done with 5-10 mole dye/mole protein at room temperature for 1-2 h to yield conjugated protein. Poor conjugation levels can be improved by increasing the reaction time and/or increasing the dye level.

With thiol-reactive dyes, the choices are somewhat limited to either maleimide or iodoacetamide derivatives. Maleimides are preferred over iodoacetamides due to their highly efficient and specific reaction with thiols (Gregory, 1955). As shown in Figs. 2 and 3, the reaction of alpha-actinin with F-5M was much faster and more efficient than 5-IAF under the same reaction conditions. A pH of 8.0 was used so that one could compare the reactivity of the two compounds. The maleimide reaction readily occurs at more neutral pH values. Some crossreactivity of maleimide with amines can occur at pH 8.0 (Brewer and Riehm, 1966) but this is

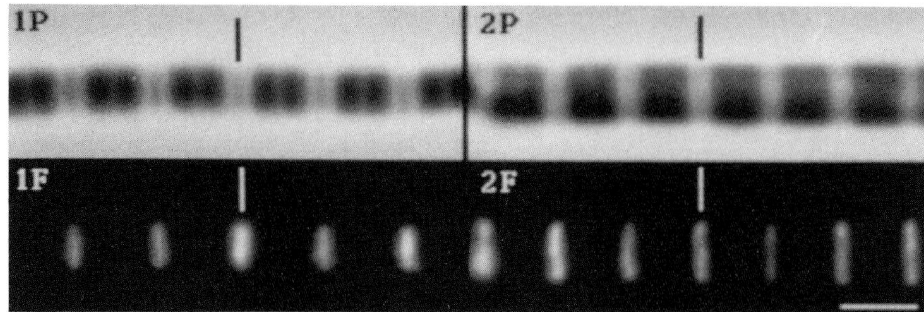

Figure 6. Labeling of rabbit skeletal myofibrils with F-5M alpha-actinin. Myofibrils were incubated with 0.1 μM alpha-actinin for 1h followed by sedimentation and resuspension in buffer without F-5M alpha-actinin, fixation and mounting on slides. Both phase-contrast (P) and fluorescein-fluorescence (F) images are shown. Myofibrils at rest (1) and short (2) sarcomere length are shown. Vertical line denotes Z-line and scale bar is 2 μm.

not likely at the low levels of maleimide used for labeling. An approach that we have used is to employ the same chemistry to couple different reporter groups to the same protein. For example, we have used maleimide derivatives of fluorescein and Texas Red to label myosin subfragment 1 (S1) (Swartz *et al.*, 1990). The coupling efficiencies are usually very similar between the different reporter groups so that the dye/protein ratios are similar. This is highly advantageous because it affords the researcher many different approaches for double labeling of biological specimens. For example, under appropriate conditions, S1 will bind to either the overlap or I-band region of myofibrils depending on the calcium level. We have used a Texas Red and fluorescein S1 to demonstrate this property in the same myofibril by sequential treatment with one color of S1 followed by the other under different conditions (Swartz *et al.*, 1990). One can also prepare biotin conjugates using the same chemistry. The advantage of having a biotin conjugate is that this can be used with avidin/streptavidin gold conjugates for electron microscopy. Also, the biotin conjugate can be detected with enzyme or fluorophore conjugates of avidin/streptavidin allowing a great latitude in the detection system. We have used the enzyme conjugates to advantage to develop biological function assays described below. One difficulty with the multilabel/same chemistry approach is that a maleimide Texas Red is no longer commercially available. Thus, we have had to modify our procedures to use iodoacetamide derivatives of Texas Red, fluorescein and biotin.

The type of reaction employed can be very protein specific. For example, actin can be specifically labeled with an amine-reactive dye at lys-61 and a thiol-reactive dye at cys-374 (Miki and dos Remedios, 1988). The specific labeling occurs because the residues are likely rendered highly reactive by their micro-environment within the protein. An extensive reference list of protein-specific labeling can be found in Haugland (1992). Another difficulty is that the most reactive residue(s) may be important in biological function. An example of this is S1 which contains two highly reactive thiols which are indirectly involved in ATPase activity. This problem can be dealt with by reversible blocking of the reactive thiols followed by reaction at other, less reactive, thiols and subsequent unblocking (Swartz *et al.*, 1990). Thus, one may have to be creative in the method employed to label the protein and the protocol may be highly protein specific.

Step 2: Pilot studies

The availability of the protein dictates to some extent the amount of development of the conjugation protocol. A simple approach for investigating reaction conditions is the use of SDS-PAGE/UV illumination to follow the time course and efficacy of the reaction as it does not require milligram quantities of protein. A few hundred micrograms of protein may suffice for following the time course of the reaction since one only needs to load 0.2-1 μg protein per lane to follow the reaction. Variables that influence the conjugation reaction include: pH, temperature, reactive dye concentration, protein concentration and time.

pH: The pH conditions are somewhat limited by the stability of the protein at high pH. The preference is for neutral to slightly alkaline conditions. Thiol-reactive dyes are generally used in the pH 7-8.5 range with maleimides in the pH 7 range and iodoacetimides in the 8-8.5 range. For example, to increase the conjugation efficiency of alpha-actinin with 5-IAF, the pH could be raised to 8.5 with the same incubation time as described above. Amine-reactive dyes are used in the pH 8-9.5 range with the succinimidyl ester being used in the pH

8-9.0 range. The buffers used should not be reactive with the dye as discussed above. For the thiol-reactive dyes, the Good's buffers: 3-N(morpholino)propanesulfonic acid (MOPS), piperazine-N,N'-bis(ethanesulfonic acid) (PIPES), N-2-hydroxyethylpiperazine-N'-2-ethanesulfonic acid (HEPES), etc.) are preferred with phosphate as another alternative.

Temperature: The reaction will be more favored the higher the temperature but one is limited by the stability of the protein, and the possibility of proteolysis of the protein during the reaction due to minor protease contamination. As a general rule, 0-25°C is the functional range of temperature one can use. Lower temperatures are preferred but the reaction will run slower under these conditions. This can be overcome by increasing other parameters which favor the reaction. One must be aware that temperature can influence the exposure of the amino acid side chains for reaction due to temperature-induced changes in protein conformation.

Time: Longer reaction times will favor higher conjugation efficiencies in most cases. For the amine-reactive dyes, the protein amino acid side chains will be competing with hydroxyl ions for the dye. A general time frame is 1-24 h. An easy approach is to start the reaction in the afternoon and let it proceed overnight.

Concentration of dye and protein: The concentration of dye used for the reaction is dictated by the efficiency of the reaction and the final dye/protein desired. A good labeling ratio is 1 mole dye/mole protein (discussed below). With this in mind, a concentration equal to 1-3 fold the protein concentration is recommended for highly reactive dyes such as the maleimides. For less reactive dyes, such as the amine-reactive dyes, 2-10 fold the protein concentration is used. The dyes are generally poorly soluble in water and are dissolved in solvents such as dimethyl sufoxide or dimethyl formamide at final concentrations in the 10-30 mM range. It is not recommended to use dimethyl sulfoxide with sulfonyl chlorides (Boyle, 1966). Once dissolved, the true concentration of the dye can be estimated spectrophotometrically using established, solvent specific, extinction coefficients. The handbook of Haugland (1992) has an extensive list of the extinction coefficients for most dyes. If the dye does not have a good spectral handle (e.g., maleimido-biotin), then you can estimate the concentration from the mass of dye dissolved in solvent. It is a good practice to pre-dilute the dye in the reaction buffer using 5-10% equivalent volume of the protein solution. This is then added to the protein solution rapidly. This approach makes it more likely that you will get uniform labeling of the protein molecules than if you added the dye directly from the stock solution. The protein concentration for the reaction depends on the type of reaction employed. Lower protein concentrations can be used with the thiol-reactive dye than the amine-reactive dyes because there are fewer competing reactions. The protein concentration should be in the 1-5 mg/ml range for reactions. Those using amine-reactive dyes should use the high end of the concentration range to favor conjugation of the protein over hydroxyl ions while the low end of the range work well for thiol-reactive dyes, especially the maleimides.

For the thiol-specific dyes, intracellular proteins should be treated with reducing agents prior to the reaction to ensure that there are no oxidized thiols in the protein. This can be done by incubating the protein with high levels of either DTT (10 mM) or 2-ME (0.1-0.5%) at pH 7-8 and room temperature in a buffer containing EDTA or EGTA and followed by desalting or dialysis to remove the reducing agent. Column desalting is recommended as it is much faster than dialysis. The buffer for the reaction with thiol-specific dyes should be degassed and contain EDTA or EGTA to minimize heavy-metal catalyzed oxidation of thiols.

Step 3: Label protein and subsequent clean-up

Once you have developed conditions which result in good labeling of the protein, you can scale up to label a larger amount to use for functionality tests and for your imaging application. The few days used developing a protocol are well spent if the reaction is successful especially considering the time likely spent isolating the protein. After you have labeled the protein, the unreacted dye can be removed by dialysis, desalting or a combination of the two. If the protein can be precipitated with ammonium sulfate, this will afford some removal of unreacted dye. However, some dyes will salt out with ammonium sulfate (e.g. Texas Red dyes). Some unreacted Texas Red dyes bind quite strongly to Sephadex G-25 so desalting on G-25 is a highly effective method for removing these dyes. A simple assay to confirm removal of unreacted dye is to analyze the protein with SDS/PAGE followed by UV illumination (see Fig. 2). There should be no fluorescence at the dye front if all unconjugated dye has been removed.

Step 4: Determine the labeling ratio

A good labeling ratio is 1 mole dye/mole protein because one has some confidence that subsequent functionality test are truly measuring the labeled protein population and not the unlabeled population. For example, functionality test with a protein labeled at ratios below 0.5 may be difficult to interpret. Higher labeling ratios may be needed to improve sensitivity; however, higher labeling ratios are much more likely to modify biological function. Estimation of the labeling ratio is done by measuring the protein and dye concentration of the same sample after removal of unreacted dye. The protein concentration can be determined using colorometric protein assays (Lowry, BCA, and Biuret) and if possible, use the unlabeled protein as the standard. It is best to use a sensitive assay read at wavelengths which have minimal overlap with the fluorescence dye. The dye concentration can be estimated spectrophotometrically using published extinction coefficients. The absorbance of most fluorescent dyes is highly solvent specific so it is best to use the solvent for which the extinction coefficient was determined. Protein conformation can also influence the spectral properties of the dye so one may need to denature the protein with urea or guanidine HCl to get an accurate measure of the dye concentration. Once both dye and protein concentration are known, one simply determines the ratio to the two concentrations to determine the labeling ratio.

Step 5: Biological activity tests

The fidelity of the conclusions one can draw from studies with labeled protein is directly related to the rigors of the biological activity tests employed. The assays used are going to be highly protein specific and dictated by the literature available on the function of the protein. In some cases, protocols can be developed which isolate functional labeled protein. For example, labeled tubulin or actin can be functionally isolated by repeated cycles of polymerization/depolymerization. From this one can conclude that the labeled proteins polymerize reversibly but you cannot conclude that they interact with other proteins in the same way as the unlabeled protein unless you assay for this. We have used reversible actin binding of labeled myosin subfragment 1 (S1) as a method to isolate functional labeled S1. The protein isolated in this fashion had ATPase activities very similar to the unlabeled protein (Swartz *et al.*, 1990). The F-5M alpha-actinin prepared in the current study appears to bundle actin as well as unlabeled actin and locate to the Z-line suggesting that it functions just as unlabeled alpha-actinin. However, we cannot conclude that it binds to proteins other than actin (if there are any) the same as unlabeled alpha-actinin. The more proteins that your labeled protein interacts with, the more complicated the activity assays become.

One approach that we have employed is to use a solid phase competition assay to monitor the binary interaction of a labeled protein with its binding partner. Specifically, we have employed this approach with a labeled subunit of the troponin complex (Swartz *et al.*, 1994). Fluorescein, rhodamine and biotin conjugates of troponin C (TnC) were prepared and assayed for their ability to bind to troponin I coated microtitre plates under high and low calcium conditions. Different levels of the fluorescently-labeled proteins or unlabeled protein were incubated in troponin-I coated wells with a fixed, saturating level, of biotin TnC. The wells were washed and biotin-TnC level was determined by color development with streptavidin-HRP. The assays showed that both unlabeled and fluorescently-labeled TnC displaced biotin-TnC but that the unlabeled TnC competed better than labeled TnC especially at low calcium. This approach has some advantages in that it can be very sensitive and one need not know the properties of the biotin conjugate. What you are interested in is how similar the unlabeled and fluorescently-labeled proteins displace the biotin conjugate.

Step 6: Use the labeled protein

After you have demonstrated that the protein has the biological activity that you are interested in, you can use it to investigate the problem of interest. In the example presented in this study, labeled alpha-actinin was incorporated into skeletal myofibril Z-lines. We have used this in conjunction with labeled S1 to understand the structure of highly shortened myofibrils which are not readily amendable to interpretation from phase-contrast images (Swartz *et al.*, 1993). This was a structural application of the probes instead of a structure/function application. We have also used the labeled S1 as a functional probe to investigate how calcium regulates the binding of myosin to the thin filament (Swartz *et al.*, 1990). We are currently using labeled alpha-actinin as a functional probe to study the dynamics of alpha-actinin exchange at the Z-line.

Another issue along with use of the protein is how to store the labeled protein. The labeled proteins are stored in the dark with reducing agents to prevent bleaching of the fluorophores. Other conditions of storage are protein dependent. The labeled alpha-actinin was stored as ammonium sulfate pellets in the dark at 4°C. Other preparations of these same protein have been stored for over a year under these conditions and still work well for imaging applications. Labeled proteins like actin and tubulin can be stored in liquid nitrogen but activity is lost over storage. Thus, one will have to experiment to determine the best conditions for storage of each specific protein.

Conclusions

In the current work, skeletal muscle alpha-actinin was used to show some of the differences in the thiol-reactive dyes in terms of their reaction efficiency and speed. The ability of the F-5M conjugate of alpha-actinin to bundle actin filaments was also used for functional characterization.

This conjugate was subsequently employed in an imaging application to determine where it was bound in skeletal muscle myofibrils.

The covalent labeling of proteins for imaging

applications is dependent upon the specific objectives and the protein to be labeled. It is a straight forward process of deciding the type of protein modification, doing pilot experiments to define conditions for labeling, labeling enough protein to allow for functionality testing, estimation of dye/protein and performing some trial imaging applications. Once you have established a protocol that gives functional protein that has sufficient signal for your imaging application, you can repeat the protein labelling on a larger scale and have plenty of labeled functional protein for more extensive experimentation.

Acknowledgements

The author would like to thank Dr. Marion Greaser for helpful suggestions on the manuscript and the Indiana University Medical School for support.

References

Aronson JF (1965) The use of fluorescein-labeled heavy meromyosin for the cytological demonstration of actin. J Cell Biol **26**: 293-306.

Boyle RE (1966) The reaction of dimethyl sulfoxide and 5-dimethylaminonaphthalene-1-sulfonyl chloride. J Org Chem **31**: 3880-3882.

Brewer CF, Riehm JP (1967) Evidence for possible nonspecific reactions between N-ethylmaleimide and proteins. Anal Biochem **18**: 248-255.

Coons AH, Creech HJ, Jones RN, Berlinger E (1942) The demonstration of pneumococcal antigen in tissues by the use of fluorescent antibody. J Immunol **45**: 159-170.

Cubitt AB, Heim R, Adams SR, Boyd AE, Gross LA, Tsien RY (1995) Understanding, improving and using green fluorescent proteins. Trends Biochem Sci **20**: 448-455.

Deerinck TJ, Martone ME, Lev-Ram V, Green DPL, Tsien RY, Spector DL, Huang S, Ellisman MH (1994) Fluorescence photoxidation with eosin: A method for high resolution immunolocalization and in situ hybridization detection for light and electron microscopy. J Cell Biol **126**: 901-910.

Ellman GL (1959) Tissue sulfhydryl groups. Arch Biochem Biophys **82**: 70-77.

Feramisco JR, Burridge K (1980) A rapid purification of alpha-actinin, filamin and a 130,000-dalton protein from smooth muscle. J Biol Chem **255**: 1194-1199.

Fritz JD, Swartz DR, Greaser ML (1989) Factors affecting polyacrylamide gel electrophoresis and electroblotting of high molecular weight myofibrillar proteins. Anal Biochem **180**: 205-210.

Gregory JD (1955) The stability of N-ethylmaleimide and its reaction with sulfhydryl groups. J Am Chem Soc **77**: 3922-3923.

Hartig PR, Bertrand NJ, Sauer K (1977) 5-iodoacetamidofluorescein-labeled chloroplast coupling factor 1: Conformational dynamics and labeling-site characterization. Biochemistry **16**: 4275-4282.

Haugland RP (1992) Handbook of Fluorescent Probes and Research Chemicals. Molecular Probes Inc. Eugene, OR. pp 5-35.

Hermanson GT (1996) Bioconjugate Techniques. Academic Press, San Diego, CA.

Hill HD, Straka JG (1988) Protein determination using bicinchoninic acid in the presence of sulfhydryl reagents. Anal Biochem **170**: 203-208.

Jue R, Lambert JM, Pierce LR, Traut RR (1978) Addition of sulfhydryls groups to *Escherichia coli* ribosomes by protein modification with 2-iminothiolane (methyl 4-mercaptobutyrimidate). Biochemistry **17**: 5399-5404.

Means GE, Feeney RE (1971) Chemical Modification of Protein. Holden-Day, San Francisco.

Meyer RK, Aebi U (1990) Bundling of actin filaments by α-actinin depends on its molecular length. J Cell Biol **110**: 2013-2024.

Miki M, Dos Remedios GC (1988) Fluorescence quenching studies of fluorescein attached to lys-61 or cys-374 in actin: Effects on polymerization, myosin subfragment-1 binding and tropomyosin-troponin binding. J Biochem **104**: 232-235.

Milligan RA, Whittaker M, Safer D (1990) Molecular structure of F-actin and location of surface binding sites. Nature **348**: 217-221.

Pardee JD, Spudich JA (1982) Purification of muscle actin. Meth Enzymol **85**: 164-181.

Safer D, Bolinger L, Leigh JS Jr (1986) Undecagold clusters for site-specific labeling of biological macromolecules: Simplified preparations and model applications. J Inorganic Biochem **26**: 77-91.

Sanger JW, Mittal B, Sanger JM (1984) Analysis of myofibrillar structure and assembly using fluorescently-labeled contractile proteins. J Cell Biol **98**: 825-833.

Suzuki A, Goll DE, Singh I, Allen RE, Robson RM, Stromer MH (1976) Some properties of purified skeletal muscle α-actinin. J Biol Chem **251**: 6860-6870.

Swartz DR (1989) Structural and Functional Studies of the Bovine Myofibril. Doctoral Thesis. University of Wisconsin-Madison.

Swartz DR, Greaser ML, Marsh BB (1990) Regulation of subfragment-1 binding in isolated rigor myofibrils. J Cell Biol **111**: 2989-3001.

Swartz DR, Greaser ML, Marsh BB (1993) Structural studies of rigor bovine myofibrils using fluorescence microscopy. II. Influence of sarcomere length on the binding of myosin subfragment-1, alpha-actinin and G-actin to rigor myofibrils. Meat Sci. **33**: 157-190.

Swartz DR, Greaser ML, Moss RL (1994) A simple solid phase assay for troponin subunit interactions. Biophys J **66**: A310.

Wang Y-L, Taylor DL (1980) Preparation and characterization of a new molecular cytochemical probe: 5-iodoacetamidofluorescein-labeled actin. J Histochem Cytochem **28**: 1198-1206.

Wang Y-L, Taylor DL (1989) Fluorescence microscopy of living cells in culture. Parts A and B. Academic Press, Inc. New York.

Discussion with Reviewers

M. Malecki: In studies on incorporation of proteins, not only would you record an increase of fluorescence intensity due to accumulation of protein derivatized with fluorochrome, but also you would experience reduction in that intensity as the result of fading and further exchange with underivatized molecules. Would you find it feasible to quantitate the final number of incorporated molecules based upon fluorescence?
Author: Yes, I would find it feasible to quantify the number of incorporated molecules based upon fluorescence intensity. One does not obtain an absolute value for number of molecules but monitors a relative change in fluorescence intensity. The competition between unlabeled and labeled molecules for incorporation can be a problem but the experiments are usually done with the labeled molecule in great excess over unlabeled such that this in not a problem. You can use the opposite approach of getting maximal incorporation of labeled molecules followed by wash-out of labeled molecule with unlabeled by following the decrease in fluorescence intensity. Photobleaching is an over-riding experimental problem when trying to measure fluorescence intensity. A few ways of dealing with it are minimizing exposure time to excitation light, adding oxygen or free-radical scavengers to the medium for in vitro experiments, and characterizing the bleaching kinetics for your system then correcting intensity values using the kinetic constants obtained from the bleaching time-course. For the systems that I work in, I do not focus using the fluorescence image, rather, I focus using phase contrast then obtain and image in the fluorescence channel only using one exposure time. Thus, bleaching complications are minimized.

M. Malecki: In your experience, what is the most efficient, but least toxic anti-fading agent applicable for studies on living systems?
Author: There is no readily accepted agent to minimize fading problems in living systems since oxygen is required for the cells. Rather, the approach is to optimize the imaging system sensitivity such that minimal exposure to excitation light is used. This is accomplished by minimizing exposure during focus and image acquisition. Currently available sensitive cameras (both video and CCD) allow one to obtain high quality images with short exposure times. Selection of more photo-stable dyes also facilitates imaging in live cells.

M. Malecki: How would you quench formation of free radicals to reduce phototoxicity for living systems?
Author: The cell itself has free-radical scavengers in the form of reduced glutathione and ascorbate but these could be depleted during long exposure to excitation light. Again, the best approach is to use minimal exposure excitation light and photo-stable dyes.

J.R. Lakowicz: Many authors have noticed that the fluorescence intensity of labeled proteins does not increase linearly with the number of covalently attached groups. Have you noticed such effects in your labeling of alpha-actinin? If you have noticed such effects, which probes seem to show the greatest amount of self quenching?
Author: Yes I have observed this most noticeably upon over-conjugation of alpha-actinin with Texas Red. I have not noticed it as readily with fluorescein. Part of this self quenching may be because the probe becomes buried in hydrophobic regions of the protein thus changing its excitation and emission properties and there is the possibility of inner-filter effects when many dye molecules are confined to the small volume of the protein. We try to use dye to protein ratios of 1 for most proteins that we use and use low amounts of dye for the conjugation reaction so that we do not over-label the protein. High labeling ratios are also much more likely to alter biological function.

J.R. Lakowicz: Many probes have been reported to photobleach rapidly in microscopy. What types of fluorophores have you found to photobleach most rapidly in microscopy? What types of probes have you found to be most stable in microscopy?
Author: From my experience, fluorescein and lower excitation wavelength probes are the most susceptible to photobleaching while the longer wavelength dyes, such as Texas Red and the Cye dyes are less susceptible whether antifade agents are added or not.

J.R. Lakowicz: What would you consider to be the characteristics of ideal fluorescent probes?
Author: The ideal fluorescence probe would be readily reactive with cysteines (preferably a maleimide derivative), have excitation above 340 nm so that special optics would not be needed, and be photostable so that bleaching is minimal.

Scanning Microscopy Supplement 10, 1996 (pages 285-294)
Scanning Microscopy International, Chicago (AMF O'Hare), IL 60666 USA
0892-953X/96$5.00+.25

CYTOSKELETON ARCHITECTURE OF C6 RAT GLIOMA CELL SUBCLONES WHOLE MOUNT ELECTRON MICROSCOPY AND IMMUNOGOLD LABELING

Wolfgang Bohn[1*], Dörte Etzrodt[1], Roland Foisner[2], Gerhard Wiche[2], and Peter Traub[3]

[1]Heinrich-Pette-Institut, Hamburg, Germany; [2]Institute of Biochemistry and Molecular Cell Biology, University of Vienna, Austria; [3]Max-Planck-Institute of Cell Biology, Ladenburg, Germany

(Received for publication August 6, 1995 and in revised form March 21, 1996)

Abstract

Whole mount electron microscopy of extracted cells combined with immunogold labeling techniques can be used to characterize the cytoskeletal architecture of cultured cells. As shown with subclones of the C6 rat glioma cell line, heavy metal shadowing was suitable for getting basic information concerning the arrangement of the various filament types within the networks. Pure carbon shadowing combined with immunogold double labeling proved to be optimal to identify linkages between filaments, to localize filament associated proteins and to follow the arrangement of filaments in dense arrays such as lamellipodiae and cell margins. Thin connecting filaments which interact with actin as well as with vimentin filaments and can be labeled with antibodies to the intermediate filament associated protein plectin may play a major role in the structural organization of the cytoskeleton of these cells.

Key Words: Connecting filaments, cytoskeleton, immunogold labeling, intermediate filaments, plectin, vimentin, whole mount.

*Address for correspondence:
Wolfgang Bohn
Heinrich-Pette-Institute
Martinistrasse 52
D-20251 Hamburg, Germany
Fax Number: +49-40-464709

Introduction

Microscopic techniques, most of all immunofluorescence but also electron microscopy have made important contributions to the current image of the cytoskeleton in mammalian cells. Intermediate filaments, microtubules, and actin filaments compose the prominent structures of this cytoskeleton and are arranged in a complex three-dimensional network. In addition, there is a growing number of filament associated proteins, which either control filament assembly and/ or stability or regulate their interaction with cellular constituents, such as proteins which cap or fragment actin filaments [35], proteins which attach to microtubules (MAPs)[21] and those showing affinity for intermediate filament (IF) proteins [12]. There is evidence that the various filament systems do not act independently but are structurally linked [38], either directly or indirectly via crosslinking proteins, and that this interaction is a prerequisite for the cell to properly react to an external stimulus [23]. Thus alterations in structure and organization of one of the interacting components would have a significant influence on the structural arrangement and function of the others, a suggestion which corresponds to observations, that substances which act quite specifically on one filament system simultaneously change the arrangement of others [3]. Similar observations have been made in studies on integrins. By interaction of the transmembrane integrins with the corresponding ligands mechanical force is transmitted into the cell, a process which induces binding of the integrin molecule to submembraneous actin filaments and leads to activation of signal transduction pathways [28, 45]. The interaction of actin filaments with these plasma membrane proteins is accompanied by selection and anchoring of certain cytoplasmic proteins (α-actinin, talin, focal adhesion kinase) into larger complexes [15, 31]. This process of force transmission seems to be dependent on the integrity of all of the major filament systems [23]. Thus, the structural interaction of the major filament systems seems to influence not only the cell shape, but also specific functions, such as the transmission of signals from the extracellular matrix into the cell, or the maintenance of junctional complexes in lateral membrane areas of epithelial cells which is essential to stabilize their polarized organization. At the same time, the data point to special arrangements and local differences in the composition of the cytoskeleton in the various cell compartments which is essential for the cell to fulfill specific functions.

Based on the suggestion that a specific structural organization of the cytoskeleton reflects a specific function, a major question concerns the localization and arrangement of the numerous cytoskeletal proteins and how their structural interactions change in complex dynamic processes.

Visualization of Cytoskeletons in Electron Microscopy

Electron microscopy combined with immunocytochemical procedures can give some answers to these questions, with certain limitations [36, 38]. Filaments of the cytoskeleton are arranged in a three-dimensional network. Thus it is reasonable to choose electron microscopic methods which will reveal the three-dimensional orientation. This is realized with whole mount cytoskeletons obtained from cells permeabilized or extracted with a non-ionic detergent and then either quick frozen/deep etched or critical point dried and shadowed with metal and carbon.

Alternatively proteins may be identified by immunolabeling on ultrathin sections of non-extracted cells (post-embedding labeling). However, with non-muscle cells localization of cytoskeletal proteins on ultrathin sections is made difficult by the anisotropic organization of the cytoskeletal network and the limited thickness of the section. In addition, labeling is largely confined to the surface of the section. Labeling on resinless sections seems to be an alternative approach, but it is still limited by the fact that only a very small area of the cell can be visualized [46].

When applying the whole mount technique one has to be aware of significant losses of proteins during extraction and possible rearrangements. To make the findings obtained with extracted cells more reliable different approaches may be followed. First, the extraction procedures can be varied, providing a better definition of the conditions leading to the appearance of certain structures. Secondly, the cell system can be manipulated, for instance with certain cytoskeleton specific drugs. This may show that a specific cytoskeletal organization depends on specific physiological state of the cell. Thirdly, the residual cytoskeleton itself can be manipulated to define conditions that maintain a typical structural arrangement at this stage. In addition, the selection of cell mutants that are deficient in certain cytoskeletal proteins has proven to be a useful approach to understand the function and dynamics of individual cytoskeletal proteins [1].

Whole Mount Cytoskeletons

Three main aspects have to be considered in the whole mount technique when it is applied to the cytoskeleton, namely detergent solubilization, fixation, and shadowing. An excellent survey of the various methods in existence has been given by Hartwig [16].

Cytoskeletons of C6 cells grown on glass coverslips were either obtained by solubilization with the non-ionic detergent NP40 (1%) in a KCl (80 mM) imidazole (20 mM) buffer at pH 7.0, supplemented with $MgCl_2$ (2 mM) and the calcium chelating agent ethylene glycol-bis-(β-aminoethylether)N,N,N',N'tetra-acetic acid (EGTA) (2 mM) at 40°C [6]. The remaining cytoskeleton consists mainly of actin, intermediate filaments and microtubules.

Stabilization of the cytoskeleton with a chemical agent is an essential point, when immunolabeling procedures will be applied that enclose prolonged incubations with antibodies and repeated steps of washing. The cytoskeleton of C6 cells was fixed with a mixture of freshly prepared paraformaldehyde (1%) and a low concentration of glutaraldehyde (0.025%) prior to labeling. This low concentration of glutaraldehyde proved to be sufficient to stabilize the structure and had the advantage of minimizing the fixative dependent background staining. Preservation of ultrastructure was not improved at this stage by increasing the glutaraldehyde concentration. The samples were postfixed with 0.5% glutaraldehyde after labeling. Higher concentrations of glutaraldehyde may disrupt microfilaments, as does also osmium tetroxide [8, 26, 37].

After fixation the C6 cytoskeleton was dehydrated through a graded series of ethanol and critical point (CP) dried. Alternatively, the cytoskeletons may be subjected to quick freezing and deep etching, which has been shown to give superior images of cytoskeleton structures at high resolution [9, 17, 20, 24, 27]. CP-drying, which was used throughout in our studies on C6 cytoskeletons, has also been applied with success [10, 11, 29, 32]. The reliability of both methods has been discussed extensively. In some studies a higher variation in filament thickness was noted in CP dried probes as compared with rapidly frozen samples [18]. Others indicated that both methods provide a similar cytoskeletal ultrastructure and when properly handled CP drying was considered to be a reliable method yielding consistent results [7]. In critical point drying, problems seem to arise when residual water is present in CO_2, which leads to a lateral aggregation of filaments [7].

After drying or deep etching the cytoskeletons are

normally shadowed with a layer of heavy metal. Cytoskeletons of C6 cells were either rotary shadowed with platinum/carbon or with pure carbon. For platinum/carbon shadowing the probes were mounted on a stage cooled with liquid nitrogen and shadowing was done at low pressure (10^{-7} bar) to improve the resolution of the metal film [22]. Immunogold-labeled probes were generally shadowed with pure carbon which provided better identification of the label because of increased contrast between the gold label and the cytoskeletal structure. In order not to decrease the filament contrast itself in these probes the cytoskeletons were generally mounted on grids without a supporting film [22].

IF-Protein Expression in C6 Rat Glioma Cells

As mentioned above cytoskeletons obtained by extraction of cells mainly consist of actin filaments, microtubules and intermediate filaments. In mammalian cells the proteins of intermediate filaments (IFs) normally represent a major cytoskeleton component, of which the expression is tissue-specific and developmentally regulated. The cytoplasmic IF-proteins probably originate from nuclear lamins as a common ancestor [25]. Those present in vertebrate cells can be divided into four groups, the vimentin-like and the neuronal IF-proteins, the cytokeratins, and the lens IF-proteins. Vimentin is the characteristic IF-protein found in terminally differentiated cells of mesenchymal origin. However, due to unknown reasons, cells frequently start to express vimentin in addition to their tissue specific IF-protein when explanted from tissue, indicating that vimentin expression is dependent on environmental conditions. Thus vimentin is the most frequent IF-protein in cultured cells and has been used as a model system to characterize the biology of IF-proteins. However, despite extensive knowledge of its molecular structure it is still a matter of question if it fulfills a specific function [41]. Vimentin seems to be linked to the nuclear lamina, and probably interacts with the microtubule and actin filament system, and with components in the plasma membrane [25]. Biochemical data indicated that the intermediate filament associated protein plectin binds to vimentin and may be involved in crosslinking cytoplasmic filament systems [12].

In contrast to the actin filament system and microtubules the presence of cytoplasmic IF-proteins is not obligatory, which is well documented by the existence of some IF-protein deficient cell lines [19, 40, 42]. IF-deficient cells have been used in studies on vimentin network formation performed by immunofluorescence either after transfection with an appropriate vector system or after injection of purified vimentin proteins. Absence of cytoplasmic IF-proteins is a rather rare event and it has proved to be difficult to establish subclones with stable phenotypes from these adherent cell lines.

The C6 rat glioma cell line, which was cloned from a chemically induced rat brain tumor, was found to consist of a mixed population of cells which either contain vimentin (80% of the cells) or completely lacked any cytoplasmic IF-proteins [34]. These latter cells showed a normal content of lamins A, B, and C [30]. In IF-deficient cells the absence of vimentin was found to result from a block at the transcriptional level. As stable subclones could be established with both phenotypes, this cell system seemed to be useful not only to follow IF-protein assembly, but to use IF-protein expression in these cells as a tool for analyzing principal mechanisms in the organization of the cytoskeleton by use of ultrastructural and immunocytochemical methods.

Cytoskeleton of C6 Cells

The cytoskeleton of the C6 subclone containing the intermediate filament protein vimentin revealed a wide spaced filament network (Fig. 1A). The prominent filaments measured 10 nm and 13 nm in width, which corresponds to that of actin and intermediate filaments taking into account a 3 nm metal layer. In between individual thin connecting filaments of about 6 nm were detectable. Similar filament types were visible in carbon shadowed cytoskeletons (Fig. 1B). In addition filaments of about 30 nm were present corresponding in size to microtubules. In IF-deficient C6 cells filament bundles and thin connecting filaments were the prominent structures (Fig. 1C). The cytoskeleton of IF-deficient cells was more fragile and more difficult to handle than that of the vimentin containing cells. To unequivocally identify the arrangement of vimentin and actin in this cytoskeleton and to localize sites of interaction we identified these filament types by immunogold labeling [4].

Immunogold-Labeling of Filaments

As shown above, the samples obtained after extraction of cells with non-ionic detergents represent a filamentous network which mainly consists of actin filaments, intermediate filaments, and microtubules which can be distinguished on the basis of their substructure and width. These filaments represent structures built of identical subunits assembled in a specific way. Antibodies developed against these proteins are directed against these subunits. Thus, in labeling experiments, one would expect a continuous labeling along these filaments as was demonstrated with antibodies to tubulin [2] and is shown with antibodies to actin and vimentin in

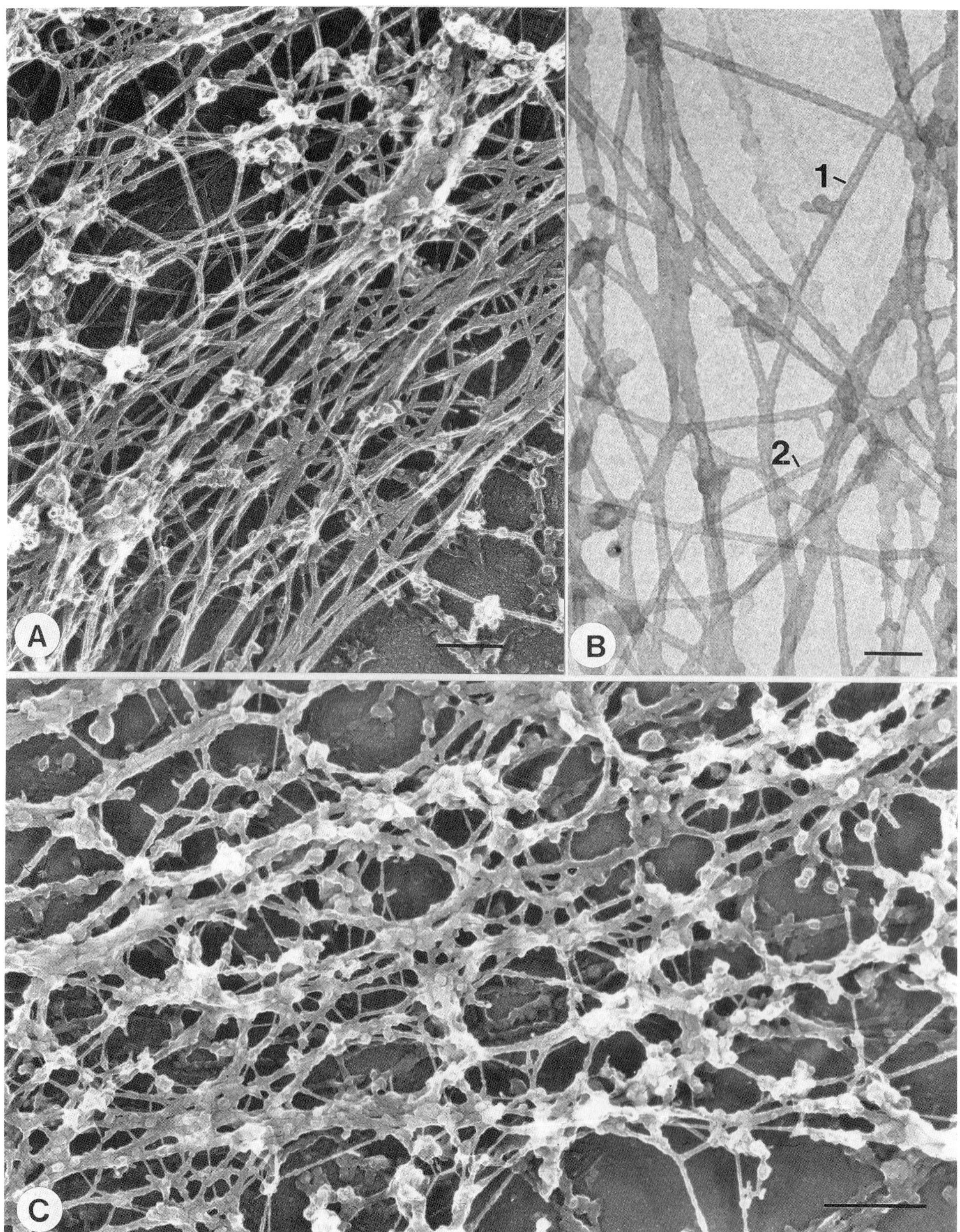

Figure 1: Cytoskeletons of C6 subclones. (A and B) Cytoskeletons of vimentin positive C6 cells; (C) Cytoskeleton of an intermediate filament deficient C6 cell. (A and C) rotary shadowing with platinum/carbon;(B) rotary shadowing with carbon. Bars in (A) = 0.2μm; (B) = 0.1μm; (C) = 0.5μm.

Fig. 2A. Despite the fact that in our labeling studies antibodies were used in excess, the amount of background label was very low. There was no random distribution of gold particles at the surface of the glass coverslip and the thin connecting filaments as well as structures corresponding to microtubules were clearly unlabeled. On the other hand the labeling density along the actin and vimentin filaments was very high and individual vimentin filaments could easily be followed in the dense filamentous networks in lammellopodiae (Fig. 2 C and D). Lower density of gold label along these filaments or a labeling gap may indicate a change in the composition of the filament as it may result from the presence of filament associated proteins. In the IF-deficient subclone antibodies to actin labeled the filament bundles but were absent on the thin connecting filaments (Fig. 2B).

In contrast to the major cytoskeletal proteins, other proteins do not assemble into filaments of repetitive subunits but are present as individual molecules or oligomeric structures. In principle, they cannot easily be identified on the basis of a typical morphology as it is possible to do with the major filament systems. Their localization relies even more on the use of immunocytochemical techniques. The proteins may be randomly distributed throughout the network. Consequently, the label may be dispersed throughout the network. This for instance is shown in immunogold labeling studies on membrane skeletons of platelets [18]. After labeling with antibodies to spectrin, gold conjugates were randomly associated with thin connecting filaments in these skeletons, and the gold labeling of actin-binding-protein complexes was also randomly distributed and could not be assigned to a specific ultrastructure. Similar problems arose when localizing plectin in the cytoskeleton of C6 subclones.

Plectin Localization in C6 Subclones

Plectin is a 466 KD protein, which originally has been identified as a major component of the salt resistant fraction of detergent solubilized C6 cells [43]. The protein contains a N-terminal globular domain, a central α-helical coiled-coil, and a C-terminal globular domain with a prominent sixfold tandem repeat organization [44]. Rotary shadowed probes of purified plectin show dumb-bell shaped (probably homotetrameric) molecules consisting of 9 nm globular domains linked by a 200 nm rod of 2 nm width [13, 14]. The high affinity of plectin for intermediate filament proteins, fodrin, MAPs and its widespread occurrence suggested that this protein plays a major role in linking cytoskeletal structures [12] mainly in interaction with the intermediate filament system. Based on these data, we investigated how plectin behaves in cells lacking cytoplasmic intermediate filament proteins. Plectin was labeled with antibodies which recognize an epitope in the rod domain of the dumb-bell shaped molecule [13]. In vimentin containing cells the plectin label was present throughout the filament network and frequently found in association with thin connecting filaments (Fig. 3A). Double labeling with antibodies to plectin and actin verified that these thin connecting filaments, the filaments with which the plectin label was associated, did not consist of actin (Fig. 3B). The label was localized at sites where these filaments made contact with intermediate filaments or in the center of these connecting filaments. Obviously, the intermediate filaments changed their direction at sites that were in contact with the thin connecting filaments. Plectin was also present in the IF-deficient subclone bound to actin containing structures. Again, the plectin label was also found in association with thin connecting filaments (Fig. 3C). The association with actin in vimentin negative cells could be substantiated by the fact that the plectin label was lost from these but not from the vimentin containing cytoskeletons after treatment with heavy meromyosin which specifically associated with actin filaments. These data support the suggestion that plectin plays a major role in crosslinking cytoskeletal filament systems and probably in forming thin connecting filaments [33].

Conclusion

The data substantiate the usefulness of whole mount electron microscopy combined with immunogold labeling for the identification of the structural arrangement of proteins within the cytoskeleton. With antibodies to actin and vimentin the immunogold conjugates continuously decorated the corresponding filaments without giving any background labeling. Any presence of gold particles at the surface of the glass coverslip outside the cytoskeleton would either indicate inappropriate labeling conditions. Alternatively the protein to be identified may be lost from the cytoskeleton during extraction and dispersed throughout the network. In this case it would be uncertain if gold particles present on specific structures reflect the original localization of this protein.

Absence of background label especially is a prerequisite for a reliable interpretation of those labeling patterns which do not follow the orientation of specific structures such as filaments. In the case of plectin the reliability of the labeling pattern is substantiated by the fact that it was associated with thin connecting filaments.

On labeled probes the structural resolution is deteriorated by the attached antibodies and immunogold conjugates. Thus shadowing with carbon proved to be sufficient for stabilization of the delicate structures and

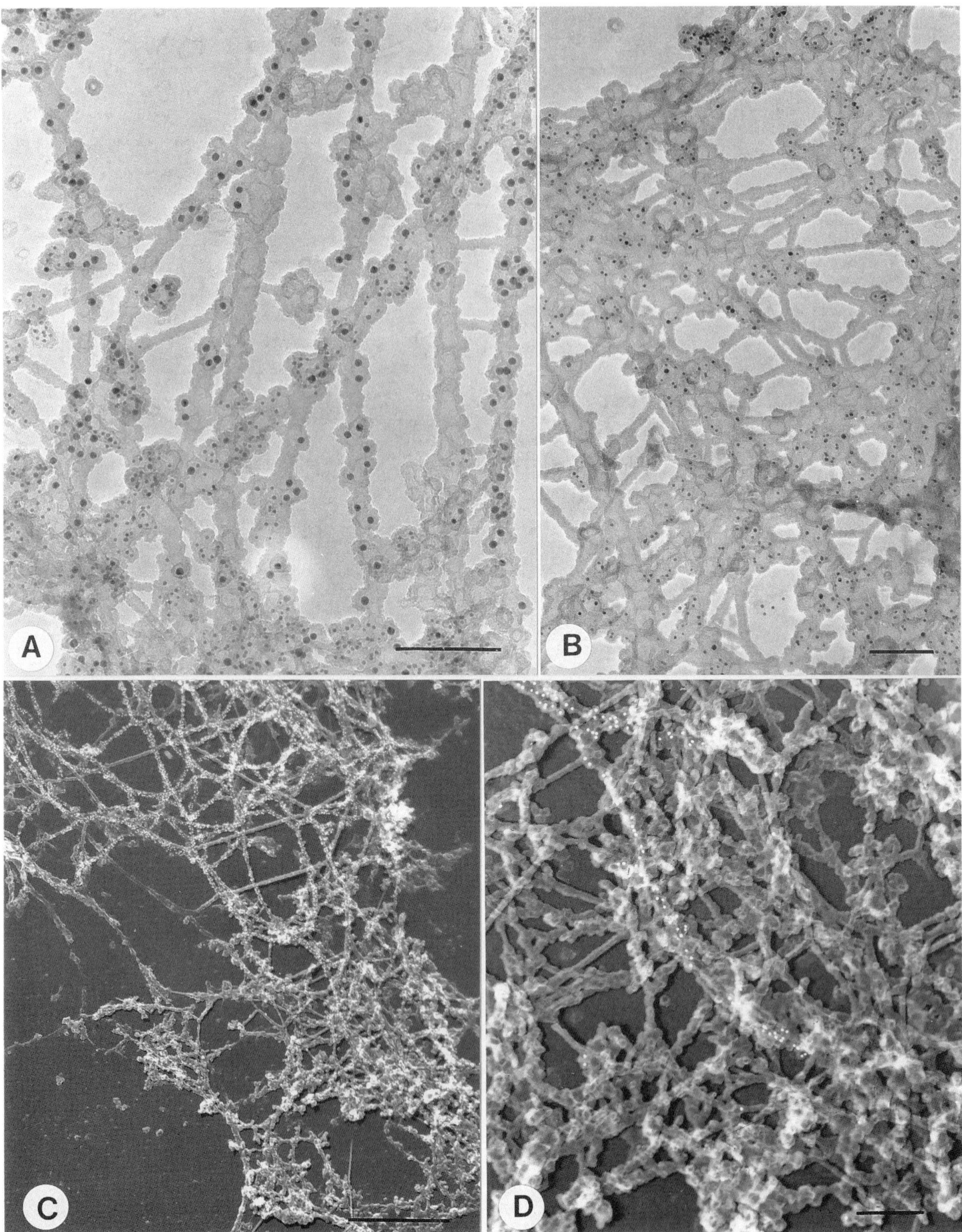

Figure 2: (A) C6 cytoskeleton labeled with rabbit antibodies to actin and a goat anti-rabbit (5nm) immunogold conjugate and a monoclonal antibody to vimentin and a goat anti mouse (10nm) gold conjugate. (B) Cytoskeleton of a intermediate filament deficient cell labeled with antibodies to actin as in (A). (C) and (D) C6 cytoskeleton labeled with a monoclonal antibody to vimentin as in (A). (D) Enlarged detail of (C). Bars: (A) and (B) = 0.1μm; (C) = 1μm; (D) = 0.2μm.

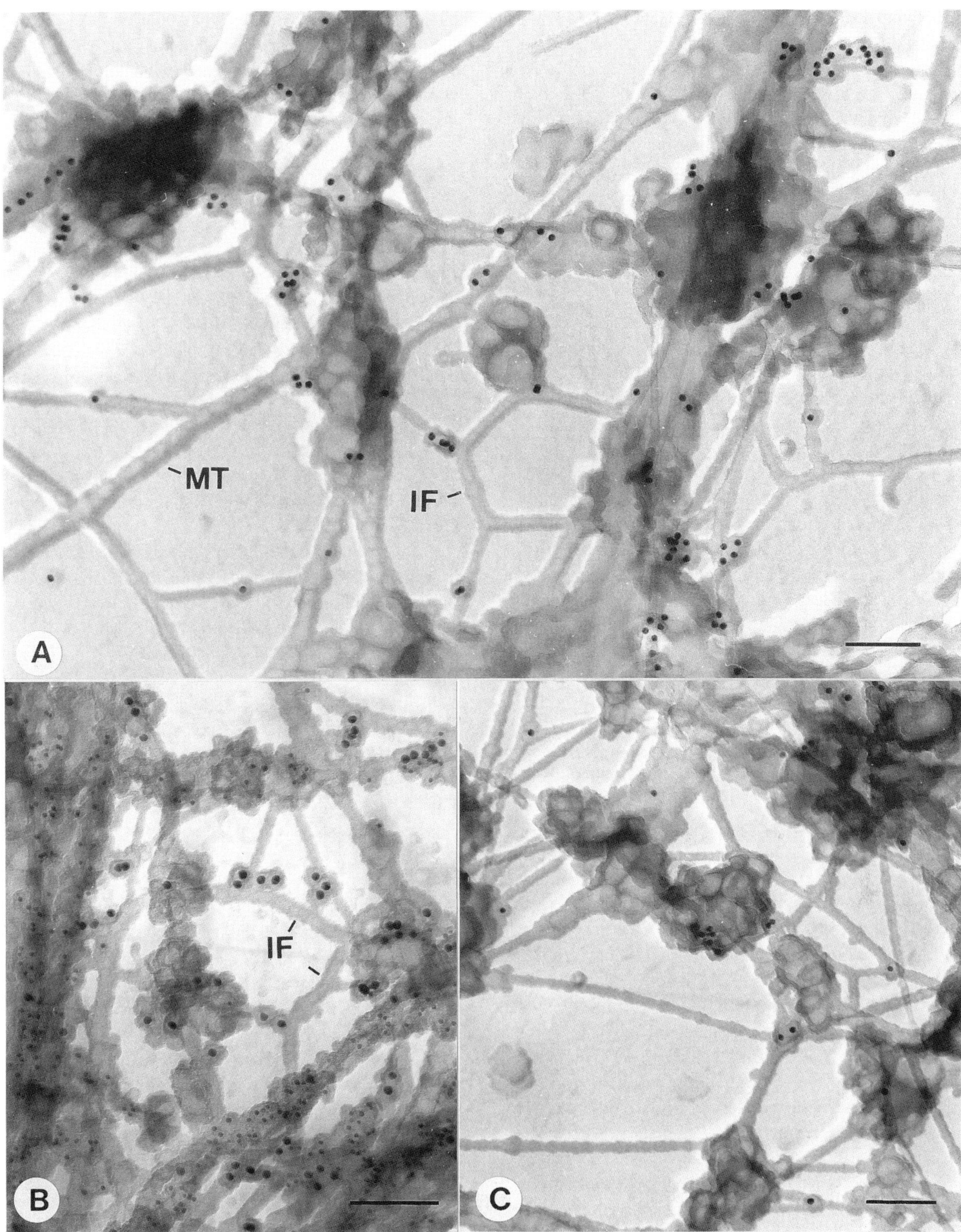

Figure 3: (A) Cytoskeleton of a vimentin positive cell labeled with a monoclonal antibody to plectin and a goat anti-mouse (10nm) immunogold conjugate. (B) Cytoskeleton of a vimentin positive cell labeled with an antibody to plectin (as in A) and antibodies to actin (as in Figure 2). (C) Cytoskeleton of a intermediate filament deficient cell labeled with an antibody to plectin as in (A). Bars in (A), (B), and (C) = 0.1μm.

had also the advantage to retain the contrast between the gold label and the biological structure.

Acknowledgements

The authors wish to acknowledge the partial support of this research by Gemeinnützige Hertie-Stiftung, Frankfurt/Main, Germany. The Heinrich-Pette-Institut is financially supported by Freie Hansestadt Hamburg und Bundesministerium für Gesundheit.

References

1. Andre E, Brink M, Gerisch G, Isenberg G, Noegel A, Schleicher M, Segall JE, Wallraff E (1989) A *Dictyostelium* mutant in severin, an F-actin fragmenting protein, shows normal motility and chemotaxis. J Cell Biol **108**: 985-995.

2. Basgall EJ, Soong MM, Tompkins WAF (1988) Visualization of cytoskeletal elements and associated retroviral antigens by immunogold transmission electron microscopy of detergent extracted cells. In: Biotechnology and Bioapplications of Colloidal Gold. Albrecht RM, Hodges GM, eds. Scanning Microscopy International, AMF O'Hare (Chicago), IL, pp 161-166.

3. Blose SH, Meltzer DI, Feramisco JR (1984) 10nm filaments are induced to collapse in living cells microinjected with monoclonal and polyclonal antibodies against tubulin. J Cell Biol **98**: 847-858.

4. Bohn W, Röser K, Hohenberg H, Mannweiler K, Traub P (1993) Cytoskeleton architecture of C6 glioma cell subclones differing in intermediate filament protein expression. J Struct Biol **111**: 48-58.

5. Bohn W, Wiegers W, Beuttenmüller M, Traub P (1992) Species-specific recognition patterns of monoclonal antibodies directed against vimentin. J Exp Cell Res **201**: 1-7.

6. Bohn W, Rutter G, Hohenberg H, Mannweiler K, Nobis P (1986) Involvement of actin filaments in budding of measles virus: Studies on cytoskeletons of infected cells. Virology **149**: 91-106.

7. Borrelli MJ, Koehm S, Cain CA, Tompkins WAF (1985) The effect of preparative protocol on the cytoskeleton ultrastructure observed in extracted whole-mount BHK cells. J Ultrastruct Res **91**: 57-65.

8. Bridgman PC, Reese TS (1984) The structure of cytoplasm in directly frozen cultured cells. I Filamentous meshworks and the cytoplasmic ground substance. J Cell Biol **99**: 1655-1668.

9. Bridgman PC, Dailey ME (1989) The organization of myosin and actin in rapid frozen nerve growth cones. J Cell Biol **108**: 95-109.

10. Capco D, Penman S (1983) Mitotic architecture of the cell: The filament networks of the nucleus and cytoplasm. J Cell Biol **96**: 896-906.

11. Carmo-Fonseca M, Cidadao AJ, David-Ferreira JF (1987) Filamentous cross-bridges link intermediate filaments to the nuclear pore complexes. Eur J Cell Biol **45**: 282-290.

12. Foisner R, Wiche G (1991) Intermediate filament-associated proteins. Curr Opin Cell Biol **3**: 75-81.

13. Foisner R, Feldman B, Sander L, Wiche G (1991) Monoclonal antibody mapping of structural and functional plectin epitopes. J Cell Biol **112**: 397-405.

14. Foisner R, Leichtfried FE, Herrmann H, Small JV, Lawson D, Wiche G (1988) Cytoskeleton-associated plectin: *in situ* localization, *in vitro* reconstitution, and binding to immobilized intermediate filament proteins. J Cell Biol **106**: 723-733.

15. Geiger B, Avnur Z, Kreis TE, Schlesinger J (1984) The dynamics of cytoskeletal organization in areas of cell contact. In: Cell and Muscle Motility, Vol 5: The Cytoskeleton. Shay JW, ed. Plenum Press, New York, pp 195-234.

16. Hartwig JH (1992) An ultrastructural approach to understanding the cytoskeleton. In: The Cytoskeleton, A Practical Approach. Carraway KL, Carraway CAC, eds. Oxford University Press, New York, pp 23-45.

17. Hartwig JH (1992) Mechanisms of actin rearrangements mediating platelet activation. J Cell Biol **118**: 1421-1442.

18. Hartwig JH, DeSisto M (1991) The cytoskeleton of the resting human blood platelet: structure of the membrane skeleton and its attachment to actin filaments. J Cell Biol **112**: 407-425.

19. Hedberg KK, Chen LB (1989) Absence of intermediate filaments in a human adrenal cortex carcinoma-derived cell line. Exp Cell Res **163**: 509-517.

20. Heuser JE, Kirschner MW (1980) Filament organization revealed in platinum replicas of freeze dried cytoskeletons. J Cell Biol **86**: 212-234.

21. Hirokawa N (1994) Microtubule organization and dynamics dependent on microtubule-associated proteins. Curr Opin Cell Biol **6**: 74-81.

22. Hohenberg H, Bohn W, Rutter G, Mannweiler K (1988) Plasma membrane antigens detected by replica technique. In: Biotechnology and Bioapplications of Colloidal Gold. Albrecht RM, Hodges GM, eds. Scanning Microscopy International, AMF O'Hare (Chicago), IL, pp 195-204.

23. Ingber DE (1993) Cellular tensegrity: defining new rules of biological design that govern the cytoskeleton. J Cell Science **104**: 613-627.

24. Lawson D (1984) Distribution of epinemin in colloidal gold-labelled quick-frozen, deep-etched cytoskeletons. J Cell Biol **99**: 1451-1460.

25. Klymkowski MW (1995) Intermediate fila-

ments: new proteins, some answers, more questions. Curr Opin Cell Biol **7**: 46-54.

26. Lehrer SS (1981) Damage of actin filaments by glutaraldehyde. Protection by tropomyosin. J Cell Biol **90**: 459-466.

27. Nakata T, Hirokawa N (1987) Cytoskeletal reorganization of human platelets after stimulation revealed by the quick-freeze deep-etch technique. J Cell Biol **105**: 1771-1780.

28. Nebe B, Rychly J, Knopp A, Bohn W (1995) Mechanical induction of ß1-integrin-mediated calcium signaling in a hepatocyte cell line. J Exp Cell Res **218**: 479-484.

29. Niederman R, Amrein PC, Hartwig J (1983) Three-dimensional structure of actin filaments and of an actin gel made with actin-binding protein. J Cell Biol **96**: 1400-1413.

30. Paulin-Levasseur M, Scherbarth A, Giese G, Röser K, Bohn W, Traub P (1989) Expression of nuclear lamins in mammalian somatic cells lacking cytoplasmic intermediate filament proteins. J Cell Sci **92**: 361-370.

31. Rinnerthaler G, Geiger B, Small JV (1988) Contact formation during fibroblast locomotion: involvement of membrane ruffles and microtubules. J Cell Biol **106**: 747-760.

32. Ris H (1985) The cytoplasmic filament system in critical point dried whole mounts and plastic embedded sections. J Cell Biol **100**: 1474-1487.

33. Roberts TM (1987) Fine (2-5nm) filaments: New types of cytoskeletal structures. Cell Motil Cytoskel **8**: 130-142.

34. Röser K, Bohn W, Giese G, Mannweiler K (1991) Subclones of C6 rat glioma cells differing in intermediate filament protein expression. Exp Cell Res **197**: 200-206.

35. Rozycki MD, Myslik JC, Schutt CE, Lindberg U (1994) Structural aspects of actin binding proteins. Curr Opin Cell Biol **6**: 87-95.

36. Shiomura Y, Hirokawa N (1987) Colocalization of microtubule-associated protein IA and microtubule associated protein 2 on neuronal microtubules *in situ* revealed with double label immunoelectron microscopy. J Cell Biol **104**: 1575-1578.

37. Small JV, Langanger G (1981) Organization of actin in the leading edge of cultured cells: Influence of osmium tetroxide and dehydration on the ultrastructure of actin meshworks. J Cell Biol **91**: 695-705.

38. Small JV (1989) Immunocytochemistry of contractile and cytoskeletal proteins. In: Electron Microscopy of Subcellular Dynamics. Plattner H, ed. CRC Press, Boca Raton, FL, pp 293-308.

39. Tint IS, Hollenbeck PJ, Verkhovsky AB, Surgucheva IG, Bershadsky AD (1991) Evidence that intermediate filament reorganization is induced by ATP-dependent contraction of the actomyosin cortex in permeabilized fibroblasts. J Cell Sci **98**: 375-384.

40. Traub UE, Nelson WJ, Traub P (1983) Polyacrylamide gel electrophoretic screening of mammalian cells cultured *in vitro* for the presence of the intermediate filament protein vimentin. J Cell Sci **62**: 129-147.

41. Traub P, Shoeman RL (1994) Intermediate filament proteins: cytoskeletal elements with a gene-regulatory function? Int Rev Cytol **154**: 1-103.

42. Venetianer A, Schiller DL, Magin T, Franke WW (1983) Cessation of cytokeratin expression in a rat hepatoma cell line lacking differentiated functions. Nature **305**: 730-733.

43. Wiche G, Herrmann H, Leichtfried F, Pytela R (1982) Plectin: a high molecular- weight cytoskeletal polypeptide component that copurifies with intermediate filaments of the vimentin type. Cold Spring Harb Symp Quant Biol **1**: 475-482.

44. Wiche G, Becker B, Luber K, Weitzer G, Castanon MJ, Hauptmann R, Stratowa C, Stewart M (1991) Cloning and sequencing of rat plectin indicates a 466-kD polypeptide chain with a three-domain structure based on a central alpha-helical coiled coil. J Cell Biol **114**: 83-99.

45. Wang N, Butler JP, Ingber DE (1993) Mechanotransduction across the cell surface and through the cytoskeleton. Science **260**: 1124-1127.

46. Wolosewick JJ (1988) To resin or not to resin in immunocytochemistry. In: Biotechnology and Bioapplications of Colloidal Gold. Albrecht RM, Hodges GM, eds. Scanning Microscopy International, AMF O'Hare (Chicago), IL, pp 155-160.

Discussion with Reviewers

P. Bell: Considering that your extraction is done under conditions that do not stabilize most of the components of the cytoskeleton, it is not surprising that your preparations are very heavily extracted. Therefore, I question your use of the term "cytoskeleton whole mounts" to refer to your preparations. Would not the term "detergent-extracted whole mount" be more appropriate since most of the cytoskeleton is gone?

Authors: In general, cytoskeletons which were obtained with the extraction procedure described were very dense (see Fig. 1A), but also showed differences in composition dependent on the subcellular compartment. The arrangement of immunogold labeled cytoskeletal structures, especially that of plectin, could best be demonstrated with micrographs showing areas of lower density. This was the basis for selecting the micrographs. Biochemical studies concerning the solubility of plectin in vimentin positive versus vimentin deficient cells substantiated the microscopical data [47].

P. Bell: Have you tried any of the methods that can improve the preservation of cytoskeletal components, such as protein crossslinking and stabilizing buffers?
Authors: So far we did not include stabilizing or crosslinking substances in our extraction procedure. We admit that especially the use of crosslinking agents as suggested by you and your colleagues has to be tested to determine if structures which connect the various filament systems can better be maintained. At least the thin connecting filaments seemed to be very sensitive to mechanical stress.

P. Bell: Did you encounter any problems with collapse or destruction of structures when you mounted preparations on uncoated grids ?
Authors: Head over drying of the grid attached cytoskeletons minimized mechanical stress during drying.

P. Bell: Under the best of circumstances there are variations in the diameter (thickness) of filamentous structures coated evaporatively with carbon or metal. Can you provide quantitative and statistically relevant data to support your statement that actin and intermediate filaments can be distinguished on the basis of size alone.
Authors: In individual preparations the various filament types can roughly be identified on basis of relative differences in thickness. However, we agree that the thickness of the carbon coat varies considerably in successive experiments. That is why we introduced immunogold labeling procedures to unequivocally follow the arrangement of individual filaments and to identify structural linkages between the various filament types.

Additional Reference

47. Foisner R, Bohn W, Mannweiler K, Wiche G (1995) Distribution and ultrastructure of plectin arrays in subclones of rat glioma C6 cell differing in intermediate filament protein (vimentin) expression. J Struct Biol **115**: 304-317.

Scanning Microscopy Supplement 10, 1996 (pages 295-307) 0892-953X/96$5.00+.25
Scanning Microscopy International, Chicago (AMF O'Hare), IL 60666 USA

FREEZE-DRIED HUMAN LEUKOCYTES STABILIZED WITH URANYL ACETATE DURING LOW TEMPERATURE EMBEDDING OR WITH OSO_4 VAPOR AFTER EMBEDDING

L. Edelmann[1*] and A. Ruf[2]

[1]Medizinische Biologie, Universität des Saarlandes, Homburg-Saar
[2]Klinikum Karlsruhe, Medizinisch-Diagnostisches Institut, Karlsruhe, Germany

(Received for publication July 5, 1996, and in revised form September 23, 1996)

Abstract

Two new simple stabilization procedures for freeze-dried biological material are introduced which are compatible with low temperature embedding (LTE) in Lowicryl. The first method uses a Lowicryl K11M/HM20 mixture supplemented with 0.3% uranyl acetate for LTE. For the second method polymerized Lowicryl blocks containing the freeze-dried material are exposed to OsO_4 vapor which penetrates into the Lowicryl block and stabilizes the embedded specimen. The quality of structural preservation is demonstrated with human leukocytes.

Key Words: Freeze-drying, low temperature embedding, Lowicryl HM20, K11M, uranyl acetate, OsO_4 vapor, lymphocytes, polymorphonuclear leukocytes.

*Address for correspondence:
Ludwig Edelmann
Medizinische Biologie
Universität des Saarlandes
D-66421 Homburg-Saar, FRG
Telephone Number: +49-6841-166254
FAX Number: +49-6841-166256

Introduction

First results obtained after freeze-drying (FD) and low temperature embedding (LTE) in the new cryosorption freeze-drying unit (CFD) of Leica (Wien, Austria) have been published in earlier papers (Edelmann, 1994a,b; Sitte *et al.*, 1994). It has been shown that good structural preservation of biological material can be obtained without the use of any chemical fixation during the preparation procedures (Edelmann, 1994a). However, depending on the type of cells and resins used labile subcellular components may be extracted from thin sections of the freeze-dried embedded material during wet cutting (Edelmann, 1994a). In order to stabilize freeze-dried specimens one may expose them to OsO_4 vapor before infiltration in a resin (see e.g., Linner *et al.*, 1986). But freeze-dried and osmicated (black) material is not well suited for low temperature embedding (LTE) in Lowicryl which requires UV polymerization.

In this paper two new simple stabilization procedures for freeze-dried biological material are introduced which are compatible with LTE in Lowicryl: (1) Uranyl acetate may be used during infiltration of the freeze-dried material with a Lowicryl K11M/HM20 mixture and (2) polymerized Lowicryl blocks containing freeze-dried specimens may be exposed to OsO_4 vapor leading to optimally stabilized preparations as seen after ultrathin sectioning and staining of the specimens. The quality of structural preservation provided by the new procedures is demonstrated with ultrathin sections of human leukocytes.

Materials and Methods

Preparation of mononuclear leukocytes (mainly lymphocytes)

Mononuclear cells were isolated from heparinized blood as described by Böyum (1968). Briefly, the anticoagulated blood was diluted 1:1 with phosphate buffered saline, layered onto Ficoll-Paque (Pharmacia, Uppsala, Sweden) and centrifuged at 400 g for 30 min. Then the mononuclear cell fraction was collected and diluted 1:4 with modified Hank's solution (mHBSS) containing (in mM) NaCl 137, KCl 5.4, Na_2HPO_4 0.34, KH_2PO_4 0.44, glucose 5.5, $NaHCO_3$ 4.2, and 20 mg/ml albumin at pH 7.4. The cell suspension was centrifuged at 130 g for 10 min. The supernatant was removed and the cell pellet was resuspended in mHBSS. In order to wash the cells free of residual Ficoll-Paque the latter steps were repeated two times. Afterwards, the cells were resuspended in platelet poor plasma and centrifuged at 400 g for 15 min. The supernatant was removed and the cell suspension was cryofixed (see below).

Preparation of polymorphonuclear (PMN) leukocytes (mainly neutrophils, few eosinophils)

The cells were isolated from heparinized blood as recently described (Ruf *et al.*, 1992). Briefly, the anticoagulated blood was diluted 2:1 with hydroxyethylstarch (4% w/v; molecular weight 450,000; Fresenius Immuno GmbH, Heidelberg, Germany), and kept at 4°C for 15 min. The buffy coats were collected and centrifuged at 190 g for 5 min. The pellets were resuspended in mHBSS. Then residual red blood cells were lysed by addition of ice-cold distilled water. After 20 sec, isotonicity was reconstituted with a 3.5% w/v NaCl solution. The cell suspensions were again centrifuged at 190 g for 5 min and the pellets were resuspended with mHBSS. The suspensions were layered onto Percoll (Sigma, Deisenhofen, Germany) that was diluted with isotonic saline so that a final density of 1.06 g/ml was achieved, and centrifuged at 190 g for 7 min. The leukocyte pellet was resuspended and washed two times with mHBSS. The viability of the cells examined by trypan blue exclusion was 98% or higher. The cells were counted with a K1000 cell counter (Sysmex, Hamburg, Germany) and stored at a concentration of $6x10^8$ leukocytes/ml until use. The entire procedure and the storage took place at 4°C. To reconstitute the functional integrity of the cells, the suspension was warmed to 37° for 10 min. Immediately before cryofixation the warmed cell suspension was diluted 3:2 with platelet-poor plasma.

Cryofixation

Cell suspensions were cryofixed in plastic spacers in the Reichert Leica MM 80 cryofixation unit as described by Sitte *et al.* (1987). The spacers were rings of anti-adhesive plastic with an inner diameter of 2.4 mm and a thickness of 0.3 mm. 3 to 4 spacers were placed on a foam rubber support of a specimen carrier and each filled with about 1.5 μl of cell suspension. The loaded carrier was transferred and fixed upside down to the MM 80. Cryofixation of lymphocytes took place immediately afterwards. Since the polymorphonuclear cells were not concentrated to the same degree as the lymphocytes it was necessary to leave the carrier for 1 min fixed upside down at the MM 80 before cryofixation. This procedure causes a concentration of cells in the lower part of the suspension which is frozen first.

Figure 1A-F: (*next three pages*) Ultrathin sections of human lymphocytes after FD and LTE. FD of **1A**, **1B**, **1E**, **1F**: 1day at -100°C, temperature increase 0.9°C/h up to -10°C (100h), 10h -10°C. FD of **1C**, **1D**: temperature increase 0.2°C/h from -90°C to -30°C (300h), temperature increase 1°C/h from -30°C to -10°C (20h), 10h -10°C. **1A:** HM20, without poststaining. **1B:** HM20, uranyl acetate and lead citrate staining. The white areas on the right side of the cells are due to differential compression of cells and extracellular matrix during sectioning (from right to left). Note the extraction artifacts in the nuclei (heterochromatin, arrows). Freezing plane: lower right corner. **1C:** K11M/HM20, procedure P1, without poststaining. Freezing plane: lower right corner. **1D:** K11M/HM20, procedure P1, poststaining with uranyl acetate and lead citrate. The freezing plane is at a distance of about 4 μm from the right side. **1E:** HM20, procedure P2, without poststaining. **1F:** HM20, procedure P2, poststaining with uranyl acetate and lead citrate. Freezing plane: lower right corner. Figures **1B** and **1F** were obtained from the same Lowicryl block.

Freeze-drying

Transfer of the cryofixed samples into the CFD and freeze-drying was carried out as described earlier (Edelmann, 1994a). Periods of drying temperatures between -100°C and -10°C are given in the figure legends under Results.

Stabilization procedures P1 and P2 after freeze-drying

P1: After FD the specimens were infiltrated inside the CFD as described in the instruction manual of the CFD with a Lowicryl K11M/HM20 mixture at a specimen stage temperature of -25°C. This mixture was prepared by dissolving 0.4% uranyl acetate in Lowicryl K11M (without shaking this amount of uranyl acetate is dissolved over night in K11M) and by mixing 3 parts of this K11M solution with 1 part of pure HM20 (in a new syringe) yielding a Lowicryl solution with 0.3% uranyl acetate. (Note that uranyl acetate cannot be dissolved in pure HM20). 1 day after start of infiltration the UV lamp was switched on for 1 day. Afterwards the specimen stage was warmed to room temperature and removed from the CFD. The polymerized blocks were then ready for sectioning.

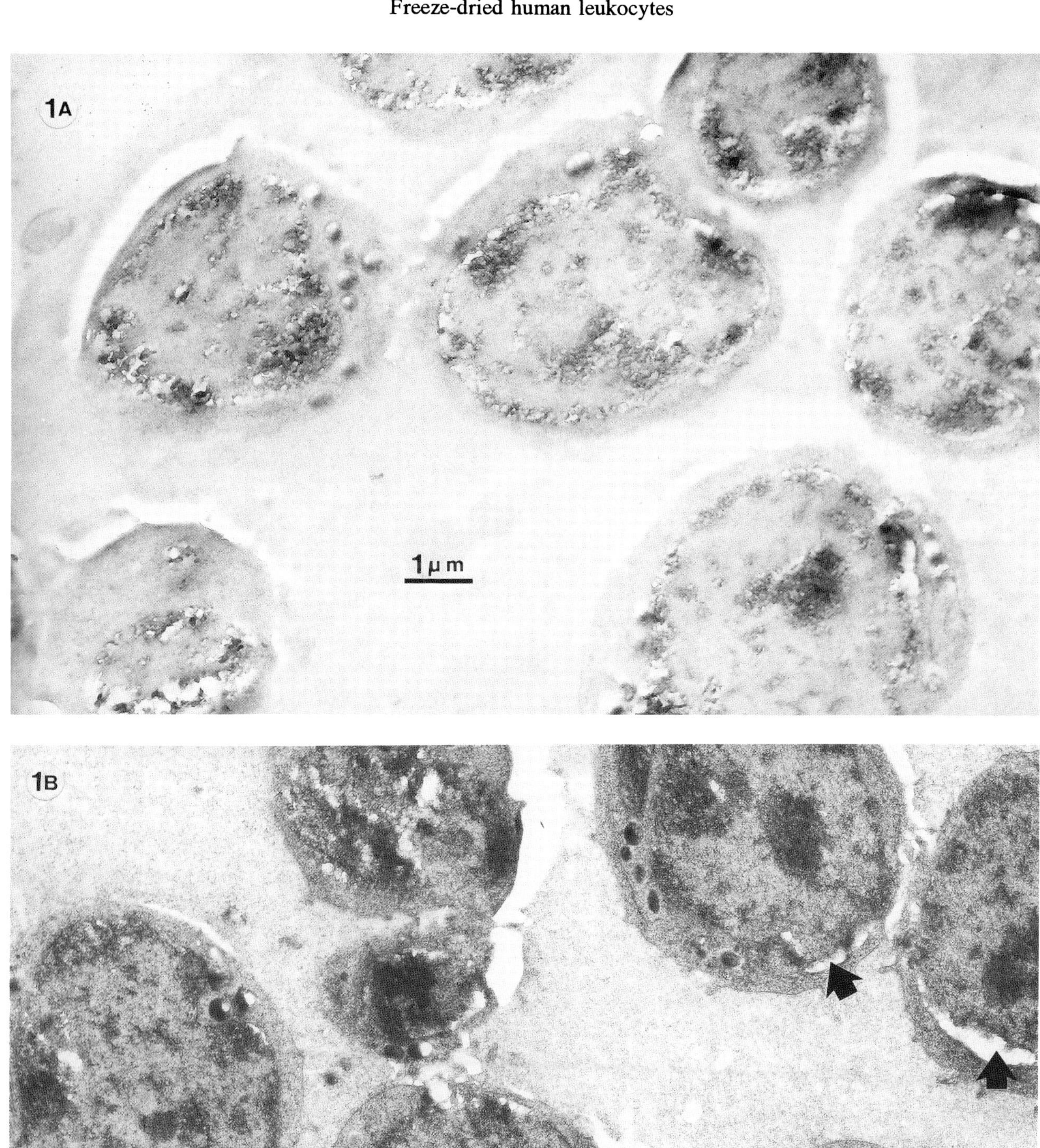
1A
1 μm
1B
1μm

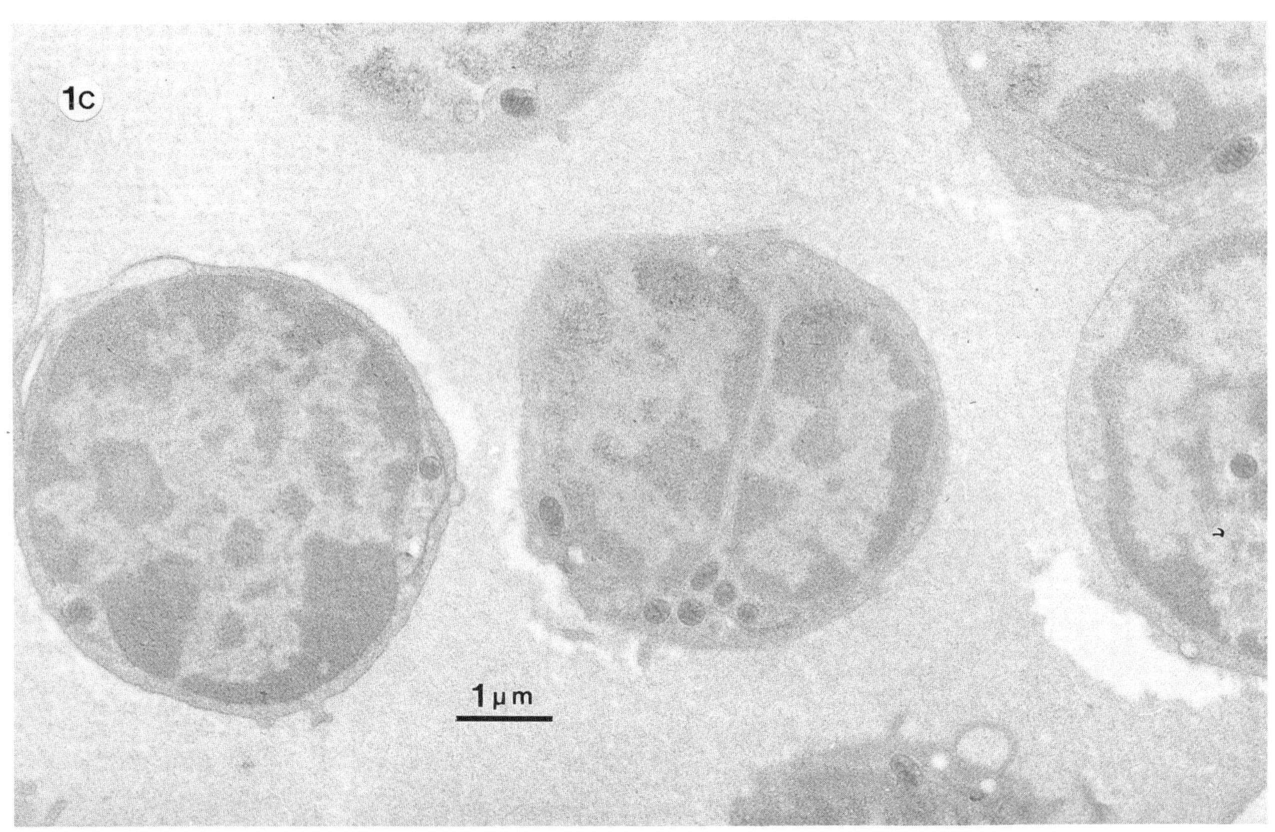
1C
1 μm

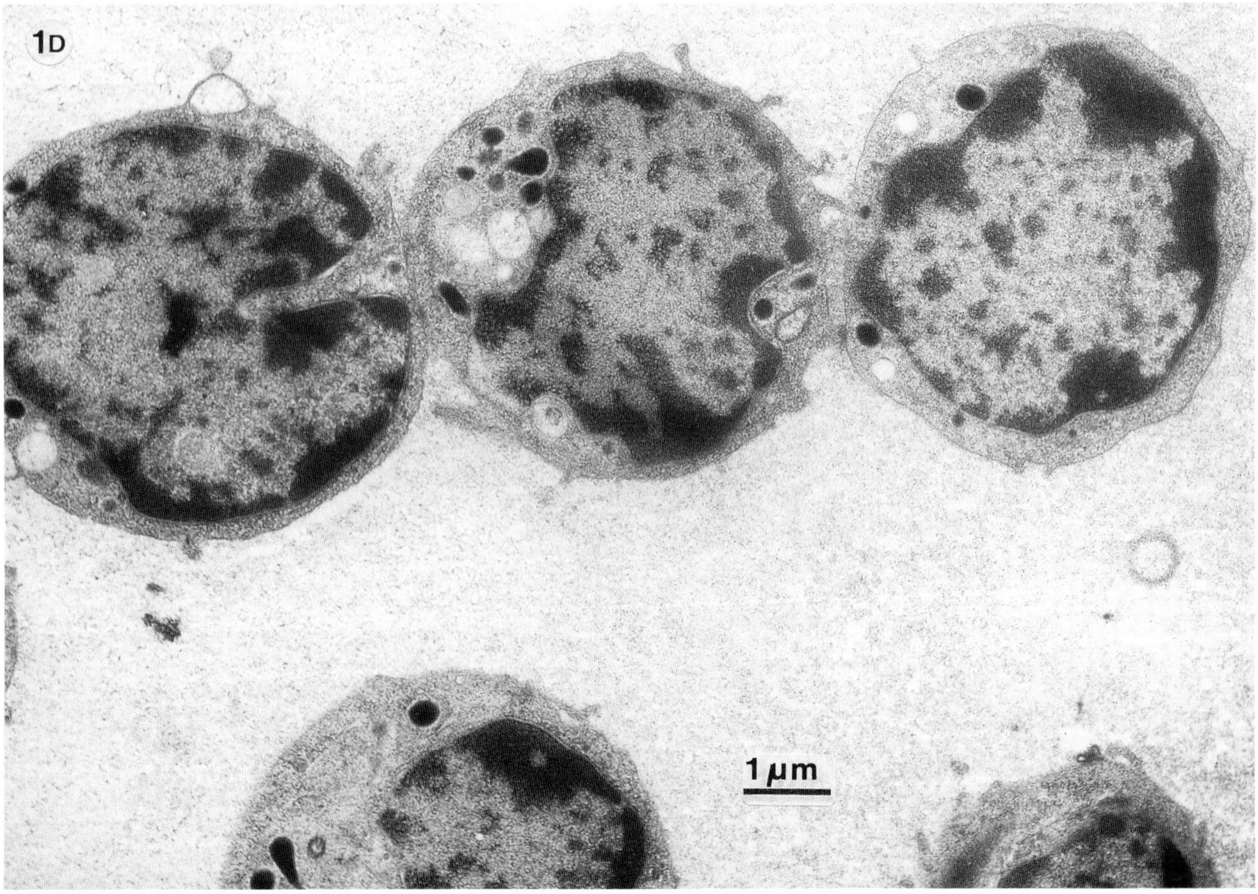

1D
1 μm

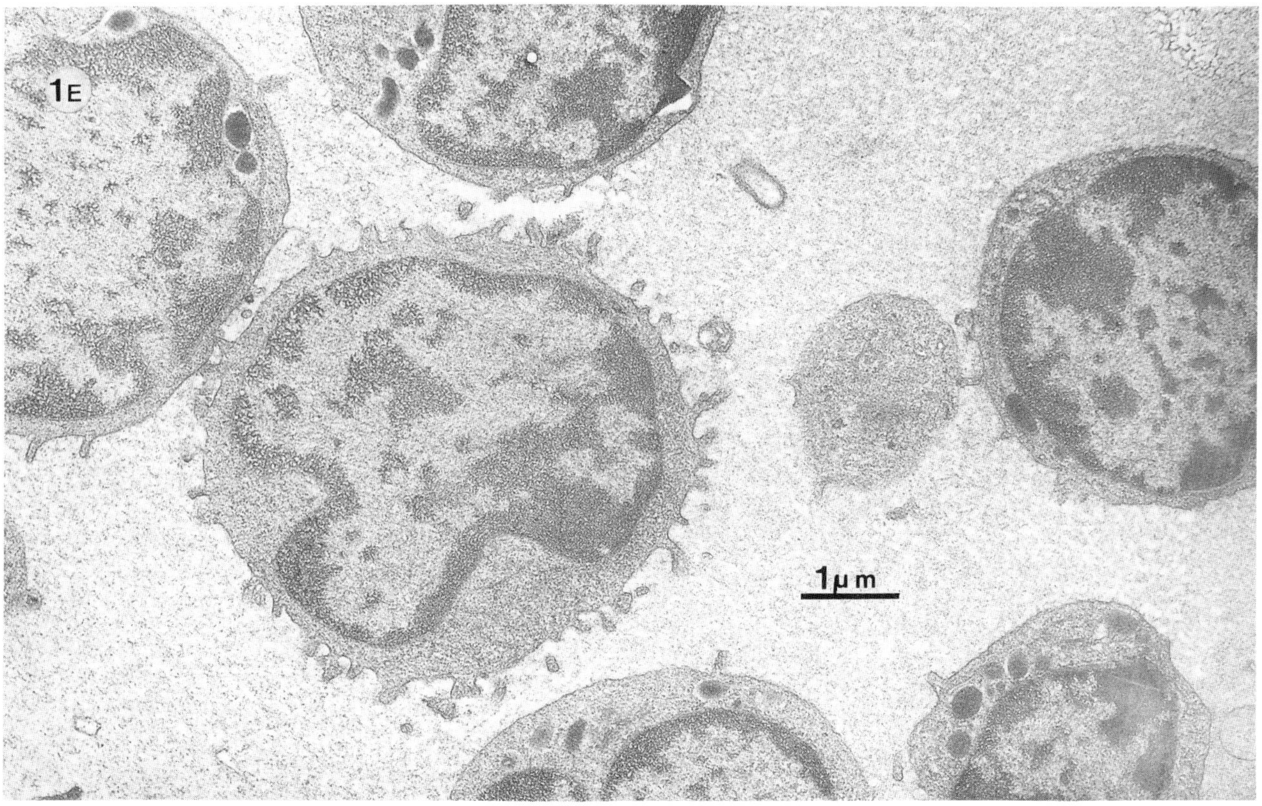

1E

1 μm

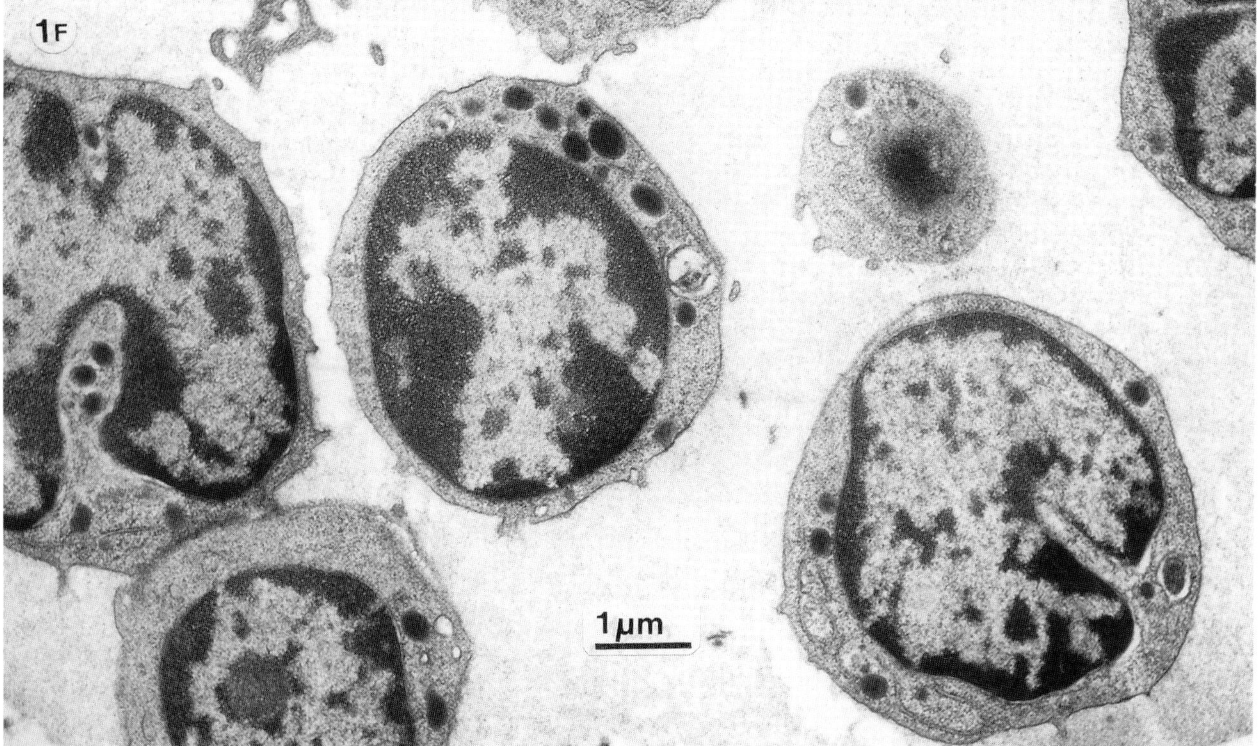

P2: After FD and LTE at -25°C (infiltration of HM20 : 10 h, UV polymerization: 24 h) the polymerized Lowicryl blocks were placed for 30 min into 0.5 ml Eppendorf capsules together with a small amount (< 0.1 g) of OsO_4 crystals (as a rule the Lowicryl blocks were first sectioned until the embedded biological material appeared at the surface of the block). After exposure of the polymerized blocks to OsO_4 vapor the blocks appear dark brown and turn black after a few hours without further exposure.

Transmission electron microscopy

Ultrathin sections (≤ 80 nm) were obtained by using a Diatome (Biel, Switzerland) diamond knife (35°). The sections were either examined without poststaining or after staining with uranyl acetate and lead citrate for short periods (≤ 10 sec). The sections were photographed in a Zeiss (Oberkochen, Germany) EM 902 with energy filter at 80 kV. All figures were obtained on identical photographic paper [Agfa (Agfa-Gevaert, Leverkusen, Germany), Brovira, grade soft].

Results

Freeze-dried lymphocytes are shown in Figs. 1A-F, freeze-dried PMN leukocytes in Figs. 2A-F. Unstained and stained ultrathin sections of HM20 embedded preparations not stabilized with P1 or P2 show severe artifacts (Figs. 1A, 1B, 2A, 2B). These artifacts are distortions of nuclear material, partial separation of the cells from the surrounding Lowicryl, and extraction of cellular granules of PMN leukocytes. Similar or even more pronounced artifacts are obtained after embedding in pure K11M or in a K11M/HM20 mixture (not shown). The effect of P1 is shown in Figs. 1C, 1D, 2C, 2D. Lymphocytes appear well preserved and faintly stained without poststaining (1C) or intensely stained after poststaining (1D). PMN leukocytes are not preserved in a satisfying manner (2C, 2D). Both sections (without and after poststaining) reveal extraction of cellular granules. To be noted is however the better preservation of nuclear material in uranyl acetate treated cells and the degree of separation of the cells from the surrounding Lowicryl is reduced (compare with Figs. 2A, 2B). The effect of OsO_4 treatment of the Lowicryl blocks is demonstrated in Figs. 1E, 1F (HM20, no former uranyl acetate treatment) and Figs. 2E, 2F (K11M/HM20, after uranyl acetate treatment). All sections either without or after poststaining show a well preserved ultrastructure. Without poststaining the contrast of subcellular structures and in particular of membranes of OsO_4 treated cells is more intense than that of uranyl acetate treated cells (compare Figs. 1E with 1C and 2E with 2C). After poststaining the contrast is very much increased (Figs. 1F, 2F). PMN leukocytes embedded in pure Lowicryl (HM20, K11M or mixture) and treated with OsO_4 vapor according to P2 (not shown) are similarly preserved as those shown in Figs. 2E and 2F.

Figure 2A-F: (*next three pages*) Ultrathin sections of human PMN leukocytes. FD of **2A**, **2B** same as for **1A**, **1B**, **1E**, **1F**. FD of **2C-F** same as for **1C**, **1D**. **2A**: HM20, without poststaining. **2B**: HM20, poststaining with uranyl acetate and lead citrate. **2C:** K11M/HM20, procedure P1, without poststaining. **2D:** K11M/HM20, procedure P1, poststaining with uranyl acetate and lead citrate, freezing plane: upper left corner. Figures **2A-D** show the loss of cellular granules. **2E**: K11M/HM20, procedures P1 and P2, without poststaining. Freezing plane: lower edge. **2F**: K11M/HM20, procedures P1 and P2, poststaining with uranyl acetate and lead citrate. Freezing plane: lower left corner. **2E** and **2F** show sections from the same Lowicryl block.

Discussion

The purpose of this contribution is the introduction of two simple new methods for preparing freeze-dried embedded material for transmission electron microscopy in such a way that ultrastructural details are preserved and visualized similarly to those obtained e.g., by the freeze-substitution (FS) technique. In the first method uranyl acetate is used as a stabilization agent. From FS experiments it is well known that many cellular components and in particular lipids can be stabilized during dehydration of a cryofixed specimen in an organic solvent supplemented with uranyl acetate (see e.g., Humbel, 1984; Voorhout *et al.*, 1991). In the FD technique the dehydration is carried out without the use of an organic solvent and the stabilization - if necessary - must occur afterwards. Since many experiments with human leukocytes showed that these cells are not optimally preserved after FD, LTE in Lowicryl, and ultrathin sectioning (Figs. 1A, 1B, 2A, 2B) it was expected that one may improve the preservation of freeze-dried cells by using uranyl acetate in the embedding medium. Two reasons led to the finally adopted and described method P1: (1) A small amount (< 0.5 %) of uranyl acetate can be dissolved in Lowicryl K11M but not in HM20, (2) Ultrathin sections (≤ 100 nm) of freeze-dried preparations embedded in K11M take up a considerable amount of water and disintegrate on the water of the knife through. This phenomenon could be largely

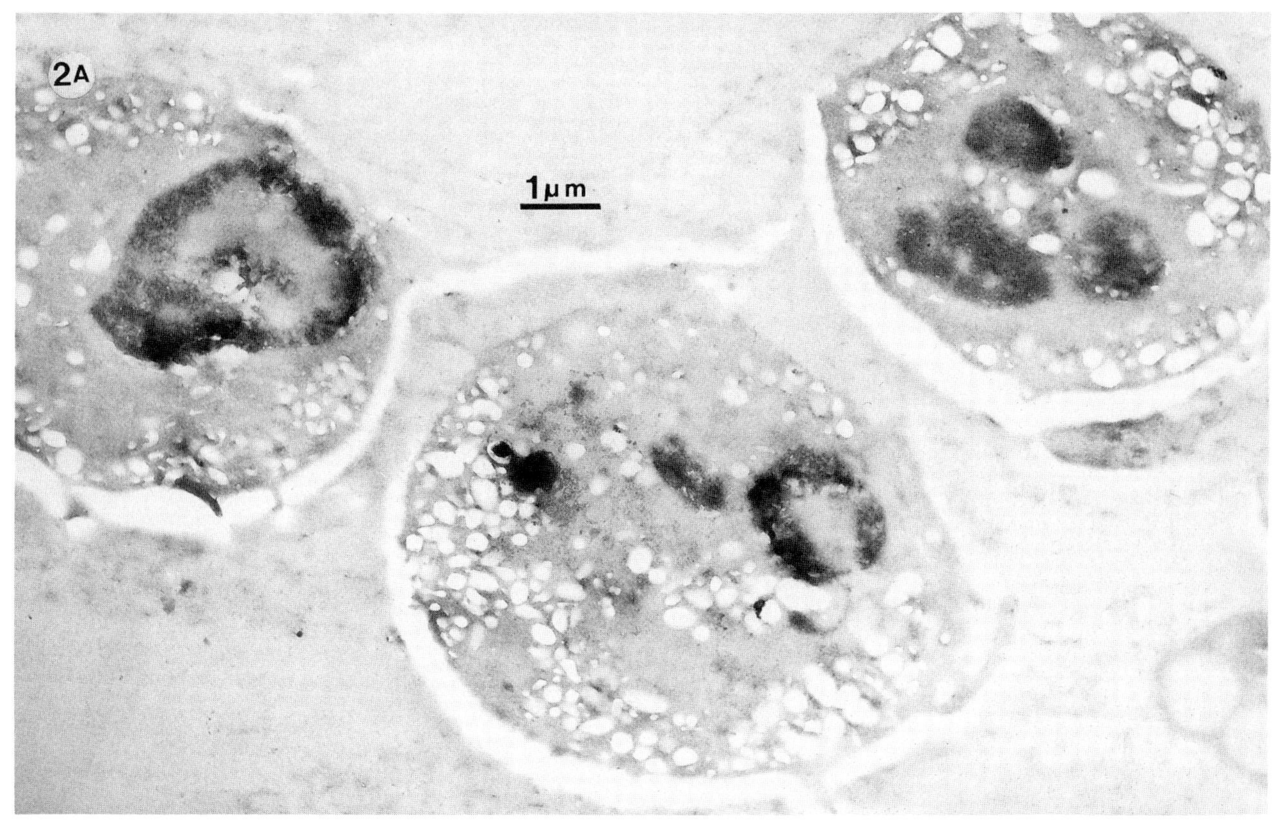
2A
1μm

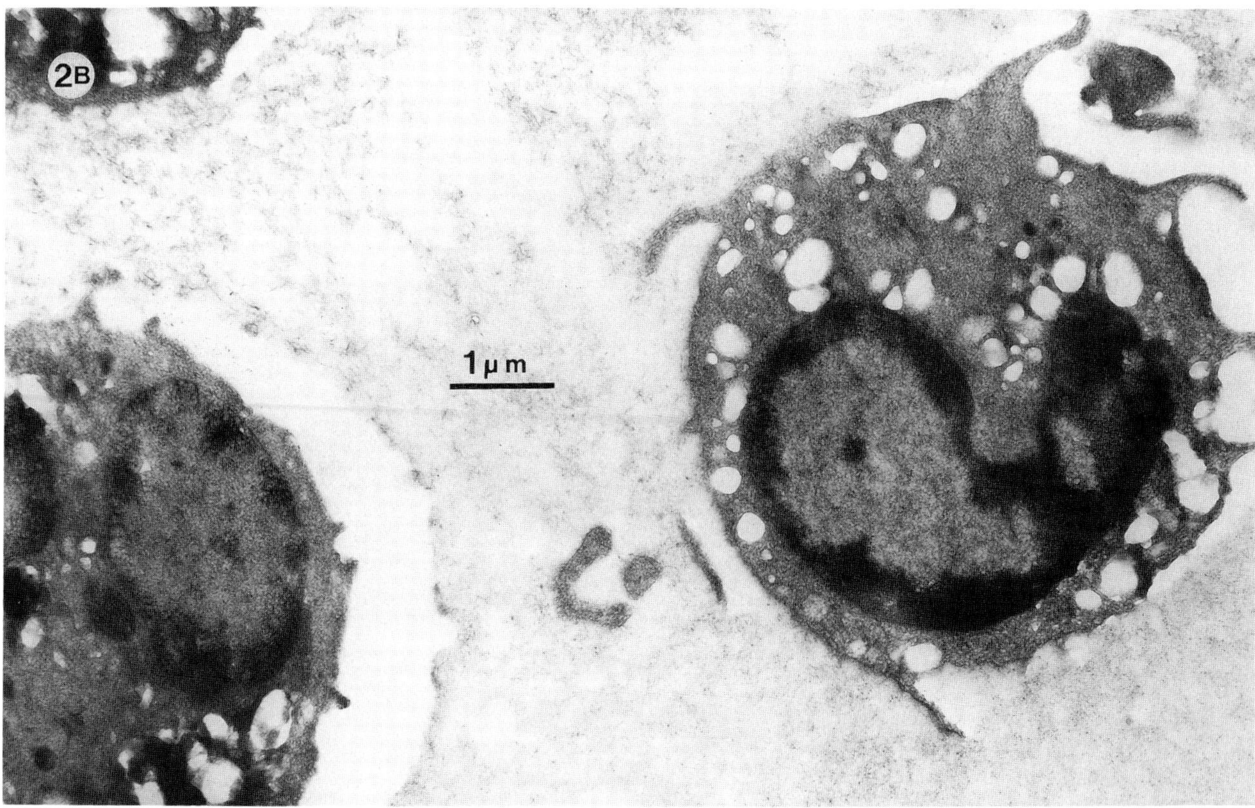
2B
1μm

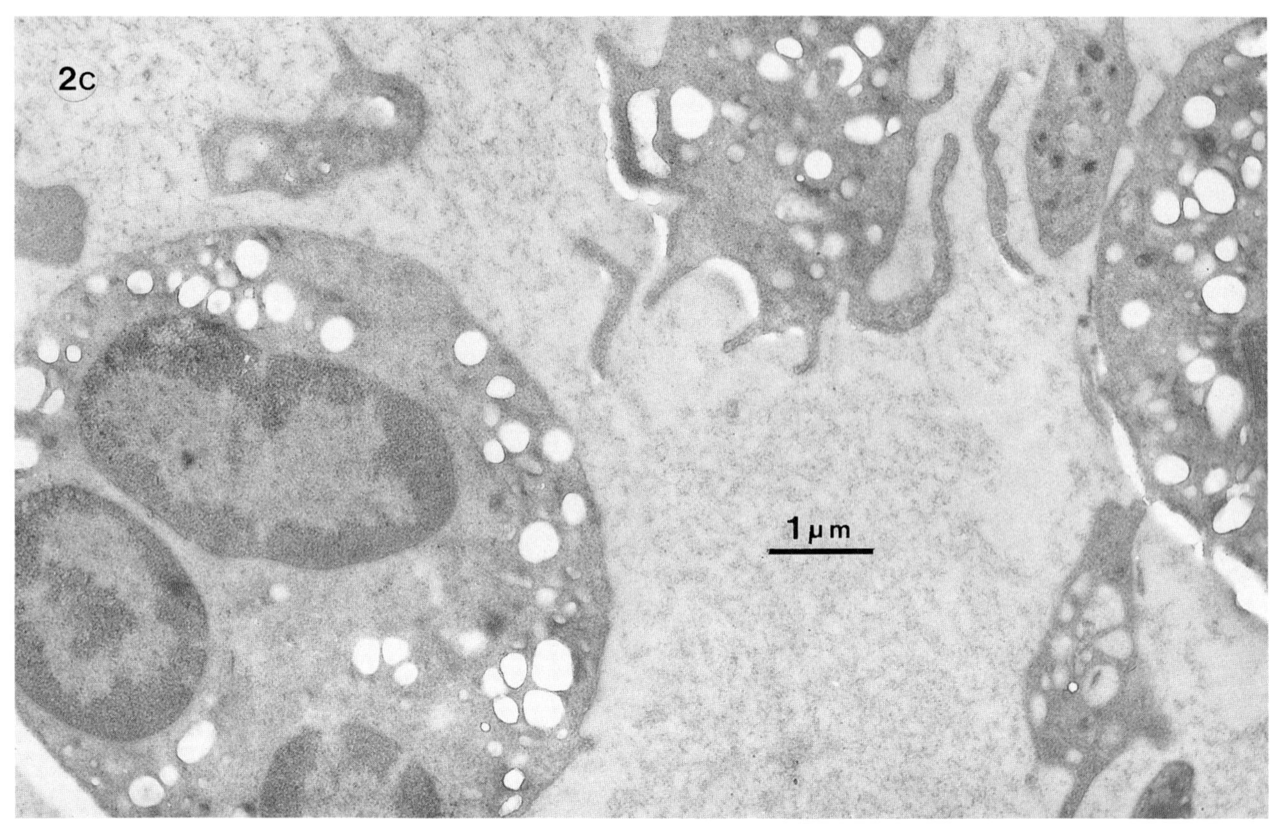
2C
1 μm

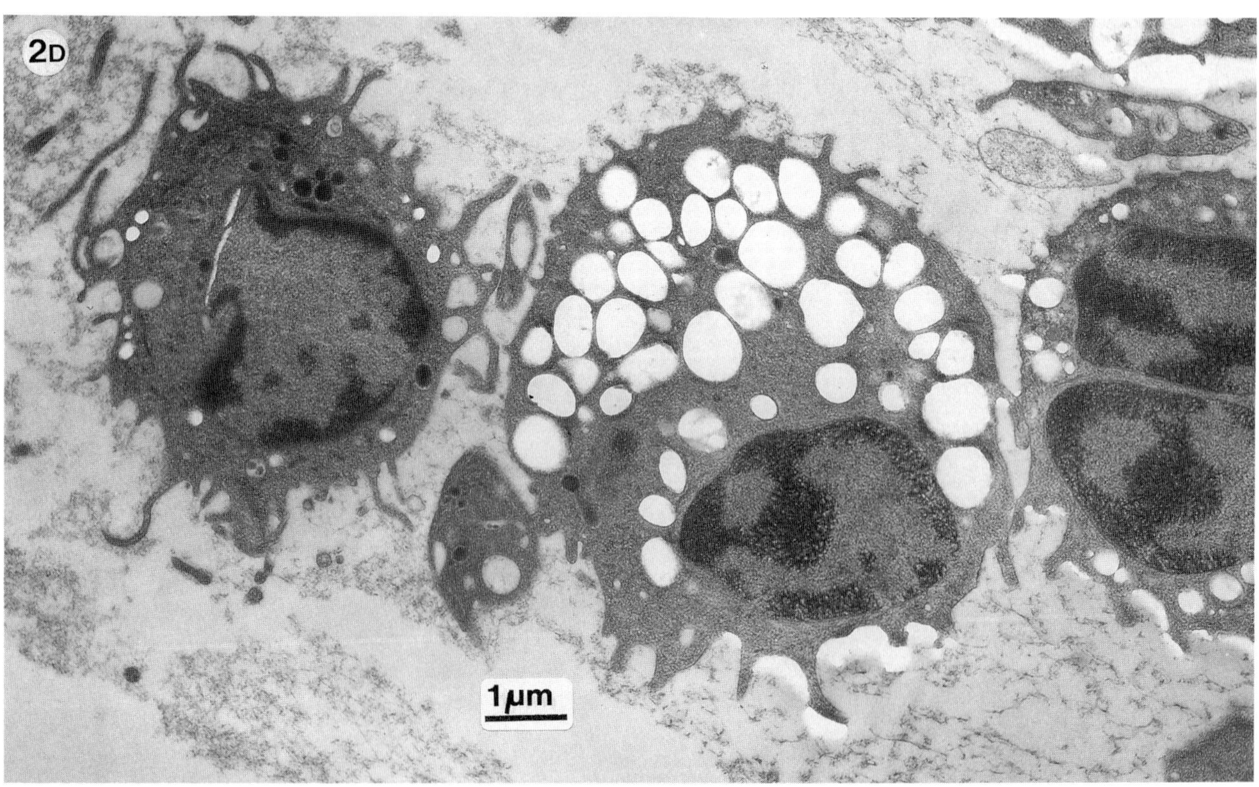
2D
1μm

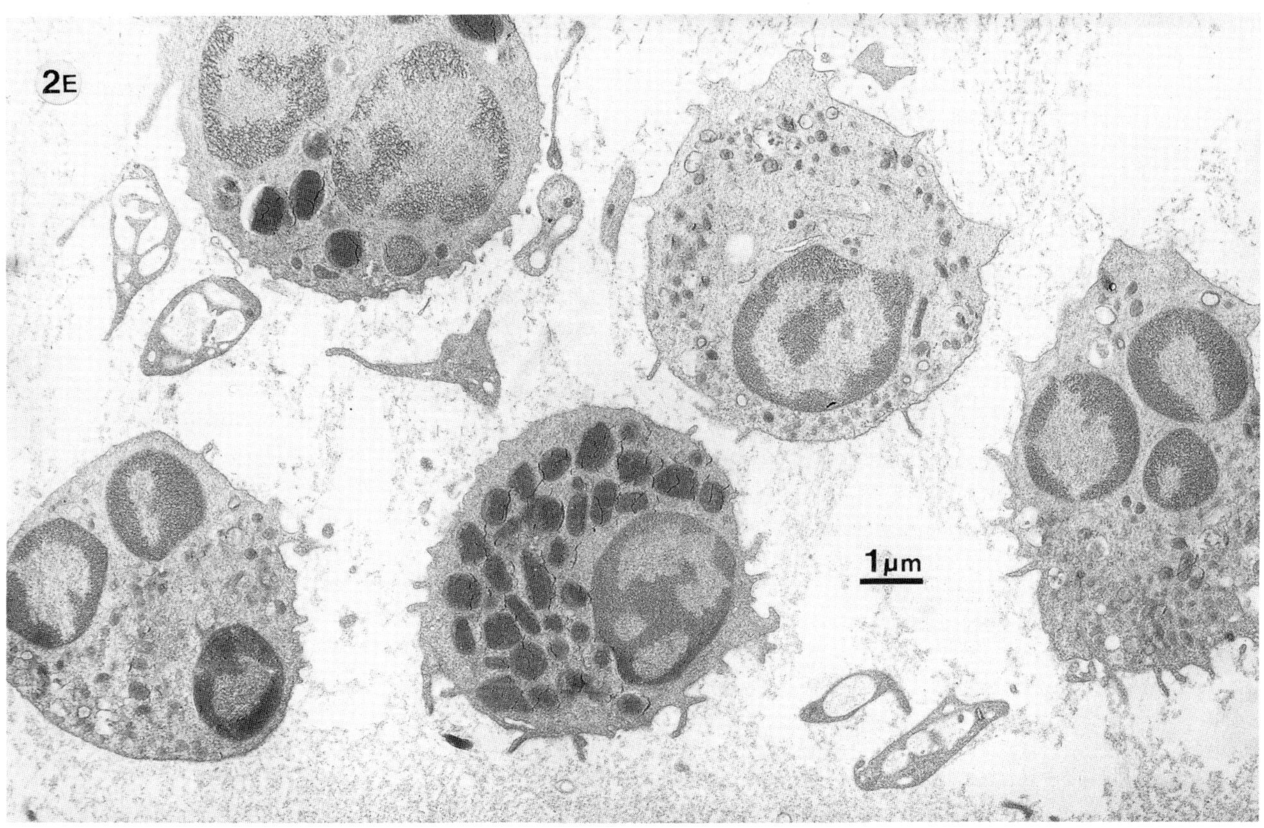

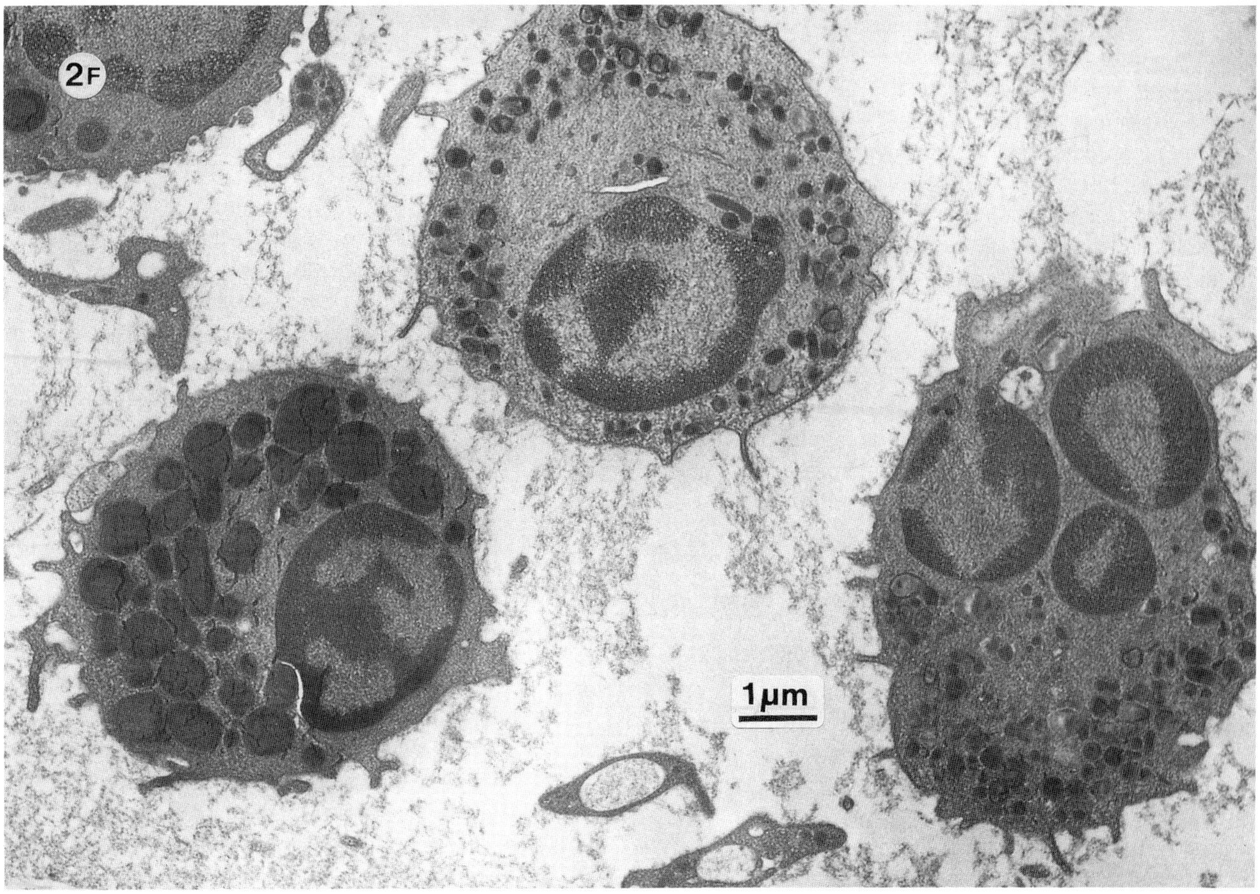

2F
1 μm

reduced by using a K11M/HM20 mixture (supplemented with uranyl acetate according to P1). With this embedding medium we obtained stable ultrathin sections of well preserved human lymphocytes (Figs. 1C, 1D). However, other leukocytes were not stabilized in a satisfying manner (Figs. 2C, 2D). Fortunately subsequent experiments with OsO_4 vapor were successful: Lowicryl blocks prepared according to P1 were exposed to OsO_4 vapor (P2) and we obtained well preserved freeze-dried and low temperature embedded PMN leukocytes (Figs. 2E, 2F). Later it was found that similar results were also obtained with method P2 alone (not shown, for lymphocytes in pure HM20 see Figs. 1E, 1F).

From these positive results we conclude: during FD and also during subsequent LTE in Lowicryl a well organized but labile ultrastructure of the biological material is maintained. The freeze-dried ultrastructure can be stabilized at least partly during infiltration in a K11M/HM20 medium supplemented with uranyl acetate (P1). In addition, the freeze-dried and embedded material can be stabilized (more effectively) by exposing the polymerized Lowicryl block to OsO_4 vapor (P2). Apparently the whole labile Lowicryl-biological material-complex with its critical internal phase boundaries is crosslinked by OsO_4. Despite the fact that P2 is a more powerful and more general applicable method than P1 both methods may be used for future studies. E.g., since lymphocytes are well preserved by both methods it would be worthwhile to compare immunocytochemical results obtained from the different preparations (treated with different chemical fixatives).

One may ask whether the biological material prepared according to the described procedures is maintained in a state closer to the living state than e.g., freeze-substituted and embedded material or more specifically: is the antigenicity of macromolecules better retained in the freeze-dried preparation? Ultimately these questions have to be answered by comparative studies. At least two facts, however, speak in favor of the freeze-dried material:

(1) During FD the biological material is not exposed to dehydrating liquids (e.g., acetone, alcohol) which may alter cellular macromolecules and extract soluble substances. Although the used resin may have an influence on the biological material the already dehydrated and hence partly stabilized specimen is certainly less influenced and modified by the liquid resin than a specimen which is first exposed to a dehydrating organic solvent and then to a resin.

(2) Freeze-dried biological material - even if it is stabilized according to the described procedures - exhibits a very different behavior towards the electron microscopic stains uranyl acetate and lead citrate than freeze-substituted material: the freeze-dried specimen is heavily stained after very short exposure of sections to the stain (< 10 sec) whereas such an intense staining is not observed in sections of freeze-substituted material. Apparently more reactive sites are preserved in the freeze-dried specimen.

Figure 3: *(facing page)* 0.2 μm thick dry-cut unstained sections of human lymphocytes (**3A**) and PMN leukocytes (**3B**) after FD and LTE in pure HM20 (same Lowicryl blocks as for Fig. **1A** and Fig. **2A**).

Finally, Fig. 1B should be discussed: it is interesting to note that the structural preservation of the unstabilized preparation is worst in the area of best cryofixation. As discussed in earlier papers (Edelmann, 1994b; Sitte *et al.*, 1994) it is assumed that in the region of best cryofixation (vitrification) a relative strong protein-water interaction is preserved leading to a reduced rate of water removal during FD and hence to less dry areas of the biological material. If more water molecules are retained in certain areas polymerization of the Lowicryl may then be incomplete. A similar water related phenomenon may be the reason for the distortions observed in Figs. 2A-D (extraction of granules, separation of cells from the surrounding Lowicryl). One may speculate that prolonged FD at rather high temperature may reduce the water content in the critical areas and that then the cells are better stabilized even in pure Lowicryls. In the present study we have used two different FD procedures (see legend of Figs. 1 and 2) which differ mainly in the speed of water removal in the temperature range between -90°C and -30°C. Both procedures yield similar results and there is probably no reason to choose the slow procedure when freeze-drying leukocytes. The final degree of dehydration obtained by both procedures is similar because it is determined by the FD time at the highest temperature (10h at -10°C). If the more tightly bound remaining water should be reduced it would be necessary to increase the FD time at -10°C or even higher temperatures before embedding at low temperature. Whether such a strategy causes morphological changes of the freeze-dried material remains to be determined.

Acknowledgements

We would like to thank Frans A. Prins, Department of Pathology, University of Leiden, The Netherlands, for the photographs obtained by Reflection Contrast Microscopy (see Discussion with Reviewers) and Mrs. B. Reiland for typing the manuscript.

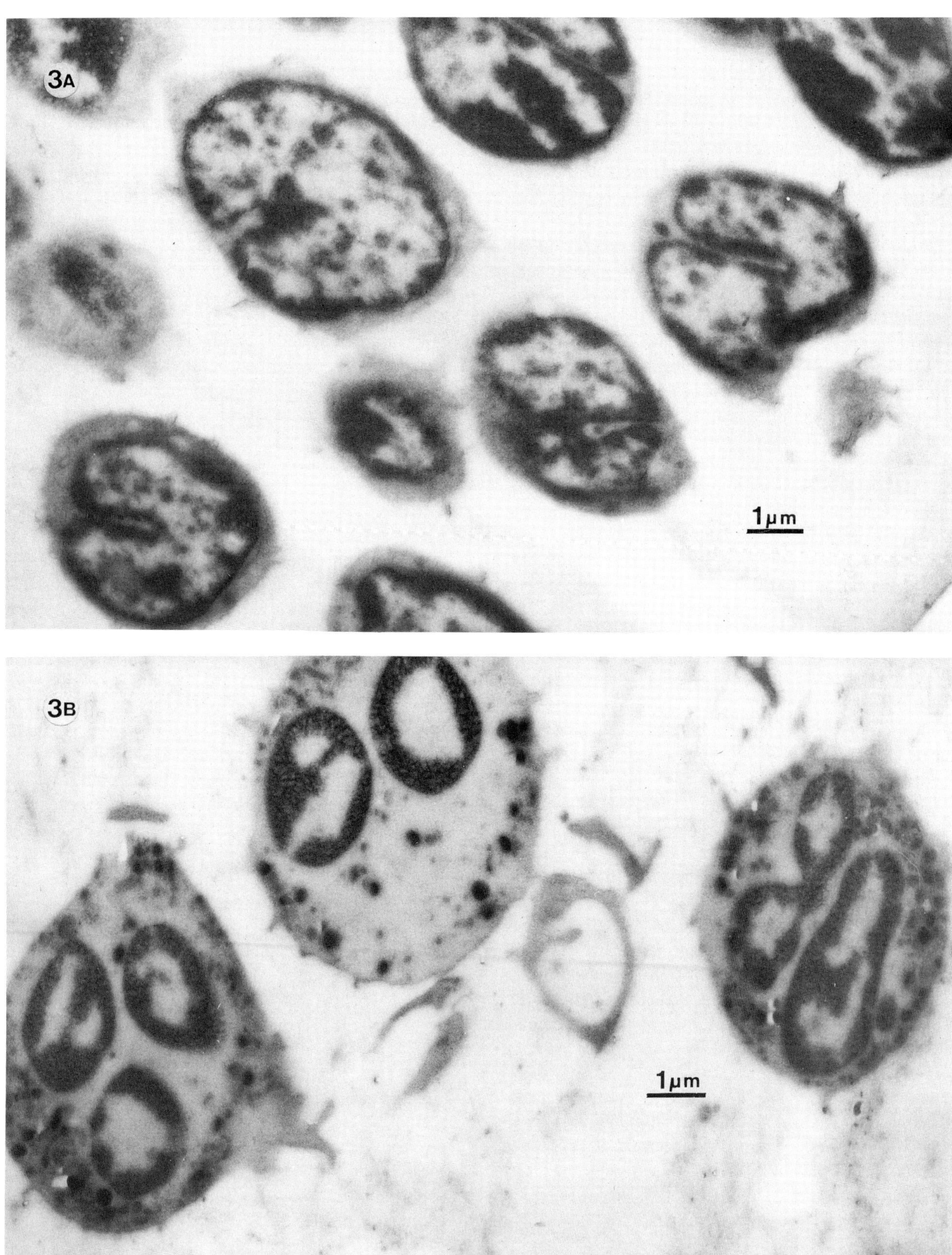
3A
1μm
3B
1μm

References

Böyum A (1968) Isolation of mononuclear cells and granulocytes from human blood. Scand J Clin Lab Invest **21** (Suppl 97): 77-90.

Edelmann L (1994a) Low temperature embedding of chemically unfixed biological material after cryosorption freeze-drying. Scanning Microscopy **8**: 551-562.

Edelmann L (1994b) Optimal freeze-drying of cryosections and bulk specimens for X-ray microanalysis. In: The Science of Biological Microanalysis. Roomans GM, Gupta BL, Leapman RD, von Zglinicki T (eds). Scanning Microscopy Suppl **8**: 67-81.

Humbel BM (1984) Gefriersubstitution - Ein Weg zur Verbesserung der morphologischen und zytologischen Untersuchung biologischer Proben im Elektronenmikroskop (Freeze-substitution - a way to improve morphological and cytological investigations of biological specimens in the electron microscope). Doctoral Thesis 7609, Eidgenössische Technische Hochschule, Zürich, Switzerland.

Linner JG, Livesey SA, Harrison DS, Steiner AL (1986) A new technique for removal of amorphous phase tissue water without ice crystal damage: A preparative method for ultrastructural analysis and immunoelectron microscopy. J Histochem Cytochem **34**: 1123-1135.

Ruf A, Schlenk RF, Maras A, Morgenstern E, Patscheke H (1992) Contact-induced neutrophil activation by platelets in human cell suspensions and whole blood. Blood **80**: 1238-1246.

Sitte H, Edelmann L, Neumann K (1987) Cryofixation without pretreatment at ambient pressure. In: Cryotechniques in Biological Electron Microscopy. Steinbrecht RA, Zierold K (eds). Springer Verlag, Berlin. pp 87-113.

Sitte H, Edelmann L, Hässig H, Kleber H, Lang A (1994) A new versatile system for freeze-substitution, freeze-drying and low temperature embedding of biological specimens. In: The Science of Biological Microanalysis. Roomans GM, Gupta BL, Leapman RD, von Zglinicki T (eds). Scanning Microscopy Suppl **8**: 47-66.

Voorhout W, van Genderen J, van Meer G, Geuze H (1991) Preservation and immunogold localisation of lipids by freeze-substitution. In: The Science of Biological Specimen Preparation for Microscopy and Microanalysis 1990. Edelmann L, Roomans GM (eds). Scanning Microscopy Suppl **5** S17-S26.

Discussion with Reviewers

Reviewer I: What is your experience with cutting of dry sections from this material?
Authors: When using a diamond knife (35°) and an ionization device (Michel *et al.*, 1992) it poses no difficulties to produce 0.2 μm thick (blue) sections from Lowicryl (HM20 or K11M) embedded material, to place them on Formvar coated grids and to investigate them in the electron microscope. Unstained sections of freeze-dried lymphocytes and PMN leukocytes embedded in *pure* HM20 are shown in Figs. 3A and 3B. These figures show a well preserved ultrastructure of both types of cells, confirming the above mentioned statement that the ultrastructure is preserved after FD and LTE. The distortions and extractions seen e.g., in Figs. 1A, 2A are absent in dry cut sections.

Reviewer I: Have you done any studies at the light microscopy level?
Authors: Figs. 4A and 4B show photographs of freeze-dried and Lowicryl embedded lymphocytes (method P1). These figures have been obtained by Frans Prins, Leiden, after toluidine blue and acridine orange (DNA) staining of two consecutive ultrathin sections in combination with Reflection Contrast Microscopy (Leica DMR Microscope equipped for epi-illumination and adapted for reflection contrast microscopy; Leica Microscopy and Systems, Wetzlar, Germany). Such histological staining can be used as counterstaining for immunohistochemistry. Immunostaining studies with the freeze-dried, embedded material are planned. For literature on reflection contrast microscopy see e.g., Prins *et al.*, 1993, 1996; Ploem *et al.*, 1995).

Additional References

Michel M, Gnägi H, Müller M (1992) Diamonds are a cryosectioner's best friend. J Microsc **166**: 43-56.

Ploem JS, Cornelese-ten Velde I, Prins FA, Bonnet J (1995) Reflection-contrast microscopy: an overview. Proc Royal Microsc Soc **30**: 185-192.

Prins FA, van Diemen-Steenvorde R, Bonnet J, Cornelese-ten Velde I (1993) Reflection contrast microscopy of ultrathin sections in immunocytochemical localization studies: a versatile technique bridging electron microscopy with light microscopy. Histochemistry **99**: 417-425.

Prins FA, Bruijn JA, DeHeer E (1996) Applications in renal immunopathology of reflection contrast microscopy, a novel superior light microscopical technique. Kidney Int **49**: 261-266.

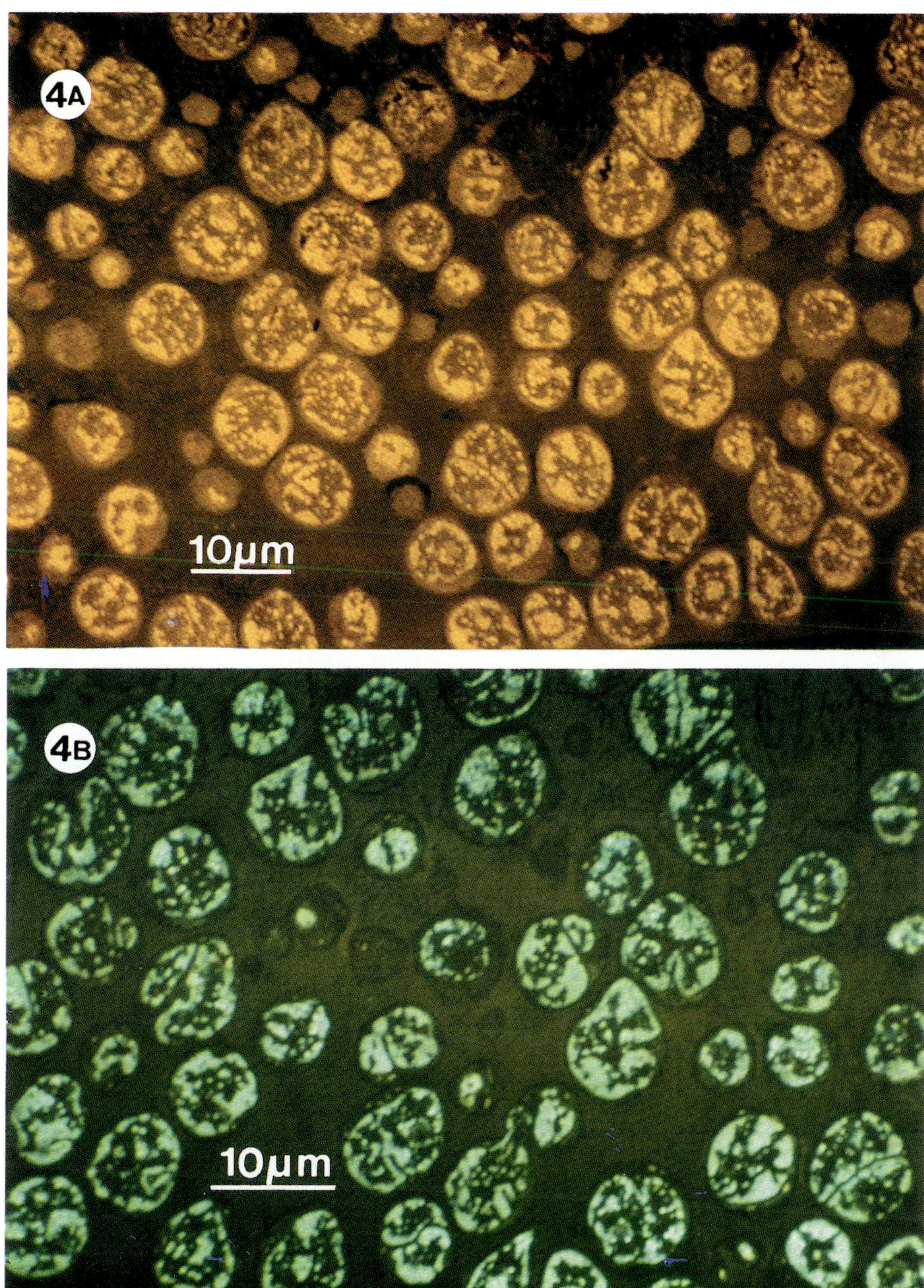

Figure 4: Sequential ultrathin ($\leq$ 80 nm) sections of human lymphocytes visualized after staining with toluidine blue (**4A**) and acridine orange (**4B**) by reflection contrast microscopy (Leica DMR Microscope equipped for epi-illumination and adapted for reflection contrast microscopy; Leica Microscopy and Systems, Wetzlar, Germany). Note that in reflection microscopy stained specimens exhibit the complementary color of the conventional transmitted light image. Photographs courtesy of F.A. Prins, Leiden.

Scanning Microscopy Supplement 10, 1996 (pages 309-325) 0892-953X/96$5.00+.25
Scanning Microscopy International, Chicago (AMF O'Hare), IL 60666 USA

LABELING WITH NANOGOLD AND UNDECAGOLD: TECHNIQUES AND RESULTS

James F. Hainfeld

Brookhaven National Laboratory, Biology Department, Upton, NY

(Received for publication October 3, 1995 and in revised form August 15, 1996)

Abstract

A significant new development in gold labeling for microscopy has been achieved through the use of gold cluster compounds that are *covalently* attached to antibodies or other probe molecules. These unique gold probes are smaller than most colloidal gold conjugates and exhibit improved penetration into tissues, higher labeling densities, and allow many new probes to be made with peptides, nucleic acids, lipids, drugs, and other molecules. A new fluorescent-gold conjugate is useful for examining localization by fluorescence microscopy, then visualizing the same label at the ultrastructural level in the electron microscope.

Key Words: Gold, Nanogold, Undecagold, colloidal gold, immunocytochemistry, labeling, silver enhancement, fluorescence labeling, FluoroNanogold, gold lipids, electron microscopy

*Address for correspondence:
James F. Hainfeld
Brookhaven National Laboratory
Biology Department
Upton, NY 11973

Telephone Number: 516-344-3372
FAX Number: 516-344-3407
E-mail: hainfeld@genome1.bio.bnl.gov

Introduction

The purpose of this review is to provide an introduction to the gold clusters (Nanogold, Undecagold, and FluoroNanogold), covering their properties and coupling chemistry. Next, examples of their use in labeling of specific sites on biomolecules for high resolution structural studies will be given; results using gold clusters in immunolabeling will also be given. As a practical guide, specific protocols will be discussed or referenced for using the clusters, both as labeling reagents, and as immunolabels.

Colloidal gold

Traditionally, colloidal gold has been used as the favored label for electron microscopy. The technology for making colloidal golds of fairly precise sizes, and adsorbing antibodies and lectins, has led to many useful applications. There are, however, some limitations and shortcomings of this technology, such as: [1] The adsorbed macromolecule (e.g., antibody), is not covalently attached, and is found to desorb to some extent (Kramarcy and Sealock, 1990), [2] Many molecules (e.g., many small ones) do not make stable conjugates with colloidal golds, [3] Because the gold is "sticky", the protein conjugates are sometimes aggregated, especially for small colloidal gold sizes (Hainfeld, 1990), and [4] penetration into tissues is frequently a problem (Takizawa and Robinson, 1994), due either to the large size of the gold, or for the smaller sizes, due to the aggregation.

Gold clusters

Gold clusters are a different approach that circumvents these problems. These are gold compounds, with defined structures, with organic linking groups for chemically attaching other molecules, as opposed to colloidal gold, which is a sphere of gold atoms with an ionic/hydrophobic surface. An example is Undecagold, which has a core of eleven gold atoms, and covalently attached phosphorus atoms, which, in turn, have organic moieties attached. The structure of Undecagold has been solved by x-ray diffraction (McPartlin *et al.*, 1969), and is shown diagramatically in Fig. 1 and in an electron micrograph in Fig. 2. A larger gold cluster that

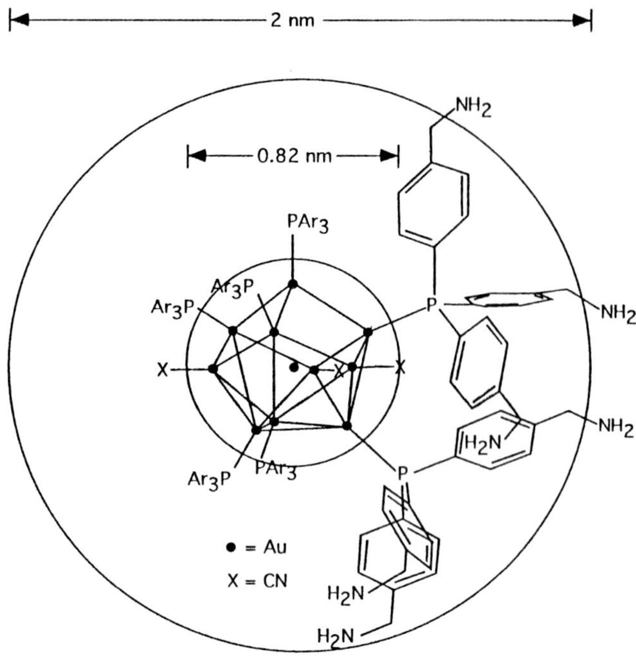

Figure 1. Diagram of the undecagold gold structure.

is 1.4 nm in gold core diameter has been made (Fig. 3), and presumably contains 67 gold atoms (Hainfeld and Furuya, 1992). This cluster has been termed "Nanogold", and its atomic structure has not yet been solved by crystallography.

The organic moieties may be changed so that different end groups are present, e.g., amines, carboxyls, or reactive linking groups, such as the maleimido group that reacts with thiols. Although there are multiple organic groups surrounding the cluster, it is possible to synthesize and purify a product that contains only one of the desired groups on the cluster's surface (Reardon and Frey 1984; Safer *et al.*, 1986). An example of using the monomaleimido-undecagold to covalently link to the hinge sulfhydryl of an Fab' antibody fragment is shown in Fig. 4 (Hainfeld, 1987, 1989).

The ability to synthesize different groups on the outer shell of the cluster gives a tremendous flexibility and control over the desired properties of the particles. For example, multiple amino groups give it a high positive charge, or one amino group might be used to link to one specific site. The organic groups may also be changed to affect solubility properties; a cluster can be made with all phenyl groups, for example, which would make it soluble in organic solvents or membranes.

Covalent reactivity

A useful group of clusters has been those with single, preformed, single reactive groups on the surface,

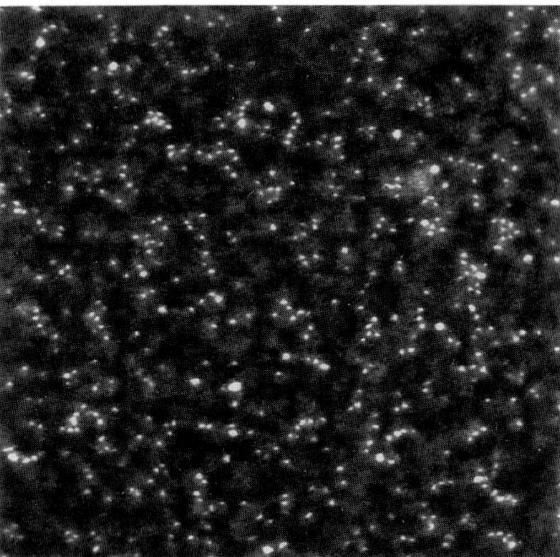

Figure 2. Darkfield field emission scanning transmission electron micrograph (STEM) of undecagold clusters on a thin 2 nm carbon film. Each bright dot is an undecagold core. Full width 128 nm.

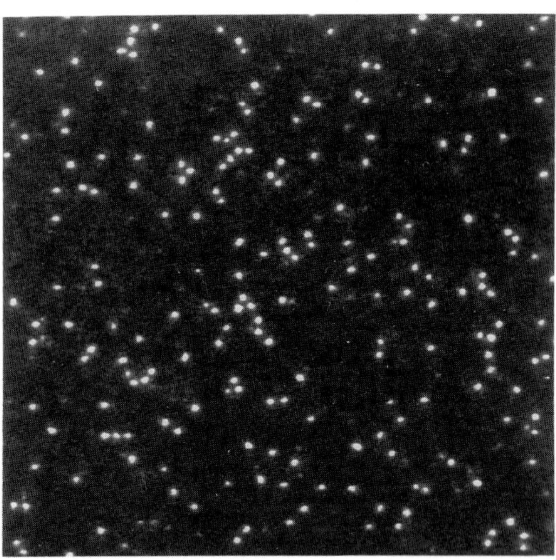

Figure 3. Darkfield STEM of 1.4 nm Nanogold clusters. Full width 128 nm.

so that the gold particle can be attached to a specific site on another molecule. The maleimide group is one example, shown above (Fig. 4) for the Fab', where the gold then reacts very specifically with free thiols, typically with a cysteine of a protein. Another specific linker is the N-hydroxysuccinimide ester, shown in Fig. 5, which reacts with free amino groups, typically lysines and the α-amino terminus of proteins. An amino gold can be reacted with carbohydrates (Fig. 6), such as on glycoproteins (Lipka *et al.*, 1983), or with the 2',3' diol

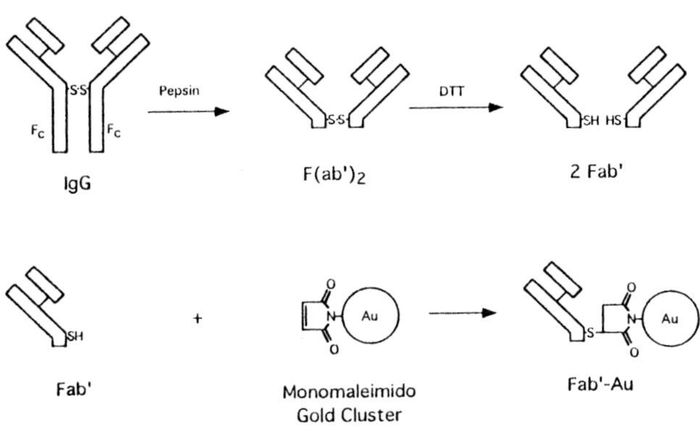

Figure 4. Coupling scheme for attaching gold clusters to the hinge sulfhydryl of Fab' fragments.

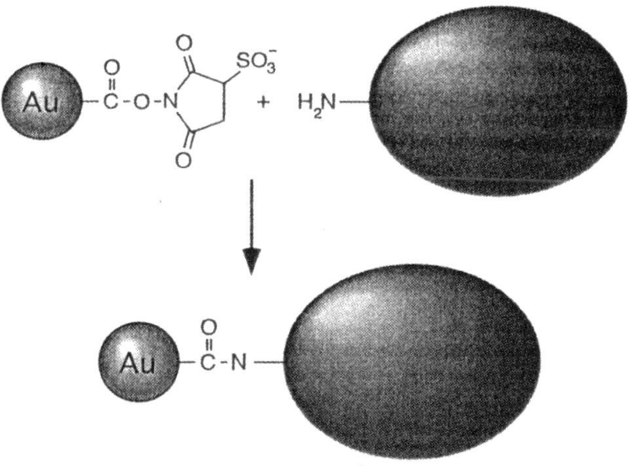

Figure 5. Reaction coupling amino groups onto N-hydroxysuccinimide (NHS) gold cluster.

---

of RNA (Skripkin *et al.*, 1993). Gold clusters can also be coupled to lipids, to make membrane labels (Fig. 7). The variety of stable covalent conjugates that can be designed certainly extends the usefulness of gold as a label.

Gold cluster immunoprobes

Gold clusters may be covalently attached to the hinge sulfhydryls of IgG or Fab' fragments using maleimido-gold. Attachment to antibody amino groups (using N-hydroxysuccinimide ester-gold) has also been done and the antibodies also show good immunoreactivity. It is also possible to couple the gold to the carbohydrate moiety. There are a number of advantages of using gold cluster immunoprobes over those made with colloidal gold: [1] They are smaller; one reason is the small size of the gold, another is that Fab' fragments are about 1/3 of the size of IgG, making the whole probe substantially smaller. This has several important

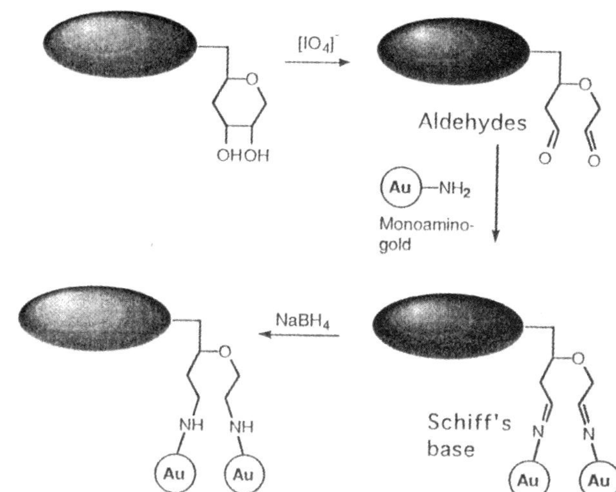

Figure 6. Reaction for covalently coupling gold clusters to carbohydrates.

---

benefits: [a] better penetration into tissues (40 μm has been reported (Sun *et al.*, 1995), whereas colloidal gold is typically < 0.5 μm) (Takizawa and Robinson, 1994), [b] more quantitative staining of antigens, since there is less steric hindrance, and [c] the small size gives higher resolution. [2] They are more stable conjugates, being covalent, whereas antibodies that "fall off" colloidal gold then compete for antigens, reducing gold labeling (Kramarcy and Sealock, 1990). [3] They are not aggregated, as some fraction of colloidal gold conjugates are, since colloidal gold is "sticky". They may be purified by gel filtration column chromatography to ensure only single Fab' fragments with gold are present, for example. [4] The gold size is very uniform, since they are compounds, whereas small colloidal gold preparations show high variability (Hainfeld, 1990); this can be important in high resolution work. [5] They actually give a better signal at low electron microscopy magnifications (with silver enhancement) than large colloidal gold, due to the higher staining density (Takizawa and Robinson, 1994).

Except for high resolution, molecular level electron microscopy (EM) work, the small size of the clusters (0.8 nm and 1.4 nm) in general leads to a signal that is too weak for good visibility, so for most general EM work and light microscopy, silver enhancement is a necessity. This is a simple procedure, and is described later.

Materials

Undecagold, Nanogold, and FluoroNanogold labeling reagents and conjugates are available commercially from Nanoprobes, Inc. (25 E. Loop Rd., Ste.

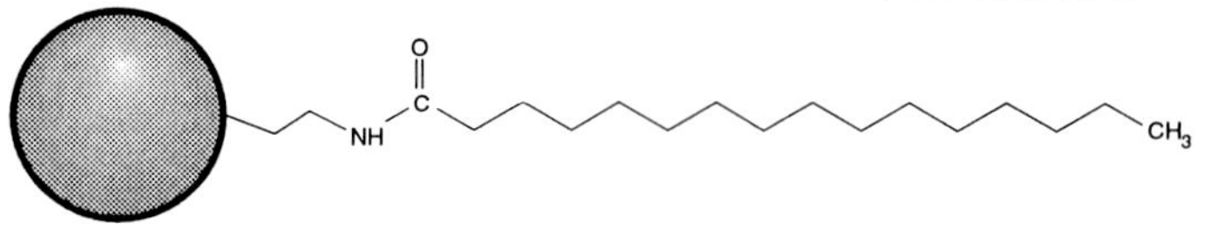

Figure 7. Gold-lipid conjugates, showing fatty acid and phospholipid derivatives.

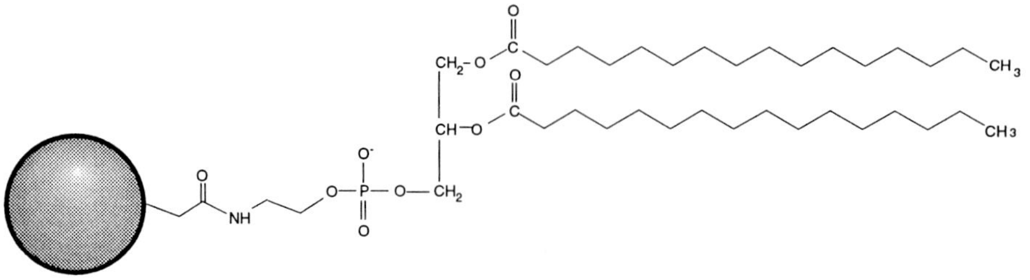

124, Stony Brook, NY 11790, Telephone Number: 516-444-8815). A suitable gel filtration column for separating free gold clusters from labeled proteins (of higher molecular weight) is Superose 12 (Pharmacia LKB Biotechnology, Inc., Piscataway, NJ 08854, Telephone Number 800-526-3593). For separating gold clusters from smaller molecules, GH25 column material is recommended (Amicon Inc., 72 Cherry Hill Dr., Beverly, MA 01915, Telephone Number: 800-426-4266).

Method of gold cluster labeling a protein

Making gold clusters for labeling requires organic synthesis of phosphines and ligands that are not commercially available; after these are made, the clusters must be formed, then purified (Safer *et al.*, 1986; Hainfeld, 1989). For monofunctional clusters, further purification, such as an ion exchange gradient is required. Next, to make the gold reactive, its shell must be activated with the appropriate crosslinker, such as a maleimide or N-hydroxysuccinimide ester. This must again be purified (by column chromatography) from the activating linker. Since most reactive groups hydrolyze in aqueous solution, and become inactive, the cluster must be used within a few hours (half lives are typically 4-16 hours). Fortunately, these clusters are now available commercially in activated form; these have been lyophilized so that they are stable and their hydrolysis only starts after they are reconstituted with water.

The steps for labeling a protein are therefore greatly simplified, and are reduced to the following:

[1] Dissolve lyophilized activated gold clusters (e.g., monomaleimido-Nanogold) by adding 1 ml water. Buffer is already in the gold and this reconstitutes to phosphate buffered saline (PBS).

[2] Mix with about 0.2 mg of protein to be labeled (the molar ratio of gold reagent to protein is typically 5:1). If reacting with amines, no protein pretreatment is necessary. For reacting with cysteine, proteins may have free -SH groups natively available, in which case no treatment is necessary. However, if these are oxidized into disulfides, they must first be reduced (with typically 50 mM mercaptoethylamine (MEA) for 1 hour), then separated by gel filtration from the reducing agent [MEA and dithiothreitol (DTT) contain thiols which would react with the gold]. If making Fab' or IgG conjugates, these must be prepared (For IgG, just reduce with mercaptoethanolamine, to expose hinge sulfhydryls, and column purify; for Fab', digest with pepsin to get $F(ab')_2$, then reduce and column purify to isolate the Fab' fragment (note: the Fab fragment is not used since it does not contain the hinge sulfhydryl).

[3] After 1-16 hrs, gel filter product (on e.g., Pharmacia Superose 12, in phosphate buffered saline, PBS), to separate labeled protein from unreacted gold.

Purification of the conjugate and quantitation

Step 3 above may be illustrated for Fab'-Au: after reacting Fab' with maleimido-Undecagold overnight at 4°C (although 1 hour is adequate for the maleimide-thiol reaction), the mixture was applied to a 1 x 30 cm Pharmacia Superose 12, in PBS, pH 7.4, running at 0.5 ml/min. The elution profile is shown in Fig. 8a. The first peak (here the tallest) is the Fab'-Au_{11}, and the second peak the unreacted, excess Au_{11}. Fig. 8b shows the spectrum of the first peak. Most proteins, including Fab', have a spectral peak at 280 nm, with no absorption above about 300 nm (they are colorless). Gold clusters have a yellow or brown color, and the spectrum of the second peak (excess Au_{11}) is shown in Fig. 8c. When gold is on the protein (first peak), the spectrum is (to first approximation) a sum of the two spectra. Here, an increase of absorbance in the 280 nm range from the protein is evident.

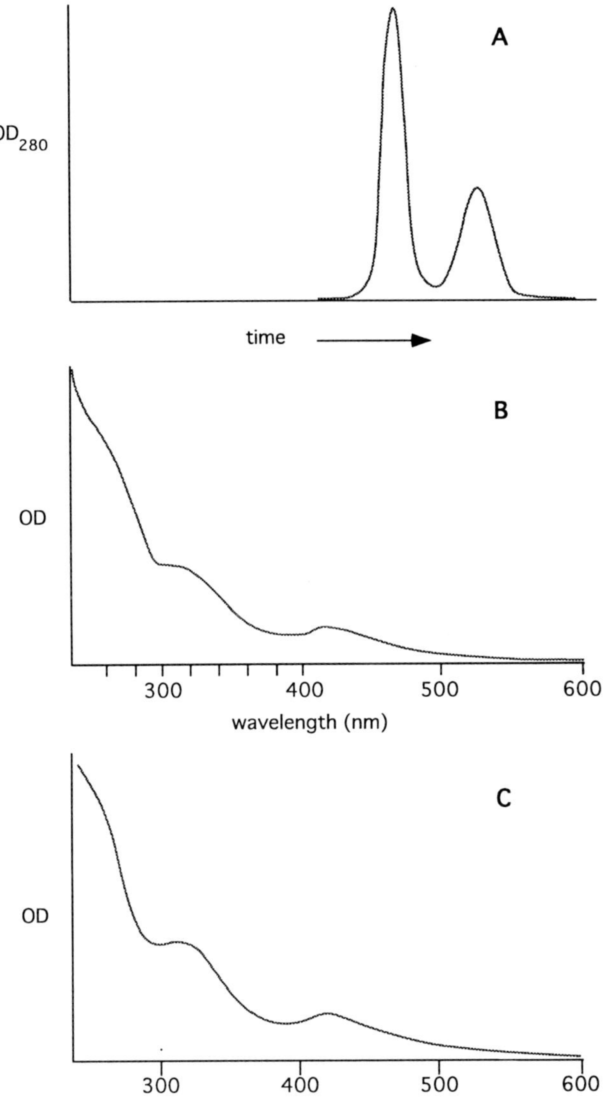

Figure 8. Gel filtration liquid chromatography of maleimide gold- Fab' reaction to isolate products. (**A**) shows the time course of products off a Superose 12 column (Pharmacia), the first and largest peak being the Fab'-Au_{11} conjugate. The second peak is unreacted gold cluster. (**B**) Spectrum of Fab'-Au_{11} conjugate peak, showing additive optical density of the two components. (**C**) Spectrum of the second peak, excess Au_{11}.

Carrying this a step further, quantitation of the labeling may be done from the spectrum; i.e., how many gold clusters are attached to each protein molecule, or what is the percentage of labeling. By knowing the spectra of the two components (and assuming that they don't alter each other when mixed), one can calculate the ratio (and amount) of each component given the composite spectrum. If the extinction coefficients are known for the two materials (gold and antibody) at two wavelengths, then a simple simultaneous equation is solved. For Fab' labeling, the equation is:

$$\text{Au/Fab ratio} = (7.5 \text{x} OD_{420})/(4.7 \text{x} OD_{280} - 16.8 \text{x} OD_{420}) \quad (1)$$

This is more clearly explained in Hainfeld (1989). This is a very useful way of assaying the success of the labeling reaction without further work such as microscopy or other tests.

High resolution labeling of isolated molecules and complexes

One application of gold cluster labeling is to react the gold at particular sites on a macromolecule, then to look with high resolution electron microscopy to map the site directly, alternatively, a substrate or Fab' fragment may be labeled, then bound to the target molecule, and its position visualized. For the highest resolution, it is advantageous to see the gold particle directly, without silver enhancement. For this work, the electron microscope must be operated at 30,000 x or higher magnification, and samples must be relatively thin, without the usual high density stains, all which would obscure visibility of the small gold clusters.

STEM Molecular Applications

The high resolution scanning transmission electron microscope (STEM) has been shown to be useful at this level, due to the high visibility of the gold clusters in darkfield on freeze dried isolated molecules adsorbed to a thin carbon film. With this microscope, undecagold clusters are clearly visible (as are even single heavy atoms), whereas with a commercial TEM, undecagold is not usually clearly seen, except by image analysis of ordered arrays. An example is the labeling of a Fab' fragment to the C-terminus of the Aα chain, shown in Fig. 9. Since this is a review, the reader is referred to published articles concerning:

* labeling Fab' with undecagold and Nanogold (Hainfeld, 1987, 1989; Hainfeld and Furuya, 1992).

* Fab'-Au_{11} localization on the phosphorylase kinase molecule to map a specific subunit (Wilkinson *et al.*, 1994),

* gold labeling of the mobile substrate-carrying arm in the pyruvate dehydrogenase complex in order to determine its position and range of motion (Yang *et al.*, 1994).

* gold labeling the substrate, dihydrofolate reductase (dhfr), of GroEL in order to determine the location where ATP-dependent folding occurs (Braig *et al.*, 1993).

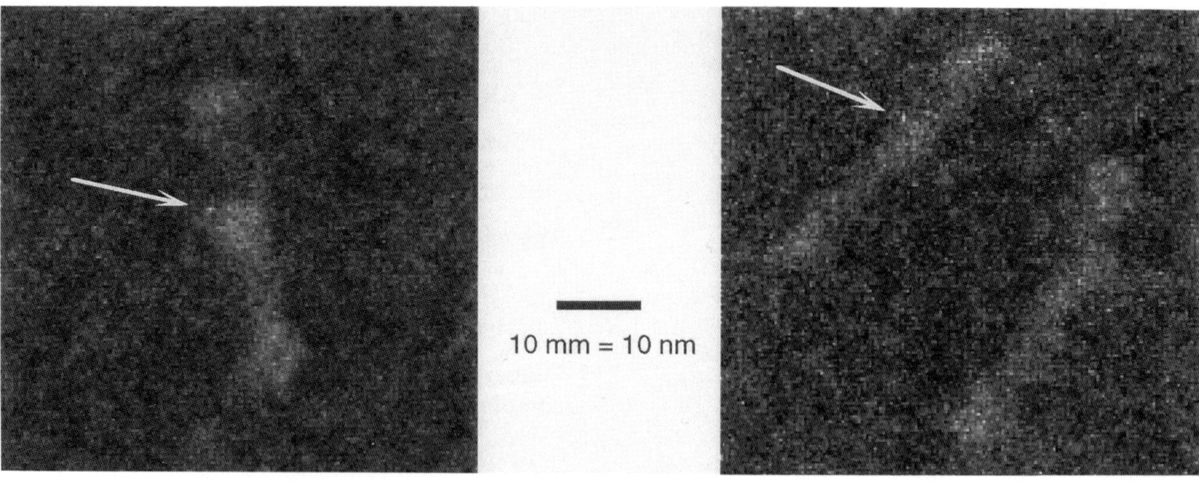

Figure 9. Darkfield field emission STEM micrographs of Fab'-Au_{11} bound to C-terminus of Aα chain of fibrinogen showing its putative molecular location (done in collaboration with G. Matsueda).

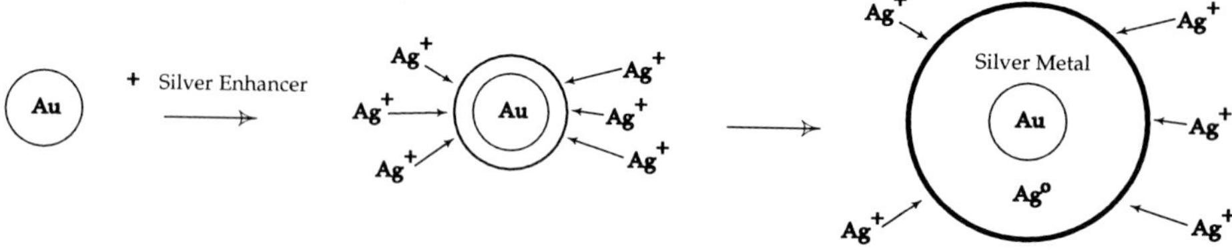

Figure 10. Schematic of the silver enhancement of gold process.

* direct labeling of tRNA and its use in mapping the ribosome binding site (Hainfeld *et al.*, 1991, 1993; Blechschmidt *et al.*, 1993)

The Brookhaven STEM is an NIH Biotechnology Resource, and is freely available to scientists with suitable molecular labeling studies. Inquiries should be directed to the author at: Brookhaven National Laboratory, Biology Dept., Upton, NY 11973, Telephone Number 516-344-3372.

TEM Molecular Applications

Although the Undecagold cluster is more difficult to see directly in conventional TEMs, the 1.4 nm Nanogold cluster is easily visualized in thin samples ($<$ about 80 nm), and in frozen hydrated samples (Boisset *et al.*, 1992). The clusters give their best signal with low values of defocus, about 0.5 μm, and can be seen at approximately 10 electrons/Å^2. Undecagold damages with beam dose (Wall *et al.*, 1982), and its signal becomes weaker, whereas Nanogold is very beam resistant (Hainfeld and Furuya, 1992) and can be clearly seen on a thin carbon film, by going to a very high magnification (100,000 x) with a high beam current. For high resolution biological applications, the beam must be kept low to prevent specimen damage, so the usual method has been to use ordered arrays or many single particles that are later computer aligned and averaged to improve the signal-to-noise ratio. A number of interesting projects have now been completed using these gold clusters to identify important biological sites or functions. The reader is referred to some of these works:

* Cys-374 of F-actin was labeled with undecagold

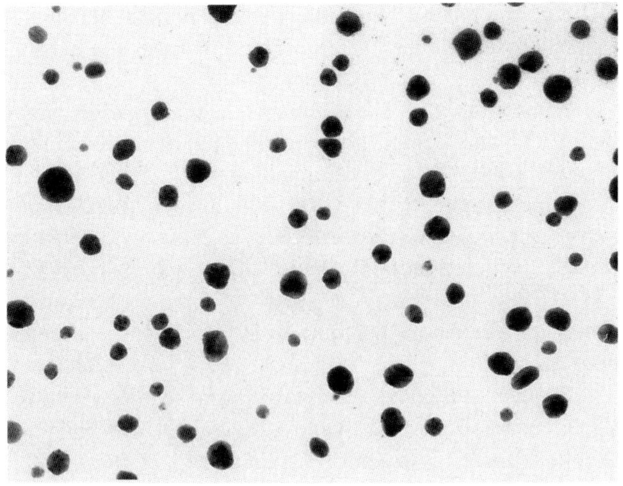

Figure 11. Electron micrograph of silver enhanced Nanogold. Particles are in the 20 nm range. Full width 530 nm.

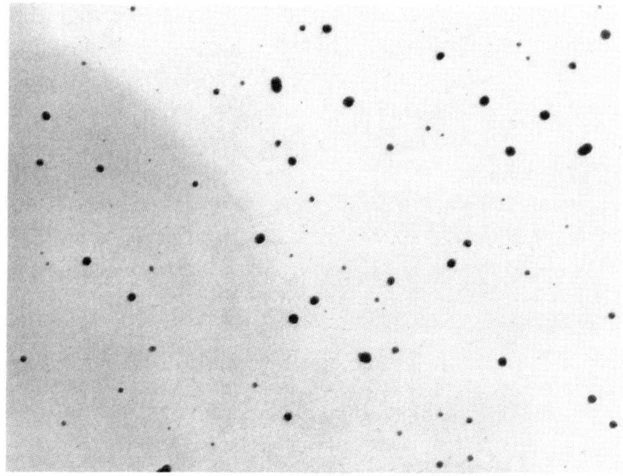

Figure 12. Electron micrograph of silver enhanced Undecagold. Particles are in the 10 nm range. Full width 530 nm.

----

and its position visualized by image processing fibers in ice (Milligan *et al.*, 1990).

* A thiol on calmodulin was labeled with Nanogold, then it was bound to the calcium release channel to determine the calmodulin binding site. This was done by image processing isolated calcium release channel complexes by cryo-EM (Wagenknecht *et al.*, 1994).

* The γ subunit of the F1 ATPase complex was localized by labeling with Nanogold and image processing of cryo-EM molecules (Wilkens and Capaldi, 1992).

* The method of insulin blocking the proteosome in protein unfolding was studied by labeling a thiol on the insulin B-chain (Wenzel and Baumeister, 1995).

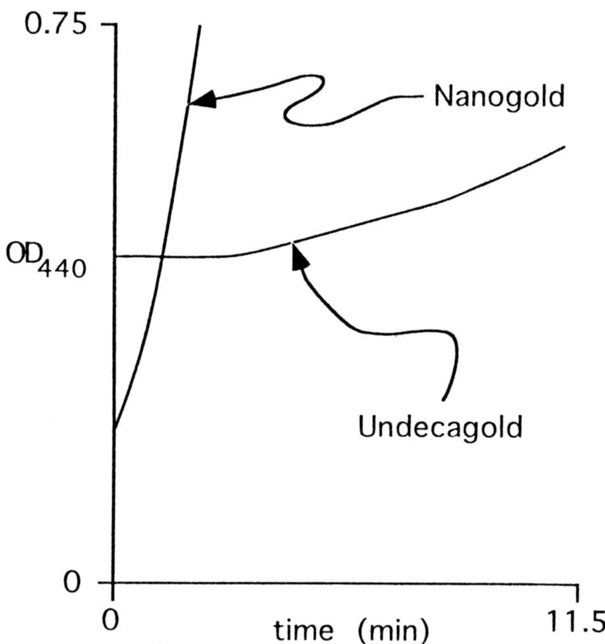

Figure 13. Graph showing the development progression with time of silver enhancement of Nanogold and undecagold. Absorbance measurements were taken of a solution in a cuvette in a spectrophotometer. Nanogold concentration was 2.2 x 10^{-6} M, and that of undecagold, 1.2 x 10^{-5} M. The undecagold, even though more concentrated, shows less development with a clear delay in starting development of about 5 min.

----

* Active site thiols in α-2 macroglobulin were visualized by Nanogold labeling, cryo-EM and image processing (Boisset *et al.*, 1992, 1994).

* Undecagold labeled cytochrome oxidase crystals were analyzed by glucose/uranyl acetate embedding and image processing to identify the subunit III site (Crum *et al.*, 1994).

* Nanogold was visualized in unstained Lowicryl sections of immunolabeled red blood cells (Hainfeld and Furuya, 1992).

Silver enhancement

An important discovery was that the silver metal deposition process, similar to photography, could be applied to gold particles. In photography, silver halide crystals are in the emulsion, and lead to silver grains after light activation and chemical reduction. Silver enhancement of gold particles is slightly different in that the source of silver is supplied in the developer solution along with the reducing agent. Silver ions are reduced to insoluble silver metal on interaction with the gold metal surface (which then becomes a silver metal surface, Fig. 10). This is a specific interaction nucleated by the gold particle, and can amplify the signal substantially.

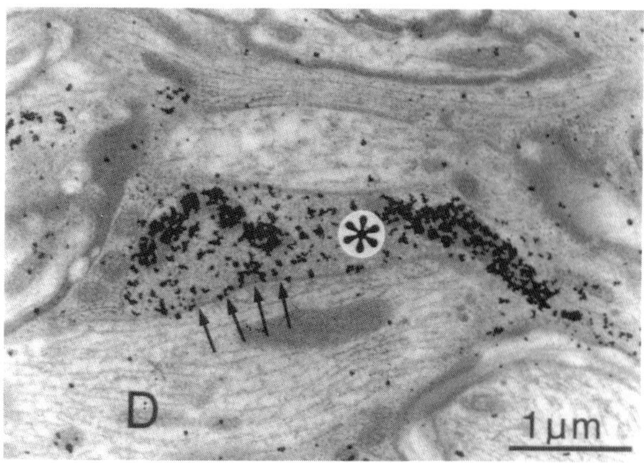

Figure 14. Transmission electron micrograph showing GABA-containing terminals (asterisks) forming symmetrical synaptic specializations (arrows) with dendrites (D) in the thoracic spinal cord of the rat. After embedding in Durcupan (Fluka) and sectioning, immunolocalization was done by incubating with GABA antiserum (Incstar, 1:2000, 4°C, 18 hours), Nanogold (goat anti-rabbit, 1:40, room temperature, 90 min), and intensified with HQ Silver (Nanoprobes) for 6 min. Counterstaining was with lead citrate. There is excellent structural preservation and the silver enhanced gold particles are about 20 nm in size. Besides the main GABA-containing region heavily stained in the center (arrows), smaller GABA-positive structures are also seen where there are multiple gold/silver particles, e.g., above and to the left of the central region. This work was done by Dr. Sarah Bacon, Oxford University, Department of Pharmacology, U.K.

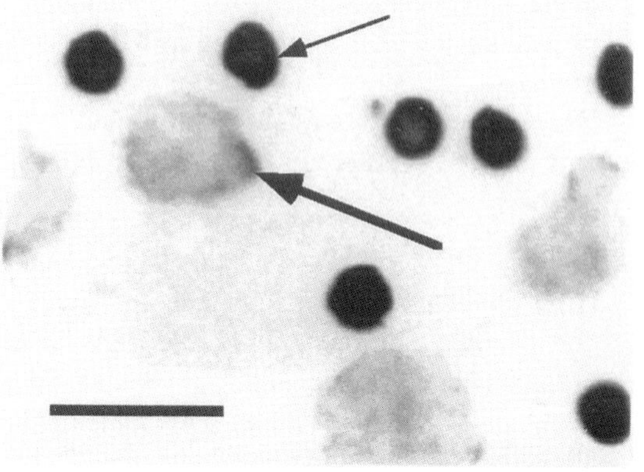

Figure 15. Light micrograph of human red blood cells (RBCs, small arrow) and HeLa cells (large arrow) incubated with anti-RBC IgG-Nanogold, then silver enhanced. The red blood cells turned totally black. Bar = 18 μm.

Developers can autonucleate after a period of time (a much slower process) and may contribute some background.

Micrographs of silver enhanced Nanogold is shown in Fig. 11 and Undecagold in Fig. 12. Even with their organic shell, they develop; however, undecagold is substantially slower, and has a delay of about 5 min before appreciably starting, whereas Nanogold develops rapidly, and immediately starts off at a high rate (Fig. 13). Gold clusters may be grown to sizes of 2-80 nm or more, but their size distribution is less uniform after the process.

Several studies report quantitative results about the silver development of Nanogold and its optimization (Hainfeld and Furuya, 1995a; Burry *et al.*, 1992; Takizawa and Robinson 1994, Stierhof *et al.*, 1995; Burry, 1995).

Osmium

Osmium tetroxide is commonly used as a counterstain. However the osmium can oxidize the silver back into solution, thus removing the silver grain either partially or completely. Burry studied this process and found a moderate level of osmium (0.1% for 30 min) gave good staining in tissues, and did not appreciably affect the silver size of the developed Nanogold (Burry *et al.*, 1992). An alternative solution was found by Sawada and coworkers, who gold toned the silver particles, which leaves a gold coating that is unreactive with osmium (e.g., 1% OsO_4 for 2 hours; Sawada and Esaki, 1994).

TEM Results in Tissue Studies: Immunolocalization with Nanogold

An excellent application of gold cluster labeling is immunohistological studies, for the reasons given above under "Gold Cluster Immunoprobes". In short, they generally give better staining than with colloidal gold probes due to their small size, high immunoreactivity and stability, and superior penetration. Several comparative studies have been done at this point to demonstrate these differences (Vandré and Burry, 1992; Takizawa and Robinson, 1994). Due to the thickness of tissue sections and counterstaining with OsO_4, lead citrate, uranyl acetate, or other stains, it is a requirement that the gold is silver enhanced. Because Nanogold develops more quickly and robustly than Undecagold, Nanogold is used in most of these applications. An example of the excellent localization, heavy antigen staining, and structural preservation is shown in Fig. 14. The use of Nanogold immunoconjugates in virtually all methods of tissue preparation (pre- and post-embedding, frozen sections and resin embedded) has now been documented.

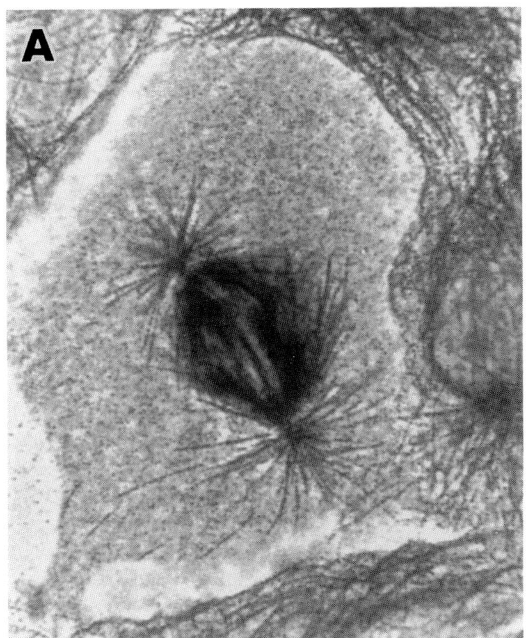

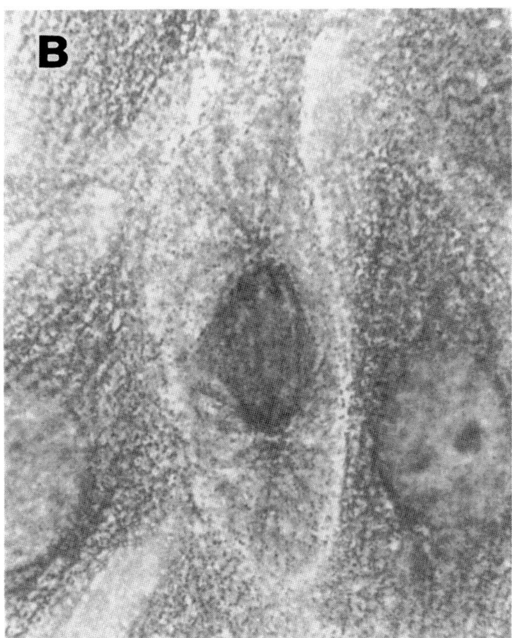

Figure 16. (A) Light micrograph of spindle microtubules labeled with a monoclonal anti-tubulin antibody, followed by Nanogold-Fab', and silver enhanced. LLC-PK cells were grown in monolayer culture, fixed with glutaraldehyde (0.7%, 15 min), permeablized with 0.1% saponin, incubated with anti-tubulin antibodies (1:250, Amersham) for 1 hour at 37°C, rinsed, then incubated with anti-mouse Nanogold-Fab' (1:50, Nanoprobes) for 1 hr at 37°C. Samples were post-fixed, silver enhanced with a N-propyl gallate enhancer for about 9 min. Cells were mounted in Mawwol and examined by bright field microscopy. Intense staining of microtubules was observed. Full width 95 μm. (B). Same preparation as described in A, except that AuroProbe One goat anti-mouse was used (1:50, Amersham). Only weak staining was observed. Full width 95 μm. This work was done by Drs. Dale Vandré and Richard Burry, Department of Cell Biology, Neurobiology, and Anatomy, The Ohio State University.

The reader is referred to the following original papers that use Nanogold for immunoelectron microscopy for detailed experimental protocols and results:

Preembedding

* Phosphoproteins associated with the mitotic spindle were localized; embedding in Epon (Vandré and Burry, 1992).

* Metabotropic glutamate receptor localization in neurons; freeze substitution and Lowicryl embedding (Baude *et al.*, 1993)

* Improved methods for *in vitro* cells; localization of synaptophysin in rat PC12 cells; embedding in Epon (Tao-Cheng and Tanner, 1994).

* Study of silver enhancement in pre-embedding immunocytochemistry; embedding in Epon (Burry *et al.*, 1992).

Localization of glutamic acid decarboxylase in the cerebellum; embedding in Spurr's resin (Gilerovitch *et al.*, 1995).

* Immunocytochemical localization of the α1 and β2/3 subunits of the $GABA_A$ receptor; freeze substitution and Lowicryl embedding (Nusser *et al.*, 1995a).

* Study of neural connections with neurobiotin and biocytin; embedding in Epon/Araldite (Sun *et al.*, 1995).

* Immunolocalization of *Drosophila* phosphatidylinositol transfer protein; embedding in Epon (Suzuki and Hirosawa, 1994).

* Quantitative immunogold method to determine densities of $GABA_A$ receptors on cerebellar cells; embedding in epoxy resin (Nusser *et al.*, 1995b).

* Anti-laminin and anti-fibronectin in rat testis localization using Nanogold, silver enhancement, gold toning, and osmium fixing; embedding in Epon 812 (Sawada and Esaki, 1994)

* Study of developers, buffers, and osmium effects; introduction of buffered n-propyl gallate as a better developer (Burry, 1995).

Postembedding

* Labeling calcium-ATPase in sarcoplasmic reticulum; embedding in LR White (Krenács and Dux, 1994).

* Quantitative immunogold method to determine densities of $GABA_A$ receptors on cerebellar cells; slam-

freezing, freeze-substitution, and Lowicryl embedding (Nusser *et al.*, 1995b)

* Metabotropic glutamate receptor localization in neurons; freeze substitution and Lowicryl embedding (Baude *et al.*, 1993)

* Immunocytochemical localization of the α_1 and $\beta_{2/3}$ subunits of the $GABA_A$ receptor; freeze substitution and Lowicryl embedding (Nusser *et al.*, 1995a).

* Comparison of embedding media (Epon 812, Durcupan ACM, Lowicryl K4M, LR-White and LR-gold; discussion of microwave treatment (Krenács and Krenács, 1995).

Ultra-thin cryosections

* Localization of lactoferrin and myeloperoxidase in neutrophils; 90 nm cryosections were labeled then embedded in Epon and thin-sectioned (Takizawa and Robinson, 1994).

Low magnification TEM and double labeling

Another interesting finding was that the silver enhanced Nanogold immunoprobe was more easily visible at low magnifications (about 10-20,000 x) than other larger colloidal golds, due to the much higher labeling density followed by the very visible silver size (approximately 20 nm) (Takizawa and Robinson, 1994). These authors also demonstrated that silver enhanced Nanogold (approximately 25 nm) could be clearly used as a double immunolabel with 10 nm colloidal gold.

Light microscopy

Immunolabeling with Nanogold-Fab' conjugates gives well labeled antigens and leads to clear visibility of the structures in the light microscope after silver enhancement (Fig. 15). Better penetration into tissues than with colloidal gold immunoprobes (even 1 nm colloidal gold) produces better visibility at the light microscope level (Fig. 16) and also at the EM level (Vandré and Burry, 1992).

An advantage of silver enhanced gold labeling at the light microscope level is the brown color produced that does not interfere with most other stains typically used. Many cells autofluoresce, and since the silver enhanced gold has equal or better sensitivity, it may be used in such cases, or as a dual label with fluorescent probes.

Staining blots

The high sensitivity of silver enhanced Nanogold makes it useful for immunoblot detection (Fig. 17). Sensitivity goes to the 0.1 pg range (7×10^{-19} moles), making it one of the most sensitive methods available (Hainfeld and Furuya, 1995a); it is in the same range as chemiluminescence, but requires about 40 min development and produces a permanent record, whereas chemiluminescence generally takes 24 hrs for this level, and requires film and film processing/printing. An example of its use in a blot of a gel was described by Wilkens and Capaldi (1992).

Staining gels

The use of gold to detect bands on gels was described for colloidal gold, but necessitated making a blot first, due to the limited penetration of colloidal gold into the gel matrix. Nanogold or undecagold labeled proteins run on gels similar to the native proteins, or in some cases, shifted up by the weight of the gold (15,000 or 5,000, respectively, Weinstein *et al.*, 1989, Hainfeld and Furuya, 1995a). When the gel is exposed to the silver enhancing solution (a different formulation than the standard silver stain for gels), a black-brown band develops in < 5 min, giving more rapid detection than with other gel stains (Coomassie blue, etc.); only the gold labeled proteins are stained.

New Developments

***In situ* hybridization (ISH)**

Testing of Nanogold-streptavidin has recently shown it to give a better signal than alkaline phosphatase or avidin-biotin (S-ABC)-peroxidase/DAB for ISH, with sensitivity down to the single copy or very low copy level (G. Hacker, personal communication). The increased sensitivity without background is probably due to the higher binding constant of streptavidin as compared to antibodies, and the efficient and high amplification silver enhancement yields with Nanogold.

Gold Lipids

Since the gold clusters can be reacted with most molecules, one can imagine a number of novel conjugates. One is to link them to fatty acids or phospholipids, thus producing a gold on one end, and a long alkyl chain on the other (Fig. 7). Since the gold is water soluble (hydrophilic), this yields an ampipathic molecule that should behave similar to the original lipids, namely, they should insert into membranes, for example. It should be possible to use them in synthetic liposomes to follow these structures microscopically.

FluoroNanogold

Another recently developed conjugate is the covalent combination of a fluorophore with a gold cluster (Powell *et al.*, 1994), also attaching a Fab' antibody fragment (Fig. 18). This has the interesting property that the fluorescence is not appreciably quenched by the gold particle; earlier work attempting dual gold-fluorescent probes with colloidal gold showed extensive quenching by the colloidal gold particle, and so further use was halted. By combining two probes into one, live cells, for example, can be viewed in the confocal microscope and a label visualized. When the distribution is optimal, these same cells can be processed for EM to determine

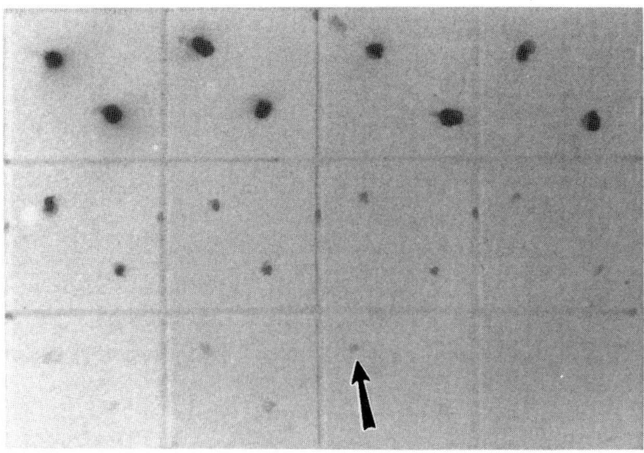

Figure 17. Sensitive immunodot blot obtained with Nanogold anti-mouse Fab' to a mouse IgG target showing 0.1 pg detection (arrow), corresponding to 6.7 x 10^{-19} moles of target. 1 μl of mouse IgG (antigen target) was spotted onto nitrocellulose containing the following amounts (each box contains a duplicate spot): top row (of spots): 10 ng, 2.5 ng, 1 ng, 0.25 ng; middle row: 100 pg, 25 pg, 10 pg, 2.5 pg; bottom row: 1 pg, 0.25 pg, 0.1 pg, buffer blank. Membrane was blocked with 4% BSA. Goat anti-mouse Fab'-Nanogold was incubated 2 hr, washed and developed 2 x 15 min with LI Silver. Buffer and non-specific antibody/antigen controls were blank.

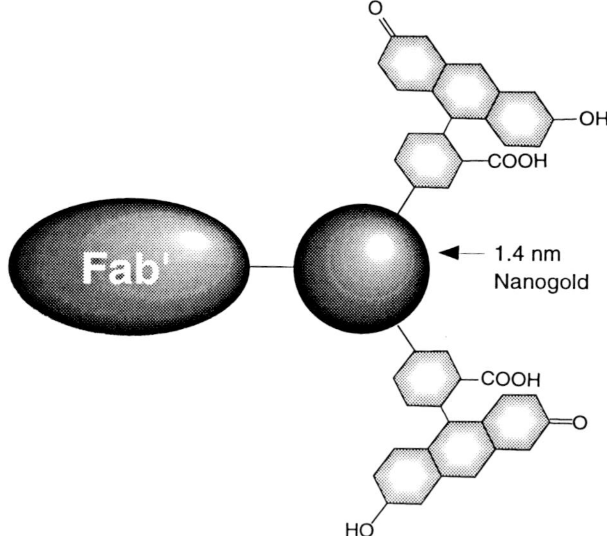

Figure 18. Diagram of FluoroNanogold.

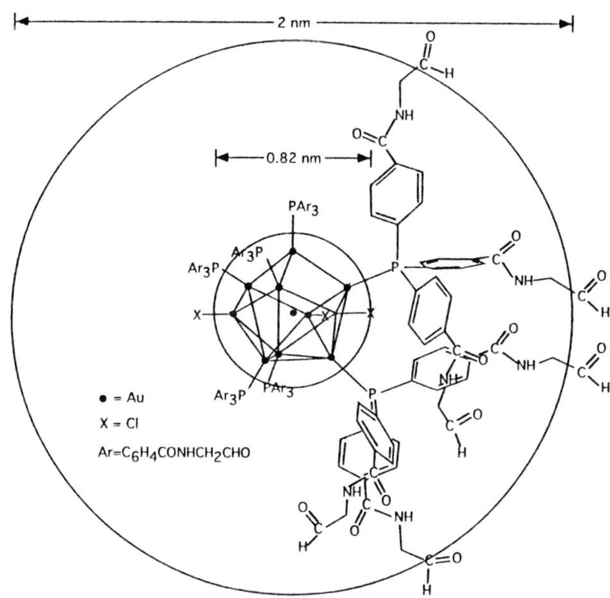

Figure 19. Diagram of polyaldehyde Undecagold.

----

the ultrastructural organization.

Polyaldehyde gold

It is possible to synthesize gold clusters with multiple reactive groups, which might be useful for attaching multiple peptides, antibody fragments, DNA strands. Undecagold has 21 organic "arms" on its surface (Fig. 1), which can then be made reactive for linking. This has been done using the aldehyde functionality (Fig. 19), which can then lead to singly labeled or oligomeric products (Hainfeld and Furuya, 1995b).

Radioactive gold clusters

Natural gold (^{197}Au) can be made radioactive (^{198}Au and ^{199}Au) by neutron bombardment in a nuclear reactor. Only a small percentage of the gold is converted (about 0.03%), and if all gold atoms are to be radioactive, pure ^{199}Au can be made by irradiating ^{198}Pt. These golds have approximately a 3-day half life, emit an intermediate β, as well as a an imageable γ. The β can kill cells, and has an average range of 100-460 μm. One possible application is to attach gold clusters to anti-tumor antibodies and thus target the dose for selective cell killing (Hainfeld *et al.*, 1990; Hainfeld, 1995). Radioactive gold is one of about 10 radionuclides deemed to be best suited for this type of work. The major problems seem to be that intravenous injection of these conjugates have uptake in other tissues which limits the dose to tumor to unsatisfactory levels. A more tractable application, for the moment, is a topical application, such as treatment of superficial bladder carcinoma, where the conjugate can be introduced into the bladder for about 0.5 hr, then lavaged, leaving the specifically targeted gold to do its work. A graph showing the

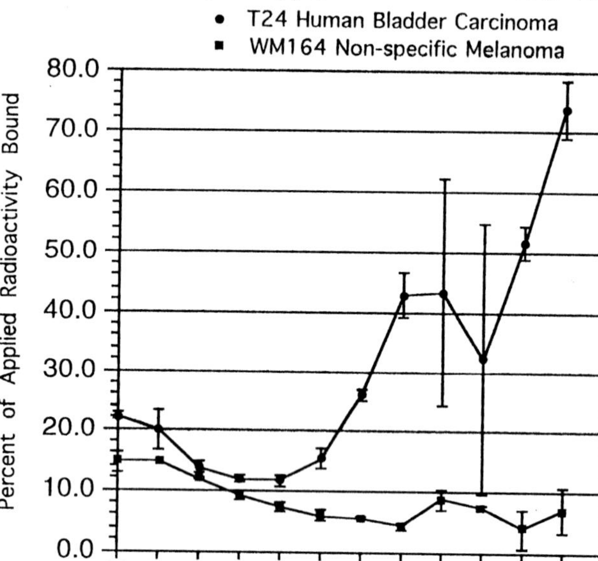

Figure 20. Graph showing the targeting of $^{198}Au_{11}$ clusters conjugated to a monoclonal antibody to human bladder carcinoma cells (MAb 48-127, gift from Y. Fradet), with minimal uptake by a non-specific cell line, WM164. Maximal tumor to non-tumor ratio was 13.6.

---

targeting of $^{198}Au_{11}$ clusters conjugated to a monoclonal antibody to human bladder carcinoma cells, with minimal uptake by a non-specific cell line is shown in Fig. 20.

Conclusion

Gold cluster chemistry has opened new areas of applications in cell biology, medicine, and material science. Many novel conjugates can be formed with stable covalent links and well defined products. The properties of the gold clusters often give them important advantages over colloidal gold counterparts. The gold provides a versatile and highly sensitive reporter group that can be seen by electron microscopy, or with silver enhancement, light microscopy, and the unaided eye.

Acknowledgements

The author would like to thank N.I. Feng for excellent technical assistance, J.S. Wall, M. Simon, B. Lin, and F. Kito for STEM microscopy, F.R. Furuya and R.D. Powell for significant contributions, and R. Burry and S. Bacon for use of their micrographs. Support of this work was through NIH #RR017777 and US DOE, OHER.

References

Baude A, Nusser Z, Roberts JDB, Mulvihill E, Mclihinney RAJ, Somogyi P (1993) The metabotropic glutamate receptor (mGluR1a) is concentrated at perisynaptic membrane of neuronal subpopulations as detected by immunogold reaction. Neuron **11**: 771-787.

Blechschmidt B, Jahn W, Hainfeld JF, Sprinzl M, Boublik M (1993) Visualization of a ternary complex of the *Escherichia coli* Phe-tRNA^{Phe} and Tu-GTP from *Thermus thermophilus* by scanning transmission electron microscopy. J Struct Biol **110**: 84-89.

Boisset N, Grassucci R, Penczek P, Delain E, Pochon F, Frank J, Lamy JN (1992) Three-dimensional reconstruction of a complex of human alpha 2-macroglobulin with monomaleimido Nanogold (Au1.4nm) embedded in ice. J Struct Biol **109**: 39-45.

Boisset N, Penczek P, Pochon F, Frank J, Lamy J (1994) Three-dimensional reconstruction of human alpha 2-macroglobulin and refinement of the localization of thiol ester bonds with monomaleimido nanogold. Ann N Y Acad Sci **737**: 229-244.

Braig K, Simon M, Furuya F, Hainfeld JF, Horwich AL (1993) A polypeptide bound by the chaperonin groEL is localized within a central cavity. Proc Natl Acad Sci USA **90**, 3978-3982.

Burry RW (1995) Pre-embedding immunocytochemistry with silver-enhanced small gold particles. In: Immunogold Silver Staining: Methods and Applications. M.A. Hayat (ed). Academic Press, San Diego, CA. pp 217-230.

Burry RW, Vandré DD, Hayes DM (1992) Silver enhancement of gold antibody probes in pre-embedding electron microscopic immunocytochemistry. J Histochem Cytochem **40**: 1849-1856.

Crum J, Gruys KJ, Frey TG (1994) Electron microscopy of cytochrome c oxidase crystals: labeling of subunit III with a monomaleimide undecagold cluster compound. Biochemistry **46**: 13719-13726.

Gilerovitch HG, Bishop GA, King JS, Burry RW (1995) The use of electron microscopic immunocytochemistry with silver-enhanced 1.4-nm gold particles to localize GAD in the cerebellar nuclei. J Histochem Cytochem **43**, 337-343.

Hainfeld JF (1987) A small gold-conjugated antibody label: Improved resolution for electron microscopy. Science **236**: 450-453.

Hainfeld JF (1989) Undecagold-antibody method. In: Colloidal Gold: Principles, Methods, and Applications. vol. 2. Hayat MA (ed). Academic Press, San Diego, CA. pp 413-429.

Hainfeld JF (1990) STEM analysis of Janssen Auroprobe One. In: Proc XIIth Int Congress Electron Microsc. Bailey GW (ed). San Francisco Press, San

Francisco, pp 954-955.

Hainfeld JF (1995). Gold, electron microscopy, and cancer therapy. Scanning Microsc **9**: 239-256.

Hainfeld JF, Furuya FR (1992) A 1.4-nm gold cluster covalently attached to antibodies improves immunolabeling. J Histochem Cytochem **40**: 177-184.

Hainfeld JF, Furuya FF (1995a) Silver enhancement of Nanogold and undecagold. In: Immunogold Silver Staining: Methods and Applications. Hayat MA (ed). Academic Press, San Diego, CA. pp 71-96.

Hainfeld JF, Furuya FF (1995b) Aldehyde gold clusters for molecular labeling. In: Proc 53rd Ann Mtg Microsc Microanal, Bailey GW, Ellisman MH, Hennigar RA, Zaluzec NJ (eds). Jones and Begell Publishing, New York, NY. pp 858-859.

Hainfeld JF, Foley CF, Srivastava SC, Mausner LF, Feng NI, Meinken GE, Steplewski Z (1990) Radioactive gold cluster immunoconjugates: Potential agents for cancer therapy. Nucl Med Biol **17**: 287-294.

Hainfeld JF, Sprinzl M, Mandiyan V, Tumminia SJ, Boublik M (1991) Localization of a specific nucleotide in yeast tRNA by scanning transmission electron microscopy using an undecagold cluster. J Struct Biol **107**, 1-5.

Hainfeld JF, Furuya FR, Carbone K, Simon M, Lin B, Braig K, Horwich AL, Safer D, Blechschmidt B, Sprinzl M, Ofengand J, Boublik M (1993) High resolution gold labeling. In: Proc 51st Ann Mtg Microsc Soc Amer. Bailey GW, Rieder CL (eds). San Francisco Press, San Francisco. pp 330-331.

Kramarcy NR, Sealock R (1990) Commercial preparations of colloidal gold-antibody complexes frequently contain free active antibody. J Histochem Cytochem **39**, 37-39.

Krenács T, Dux L (1994) Silver-enhanced immunogold labeling of calcium-ATPase in sarcoplasmic reticulum of skeletal muscle. J Histochem Cytochem **42**, 967-968.

Krenács T, Krenács L (1995) Comparison of embedding media for immunogold-silver staining. In: Immunogold Silver Staining: Methods and Applications. Hayat MA (ed). Academic Press, San Diego, CA. pp 57-70.

Lipka JJ, Hainfeld JF, Wall JS (1983) Undecagold labeling of a glycoprotein: STEM visualization of an undecagoldphosphine cluster labeling the carbohydrate sites of human haptoglobin-hemoglobin complex. J Ultrastruct Res **84**: 120-129.

McPartlin M, Mason R, Malatesta I (1969) Novel cluster complexes of gold(0)-gold(1). J Chem Soc Chem Commun 334.

Milligan RA, Whittaker M, Safer D (1990) Molecular structure of F-actin and location of surface binding sites. Nature **348**, 217-221.

Nusser Z, Roberts JDB, Baude A, Richards JG, Sieghart W, Somogyi P (1995a) Immunocytochemical localization of the $\alpha 1$ and $\beta 2/3$ subunits of the $GABA_A$ receptor in relation to specific GABAergic synapses in the dentate gyrus. Eur J Neurosci **7**: 630-646.

Nusser Z, Roberts JDB, Baude A, Richards JG, Somogyi P (1995b). Relative densities of synaptic and extrasynaptic $GABA_A$ receptors on cerebellar granule cells as determined by a quantitative immunogold method. J Neurosci **15**, 2948-2960.

Powell RD, Hainfeld JF, Churchill MEA, Belmont ASI (1994) Combined fluorescent and gold nucleic acid probes. In: Proc 52nd Ann Mtg Microsc Soc Amer. Bailey G, Garratt-Reed AJ (eds). San Francisco Press, San Francisco. pp 176-177.

Reardon JE, Frey PA (1984) Synthesis of undecagold cluster molecules as biochemical labeling reagents. 1. Monoacyl and mono[N-(succinimidooxy)succinyl] undecagold clusters. Biochemistry **23**: 3849-3856.

Safer D, Bolinger L, Leigh JS (1986) Undecagold clusters for site-specific labeling of biological macromolecules: simplified preparation and model applications. J Inorg Biochem. **26**: 77-91.

Sawada H, Esaki H (1994) Use of Nanogold followed by silver enhancement and gold toning for preembedding immunolocalization in osmium-fixed, epon-embedded tissues. J Electr Microsc, **43**, 361-366.

Skripkin E, Yusupova G, Yusupov M, Kessler P, Ehresmann C, Ehresmann B (1993) Synthesis and ribosome binding properties of model mRNAs modified with undecagold cluster. Bioconj Chem **4**, 549-553.

Stierhof Y-D, Hermann R, Humbel BM, Schwarz H (1995) Use of TEM, SEM, and STEM in imaging 1 nm colloidal gold particles. In: Immunogold Silver Staining: Methods and Applications. Hayat MA (ed). Academic Press, San Diego, CA. pp 97-118.

Sun XJ, Tolbert LP, Hildebrand JG (1995) Using laser scanning confocal microscopy as a guide for electron microscopic study: a simple method for correlation of light and electron microscopy. J Histochem Cytochem **43**, 329-335.

Suzuki E, Hirosawa K (1994) Immunolocalization of a *Drosophila* phosphatidylinositol transfer protein (rdgB) in normal and rdgA mutant photoreceptor cells with special reference to the subrhabdomeric cisternae. J Electron Microsc **43**: 183-189.

Takizawa T, Robinson JM (1994). Use of 1.4-nm immunogold particles for immunocytochemistry on ultrathin cryosections. J Histochem Cytochem **42**, 1615-1623.

Tao-Cheng J-H, Tanner VA (1994) A modified method of pre-embedding EM immunocytochemistry which improves specificity and simplifies the process for *in vitro* cells. In: Proc 52nd Ann Mtg Microsc Soc Amer. Bailey G, Garratt-Reed AJ (eds). San Francisco

Press, San Francisco, 306-307.

Vandre DD, Burry RW (1992) Immunoelectron microscopic localization of phosphoproteins associated with the mitotic spindle. J Histochem Cytochem **40**, 1837-1847.

Wagenknecht T, Berkowitz J, Grassucci R, Timerman AP, Fleischer S (1994) Localization of calmodulin binding sites on the ryanodine receptor from skeletal muscle by electron microscopy. Biophys J **67**, 2286-2295.

Wall JS, Hainfeld JF, Bartlett PA, Singer SJ (1982) Observation of an undecagold cluster compound in the scanning transmission electron microscope. Ultramicroscopy **8**: 397-402.

Weinstein S, Jahn W, Hansen H, Wittmann HG, Yonath A (1989) Novel procedures for derivatization of ribosomes for crystallographic studies. J Biol Chem **264**, 19138-19142.

Wenzel T, Baumeister W (1995) Conformational constraints in protein degradation by the 20S proteasome. Nature Struct Biol **2**, 199-204.

Wilkens S, Capaldi RA (1992) Monomaleimidogold labeling of the γ subunit of the *E. coli* F_1 ATPase examined by cryoelectron microscopy. Arch Biochem Biophys **229**, 105-109.

Wilkinson DA, Marion TN, Tillman DM, Norcum MT, Hainfeld JF, Seyer JM, Carlson GM (1994) An epitope proximal to the carboxyl terminus of the α-subunit is located near the tips of the phosphorylase kinase hexadecamer. J Mol Biol **235**, 974-982.

Yang Y-S, Datta A, Hainfeld JF, Furuya FR, Wall JS, Frey PA (1994) Mapping the lipoyl groups of the pyruvate dehydrogenase complex by use of gold cluster-labels and scanning transmission electron microscopy. Biochemistry **33**: 9428-9437.

Discussion with Reviewers

R.M. Albrecht: Do non-covalent interactions between the gold-organic clusters and antibodies, ligands, etc. occur and is this ever a consideration when conjugating them to molecular species or when labeling with the conjugates?

Author: For the most part the answer is no. One can take gold clusters that do not have the activated linking arm and incubate them with the antibody, or other target, then separate the products by column chromatography or polyacrylamide gel electrophoresis. Peaks from columns can easily be quantitated for gold cluster by the UV-visible absorption, e.g., at 420 nm, where most proteins don't absorb. Gold in gels may be detected by silver enhancement. Typically, virtually no gold is found with the protein, unless it is specifically covalently linked. The linkers used, e.g., maleimide and N-hydroxysuccinimide ester are very specific for thiols and amines, respectively, and this can be similarly demonstrated. In a few unusual cases, gold clusters have been found to bind to certain macromolecules. However, this may usually be reversed by using a higher ionic strength. Another alternative is to make the gold with different surface groups, e.g., with a sugar- or PEG-like coating. The converse is also true: gold clusters can be made to intentionally non-covalently bind to certain substrates, e.g., by making them highly charged (plus or minus) or hydrophobic. The usual clusters are made with N-methyl benzamide groups which confer high water solubility, but have no charge.

R.M. Albrecht: What is the overall size, gold plus organic shell, of the nanoprobes?

Author: For Undecagold, the core contains 11 gold atoms in a 0.82 nm sphere, and the covalently attached phosphine ligand shell makes the total diameter 2.0 nm. This structure has been solved by x-ray crystallography. For Nanogold, the core contains approximately 67 gold atoms in a 1.4 nm sphere, and with the ligand shell, it is 2.7 nm in diameter. This has not been solved by x-ray, but in the EM we have observed small crystals. When labeling a target molecule, the distance measured would be from the linking site to the center of the gold cluster, giving about 1.4 nm resolution.

R.M. Albrecht: In our experience the conjugation of 3 nm colloidal gold to Fab fragments is useful as it produces a 1:1 gold:Fab ratio; the probe has a valence of 1 and is readily detectable. However, not all Fab fragments can be successfully conjugated to 3 nm gold via the non-covalent, hydrophobic route. Do the undecagold-Fab' or nanogold-Fab' conjugates (via the sulfhydryl link) generally work well with most Fab' without a drop in antibody affinity?

Author: Yes. The hinge sulfhydryl on Fab' fragments is very accessible, and labels very well across all species. The maleimide linking group is very specific and reactive with free thiols, and forms a stable covalent bond. Coupling is done under mild conditions (e.g., pH 7, phosphate buffer, 150 mM NaCl, 4°C), so that the antibody is not denatured. Since the hinge site is at the opposite end of the molecule to the antigen binding hypervariable region, this is an ideal place for the gold, so as not to compromise immunological activity. Some quantitative measurements of antibody activity after gold cluster labeling were made using a radioimmunoassay with mouse monoclonals and showed immunoreactivities of 76-83% (Hainfeld, 1995).

R.M. Albrecht: Probes synthesized to contain one undecagold or one nanogold cluster per Fab' or per other molecule or active molecular fragment are desirable for quantitative studies. In cases where silver enhancement is necessary to detect the undecagold or nanogold is the enhancement procedure sufficiently uniform and controllable to permit accurate quantitative analysis?

Author: For quantitation, one would like to have no particles so underdeveloped that they are below the detection limit, or so overdeveloped that they coalesce. Nanogold may be developed to produce ~15 nm particles which are not generally overlapping, but their size varies more than a well prepared 15 nm colloidal gold. Frequently a higher density of labeling is seen with Nanogold than in a parallel experiment with colloidal gold (e.g., Vandré and Burry, 1992), which raises some questions about quantitating antigens with colloidal gold probes (which may not penetrate as well, and may have antibody that has dissociated from the gold and competes for sites). In high resolution STEM or TEM of silver enhanced Nanogold and undecagold, there are still some gold clusters that would be below the usual detection level in tissue work, so quantitation will be only approximate. A well controlled study is needed to thoroughly answer this question. Undecagold develops more slowly than Nanogold, and with more variability, and with less final product for the same concentration of gold particles (Hainfeld and Furuya, 1995). It would appear that undecagold is less suitable for quantitative work when using silver enhancement.

R.M. Albrecht: Where the gold particles are sufficiently smaller than the antibody or ligand or the active fragments of antibody or ligand, we have seen, via correlative HR-SEM and SFM, that the colloidal gold particles intercalate themselves within the molecule rather than sticking to the "outside" of the molecule. What is the structure of Fab'-undecagold and Fab'-nanogold conjugates? Does attaching undecagold or nanogold via the free amino route ever tend to "blanket" the molecule and reduce activity or penetration?

Author: Contrary to your results with colloidal gold, undecagold and Nanogold appear at the periphery of many molecules, and specifically for Fab', which has a ellipsoid shape (5.0 x 4.0 x 3.0 nm), it appears at one end (see Fig. 4 in Hainfeld, 1987). Another interesting result demonstrating the site specific attachment of the gold clusters was shown for Nanogold, where it was reacted with the hinge sulfhydryls in IgG, and high resolution images showed it to be at exactly the expected position at the vertex of the "Y" shaped molecules (Fig. 7a in Hainfeld and Furuya, 1992). Nanogold synthesized with two reactive arms reacted with 2 Fab's and the gold was found in the center of this construct, as expected (Fig 7b, same paper).

Your second question pertains to amino labeling; since proteins can have many lysines, do gold clusters that react with primary amines (e.g., NHS-Nanogold) overly coat the molecule and reduce activity/penetration? This is not an actual problem, since the microenvironment (pK_a) of the amines varies and some are more reactive. One limits the amount of NHS-Nanogold, and the reaction time, so that typically one gold cluster is attached (if that is what's desired). From the column purification and UV-visible spectrum, the stoichiometry can easily be determined (of gold to protein). In practice, overloading proteins with gold is not a problem.

R.M. Albrecht: It is suggested that penetration of undecagold probes (2 nm diameter gold+organic) and nanoprobes may be better than probes attached to 1-3 nm colloidal gold particles. Given that in both cases the marker, undecagold or colloidal gold, is considerably smaller that the molecule to which it is conjugated, what do you feel is the most likely explanation for the increased penetration? In the Vandré paper whole antibody molecules were used for the primary antibody. Increased labeling seen with the second antibody-nanoprobe conjugate as compared to second antibody-1 nm colloidal gold conjugate was attributed to the presence of un(gold)labeled 2nd antibody in the prep. The Takizawa and Robinson (1994) study showed that nanoprobes attached to Fab' gave greater intensity of labeling, than 1 nm colloidal gold conjugated to whole antibody. Not surprising since the total probe size (antibody fragment + gold) of the Fab'-nanoprobe is very much smaller than the whole antibody-1 nm colloidal gold. Again this study employed gold labeled second antibody and in both cases the primary antibody was whole antibody. So in any case the presence of specific second antibody (either as whole ab or as Fab' antibody fragments) is limited to wherever the primary antibody, a whole antibody, can access. Have there been any comparative studies where the primary antibody prepared as an Fab' conjugated to 1-3 nm colloidal gold was compared to primary antibody prepared as an Fab' conjugated to nanogold and/or undecagold -- with care taken to insure that neither prep had free un(gold)conjugated Fab'?

Author: Fab'-3 nm colloidal gold conjugates (goat anti-mouse) were freshly prepared and chromatographed, and compared with Fab'-Nanogold (using the same antibody) on immunoblots targeting mouse IgG, followed by silver intensification. The Nanogold probe gave a factor of 10 greater sensitivity; this result was consistently found with other antibodies. Since accessibility was not a problem with this assay, one simple explanation is that the Nanogold conjugates retain better immunoreactivity.

When a 1 nm colloidal gold antibody conjugate was

analyzed by high resolution STEM microscopy, substantial aggregation was observed (Hainfeld, 1990). The STEM is useful since both antibody and gold are clearly visible, and mass measurement may be used to identify one antibody. Since Nanogold conjugates have no tendency to aggregate, and are column purified as monomers, it appears that typical colloidal gold conjugates may have limited penetration due to the fraction of the material that is aggregated. The STEM study also revealed a much wider gold size distribution for the "1 nm" colloidal gold; it varied from 1 to 3 nm.

The Takizawa and Robinson (1994) study compared penetration of 1.4 nm (Nanogold)-Fab', 1.4 nm (Nanogold)-IgG, and 1 nm colloidal gold IgG in 1-2 μm cryosections that were then cross-sectioned. They found 1.4 nm (Nanogold)-Fab' > 1.4 nm (Nanogold)-IgG > 1 nm colloidal gold-IgG (their Fig. 8). In other words, better penetration was not only due to the use of Fab' rather than IgG.

A last point concerns the actual size of the gold conjugates. Nanogold has a 0.6 nm organic shell around it and needs no further stabilization. It has a single Fab' covalently attached. Colloidal gold needs to be stabilized with antibody, BSA, PEG, or other substance. BSA is standardly used, which is 68 kD, larger than an Fab' (50 kD). Hence a colloidal gold may have other molecules bound that are 4-5 nm in size. This is about 7 times the size of the Nanogold shell, and could easily double the size of a Fab'-3 nm colloidal gold probe, thus making it an overall larger probe. Baschong and Wrigley (1990) discuss in detail experience in making 1.5-2.6 nm colloidal gold-Fab conjugates. They found that 2.6 nm colloidal gold adsorbed from 1 to 6 active Fab molecules per gold particle (their Fig. 4). By titrating BSA, they could prepare conjugates that contained predominantly one active Fab molecule (but also had adsorbed BSA molecules).

J. Beesley: In Figs. 11 and 12, especially 11, there appears to be a considerable number of small particles (approximately 1/10 size of the silver-enhanced particles) in the background. Would the author please comment on what these are and how to reduce them?
Author: My interpretation is that these are gold particles that have developed less than the larger particles. The silver enhancement of gold is already known to produce some variability in size, but in many examples, the development of Nanogold has produced strikingly uniform particle sizes. See, for example, Krenács and Krenács (1995), Fig. 4 on page 65. They found that more uniform particles developed with post-embedding immunostaining using Lowicryl K4M and LR-White than with Epon. Development for 3 min on a Lowicryl K4M section using a silver acetate developer (without gum arabic) gave extremely uniform 8-10 nm particles with no small (or large) particles (their Fig. 4). One might well use their methods to also achieve very uniform development.

Silver growth characteristics will also depend on the developer used. R. Burry and colleagues compared the growth time course and particle size distribution of silver enhancement using his N-propyl gallate developer formulation, and found that Nanogold gave a tighter size distribution than 1 nm colloidal gold (Burry *et al.*, 1992, see Fig. 5).

J. Wroblewski: There is a great need to obtain highly specific probes and good detection systems for *in situ* hybridization on the ultrastructural level. Are there any other Au-based, sensitive detection systems than Nanogold-streptavidin that can be used in electron microscopical studies?
Author: Colloidal gold-streptavidin conjugates have been used, but are somewhat unstable and give poorer results (lower signal/more background) than Nanogold-streptavidin (Hacker *et al.*, 1996). A recent improvement in the sensitivity to routine single copy detection has been achieved by combining the CARD system (catalyzed reporter deposition, where biotin is enzymatically amplified by peroxidase acting on the substrate biotinyl-tyramide) followed by Nanogold-streptavidin (Hacker *et al.*, 1996). Another alternative to streptavidin is to use a Nanogold-anti-biotin conjugate, but in our tests this does not perform as well as Nanogold-streptavidin. Also, DNA/RNA probes with digoxigenin or fluorescein (or other antigens) can be probed with gold-antibody conjugates to digoxigenin or fluorecein.

J. Wroblewski: Can Nanogold-streptavidin be used on resin-embedded material, and which resins are most suitable for this technique?
Author: No exact study of this question has been done, to my knowledge. However, rather extensive work was done evaluating Nanogold-antibody conjugates on resin-embedded material (i.e., incubation of gold with embedded material that had been sectioned) (Krenács and Krenács, 1995). Nanogold was successfully used with Epon 812, Araldite (Durcupan ACM), LR-White, LR-Gold, and Lowicryls. Epoxides needed to be treated with sodium (m)ethoxide, and acrylate (Lowicryl and LR-white/gold) sections were microwaved for best results. Further technical details may be gleaned from other articles referred to above under the heading "Pre-embedding" and "Postembedding".

Additional References

Baschong W, Wrigley NG (1990) Small colloidal

gold conjugated to Fab fragments or to immunoglobulin G as high-resolution labels for electron microscopy: a technical overview. J Electron Microsc Techn **14**: 313-323.

Hacker GW, Zehbe I, Hainfeld J, Sällström J, Hauser-Kronberger C, Graf A-H, Su H, Dietze O, Bagasra O (1996) High-performance Nanogold™ *in situ* hybridization and *in situ* PCR. Cell Vision **3**: 209-215.

Scanning Microscopy Supplement 10, 1996 (pages 327-344)
Scanning Microscopy International, Chicago (AMF O'Hare), IL 60666 USA
0892-953X/96$5.00+.25

IMMUNOCYTOCHEMISTRY BY ELECTRON SPECTROSCOPIC IMAGING USING WELL DEFINED BORONATED MONOVALENT ANTIBODY FRAGMENTS

M.M. Kessels, B. Qualmann and W.D. Sierralta*

Max-Planck-Institut für experimentelle Endokrinologie, Hannover, FRG

(Received for publication June 11, 1996, and in revised form December 31, 1996)

Abstract

Contributing to the rapidly developing field of immunoelectron microscopy a new kind of markers has been created. The element boron, incorporated as very stable carborane clusters into different kinds of peptides, served as a marker detectable by electron spectroscopic imaging (ESI) - an electron microscopic technique with high-resolution potential.

Covalently linked immunoreagents conspicuous by the small size of both antigen recognizing part and marker moiety are accessible by using peptide concepts for label construction and their conjugation with Fab' fragments. Due to a specific labeling of the free thiol groups of the Fab' fragments, the antigen binding capacity was not affected by the attachment of the markers and the resulting immunoprobes exhibited an elongated shape with the antigen combining site and the label located at opposite ends. The labeling densities observed with these reagents were found to be significantly higher than those obtained by using conventional colloidal gold methods.

Combined with digital image processing and analysis systems, boron-based ESI proved to be a powerful approach in ultrastructural immunocytochemistry employing pre- and post-embedding methods.

Key Words: Immunocytochemistry, electron spectroscopic imaging, boron, immunoglobulin fragments, peptides, small sized markers, carboranes, antigen localization, electron energy loss spectroscopy, energy-filtered transmission electron microscopy.

*Address for correspondence
W.D. Sierralta
Max-Planck-Institut für experimentelle Endokrinologie
Postfach 610309
D-30603 Hannover, FRG
Telephone/FAX Number: +49 (511) 5359-156
E-mail: 106002.547@compuserve.com

Introduction

The ultrastructural study of events and functions in biological specimens provides a better comprehension of their cell biology. For this reason, efforts are continuously being made to improve the spatial resolution of the techniques used in electron microscopy for the analysis of biological systems. The improvements in immunoelectron microscopy since the introduction of ferritin as an electron dense label for immunoreagents (Singer, 1959) illustrate the steady search for optimal probes of high resolving power. The application of colloidal gold particles as labels for immunoelectron microscopy (Faulk and Taylor, 1971) increased dramatically the resolution, specificities and resolving power attainable in immunocytochemistry. Although the current immunogold labeled reagents are ideal for routine applications, their relatively large overall size does not support their use if higher resolution is needed as in the tagging of vicinal domains in single macromolecules or of subunits in oligomeric complexes. Another disadvantage of colloidal gold-labeled immunoreagents is the negative charge carried by the particles, causing electrostatic repulsions with components of the specimen and resulting in lower labeling efficiency than expected (van de Plas and Leunissen, 1993).

In order to improve the performance of the electron dense labels, several groups have produced ultrasmall markers. Tetramercury (Lipka *et al.*, 1979), tetrairidium (Furuya *et al.*, 1988), undecagold (Hainfeld, 1987) and tungstate clusters (Keana and Ogan, 1986) have been described and may be covalently bound to immunoglobulins or their fragments. These compounds differ in charge and are in part not stable under the electron beam. All of them are virtually undetectable in conventional transmission electron microscopy, restricting their application to studies of periodical structures with the help of image averaging and processing. Only the 1.4 nm gold cluster developed by Hainfeld and Furuya (1992) has been successfully applied in immunocytochemistry. Post-incubation size-enlargement by coating with metallic silver ("silver-enhancement") (Holgate *et al.*, 1983; Danscher and Nørgaard, 1983) allows for its easy visualization in transmission electron microscopes and even in light microscopes.

All the labels described above belong to a group that may be detected in transmission electron microscopes because elements with a high atomic number strongly scatter elastically electrons of the beam. The probability and strength of this elastic Rutherford scattering, based on the deflection of beam electrons by the positively charged atomic nucleus, strongly depend upon the sample thickness and the atomic number of the elements present.

A different approach is used in the ultrastructural detection of specific elements by their inelastic scattering properties via electron spectroscopic imaging (ESI) and electron energy loss spectroscopy (EELS) (Ottensmeyer, 1982; 1984; Colliex, 1986; Reimer, 1991). Here, irradiated electrons interact with the shell electrons of the sample, and - by transferring energy - cause excitation or even ionization of the specimen atoms. A part of the electrons in the transmitted beam has therefore lost a defined amount of energy, which allows to draw conclusions about the kind and state of the atoms present in the specimen. To split the electrons according to their energy content, the transmission electron microscopes must be equipped with energy filters (Castaing and Henry, 1962; Henkelman and Ottensmeyer, 1974): The electrons leaving the specimen are deflected on circular paths by the Lorentz force of a magnetic field; the deflection radii are proportional to the energy content of the electrons.

An electron energy loss spectrum represents graphically the intensity distribution of electrons according to their energy content. The presence of an element occurring in sufficient quantities in a specimen can be demonstrated on the basis of its specific absorption edge. The positions of these absorption edges correspond to the characteristic binding and therefore ionization energies of a specific element according to the energy level (K-shell, L-shell etc.). These edges are superimposed on the offset of an intense, broad peak (plasmon peak) which includes electrons having created collective excitations, inter- and intraband transitions. For this reason, the spectrum or the elemental map must be cleared from the relatively large amount of the non-element-specific, intense background to obtain a netto element signal.

The representation of the spatial distribution of electrons of a specific selected energy loss in the form of two-dimensional images is called electron spectroscopic imaging. To obtain the net elemental distribution for a chosen element it is necessary to record, in addition to the element-specific picture just above the absorption edge, at least one reference image below the edge onset. This permits the extrapolation of the non-element-specific background.

Energy filtered transmission electron microscopic (EFTEM) methods have been rapidly developing into very powerful tools for elemental microanalysis in biological specimens during the last years. Although mainly used for detections of naturally occurring (Ottensmeyer and Andrew, 1980; Arsenault and Ottensmeyer, 1983; Leapman and Andrews, 1991) or exogenously supplied elements (Köpf-Maier and Martin, 1989; Wagner and Chen, 1990) - often as precipitates - these techniques may also be used in immunocytochemistry to detect reagents labeled with appropriate marker molecules. Outstanding features of this method are the high spatial resolution and sensitivity potentials achievable. In commercial instruments, ESI allows for a spatial resolution of 0.5 nm and provides enough sensitivity to detect 50 accumulated atoms of a light element like phosphorus (Adamson-Sharpe and Ottensmeyer, 1981).

Marker molecules for ESI-immunocytochemistry must fulfill the following requirements: The elemental marker should not be a component of the investigated cells or tissues and should exhibit a strong, sharp absorption edge in the electron energy loss spectrum sufficiently distinct from those of other elements present. Furthermore, the detection limit must be exceeded, ensured by a high topical density of the element chosen.

Organo-boron compounds appear to be particularly well suited for an application as labels in ESI-based immunocytochemistry. Beside its absence in most biological specimens and its sharp absorption edge in EELS not particularly overlapping with those of other elements normally present, boron has the property to form stable and defined covalent clusters which can be incorporated into organic molecules (Plešek, 1992; Hawthorne, 1993; Morin, 1995).

In 1989, a first experiment with boron-labeled immunoreagents was described by Bendayan *et al.* (1989). They employed a heterogeneous, positively charged marker with varying boron content, a boronated polylysine (Barth *et al.*, 1986; Alam *et al.*, 1989) linked to protein A, for the indirect detection of antigenic sites. Their approach resulted in strong ESI signals, with a reduced spatial resolution, limited by the relatively large size of the probe (approximately 15 nm - not considering the sizes of the primary reagents used in the indirect approach). These immunoreagents did not allow for high resolution analyses with ESI, compelling compact, small-sized marker molecules. A size-reduction should additionally increase tagging efficiencies, because steric hindrances are lowered and infiltrations of section's surface depths are facilitated (Horisberger, 1981; Yokota, 1988).

A direct boronation of antibodies through multiple attachment of small, boron-containing molecules (Mizusawa *et al.*, 1982; Alam *et al.*, 1985) is limited

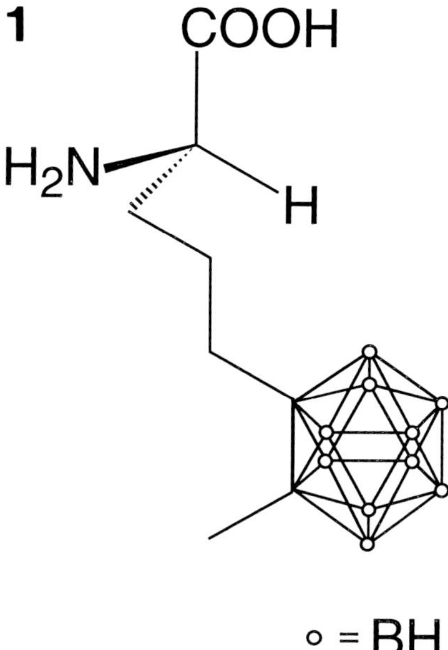

Figure 1. Stereochemical structure of the carborane cluster containing amino acid used for the different peptide syntheses. L-5-(2-methyl-1,2-dicarba-closo-dodecaborane(12)-1-yl-)2-amino pentanoic acid (L-MeCBA **1**) was produced by the bislactim ether method (Schöllkopf *et al.*, 1981) following the concept of auxilliars, i.e., a glycine building block is incorporated into a molecule containing a group which sterically controls the alkylation leading enantioselectively to the desired amino acid derivative.

---

because of the strong effects on immunoreactivity and solubility of the proteins. Therefore, research was recently more focussed on the grafting of boron-loaded polymers to single sites of carrier molecules. However, the difficulties caused in this procedure by the inherent heterogeneity of the boron-containing labels like poly-lysine (Barth *et al.*, 1986; 1989), poly-ornithine (Tamat *et al.*, 1989), dextranes (Abraham *et al.*, 1989; Pettersson *et al.*, 1989) and starburst dendrimers (Barth *et al.*, 1994) have stimulated the design and synthesis of more defined molecules, such as oligomeric peptides containing carboranyl amino acids (Varadarajan and Hawthorne, 1991; Kane *et al.*, 1993; Leusch *et al.*, 1994).

Development of Probes for EFTEM

Immunocytochemistry by electron spectroscopic imaging requires the development of appropriate immunoreagents consisting of a small and homogeneous marker moiety and an efficient antigen recognizing part. In this context the use of Fab' fragments of antibodies is advantageous due to their small size amounting to 5 nm x 4 nm x 3 nm (Hainfeld, 1987). These fragments are able to reach even antigenic sites of which the accessibility is diminished by other constituents of the specimen or by the tertiary structure of the given target protein, they easily penetrate the depths of the section surface and can also be used as valuable reagents for pre-embedding methods. Since the dimensions of the antibody fragments are quite low, the achievable lateral resolution is high. The Fab' fragments exhibit one single antigen combining site, i.e., they are monovalent, a basic requirement for examinations of some special problems.

Additionally, Fab' fragments provide the possibility of a regiospecific derivatization with the marker molecule, since the thiol groups of the former hinge region of the IgG (immunoglobulin G) molecule are accessible and allow for a SH-specific labeling at a domain opposite to the antigen combining site (Kato *et al.*, 1976; Yoshitake *et al.*, 1982; Padlan, 1994). Thus, the immunoreactivity is not affected by the marker compound. This is in strong contrast to labeling approaches using all the electron-rich moieties of the protein for covalent linkages.

In the design of the marker molecules the following special demands must be satisfied:

[1] a fully accessible, reactive handle permitting the thiol-selective conjugation to the antibody fragment has to be incorporated,

[2] the marker molecule should carry a high boron load spread over a minimized volume to exceed the ESI detection limit and to achieve a high signal to noise ratio,

[3] the label should be homogeneous in structure, boron content and size,

[4] a fluorescent group should be incorporated to quantify the conjugation reaction,

[5] the compound has to be soluble in a solvent miscible with water or even in water itself.

Since using amino acids as well characterized multifunctional building blocks is an appropriate approach, we applied peptide chemistry to fulfill these requirements.

First of all, an amino acid containing the element boron was necessary. R,S-5-(2-methyl-1,2-dicarba-closo-dodecaborane(12)-1-yl-)2-amino pentanoic acid (D,L-MeCBA) (Varadarajan and Hawthorne, 1991) is an attractive compound for our application, since 10 boron atoms are condensed in a cluster of high stability and steric hindrances are comparatively low. This carboranyl amino acid has been successfully used in peptide syntheses for different purposes (Varadarajan and Hawthorne, 1991; Kane *et al.*, 1993; Leusch *et al.*, 1994).

First experiences with linear racemic peptides

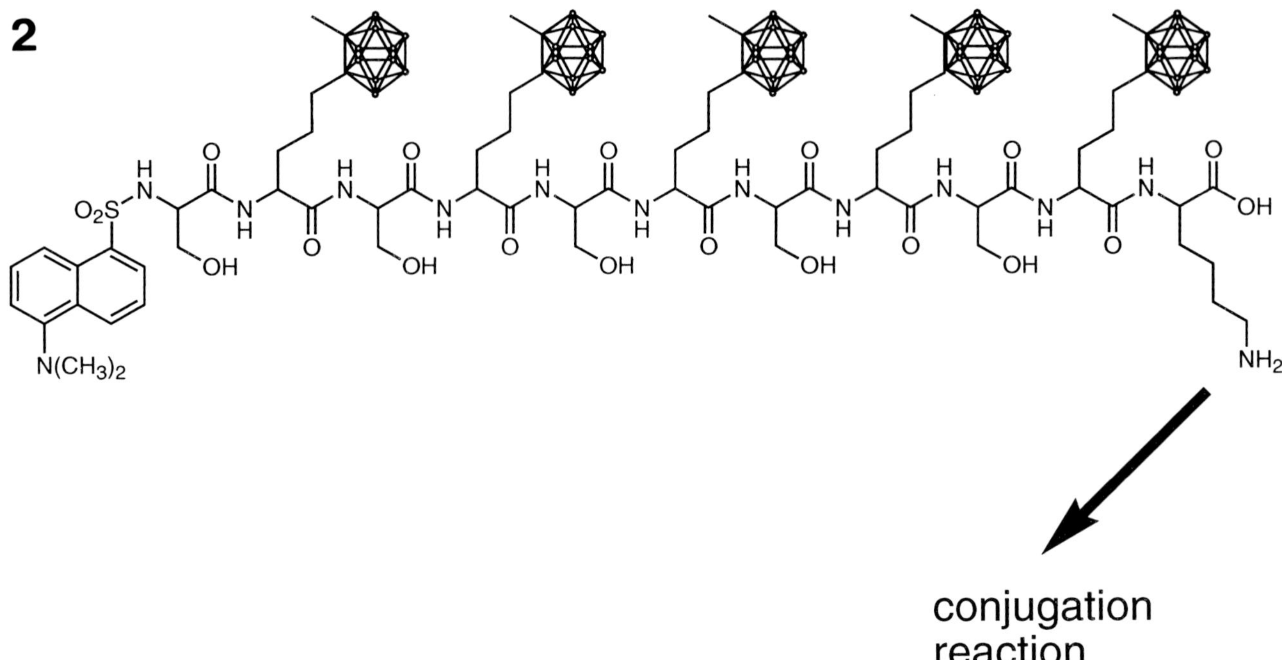

Figure 2. Structure of the linear, N-terminally dansylated peptide **2** consisting of 11 amino acids (5 MeCBAs **1**, 5 serines and a lysine), which has been successfully linked to Fab' antibody fragments and detected by ESI using the incorporated boron as marker element.

(Leusch *et al.*, 1994) encouraged us to work on the further development of boron-rich marker molecules. Two routes were followed to improve the performance of the peptides: Since an enantioselective synthesis allows for a more defined and condensed secondary structure of the marker, we developed an efficient synthesis of the pure L-enantiomer of the carborane-containing amino acid MeCBA **1** (Fig. 1). The synthesis is short, convenient and proceeds in high overall yield (68%). The carboranyl amino acid MeCBA **1** is obtainable in high enantiomeric purity (enantiomeric excess > 98%) and in amounts sufficient for peptide syntheses (Kessels and Qualmann, 1996). The second way to enhance the signals of the marker molecules consists in the incorporation of a greater number of carboranyl amino acids.

The linear peptides produced followed partly the concept of Varadarajan and Hawthorne (1991). They contain a dansyl group as a fluorescence marker, a lysine moiety as the later coupling site and a different number of MeCBAs. Diverging from Varadarajan and Hawthorne (1991), alternating serines were incorporated to improve the solubility of the marker peptides and a segment condensation synthesis in solution was used (Leusch *et al.*, 1994; Kessels, 1996; Kessels *et al.*, 1996). In general, the extremely high hydrophobicity of the carborane cage often represents a problem; the side chain of carboranylalanine, for example, is reported to be 1000 times more hydrophobic than that of valine (Fauchère *et al.*, 1980). To decrease the hydrophobicity of the compounds we incorporated polar groups like hydroxyl and amino functions.

The all-L-peptide **2** produced (Fig. 2) represents a small and compact marker molecule. The maximal possible length extension assuming a fully stretched conformation is approximately 3.5 nm. The possibility of forming a more condensed secondary structure in solution ought to be considered, especially in the case of enantiomerically pure compounds.

The peptide proved to be soluble in polar organic solvents like dimethylformamide and different alcohols and in aqueous organic mixtures allowing for coupling to the antibody fragments Fab'. Conjugations were performed with heterobifunctional crosslinkers of the N-hydroxysuccinimidylmaleimidyl type linking the free thiol groups of Fab' with the amino group of the lysine anchor present in the marker peptide (Kessels *et al.*, 1996). The efficiency of conjugation exhibited a strong dependence on the length of the spacer moiety incorporated in the different crosslinkers used (Kessels, 1996), probably due to steric hindrances. The coupling reactions were verified by fluorescence examination. The boron content of the peptide **2**-immunoprobes was impressively demonstrated by electron energy loss spec-

3

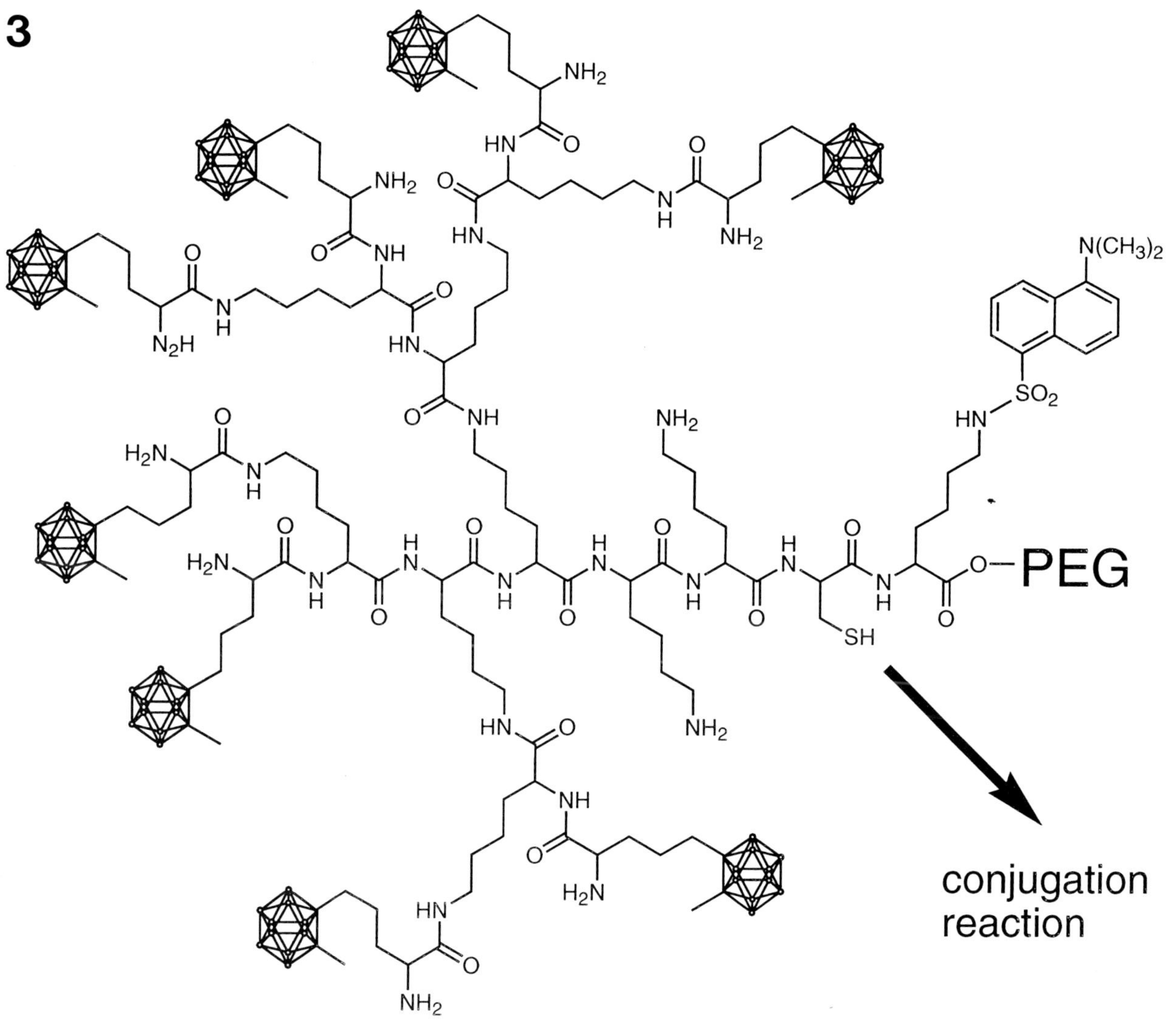

PEG

conjugation reaction

Figure 3. Molecular structure of the dendritic peptide construct **3** containing 8 carboranyl amino acids attached at the outer surface of a poly(α,ϵ-L-lysine) core and a linear tail carrying the coupling site for Fab' conjugation, the dansyl group and a polyethylene glycol chain (PEG) spaced from the dendrimer moiety by 2 lysines. An identical peptide, only lacking the PEG-tail, was also produced. The peptides **2** (Fig. 2) and **3** are shown in the same size.

---

troscopy (Kessels *et al.*, 1996).

A further synthesized elongated peptide (8 incorporated carboranyl amino acids, 21 steps) exhibited a poor solubility in aqueous organic mixtures and the coupling ratios were low (Kessels, 1996). These immunoreagents were not promising - obviously a limitation of the linear peptide concept had been reached.

Furthermore, there are some common disadvantages of the linear peptide syntheses, such as the very laborious, multi-step procedures (even though we used a segment condensation approach) leading to low overall yields (even assuming, e.g., a yield of 90% in each individual step the overall yield of a 15-step synthesis (peptide **2**) would nevertheless be only 20% and that of a 21-step procedure only 11%, respectively). Another disadvantage is the necessary, time-consuming and increasingly difficult characterization of each synthetic intermediate by analytical means like thin layer chromatography, nuclear magnetic resonance spectroscopy and mass spectrometry.

To bypass the limitation and the inconveniences we left the linear approach and generated spherical dendritic peptides carrying the carboranyl amino acids at the outer surface (Qualmann *et al.*, 1996b). We developed a solid phase peptide synthesis (SPPS; Merrifield, 1963) employing the orthogonal concept of fluorenyl-methyloxycarbonyl-SPPS (Atherton *et al.*, 1981). The coupling difficulties encountered in solid phase and solution

Figure 4. Schematic presentation of the different functional moieties incorporated into the dendritic peptide **3** and their arrangement.

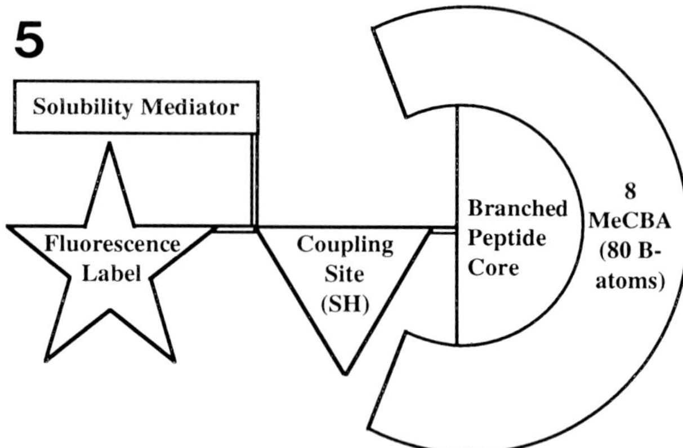

Figure 5. Electron energy loss spectrum of immunoreagents bearing peptide **3** adsorbed to nitrocellulose-coated gold grids. Notice the sharp boron K-shell absorption edge at 196 eV rising from the steep declining background curve. Instrument settings: objective aperture 90 μm, spectrometer entrance aperture 100 μm, slit width 1-2 eV, beam current 30 μA, magnification 30000x, scan speed 1.6 eVs^{-1}.

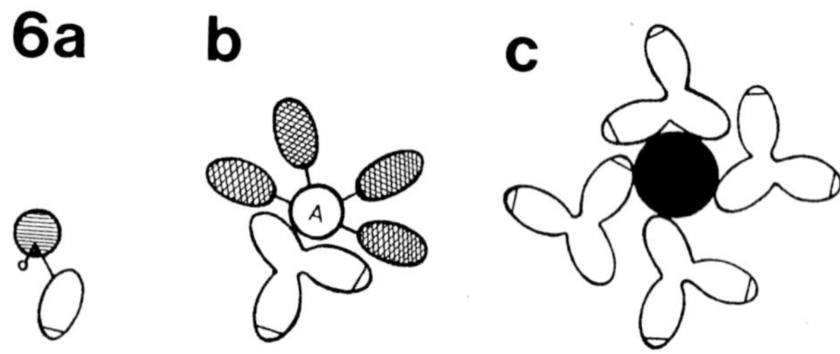

Figure 6. Comparison of sizes and shapes of different immunoreagents. Schematic representations. **a**. Fab' fragment conjugated with the boronated dendritic peptide **3**. **b**. Indirect immunoprobe employed by Bendayan *et al.* (1989), i.e., an antibody detected by protein A, which has been linked with several boronated polylysines. **c**. Conventional colloidal gold (Au_{6nm}) immunoreagent.

Figure 7 (*on facing page*). Comparison of conventional immunogold technique with boron electron spectroscopic imaging. **a, b**. Labeling of the secretory vesicles (sv) in somatotrophic cells of porcine anterior pituitary with goat anti-GH, detected by rabbit anti goat-Au_{10nm}. *N* nucleus, *arrow* plasma membrane with adjacent non-somatotroph. **c**. ESI with anti-GH-Fab'-peptide-**2**-conjugate: one of the two background images taken below the boron absorption edge (*178 eV*), an image just above the boron absorption edge (*201 eV*), a conventional zero loss image (*0 eV*) and the net boron distribution (*B-distribution*) calculated by the three-window method (150 eV, 178 eV, 201 eV). With kind permission from Kessels *et al.* (1996), Fig. 3; © Springer Verlag, Heidelberg, FRG.

synthesis of carboranyl amino acid-containing oligopeptides derive from steric hindrances particularly of the N-terminus of the carboranyl amino acids due to the bulky carborane moiety (Varadarajan and Hawthorne, 1991; Kane *et al.*, 1993; Leusch *et al.*, 1994). The attachment of the carboranyl amino acids to the terminal amino groups of a poly(α,ϵ-L-lysine) dendrimer (Denkewalter and Kolc, 1981; Tam, 1988) in a final acylation step overcame these limitations. The amino functions of the molecule, whose quantity grows exponentially with each step of the dendrimer synthesis, are spherically distributed, thus steric hindrances are diminished and acylation occurs quantitatively (Posnett *et al.*, 1988; Qualmann *et al.*, 1996b).

Due to the globular shape of the marker molecules the topical boron concentration is very high. The maxi-

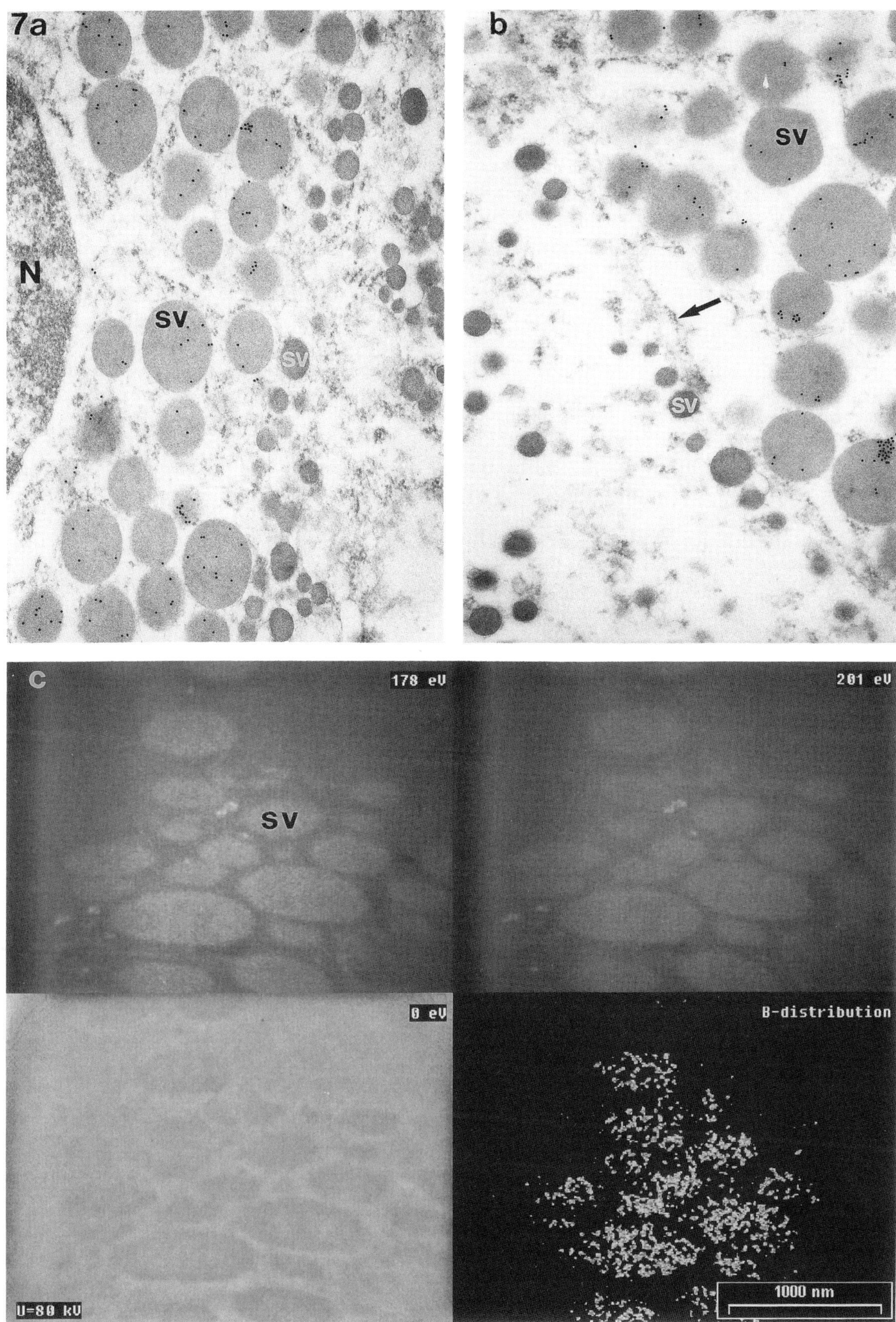
7a
N
SV
SV
b
SV
SV
c
178 eV
SV
201 eV
0 eV
U=80 kV
B-distribution
1000 nm

mal extension of the 80 boron atoms can theoretically reach 4 nm. In fact, for a poly(α,ϵ-L-lysine) dendrimer of the same size terminally acylated with a further generation of lysines instead of carboranyl amino acids a diameter of 2.28 nm in solution is described (Aharoni *et al.*, 1981).

The marker peptides prepared consist of two parts (Fig. 3): the globular, branched core of lysines with 8 terminal amino groups each carrying a carboranyl amino acid - the marker moiety - and a C-terminal, linear peptide chain containing the coupling site (a thiol function of an incorporated cysteine) and a fluorescence label (a dansylated lysine like in the linear peptides) both spaced from the voluminous hepta(α,ϵ-L-lysine) core by two more lysines. Their highly polar amino functions should also enhance the hydrophilicity of the molecule.

A further increase of the hydrophilicity was obtained by the derivatisation of the C-terminus with a polyethyleneglycol (PEG) tail. This solubility mediator proved to be necessary since the marker peptide lacking the polyether tail was only poorly soluble in aqueous solutions. Probably due to its low solubility the conjugation efficiency was as low as that observed for the elongated linear peptide containing 8 carboranyl amino acids. In contrast, the PEG-bearing analog **3** (Figs. 3 and 4) exhibited good solubilities in a wide variety of solvents (e.g., chloroform, methanol, water) and reacted smoothly in Fab'-conjugations with different types of crosslinkers employed (Qualmann *et al.*, 1996a,b). No disturbing precipitates were formed and the associated loss of antibodies was diminished, ensuring the use of even monoclonal antibody fragments usually available in only moderate quantities (Parham, 1983). Furthermore, the addition of organic solvents to dissolve the marker peptide can be omitted reducing the risk of decreasing the antigen binding capacity of the Fab' molecules.

The quality of the immunoprobes was again checked by fluorescence measurements and examinations using an energy filtered transmission electron microscope to detect the marker element boron. An electron energy loss spectrum of an immunoreagent bearing the marker peptide **3** is shown in Fig. 5 exhibiting the sharp boron absorption edge at an energy loss of 196 eV.

The marker compound was proven to be remarkably stable during beam irradiation: no losses of signal intensity in successively recorded electron energy loss spectra were detected when a given area of the specimen was exposed to a 30 μA electron beam for more than five minutes (Qualmann *et al.*, 1996a). This is in striking contrast to observations of Huxham *et al.* (1992). They examined borate adsorbed to polystyrene beads and described a rapid loss of boron signals: in their case after only 3 minutes of exposure a complete loss of signals was seen. In contrast, immunoprobes containing boron in the form of chemically and thermically very stable and covalently attached closo-carborane cages (Wade, 1976) are not affected by irradiation, thus proving to be appropriate for immunocytochemistry.

(*Figure 8 on facing page*)

Figure 8. Uptake of BSA by ileal enterocytes of newborn piglets demonstrated by tagging with the Fab'-boronated peptide conjugate and ESI-examination (three-window-method - 150 eV, 178 eV, 201 eV). Differences in labeling of BSA-containing vesicles obtained with the BSA-specific boronated immunoprobe (a, c) and a non-related control conjugate (b). **a**. Overview consisting of the background images taken below the boron absorption edge (*178 eV*), the boron specific image just above the edge (*201 eV*), the zero loss image (*0 eV*) and the net boron distribution. Labeling performed with anti-BSA-Fab' fragments conjugated with peptide **3**. The sizes and shapes of the vesicles are delineated even by the BSA-tagging alone, permitting the examination of antigen localization just by comparison of the calculated boron distribution and the zero loss image. The frame (in a) marks the part of the vesicle shown beyond in 8fold enlargement (c) The grayscale signal intensities range from white (highest intensity) over light gray to dark gray). **b**. ESI-micrograph demonstrating the background noise detected over a vesicle area in control sections incubated with a boronated anti-GH immunoprobe. The images b and c are shown at the same magnification. With kind permission from Qualmann *et al.* (1996a); © Blackwell Science Ltd, Oxford, UK.

In summary, defined, boron-rich marker peptides of small, homogeneous size were generated in good purity. Since standard procedures of solid phase peptide synthesis are used in the case of the dendritic peptide **3**, the synthesis is reliable, easy to control by the analytical means of SPPS and proceeds in high overall yields.

In general, the immunoreagents are of elongated shape carrying the covalently linked marker at the opposite end of the antigen combining site, thus immunoreactivity is not affected sterically by the marker peptides and the lateral extension of the immunoprobe just depends on the diameter of the Fab' fragment (approximately 4 nm). In contrast, other immunoreagents previously described show much larger spatial extensions: in conventional immunogold labeling the diameter of the colloidal gold is greatly enlarged by the antibody coat and in the boron/ESI method of Bendayan *et al.* (1989) the heterogeneously linear boronated polylysines linked to the protein A, used as indirect label, contribute substantially to the overall size of the marker even assuming a rather globular structure (Figure 6).

All the immunoreagents produced have been puri-

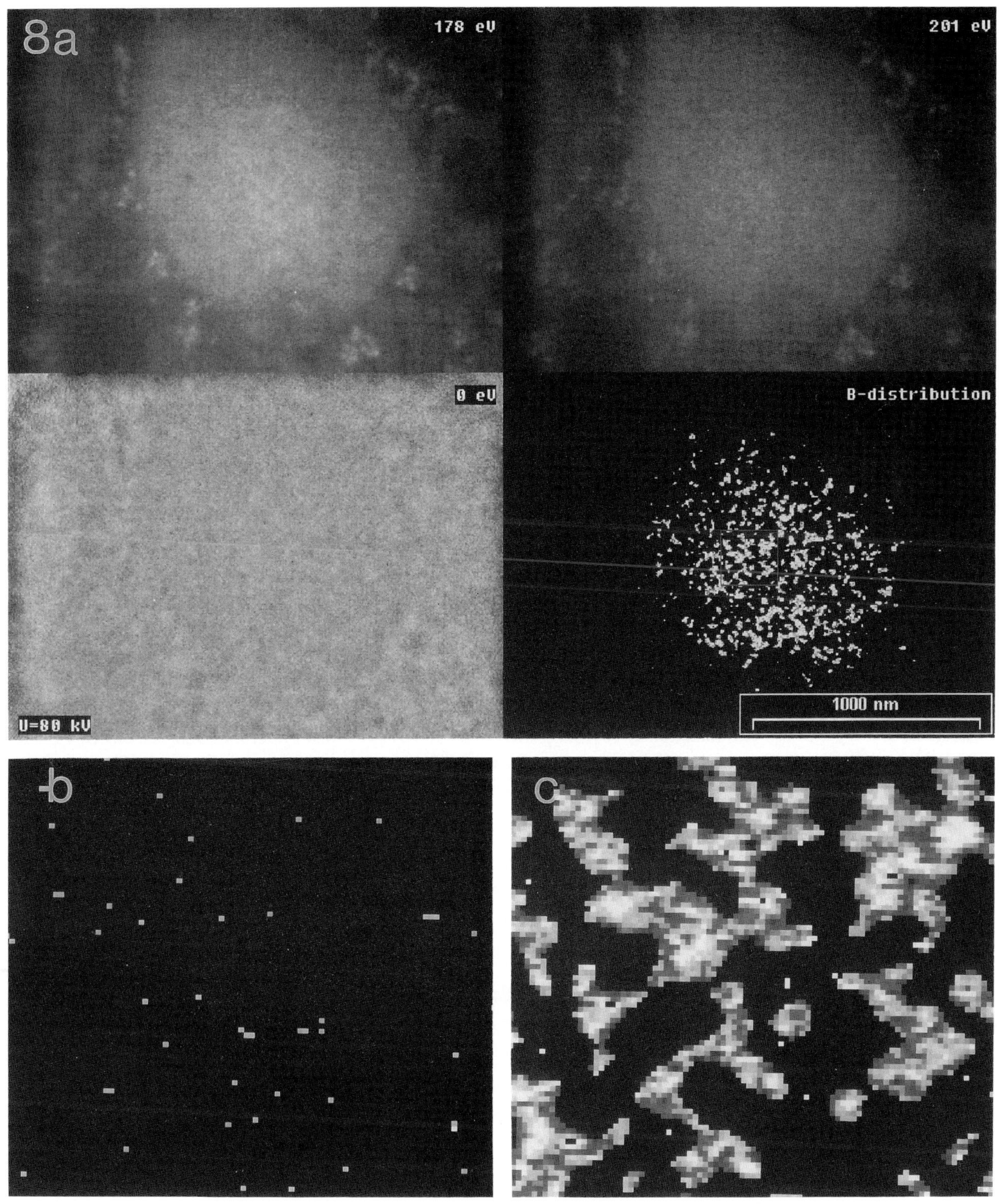

fied from unreacted low molecular weight components of the conjugation reaction, concentrated and stored at 4°C in the presence of 1 mM azide as preservative and 0.5% (w/v) bovine serum albumin and ovalbumin as expander, respectively. No loss in immunoreactivity and no aggregation was observed even during prolonged storage of several months (Kessels *et al.*, 1996; Qualmann *et al.*, 1996a; Qualmann, 1996).

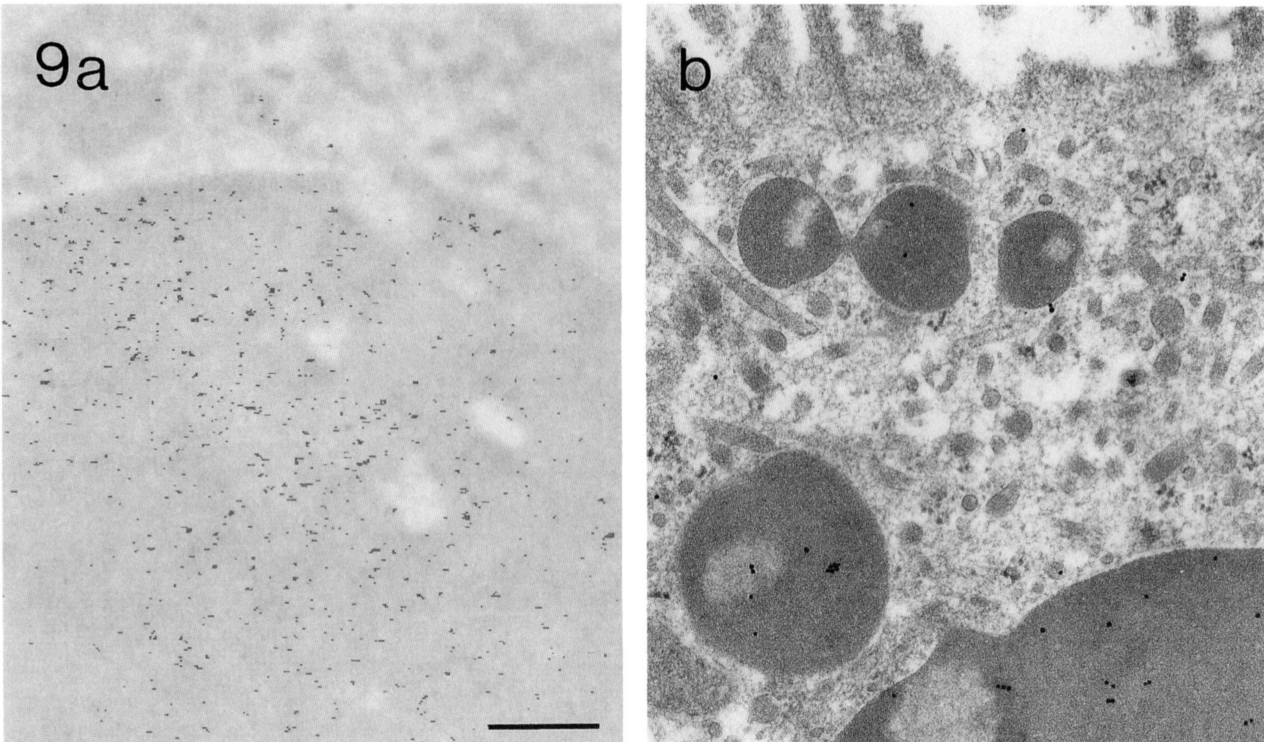

Figure 9. Assignment of BSA-labeling to transcytotic vesicles in an epithelial cell of ileum taken from newborn piglets after the administration of bovine serum. **a**. Superimposition of calculated net boron signals (shown in here in black, but in green in the original publication) obtained by incubation with anti-BSA-Fab'-peptide **3** on the corresponding zero loss image of endocytotic vesicles (unstained, 40 nm thin section). **b**. Conventional direct immunogold labeling ($F(ab')_2$-Au_{12nm}) of BSA in endocytotic vesicles of different size. Section (80 nm) contrasted with OsO_4, uranyl acetate and lead citrate. Scale bar: 400 nm. With kind permission from Qualmann *et al.* (1996a); © Blackwell Science Ltd, Oxford, UK.

Requirements in the Preparation of Specimens for Immunocytochemistry

Retention of antigenicity in sufficient amounts for good antigen-antibody interactions requires mild fixation. The choice of embedding resin also has a great influence on the analysis by immunocytochemistry, especially if post-embedding is performed. The suitability of Epon for pre-embedding experiments and of Lowicryl-embedded tissue for post-embedding immunocytochemistry with ESI has been reported by Bendayan *et al.* (1989). For our studies with boron-labeled antibody fragments, Spurr resin (Spurr, 1969) was found suitable even for post-embedding ESI immunocytochemistry because sections of this plastic allows for the reaction of antigens at the depths of the surface with immunoreagents as confirmed by conventional immunogold labeling. As shown in the Figures 7, 9 and 10, a satisfactory retention in antigenicity after fixation, dehydration and embedding procedures was secured.

The resin showed practically no interference with the detection of boron; the high purity of the plastic components, free of traces of elements like sulphur and phosphorus, resulted in a very low background (see Figure 10, for example).

A crucial factor in the performance of energy-filtered transmission electron microscopic methods is the thickness of the sample, because the probability of multiple scattering events decreasing the signal to noise ratio augments with the thickness of the probe under the electron beam. The energy loss of an electron is only specific for an element if it is attributed to a single scattering event in the specimen. Therefore, if the percentage of multiple scattering events should not exceed 15%, the samples should be no thicker than about 0.3 mean free paths for total electron scatter; this means a thickness of about 30 nm for sections of biological material (Ottensmeyer, 1986). Despite their thinness, these sections of polymerized Spurr resin exhibited reasonable stability under the electron beam with low tendency to tear or to drift. Sections of material embedded in Spurr resin stay adherent on 600 mesh-

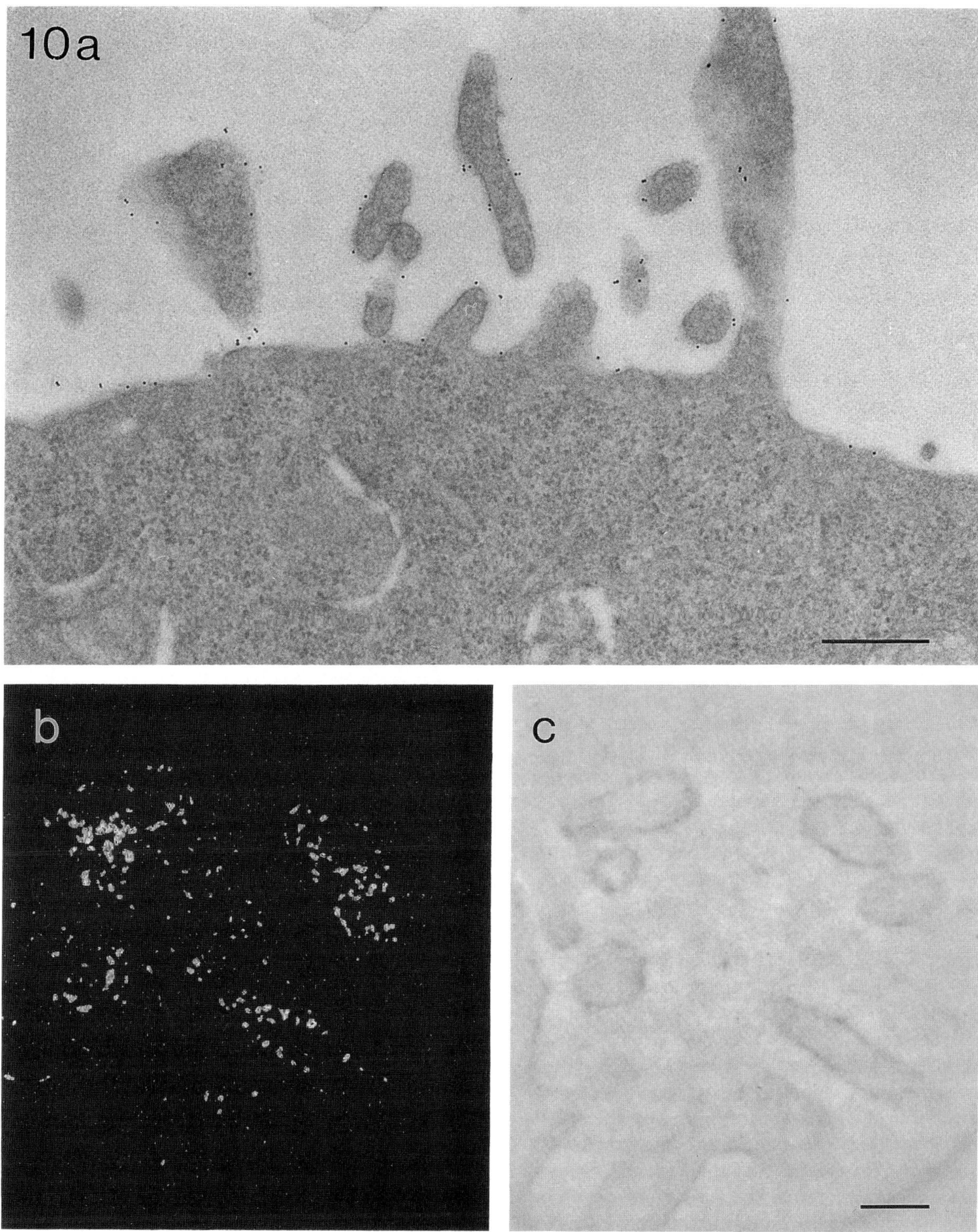

Figure 10. Labelling of high affinity IgE receptors on membranes of rat basophilic leukemia cells (line RBL-2H3, cells were incubated with the immunoreagent at 4°C in the presence of azide to prevent internalization). **a**. Immunogold labelling using 6 nm gold coated with rat monoclonal antibody J17 (contrasted 100 nm section). **b**. Grayscale representation of the net boron distribution in an ultrathin section of cells incubated with a boron-containing immunoconjugate of J17. **c**. Zero-loss image corresponding to the area shown in **b**. Scale bars: 400 nm (**a**), 150 nm (**b,c**).

grids allowing incubations on aqueous solutions and avoiding supporting films which will undesiredly increase the path length of beam electrons through the specimen. An increase in stability may be obtained by subsequent carbon coating, but this significantly lowers the signal to noise ratio.

Additionally, contrasting of the sections with OsO_4, uranyl acetate, lead citrate or other heavy metal stains should be omitted otherwise a reduction of the element-specific signal is obtained.

Hardware Requirements for ESI-Immunocytochemistry

In order to detect specific elements by electron spectroscopic imaging, the microscopes have to be upgraded with an imaging electron spectrometer. Two different types of arrangements of energy filters have been realized in transmission electron microscopes commercially available at present:

[1] the addition of a post-column spectrometer to a conventional transmission electron microscope (cTEM) as accessory,

[2] the integration of an in-column spectrometer in the magnification system of specially constructed TEMs.

The work described here was carried out with a ZEISS 902 (Zeiss, Oberkochen, FRG), which is equipped with an imaging spectrometer of the prism-mirror-prism type (Castaing and Henry, 1962) and an energy selecting slit. Electron energy loss spectra were recorded with the electron detector of the microscope connected to a digital compensation X/Y-plotter. The effective diameter of the selected area analyzed is restricted by the spectrometer entrance aperture and the magnification.

In ultrathin sections, only a relatively small portion of electrons is inelastically scattered (approx. 1.5%; Ottensmeyer, 1982) because the ionization cross section is very small even for elements with a low atomic number. Therefore, EFTEM techniques require greater electron doses than usual, which may result in radiation damage, instabilities and even mass losses in the specimen. If the acceleration voltage is raised, the impulse- and thus the energy-transfer during interaction with matter is lower, in consequence, the mean free path length increases. The signal to noise ratio augments, because the probability of multiple scattering events is diminished. In the ZEISS 902, the acceleration voltage is factory-limited to 80 kV to prevent the electrical collapse of the filter. For operations at higher voltage, the electrostatic mirror must be replaced by a fully magnetic deflection system, as realized in the ZEISS 912 with an omega filter (Zanchi *et al.*, 1975) design.

The efficient collection of the scattered electrons is of uppermost importance in order to keep the total exposure time and thus radiation-induced alterations in the specimen under the electron beam low. A large objective aperture allows the capture of scattered electrons in a great area.

Fundamental limitations of the photographic materials, like their relatively low sensitivity and non-linear responses as well as the tedious chemical development have stimulated the search of appropriate alternative recording systems in electron microscopy (Krivanek and Mooney, 1993; De Ruijter, 1995). The fast development of electronic media in the last years has been of considerable benefit for ESI. Thus, the detection of the low level of electrons inelastically scattered by the lighter elements present in the specimen makes the use of electronic cameras in almost all the cases mandatory.

Besides a high sensitivity, the high-quality electronic cameras currently in use possess a good spatial resolution, with less or no geometric distortion, are able to acquire the images fast, react to changes in the input intensity rapidly and last, but not least give signal outputs proportional to the incoming energy currents.

Two sorts of electronic cameras are currently in use: the silicon intensified target (SIT) vidicon cameras and the solid-state slow-scan charge-coupled devices (SCCD). In the first type, a scintillator on top of the electronic tube transduces the electrons into photons and a photoconductive layer under the scintillator changes its resistance in proportion to the total number of photons absorbed. An electron beam produced inside the tube scans the faceplate and produces an interlaced analogue signal compatible with video processor devices, like monitors, frame grabbers, recorders.

In the SCCD, discrete microscopic wells are built on a chip, each well representing a pixel. The wells accumulate charge in proportion to the photons collected. After a given time, the electrons accumulated in a row of pixels are shifted to an on-chip output amplifier. The charge accumulated in the chip is read out by the orderly transference of all rows. As in the case of vidicon tubes, the SCCD-analogue images are further amplified and digitized by separate electronics, resulting in images with pixel intensities expressed in digital numbers, allowing an uncomplicated processing (Center for Light Microscopic Imaging and Biotechnology, CLMIB, 1995).

The SCCD have many advantages over the SIT cameras, especially as far as sensitivity, resolution, speed, noise, response time and uniformity are concerned. Of uppermost importance is the low level of "shading" (nonuniformity in the camera, especially in the corners of the image scanned) that SCCD exhibit as compared to the SIT cameras. Probably the major advantage of the SIT camera is its lower price.

Great success in the recording and analysis of element-specific images in EFTEM has been achieved by the application of digital image analysis. Recording times of a few seconds when information is digitalized and integrated contrast sharply with exposure times of up to 100 seconds needed for photographic film material. Computer assistance is an essential component in data acquisition, data processing and displaying of results.

To obtain net elemental maps, the relatively strong background signal has to be removed. The photographically recording and subtraction of elemental maps can be done by either the rather imprecise, subjective and less reproducible photographic approach or by employing an image processor. The computer-assisted use of two or more reference images and the application of different kinds of complex algorithms allows to extrapolate the background signal underneath the ionization edge with great accuracy. For the calculation of the boron net distribution on sections of material from biological specimens embedded in resin, we employed a three-window-method. From two images recorded at energy losses of 150 eV and 178 eV (width of the energy selecting slit was approx. 15 eV) below the boron K edge, the best fitting parameters for the background model curve are examined for each of the 765 x 574 pixel and extrapolated for the third image recorded at 201 eV above the edge; the difference between the extrapolated background curve and the recorded data at an energy loss of 201 eV represents the net element signal. Our microscope is upgraded with a Dage SIT camera and connected to the Image Analysis System ESI Vision from AnalySIS, Münster (FRG). The newer version (2.1) of this analysis program will additionally permit the use of multi-window methods for background calculation.

Immunocytochemical Applications

The validity of the new approach has been proven by labeling different antigens employing either pre- or post-embedding methods. A first set of experiments was performed using polyclonal goat antibodies directed against porcine growth hormone (GH). Fab' fragments prepared from them were labeled with the boronated peptides and employed for a direct post-embedding detection of the antigen in ultrathin sections of porcine pituitary embedded in Spurr resin.

The pituitary gland contains different kinds of secretory cells, which can be distinguished by examining their morphology. Somatotrophs stand out for their large and rather low electron dense secretory vesicles filling wide areas of the cytoplasm. Conventional immunogold labeling was performed using anti-goat IgG-Au_{10nm} as secondary antibody to examine the subcellular distribution of the growth hormone. The labeling was practically restricted to secretory vesicles of somatotrophs (Fig. 7); other parts of the cells like cytoplasm and nuclei were not labeled. Secretory vesicles containing hormones other than GH exhibited no gold tags either.

As the antigen GH occurs in large amounts in these organelles, this system served as an excellent biological system to allow the reliable development of the ESI-immunocytochemical approach based on the newly created boron-rich marker molecules. Since high accuracy is desired, we used the three-window-method for our ESI examinations. Fig. 7c shows the GH-localization by means of ESI-immunocytochemistry, an example of the labeling using the linear peptide **2** as marker is shown. Similar results were obtained using peptide **3** conjugates (Kessels, 1996). The net distribution of the marker element boron was assessed by recording images below and just above the boron edge. The higher brightness of the presented 178 eV background image compared to the 201 eV image reflects the steep declining of the recorded intensities at higher energy losses out of which the boron edge is rising as it can be observed in the detail of an electron energy loss spectrum shown in Fig. 5. Using the three-window method an intense labeling of GH-secretory vesicles was observed. In many cases the signal density was high enough to reveal the size and shape of the tagged organelles. Other constituents of the cells and secretory vesicles of other cell types exhibited no boron signals, demonstrating the specificity of the labeling procedure.

A second experimental model was used to confirm the validity of the approach: the transepithelial passage of proteins from the lumen of the small gut through the ileal enterocytes into the circulation of newborn piglets (Nelson, 1932; Brambell, 1959). We employed the heterologous bovine serum albumin (BSA) as a model protein to follow this process (Sierralta *et al.*, 1994). The system was tested by conventional immunocytochemistry using a direct reagent (12 nm gold coated with $F(ab')_2$ fragments of polyclonal goat antibodies against BSA). Ileal endocytotic vesicles filled with the antigen were observed in the epithelial cells, the vesicles were very heterogeneous in size (Figs. 8 and 9).

The labeling with immunoconjugates bearing boronated peptides **2** and **3**, respectively, analysed by ESI also localized the antigen in the endocytotic vesicles (Figs. 8 and 9; Qualmann, 1996; Qualmann *et al.*, 1996a). Other compartments of the ileal cells exhibited no boron signals. Corresponding to the density of antigens present the sizes and shapes of the vesicles were again delineated even by the net boron signals alone.

A very exact examination of the localization of calculated net boron signals can be performed by superimposing them on the conventional zero-loss image

(Fig. 9a) or on images at other energy losses generally providing a higher contrast of morphological structures (e.g., below the carbon edge at an energy loss of 250 eV). In general, zero-loss images recorded in ESI examinations of ultrathin sections of biological materials display low contrast, due to their thinness, due to the omitted contrasting step and due to the fact that the images are collected with wide objective apertures to ensure an efficient data collection (e.g., we used an objective aperture of 90 μm instead of 60 μm as chosen in standard cTEM applications) leading to very light images which are no longer optimally recorded by the camera. Therefore, for graphical presentations zero-loss images have as a rule to be digitally remastered.

In a subsequent set of experiments the immunoboron approach was extended to pre-embedding methods. High affinity membrane receptors for IgE present on basophils and mast cells (Metzger and Ishizaka, 1982; Kinet, 1989) are able to induce secretions of mediators leading to allergic and inflammatory symptoms. We localized the IgE receptor protein FCϵR I employing the rat basophilic leukemia cell line RBL-2H3 first by conventional immunogold methods. Colloidal gold of 6 nm in diameter coated with the monoclonal rat antibody J17 (Ortega *et al.*, 1988) showed the IgE-receptor protein to be localized under the given conditions at the plasma membrane (Fig. 10a). The antibody is directed against an extracellular domain (Ortega *et al.*, 1988).

These experiments were followed by subcellular examinations based on the immunoreaction with antibody fragments labeled with the boronated peptide **3** (Qualmann, 1996; Kessels, 1996). Boron signals were detected at the surface of the plasma membrane and the extensions of the cells. Due to the high labeling efficiency of the antigen which occurs in approximately 270000 molecules per cell (Mao *et al.*, 1992), the membranes were often drawn by the net boron signals alone (Fig. 10b). Additionally, the ultrathin sections of the pre-embedded material showed very well the low level of interference introduced by Spurr resin in the boron detection by ESI. The background signals recorded in areas of pure resin were low (Fig. 10b).

In all cases several control experiments were run to assess the background noise by omitting the boronated probe or by using boron-labeled antibodies directed against antigens non-occurring in the material under investigation. The results were comparable, unspecific labeling was not observed (see Figure 8b, for example).

The high signal densities obtained were impressively seen independent of the experimental model used. The labeling efficiencies were estimated to be more than one order of magnitude higher than those observed with the immunogold labeling method (Kessels *et al.*, 1996). The very small, boronated immunoreagents we employed thus confirmed the observations of Horisberger (1981) and Yokota (1988), who demonstrated the crucial influence of the spatial extension of immunoreagents on the achievable tagging frequency.

The exploitation of the high-resolving power of ESI using novel marker compounds like those described here, i.e., the full spatial resolution of two or more neighbored or even overlapping marker molecules, will be left to future experiments. Preliminary results show that with our present picture acquisition system a magnification of at least 150000x should be used to obtain pixel areas small enough (approximately 0.5 nm and smaller) to image accurately the intensity distribution of boron signals predicted from the peptide structures, i.e. to display the markers with sufficient precision.

In general, ultrasmall immunoreagents will allow for successful studies on interactions between macromolecular domains and between constituents of supramolecular complexes by means of high-resolution electron microscopy. Since the achievable resolution of ESI is high and can be further enhanced by image analysis systems (Adamson-Sharpe and Ottensmeyer, 1981), immunocytochemistry based on ESI/EELS techniques may become of great importance. Furthermore, immunocytochemistry by ESI-detection benefits from the separation of morphological data and marker localization data opening the gate for multiple immunostaining not distinguished by different sizes of not further characterized electron-dense spots but elementally by ESI-examinations of markers consisting of different marker elements. Using modern digital superimposure techniques electron microscopy will become "colored".

The actual focussing on the use of boron is due to its now well known chemistry (Stock, 1933; Lipscomb, 1977) and appeared very promising since in addition, the study of the distribution and subcellular localization of administered boronated compounds represents an important aspect for boron neutron capture therapy of cancer (Locher, 1936; Barth *et al.*, 1990). Also in this context, the development of modified head groups with even higher boron load and better solubility to be used in our dendritic peptide concept represents a promising project.

Despite the exclusive use of boron so far, this element should not remain the only one being potentially incorporable into immunoprobes; several other elements have been successfully detected by ESI/EELS over the last years (Costa *et al.*, 1978; Jeanguillaume, 1987; Köpf-Maier and Martin, 1989; Wagner and Chen, 1990; Fehrenbach *et al.*, 1994). It will be interesting to see what kind of marker elements will be used next.

Acknowledgements

The authors wish to express their gratitude to Prof. P.W. Jungblut, Max-Planck-Institut für experimentelle Endokrinologie, Hannover, FRG, for his encouragement. We are grateful to Prof. L. Moroder and Mr. H.-J. Musiol, Max-Planck-Institut für Biochemie, Martinsried, FRG, for their invaluable help in peptide construction. We also thank Dr. F. Klobasa, Institut für Tierzucht und Tierverhalten (FAL), Mariensee, FRG, and Prof. I. Pecht, Weizmann Institute of Science, Rehovot, Israel, for the antibodies provided and Mrs. V. Ashe for linguistic assistance.

References

Abraham R, Müller R, Gabel D (1989) Boronated antibodies for neutron capture therapy. Strahlenter Onkol **165**: 148-151.

Adamson-Sharpe KM, Ottensmeyer FP (1981) Spatial resolution and detection sensitivity in microanalysis by electron energy loss selected imaging. J Microsc **122**: 309-314.

Aharoni SM, Crosby III CR, Walsh EK (1982) Size and solution properties of globular tert-butyloxycarbonyl-poly(α,ϵ-L-lysine). Macromolecules **15**: 1093-1098.

Alam F, Soloway AH, McGuire JE, Barth R, Carey WE, Adams DM (1985) Dicesium N-succinimidyl 3-(undecahydro-closo-dodecaboranyldithio)propionate, a novel heterobifunctional boronating agent. J Med Chem **28**: 522-525.

Alam F, Soloway AH, Barth RF, Mafune N, Adams DM, Knoth WH (1989) Boron neutron capture therapy: Linkage of a boronated macromolecule to monoclonal antibodies directed against tumor-associated antigens. J Med Chem **32**: 2326-2330.

Atherton E, Logan CJ, Sheppard RC (1981) Peptide synthesis. Part 2. Procedures for solid phase synthesis using N^{α}-fluorenylmethoxycarbonylanino-acids on polymamide support. Synthesis of substance P and of acyl carrier protein 65-74 decapeptide. J Chem Soc Perkin Trans **1**: 538-546.

Arsenault AR, Ottensmeyer FP (1983) Quantitative spatial distributions of calcium, phosphorus, and sulfur in calcifying epiphysis by high resolution electron spectroscopic imaging. Proc Natl Acad Sci USA **80**: 1322-1326.

Barth RF, Alam F, Soloway AH, Adams DM, Steplewski Z (1986) Boronated monoclonal antibody 17-1A for potential neutron capture therapy of colorectal cancer. Hybridoma **5**: S43-S50.

Barth RF, Mafune N, Alam F, Adams DM, Soloway AH, Makroglou GE, Oredipe OA, Blue TE, Steplewski Z (1989) Conjugation, purification and characterization of boronated monoclonal antibodies for use in neutron capture therapy. Strahlenther Onkol **165**: 142-145.

Barth RF, Soloway AH, Fairchild RG (1990) Boron neutron capture therapy of cancer. Cancer Res **50**: 1061-1070.

Barth RF, Adams DM, Soloway AH, Alam F, Daby MV (1994) Boronated starburst dendrimer-monoclonal antibody immunoconjugates: evaluation as a potential delivery system for neutron capture therapy. Bioconjugate Chem **5**: 58-66.

Bendayan M, Barth RF, Gingras D, Londoño I, Robinson PT, Alam F, Adams DM, Mattiazzi L (1989) Electron spectroscopic imaging for high-resolution immunocytochemistry: Use of boronated protein A. J Histochem Cytochem **37**: 573-580.

Brambell FWR (1959) The transmission of immunoglobulins from the mother to the foetal and newborn young. Proc Nutr Sci **23**: 35-41.

Castaing R, Henry L (1962) Filtrage magnetique des vitesses en microscopie electronique. (Magnetic speed filtering in electron microscopy). C R Acad Sci Paris **255**: 76-78.

Center for Light Microscopic Imaging and Biotechnology (CLMIB) (1995) A guide to selecting electronic cameras for light microscope-based imaging. American Laboratory **4**: 25-40.

Colliex C (1986) Electron-energy loss spectroscopy analysis and imaging of biological specimens. Ann N Y Acad Sciences **483**: 311-325.

Costa JL, Joy DC, Maher DM, Kirk KL & Hui SW (1978) Fluorinated molecule as a tracer: Difluoroserotonin in human platelets mapped by electron energy-loss spectroscopy. Science **200**: 537-539.

Danscher G, Nørgaard JOR (1983) Light microscopic visualization of colloidal gold on resin-embedded tissue. J Histochem Cytochem **31**: 1394-1398.

De Ruijter WJ (1995) Imaging properties and application of slow-scan charge-coupled device cameras suitable for electron microscopy. Micron **26**: 247-275.

Denkewalter RG, Kolc J, Lukasavage WJ; Allied Corp (1981) Macromolecular highly branched homogeneous compound based on lysine units. US-A 428972, Chem Abstr (1985) **102**: 79324q.

Fauchère JL, Kim Quang Do, Jow PYC, Hansch C (1980) Unusually strong lipophilicity of "fat" or "super"amino-acids, including a new reference value for glycine. Experientia **36**: 1203-1204.

Faulk WP, Taylor GM (1971) An immunocolloid method for the electron microscope. Immunochemistry **8**: 1081-1083.

Fehrenbach H, Schmiedl A, Brasch F, Richter J (1994) Evaluation of lanthanide tracer methods in the study of mammalian pulmonary parenchyma and cardiac muscle by electron spectroscopic imaging. J Microsc **174**: 207-223.

Furuya FR, Miller LL, Hainfeld JF, Christopfel WC, Kenny PW (1988) Use of $Ir_4(CO)_{11}$ to measure the lengths of organic molecules with a scanning transmission electron microscope. J Am Chem Soc **110**: 641-643.

Hainfeld JF (1987) A small gold-conjugated antibody label: improved resolution for electron microscopy. Science **236**: 450-454.

Hainfeld JF, Furuya FR (1992) A 1.4-nm gold cluster covalently attached to antibodies improves immunolabeling. J Histochem Cytochem **40**: 177-184.

Hawthorne MF (1993) Die Rolle der Chemie in der Entwicklung einer Krebstherapie durch die Bor-Neutroneneinfangreaktion. (The role of chemistry in the development of boron neutron capture therapy of cancer) Angew Chem **105**: 997-1033; Also published as: The role of chemistry in the development of boron neutron capture therapy of cancer. Angew Chem Int Ed Engl **32**: 950-984.

Henkelman RM, Ottensmeyer FP (1974) An energy filter for biological electron microscopy. J Microsc **102**: 79-94.

Holgate CS, Jackson P, Cowen PN, Bird CC (1983) Immunogold-silver staining: new method of immunostaining with enhanced sensitivity. J Histochem Cytochem **31**: 938-944.

Horisberger M (1981) Colloidal gold: a cytochemical marker for light and fluorescent microscopy and for transmission and scanning electron microscopy. Scanning Electron Microsc 1981; II: 9-31.

Huxham IM, Gaze MN, Workman P, Mairs RJ (1992) The use of parallel EEL spectral imaging and elemental mapping in the rapid assessment of anti-cancer drug localization. J Microsc **166**: 367-380.

Jeanguillaume C (1987) Electron energy loss spectroscopy and biology. Scanning Microsc **1**: 437-450.

Kane RR, Pak RH, Hawthorne MF (1993) Solution-phase segment synthesis of boron-rich peptides. J Org Chem **58**: 991-992.

Kato K, Fukui H, Hamaguchi Y, Ishikawa E (1976) Enzyme-linked immunoassay: conjugation of the Fab' fragment of rabbit IgG with β-D-galactosidase from *E. coli* and its use for immunoassay. J Immunol **116**: 1554-1560.

Keana JFW, Ogan MD (1986) Functionalized Keggin- and Dawson-type cyclopentadienyltitanium heteropolytungstate anions: small, individually distinguishable labels for conventional transmission electron microscope. 1. Synthesis. J Am Chem Soc **108**: 7951-7957.

Kessels MM (1996) Immuncytochemie auf der Basis von Electron Spectroscopic Imaging. Design, Synthese und Anwendung Bor-haltiger Markermoleküle (Immunocytochemistry based on electron spectroscopic imaging. Design, synthesis, and application of boron-containing marker molecules). Doctoral Thesis, University of Hannover, FRG.

Kessels MM, Qualmann B (1996) Facile enantioselective synthesis of (S)-5-(2-methyl-1,2-dicarba-closo-dodecaborane(12)-1-yl)-2-aminopentanoic acid (L-MeCBA) using the bislactim ether method. J prakt Chem **338**: 89-91.

Kessels MM, Qualmann B, Klobasa F, Sierralta WD (1996) Immunocytochemistry by electron spectroscopic imaging using a homogenous boronated peptide. Cell Tissue Res **284**: 239-245.

Köpf-Maier P, Martin R (1989) Subcellular distribution of titanium in the liver after treatment with the antitumor agent titanocene dichloride. Virch Arch B [Cell. Pathol] **57**, 213-222.

Kinet JP (1989) Antibody-cell interactions: Fc receptors. Cell **57**: 351-354.

Krivanek OL, Mooney PE (1993) Applications of slow-scan CCD cameras in transmission electron microscopy. Ultramicroscopy **49**: 95-108.

Leapman RD, Andrews SB (1991) Biological electron energy loss spectroscopy: The present and the future. Microsc Microanal Microstruct **2**: 387-394.

Leusch A, Jungblut PW, Moroder L (1994) Design and synthesis of carboranyl peptides as carriers of 1,2-dicarbadodecacarborane clusters. Synthesis: 305-308.

Lipka JJ, Lippard SJ, Wall JS (1979) Visualization of polymercurimethane-labeled fd bacteriophage in the scanning transmission electron microscope. Science **206**: 1419-1421.

Lipscomb WN (1977) Die Borane und ihre Derivate - Nobel-Vortrag (The boranes and their derivatives - nobel lecture). Angew Chem **89**: 685-696.

Locher GL (1936) Biological effects and therapeutic possibilities of neutrons. Am J Roentgenol Radium Ther **36**: 1-13.

Mao SY, Pfeiffer JR, Oliver JM, Metzger H (1992) Effects of subunit mutation on the localization to coated pits and internalization of cross-linked IgE-receptor complexes. J Immunol **151**: 2760-2774.

Merrifield RB (1963) Solid phase peptide synthesis. I. The synthesis of a tetrapeptide. J Am Chem Soc **85**: 2149-2154.

Metzger H, Ishizaka T (eds) (1982) Symposium: transmembrane signalling by receptor aggregation: the mast cell receptor for IgE as a case of study. Fed Proc **41**: 7-34.

Mizusawa E, Dahlman HL, Bennett SJ, Goldenberg DM, Hawthorne MF (1982) Neutron-capture therapy of human cancer: In vitro results on the preparation of boron-labeled antibodies to carcinoembryonic antigen.

Proc Natl Acad Sci USA **79**: 3011-3014.

Morin C (1995) The chemistry of boron analogues of biomolecules. Tetrahedron **50**: 12521-12569.

Nelson JB (1932) The maternal transmission of vaccinal immunity in swine. J Exp Med **56**: 835-840.

Ortega E, Schweitzer-Stenner R, Pecht I (1988) Possible orientational constraits determine secretory signals induced by aggregation of IgE receptors on mast cells. EMBO J **7**: 4101-4109.

Ottensmeyer FP (1982) Scattered electrons in microscopy and microanalysis. Science **215**: 461-466.

Ottensmeyer FP (1984) Electron spectroscopic imaging: parallel energy filtering and microanalysis in the fixed-beam electron microscope. J Ultrastruct Res **88**: 121-134.

Ottensmeyer FP (1986) Elemental mapping by energy filtration: advantages, limitations, and compromises. Ann N Y Acad Sci **483**: 339-350.

Ottensmeyer FP, Andrew JW (1980) High-resolution microanalysis of biological specimens by electron energy loss spectroscopy and by electron spectroscopic imaging. J Ultrastruct Res **72**: 336-348.

Padlan EA (1994) Anatomy of the antibody molecule. Molec Immunol **31**: 169-217.

Parham P (1983) On the fragmentation of monoclonal IgG1, IgG2a, and IgG2b from BALB/c mice. J Immunol **131**: 2895-2902.

Pettersson ML, Courel MN, Girard N, Abraham R, Gabel D, Thellier M, Delpech B (1989) Immunoreactivity of boronated antibodies. J Immunol Methods **126**: 95-102.

Plešek J (1992) Potential applications of the boron cluster compounds. Chem Rev **92**: 269-278.

Posnett DN, McGrath H, Tam JP (1988) A novel method for producing anti-peptide antibodies. J Biol Chem **263**: 1719-1725.

Qualmann B (1996) Hochauflösende Methoden der Immunelektronenmikroskopie. Abbildende Elektronen-Energieverlust-Spektroskopie zur Antigendetektion mittels Bor-enthaltender Markermoleküle (High-resolution methods in immunocytochemistry. Electron spectroscopic imaging for antigen detection using boron-containing marker molecules). Doctoral Thesis, University of Hannover, FRG.

Qualmann B, Kessels MM, Klobasa F, Jungblut PW, Sierralta WD (1996a) Electron spectroscopic imaging of antigens by reaction with boronated antibodies. J Microsc **183**: 69-77.

Qualmann B, Kessels MM, Musiol HJ, Sierralta WD, Jungblut PW, Moroder L (1996b) Synthese Borreicher Lysindendrimere zur Proteinmarkierung in der Elektronenmikroskopie (Synthesis of boron-rich lysine dendrimers as protein labels in electron microscopy) Angew Chem **108**: 970-973. Also published as: Synthesis of boron-rich lysine dendrimers as protein labels in electron microscopy. Angew Chem Int Ed Engl **35**: 909-911.

Reimer L (1991) Energy-filtering transmission electron microscopy. In: Advances in Electronics and Electron Physics. Hawkes PW (ed). Academic Press Inc., Boston. **81**: 43-126.

Schöllkopf U, Groth U, Deng C (1981) Enantioselektive Synthese von (R)-Aminosäuren unter Verwendung von L-Valin als chiralem Hilfsstoff (Enantioselective synthesis of (R)-amino acids using L-Valin as chiral agent) Angew Chem **93**: 793-795. Also published as: Enantioselective synthesis of (R)-amino acids using L-Valin as chiral agent. Angew Chem Int Ed Engl **20**: 798-800.

Sierralta WD, Qualmann B, Klobasa F (1994) Passage of a heterologous protein through ileal enterocytes of newborn piglets: Immunolabelling of bovine serum albumin at the light- and electron microscopic level. J Vet Med A **41**: 421-430.

Singer SJ (1959) Preparation of an electron-dense antibody conjugate. Nature **183**: 1523-1524.

Stock A (1933) The Hydrides of Boron and Silicon. Cornell University Press, Ithaca, New York. pp 1-250.

Spurr AR (1969) A low-viscosity epoxy resin embedding medium for electron microscopy. J Ultrastruct Res **26**: 31-43.

Tam JP (1988) Synthetic peptide vaccine design: synthesis and properties of a high-density multiple antigenic peptide system. Proc Natl Acad Sci USA **85**: 5409-5413.

Tamat SR, Patwardhan A, Moore DE, Kabral A, Bradstock K, Hersey P, Allen BJ (1989) Boronated monoclonal antibodies for potential neutron capture therapy of malignant melanoma and leukaemia. Strahlenther Onkol **165**: 145-147.

van de Plas P, Leunissen JLM (1993) Ultrasmall gold probes: characteristics and use in immuno(cyto)chemical studies. In: Antibodies in Cell Biology. Asai DJ (ed). Academic Press, San Diego. pp 241-257.

Varadarajan A, Hawthorne MF (1991) Novel carboranyl amino acids and peptides: Reagents for antibody modification and subsequent neutron-capture studies. Bioconjugate Chem **2**: 242-253.

Wade K (1976) Structural and bonding patterns in cluster chemistry. Adv Inorg Radiochem **18**: 1-66.

Wagner RC, Chen SC (1990) Ultrastructural distribution of terbium across capillary endothelium: detection by electron spectroscopic imaging and electron energy loss spectroscopy. J Histochem Cytochem **38**: 275-282.

Yokota S (1988) Effect of particle size on labeling density for catalase in protein A-gold immunocytochemistry. J Histochem Cytochem **36**: 107-109.

Yoshitake S, Imagawa M, Ishikwa E, Nijtsu Y, Urushizaki I, Nishiura M, Kanazawa R, Kurosaki H, Tachibana S, Nakazawa N, Ogawa H (1982) Mild and efficient conjugation of rabbit Fab' and horseradish peroxidase using a maleimide compound and its use for enzyme immunoassay. J Biochem **92**: 1413-1424.

Zanchi G, Perez JP, Sevely J (1975) Adaption of a magnetic filtering device on a one megavolt electron microscope. Optik **43**: 495-501.

Discussion with Reviewers

F.P. Ottensmeyer: An improvement I would urge is in the presentation, or perhaps even in the acquisition of the series of energy loss images. One is not left with the impression that anything logical is happening that would permit one to arrive at the result of the B-distribution: The images at 178 eV are brighter than the one on the boron edge at 201 eV. This could be due to photographic processing or possibly an automatic gain adjustment of the camera of the microscope. Thus it appears that some magic has occurred to get the result. Please comment.

Authors: We agree that the fact that the reference images recorded at lower energy losses (150 eV, 178 eV - only the latter are presented in the manuscript) are brighter than the boron-specific one (201 eV) might be irritating to the reader. However, by looking at electron energy loss spectra it becomes clear that this is a common situation in most of the cases because of the steep declining of intensity with raising energy losses; e.g., please consider figure 5: though it is not the purpose of this figure, the high intensity of images recorded at an energy loss of 178 eV can easily be estimated by visual extrapolation. The boron signal presented in the figure is, however, clearly distinct from the backgroud curve, which can be easily extrapolated visually.

As explained in the text, in the case of ESI recordings this extrapolation of the background curve is done accurately for each pixel: using two reference images recorded at 150 eV and 178 eV, the well established three-window-method was used to compute the slope of the background curve to be substracted from the values recorded just above the boron edge. Positive differences mean presence of an onset of the curve due to boron presence, no differences mean no signal.

We shortly described the course of electron energy loss spectra in our manuscript and also explained the three-window-method (see Hardware requirements for ESI-Immunocytochemistry).

Scanning Microscopy Supplement 10, 1996 (pages 345-348)
Scanning Microscopy International, Chicago (AMF O'Hare), IL 60666 USA
0892-953X/96$5.00+.25

SPECIMEN PREPARATION OF THE HUMAN CEREBELLAR CORTEX FOR SCANNING ELECTRON MICROSCOPY USING A *t*-BUTYL ALCOHOL FREEZE-DRYING DEVICE

Teruyuki Hojo*

Department of Anatomy and Anthropology, School of Medicine, University of Occupational and Environmental Health, Kitakyushu City 807, Japan

(Received for publication August 7, 1995, and in revised form January 2, 1996)

Abstract

A freeze-drying device was applied to *t*-butyl alcohol substituted nerve cells, fibers and synaptic terminals. Ten percent formalin-fixed human cerebellar cortex specimens were postfixed in 2.5 % glutaraldehyde in 0.1 M cacodylate buffer solution and were rinsed three times in 5 % sucrose solution in the same buffer. After the postfixed specimens were dehydrated using a graded series of ethanols and then transferred into a graded series of *t*-butyl alcohols (freezing point 25.4°C), the *t*-butyl alcohol substituted specimens were freeze-dried at 15°C and at high vacuum ($5x10^{-2}$ Torr). The freeze-dried specimens were sputter coated with gold. Scanning electron microscopy revealed synaptic terminals on the surfaces of a Golgi cell and a small flat polygonal cell. Rough somatic surfaces of granule cells were also observed.

Key Words: *t*-butyl alcohol freeze-drying device, Golgi cell, granule cell, flat polygonal cell, climbing fibers, synaptic terminals.

*Address for correspondence:
T. Hojo
Department of Anatomy and Anthropology
School of Medicine
University of Occupational and Environmental Health,
Kitakyushu City 807, Japan
Telephone number: 81-93-691-7232
FAX number: 81-93-691-8544
E-mail: hojo@med.uoeh-u.ac.jp

Introduction

Scanning electron microscopy (SEM) has provided a method for the analysis of the three-dimensional microfeatures of nerve cells and the interrelationship between nerve cells without the need for serially sectioned specimens (e.g., Castejón, 1988, 1992, 1993, 1994; Hojo, 1994; Reese *et al.*, 1985). In comparison to studies of others, the author has been able to analyze relatively large areas of three-dimensional (3-D) microfeatures of the human cerebellar cortex. In a previous SEM study, micrographs of large areas of cerebellar cortex, freeze-dried using a *t*-butyl alcohol freeze-drying device could be used to clarify the shape of the Purkinje cells and the interrelationship between nerve cells and various fibers (Hojo, 1994). Also, in the present study, the synaptic terminals on nerve cells were examined in specimens prepared in the *t*-butyl alcohol freeze-drying device, but now the specimens were prepared by double fixation and sucrose buffer rinse.

For the study of the cerebellar cortex, light microscopy of Golgi impregnated tissue is useful, but this technique reveals an incomplete cellular outline or silhouette of the relationships of cells, and synaptic terminals on the cell body surfaces can be examined only at low resolution. Light microscopy and transmission electron microscopy are also useful for examining sectioned specimens to demonstrate the interior of neurons (e.g., Chan-Palay and Palay, 1970), but better preparation methods are needed for larger specimens in order to show the microstructure of the outer surface of entire large cells and neurons, such as round synapses and cell somas of the Purkinje cells and Golgi cells (e.g., Castejón, 1988, 1992, 1993, 1994; Hojo, 1994; Reese *et al.*, 1985).

The improved preparation method used in the present study consists of two steps. First, 10 % formalin fixed specimens were postfixed with 2.5 % glutaraldehyde in cacodylate buffer, rinsed in the same buffer, and then dehydrated. Second, a *t*-butyl alcohol freeze-drying

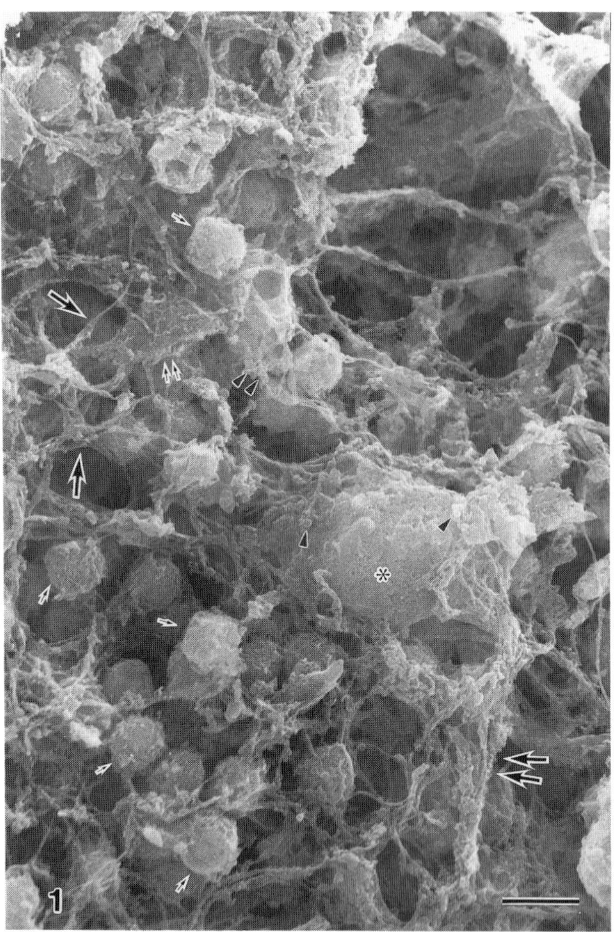

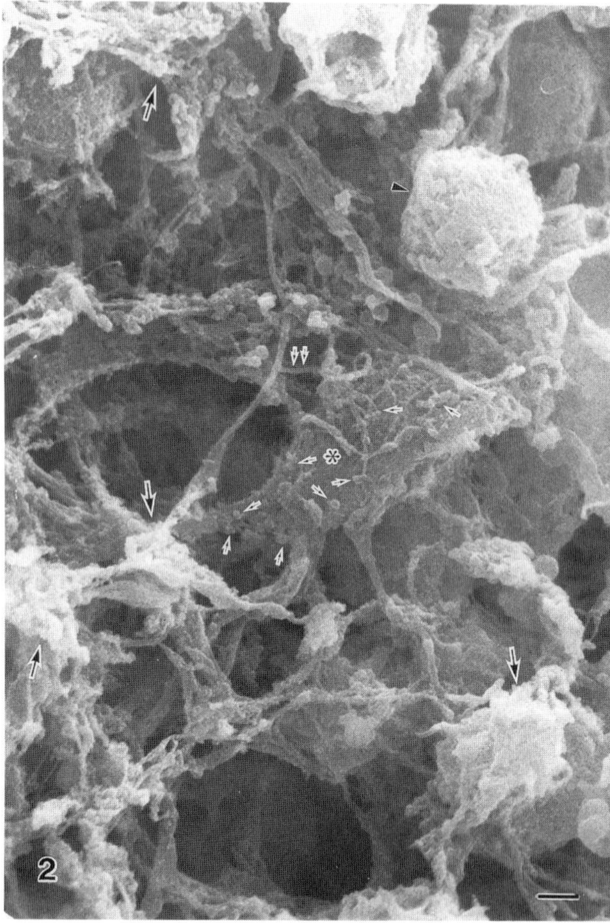

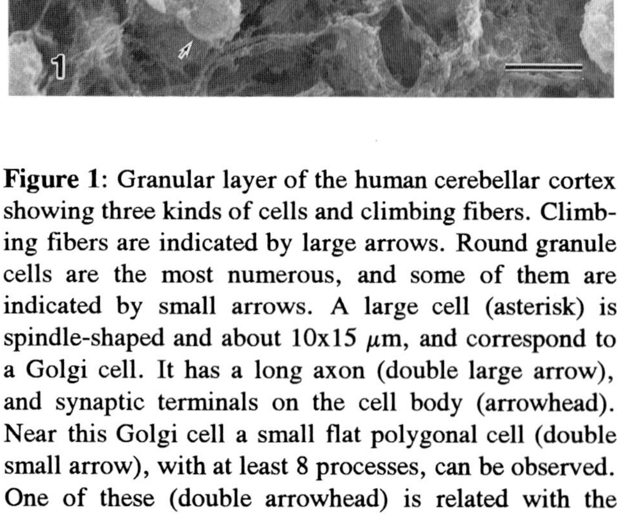

Figure 1: Granular layer of the human cerebellar cortex showing three kinds of cells and climbing fibers. Climbing fibers are indicated by large arrows. Round granule cells are the most numerous, and some of them are indicated by small arrows. A large cell (asterisk) is spindle-shaped and about 10x15 μm, and correspond to a Golgi cell. It has a long axon (double large arrow), and synaptic terminals on the cell body (arrowhead). Near this Golgi cell a small flat polygonal cell (double small arrow), with at least 8 processes, can be observed. One of these (double arrowhead) is related with the Golgi cell. Bar=5 μm.

Figure 2: Higher magnification of a small flat polygonal cell (asterisk), about 6x3 μm. Many synaptic terminals (small arrows) on the cell body as well as on the processes can be seen. This cell is connected with granule cells, and is apparently forming connections with a climbing fiber (double small arrow). Dendritic contacts of the climbing fibers with the granule cells (large arrows) can be observed. One of the granule cells (arrowhead) shows a rough somatic surface. The dark empty spaces were previously occupied by the neuroglial cell layer. Bar=2 μm.

device was used to prepare the human cerebellar cortex at relatively high temperature and at high vacuum as described earlier (Hojo, 1994).

This freeze-drying technique results in better specimen preservation at suitable temperatures and at high vacuum compared to conventional CO_2 critical point drying (Akahori *et al.*, 1988; Inoue and Osatake, 1988). In the present study further improvements in preserving nerve cell bodies, processes, fibers and synaptic terminals, compared to my previous study and studies of other authors are presented. Furthermore, the SEM used in the present study has a resolution of 5 nm and can examine nerve cells with processes and fibers also in large specimens.

Materials and Methods

In the present study the specimen was a 10 % formalin-fixed human cerebellar cortex (from an 82 year old male) which was postfixed with 2.5 % glutaraldehyde in 0.1 M cacodylate buffer, pH 7.4, for 36 hours. This method does not include maceration with alkalis like that of Reese *et al.* (1985). Instead of being rinsed in water, the specimens were rinsed three times in 5 % sucrose solution in the same buffer for 30 minutes and then kept in the same solution in the same buffer for 24 hours.

After the specimens were dehydrated using a graded series of ethanols and then transferred into a graded series of *t*-butyl alcohols (freezing point 25.4°C), the *t*-butyl alcohol substituted specimens were freeze-dried at 15°C and at high vacuum ($5x10^{-2}$ Torr) using the *t*-butyl alcohol freeze-drying device. After drying, the specimens were mounted with silver paste on aluminum stubs and sputter coated with a 10 nm film of gold. They were then examined in an SEM (ABT SX 40-A, Akashi Beam Technology, Tokyo, Japan). This SEM has a theoretical resolution of 5 nm at 2 mm working distance.

Results and Discussion

This study is the second application of freeze-drying to *t*-butyl alcohol substituted specimens. In the specimen preparation method used in the present study, *t*-butyl alcohol substituted specimens are freeze-dried in a high vacuum chamber. In the present study the double fixation and sucrose buffer rinse were used instead of 10 % formalin fixation only and water-rinsing of the specimens. As reported in my previous study, the freezing-point of *t*-butyl alcohol is relatively high (25.4°C), and the procedure of freeze-drying using *t*-butyl alcohol causes little damage to the specimens. The specimen preparation method of double fixation and sucrose buffer rinsing used in the present study revealed synaptic terminals and nerve fibers more clearly than the preparation method of 10% formalin fixation with water-rinsing used in my previous study. These improvements in the preparation method make it easier to get a clear image of the neurons than that by Reese *et al.* (1985) using alkali maceration.

Three kinds of cells were found in the granular layer. Most cells were granule cells; these cells were round and had a mean diameter of about 3 μm (Figure 1). Moreover, a larger spindle-shaped cell and a flat polygonal cell were observed. The larger spindle-shaped cell, about 10 μm x 15 μm, may be a Golgi cell. One of the processes of the Golgi cell was connected with one of the processes of the small flat polygonal cell, which measured about 6 μm x 3 μm (Figure 1).

On the Golgi cell body, synaptic terminals were observed (Figure 1). The flat polygonal cell had at least 8 processes. One of its processes was connected with the Golgi cell. Some processes as well as the cell body of the flat polygonal cell had many round synaptic structures (Figure 2). The small flat polygonal cell was contrasted with the large Golgi cell, and was connected with granule cells and a climbing fiber (Figure 2). These findings favor the idea that the flat polygonal cell may be an accessory cell to the Golgi cell, and to the best of our knowledge has not yet been reported by other authors (e.g., Carpenter, 1991; Castejón, 1988, 1992, 1993, 1994; Chan-Palay and Palay, 1970; Reese *et al.*, 1985).

Moreover, the dendritic contacts of the climbing fibers with the granule cells of the human cerebellar cortex as shown in the present study were similar to those shown in Teleost fish by Castejón (1988, 1992, 1993, 1994) (Figure 2). The granule cell showed the rough somatic surface as described by Castejón (1992, 1993).

The SEM in the present study has almost the same resolution (5 nm at 2 mm working distance) as that of Reese *et al.* (1985) and Castejón (1988, 1992, 1994). Both my present and previous study (Hojo, 1994) showed as many fibers and synaptic terminals on surfaces of various cell bodies as Castejón's studies (1988, 1992, 1993, 1994). The results of the present study seem to show that the method described here is an improvement over my previous method (Hojo, 1994) and may be easier than other methods, e.g., using alkali maceration (Reese *et al.*, 1985) in preparing the cerebellar cortex for examination by SEM.

References

Akahori H, Ishii H, Nonaka I, Yoshida H (1988) A simple freeze-drying device using *t*-butyl alcohol for SEM specimens. J Electron Microsc **37**: 351-352.

Carpenter MB (1991) Core text of neuroanatomy 4th ed, Williams & Wilkins, pp. 227-229.

Castejón OJ (1988) Scanning electron microscopy of vertebrate cerebellar cortex. Scanning Microsc **2**: 569-597.

Castejón OJ (1992) Conventional and high resolution scanning electron microscopy of outer and inner surface features of cerebellar nerve cells. J Submicrosc Cytol Pathol **24**: 549-562.

Castejón OJ (1993) Sample preparation techniques for conventional and high resolution scanning electron microscopy of the central nervous system. The cerebellum as a model. Scanning Microsc **7**: 725-740.

Castejón OJ (1994) Conventional and high resolution scanning electron microscopy and cryofracture techniques as tools for tracing cerebellar short intracortical circuits. Scanning Microsc **8**: 315-324.

Chan-Palay V, Palay SL (1970) Interrelations of basket cell axons and climbing fibers in the cerebellar cortex of the rat. Z Anat Entwickl-Gesch **132:** 191-227.

Hojo, T (1994) An experimental scanning electron microscopic study of human cerebellar cortex using a *t*-butyl alcohol freeze-drying device. Scanning Microsc **8**: 303-313.

Inoue T, Osatake H (1988) A new drying method of biological specimens for scanning electron microscopy: The *t*-butyl alcohol freeze-drying method. Arch Histol Cytol **51:** 53-59.

Reese BF, Landis DMD, Reese TS (1985) Organization of the cerebellar cortex viewed by scanning electron microscopy. Neuroscience **14:** 133-146.

Discussion with Reviewers

O.J. Castejón: Most of the techniques employed thus far preservation of nerve tissues for scanning electron microscopy selectively removes the neuroglial cell layer, (Castejón, 1993). The *t*-butyl alcohol freeze-drying device used in your study also selectively removed the granule cell neuroglial cell layer exposing the outer surface of nerve cells and synaptic connections of afferent fibers. Do you have an explanation for such useful artifact?
Author: Thank you for your suggestion. The *t*-butyl alcohol freeze-drying device selectively removed the outer surface of nerve cells and synaptic connections of afferent fibers. As you noted in your paper (Castejón, 1993), the narrow extracellular space between the outer surface of nerve cells and that of synaptic connections of afferent fibers may be enlarged during the *t*-butyl alcohol freeze-drying preparation. And the enlarged space may easily be removed.

O.J. Castejón: Which could be the alterations induced by the aging process in your sample, an 82 years old male?
Author: In this paper I did not compare this old male sample with other young samples. But the number and the shape of granule cells may be changed with the advance of age.

O.J. Castejón: Could you identify afferent mossy fibers in your preparations?
Author: I have not yet identified afferent mossy fibers in my preparations.

Scanning Microscopy Supplement 10, 1996 (pages 349-358)
Scanning Microscopy International, Chicago (AMF O'Hare), IL 60666 USA 0892-953X/96$5.00+.25

X-RAY MICROSCOPY: PREPARATIONS FOR STUDIES OF FROZEN HYDRATED SPECIMENS

A. Osanna[1*], C. Jacobsen[1], A. Kalinovsky[1,2], J. Kirz[1], J. Maser[1] and S. Wang[1]

[1]Department of Physics, [2]Electrical Engineering Dept., S.U.N.Y. at Stony Brook, Stony Brook, NY

(Received for publication September 21, 1995 and in revised form June 6, 1996)

Abstract

X-ray microscopes provide higher resolution than visible light microscopes. Wet, biological materials with a water thickness of up to about 10 μm can be imaged with good contrast using soft X-rays with wavelengths between the oxygen and carbon absorption edges (at 24 and 43 Å). The Stony Brook group has developed and operates a scanning transmission X-ray microscope (STXM) at the National Synchrotron Light Source (NSLS) at Brookhaven National Laboratory. The microscope is used for imaging with a current resolution of 50 nm, and for elemental and chemical state mapping.

Radiation damage imposes a significant limitation upon high resolution X-ray microscopy of room temperature wet specimens. Experience from electron microscopy suggests that cryo techniques allow vitrified specimens to be imaged repeatedly. This is due to the increased radiation stability of biological specimens in the frozen hydrated state. Better radiation stability has been shown recently with a cryo transmission X-ray microscope developed by the University of Göttingen, operating at the BESSY storage ring in Berlin, Germany. At Stony Brook, we are developing a cryo scanning transmission X-ray microscope (CryoSTXM) to carry out imaging and spectro-microscopy experiments on frozen hydrated specimens. This article will give an outlook onto the research projects that we plan to perform using the CryoSTXM.

Key Words: X-ray microscopy, soft X-rays, imaging, mapping, water window contrast, hydrated specimens, frozen-hydrated specimens

*Address for correspondence:
Angelika Osanna
Dept. of Physics, S.U.N.Y.
Stony Brook, NY 11794-3800
Phone number: (516) 632-8056
FAX number: (516) 632-8101
e-mail: osanna@xray1.physics.sunysb.edu

Introduction

X-ray microscopes provide intermediate resolution between visible light and electron microscopy. Many biological structures such as the cytoskeleton, synapses and organelles are at or below the resolution limit of the visible light microscope. These structures can be studied in thin sections using electron microscopes or in whole cell mounts using soft X-ray microscopy.

Soft X-rays with energies between 250 and 530 eV (50 to 24 Å) provide "natural" contrast for hydrated organic materials. The basic contrast mechanism involves photoelectric absorption of X-rays in matter. Fig. 1 shows the penetration depth for electrons and X-rays in water. The "carbon edge" and the "oxygen edge" are X-ray *absorption edges*. Absorption edges arise because absorption of X-rays leads to the ejection of electrons and the ionization of atoms. All materials show absorption edges in their X-ray spectrum; for soft X-rays they are mostly at the binding energies for K and L electrons of low-Z elements.

The spectral region between 285 and 530 eV is called the *water window*. Here, water is much more transparent than carbon-based molecules like protein or DNA (see Fig. 1). Soft X-ray microscopes operate chiefly in the water window spectral region:

* It allows biological samples to be imaged in their natural, i.e. wet environment.

* High intrinsic contrast is obtained without staining; however, gold or luminescent labels are used for special studies.

* It allows imaging of samples in up to 10 μm water or ice; therefore, there is no need for sectioning of thick specimens.

Fig. 1 also gives the mean free paths of electrons for elastic and inelastic scattering in protein and water. In the energy range of typical transmission electron microscopes, the mean free path is typically smaller than the penetration distance for soft X-rays used in X-ray microscopes. Longer penetration depth means less absorption of the incident radiation per penetrated sample thickness. Whereas in electron microscopy contrast is generated by the elastically scattered electrons, contrast in X-ray microscopes is due to photoab-

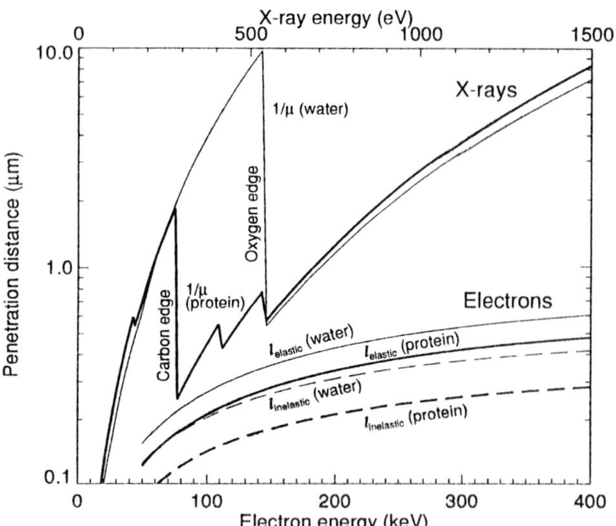

Figure 1. Penetration distance for X-rays and mean free paths l for electrons in water and protein.

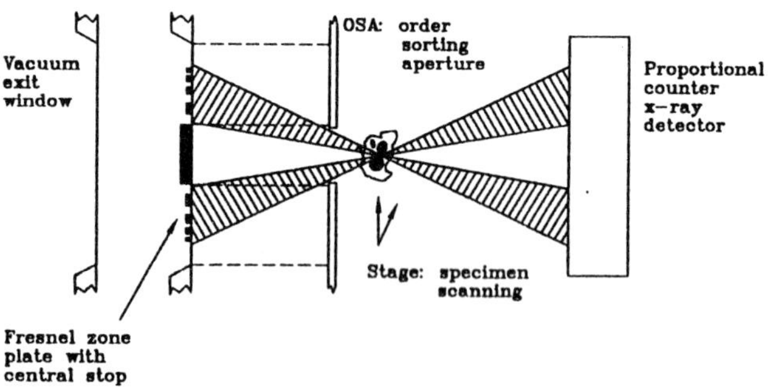

Figure 2. Schematic of the microscope. X-rays are incident from the left.

----

sorption of incident X-rays in the sample. Elastic and inelastic scattering is negligible at soft X-ray energies. Minimum radiation doses for sufficient contrast in biological specimens in electron and X-ray microscopes were first calculated and compared by D. Sayre *et al.* [12]. Detailed calculations by Jacobsen and Williams based on [8] and [12] show that frozen samples in a thick waterlayer (>500 nm) can be expected to be imaged with lower dose in X-ray microscopes, even when phase contrast electron microscopy is considered (Jacobsen C, Williams S, Contrast and dose for ice-embedded biological specimens in electron and X-ray microscopy, in preparation). For thin sections, however, electron microscopes provide a lower radiation dose.

The Scanning Transmission X-ray Microscope

Our group at Stony Brook has developed a Scanning Transmission X-ray Microscope (STXM) which operates

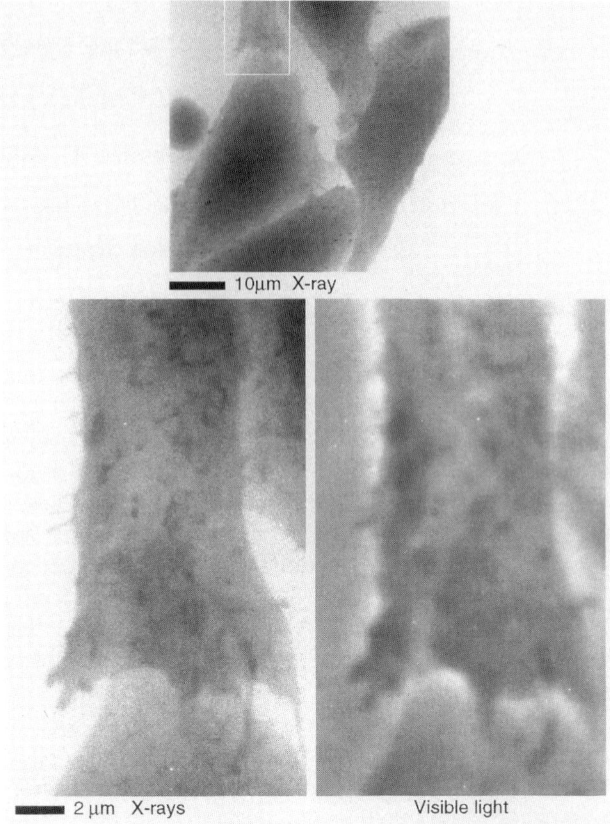

Figure 3. X-ray and optical micrograph of a whole wet cultured Chinese hamster ovarian fibroblast fixed in 1% glutaraldehyde. The optical micrograph was taken at first using a Zeiss 63 x N.A.=1.4 oil immersion objective lens in phase contrast. The cell was then imaged using the X1A Scanning Transmission X-ray Microscope. Figure courtesy J. Fu, C. Jacobsen, A. Osanna (Stony Brook Physics), W. Mangel, W. McGrath (Brookhaven Biology).

----

at the X1A beamline of the National Synchrotron Light Source (NSLS) at Brookhaven National Laboratory [7, 15, 9, 6]. A spherical grating with a resolving power of up to $\lambda/\Delta\lambda \sim 1800$ is used to select 16 - 50 Å (250 - 750 eV) monochromatic X-rays from an undulator source. The resolving power will be increased to $\lambda/\Delta\lambda \sim 4000$ with a beamline upgrade in 1996. Fig. 2 shows a schematic of the microscope. The monochromatized X-rays are focused by a Fresnel zone plate. The specimen is mounted on a scanning stage and placed in the focal spot by adjusting the position of the zone plate. A transmission image is recorded by scanning the sample through the focal spot. The transmitted radiation is detected with a gas flow proportional counter. The recorded data are used to form the image and are stored on a VAX computer. They can be used directly for

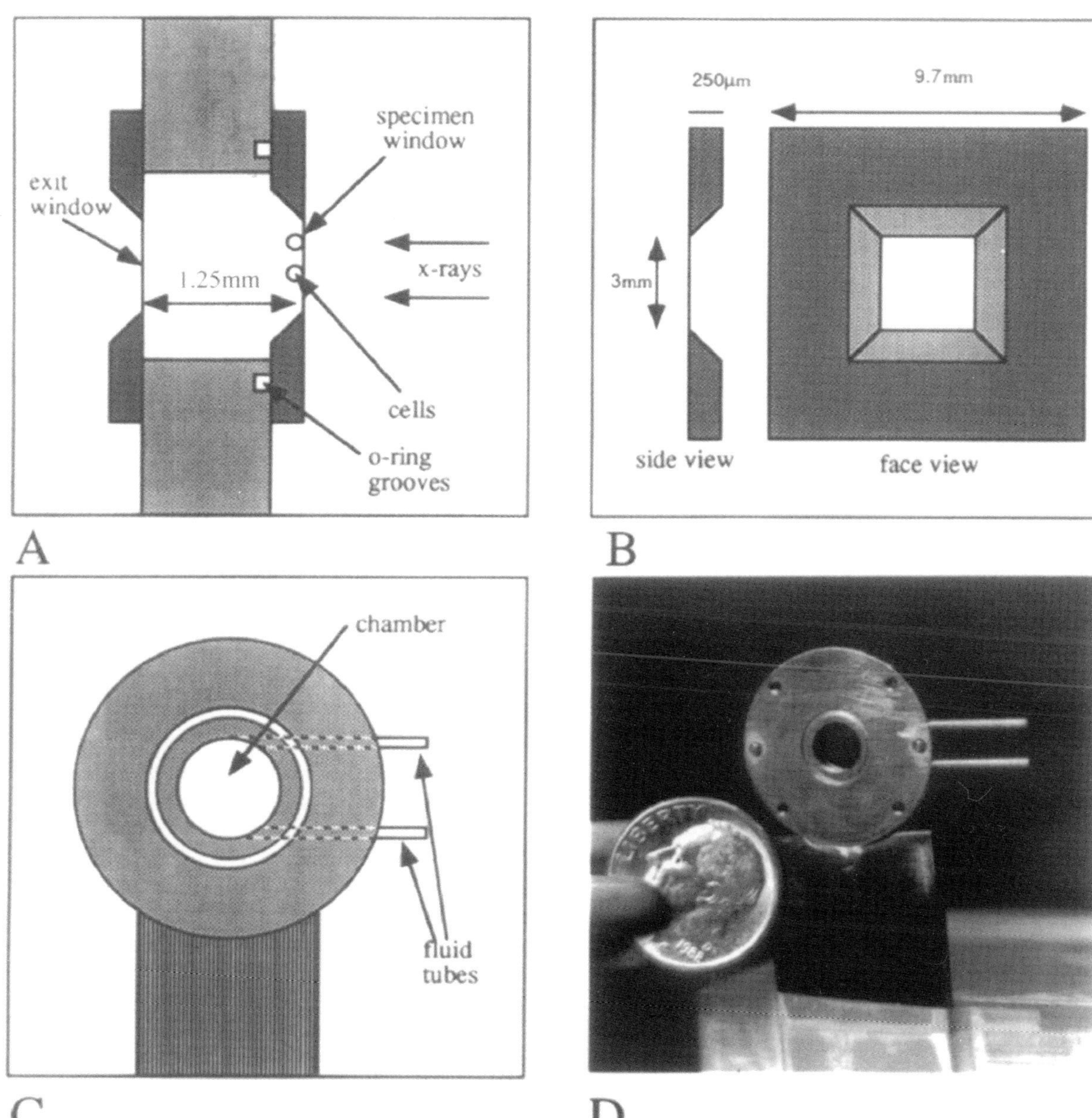

Figure 4. The CalTech wet specimen chamber (from [11]).

image processing. At present, the microscope operates in an ambient pressure environment. To avoid the strong absorption of X-rays in air, the sample area is flooded with helium.

The resolution of STXM images is primarily determined by the outermost zone width of the zone plate used as a probe-forming optic [4]. The resolution of presently ~ 50 nm is superior to that of a far-field optical microscope (see Fig. 3). 30 nm zone plates have been produced at the University of Göttingen and are also now being fabricated by us with collaborators at Lucent Technologies (Bell Laboratories).

STXM allows two basic modes of operation: the *imaging mode* and the *spectral mode.*

* In the *imaging mode,* the sample is scanned in x and y to obtain a transmission image. The wavelength is kept fixed.

* In the *spectral mode,* the wavelength is changed continuously to acquire an X-ray spectrum of a single point in the specimen. Because zone plates are chromatic optics (i.e., the focal length is determined by the wave length of the X-rays), the zone plate has to be moved

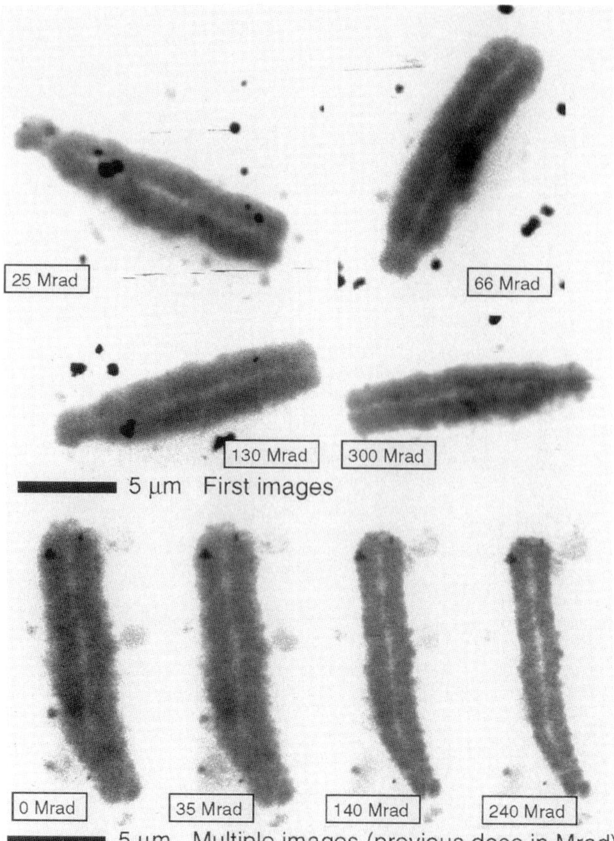

Figure 5. Wet *Vicia faba* chromosomes lightly fixed in 0.2% glutaraldehyde. Multiple images of the same chromosome show degradation due to radiation damage; however first images of previously unexposed chromosomes give accurate mass and diameter measurements up to ~200 MRad. See [14, 16].

----

along the beam axis during the scan to keep the sample in focus. By taking several spectra of different sample areas, elemental and chemical properties of the specimen can be determined. In particular, one can combine these two methods and use "chemical contrast" imaging. This makes use of the fact that the spectra of many organic materials show fine structures near an absorption edge. These X-ray Absorption Near-Edge Structures (XANES) are due to e.g. double bonds between C and other atoms. They make it possible to separate different organic constituents like DNA or protein in the specimen by choosing wavelengths for imaging at characteristic peaks. Details about XANES microscopy done at the X1A beamline can be found in [17, 18, 19, 20].

Sample Preparation for X-ray Microscopy

The STXM accommodates three different kinds of sample holders: silicon nitride (SiN) windows, the

Figure 6 on facing page

Figure 6. DNA and protein map for air-dried bull sperm cells. Fig. 6A shows the Carbon XANES spectra for DNA and protamine 1 and 2. The vertical lines indicate the STXM imaging wavelengths. Fig 6B gives the STXM pictures (a, d) and the calculated DNA (c, f) and protein (b, d) distributions for two individual sperm cells. The bar length is 2μm; both pictures have a resolution of 50nm. See [17, 19, 20].

CalTech wet specimen chamber (Fig. 4), and electron microscope grids.

Fig. 4B shows a SiN window. It consists of a silicon frame with a 1000 Å thick SiN membrane. Cells grow very well on SiN windows, and they can be imaged dried or in a wet state after loading them into the wet chamber.

The CalTech wet specimen chamber [11] uses two SiN windows, on one of which the cells have been grown (see fig. 4). The volume inside the chamber is initially filled with a buffer solution that is drained immediately before imaging. Thus, the cells remain in a humid atmosphere. Without exposure to X-rays, the cells can be kept alive for several hours if the chamber is flushed and drained every 20 to 30 minutes.

Electron microscope grids are used for dry samples and sections. They allow loading the sample into both a TEM and STXM and thus do comparative imaging.

Generally, no additional labeling or staining for contrast is necessary in X-ray microscopes. However, labeling with fluorescent labels like biotinylated terbium is done for luminescence microscopy [10]; gold sphere labeling is used for contrast enhancement in darkfield imaging [1].

Radiation Damage

The dominant interaction of X-rays in matter is the photoelectric absorption. In this process energetic electrons are produced which further ionize atoms and molecules in the specimen. This and the generation of free radicals causes *radiation damage* to the specimen which will ultimately limit the resolution of the microscope in use.

Several groups have investigated radiation damage by X-rays. In Stony Brook, Williams *et al.* have studied radiation damage to metaphase chromosomes as measured by mass loss [16]. First images of previously unexposed, wet, lightly fixed *vicia faba* chromosomes give accurate mass and diameter measurements up to a cumulative dose of about 200 MRad. Multiple images of the same chromosome, however, show degradation already at 35 MRad. (Fig. 5) Unfixed, wet specimens

Linear Absorption Coefficient (inverse micron)
X-ray wavelength (nm)
0
1
2
3
4
5
4.10
4.20
4.30
4.40
4.50
DNA
Protamine 1
Protamine 2

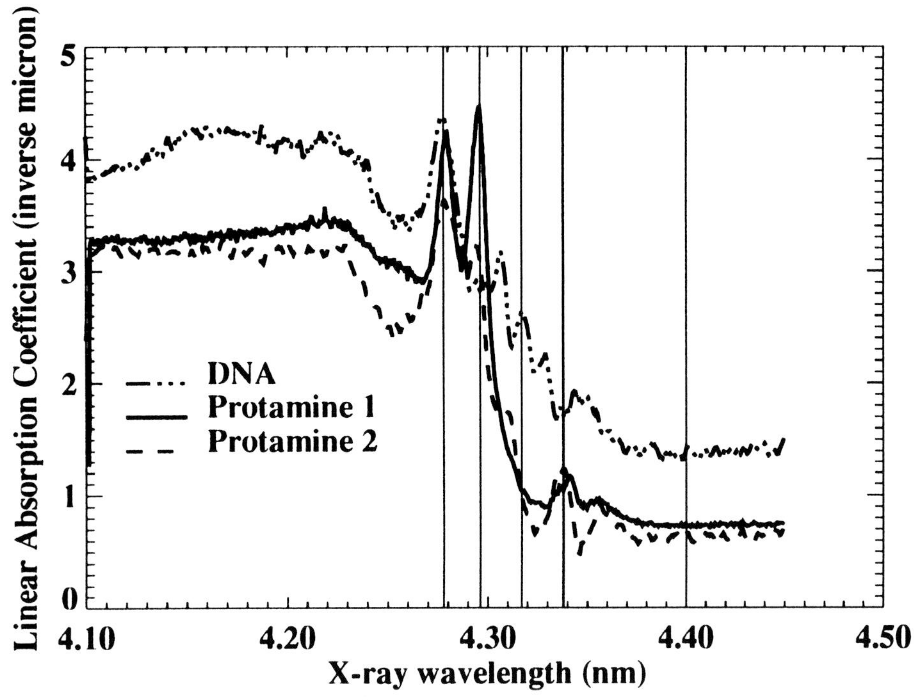

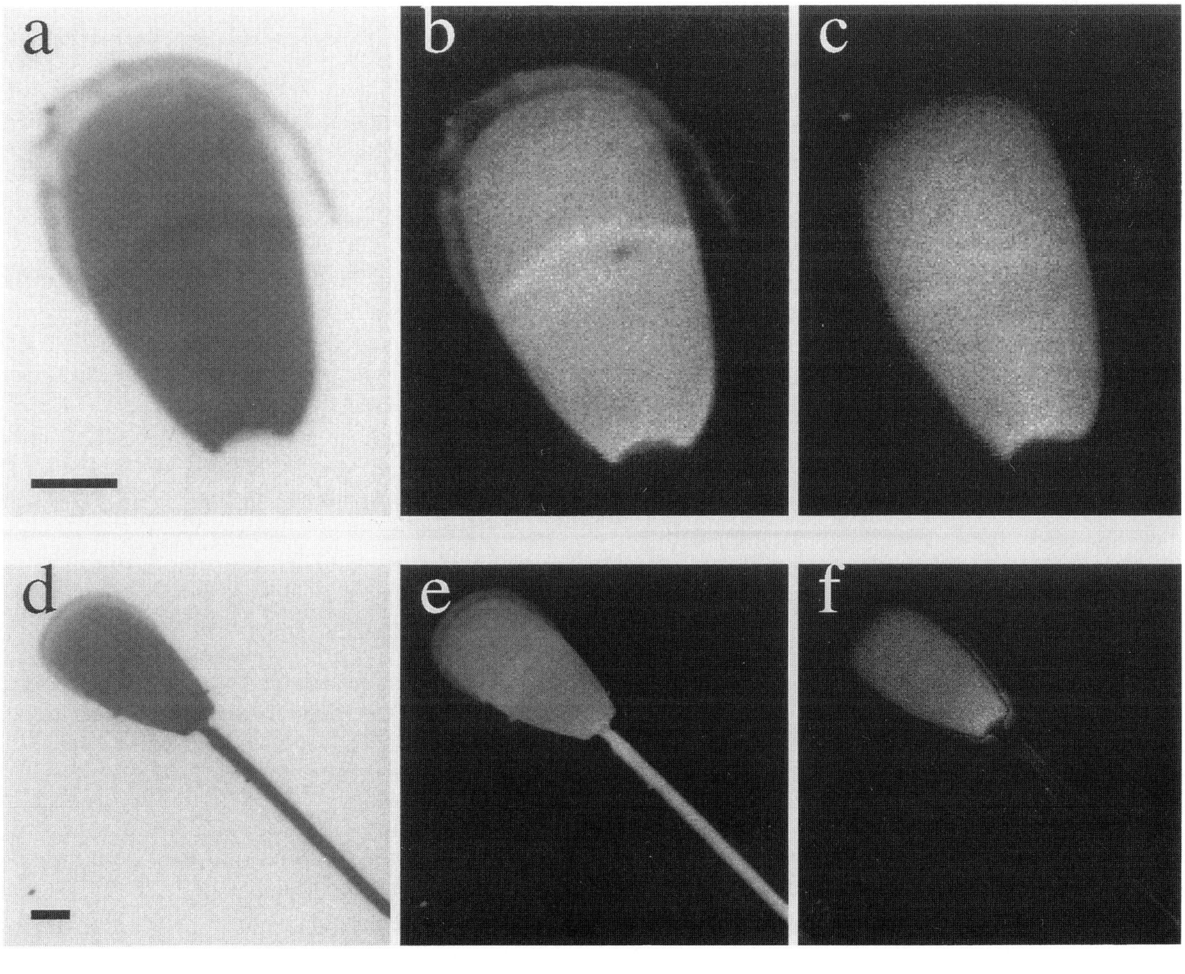

a
b
c
d
e
f

show even more severe radiation damage at comparable doses. The mass loss of wet chromosomes has been determined as a function of the incident radiation dose and the data have be fitted with an exponential mass loss law. It has been found that ionic strength and choice of fixation influence the mass loss under irradiation. Details of the numerical work go beyond the scope of this paper and can be found in [16].

Radiation damage influences the operation of STXM in two different ways. On the one hand, it limits the resolution by not allowing for long exposure times to obtain good photon statistics. On the other hand, one would often like to image a sample repeatedly, for example to get information about the 3D structure (tomography) or to do XANES microscopy.

The Cryo Scanning Transmission X-ray Microscope

To reduce the effects of radiation damage we are planning to image frozen hydrated samples. Cryo methods have been employed in transmission electron microscopy for about 10 years with good results (see for example [2] for an overview). A cryo X-ray microscope has recently been developed by the University of Göttingen. First results [13] show excellent structural preservation of frozen hydrated specimens for radiation doses up to three orders of magnitude higher than what has been possible with room temperature samples.

J. Maser *et al.* at Stony Brook have been developing a Cryo Scanning Transmission X-ray Microscope (CryoSTXM) which is currently being commissioned. A description of the technical features can be found in [9]. CryoSTXM is using an optical arrangement similar to that of STXM (see fig. 2) and will allow imaging and spectral mode operation just like its room-temperature predecessor. To improve thermal insulation and reduce thermal drifts, it has been designed as a vacuum system. It will allow us to use the full water window spectral region for imaging, including the oxygen and nitrogen edge (air leaks in the helium enclosure of STXM have prevented this up to now). Sample preparation will be done by standard cryo methods like plunge-freezing without the necessity of cryo sectioning for whole-cell specimens.

STXM Results and Cryo Plans

While the STXM has been used for studies both of material science and of biological specimens, efforts with cryoSTXM will concentrate on extending the biological studies outlined below. We now describe the current research projects that will be continued with the cryo microscope.

Metaphase chromosome structure

Vicia faba chromosomes have been imaged using water window contrast. Besides the radiation damage studies described above, Williams *et al.* have combined STXM carbon mass measurements with chemical measurements of the total DNA content to find that metaphase chromosomes from four different species all contain 38% DNA by mass [7, 16]. This is in contrast to previous results which gave variable and species-dependent DNA mass fractions. Previous studies have determined the DNA mass through chemical methods which cannot distinguish between different chromosome types in bulk specimens and where the sample is also likely to contain non-chromosome mass. From STXM images Williams *et al.* were able to sort chromosomes by morphology and thus measure the DNA mass from a pure sample. The results of Williams *et al.* suggest a common mechanism for DNA packing at the molecular level, while the differences in mass per unit length and total chromosome length suggest DNA-dependent mechanisms for higher-level chromosome organization. Further studies using chemical state mapping and tomographic imaging are planned with CryoSTXM.

Mapping of protamine and DNA in sperm

Work by Zhang *et al.* has established [17, 19, 20] that the protein-to-DNA ratio is constant among different mammalian sperm cells; however, the protamine-1 and -2 contents vary strongly. Protamine-1 and -2 together constitute 95-99% of the protein present in mammalian sperm. Sperm cells of four different mammalian species have been imaged and the DNA and protein distribution have been determined by imaging at X-ray energies corresponding to distinct peaks of protamine-1 and -2 in their XANES spectra. Fig. 6a shows the spectra for DNA and protein; Fig. 6b shows the results for air-dried bull sperm cells. In human sperm, infertility may be related to protamine-2 deficiency. The protamine-2 content of human sperm has not been determined yet since the dry mass of human sperm heads is too high to view them at the carbon edge; the much larger absorption length of X-rays at the oxygen edge will make this measurement possible in CryoSTXM.

Malaria parasite structure

Malaria is still the largest killer of humans on earth. Searches for a vaccine have been largely unsuccessful so far. Fig. 7 shows the life cycle of *Plasmodium falciparum* in red blood cells. Our studies are aiming at more information about the infection process and the intracellular stages of the parasite. Both light and electron microscopy require some form of staining for contrast and (for electron microscopy) sectioning to reduce the

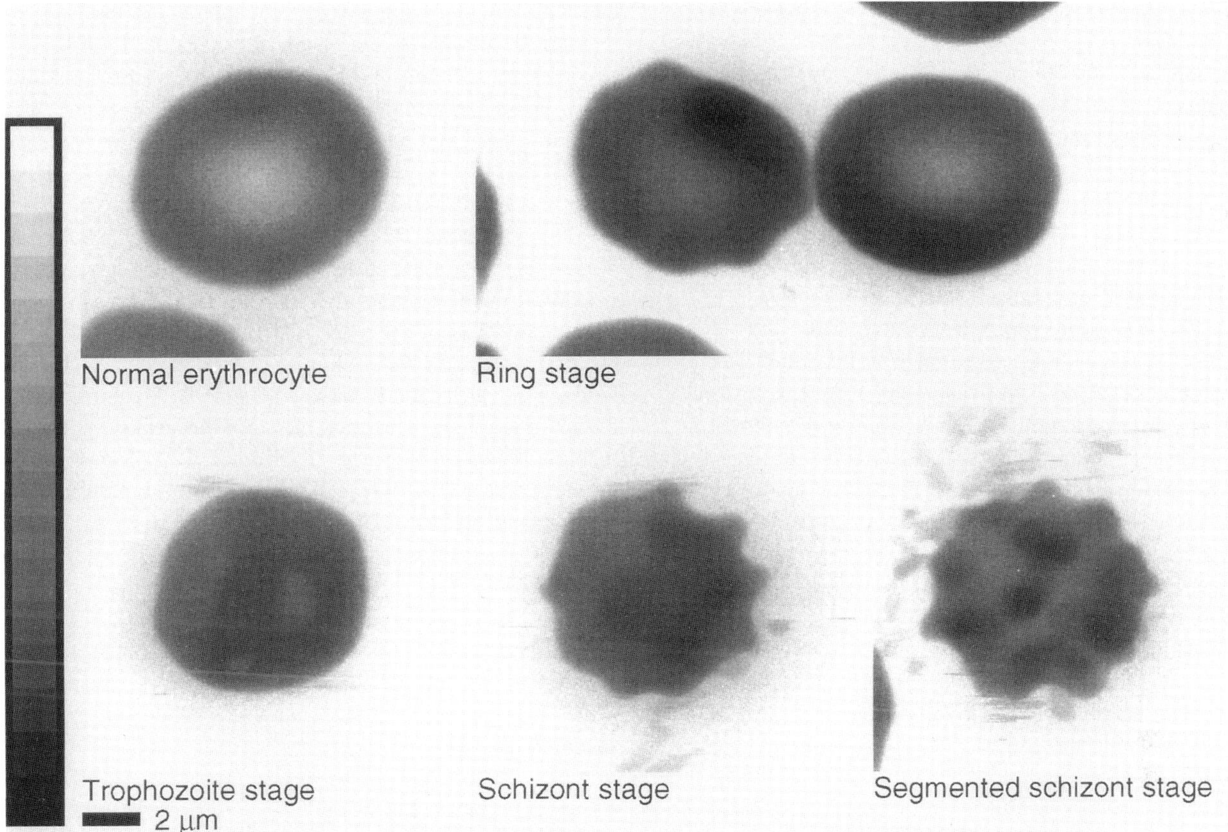

Figure 7. The life cycle of the malaria parasite in human red blood cell as seen in STXM. Figure courtesy W. Mangel, W. McGrath, S. Williams (Brookhaven Biology), C. Jacobsen, J. Kirz, X. Zhang (Stony Brook Physics)

---

specimen thickness. This has been an issue in the discussion about the evidence for a parasitophorous duct in confocal microscope observations. STXM provides better resolution than confocal light microscopy (which has a resolution of ~200nm) without any sectioning or staining. While we have obtained some images which are suggestive of duct structures, part of the controversy over confocal and electron microscope observations has centered on glutaraldehyde fixation artifacts; it should be possible to avoid such complications with cryoSTXM.

STXM and cryoSTXM both allow for luminescence and darkfield microscopy which are described in [5, 10] (luminescence) and [1] (darkfield). First results on tomography using STXM have been reported in [3].

Conclusion

Soft X-ray microscopy offers new opportunities for quantitative imaging and elemental and chemical state mapping of whole, hydrated cells. Further studies using frozen hydrated specimens will become possible with the development of cryo X-ray microscopes, such as the cryo scanning transmission X-ray microscope of the Stony Brook group. Applications include multiple imaging and tomography of specimens like chromosomes, sperm cells and malaria-infected red blood cells.

Acknowledgements

We want to thank H. Ade, C. Buckley and M. Rivers for their continuing contributions to the work at the X1A beamline, and other past and present members of the Stony Brook group, including H. Chapman, J. Fu, S. Wirick and X. Zhang. We also thank to T. Oversluizen, J. Miao and B. Winn for their work with the beamline upgrade, to E. Anderson for the currently used zoneplates and S. Spector for fabricating new zoneplates and test patterns. This work was supported by the Office of Health and Environmental Research, Department of Energy, under contract FG02-89ER60-858, and by the National Science Foundation under grants BIR-9316594 and BIR-9112062, and Presidential Faculty Fellow award RCD-9253618 (CJ).

References

1. Chapman H, Fu J, Jacobsen C, Williams S (1996) Darkfield X-ray microscopy of immunogold-labeled cells. J Soc Microsc America **2**: 53-62.

2. Dubochet J, Adrian M, Chang J-J, Homo J-C, Lepault J, McDowall AW, Schultz P (1988) Cryo-electron microscopy of vitrified specimens. Q Rev Biophys **21**: 129-228.

3. Haddad WS, McNulty I, Trebes JE, Anderson E, Levesque RA, Yang L (1994) Ultrahigh-resolution X-ray tomography. Science **266**: 1213-1215.

4. Jacobsen C, Kirz J, Williams S (1992) Resolution in soft X-ray microscopes. Ultramicroscopy **47**: 55-79.

5. Jacobsen C, Lindaas S, Williams S, Zhang X (1993) Scanning luminescence X-ray microscopy: imaging fluorescent dyes at suboptical resolution. J Microsc **172**: 121-129 .

6. Jacobsen C, Williams S, Anderson E, Browne MT, Buckley CJ, Kern D, Kirz J, Rivers M, Zhang X (1991) Diffraction-limited imaging in a scanning transmission X-ray microscope. Opt Comm **86**: 351-364.

7. Kirz J, Jacobsen C, Howells M (1995) Soft X-ray microscopes and their biological applications. Q Rev Biophys **28**: 33-130.

8. Langmore JP, Smith MF (1992) Quantitative energy-filtered electron microscopy of biological molecules in ice. Ultramicroscopy **46**: 349-373.

9. Maser J, Chapman H, Jacobsen C, Kalinovsky A, Kirz J, Osanna A, Spector S, Wang S, Winn B, Wirick S, Zhang X (1995) The scanning transmission X-ray microscope at the NSLS: From XANES to Cryo. In: X-ray Microbeam Technology and Applications, vol. 2516, Yun W (ed) Society of Photo-Optical Instrumentation Engineers (SPIE) Proceeding, Bellingham, Washington, pp. 78-79.

10. Moronne M, Larabell C, Selvin P, von Brenndorff AI (1995) Development of fluorescent probes for X-ray microscopy. In: Proceedings of the 52nd Annual Meeting of the Microscopy Society of America, Bailey GW, Garratt-Reed AJ (eds) San Francisco Press, San Francisco, pp. 48-49.

11. Pine J, Gilbert J (1992) Live cell specimens for X-ray microscopy. In: X-ray Microscopy III (Springer Series in Optical Sciences Vol. 67) Michette AG, Morrison GR, Buckley CJ (eds) Springer Verlag, Berlin, pp. 384-387.

12. Sayre D, Kirz J, Feder R, Kim DM, Spiller E (1977) Transmission microscopy of unmodified biological materials: comparative radiation dosages with electrons and ultrasoft X-ray photons. Ultramicroscopy **2**: 337-349.

13. Schneider G, Niemann B, Guttmann P, Rudolph D, Schmahl G (1995) Cryo X-ray microscopy. Synch Rad News **8:3**: 19-28.

14. Williams S, Jacobsen C, Kirz J, Lamm SS, Van't Hof J, Zhang X (1995) Metaphase chromosome DNA mass fraction is independent of species. In: Proceedings of the 52nd Annual Meeting of the Microscopy Society of America, Bailey GW, Garratt-Reed AJ (eds) San Francisco Press, San Francisco, pp. 46-47.

15. Williams S, Jacobsen C, Kirz J, Maser J, Wirick S, Zhang X, Ade H, Rivers M (1995) Instrumentation developments in scanning soft X-ray microscopy at the NSLS. Rev Sci Instrum **66**: 1271-1275.

16. Williams S, Zhang X, Jacobsen C, Kirz J, Lindaas S, van't Hof J, Lamm SS (1993) Measurements of wet metaphase chromosomes in the scanning transmission X-ray microscope. J Microsc **170**: 155-165.

17. Zhang X (1995) Development and Application of Quantitative X-ray Microscopy with Chemical Sensitivity. Doctoral Thesis, S.U.N.Y. Stony Brook, New York 11794.

18. Zhang X, Ade H, Jacobsen C, Kirz J, Lindaas S, Williams S, Wirick S (1994) Micro-XANES: chemical contrast in the scanning transmission X-ray microscope Nuclear Instruments and Methods in Physics Research A **347**: 431-435.

19. Zhang X, Balhorn R, Jacobsen C, Kirz J, Williams S (1995) Mapping DNA and protein in biological samples using the scanning transmission X-ray microscope. In: Proceedings of the 52nd Annual Meeting of the Microscopy Society of America, Bailey GW, Garratt-Reed AJ (eds) San Francisco Press, San Francisco, pp. 50-51.

20. Zhang X, Balhorn R, Mazrimas J, Kirz J (1996) Mapping and measuring DNA to protein ratios in mammalian sperm head by XANES imaging. J Struct Biol **116**: 335-344.

Discussion with Reviewers

J. Hainfeld: The X-ray C and O absorption edges are very dramatic as shown in Fig. 1. Have you tried two image differencing, one above the O edge and one below the O edge, for example, as a way to increase contrast even further over "natural" contrast?

Authors: "Two image differencing" is the basic principle of elemental and chemical state mapping that is done routinely in STXM (see section on description of the microscope). Imaging above and below an absorption edge like the carbon edge is the simplest version of this method that usually uses X-ray Absorption Near-Edge Structures (XANES). It is used for example to image samples that are known to contain protein but where a protein standard is not available. While imaging across the carbon edge is easily performed and yields nice results, the method has not yet been used at the oxygen edge, the reason being residual air in the X-ray path of STXM. As mentioned in the description of our cryo version of STXM, mapping across the oxygen and also the nitrogen edge will be done in the vacuum environment of cryoSTXM. However, for wet specimens the oxygen edge is dominated by water and may be less informative.

J. Hainfeld: In Fig. 3 you compare a STXM and a light microscope cell image, and the light image is obviously of lower quality. However, thick samples, e.g., cells, suffer from out-of-focus information being imaged, resulting in image degradation, whereas the STXM image is produced from a highly collimated source. Could you comment on this effect and would a confocal image be better and perhaps a better comparison?
Authors: The light microscope image shown may indeed suffer from out-of-focus information from the full thickness of the specimen. High depth of focus is in fact an advantage of X-ray microscopy (our STXM has a depth of focus of $\approx 2\mu$m at a spatial resolution of 50nm and a wavelength of 40Å). Confocal microscopy could be used to deliver ~200 nm resolution, 500 nm thick optical sections of fluorescently labeled features, which would provide useful complementary information to the X-ray microscope's 50 nm resolution image of total organic mass.

J. Hainfeld: The very high doses sustained by x-radiation in some STXM images result in damage from free radicals. Have you tried to bathe samples in free radical scavengers and does it help?
Authors: We have indeed used free radical scavengers; Williams *et al.* report in [21] that bathing glutaraldehyde-fixed *V. faba* chromosomes in dithiothreitol or ethanol reduces the amount of radiation damage incurred during imaging by a factor of 4 and 10, respectively. Despite this improvement in radiation resistance, we have chosen to concentrate on cryo methods because it offers even more radiation damage resistance and allows one to avoid the artifacts which sometimes accompany glutaraldehyde fixation.

J. Hainfeld: The cryoSTXM has been shown to slow the beam damage of biological specimens. Are these studies at liquid nitrogen temperature and is this where you propose to operate? Would it pay to go to liquid helium temperatures; what is the damage-temperature profile?
Authors: The Göttingen X-ray microscope is operating at liquid nitrogen temperature, and so will the Stony Brook microscope. Even though there may be advantages from operating at liquid helium temperature, we have chosen to begin with using the commercially-developed technologies available for reaching liquid nitrogen temperatures. No damage-temperature profile for cryo X-ray microscopy is available yet.

J. Hainfeld: In reference to the protamine/DNA work, do DNA and protein damage differently in the X-ray beam?
Authors: The mapping of DNA and protein in sperm was done with dried specimens. The studies of Williams *et al.* reported in [16] show no measurable mass loss at doses up to 2.4 Grads. No difference in radiation sensitivity of protein and DNA has been observed.

J.A.N. Zasadzinski: Are there samples that can be analyzed better with the current generation of X-ray microscopes than with currently available electron microscopy techniques - especially when cryo techniques are required?
Authors: In this paper, we have summarized several studies which required the ability to study unsectioned specimens: total mass measurements of chromosomes, protamine mapping in sperm, and a search for a duct structure which extends in three dimensions through malaria-infected erythrocytes. X-ray microscopy is well suited to studying samples for which 0.1 to 0.3 μm thick sections provide incomplete information, or for which chemical state maps of low-Z elements are required.

J.A.N. Zasadzinski: When imaging large samples such as whole cells, what is the practical resolution, and what limits the resolution?
Authors: It is difficult to determine the resolution from images of unknown objects which is what we are looking at with wet samples (they cannot be imaged in a transmission electron microscope before using STXM). Power spectra taken from glutaraldehyde-fixed specimens suggest rendering information close to the 50nm level (for details, see for example [6]). We have not observed any degradation of resolution due to multiple scattering and refractive index variations in large specimens.

J.A.N. Zasadzinski: How many X-ray microscopes do you envision as being necessary for biological research in the US? - One for every university - one for each state or one in the country?
Authors: Due to the fact that X-ray microscopes need a synchrotron to operate, there will not be "one STXM per biology lab". But then, not every laboratory can afford a high voltage-TEM, and yet they are regarded to provide useful information to the community. At the Stony Brook STXM we have demands for much more beamtime than we have to offer. Furthermore, X-ray microscopy is under development at more than a dozen

synchrotron centers worldwide.

G.M. Roomans: Will ice crystals be a problem in obtaining images of frozen-hydrated specimens by X-ray microscopy?

Authors: Cryo electron microscopy indicates that in flash frozen samples there is no sign of ice damage. STXM is insensitive to ice crystals smaller than ~10 nm, a requirement that is satisfied by the usual plunge-freezing methods. The Göttingen results indicate that sufficient vitrification of thick samples is possible.

Additional References

[21] Williams S, Jacobsen C, Kirz J, Zhang X, Van't Hof J, Lamm SS (1992). Radiation damage to chromosomes in the scanning transmission X-ray microscope. In: Soft X-ray Microscopy, vol. 1741, Jacobsen C, Trebes J (eds) Society of Photo-Optical Instrumentation Engineers (SPIE), Bellingham, Washington,

Scanning Microscopy Supplement 10, 1996 (pages 359-373) 0892-953X/96$5.00+.25
Scanning Microscopy International, Chicago (AMF O'Hare), IL 60666 USA

IN VITRO SYSTEMS AND CULTURED CELLS AS SPECIMENS FOR X-RAY MICROANALYSIS

Godfried M. Roomans*, Jarin Hongpaisan, Zhengzhu Jin, Ann-Christin Mörk and Ailing Zhang

Department of Human Anatomy, University of Uppsala, Box 571, 75123 Uppsala, Sweden

(Received for publication November 29, 1995 and in revised form April 19, 1996)

Abstract

In vitro systems and cultured cells are recognized as useful systems in many areas of biomedical research, including X-ray microanalysis. To be reliable, in *an vitro* system should have an elemental composition close to that of the tissue *in situ*, react in the same way to stimuli, and retain the *in situ* regulation of ion transport. In the present paper, four of the most commonly used *in vitro* systems will be reviewed: incubated tissue slices (liver and pancreas), isolated glands (submandibular gland acini, sweat glands), primary cell cultures (sweat glands, endometrium), and cell lines (the colon cancer cell line T84, immortalized sweat gland cells). Incubation of tissue slices of liver in Krebs-Ringers buffer caused a significant increase in Na and Cl and a decrease in K. Initially, these changes were also observed in the pancreas, but here the values gradually returned to normal. Isolated submandibular gland acini, and isolated sweat gland ducts and coils react in a similar way to stimulation as their *in situ* counterparts. In primary cultures of coil cells, however, part of the cell population acquires different ion transport characteristics. Technically simplest is the use of cell lines originating from cancer cells (e.g., the T84 cell line) and immortalized cell lines. X-ray microanalysis not only confirms data on ion transport obtained with other techniques, but adds the possibility to investigate the presence of subpopulations within a culture.

Key Words: X-ray microanalysis, specimen preparation, *in vitro* systems, cell culture, diffusible ions, cryotechniques, chloride transport, epithelia.

*Address for correspondence:
Godfried M. Roomans
Department of Human Anatomy, University of Uppsala, Box 571, 75123 Uppsala, Sweden
Telephone number: +46-18-174114
FAX number: +46-18-551120
e-mail: godfried.roomans@anatomi.uu.se

Introduction

The usefulness of *in vitro* systems and cultured cells has been recognized in many areas of biomedical research, and it is not surprising that this type of specimens is increasingly being used in X-ray microanalysis. In general, *in vitro* systems and cultured cells allow a wider range of experimental conditions to be applied to the system, in comparison to *in situ* experiments. It has also been argued (Roomans, 1991; Hongpaisan *et al.*, 1994; Hongpaisan and Roomans, 1995) that especially in X-ray microanalysis of tissue in human pathology, the use of *in vitro* systems and cell cultures would have advantages: since it is difficult to optimize the freezing conditions in a clinical setting, *in vitro* systems would make it possible to separate taking the biopsy from the freezing process, and to perform both in an optimal way.

In the present paper, we will review four of the most commonly used *in vitro* systems: incubated tissue slices, isolated glands, primary cell cultures, and cell lines. To be useful in X-ray microanalysis, the *in vitro* systems should ideally have an elemental composition similar to the *in situ* systems from which they are derived, and they should react in a similar way to physiological stimuli.

Incubated tissue slices are routinely used in many physiological and biochemical studies (e.g., brain slices, see Ballyk and Goh, 1992 and McIlwain, 1987) but have been rarely used in X-ray microanalytical studies. The tissue slices are incubated in an aerated physiological buffer (Krebs-Ringers buffer, artificial cerebrospinal fluid) at physiological temperature, and left for a period of minutes to hours under these conditions. It could be shown (Hongpaisan *et al.*, 1994; Hongpaisan and Roomans, 1995) that even a brief exposure to a physiological buffer introduced significant changes in the elemental composition of the cells. In some cases (pancreas, submandibular gland) the situation was stabilized by prolonged incubation, in other cases (brain tissue) the situation deteriorated during incubation with further increase of Na and Cl and decrease of K. Isolated glands are used in physiological studies, but have so far not been used for X-ray microanalysis.

Primary cell cultures have been used in a number of studies (Wroblewski and Roomans, 1984): one of the first such studies used cultured vascular smooth muscle cells (James-Kracke *et al.*, 1980), but also epithelial cells, such as respiratory epithelium (Sagström *et al.*, 1992) and sweat gland epithelium (Mörk *et al.*, 1995), and endocrine cells such as thyroid follicular cells and endocrine pancreatic cells (Wróblewski and Wroblewski, 1994) have been used. Often it has been noted that the primary cultures have higher concentrations of Na and Cl, and lower concentrations of K than the corresponding *in situ* system (Wróblewski and Wroblewski, 1994) and the situation may deteriorate with increasing passage number (Sagström *et al.*, 1992).

Cell lines (fibroblasts, cancer cells) have been used on numerous occasions (reviewed, e.g., in Von Euler *et al.*, 1993, Hongpaisan *et al.*, 1994 and Warley, 1994). If the analysis is not carried out on sections (Zierold and Schäfer, 1988) but rather on entire cells, a washing step has to be introduced to remove the culture medium and the possibility that the washing procedure may result in changes in elemental concentration has been a cause of some concern (Wroblewski and Roomans, 1984; Abraham *et al.*, 1985; Borgmann *et al.*, 1994). For X-ray microanalysis, cells may be cultured on a variety of substrates (reviewed, e.g., by Warley, 1994) that may be classified into two groups: ultrathin or thick. In the former case, the specimens can be quantitatively analyzed as thin sections, in the latter case, there may be a problem due to the presence of the substrate, which may contribute to the spectrum. There is some advantage in culturing cells on an ultrathin substrate (Von Euler *et al.*, 1993) because a better signal can be obtained in the (scanning) transmission electron microscope with a higher accelerating voltage, and because of the more straightforward quantitative procedure, but not all cells can be cultured in this way and good results may also be obtained with cells grown on a thick substrate and analyzed in the scanning electron microscope, particularly if the accelerating voltage is adjusted to give optimal results (Borgmann *et al.*, 1994).

The experiments described in this paper were carried out on epithelial cells. Since epithelia form the body's boundary and therefore regulate and control the passage of substances (including ions and water) into and out of the organism, regulation of epithelial ion transport is an important biological problem. In this study, in particular the aspect of chloride transport in epithelial cells is considered. Most epithelial cells have, or are assumed to have, an apical chloride channel responsible for chloride efflux. In the duct of the sweat gland, the apical cell membrane has an inwardly-directed chloride channel whereas the basolateral membrane has an outwardly directed chloride channel. Together these channels are responsible for reabsorption of Cl^- from the primary sweat. Some chloride channels are regulated by Ca^{2+} ions, and are activated by substances that increase the intracellular Ca^{2+} level. Other chloride channels are activated by cAMP. In the genetic disease cystic fibrosis, the cAMP-activated channels are defective, whereas the Ca^{2+} activated channels appear to function in a normal way. In some epithelial cells, both types of Cl^- channel may occur, whereas in other epithelial cell types, either cAMP- or Ca^{2+}- activated Cl^- channels are present. Whether this regulation of chloride secretion is retained in different *in vitro* systems is one aspect to be considered in the present paper.

Materials and Methods

Incubated slices of liver and pancreas

For incubation experiments on liver and pancreas, Sprague-Dawley rats (8-9 weeks old, 190-220 g) were used. The animals were anesthetized with pentobarbital (45 mg/kg body weight). In some animals, the tissue was frozen *in situ* with liquid-nitrogen cooled brass clamps. In other animals the tissue was dissected to 1-1.5 mm thick slices and incubated in a Krebs-Ringers buffer (KRB), containing 140 mM NaCl, 5 mM KCl, 1.5 mM $CaCl_2$, 5 mM HEPES (2-hydroxyethylpiperazine-2'-ethanesulfonic acid), and 1 mM $MgCl_2$ supplemented with 5 mM D-glucose at pH 7.4. In some experiments, NaCl was exchanged for sodium gluconate or potassium gluconate. The tissue was superfused with the incubation fluid at a rate of 2 ml/min for 2 h at 36 ± 1°C and oxygenated with 95% O_2 and 5% CO_2; samples were taken at half-hour intervals and frozen in liquid propane cooled by liquid nitrogen.

For X-ray microanalysis either 16-μm thick (Wróblewski *et al.*, 1978, 1987; McMillan and Roomans, 1990) or 4 μm-thick (Wroblewski *et al.*, 1983) cryosections were cut on a conventional cryostat, mounted on a carbon support (the thinner sections were mounted over a Formvar (Merck, Darmstadt, Germany) film covered hole in the support), freeze-dried, and coated with a thin carbon layer to prevent charging in the electron microscope. Sections were cut a few cell layers away from the dissected edge of the tissue block.

The 16 μm-thick cryosections were analyzed in a Philips 525 scanning electron microscope in the secondary electron mode at 20 kV with a LINK AN10000 (Oxford Instruments, Oxford, UK) energy-dispersive spectrometer system. The 4 μm-thick sections were analyzed in a JEOL 1200EX TEMSCAN in the scanning transmission mode at 100 kV with a Tracor 5500 energy-dispersive spectrometer system. All analyses were carried out with a stationary beam (probe size 100 nm). Quantitative analysis was carried out based on the ratio of characteristic counts to background intensity in the same energy region (P/B-ratio) (Roomans, 1988; Von Euler *et al.*, 1992). P/B-ratios obtained on the samples were compared with those obtained on standards consisting of a gelatin/glycerol matrix containing mineral salts in known concentrations (Roomans, 1988). Data from 4 μm-thick sections were not significantly different from those obtained on 16 μm-thick sections and the results of these two specimen types were pooled.

Isolated submandibular gland acini

Sprague-Dawley rats (male, 200 g, about 5 weeks old) were used for these experiments. Animals had access to food and water *ad libitum* until the moment of the experiment. The animals were anesthetized with pentobarbital sodium (30 mg/kg body weight). The submandibular glands were dissected free, removed and placed in a small volume of Dulbecco's Modified Eagle Medium (DMEM; Gibco, Paisley, UK), which had previously been aerated with 95%O_2/5%CO_2 for 30 minutes at 37°C; the pH of this medium was adjusted to 7.4 with 1M NaOH before use. The submandibular glands were then chopped into about 1 mm^3 pieces, and digested in DMEM containing 1000U/15ml collagenase (Worthington Biochemical Corporation, Freehold, New Jersey, USA) and 11mg/15ml hyaluronidase (Sigma, St. Louis, Missouri, USA). The pH of the digestion medium was adjusted to 7.4 at 0, 5, 15, 30, 40 and 50 minutes. To promote dissociation of gland fragments, the digestion medium with the fragments was sucked through a glass pipette with a broken tip at 30 and 40 minutes, and through a 10 ml plastic pipette at 50 minutes. After digestion, the suspension was filtered through a nylon mesh and the fragments were washed 3 times with DMEM containing 2% albumin. The fragments were then allowed to sediment for 4-5 minutes, the supernatant was removed, and finally the fragments were resuspended in DMEM containing 2% albumin. All the steps of isolation were performed under 95%O_2/5%CO_2 at 37°C. To check the effect of the various steps of the isolation procedure, gland fragments were frozen by slam-freezing against a liquid-nitrogen cooled, gold-coated copper block (CF100, Life Cell, Houston, TX) after collagenase digestion for 15 and 50 minutes, as well as after 50 minute digestion followed by disaggregation. As controls, glands were dissected and frozen without any treatment. The frozen samples were transferred into liquid nitrogen in which they were kept until cryosectioning. Ultrathin cryosections (0.2 μm or less) were cut with an LKB (Uppsala, Sweden) Cryo Nova cryoultramicrotome at a specimen temperature of -140°C and a knife temperature of -135°C. The sections were collected dry onto copper grids and sandwiched between two grids. The grids had been covered with a Formvar film and then coated with a thin carbon layer before they were used to carry sections. The sections were freeze-dried at -80°C and 10^{-5} torr in an external freeze-drier. X-ray microanalysis was performed at 100 kV in the transmission (TEM) mode of a Philips 400 electron microscope (spot size about 0.2 μm) with a LINK QX200 energy-dispersive X-ray microanalysis system. Quantitative analysis was carried out based on the peak-to-continuum ratio after correction for extraneous background (Roomans, 1988) and by comparing the spectra from the cells with those from a standard. Only one spectrum was acquired from each cell. Spectra were acquired for 100 seconds.

Primary cultures of sweat glands

The coil or duct part, respectively, was cultured in 25 cm^2 tissue culture flasks (Costar, Cambridge, MA) containing 800 μl of the culture medium (1% fetal calf serum added to the pre-culture medium). When cellular outgrowth was seen, several ml of the culture medium was added. After 10-14 days the cells were incubated with 1 ml dispase (Boehringer Mannheim, Germany) for 30-45 minutes. Sheets of cells detached from the floor of the culture flask were allowed to recover in culture medium. For X-ray microanalysis the cells were seeded out on 75 mesh titanium grids (Agar Scientific, Stansted, UK). The grids had been covered with a Formvar (Merck) film coated with a thin carbon layer. The grids were sterilized under ultra-violet light before use. The cells were allowed to attach and spread for 3-7 days at 37 °C in a humidified atmosphere of 5% CO_2/95% air in a culturing chamber. In stimulation experiments, the cells were exposed to an agonist (as stated in the Results section). As controls, unstimulated cells, or cells exposed to buffer only were used; there was no significant difference between these types of controls. The incubation was stopped by a 4 second rinse in 0.15 M ammonium acetate to remove the salt-rich Krebs-Ringer solution, which would otherwise disturb the analysis. After the rinsing, the grids were blotted on filter paper and frozen in liquid propane cooled by liquid nitrogen and allowed to freeze dry at -80 °C in vacuum overnight. The freeze-dried grids were coated with a conductive carbon layer before analysis. X-ray microanalysis was performed at 100 kV in the TEM mode of a Philips 400 electron microscope as described above.

Primary cultures of endometrium

Female NMRI (Naval Medical Research Institute) mice (4-5 weeks old), obtained from B&K, Sollentuna, Sweden, were used for this part of the study. The animals were kept in separate cages until defecation habits became normal, which usually took 1 week.

Primary cultures of epithelial cells were isolated and cultured using a method modified from Glasser *et al.* (1988). Briefly, the uterine horns were rapidly removed, cut lengthwise, and placed (10 uterine horns from 5 mice) in 4 ml sterile Hank's balanced salt solution without Ca^{2+} and Mg^{2+} (HBSS -Ca^{2+} -Mg^{2+}) (Gibco) at room temperature. They were transferred to a 35 mm Petri-dish with 2 ml sterile enzyme solution containing 0.5% trypsin and 2.5% pancreatin which were prepared in HBSS (-Ca^{2+} -Mg^{2+}) and incubated for 60 min at 4°C and for 60 min at room temperature. The enzyme solution containing part of uterine epithelial cells was collected, centrifuged, and the supernatant was discarded. The uterine horns were washed using 3 ml HBSS on a vortex mixer. The washing solution was collected and used to suspend the epithelial cell pellet. The washing was repeated once. The epithelial pellet was washed again 2 more times using HBSS by centrifugation at 1200 rpm. The pellet was finally resuspended in 5 ml normal culture medium by gentle aspiration. The cell plaques were allowed to settle by gravity (10-15 min). The supernatant, approximately 4.5 ml, was discarded, and the remaining cells were resuspended in the culture medium (50 μl medium per uterine horn). A drop of the cell suspension (40-50 μl) was seeded out on a titanium grid as described above. The cells were allowed to attach and spread in a humidified chamber overnight at 37°C in an atmosphere of 5% CO_2/95% air. Then 2.5 ml culture medium was added to each culture dish. The medium was changed every 2 days and the cells were cultured for 6 days. The cells were washed in Krebs-Ringer buffer (KRB), pH 7.4. Experiments were carried out by exposing cells to stimulants or inhibitors of ion transport dissolved in KRB, and the incubation was terminated by a quick rinse in KRB solution followed by a quick rinse in the distilled water in order to remove the salts from the KRB solution. The grids were then blotted dry on a filter paper. The whole rinsing procedure did not take more than 10 seconds. The grids with the cells were frozen in liquid nitrogen and freeze-dried at -80°C at 10^{-5} torr overnight. For X-ray microanalysis, the specimens were viewed in the STEM mode of a Hitachi 7100 electron microscope and analyzed at an accelerating voltage of 100 kV with an Oxford Instruments (Oxford, UK) ISIS X-ray microanalysis system. Quantitative analysis was carried out as described above.

Immortalized sweat gland cell lines.

A human eccrine sweat gland cell line NCL-SG3 (Lee and Dessi, 1989) was used in the study. NCL-SG3 cells were cultured in William's E medium (Gibco) supplemented with penicillin (100 IU/ml), streptomycin (100 μg/ml), L-glutamine (2 mM), insulin (10 μg/ml), transferrin (10 μg/ml), hydrocortisone (5 ng/ml), epidermal growth factor (10 ng/ml) and sodium selenite (10 ng/ml). Cells grown on plastic were trypsinized with 0.05% trypsin in 0.02% ethylenediaminetetraacetic acid (EDTA) according to standard methods. As soon as the cells disengaged from the culture dish they were flushed with culture medium containing 10% fetal calf serum (FCS), collected in a sterile centrifugation tube and pelleted at low speed. The cell pellet was gently agitated in fresh medium and for X-ray microanalysis, the cells were then reseeded on titanium grids as described above for the primary cultures. The cells were allowed to attach and spread for 2 days at 37°C in a humidified atmosphere of 5% CO_2/95% air in a culturing chamber (Mörk and Roomans 1993; Von Euler *et al.* 1993). Stimulation experiments were carried out as described above, the cell cultures were rinsed with 0.15 M ammonium acetate or with distilled water (there was no difference between these methods) and X-ray microanalysis was carried out in the TEM mode at 100 kV as described above.

In some experiments, cells were also cultured on other substrates (Mörk and Roomans, 1993), namely, glass or plastic coverslips, or Transwell inserts (Costar). These cells were after freezing and freeze-drying analyzed in a Philips 525 scanning electron microscope at 20 kV, using a LINK AN 10000 energy-dispersive spectrometer system. Semi-quantitative analysis was carried out (because the analytical volume presumably included the substrate) by normalizing the values for the elemental concentrations to phosphorus and/or sulfur.

Tumor cell lines

The colonic tumor cell line T84 was cultured as described previously (von Euler and Roomans, 1992), and for X-ray microanalysis, the T84 cells were grown on Formvar-covered titanium grids and analyzed as described above.

Statistics

Comparisons between two experimental groups were carried out by Student's t-test; for comparisons between more groups, analysis of variance (ANOVA) was used, followed by Student's t-test.

Results

Incubated slices of liver and pancreas

Rapid dissection of the tissue did not cause marked changes in the elemental composition of the hepatocytes or the pancreatic acinar cells, but even a brief incubation in KRB caused an increase in Na and Cl and a decrease in K (Fig. 1). For the pancreas, incubation for 2h in standard KRB resulted in a decrease of Na and an increase of K, so that after 2h incubation the elemental composition was more close to the *in situ* values. For

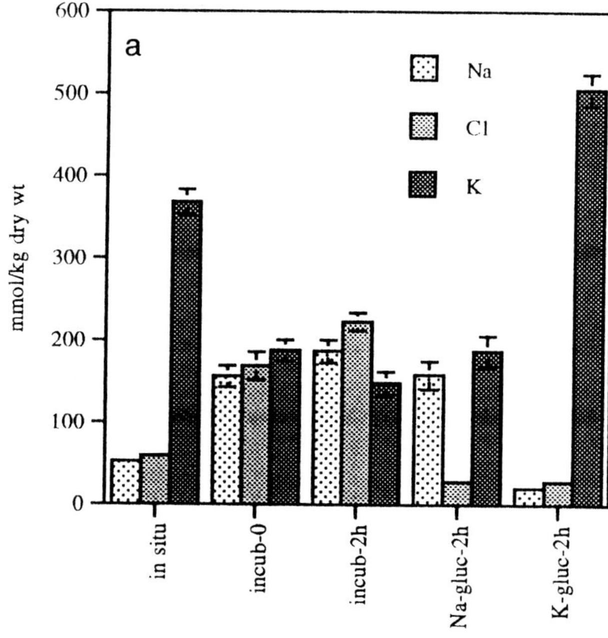

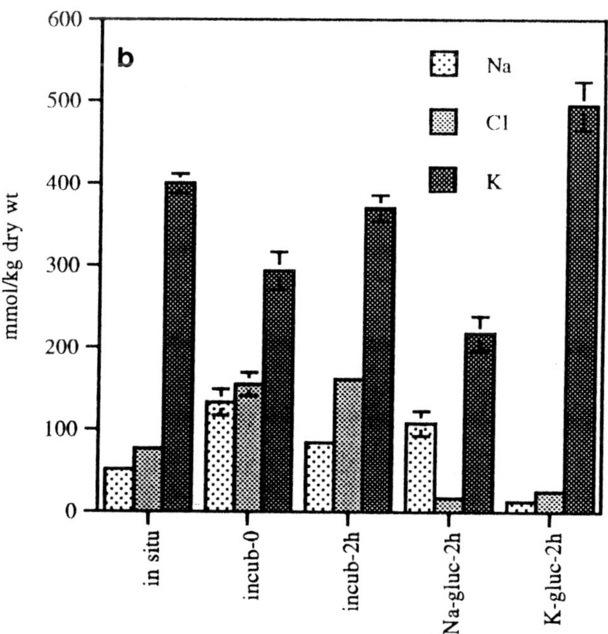

Figure 1: Effect of incubation (at 36°C) of (a) rat liver and (b) rat pancreas in standard KRB, compared to buffer in which the NaCl was replaced by Na gluconate or K gluconate, respectively. The following data are given: *in situ* concentrations, incubation for a few seconds in standard KRB (incubation-0), incubation for 2h in standard KRB (incubation-2h), incubation for 2h in Na-gluconate buffer (Na-gluconate) and incubation for 2h in K-gluconate buffer (K-gluconate). Data mainly from Hongpaisan and Roomans (1995).

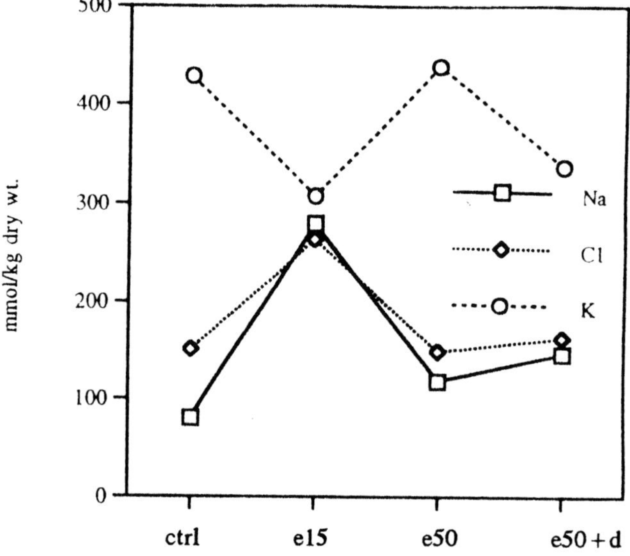

Figure 2: Concentrations of Na, Cl and K in the nucleus of submandibular gland acinar cells, determined on ultrathin cryosections, under the following conditions: *in situ* (ctrl), after 15 min collagenase treatment (e15), after 50 min collagenase treatment (e50), and after collagenase treatment and disaggregation (e50+d). Mean and standard error, data in mmol/kg dry weight.

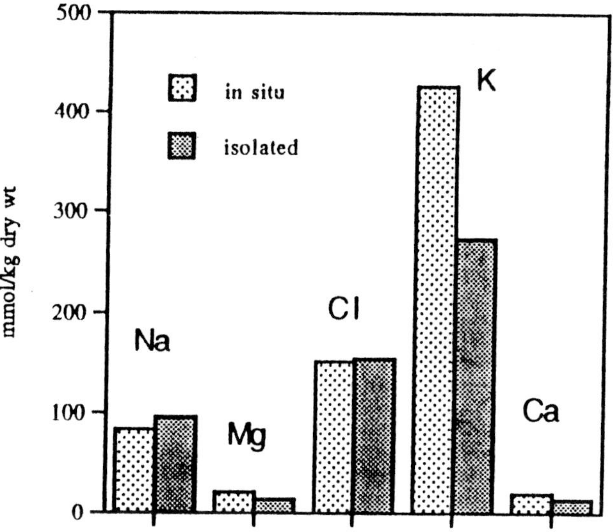

Figure 3: Nuclear elemental concentrations in a typical preparation of isolated submandibular gland acinar cells compared to data obtained from the gland *in situ*. Only for K, the isolated cells differ significantly from the *in situ* concentrations. Data in mmol/kg dry weight.

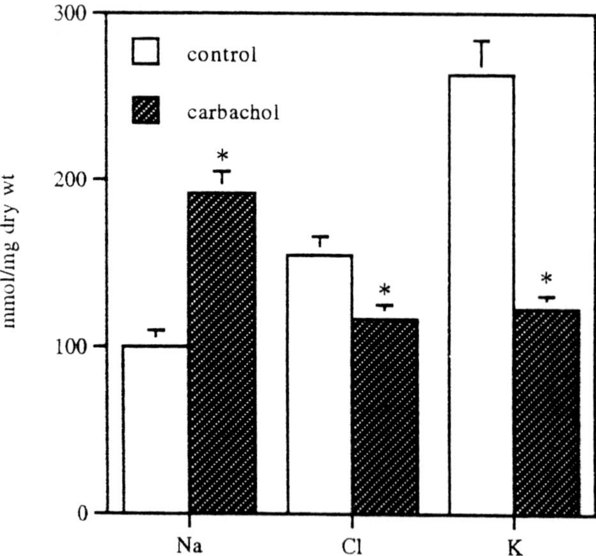

Figure 4. Effect of carbachol stimulation on the elemental concentrations in the nucleus of isolated submandibular gland acini. Data in mmol/kg dry weight, mean and standard error (number of measurements: 26 in each group), significance of differences indicated by an asterisk (*). Differences were considered significant for $p<0.05$.

liver, on the other hand, a further decrease of K and increase in Na and Cl was noted. Exchanging NaCl in the KRB for Na gluconate had only minor effects on the intracellular Na concentration. The exchange of chloride for gluconate resulted, regardless of the accompanying cation, in a decrease of the cellular Cl concentration. When NaCl was replaced by K gluconate, the intracellular K concentration increased, and intracellular Na decreased (Fig. 1). Marked oedema could be observed not only in tissues incubated in KRB, but also in the tissue incubated for 2 hours in Na gluconate or K gluconate (Hongpaisan and Roomans, 1995).

Isolated submandibular glands

While incubation with collagenase gave rise to a marked increase of the Na/K ratio of the acinar cells in comparison to the *in situ* situation, the value improved after prolonged incubation and even after the disaggregation step (Fig. 2). It was well possible to obtain an acinar preparation with elemental concentrations quite close to the in situ values (Fig. 3). Carbachol caused Cl^- and K^+ efflux and an increase in the intracellular Na^+ concentration (Fig. 4).

Isolated sweat glands

In one series of experiments, the effect of the isolation procedure on the elemental composition of sweat gland cells was investigated. Dissected sweat glands showed a Na/K ratio of about 2, which was much higher than literature data on *in situ* coils and ducts, which show Na/K ratios in the range of 0.2-0.3 (Wilson *et al.*, 1988). When the glands were incubated, the Na/K ratio increased additionally, to about 4. Addition of collagenase to the incubation medium had no significant effect on the Na/K ratio. After the glands had been incubated with collagenase, they could be separated into coil parts and duct parts. At this stage of the procedure, the sweat gland coils had a Na/K ratio of about 8, whereas the ductal part had an even higher Na/K ratio, about 16 (Fig. 5).

If at this stage, isolated ducts were stimulated with cAMP, a significant decrease of the intracellular Cl^- concentration was observed. Most of this decrease could be inhibited by the chloride channel blocker NPPB (5-nitro-2-(3-phenylpropylamino)-benzoicacid). Carbachol, on the other hand, did not cause a significant loss of Cl^- from duct cells. If, on the contrary, isolated coils were stimulated by cAMP, no significant effect on the intracellular Cl was noted. Carbachol, however, caused Cl^- efflux, which to a large part could be inhibited by NPPB (Fig. 6).

While coil cells could be routinely cultured, it proved to be very difficult to obtain primary cultures of the duct part.

Primary cultures of sweat gland cells

Stimulation with carbachol of cultured coil cells resulted in a significant decrease (about 60%) in the cellular content of Cl and a slight decrease in the cellular content of K and Na. This chloride efflux could be blocked by NPPB (Fig. 7). Also acetylcholine and the calcium ionophore A23187 stimulated chloride secretion (Mörk *et al.*, 1995). NPPB inhibited acetylcholine-induced chloride secretion. Stimulation of cultured coil cells with 5 mM cAMP caused a significant decrease in the cellular content of Cl and K. The decrease in Cl after stimulation with cAMP (about 30%) was, however, much smaller than after stimulation with carbachol (Fig. 7). In the control cells, the frequency distribution of the cellular Cl concentration displayed a bell-shape with a median concentration close to the mean concentration of Cl in the cultured cells. However, after stimulation with cAMP the frequency distribution of Cl showed two peaks, one with a maximum at the control value, and the other with a maximum at about 40-50% of the control value (Fig. 8). Hence, it appears, that only part of the cells secrete Cl^-. NPPB (but not another chloride channel blocker, diphenylamine-2-carboxylate (DPC)) inhibited cAMP-induced Cl^- secretion to a small extent.

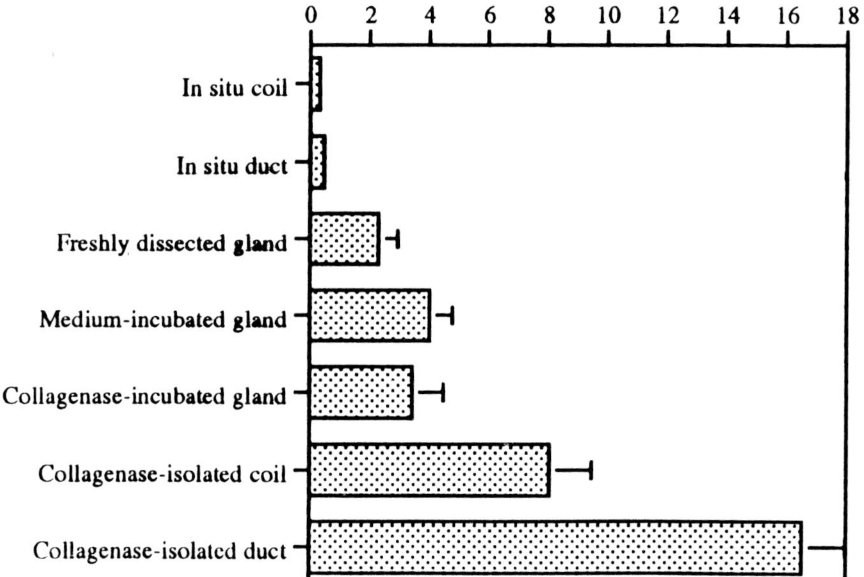

Figure 5. The Na/K ratio in the cytoplasm of cells in the human sweat gland. The data for *in situ* coil and duct were taken from Wilson *et al.* (1988). Data were obtained from freshly dissected glands (several hours after surgery), glands incubated in buffer, glands during collagenase treatment (no distinction made between coils and ducts in these cases), and isolated coils and ducts after collagenase treatment.

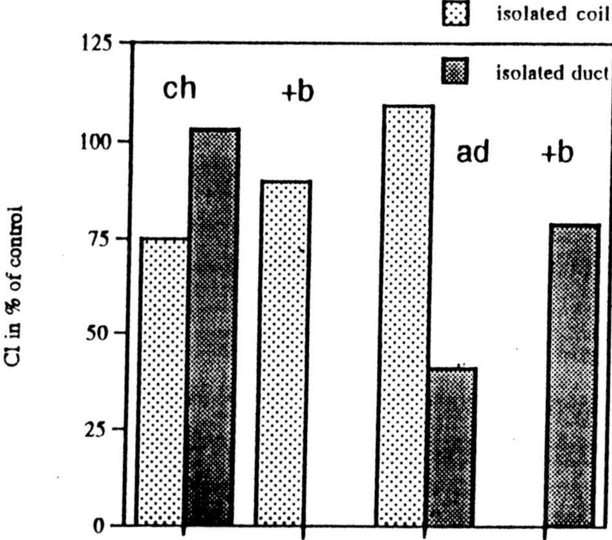

Figure 6. Intracellular Cl concentrations in sweat gland cells (data in % of unstimulated glands). Cholinergic stimulation (ch) causes a 25% loss of Cl from isolated coil cells, but no loss from isolated duct cells. In the isolated coil cells, the Cl loss can be reduced in the presence of the chloride channel blocker (+b) NPPB. β-Adrenergic stimulation (ad) causes no loss of Cl from isolated coil cells, but a substantial loss from the isolated duct cells. The Cl loss from the duct can be reduced by NPPB (+b).

Primary cultures of endometrium

The epithelial cells of the uterus had grown into continuous epithelium monolayer after 5 days in culture. Transmission electron microscopy showed that the cells had formed both apical microvilli and tight junctions (Fig. 9). Stimulation of the cells with cAMP caused a highly significant decrease of intracellular Cl and K and an increase in Na. Ouabain (10^{-4} M) caused an increase in Na and Cl and a decrease in intracellular K (Fig. 10).

Immortalized sweat gland cell lines

Stimulation of NCL-SG3 with cAMP caused a significant net Cl^- and K^+ secretion from the cells (Fig. 11). cAMP had no significant effect on the intracellular Na and Ca concentrations, although a tendency to increased Na levels could be observed. Cl^- secretion was blocked by the chloride channel blockers NPPB (Fig. 11), DPC, and 9-AC (Mörk *et al.*, 1996a). Administration of these chloride channel blockers (without cAMP) or of dimethylsulfoxide (DMSO) or ethanol in the concentration needed to dissolve the chloride channel blockers did not cause significant changes in ion content of the cells. The effect of cAMP was independent on the type of substrate used to culture the cells (Mörk and Roomans, 1993). Also carbachol caused a decrease of the intracellular Cl content, indicative of Cl^- efflux (Fig. 11). However, this was only the case for cells grown on permeable substrates (Formvar-film covered grids or Transwell inserts) but not for cells grown on impermeable substrate (glass or plastic cover slips) (Ring *et al.*, 1995).

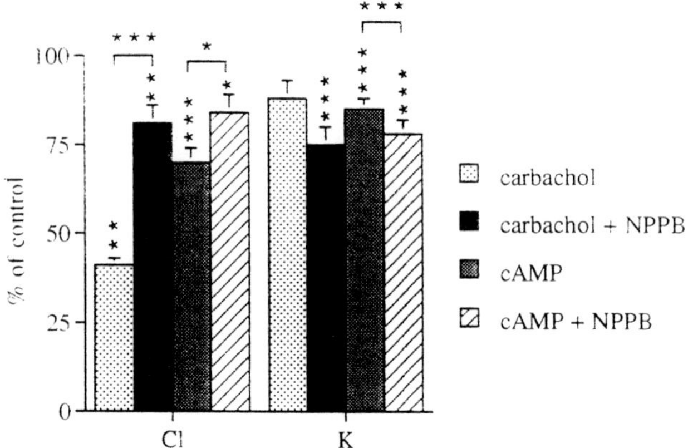

Figure 7. Primary cultures of human sweat gland coil cells (Mörk *et al.*, 1995). Carbachol stimulation results in a loss of about 60% of the intracellular Cl, and this loss can be greatly reduced by NPPB. cAMP causes only about 30% loss of Cl, also this loss can be reduced by NPPB. The significance of differences is given by asterisks (*, $p<0.05$; **, $p<0.01$; ***, $p<0.001$). Asterisks directly over the bar refer to differences between treatment and control, asterisks over brackets refer to differences between stimulated cells and cells stimulated in the presence of NPPB. The data are based on 3-5 experiments with 10-15 cells analyzed in each experiment; mean and standard error are given.

Tumor cell lines

The cell line T84 is derived from an adenocarcinoma. Both cAMP and ionomycin (which increases the intracellular Ca^{2+} concentration) cause Cl^- and K^+ efflux from T84 cells (Fig. 12).

Discussion

Many of the types of experiments described in this study could not, or only with great difficulty, be carried out in humans or animals *in situ*. The difficulties include controlling the concentration of the drug *in situ*, getting the drug to the correct location, systemic effects, and possible risks for the volunteers. As alternatives to *in situ* studies it can be attempted to use a variety of *in vitro* systems: slices, isolated glands, primary cultures, and cell lines.

The advantage of the tissue slice system is that it is technically simple. The data on the pancreas slices show that this system performs relatively well: although there are some initial changes in elemental concentration (increased Na and Cl, decreased K), the concentrations stabilize during the incubation in KRB. Similar results were obtained for the submandibular gland (Hongpaisan and Roomans, 1995). In that study it was also shown that the acinar cells in pancreas and submandibular gland react in the same way to cholinergic stimulation as the cells *in situ*. However, the liver (in common with brain tissue, Hongpaisan and Roomans, 1995) shows more extensive changes in elemental composition when slices

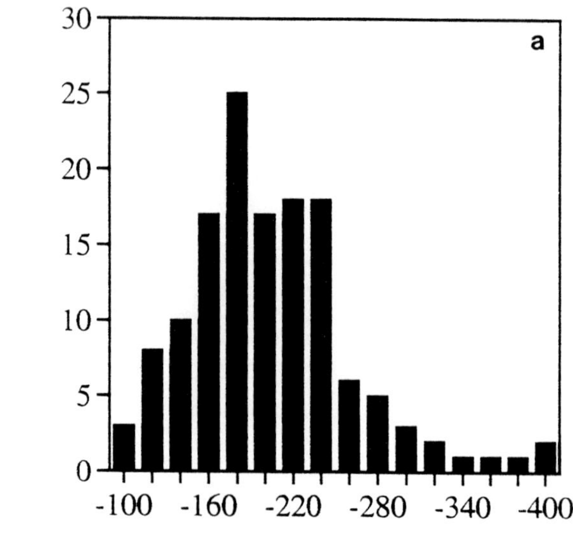

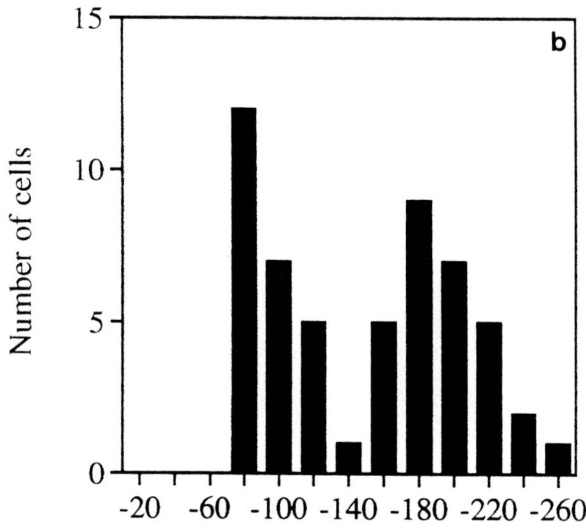

Figure 8. Primary cultures of human sweat gland coil cells (Mörk *et al.*, 1995). Frequency distribution of the Cl concentration in unstimulated cells (a) and in the cAMP-stimulated cells (b). About half of the cells lose about 60% of their Cl whereas the remaining half does not appear to lose any Cl. This results in an about 30% loss of Cl for the population as a whole as seen in Figure 7.

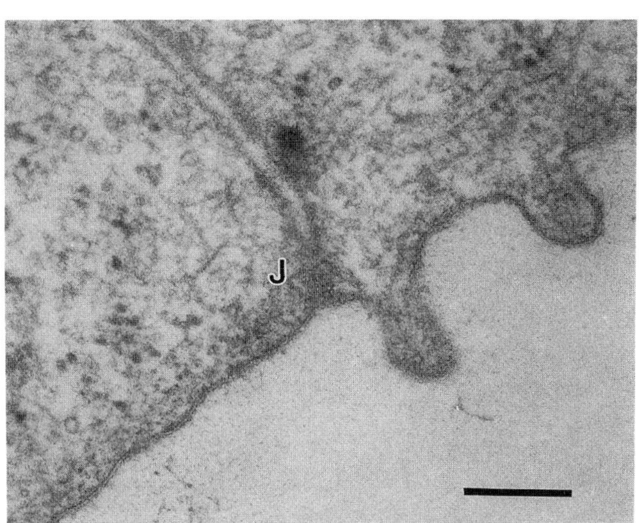

Figure 9. TEM micrograph of a primary culture of uterine epithelial cells, showing a junctional complex between cells (J). Bar is 1 μm.

are incubated in KRB buffer. The disadvantage of slices is that diffusion of oxygen and nutrients is limited, resulting in decreased activity of ATPases such as the Na^+-K^+-ATPase and the Ca^{2+}-ATPase. The resulting elemental changes are typical for this kind of damage.

It is of interest that the elemental content of the cells is preserved best when the slices are stored in a medium with a K/Na ratio closer to the intracellular ratio (instead of in a solution resembling the extracellular environment). This method has been used in some biochemical studies (Elliott, 1969) and is, e.g., used in the storage of tissue for transplantation (Wahlberg *et al.*, 1986). Similarly, Sjöstrand (1992) has suggested that it should be possible to reduce the so-called "infarct-reflux" damage to heart tissue by perfusion with K-rich instead of Na-rich fluids. Exchanging NaCl for K gluconate results in a high intracellular K/Na ratio; it is not unlikely that K^+ ions diffuse into the cell from the extracellular medium. On the other hand, chloride ions slowly diffuse out of the cells into the gluconate medium. Morphological studies show, however, that despite the "normal" intracellular K/Na ratio, significant oedema occurs even in K gluconate medium (Hongpaisan and Roomans, 1995), and it is likely that the cell membrane, under the conditions of the experiment, is not in full control of the ion fluxes.

Also during isolation of acini from the rat submandibular gland and ducts or secretory coils, respectively, from the human sweat glands, marked ionic changes take place, especially in the sweat glands. With regard to the submandibular gland acini, it appears possible to

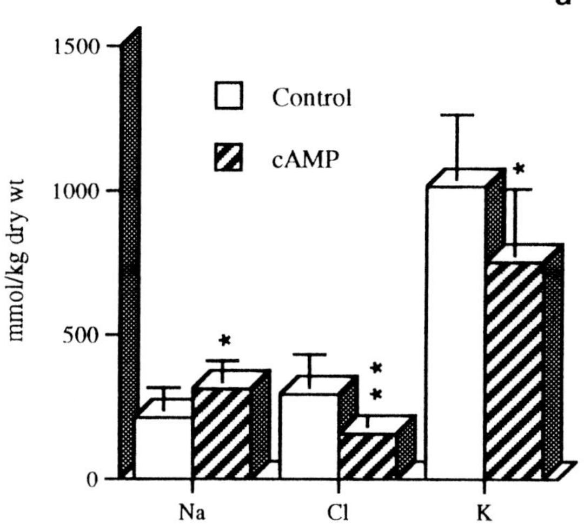

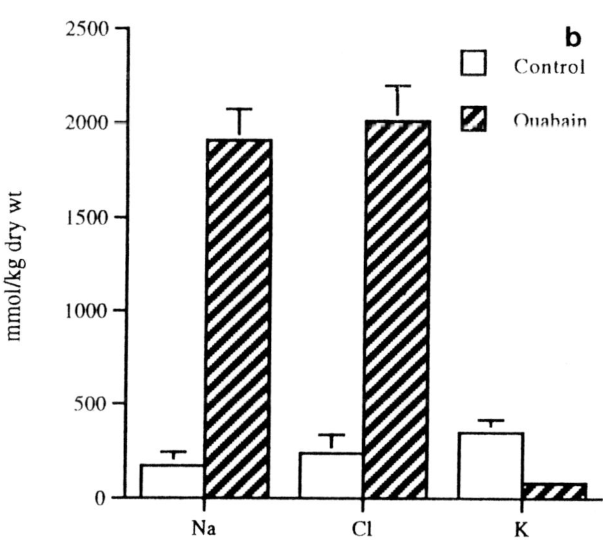

Figure 10. (a) Effect of cAMP (5 mM, 5 min), (b) effect of ouabain (50 μM, 5 min) on the elemental composition of cultured uterine epithelial cells; mean ± standard deviation (n=25), data in mmol/kg dry weight. The data are based on 25 spectra per treatment. Significant differences are indicated by * ($p<0.01$), ** ($p<0.05$) and *** ($p<0.001$).

have a cell preparation that in its ionic content resembles the *in situ* elemental content, and reacts in a similar way to a cholinergic stimulus, even though a transient increase of the Na and Cl concentration takes place during the isolation procedure. It is likely that this preparation can be relatively easily used for *in vitro* experiments. With regard to the human sweat glands, the situation is much more difficult. The increase in Na and Cl, and the decrease in K are very marked in the coil

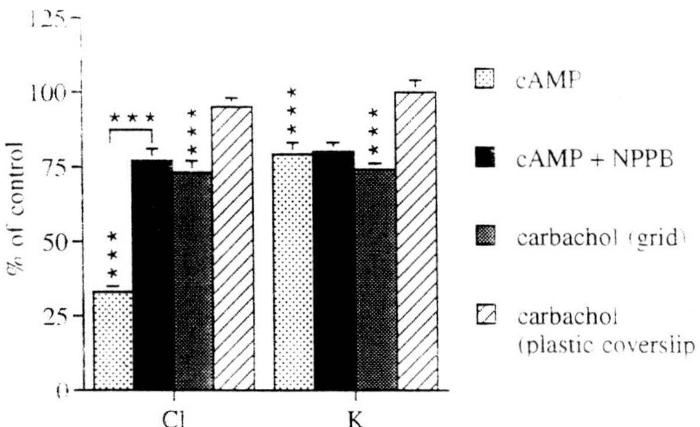

Figure 11. Immortalized human sweat gland cell line NCL-SG3 (Mörk *et al.*, 1996a; Ring *et al.*, 1995). Stimulation with cAMP results in a considerable loss of Cl (data in % of unstimulated cells). This loss can be reduced by the chloride channel blocker NPPB. Stimulation by cAMP also leads to a loss of K. On cells cultured on Formvar-film covered titanium grids, carbachol induced loss of Cl and K, but this was not the case in cells cultured on an impermeable substrate, such as a plastic coverslip. The significance of differences is given by asterisks (*, $p<0.05$; **, $p<0.01$; *, $p<0.001$). Asterisks directly over the bar refer to differences between treatment and control, asterisks over brackets refer to differences between stimulated cells and cells stimulated in the presence of NPPB. The data are based on 4-6 separate experiments, with a total of 20-80 cells for each treatment. Mean and standard error are given.

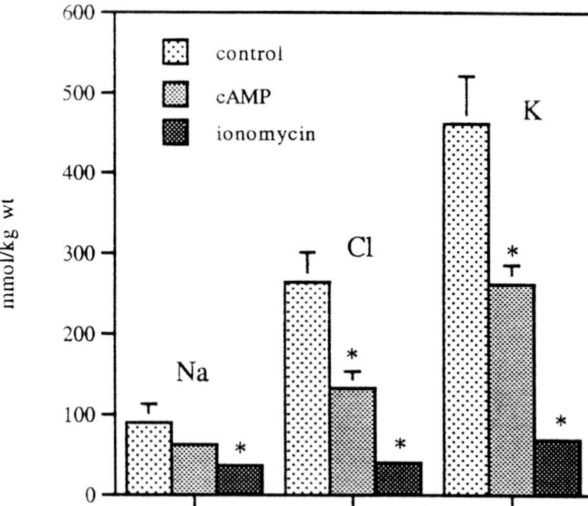

Figure 12. Effect of stimulation with cAMP and ionomycin on cultured T84 cells. Data in mmol/kg dry weight, mean and standard error (number of measurements 35 in each group, two separate experiments), significance of differences indicated by an asterisk (*). Differences were considered significant for $p<0.05$.

and even more pronounced in the duct. It appears that the changes in the duct are mostly non-reversible, and this may be the explanation for the fact that it was difficult to prepare primary cultures of duct cells. The coil cells are evidently damaged to a lesser extent, and coil cells can be more easily cultured.

The sweat gland can be functionally divided into a secretory coil, where the primary sweat is produced under cholinergic control with Ca^{2+} as a second messenger, and a duct where Cl^- and Na^+ are reabsorbed under β-adrenergic control with cAMP as a second messenger. It is of interest that the specificity of activation of chloride efflux in the different parts of the sweat gland is retained during the isolation, but that it is lost in the primary cultures of sweat gland cells, where both cAMP- and Ca^{2+} activated chloride channels are present, at least in part of the cells. Analysis of a cell population by X-ray microanalysis provides a unique opportunity to demonstrate a heterogeneity in the cell population (Von Euler *et al.*, 1993). In this particular experiment it can be seen that about half of the cells does not react to stimulation with cAMP, but that the other half of the cells reacts and loses about the same relative amount of Cl compared to cholinergic stimulation. Also in the immortalized NCL-SG3 sweat gland cell line (Lee and Dessi, 1989) which was thought to originate from duct cells, both cAMP- and Ca^{2+}-activated chloride channels have been demonstrated (Mörk *et al.*, 1996a; Ring *et al.*, 1995). Using the patch-clamp technique, Mörk *et al.* (1996a) could show that increasing the cellular Ca^{2+} concentration by calcium ionophores activated chloride channels, but that carbachol had no effect. This was in apparent contradiction with the data obtained by X-ray microanalysis. Patch-clamp experiments are, however, carried out with cells cultured on an impermeable substrate. When the X-ray microanalysis experiments were repeated on cells cultured on some types of impermeable substrate, no effect of carbachol could be seen, whereas cells cultured on permeable substrates consistently showed activation of Cl^- efflux by carbachol. These results presumably indicate that Ca^{2+}-activated chloride channels are indeed present in NCL-SG3 cells (despite the fact that these cells are of ductal origin), but that the cholinergic receptor is on the side of the cells touching the substrate and cannot be reached by cholinergic agonists when the cells are grown on impermeable substrates. When cells are grown on permeable substrates, the receptors are more accessible, and the cells can be stimulated by cholinergic agonists to secrete chloride. cAMP-activation of Cl^- efflux can, however, be obtained both in cells grown on permeable and in cells grown on impermeable substrates (Mörk and

Roomans, 1993). In principle, it is not uncommon for epithelial cells to have mechanisms for both a Ca^{2+}-activated and a cAMP-activated chloride secretion, as is shown in the experiments on T84 cells where both cAMP and ionomycin (presumably via an increase in intracellular free Ca^{2+}) cause chloride efflux. *In situ*, in general, one of the mechanisms appears to predominate, e.g., in the colonic epithelium, cAMP-induced chloride efflux predominates. Why this is not exactly the same in the cultured counterparts is a very interesting question.

Relatively little is known about ion transport mechanisms in the uterine epithelium. Matthews *et al.* (1993) have proposed that Na^+ absorption by this epithelium plays an important role in the regulation of the composition of the uterine fluid. It is well known that female patients with cystic fibrosis have a decreased pregnancy rate, which is suspected to be associated with altered quality of the cervical mucus and altered ion concentrations in the uterine fluid (Brugman and Taussig, 1984). Since it is known that the basic defect in cystic fibrosis is a defective cAMP-regulated chloride channel, it is not unreasonable to suspect that the cells of the uterine epithelium possess a cAMP-activated chloride channel. The results of the X-ray microanalysis bear this out.

Acknowledgements

Work reported in this review was supported by grants from the Swedish Medical Research Council (project 12X-07125) and the Swedish National Heart and Lung Foundation.

References

Abraham EH, Breslow JL, Epstein J, Chang-Sing P, Lechene C (1985). Preparation of individual human diploid fibroblasts and study of ion transport. Am J Physiol **248**: C154-C164.

Ballyk BA, Goh JW (1992) Elevation of extracellular potassium facilitates the induction of hippocampal long-term potentiation. J Neurosci Res **33**: 598-604.

Borgmann S, Granitzer M, Crabbé J, Beck FX, Nagel W, Dörge A (1994) Electron microprobe analysis of electrolytes in whole cultured epithelial cells. Scanning Microsc Suppl **8**: 139-148.

Brugman S, Taussig L (1994) The reproductive system. In: Cystic Fibrosis. Taussig LM, ed. Thieme-Stratton, New York, 323-337.

Elliott KAC (1969) The use of brain slices. In: Handbook of Neurochemistry, vol. II. Lajtha A, ed. Plenum Press, New York, 103-114.

Glasser SR, Julian JA, Decker GL, Tang JP, Carson DD (1988) Development of morphological and functional polarity in primary cultures of immature rat uterine epithelial cells. J Cell Biol **107**: 2409-2423.

Hongpaisan J, Roomans GM (1995) Use of post-mortem and *in vitro* tissue specimens for X-ray microanalysis. J Microsc **180**: 93-155.

Hongpaisan J, Mörk AC, Roomans GM (1994) Use of *in vitro* systems for X-ray microanalysis. Scanning Microsc Suppl **8**: 109-116.

James-Kracke MR, Sloane BF, Shuman H, Karp R, Somlyo AP (1980) Electron probe analysis of cultured vascular smooth muscle. J Cell Physiol **103**: 313-322.

Lee CM, Dessi J (1989) NCL-SG3: a human eccrine sweat gland cell line that retains the capacity for transepithelial ion transport. J Cell Sci **92**: 241-249.

Matthews CJ, McEwan GTA, Redfern CPF, Thomas EJ, Hirst BH (1993) Bradykinin stimulation of electrogenic ion transport in epithelial layers of cultured human endometrium. Pflügers Arch **422**: 401-403.

McIlwain H (1987) Tissue slice techniques in neuroscience. In: Encyclopeadia of Neuroscience, vol 2. Adelmann A, ed. Birkhäuser, Boston-Stuttgart. pp 1219-1220.

McMillan EB, Roomans GM (1990) Techniques for X-ray microanalysis of intestinal epithelium using bulk specimens. Biomed Res (India) **1**: 1-10.

Mörk AC, Roomans GM (1993) X-ray microanalysis of cAMP-induced ion transport in NCL-SG3 sweat gland cells. Scanning Microsc **7**: 1233-1240.

Mörk AC, Hongpaisan J, Roomans GM (1995) Ion transport in primary cultures from human sweat gland coils studied with X-ray microanalysis. Cell Biol Int **19**: 151-159.

Mörk AC, von Euler A, Roomans GM, Ring A (1996) cAMP-induced chloride transport in NCL-SG3 sweat gland cells. Acta Physiol Scand **157**: 21-32.

Ring A, Mörk AC, Roomans GM (1995) Calcium-activated chloride fluxes in cultures NCL-SG3 sweat gland cells. Cell Biol Int **19**: 265-278.

Roomans GM (1988) Quantitative X-ray microanalysis of biological specimens. J Electron Microsc Techn **9**: 19-44.

Roomans GM (1991) Cryopreparation of tissue for clinical applications of X-ray microanalysis. Scanning Microsc Suppl **5**: S95-S107.

Sagström S, Roomans GM, Wroblewski R, Keulemans JLM, Hoogeveen AT, Bijman J (1992) X-ray micro-analysis of cultured respiratory epithelial cells from patients with cystic fibrosis. Acta Physiol Scand **146**: 213-220.

Sjöstrand FS (1992) Information and misinformation regarding ischemia of heart muscle tissue. The cause of cell death during blood reperfusion and reactivation of heart muscle tissue after prolonged ischemia. Scanning Microsc **6**: 1041-1060.

Von Euler A, Roomans GM (1992) Ion transport in

colon cancer cell cultures studied by X-ray microanalysis. Cell Biol Int Rep **16**: 293-306.

Von Euler A, Wróblewski R, Roomans GM (1992) Correction for extraneous background in X-ray microanalysis of cell cultures. Scanning Microsc **6**: 451-456.

Von Euler A, Pålsgård E, Vult von Steyern C, Roomans GM (1993) X-ray microanalysis of epithelial and secretory cells in culture. Scanning Microsc **7**: 191-202.

Wahlberg J, Southard JH, Belzer FO (1986) Development of a cold storage solution for pancreas preservation. Cryobiology **23**: 477-482.

Warley A (1994) The preparation of cultured cells for X-ray microanalysis. Scanning Microsc Suppl **8**: 129-138.

Wilson SM, Elder HY, Sutton AM, Jenkinson DMcE, Cockburn F, Montgomery I, McWilliams SA, Bovell DL (1988) The effects of thermally-induced activity *in vivo* upon the ultrastructure and Na, K and Cl composition of the epithelial cells of sweat glands from patients with cystic fibrosis. Tissue & Cell **20**, 691-700.

Wroblewski J, Roomans GM (1984) X-ray microanalysis of single and cultured cells. Scanning Microsc 1986; IV: 1875-1882.

Wroblewski J, Müller RM, Wróblewski R, Roomans GM (1983) Quantitative X-ray microanalysis of semi-thick cryosections. Histochemistry **77**: 447-463.

Wróblewski R, Roomans GM, Jansson E, Edström L (1978) Electron probe X-ray microanalysis of human biopsies. Histochemistry **55**: 281-292.

Wróblewski R, Wroblewski J, Roomans GM (1987) Low temperature methods in biological microanalysis. Scanning Microsc **1**: 1225-1240.

Wróblewski R, Wroblewski J (1994) X-ray microanalysis of endocrine, exocrine, and intestinal cells and organs in culture: technical and physiological aspects. Scanning Microsc Suppl **8**: 149-162.

Zierold K, Schäfer D (1988) Preparation of cultured and isolated cells for X-ray microanalysis. Scanning Microsc **2**: 1775-1790.

Discussion with Reviewers

K. Izutsu: With regard to the sweat gland duct, it is stated that there is an inwardly directed Cl^- channel in the apical membrane, and an outwardly directed Cl^- channel in the basolateral membrane, and that together these channels effect Cl^- resorption from the primary sweat. If the composition of the primary sweat is isotonic, how can passive forces cause an influx at one surface, and an efflux at the other? Is the membrane potential different at the two surfaces?

Authors: The accepted model for NaCl uptake from the sweat gland duct (Quinton, 1990) involves passive diffusion of Na^+ across the apical membrane into the duct cell through an amiloride-sensitive pathway and an active extrusion of Na^+ from the cell across the basolateral membrane requiring an ouabain-sensitive Na^+-K^+-ATPase. The normal sweat duct appears to be highly conductive to Cl^- so that its movement seems to be governed by the electrochemical gradients established by active Na^+ transport, and Cl^- moves passively across both the apical and the basolateral membrane down electrochemical gradients. The chloride channel CFTR (cystic fibrosis transmembrane regulator) has been shown to be present in both apical and basolateral membranes of sweat duct cells (Cohn *et al.*, 1991; Denning *et al.*, 1992).

H.Y. Elder: I do not agree that your results obtained with the primary culture of sweat gland coil cells indicate specificity being lost during culture. In the secretory coil of human sweat glands cholinergic and adrenergic α and β pathways are well known to exist (Sato, 1977; Bovell *et al.*, 1989).

Authors: Indeed, we accept the evidence that coil cells are sensitive to β-adrenergic activation, but the references you quote show that the role of β-adrenergic activation is only minor. The effect was not sufficient to be seen in our experiments on isolated coils. In the experiment on cultured cells, the effects of the Ca^{2+}-induced (cholinergic) and the cAMP-induced (β-adrenergic) pathway appeared to be in the same order of magnitude. Possibly we should qualify our statement to say that we do not mean absolute specificity but rather the clear predominance of one system over the other.

K. Izutsu: The KRB contained no bicarbonate, yet it was oxygenated with 95% O_2 and 5% CO_2. Was the pH of this solution stable over a long time?

Authors: In the incubation experiments, the tissue slices were superfused with KRB at a rate of 2 ml/min, to mimic circulation that would provide oxygen and nutrients and clear the waste products produced by the cells. Moreover, the KRB used in this study contained HEPES (Good *et al.*, 1966), which is an effective hydrogen buffer. The pH was constant over the incubation time.

K. Izutsu: Incubation was stopped by a 4 sec rinse in ammonium acetate or in distilled water. Were the sweat glands sensitive to hypotonic conditions? Many cell types would undergo regulated volume changes under these conditions or would increase their cytosolic Ca^{2+} concentrations. Was there any evidence for any of these changes, or perhaps morphological findings of cell swelling?

K. Zierold: Have you found changes of intracellular dry

Table 1. Elements in nucleus and cytoplasm in submandibular gland acini *in situ* and at the end of the isolation procedure.

	gland *in situ*		isolated acini	
	nucleus	cytoplasm	nucleus	cytoplasm
Na	80 ± 7	74 ± 6	106 ± 7	72 ± 6 ***
P	603 ± 33	418 ± 27 ***	433 ± 14	253 ± 14 ***
Cl	151 ± 7	150 ± 7	126 ± 7	107 ± 8
K	427 ± 27	354 ± 19 *	252 ± 10	216 ± 12 *

Data given as mean and standard error, about 40 measurements in each group. Significant differences between nucleus and cytoplasm are indicated by * ($p<0.05$) and *** ($p<0.001$), respectively.

mass or water portion depending on experimental conditions?

Authors: We realize that rinsing the cells with ammonium acetate or distilled water can induce artefacts. The ammonium acetate is (about) isotonic and no volume changes would therefore be expected, but the possibility of pH changes cannot be excluded. On the other hand, washing with distilled water could induce osmotic effects. For the sweat gland cell line, we determined that there were no significant differences between the washing fluids, and the same was previously shown for fibroblasts (Roomans, 1991; text reference). It should be realized though, that we measure freeze-dried cells: any (transient) changes in water content of the cells during rinsing (or as a result of other experimental conditions) will not be apparent after freeze-drying. Damage such as bursting of the cells was not observed. No significant changes in elemental composition after rinsing with ammonium acetate or distilled water, respectively, were observed. It should also be taken into consideration that controls and experimentally treated cells were rinsed in the same way, and that minor artefacts caused by the rinsing procedure therefore would cancel out.

H.Y. Elder: Some of the cryofixation methods selected would seem to be well short of the best that could have been applied, e.g., the brass clamps used to freeze the tissue slices, or the endometrial primary cultures frozen by plunging into liquid nitrogen. Why were these methods selected in preference to proven better methods? Did you examine the ice crystal artefact sizes in the analyzed tissues? What probe diameters were employed and what was the relation of these to the phase separation artefacts found?

Authors: X-ray microanalysis of the incubated tissue slices was carried out on 16 μm freeze-dried cryosections, in which the analyzed volume is not less than 8-10 μm due to scattering of the electron beam in the tissue, even if the initial probe is narrow. Ice crystal remnants are well below this size. Also because sectioning is carried out at relatively high temperatures (-30°C), there is no need to use the most sophisticated freezing techniques. Because of their small size, cell cultures are easy to cryofix and again, in the analysis of cultured cells the analyzed volume comprises a large part of the cells, and ice crystals are small in relation to the probe volume.

H.Y. Elder: You mention oedema in the context of tissue slices. Have you measured the degree of swelling and/or correlated it with any of the imposed ionic changes?

Authors: We have so far only qualitatively shown oedema in micrographs (Hongpaisan and Roomans, 1995); a more quantitative study is under way. Quantitative data on oedema in different storage fluids on a macroscopic scale have been published by Wahlberg *et al.* (1986) (text reference).

K. Izutsu: The finding that submandibular acini have significantly altered K^+ concentrations is an interesting and important finding in that it indicates that significant changes in K^+ concentrations occur during isolation of acini. Hence, the limit of these preparations should be borne in mind when performing experiments with dispersed acini.

Authors: It appears that it is not uncommon for *in vitro* preparations to have different Na, K and Cl concentrations compared to tissue *in situ*. This is not only valid for isolated submandibular acini, but also for many, if not all, other tissues. Possibly, insufficient attention has been paid to this in the past. However, this does not mean that it is impossible to use *in vitro* preparations.

R. Wróblewski: When working with tissue slices or with cultures from dissected organs, one is always wondering what happens during the phase which we call for incubation or stabilization. Several morphological studies have shown rather extensive changes at the morphological level. Still, cells in the slices and in the organ cultures do react to different types of stimuli several days after they have been removed from the body. Could you comment on the following issues: (a) How quick is the diffusion of the stimulus in a slice of 1 mm in diameter in the tissue slices of the organs you have been investigating (depth versus time)?; (b) Do you expect any differences in the elemental content of mobile ions in the nucleus and cytoplasm after incubation for 2-3 hours in the incubation buffer?

Authors: We cannot give an exact answer to your question about the diffusion rate of the stimulus in tissue slices, since we have not measured this parameter. We analyzed cells within 240 μm of the surface; the stimulus could reach the cells either by diffusion or by propagation via gap junctions. It is evident that diffusion becomes less of a problem with thinner *in vitro* samples, and that therefore e.g., isolated submandibular gland acini are more suitable as an experimental system than a slice of submandibular gland. We think that in general, the nuclear membrane is freely permeable to Na^+, Cl^- and K^+, and that this does not change during incubation. In the isolated submandibular gland acini, where we have measured both nucleus and cytoplasm, the nucleus had (somewhat) higher concentrations of Na, Cl. K and P than the cytoplasm; this was the case both in the *in situ* gland and at the end of the incubation (Table 1).

K. Zierold: Are the intracellular elemental contents compiled in the diagrams shown, distributed homogeneously in the cells? Have you observed intracellular gradients, storage or binding sites?

Authors: Measurements at the subcellular level were only carried out in the isolated sweat glands and in the submandibular gland *in situ* and in the isolated acini. In the sweat gland we could not find significant differences in elemental concentrations between nucleus and cytoplasm, except for P, which was higher in the nucleus. Results for the submandibular gland are shown in Table 1. A more detailed study of the submandibular gland *in situ* is in press (Mörk *et al.*, 1996b).

H.Y. Elder: You quote Na/K ratios of up to 16 from the isolated sweat glands and imply that these exceptionally high ratios could be correlated with the preparative state. You indicate that such ratios are not simply indications of moribund cells since such cells could subsequently be cultured (though rarely the duct cells with the highest ratios). However, other authors (e.g., Zierold *et al.*, 1994) have taken Na/K ratios of higher than about 0.3 to indicate poor preparative technique and damaged physiological state. Could you comment further?

Authors: There is no question that a high Na/K ratio indicates a damaged physiological state. However, our data, e.g., from the submandibular gland acini, show that cells can recover from a temporary rise of the intracellular Na/K ratio, and the isolated acini have a Na/K ratio of 0.4. Also, within a population of cells with a mean Na/K ratio of 16, there are cells with an even higher and cells with a lower Na/K ratio value. It is not unlikely that the cells that manage to survive and grow, belong to the group with relatively low Na/K ratios in this population.

K. Zierold: You report that the elemental content of cells in tissue slices is preserved best when the slices are stored in a medium of high K/Na ratio. However, under these conditions a high intracellular K/Na ratio can be expected even in cells with completely destroyed membranes! Do you have any other indications for preservation of the vitality of cells incubated in a medium with high K content?

J. Beesley: The authors state that the elemental content of the cells is preserved best when the slices are stored in a medium with a K/Na ratio close to the intracellular ratio instead of a solution resembling the extracellular environment and suggest that under the conditions of the experiment it is likely that the cell membrane is not in full control of the ion fluxes. Would this condition have any adverse effect on the reaction of cells to stimuli?

Authors: We agree that under these experimental conditions the high intracellular K/Na ratio is no evidence for proper functioning of the membrane or even of the vitality of the cell. However, the fact that storage of tissue in solutions with a high K/Na ratio experimentally has been found to work well during organ transplants indicates that tissues can retain their viability in solutions with a high K/Na ratio. Whether the reaction of the tissue to stimuli would be affected by the extracellular K/Na ratio would depend on the tissue and the stimulus under investigation. Certainly all reactions to stimuli that are dependent on membrane potential or influx of Na^+ would be affected by changes in the extracellular fluid. Therefore, we would not recommend carrying out stimulation experiments in a medium with a high ratio, but it would be interesting to test whether cells that have been stored for some time in a medium with high K/Na ratio could be transferred to extracellular fluid with a low K/Na ratio, and if they would perform better than cells kept all the time in fluid with a low K/Na ratio. This will be the subject of further experiments.

K. Izutsu: The ability to demonstrate a heterogeneity in the cell population is also a unique application of the microanalysis technique. Hence, experimenters who perform studies using cell suspensions should be aware of possible complications arising from working with a heterogeneous population of cells.

Authors: We agree with this comment, and would like to add that the same is valid for e.g., Ussing chamber experiments with cell monolayers.

Additional References

Bovell DL, Elder HY, Jenkinson DMcE, Wilson SM (1989) The control of potassium ($^{86}Rb^{+}$) efflux in the isolated human sweat gland. Quart J Exp Physiol **74**: 267-276.

Cohn JA, Melhus O, Page LJ, Dittrich KL, Vigna SR (1991) CFTR: development of high-affinity antibodies and localisation in sweat gland. Biochem Biophys Res Comm **181**: 36-43.

Denning GM, Ostedgaard LS, Cheng SH, Smith AE, Welsh MJ (1992) Localisation of cystic fibrosis transmembrane conductor in chloride secretory epithelia. J Clin Invest **89**: 339-349.

Good NE, Winget GD, Winter W, Connolly TH (1966) Hydrogen ion buffers for biological research. Biochemistry **5**: 467-477.

Mörk AC, Zhang A, Martinez JR, Roomans GM (1996b) Chloride secretion in the submandibular gland of adult and early postnatal rats studied by X-ray microanalysis. Histochem Cell Biol, in press.

Quinton PM (1990) Cystic fibrosis: a disease in electrolyte transport. FASEB J **4**: 2709-2717.

Sato K (1977) The physiology, pharmacology and biochemistry of the eccrine sweat gland. Rev Physiol Biochem Pharmacol **79**: 51-131.

Zierold K, Hentschel H, Wehner F, Wessing A (1994) Electron probe X-ray microanalysis of epithelial cells: aspects of cryofixation. Scanning Microsc Suppl **8**: 117-127.

Scanning Microscopy Supplement 10, 1996 (pages 375-386)
Scanning Microscopy International, Chicago (AMF O'Hare), IL 60666 USA

0892-953X/96$5.00+.25

ASPECTS OF CRYOFIXATION AND CRYOSECTIONING FOR THE OBSERVATION OF BULK BIOLOGICAL SAMPLES IN THE HYDRATED STATE BY CRYOELECTRON MICROSCOPY

K. Richter*

University of Lausanne, Lausanne, Switzerland

(Received for publication September 17, 1996 and in revised form December 17, 1996)

Abstract

Cryoelectron microscopy allows the observation of hydrated samples at high spatial resolution, and it would be of great interest in biology to apply this method to cells and tissues. However, because of technical problems, the cryo-observation of frozen hydrated ultrathin sections of bulk material has not become an established method. The major limitations are due to the difficulty of achieving the vitrification of such material, and the structural deformation caused by ultrathin sectioning: 1. The vitrification of cells in a physiological environment requires high-pressure freezing. However, new results suggest that the pressure may alter the ultrastructure of the sample. 2. Cryosectioning compresses structures in the cutting direction about 40%. This deformation does not necessarily destroy the character of macromolecular assemblies, but since it depends on the properties of the material, internal standards cannot be used to correct for the deformation of all the structures in a cell.

Key Words: Cryofixation, cryosectioning, cryoelectron microscopy, vitrification, plunge-freezing, high-pressure freezing.

*Address for correspondence:
K. Richter
Biomedical Structure Analysis Dept. 0195
Im Neuenheimer Feld 280
D-69120 Heidelberg
Germany

Telephone Number: +49(6221)423263
FAX Number: +49(6221)423459
E-mail: k.richter@dkfz-heidelberg.de

Introduction

Biological specimens are not well suited to electron microscopy at room temperature. One major restriction is the vacuum in the microscope column, which does not allow to observe water at room temperature. For this reason, biological specimens are usually observed as dry artifacts. Operating the microscope at temperatures below 170 K, when water is stable in the vacuum, does not improve the situation. When water crystallizes to ice, the biological matrix separates as a dry network. However, if the cooling is quick enough, water is prevented from transforming into the crystalline solid, and if this highly viscous vitreous state is reached, the fully hydrated ultrastructure of the biological samples can be studied by cryoelectron microscopy (Dubochet and McDowall, 1981).

Cryofixation is much quicker than chemical fixation, ion-gradients are not washed out, the antigenicity of epitopes is not altered due to reactions with chemical fixatives, and no external mass is added as it is with the chemical crosslinkers. For these reasons, cryofixation is also employed in other electron microscopy techniques that end up with a dry specimen, e.g., freeze-substitution and freeze-drying. Since high-resolution structural preservation is sacrificed by dehydration, vitrification is not the necessary aim with these techniques: Cryofixation is satisfactory as long as displacements of the organic matter due to ice-crystal growth are beyond the other resolution limits of the method. If, on the other hand, the structures are to be observed in the hydrated state, vitrification is a prerequisite.

Cryoelectron microscopy of vitrified biological samples is well established for objects small enough to be observed directly without the need for ultrathin sectioning. These specimens are prepared as a thin layer of aqueous suspension of about 80 nm in thickness and 1 μm in diameter spanned throughout the holes of a perforated carbon film. The thin layer is reproducibly vitrified by plunge-freezing of the grid into liquid ethane at 110 K (Adrian *et al.*, 1984). Cryoelectron micrographs reveal the fully hydrated structure of embedded particles, e.g., viruses, and the image contrast corrected by the contrast-transfer function of the microscope can be interpreted as the native distribution of mass within the sample. An extensive review about the benefits of cryoelectron microscopy is given by Dubochet *et al.* (1988).

The smallest *in-vivo* system in biology is the cell. It would be of great interest to be able to look inside a cell by cryoelectron microscopy. Many questions concerning the organization of chromatin in the nucleus, the compartmentalization of the cytoplasmic gel or the extension of membranes by associated matrix-proteins are waiting to be answered. As these structures depend upon hydration for their existence, they cannot be observed in a dry specimen. However, two technical requirements hinder the employment of cryoelectron microscopy for hydrated cells: (1) the vitrification of cells in their physiological environment requires costly equipment and is difficult to achieve, and (2) Ultrathin sectioning distorts the volume of the section.

This article presents personal experiences with the cryofixation of cells and the interpretation of cutting artifacts in cryosections.

Materials and Methods

Plunge-freezing was performed with a KF80 (Reichert-Jung, Vienna, Austria) using ethane at 95 K as a cryogen. In order to reduce the heat flow from the plunger to the cryogen, the plunge rod was modified. The end-fixation for the specimen pins was replaced by a yellow plastic disposable tip of the type normally used for laboratory micro-pipettes. The plunge depth was readjusted fixing a plastic tube of convenient length underneath the original stop-face of the iron plunge rod. Parafilm was wrapped around this plastic tube to prohibit bouncing at the end of the plunge. As specimen carriers, small aluminum pins (purchased by a local electronics store), 0.65 mm in diameter and only 33 mg in weight were stuck in the small orifice of the yellow tip.

For slam-freezing, the KF80 impact freezer was employed as described by the manual. High-pressure freezing was performed with a prototype freezer, kindly provided by M. Wohlwendt (Sennwald, Switzerland).

Specimens were cut at 110 K with the Ultracut E ultramicrotome equipped with the FC4 cryochamber (Reichert-Jung). The nominal feed for sectioning was set to 60 nm. This number is rather approximate, since the cutting feed changes with thermal irregularities. Determined by the transmission in the electron microscope (9 mrad objective aperture, 190 nm mean free path for vitreous ice) the thickness of cryosections was generally about 100 nm. In rare cases much thinner sections (down to 40 nm) could be studied. Sections were cut with cryodiamond knives (Diatome, Biel, Switzerland) of either 35° or 45° knife angle. The clearance angle was set to 3°. The cutting stroke was performed manually, as slow as possible. An ion-spray (Static Line, Hauck, Biel, Switzerland) sometimes considerably improved the yield of flat sections. Its effect depends on the distance from the knife edge and seems to depend on the laboratory atmosphere. The proper distance (about 25 mm) must be found by trial and error. For most of the work discussed here this ion spray was not yet available.

Slam-frozen and high-pressure frozen samples were glued with a mixture of ethanol/propanol-2 (2:3) to the cutting support. This cryoglue is sticky at 130 K and becomes stiff below 115 K (Richter, 1994a).

Cryosections were manipulated with an eyelash to transfer them from the knife edge onto a carbon coated 600 mesh copper grid. The grids were held in place nearby the knife edge by a pair of special tweezers which are part of the tool box of the cryomicrotome. The piston used to press the sections onto the grid was also part of this toolbox. Until observation in the electron microscope, the loaded specimen grids were stored in small grid boxes (Gatan, Pleasanton, CA, USA) under liquid nitrogen.

All the way from the cryofixation device to the electron microscope, the specimen should never be warmed up above 135 K. Transfers between cryochambers were performed using small vessels to keep the samples covered by liquid nitrogen. In addition, care was taken that all tools coming in contact with the specimen were precooled.

Cryo-observation took place in a CM12 electron microscope (Philips, Eindhoven, The Netherlands) at 80 kV, using a Gatan 626 cold stage at 110 K. The microscope was equipped with a double blade anticontaminator (Marc Adrian, University of Lausanne, Switzerland). Specimens stayed clean for at least 2 h observation time. Images were registered at 17000x primary magnification on Kodak SO-163 film sheets (Eastman Kodak Co., Rochester, NY), exposed to a speed of 2 and developed in D-19 (Kodak) full strength for 12 min at 20°C. For low-dose imaging, the programmable low-dose device of the microscope was employed. The specimen area was exposed to 250 e^-/nm^2 for the registration of one micrograph.

Vitrification of Bulk Specimens

Ice-crystal formation involves two processes: germination and crystal-growth. Ice-crystal growth as an exothermic process will not cease on further cooling

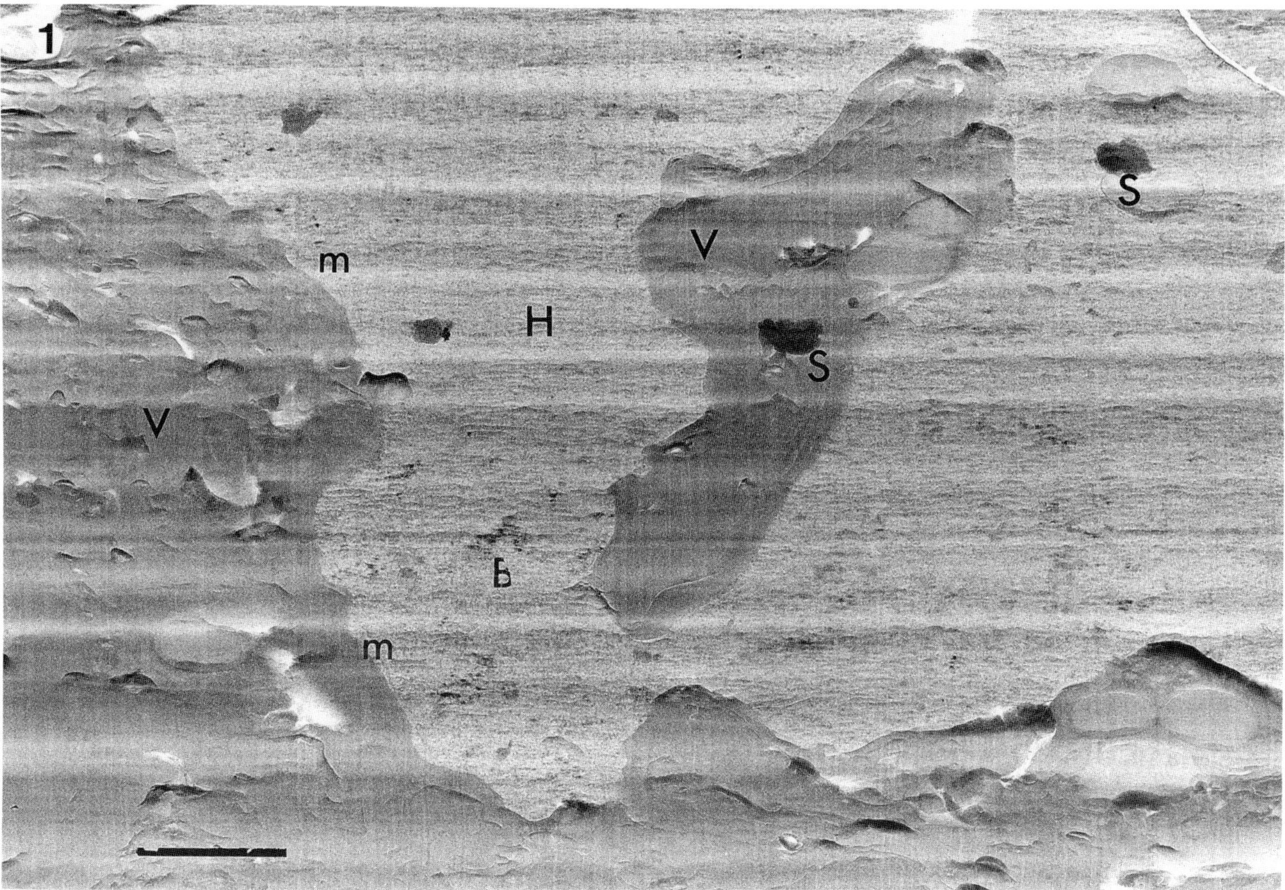

Figure 1: CV1 culture cells, pelleted in culture medium and plunge-frozen into liquid ethane. The cryosection contains vitreous and crystalline frozen regions in close neighborhood. Vitreous areas (V) are optically denser than the crystalline frozen ones (H). The crystalline state is indicated by the nested arrangement of dark spots (Bragg reflections B) which represent fractures of the ice-crystal so oriented that they diffract the electron beam beyond the aperture of the microscope. Note that both states of water are sharply separated. The separation line is the cell membrane (m). S contamination by a small piece of a cryosection. Scale bar 10 μm.

---

once it has started. Any vitrification protocol must therefore prevent the establishment of ice-crystal germs in the system. Ice-crystal germination is a stochastic process. Vitrification is easiest the smaller the volume of freezable bulk water, and the shorter the time required to pass the devitrification temperature (135 K for pure water). Vitrification depends, therefore, not only on the cooling method, but also on the sample size and the extend to which water is bound to other molecules. Biological macromolecules act as cryoprotectants, since they bind hydration water. In addition, eukaryotic cells are compartmentalized by lipid membranes which reduce the freezable volume accessible to a growing ice-crystal (Fig. 1). Despite the cryoprotective effect of biomolecules and the segmentation of the sample into a number of small compartments by membranes, vitrification of entire cells is not easily achieved. Two freezing methods have proved to be most successful: plunge-freezing into liquid ethane and high-pressure freezing.

Plunge freezing

Plunge-freezing takes place at ambient pressure. The specimen is projected via a guillotine-like frame into the liquid coolant, e.g., propane or ethane, cooled in a bath of liquid nitrogen to close to the melting temperature of the respective coolant (McDowall *et al.*, 1982). As long as the coolant moves around the specimen during the plunge, the heat transferred to it from the specimen is dissipated by convection. Cooling devices should therefore ensure that the sample moves relative to the cryogen until it is completely frozen (Le *et al.*, 1989; Ryan *et al.*, 1992). The surfaces of small pieces of mouse liver or heart were vitrified reproducibly by plunge-freezing into liquid ethane. However, the depth of the

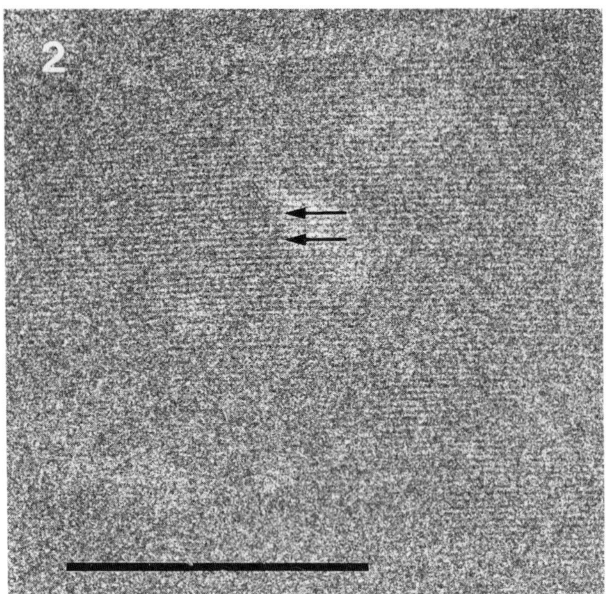

Figure 2: High-pressure frozen n-hexadecene. n-Hexadecene is a liquid at ambient pressure and temperature. After high-pressure freezing it is a crystalline solid. The spacing of the crystal planes (arrows) visible in this micrograph measures 2.1 nm. It is not known if the transformation into the solid state is caused by the pressure or the cooling. Scale bar 100 nm.

vitrification did not go beyond the first cell layer. The plunge-freezing of other test-specimens including cultured *Ptk* cells, *Saccharomyces cerevisiae* and *Chlorella* algae only exceptionally produced vitrified cells (Richter, 1992).

The plunging method was compared with slam-freezing onto a liquid nitrogen cooled copper block using mouse liver as a test specimen. One might expect improved cooling properties from this design since the heat-capacity and heat-conductivity of solids are much higher compared to liquids. However, in our hands, no vitrification was observed within the slam-frozen samples. One reason for the more effective cooling with the plunging device may be that the liquid coolant washes around the specimen, maintaining the highest possible temperature gradient at the specimen surface. The solid heat-sink, in contrast, warms up at the contact with the warm specimen. Furthermore, heat transfer is impaired by insulating oxidation layers, polishing residues, or contaminating ice on the cold mirror surface. In contrast to our experience, Shi *et al.* (1996) did succeed to vitrify the surface cell-layer of mouse liver by in-situ slam-freezing. Beside the possibility, that there are differences in the design of the slam freezers and in the protocol to use them, the tissues cannot be compared. In our experiment, the liver was dissected and small pieces cut from inside the organ. These pieces of tissue consisted mainly of epithelial hepatocytes which are big, isomorphous cells. The liver as an organ is covered by connective tissue consisting of elongated cells embedded in extensive extracellular matrix. This type of tissue might be vitrified more easily than the interior of the organ.

If cells are to be studied "close to the native state" vitrification of a surface cell-layer alone is not acceptable. Cells on the surface of a freshly prepared sample are not in a natural environment. They are subject to anemia, shortage of nutrients and water stress. Therefore, structures observed in these cells cannot be regarded as existing in a physiologically active state. As a consequence, cryoelectron microscopy of hydrated biological cells demands cooling methods that vitrify sample thickness spanning several layers of cells.

Whatever cooling-device is used, cooling is limited at a certain distance from the surface by heat-conduction through the specimen. Vitrification in deeper zones therefore requires conditions that reduce the probability for ice crystal formation. The administration of chemical cryoprotectants, e.g., sugars, salts, or dimethyl sulfoxide (DMSO), is not acceptable as long as the physiological integrity of the cells is to be preserved. An alternative method is that of cooling under high pressure.

High-pressure freezing

Since *Ice I* (the hexagonal ice that forms at ambient pressure) has a lower specific mass than liquid water it melts under the effect of hydrostatic pressure. Pressure induced melting is the underlying principle of ice-skating in winter. The melting point of water decreases with increasing pressure up to 2050 bar when conditions for the formation of high-density polymorphs of ice become favorable (Hobbs, 1974). At 2050 bar, the equilibrium melting point of water is depressed to 251 K and supercooling down to 183 K (Angell, 1982).

High-pressure-freezing involves two steps (Moor, 1987). After rapid establishment of the sample-pressure, cooling must take over before the sample is damaged by the high pressure. In commercialized high-pressure freezing machines, cooling is brought about by a jet of liquid nitrogen. Metal shields protect the sample against the impact of the jet and the turbulence which occurs when the specimen chamber is pressurized. In addition, the sample is embedded in hexadecene (Studer *et al.*, 1989). Hexadecene is supposed to solidify under pressure, thus, transferring the pressure from the metal shields to the sample. These additional layers of material between the active heat sink (i.e., the nitrogen jet) and the sample reduce the cooling efficiency of the device. Nevertheless, vitrification depths of more than 100 nm in samples as different as collagen of growing bone

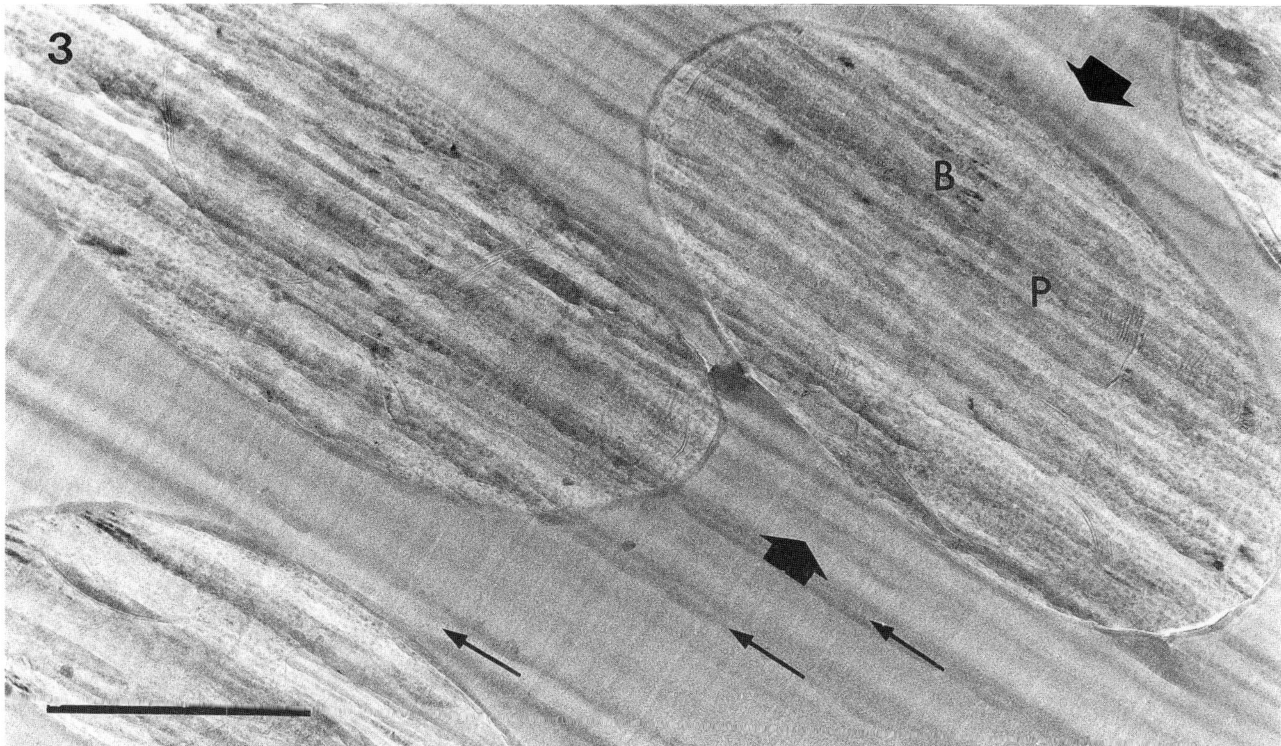

Figure 3: Chlorella algae embedded in 20% gelatin, plunge frozen into liquid ethane. In contrast to the cryoprotected embedding matrix, the cells are not vitrified. Bragg-reflections (B) indicate the hexagonal frozen state (confirmed by electron diffraction, not shown). Note the elliptical deformation of the originally round cells. They are compressed parallel to the knife marks (pair of thick arrow heads). The thickness variations perpendicular to the knife marks represent the deformation lines (arrows). P: Chloroplast. Scale bar: 1 μm.

(Studer *et al.*, 1995), liver tissue (Sartori *et al.*, 1993), and apple leafs (Michel *et al.*, 1991) can be achieved. Therefore, for the cryoelectron microscopy of bulk specimens, high-pressure freezing is the cooling method of choice.

Whereas cells frozen under high pressure appear to be well preserved (Dahl and Staehelin, 1989), the impact of the pressure may influence the ultrastructure of certain macromolecular organizations (Fig. 2). Ding *et al.* (1992) noted a loss of order in microfilaments in leaves of *Nicotiana tabacum*. The reorganization of model systems after high-pressure freezing has been reported in recent articles by Meyer *et al.* (1996) and Leforestier *et al.* (1996).

If the pressure is adjusted above 2050 bar, ice-crystals formed by high-pressure-freezing are the high-pressure morphologies *Ice II* and *Ice III/IX* (Richter, 1994a,b). In contrast to *Ice I*, it has been observed, that electron beam irradiation transforms the crystalline *Ice II* or *Ice III/IX* into a dense modification of vitreous water (Sartori *et al.*, 1996). It cannot be expected that this transformation would involve the rehydration of the organic phase which separated from the crystalline ice in eutectic concentration. Thus, the mere fact that the vitreous state is observed does not guarantee that biological structures are still hydrated in the native state. Further investigation will be necessary to evaluate this aspect of the preservation of structure in high-pressure frozen samples.

Cryo-Ultramicrotomy of Vitreous Biological Material

The most positive that can be said about cryo-sectioning is that the vitreous state permits ultrathin sectioning without pretreatment of the sample. Hutchinson *et al.* (1978) were probably the first to use the method for observing hydrated cryosections of vitrified biological bulk material. A comprehensive study of vitrified ultrathin cryosections has been reported by Chang *et al.* (1983). Cryosections are much more distorted by the cutting process as conventional plastic sections. For the interpretation of the micrographs the possibilities of structural deformation should be known.

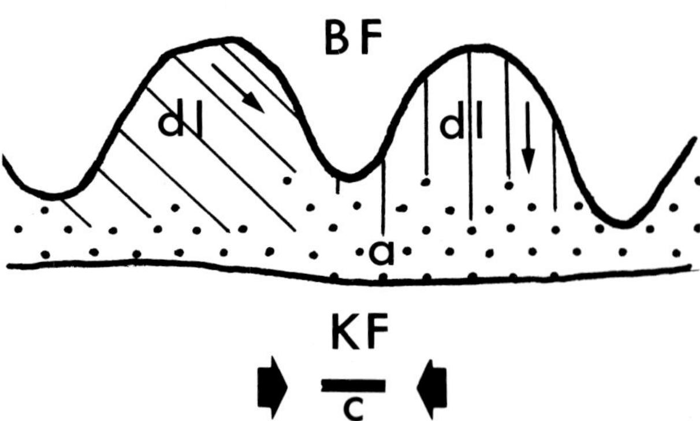

Figure 4: Schematic cross section through a cryosection in cutting direction. Cryosections have two different surfaces: The "knife face" (KF) is flat. In contrast, the relief of the former "block face" (BF) reflects the deformation lines as variations in section thickness (dl). Structures of the samples visible on cryoelectron micrographs are interpreted as projection views of the cut volume. It is not known whether the entire volume of the section contributes to the visible structural information. It is possible that part of the section near the "knife-face" loses its internal structure due to the friction on the knife. Apart from the compression in the cutting direction (thick double-arrow, c), cryosectioning also causes the tilting of parts of the sections to various degrees (thin arrows; see Fig. 5).

Sectioning and cryo-observation of bulk vitreous material

Vitreous water is metastable and transforms via cubic into hexagonal ice when warmed above the devitrification temperature (135 K for pure water; Dubochet and McDowall, 1981). In aqueous solutions where the water is partly bound in hydration shells, the devitrification temperature is higher, e.g., 171 K for 35% of gelatin (Chang *et al.*, 1983). Since biological systems are heterogeneous in nature, they should not be warmed above the devitrification temperature of pure water. The exothermic heat liberated by ice crystallization in regions where the water is relatively free may also promote the growth of ice crystals in regions of higher organic concentration. Despite the fact that the vitreous state is thermodynamically metastable, vitreous samples can be safely stored in liquid nitrogen for at least one year.

Cryo-ultramicrotomy requires an ultramicrotome equipped with a cryochamber where the temperatures of the sample and the knife can be adjusted to below 135 K. Clamping is not recommended for fixing the sample to the microtome arm, since the samples are too brittle and easily will break under the force necessary to secure them. The best way is to freeze the sample onto a support that can itself be attached to the microtome arm. Alternatively, a cryoglue can be used to fix the sample to the cutting support (Richter, 1994a).

Cryosectioning is best performed with diamond knives in combination with an electron-spray device (Michel *et al.*, 1992). Compression in the direction of cutting can be reduced by using low-angle knives. Richter (1994c) found that the deformation of lipid droplets in cryosectioned mouse hepatocytes into ellipsoids amounts to a longitudinal compression of 30% when sections are cut at a cutting angle of 38°. In contrast, compression reached 41% using the standard 45°-knife at a cutting angle of 48°. These compression values explain the observation that the sections are usually thicker than the nominal cutting feed.

Since there is no liquid which has the same surface-tension at low-temperature as water does at room-temperature, cryosections are cut with a dry knife. Lubrication of the cutting edge is replaced by an ion-spray that charges the surface of the diamond so as to repel the cut sections and thus reduce friction. The ion spray often helps to obtain flat sections when without the spray the sections would roll up during cutting. However, it could not be observed, that the use of the ion spray reduced the compression in the cutting direction. Accordingly, Shi *et al.* (1996) measured quite high compression (40 to 50 %) within their model system despite the use of a low-angle knife in combination with the ion-spray.

The sections laying on the back of the knife are lifted with an eyelash onto a carbon coated supporting grid. Care must be taken that the grid as well as the eyelash are both cooled to below the devitrification temperature. Measurement of the temperature of the eyelash or the grid will be difficult to be performed. However, a simple check can be done using ultrathin sections of the cryoglue mentioned above. The sections will stick to the eye-lash or melt on the grid when these latter are too warm. After charging the sections, the grid is placed on a polished metal surface in the cryochamber and the sections are pressed onto it with a polished metal-piston. The grid is much thicker than the sections, and therefore structures above the meshes are not damaged. Over the grid-bars the sections are brought into firm contact with the grid, which helps to reduce drift during observation in the electron microscope. Drift is a major problem for the registration of useful micrographs with a conventional transmission electron microscope (CTEM). The reason for drift seems to be the accumulation of electric

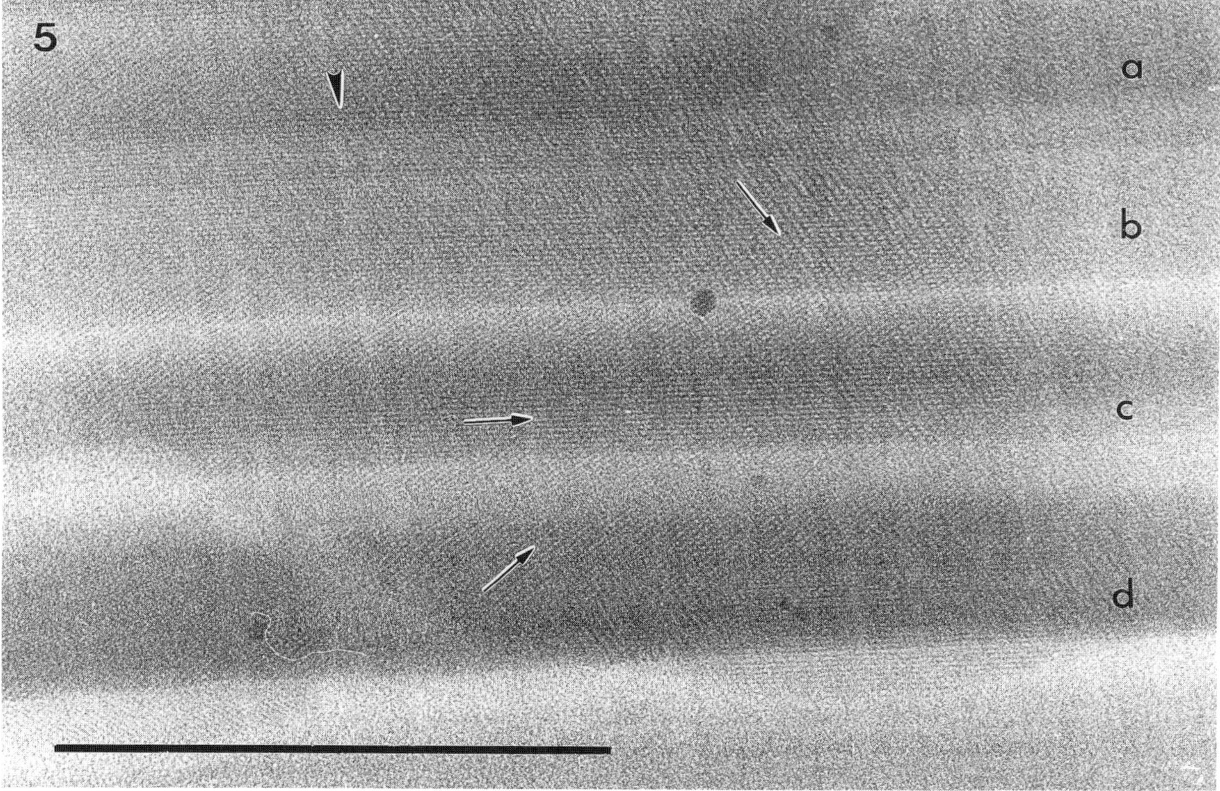

Figure 5: Cryosection of a catalase crystal. The micrograph shows part of one crystal spanning four deformation lines (a, b, c, d). Note that the crystal is seen in different views (arrows). Areas corresponding to one crystal plane are approximately delimited by the deformation lines: The crystal appears to be broken into pieces which are slightly tilted relative to one another by a deformation process that also produced the deformation lines. Arrowhead: knife mark; Scale bar 100 μm.

---

charges on the section. Accordingly, a supporting carbon film further reduces the drift problem.

Cryosections of vitreous samples are very sensitive to beam induced damage. Electron doses above 1000 e/nm^2 cause bubbling of the organic phase embedded in vitreous water. For this reason, the micrographs must be taken under low-dose conditions. In practice, the grid is searched for thin and flat sections at low magnification (400x) and lowest beam intensity acceptable for the search. Thus the electron dose deposited on the sample for the search can be kept below 10 e$^-$/nm^2. Care must be taken, not to hit the anticontaminator with the spread electron beam. Reduction of beam-intensity should therefore be controlled by the spot size (condenser 1). After adjustment of the microscope, the section is scanned micrograph by micrograph focusing in direct neighborhood of the region from which the micrograph is taken. Thus, structural details are not screened on the cryosection, but on the micrograph negatives.

Contamination with ice is another technical problem in cryoelectron microscopy. Two forms of contamination occur: (1) Condensation of atmospheric humidity which falls as hexagonal ice onto the cold specimen, and (2) layers of vitreous water which arise from water molecules in the vacuum which condense on the cold specimen in the electron microscope. The deposition of snow can be controlled by quick operation, and by taking care to prevent turbulence of the air at the interface between the cryochambers and the laboratory atmosphere. Contamination within the microscope can be reduced ensuring that the microscope vacuum is free from water. In practice, the use of a fork-type cryopump (Homo *et al.*, 1984) has been found necessary. This anti-contaminating device shields the specimen from the upper and lower parts of the column by two cooled metal-plates. Holes in the middle of the plates leave a small unprotected solid angle (70 mrad) for the transmitting electron beam. The efficiency of the anti-contaminator should be checked by using a clean carbon-coated grid as a test specimen. It is transferred on a cryostage into the electron microscope at room-temperature and then cooled down. The contamination is measured after a resting time equal to that required for normal observation. A contamination layer of water can be recognized by locally etching the water away with the focused electron beam.. The contrast provided by the hole makes it possible to estimate the thickness of the contamination. Especially thin layers can be made conspicuous by warming the grid to 140 K. At this temperature, the vitreous layer transforms to cubic ice which is identified by its powder diffraction pattern (Dubochet *et al.*, 1982).

Section morphology and deformation induced by cutting

During cutting one can already see that cryosections are compressed in the cutting direction, since they are considerably shorter than the corresponding block face. The elliptical deformation of known objects in the cryosections confirms that structures are compressed in the cutting direction (Fig. 3). This compression is related to an increase in section thickness, since the section volume cannot be reduced. The longitudinal compression therefore implies two superimposed deformation processes: decrease in length in the direction of cutting and increase in length perpendicular to the plane of the section. One might argue that turbulence occurs during cutting and destroys structural aspects of the specimen. On the other hand, inter-atomic distances are not influenced by such deformation processes. Two questions arise: (1) At what resolution is the deformation recognizable? (2) Can the degree of deformation be predicted and calculated?

Richter *et al.* (1991) have described a model of the deformation process based on the morphology of cryosections. Typical morphological features of cryosections are the deformation lines, i.e., the periodic ribbon-like increases in thickness perpendicular to the cutting direction (Fig. 3). As most of the knife-marks on cryosections are straight, it is concluded that cryosections have two different surfaces; one surface being flat and the other undulating or even broken up by the deformation lines (Fig. 4). In general, the two surfaces of a section are of different origin. When the knife separates the cut volume as a section from the specimen block, two surfaces are created. The surface on the back of the knife belonging to the newly cut section will be called "knife face" in the following. The second surface in front of the knife belongs to the specimen block. It will become the "block face" on top of the following section. Stereo views from cryosections showed that the flat surface corresponds to the "knife-face" of the section and the undulating surface to the "block-face" (Richter, 1992). During the cutting stroke, the section is deflected from its original position by the cutting angle and becomes displaced onto the back of the knife. Deflection by the knife and friction on the knife back result in a shearing stress perpendicular to the cutting direction, and a compression force parallel to it. The deformation lines may thus be due to material being periodically squeezed out from the "block face" of the section. Friction on the knife-back keeps the "knife-face" flat. The knife marks at this side will be straight, whereas those from the "block-face" will be distorted by the deformation of the surface.

The ultrathin section is a thin layer of the specimen. It consists of the cut volume and the two surfaces, and the micrograph is a two-dimensional projection of this three-dimensional structure. It is important to know how the structural information available from the micrograph can be influenced by cutting. When considering the structural information obtainable from the micrograph, one must distinguish between surface structures and projections of at least part of the volume. Comparison of tilted cryosections of catalase crystals has confirmed that the visible crystalline arrangement is a projection and not a surface-image (Richter, 1994c). Since the crystalline pattern is still visible, sectioning does not involve mass-displacements that mix up the structures within the cut volume (Fig. 5).

Cutting may break the section into chips that are tilted and stuffed together, causing the compression of structures in the μm-range, e.g., the lipid droplets. Within one chip, the ultrastructure would then still be preserved. This model was tested with the cryosectioned catalase crystals. The preservation of the crystalline order as determined by electron diffraction extended further than 3 nm. However, the diffraction patterns also revealed, that the crystalline arrangement was different from the original orthogonal type. Geometrical reconstruction of the original unit cell under the assumption, that distances are compressed in the cutting direction,

Figure 6 (*on facing page*): Comparison of two hydrated cryosections of mouse liver. Scale bar 1 μm. **a.** Vitreous state: The overall contrast is very low. Especially in the nuclear region (N) no internal structure can be seen. C: Cytoplasm; L: Lipid droplet; S: Contamination. **b.** Hexagonal frozen state: The hexagonal frozen state is indicated by the Bragg reflections (B). Note the phase-separation network (asterisks): When water freezes to ice it separates as a pure substance from the organic phase which aggregates in a non-freezable eutectic concentration. In this example, phase separation did not obscure the cell morphology. However, compacting the biological matrix into a fine network improved the contrast. Compared with Fig. 6a, the cell compartments are much easier discernible. N: Nucleus; C: Cytoplasm; M: Mitochondrion; Arrow-head: Nuclear pore; b: Cell border.

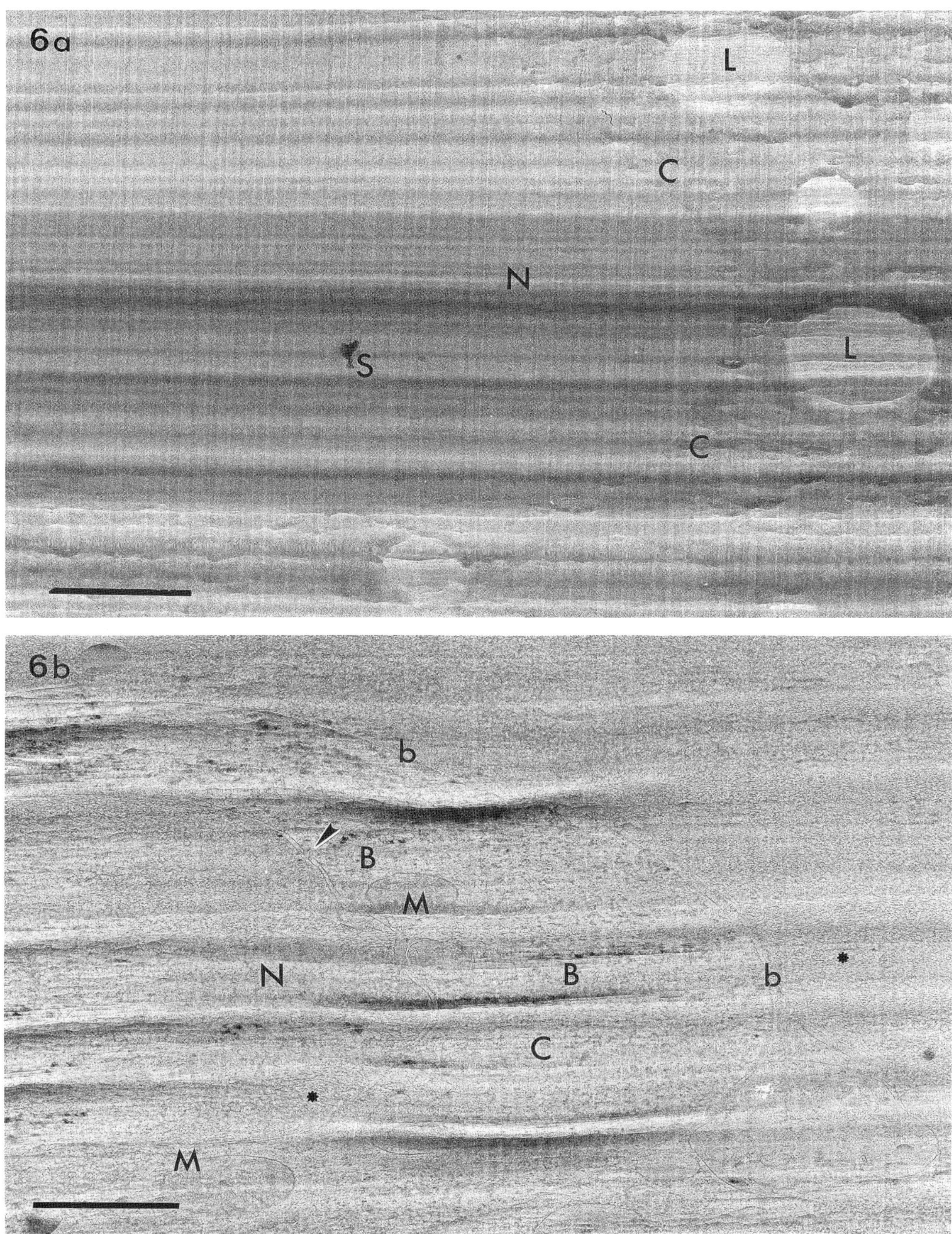
6a
L
C
N
S
L
C
6b
b
B
M
N
B
b
C
M

revealed a compression of 40%, the same as for the elliptical deformation of lipid-droplets (Richter, 1994c). The deformation lines are therefore not "chips" broken off from the section plane.

Another sample, hexagonal liquid crystals of DNA (kindly provided by F. Livolant, Université Paris Sud, France), could in contrast provide arguments for the chip-model, since the hexagonal arrangement did not appear to be compressed in the cutting direction (Richter, 1994c). However, the section itself showed the usual signs of deformation (presence of deformation lines and shortening compared with the size of the block). As this crystal is a closed pack of DNA molecules, the crystal planes will not be approached further without breaking the molecules. Cutting will therefore rather displace single molecules perpendicular to the section plane while compression in the direction of cutting permanently restores the dense package.

This leads to the conclusion that no general rule can be given for the degree of distortion that cutting induces in the different structures which are present in biological cells.

Micrographs of frozen hydrated sections are very low in contrast. The nucleus of well prepared, vitreous cells, e.g., appears nearly structureless. Only faint differences in granularity can sometimes be distinguished. One reason is the small difference in mass between water and biomolecules (McDowall *et al.*, 1982). There may, however, be another reason for this poor yield of structural information form cryosections. Because of the good structural preservation provided by cryofixation, the biomolecules are not denatured into an aggregation network, but remain finely dispersed as a hydrated gel. Accordingly, structures become visible when the water freezes to crystalline ice and the organic phase aggregates in the eutectic concentration (Fig. 6). In order to reveal the structure of hydrated macromolecular complexes by cryoelectron microscopy, very thin sections will be required. Here the present "state of the art" offers a challenge to those undertaking future research in the field of cryoelectron microscopy.

Beside the interest of high-resolution structure determination, another application in electron microscopy requires hydrated cryosections. Sun *et al.* (1995) showed, that the low-loss region of electron energy loss spectra contains information about the water content of the measured specimen region. The spatial resolution for these maps of water distribution in the cell is limited to 80 nm, which is precise enough to differentiate among most cellular compartments in eukaryotic cells.

Acknowledgement

I would like to express my thanks to Prof. Jacques Dubochet who provided me with the opportunity to learn about cryoelectron microscopy when I was working as a doctoral student in his laboratory.

References

Adrian M, Dubochet J, Lepault J, McDowall AW (1984) Cryoelectron microscopy of viruses. Nature **308**: 32-36.

Angell CA (1982) Supercooled water. In: Water - A Comprehensive Treatise, Vol. 7. Franks F (ed). Plenum Press, New York. pp 1-82.

Chang J-J, McDowall AW, Lepault J, Freeman R, Walter CA, Dubochet J (1983) Freezing, sectioning and observation artifacts of frozen hydrated sections for electron microscopy. J Microsc **132**: 109-123.

Dahl R, Staehelin A (1989) High-pressure freezing for the preservation of biological structure: Theory and practice. J Electron Microsc Techn **13**: 165-174.

Ding B, Turgeon R, Parthasarathy MV (1991) Effect of high-pressure freezing on plant microfilament bundles. J Microsc **165**: 367-376.

Dubochet J, McDowall AW (1981) Vitrification of pure water for electron microscopy. J Microsc **124**: RP3-RP4.

Dubochet J, Lepault J, Freeman R, Berriman JA, Homo J-C (1982) Electron microscopy of frozen water and aqueous solutions. J Microsc **128**: 219-237.

Dubochet J, Adrian M, Chang J-J, Homo J-C, Lepault J, McDowall AW, Schultz P, (1988) Cryoelectron microscopy of vitrified specimens. Quart Rev Biophys **21**: 129-228.

Hobbs PV (1974) Ice Physics. Clarendon Press, Oxford. pp. 60-78.

Homo J-C, Booy F, Labouesse P, Lepault J, Dubochet J (1984) Improved anti-contaminator for cryoelectron microscopy with a Philips EM 400. J Microsc **136**: 337-340.

Hutchinson TE, Johnson DE, Mackenzie AP (1978) Instrumentation for direct observation of frozen hydrated specimens in the electron microscope. Ultramicroscopy **3**: 315-324.

Le PW, Murray R, Robards W, Waites PR (1989) Counter-current plunge cooling: A new approach to increase reproducibility in the quick freezing of biological tissue. J Microsc **156**: 173-182.

Leforestier A, Richter K, Livolant F, Dubochet J (1996) Comparison of slam-freezing and high-pressure-freezing effects on the DNA cholesteric liquid crystalline structure. J Microsc **184**: 4-13.

McDowall AW, Chang J-J, Freeman R, Lepault J, Walter CA, Dubochet J (1982) Electron microscopy o frozen hydrated sections of vitreous ice and vitrified biological samples. J Microsc **131**: 1-9.

Meyer WH, Dobner B, Semmler K (1996) Macro-ripple-structures induced by different branched-chain phosphatidylcholins in bilayers of dipalmitoylphosphatidylcholine. Chem Phys Lipids **82**: 179-189.

Michel M, Hillmann T, Müller M (1991) Cryo-sectioning of plant material frozen at high pressure. J Microsc **163**: 3-18.

Michel M, Gnägi H, Müller M (1992) Diamonds are a cryosectioner's best friend. J Microsc **166**: 43-56.

Moor H (1987) Theory and practice in high-pressure freezing. In: Cryotechniques in Biological Electron Microscopy. Steinbrecht RA, Zierold K (eds). Springer, Berlin. pp. 175-192.

Richter K (1992) Cryoelectron Microscopy of Frozen Hydrated Ultra-Thin Sections from Biological Material. Doctoral Thesis, UNIL, Lausanne, Switzerland.

Richter K (1994a) A cryo-glue to mount vitreous biological specimens for cryoultra-microtomy at 110 K. J Microsc **173**: 143-147.

Richter K. (1994b) High-density morphologies of ice in high-pressure frozen biological specimens. Ultramicroscopy **53**: 237-249.

Richter K (1994c) Cutting artifacts on ultrathin cryosections of biological bulk specimens. Micron **25**: 297-308.

Richter K, Gnägi H, Dubochet J (1991) A model for cryosectioning based on the morphology of vitrified ultrathin sections. J Microsc **163**: 19-28.

Ryan KP (1992) Cryofixation of tissues for electron microscopy: A review of plunge freeze cooling methods. Scanning Microsc **6**: 715-743.

Sartori N, Richter K, Dubochet J (1993) Vitrification depth can be increased more than 10-fold by high-pressure freezing. J Microsc **172**: 55-61.

Sartori N, Bednar J, Dubochet J (1996) Electron-beam-induced amorphization of ice III or IX obtained by high-pressure freezing. J Microsc **182**: 163-168.

Shi S, Sun S, Andrews SB, Leapman RD (1996) Thickness measurement of hydrated and dehydrated cryosections by EELS. Microsc Res Techn **33**: 241-250.

Studer D, Michel M, Müller M (1989) High-pressure freezing comes of age. Scanning Microsc Suppl **3**: 253-269.

Studer D, Michel M, Wohlwend M, Hunziger EB, Buschmann MD (1995) Vitrification of articular cartilage by high-pressure freezing. J Microsc **179**: 321-332.

Sun S, Shi S, Hunt JA, Leapman RD (1995) Quantitative water mapping of cryosectioned cells by electron energy-loss spectroscopy. J Microsc **177**: 18-30.

Discussion with Reviewer

B. Andrews: The paper sends a somewhat mixed message regarding the advantages and disadvantages of different cryofixation techniques. Thus, it is variously mentioned that high-pressure freezing is required for vitrification of cells (Abstract), but that plunge freezing reproducibly vitrified liver and heart (see also Fig. 1), while metal-block freezing is dismissed altogether. Perhaps the author could clarify and expand his views, particularly with regard to metal-block freezing, which has a sound physical basis, and which many other laboratories have found to give superior results.

Author: The advantages of the solid heat sink compared with the liquid coolant are the higher heat conductivity, the increased heat capacity and the possibility to establish a temperature gradient which goes to the limits of the primary coolant (e.g., liquid helium), whereas the liquid can only be cooled down to its freezing temperature. With these three values an ideal cooling rate can be calculated. However, things are not so easy: the interface between the sample and the solid heat-sink may sensitively reduce the heat-transfer. This interface is very difficult to control for the experiment in question. Parameters often discussed in this respect are bouncing of the sample and thermal isolating contamination layers on the mirror surface. However, the sample will also shrink when cooled which may impair an initial neat contact with the mirror.

These interface-problems might better be solved with a liquid coolant. With a liquid coolant heat is essentially transported by convection, the bad coefficients for heat conductivity and heat capacity becoming less important as long as the flow of coolant around the sample is well adjusted, i.e. laminar and rapid. Also this condition is difficult to control with the various sample sizes and shapes to be frozen.

The practical approach to this problem is the experiment. In my hands, plunge-freezing into liquid ethane resulted in a good yield of vitrified liver tissue, whereas the same type of sample was not to be vitrified by slam-freezing onto a liquid nitrogen cooled copper block.

B. Andrews: The author states that plunge freezing vitrification "did not exceed the first cell layer" of liver and heart, thus implying a vitrification depth of around 20 μm. This seems rather optimistic. Please comment.

Author: To my knowledge, no ambient-pressure cooling method is capable to vitrify more than the first cell layer of a sample. It has taken long time to show that a quantity of pure bulk, liquid water (i.e. no vapor) can be cooled to an amorphous state. The crucial insight was, that this can be achieved only with really small quantities: a 1 μm thick layer of pure water cooled simultaneously from both sides by plunge freezing is about the upper limit that can reproducibly be vitrified with this method (Dubochet and McDowall, 1981). An animal cell is one order of magnitude larger in size. However, the situation is easier with cells than with pure water. Cell-water is more or less bound to organic components of the cell, what reduces the volume of freezable water. In addition, the ice-crystal front stops at cell-membranes (e.g., Fig. 1). Hence, within each compartment a new germination event must occur to freeze the entire system (this holds only for conditions where freezing damage does not yet rupture the membranes).

Freezing of the water in cells generally involves the entire compartment, which results in large, highly branched crystals of hexagonal ice (Dubochet *et al.*, 1988). Only in rare cases I could observe a crystallization-front within the cytoplasm of a cell, where the vitreous state changed first to cubic and then to hexagonal ice. The normal case is, that vitreous and hexagonal frozen regions are in direct neighborhood, separated by a membrane. The growth of an ice-crystal seems to provide enough energy to transform already vitrified parts of the compartment into ice (Fig. 1), and it is not possible to vitrify only part of the cytoplasm of a cell. Cryofixation techniques may thus be ranked by the number of cell layers they are capable to vitrify (which also depends on the type of tissue).

Measuring the vitreous zone in cross-sections through the samples, I found, that plunge-freezing can vitrify the first cell-layer of liver tissue (i.e., about 20 μm), but not the second one.

B. Andrews: Vitreous and crystalline ice appear to cryosection very differently, as can be seen from Figs. 1 and 3. In particular, it appears (Fig. 1) that vitreous specimens are more likely to show abrupt surface deformations ("crevasses"). Is this true in general, and are there any other important differences between these types of specimens regarding sectioning properties, e.g., compressibility, strength or continuity of sections and ribbons, adhesion to supports, etc.?

Author: The deformation of the section depends on the mechanical properties of the sample, which are not only determined by the state of the water, but also by the other components of the sample. Sharp deformation lines e.g. frequently occur along membranes. In addition, the actual section thickness and cutting speed have an influence. Larger feed and faster cutting strokes provoke more abrupt surface deformations.

One clear statement can be done concerning the state of the water. A crystal cannot be distorted plastically. Therefore, cutting breaks the large hexagonal ice-crystals of frozen material into small pieces. This can be deduced from the shape of spots in the electron diffraction patterns (discussed in Richter, 1994c). These spots are crescent-shaped, i.e. they are sharp in the direction of reciprocal space, and blurred throughout a small angle of directions. As the large crystal not only is broken into pieces but the pieces are also tilted within a small angular distribution, the spots of a single-crystal diffraction pattern smear into crescents reflecting the variation in tilt-angle. The small crystal-pieces are separated by crevasses which are visible in Fig. 1 as fine lines in the frozen part. Lines of this kind are missing in the vitreous part.

B. Andrews: It was surprising to see no diffractograms. Does the author not rely on diffraction as a method for determining the state of frozen water?

Author: Electron diffraction is the only objective way to demonstrate the state of the water with electron optics. Each sample observed was checked by electron diffraction. However, also morphological criteria can be applied to distinguish vitreous from frozen regions on bright-field images. Vitreous areas appear optically denser than those containing hexagonal ice. Crystalline frozen regions have a coarse appearance due to the crevasses described above. Fragments of the broken crystal oriented such that they diffract beyond the objective aperture appear as sharp dark spots on the micrograph (Fig. 1, "Bragg reflections").

The sample region for electron diffraction is defined by the round selected area aperture. In micrographs like the one of Fig. 1, it will be difficult to select separately crystalline frozen and vitreous regions to obtain the diffraction pattern. In this case morphological criteria are more convincing.

B. Andrews: The use of carbon-only films is recommended as supports for cryosections. Can the author please provide some details about these films, since it would be interesting to know how their thickness and method of preparation affect the quality and feasibility of imaging frozen-hydrated sections.

Author: As mentioned in the text, I consider the carbon-support necessary to reduce beam-induced drift of the cryosections.

The carbon film was produced according to standard procedures. Carbon was evaporated from carbon-threads in a Balzers table-evaporator at a pressure of $5x10^{-5}$ mbar, and a distance of about 7 cm to freshly broken mica (all from Balzers, Liechtenstein). The thickness of the carbon-layer was in the range of 8 to 10 nm. The film was cast onto a water surface and descended onto submerged 600 mesh copper grids. After air-drying during one day the grids were ready to use.

Scanning Microscopy Supplement 10, 1996 (pages 387-466)
Scanning Microscopy International, Chicago (AMF O'Hare), IL 60666 USA
0892-953X/96$5.00+.25

ADVANCED INSTRUMENTATION AND METHODOLOGY RELATED TO CRYOULTRAMICROTOMY: A REVIEW[‡]

H. Sitte

Fachrichtung 3.5, Medizinische Biologie, Fachbereich Theoretische Medizin, Universität des Saarlandes, D-66421 Homburg-Saar, Germany
Telephone Number: (49) 6841-166250; Fax Number: (49) 6841-68401

(Received for publication March 13, 1996 and in revised form August 8, 1996)

Abstract

This review is concerned with the considerable progress in the field of cryo-ultramicrotomy (cryofixation, cryosectioning, investigation and analysis of cryosections) during recent years. This progress includes both more efficient instrumentation and methodology. The article is mainly directed to the investigation and analysis of frozen-hydrated sections in the low dose cryo-transmission electron microscopy (TEM) and cryo-energy filtered TEM (EFTEM). A general survey is followed by an evaluation of the different relevant procedures. Both cryo-ultramicrotomy for macromolecular cytochemistry (Tokuyasu technique) and cryo-ultramicrotomy for element analysis are only shortly mentioned without discussion of the chemical and analytical approach. Because of lack of first hand experience, cryo-sectioning for X-ray microanalysis in the frozen-hydrated state according to Hall and Gupta is not included into this review. The methods and instruments required for ultrathin sectioning at low temperatures are described and discussed in detail. This concerns the preceding cryofixation, the cryosectioning itself with special emphasis to the required stability and precision of the cryo-ultramicrotome, the characteristics of the knives, the charging phenomena due to sectioning and the subsequent TEM investigation including EFTEM with electron spectroscopic imaging (ESI) and the available accessories for digital low dose registration of signals.

Key Words: Cryo-ultramicrotomy, cryo-immobilisation, high-pressure freezing, ambient-pressure freezing, vitrification, cryoprotection, cryo-stabilisation, cryotransfer, low dose imaging, energy filtering (EFTEM), electron spectroscopic imaging (ESI), beam damage, radiolysis, hybrid methods.

[‡]The paper is dedicated with some delay to my brother Prof. Peter Sitte on the occasion of his 65th birthday (December 8, 1994) with very best wishes and the highest fraternal respect.

Table of Contents

"Den Vorsatz, etwas zu verbessern
muß mancher Forscher sehr verwässern.
Die so erzeugte Wasserkraft
treibt dann die Arbeit fabelhaft"
[frei nach Eugen Roth, 1977 (translation page 460)]

Introduction

In most cases the real possibilities of new methods or instruments are considerably overestimated, a lot of difficulties and limitations overlooked or hidden. At the end it often turns out, that the power of the innovation is below the expectations and that the hurdles to be taken increase with each step forward. In the early fifties the *"ultrathin serial sections in the thickness range of 100 Å"* (see Porter and Blum, 1953; Sitte, 1955; Sjöstrand, 1953, 1954) turned out often to be *"beefsteaks"* ten times thicker, as clearly shown by Peachey (1958) or Bachmann and P. Sitte (1958, 1960). And even those *"beefsteaks"* were not collected

Table 1. Important improvements 1985 - 1995 influencing cryo-ultramicrotomy

New Method or Technology	Main Improvements	Literature	See this Review
High-Pressure Freezing (Synchronisation of pressurisation and cooling by alcohol volume at ambient temperature, 1-hexadecane as intermedium between sandwich container and object, variable thickness 0.1 to 0.6 mm of specimen container, dialysis tubes for cell suspensions)	Yield of well frozen specimens often near 100%, minimum delay between pressurisation and freezing ($\leq$ 13 msec for layers of 0.1 mm) → minimised pressure artefacts, prevention of blowing out of suspensions by use of dialysis tubes	Müller and Moor, 1984, Studer *et al.*, 1989, Michel *et al.*, 1991, Hohenberg *et al.*, 1994. See also catalogues of Balzers (HPM010) and Leica (EM-HPF)	Sections "Freezing and the Frozen State of Water" and "High Pressure Freezing", Fig. 12, Table 5
Cryo-Ultramicrotomes (through-the-wall specimen arm, shell-mounted cryochamber, continuous refilling of LN_2)	Mechanical stability at least identical to standard ultramicrotomes for ambient temperature work, better reproducible section thicknesses (thinner sections) regularly available → minimising of crevasses	Catalogues of RMC (Cryochamber RMC 21) and of Leica-Reichert (Ultracut-S/FCS and Ultracut-UCT/ FCS)	Sections "Cryo-Ultramicrotomes" and "Ultrathin Sectioning and Handling..", Figs. 13e,f, 14b, 15
Adjustable Ioniser Spike inside the GN_2 of the cryochamber	Minimising of charging phenomena (cryosections versus knife material), perfect diamond trimming and diamond knife sectioning of sugar protected material according to Tokuyasu, easier section transfer	Michel *et al.*, 1992. See also catalogue of Diatome (Static Line II)	Sections "Cryo-Ultramicrotomes" and "Ultrathin Sectioning and Handling..", Fig. 17
(a) **Low Angle Cryo-Diamond Knives** and (b) **Diamond Trimming Tools,** in combination with an adjustable ioniser	(a) Reduced section compression and formation of *"crevasses"*, smoother sections, lower section thicknesses available. - (b) Better geometry of sectioning surface by diamond trimming improves all results	Michel *et al.*, 1992. See also catalogues of Diatome	Sections "Cryo-Ultramicrotomes" and "Ultrathin Sectioning and Handling..", Figs. 17 - 21
High-Performance Cryotransfer-Systems (cold stages) for low temperature TEM work down to -180°C coupled with efficient decontamination systems (cold traps) for TEM	Minimising of frost deposition during and after cryotransfer, good temperature stability → minimum drift (< 0.1 nm/sec) → optimum resolution ($\leq$ 0.34 nm)	Catalogues of Gatan (Model 626-DH) and of Oxford (CT 3500 and Anticontamination system)	Section "Ultra-thin Sectioning and Handling ..", Fig. 22
Freeze-Drying for EDX and EELS	Freeze-drying considerably simplified and improved by cryosorption freeze-dryers	Sitte *et al.*, 1994. See also catalogue Leica (EM-CFD)	Sections ".. Element Analysis" and Discussion, Fig. 25
Slow Scan CCD Cameras for digital on-line low dose recording of TEM images or diffraction patterns or autotuning/autofocus of TEM	$\geq$ 10:1 reduction of e^- dose (beam damage), linear ratio between e^-dose and signal intensity, wide dynamic range, automatic TEM operation (focus, tomography, lens calibration, stigmator control etc.)	De Ruijter, 1995, Dierksen *et al.*, 1993, Koster and De Ruijter, 1992, Krivanek and Mooney, 1993, Tietz, 1992. See also the catalogues of Gatan and LEO (Zeiss)	Section ".. Frozen-Hydrated Ultrathin Sections", Fig. 27, Table 6
Electron Stimulable Phosphorus Image Plates for digital off-line low dose recording of TEM images or diffraction patterns	$\geq$ 10:1 reduction of e^- doses (beam damage), linear ratio between e^- dose and signal intensity, wide dynamic range, extremely large detector surface identical to photographic plates/films, compatible with all TEM models	Ayato *et al.*, 1990, Burmester, 1992, Mori *et al.*, 1988, 1990, Oikawa, 1990, Shindo *et al.*, 1990, 1991. See also catalogue Fuji FDL-5000 system	Section ".. Frozen-Hydrated Ultrathin Sections", Figs. 28 and 29, Table 6

Electron Energy Filters (EFTEM) for Electron Spectroscopic Imaging (ESI) in "*in column*" or "*post column*" design	Minimising of underfocus < 0.1 μm needed in conventional TEM to obtain sufficient contrast by "*zero loss imaging*" (elimination of inelastically scattered el⁻) → better contrast and resolution of frozen-hydrated specimens with high contents of elements with low atomic numbers	Bauer, 1988, Krivanek *et al.*., 1995, Benner *et al.*, 1994, Bihr *et al.*, 1991, Schröder, 1992, Schröder *et al.*, 1990. See also catalogues of LEO (Zeiss) and Gatan	Section ".. Frozen-Hydrated Ultrathin Sections", Fig. 30

as routinely as often claimed at that time. Pictures presented as results of daily work were often identified as extraordinary singles, selected from thousands of electron micrographs. Neither cryostats nor cryo-chambers for sectioning (see Appleton, 1974; Bernhard, 1965; Bernhard and Nancy, 1964; Bernhard and Leduc, 1967; Bernhard and Viron, 1971; Christensen, 1969, 1971; Dollhopf and Sitte, 1969; Leduc *et al.*, 1967) reached the claimed minimal temperatures without severe drawbacks and cryowork was much more complicated than announced (see *e.g.* Echlin, 1992, pp. 138-140; Robards and Sleytr, 1985, p. 504; Sitte, 1982; Tokuyasu, 1980, p. 383; Zierold, 1987, p. 145). The 100:1 profit in "*cryostabilisation*" by extremely low temperatures with the LHe-cryostat-lens at 4.2 K (Dietrich *et al.*, 1980; Knapek and Dubochet, 1980) in comparison to work at ambient temperature was shrinking to a small x 3 to x 10 stabilising factor for the observed organic crystals exposed to an electron beam at low temperature (Chiu *et al.*, "*International Experimental Study Group*", 1986) already obtainable with a liquid nitrogen (LN_2) cold stage. But, very important for scientific progress in our field - and probably in most similar new areas of high-tech research: nobody with rational concepts would have continued to invest and waste precious time and a lot of money after a clear cut statement from the leading pioneers, that there exist tremendous difficulties to obtain the desired results. Nobody would have made further effort to run cryo-apparatuses which neither reached the wanted temperatures at a reasonable time nor worked correctly. And nobody would have been interested in cryo-scopes, which only cause big problems without really preserving precious samples from beam damage. Nevertheless: we should not worry about all the mistakes and misinterpretations made (and published) in scientific enthusiasm. They were mostly at the end coupled with remarkable real progress. It should for example not hamper, that high pressure freezers vitrify only 100 μm thick samples (instead of calculated 600 μm, see Moor, 1987; Sartori *et al.*, 1993; Studer *et al.*, 1995) if it becomes evident at the same time, that ambient pressure cryo-fixation also does not reach its goal (see Sitte *et al.*, 1987a). High pressure remains still more than 10 times better. This makes the point. Therefore: progress is evident, and in cryo-ultramicrotomy and all directly associated fields of freezing and low dose electron microscopy just the recent years brought a lot of really important improvements as shown in Table 1. Of course, there exist a couple of excellent reviews covering this field (see *e.g.* Echlin, 1992; Dubochet *et al.*, 1987, 1988; Menco, 1986; Morgan, 1995; Reid and Beesley, 1991; Robards and Sleytr, 1985; Roos and Morgan, 1990; Zierold, 1987). But they now have in some respects more "*historical*" character and do not cover the present state of the art. On the other hand, also the current literature is not always based on the best methods and instruments available today. Additionally a lot of comparative work is needed to evaluate, what is really possible with different specimen categories. The entire sense of this review is therefore, to close these gaps and to encourage interested colleagues to start again and to gain from those fascinating advances.

To avoid misunderstandings, it is important to differentiate precisely between the various fields of applications, in which cryo-ultramicrotomy nowadays is used. Besides the broad application in materials sciences (not considered in this article) the main application is immuno cytochemistry (or more general: macromolecular cytochemistry). As shown in Table 2 this specific application follows a completely different protocol in comparison to all other areas of application. Chemical prefixation with aldehydes, necessary for these freeze-thawing procedures, does not affect most of the biomacromolecules in respect of their antigenicity. Since even a weak aldehyde fixation opens the cell membranes, efficient intra- and intercellular cryoprotection is possible, which simplifies freezing and sectioning tremendously. It was the incredible merit of Tokuyasu (1973) to introduce sucrose as cryoprotectant, which makes bulk tissue samples so well sectionable at low temperature, that his protocol pushed cryo-ultramicrotomy into the field of well reproducible routine work not essentially different from standard ultramicrotomy of resin blocks (for further detail see *e.g.* Griffiths, 1993; Griffiths *et al.*, 1983, 1984; Tokuyasu, 1986; Sitte *et al.*, 1988). Nevertheless the Tokuyasu technique also took and

Table 2. Comparison between the pathways and protocols for cryo-ultramicrotomy (CUM), if macro-molecular histochemistry (*e.g.*, Tokuyasu method), EDX or EELS for element analysis on ultrathin cryosections after freeze drying or direct imaging of frozen-hydrated ultrathin cryosections in the cryo-TEM is the goal. Note the considerable difference between macromolecular histochemistry (Tokuyasu technique), element analysis and frozen hydrated work.

Step	Macro-molecular Histo-chemistry	Element Analysis on Ultrathin Sections	Frozen-Hydrated Investigation
Chemical prefixation	Low molarity aldehyde mixture	—	—
Cryoprotection	2.3 M sugar (only in exceptional cases lower molarity)	—	—
Cryo-ultramicrotomy (CUM)	CUM at temperatures around -100°C with cryo-diamond knives and ioniser (easy sectioning after 2.3 M sugar protection even of surfaces around 0.5 x 0.5 mm²)	CUM at temperatures ≤ -160°C with cryo-diamond knives and ioniser (very difficult sectioning of the extremely brittle material even in surface areas below 0.1 x 0.1 mm²)	CUM at temperature of ≤ -160°C with cryo-diamond knives and ioniser (very difficult sectioning of the extremely brittle material even in surface areas below 0.1 x 0.1 mm²)
Follow-up procedures	Picking-up with droplets of concentrated sugar solution within wire loop → thawing → labelling → coating (*e.g.* methyl cellulose) → drying → TEM at +20°C	Dry transfer to grid → freeze drying ≥ 12 h at low temperature (start -80°C) → EDX or EELS in TEM/ STEM at +20°C or low temperature	Dry transfer to grid → cryo-transfer ≤ -150°C to TEM → low dose imaging with SS-CCD or image plate in EFTEM ("*zero loss mode*") at temperature ≤ -150°C

takes advantage of the progress in cryosectioning, specifically of the new cryodiamond knives together with the discharging ionisers and of course of the improved cryochambers themselves (see separate Section "*Cryo-Ultramicrotomy According to Tokuyasu*" and Table 2).

The demands for the investigation or diffraction analysis of frozen-hydrated specimens in the cryo-transmission electron microscopy (TEM) and/or for energy-dispersive X-ray analysis (EDX) or electron energy loss spectroscopy (EELS) of ultrathin sections differ considerably from the needs of Tokuyasu work for immuno cytochemistry (see again Table 2). Chemically prefixed and sugar protected (2.3 M sucrose) specimens do not need a rapid cryofixation: LN_2 does the job mostly sufficiently and sometimes even better than ethane, since vitrification is easy obtainable and clefts or ruptures in the samples are not provoked by high cooling rates. Also sectioning of the "*sugar embedded*" material even in rather old cryo-ultramicrotomes poses no severe problems, since in the range around -100°C those specimens have a nice consistency: sections with a width up to 0.5 mm (in some cases up to 1.0 mm) are easily available. The collection of sections from the dry knife with a droplet of 2.3 M sugar solution (also an invention of Tokuyasu) is as simple and elegant as spreading and deposition of the sections on grids or cover slips.

The technical and methodological progress of the recent years (see Table 1) eliminated some of the severe draw-backs of the cryo-ultramicrotomy and investigation of fresh frozen samples both for element analysis and studies in the frozen-hydrated state by TEM/scanning transmission electron microscopy (STEM)/EDX/EELS. This progress started with the development of a better and far more efficient methodology of high pressure freezing (see *e.g.* M. Müller and Moor, 1984; Moor, 1987; Studer *et al.*, 1989, 1995; Michel *et al.*, 1991), which has advanced to a routine method at least for all cell suspensions (Hohenberg *et al.*, 1994), some stable animal tissues (*e.g.*, cartilage) and a lot of plant material not accessible for ambient pressure freezing due to their high water content. Now in those fields high pressure freezing allows to vitrify rather large samples. This is of great importance since clear evidence was presented recently, that microcrystalline freezing generates artefacts (McDowall *et al.*, 1984; Studer *et al.*, 1995), although the small crystals are not resolvable in the electron micrograph and do not influence the cryo-

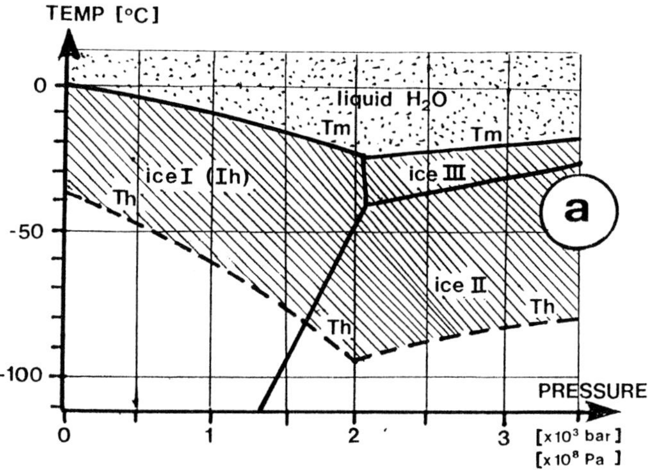

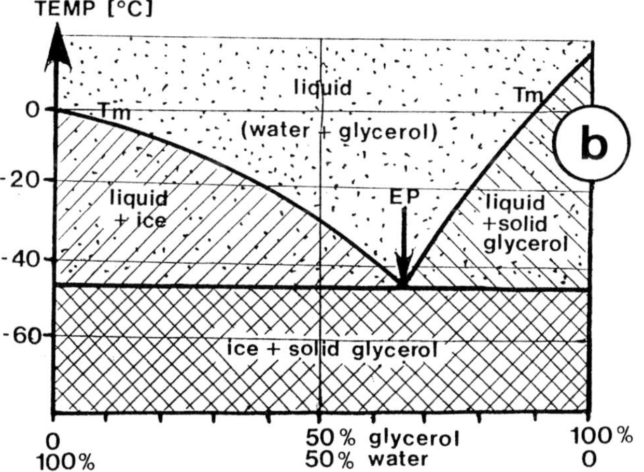

Figure 1. Equilibrium phase diagrams for freezing of "*pure*" water (a) and of the eutectic binary system "*water/glycerol*" (b). As explained in this section ("Freezing and the Frozen State of Water.."), freezing processes are essentially kinetic events: the diagrams allow therefore no conclusion about the periods of time needed to reach the state of equilibrium. (a) Dependence of melting temperature T_m, temperature T_h of "*homogenous nucleation*" (broken line) and transition from one of the polymorphic modifications I ⇔ II ⇔ III into one another, on pressure. Ice I is a synonym for hexagonal ice I_h. Ice II and ice III are like ice IV to IX high pressure modifications (IV to IX stable at pressures above 3 kbar and not considered in this review, see *e.g.*, Eisenberg and Kauzman, 1969). Cubic ice (I_c) is mostly generated by warming up amorphous (vitreous) ice (I_v) above the devitrification temperature T_d. Since both I_c and I_v are unstable modifications, they are not included in this equilibrium phase diagram. The diagram shows clearly, that T_m has a minimum slightly below -20°C at a pressure of 2.1 kbar. T_h for homogenous nucleation shows also the minimum at the same pressure. (b) Dependence of freezing point T_m in the binary system "*glycerol/water*" on the relative amount of glycerol/water: maximum depression occurs in the equilibrium at 67 % glycerol, where according to the diagram below -40°C (EP = eutectic point) a mixture of ice and solid glycerol ("*eutectic mixture*") should crystallise.

---

sectioning process at the same extent as larger crystals of hexagonal ice (I_h). Besides the high tech pressure freezing also the simpler ambient pressure freezing methods should not be overlooked: they are now further developed, better understood and open new possibilities for subsequent cryosectioning. Dramatic improvements resulted in the design and operation of cryochambers since 1990: both the new "*through-the-wall*" operation of the specimen arm of the ultramicrotome, the "*shell-mounted cryochamber*" and the nowadays available systems for electronically controlled "*continuous LN_2-refilling*" of the latest generation of cryochambers (Leica FC S and FC R, partially RMC CR 21) (see list of suppliers) reach or surpass the stability and precision of standard ultramicrotomes for ambient temperature work. Everybody struggling with a former system knows exactly, that especially stability and precision are of the greatest practical importance, since all cryosectioning artefacts are increasing with increasing section thicknesses. Irregular sectioning, that means thicker sections now and then, destroys all hopes and benefits of a stable system. The maximum precision (exclusion of outside thermal influences, further improved advance and drive systems) is therefore really needed in this field. The application of an adjustable ioniser and the use of the new cryodiamond knives for dry trimming and sectioning work act in the same direction (Michel *et al.*, 1992): they indeed make cryosectioning of fresh frozen specimens more reproducible and efficient. Additional advantages will arise for sure from cryodiamond knives with facet angles below 35°: section compression will be lowered once more noticeably, as Jésior (1986, 1989) has demonstrated already for ambient temperature sectioning. Since frozen sections do not allow any spreading operation by heat or organic solvent vapours, this is a very important measure in order to reduce this disturbing artefact.

Finally, any progress to lower the needed electron dose for documentation of biostructures in the frozen-hydrated state or to minimise beam damage by other means, should be carefully realised. This concerns both screening and selection of useful areas in the frozen-hydrated sections, focusing and digital registration of all signals excited by the electron beam. Since the expected considerable stabilisation of sensitive biological ultrastructures by low temperature (The

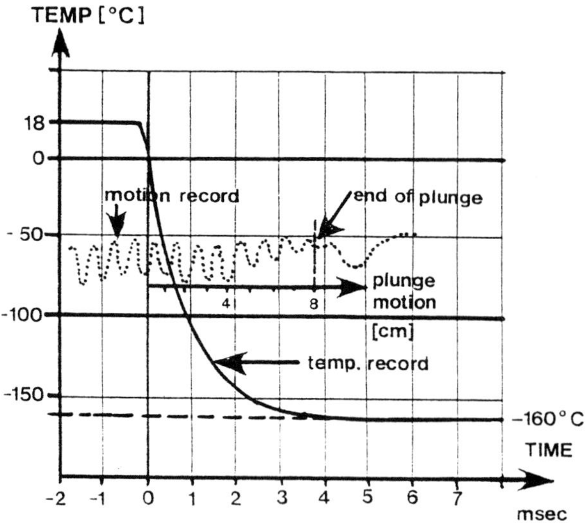

Figure 2. Typical bare thermocouple record from plunging into liquid ethane. The ethane was kept at -160°C. The motion record was realised with a photo diode on a black and white sequence moved and recorded together with the plunger. The plunge velocity was 2 m/sec. The cooling rate between 0 and 100°C was calculated according to the definition (approximately 9300°C/sec). The plunging motion in the cryogen was limited to 8 cm (modified Fig. 7 of Ryan, 1992, published with permission). Further explanation in the section on "Freezing and the Frozen State of Water...".

----

term "*cryostabilisation*" used by Dubochet *et al.* (1981) seems to me much better than the previously used term "*cryoprotection*", which provokes some confusion with cryoprotection by sucrose or glycerol) was not reproducible for the whole range of structures and procedures (Chiu *et al.*, 1986), the new possibilities of digital accumulation of signals at minimum electron doses are a precious tool for frozen-hydrated work. This low dose registration is now possible both with slow-scan charge-coupled-devices (SS-CCD-systems, see *e.g.*, De Ruijter, 1995) and with image plates (see *e.g.*, Burmester, 1992; Mori *et al.*, 1988, 1990; Oikawa *et al.*, 1990) already used in radiology and now just for the first time introduced and made commercially available for TEM-work by Fuji (FDL-5000 system). Both different systems besides their higher sensitivity have the big advantage of a precise linearity of signal intensity to the electron dose enabling an accurate quantification of the recorded signals. Finally "*Energy Spectroscopic Imaging*" (ESI) with energy filtering energy filtering TEMs (EFTEMs, see *e.g.*, Bauer, 1988; Henkelmann and Ottensmeyer, 1974; Krivanek *et al.*, 1995; Ottensmeyer and Andrew, 1980; Schröder, 1992; Schröder *et al.*, 1990) opens a possibility to reduce underfocus, to improve signal-to-noise ratio and contrast of fresh-frozen material considerably. Since all above mentioned recent progress contributes to frozen-hydrated TEM work in a highly welcome manner, the different new approaches will be described and discussed in detail in the following sections. As far as necessary and useful, earlier (already forgotten) progress will also be mentioned and integrated. Since some instruments will not be well known, a short list of suppliers is added for interested colleagues.

Table 3. Comparison of liquid cryogens for immersion cryofixation ("*plunging*") according to Sitte *et al.* (1987a) (modified). See also Bald (1984), and (Ryan) 1992. The cooling efficiency (CE) of the different cryogens was roughly calculated according to published microthermocouple measurements in comparison to the most frequently used cryogen (CE for propane = 1.0). It is evident, that melting points (T_m), boiling points (T_b) and their differences (T_b-T_m) are not the only deciding parameters for CE. "*Primary cryogens*" like boiling LN_2 are not suited at ambient pressure due to the immediate formation of a thermally insulating gas layer ("*Leidenfrost phenomenon*"). Only pressurised "*hyperbaric LN_2*" successfully serves as an excellent cryogen for high pressure freezing. All values of T_m and T_b in [°C].

Cryogen	Tm	Tb	(Tb-Tm)	CE
Ethane	-171	-89	82	1.3
Propane	-190	-42	148	1.0
Freon 13	-185	-81	104	0.8
Freon 22	-155	-41	114	0.7
Freon 12	-152	-30	122	0.5
Isopentane	-160	+28	188	0.5
LN2 cooled to Tm	-210	-196	14	0.2
Boiling LN2	—	-196	—	0.1
Hyperbaric LN2 (pressurised) at about 2 kbar	—	—	—	> 1.3

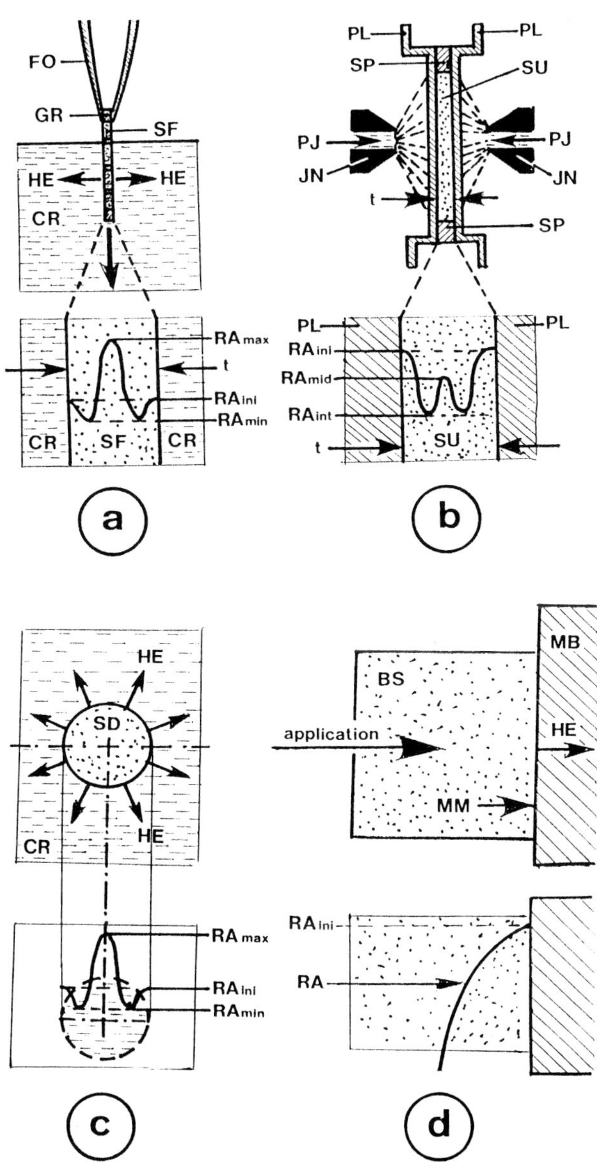

Figure 3. Cooling rate RA profiles depend on the geometry of the specimen and the uni-, bi- or multi-directional access of the liquid or solid state cryogen CR/MB. The upper schema shows the geometrical situation for heat extraction HE, the lower shows a schematical diagram of the expected RA-profile. Further explanations in the section on "Ambient Pressure Freezing". **(a)** Immersion of a "*bare grid*" GR according to Adrian *et al.* (1984) with suspension films SF between the grid bars (grid hold by forceps FO) into liquid cryogen CR (*e.g.*, ethane at -170°C). The immersion guarantees an exactly synchronous bi-directional heat extraction HE with cooling rates RA > 1 million °C/sec provided a film thickness t ≤ 0.1 μm. The RA-profile shows under such preconditions the RA_{max} (maximum cooling rate) in the centre of the film SF and two RA_{min} in the intermediate positions right and left. The initial rate RA_{ini} lies between these extremes. **(b)** Double-sided cryogen- (propane)-jet PJ according to M. Müller *et al.* (1980) cooling a suspension SU between two sandwich-planchettes PL (thickness t given by spacer element SP). Jet nozzles JN right- and left-hand in exact symmetry to the sandwich PL/SU/PL. The extreme velocity of PJ results in a very high initial rate RAini. As far as the exact symmetry and a perfect synchronisation of both jet-streams PJ are guaranteed (*e.g.*, by Balzers JFD 030 Jet Freezer), the RA-profile is similar to (a) with a peak RA_{mid} in the middle of the sandwich due to the bi-directional heat extraction. The absolute value of CR_{mid} depends on thickness t of the suspension layer SU and both thickness and material of the planchettes PL. **(c)** Cooling of a suspension droplet SD (diameter d << 100 μm) during "*spray-freezing*" according to Bachmann and Schmitt (1971) offers the best conditions, since heat is extracted into all spatial directions (see arrows HE) by the cryogen CR. The maximum rate RA_{max} results in the centre of the sphere. **(d)** Strictly uni-directional heat extraction HE through the highly polished mirror-like contact area ("*metal mirror*" MM) between the cold metal block MB and the bulk specimen BS. The initial rate RA_{ini} is extremely high due to the high heat capacity and heat conductivity of the solid state (mostly copper, cooled by LN_2). But RA decreases rapidly inside the specimen BS. The same happens in all larger bulk samples with diameters > 0.5 mm.

Freezing and the Frozen State of Water - Cryoprotection

It is indeed very encouraging for a biologist, to read the statement of two well known specialists in the field of the physicochemistry of water in an excellent review article (see Bachmann and Mayer, 1987, p. 4), that "*water has some unusual properties which have made it one of the most investigated liquids whose structure is still not fully understood*". But this comment should not be an excuse and some basics have to be understood or at least noticed and accepted before starting a cryopreparation.

As long as we work at "*normal*" ambient pressure of 1 bar, we know that water freezes (or melts) at 0°C (T_m). Unfortunately things are not as simple as expected, since freezing is a typical "*kinetic*" event. If we lower the temperature rapidly enough, *e.g.*, with a cooling rate of several 100.000°C per second, cooling is more rapid than freezing. We will therefore probably observe liquid "*supercooled*" water with a temperature below 0°C, *e.g.* -10°C, which does not seem to be in agreement with the diagram in Fig. 1a (for physicochemical details see *e.g.*, Bachmann and Mayer (1987), Eisenberg and Kauzmann (1969) or Franks (1972-1982). Under such conditions finally freezing mostly starts from an impurity (if "*pure*" H_2O is our liquid) which acts as "*nucleus*" or "*crystallisation center*". Since those nuclei are "*foreign*" particles in water, one speaks of a "*heterogeneous nucleation*". Freezing in this case is a very speedy process and leads to the well known normal hexagonal ice (I_h) - the only ice which we observe in our terrestrial environment: the ice in our deep freeze, the beautiful snow crystals, the ice of the increasingly melting glaciers and pole regions - all consisting of I_h. Provided that no "*foreign*", that means heterogeneous nucleus is present in our water, we reach by further cooling at ambient pressure a temperature T_h near -40°C, were the society of water molecules itself serves for nucleation: in this case one speaks of "*homogenous*" or "*spontaneous nucleation*". Freezing now starts everywhere in the whole liquid phase and huge amounts of small ice crystals are formed.

Things change again if pressure is applied: according to the phase diagram in Fig. 1a both melting point T_m and the temperature T_h, where homogenous nucleation starts during cooling, decrease with increasing pressure. The cross-hatched area in the diagram represents the region within which supercooled water can exist. Also T_h is a limit, which follows kinetic laws: during an extremely rapid cooling the start of spontaneous crystallisation becomes delayed. But this is not the whole story: like all other liquids also water gets increasingly viscous with decreasing temperature. Like for very viscous sugar solutions crystallisation processes in other viscous liquids are very slow. For example honey tends to crystallise but often needs several years for this event due to its high viscosity. Also pure water at -100°C has already a remarkable viscosity and finally crystallisation becomes impossible if the temperature decreases below about -135°C. We have an extremely viscous glass-like fluid. In other words: we have vitrified our water to "*amorphous*" ice I_v without a crystalline structure. This vitreous ice I_v is stable over a long time at these temperatures. It devitrifies and changes into microcrystalline cubic ice I_c, if we heat it above this "*devitrification temperature* T_d" around -135°C (for physicochemical details see Bachmann and Mayer, 1987). If we now take a look at the phase diagram in Fig. 1a, the basics of pressure freezing should be evident and understandable: at the pressure of about 2,000 bar T_m is reduced to -20°C, T_h below -90°C. Homogenous nucleation is shifted into the range of more viscous water, since viscosity is strongly dependent on temperature and not changing with pressurisation. To produce I_v at 2,000 bar we need therefore much lower cooling rates in comparison to ambient pressure.

Instead of applying pressure similar effects with respect to reduction (depression) of T_m and T_h result from suited additives (antifreezes or cryoprotectants) as visible in diagram Fig. 1b. If we add for example 50 % glycerol to pure water, freezing at ambient pressure starts at a lower temperature. The 50/50-mixture starts freezing below -20°C. Since I_h is formed, concentration of glycerol is increasing and the system follows the T_m-curve to the "*eutectic point EP*" at the eutectic temperature near -50°C and a glycerol concentration of 67%. A further decrease of temperature should result in complete crystallisation of the residual liquid if we have the patience to wait for it. The kinetic character of this process is even more expressed than during rapid cooling of unpretreated fresh samples and in reality those "*cryoprotected systems*" react extremely slowly : reality is far away from all diagrams. Probably supercooled liquids result rather than the announced eutectic crystallisation, consisting theoretically of I_h and frozen glycerol according to Fig. 1b.

The first successful cryoprotection and vitrification in biological cryopreparation used the mentioned cryoprotection by glycerol according to Fig. 1b (see Polge *et al.*, 1949; Moor, 1964). Cell suspensions and tissues were frozen for subsequent freeze fracture followed by freeze etching and freeze replication, which offered for the first time an alternative to chemical fixation, dehydration and resin embedding by an exclusively physical process (see *e.g.*, Benedetti and Favard, 1973,; Moor and Mühlethaler, 1963; Steere, 1957, 1973). Glycerol made it possible to vitrify those specimens even by quite simple freezing procedures. For the first time disturbing ice segregation, that means separation of the cytoplasmic phase with its high water content between 50 and sometimes near 100 % firstly into pure I_h and finally into an eutectic mixture could be suppressed. In the meantime a lot of different cryoprotectants for cells and tissues were checked (*e.g.*, DMSO: dimethylsulfoxide, PVP: polyvinylpyrrolidone, sucrose; for detailed information see Meryman, 1971 or Skaer, 1982). The initially used glycerol made the frozen samples at lower sectioning temperatures around or below -100°C extremely brittle. It was therefore nicely suited for freeze fracturing but terrible for cryo-ultramicrotomy at low temperatures. Contrary to glycerol the already mentioned sugar protection according to Tokuyasu (1973) is probably not suited for freeze fracture, but excellent for sectioning around -100°C. The drawback of this kind of intracellular cryoprotection is, that cell membranes (and of course also other cellular structures) have to be chemically damaged and by this treatment (chemical prefixation with aldehyde) opened. After this pre-treatment *e.g.*, by aldehyde in low concentrations, which does not affect severely the protein conformation, sugar or glycerol molecules can freely enter through the leaky membranes and a complete protection is possible. Of course such a treatment changes all equilibria of living cells totally: ions and water are moving rapidly and practical nothing is preserved in the original state as far as smaller molecules and ions are concerned. Biomacromolecules (*e.g.*, proteins, nucleic acids, macromolecular carbohydrates) retain most of their antigenicity and that is exactly the value of this pre-treatment, because cell destruction and changes in the conformation *e.g.*, of proteins resulting from the much more aggressive dehydration by organic solvents for resin embedding together with OsO_4-stabilisation of membranes are avoided.

Often cryoprotection of the extracellular (extraplasmatic) compartments or fluids without preceding chemical fixation is advantageous, since mostly severe artefacts by segregation of I_h start around the cells. In such cases extraplasmatic protection by non poisoning agents (*e.g.*, sugar in the suspension medium, see Moor and Mühlethaler, 1963; Dubochet *et al.*, 1983; McDowall *et al.*, 1984, and discussion in the next section) improves the results often considerably, even if the cryoprotectant does not penetrate or enters only slowly or in small amounts into the cytoplasmic phases.

We have to ask: what do we really gain by application of high pressure or physicochemical cryoprotection by glycerol or sucrose in freezing procedures ? There is strong evidence, that water is by several reasons one of the most speedy crystallising liquids: cooling of pure water must be extraordinarily fast to suppress formation of I_h, which is deadly for subsequent morphological studies of cells and tissues. When I_h crystallises, the increasingly concentrated residues of cytoplasm with all incorporated particles (*e.g.*, ribosomes) and structures (*e.g.*, filaments of cytoskeleton) are shifted rapidly to other places, which they never occupied before. Essentially the same occurs in extraplasmatic (extracellular) compartments and all these events change morphology dramatically. For a long time it was a strict belief in physicochemistry, that amorphous I_v can be only produced by apposition of single H_2O molecules from the gaseous phase (sublimation) on cold surfaces, as already demonstrated a long time ago (for details see *e.g.*, Burton and Oliver, 1935a,b; Eisenberg and Kauzmann, 1969; Franks, 1972-1982; Riehle and Hoechli, 1973). Recently, against this belief, amorphous I_v could be obtained directly from pure liquid H_2O (Mayer, 1985). For this direct transformation of pure liquid H_2O to I_v probably cooling rates between 10^6 and 10^7°C/sec are essential, if layers or droplets of about 1 μm size are used (see Bachmann and Mayer, 1987). According to the same authors for 10 μm layers or droplets with respect to the higher probability for heterogeneous nucleation already cooling rates two or three dimensions higher are estimated (which will be utopic). In biology, in comparison to pure water, we have somewhat better conditions: specimens do not contain 100% H_2O and have therefore some "*internal cryoprotection*" by dissolved material. Partially this internal protection is remarkable, *e.g.*, in some frost resisting plants. This improves our situation. On the other hand proper cryofixation (vitrification) is complicated by the frozen water itself, that means by its low heat conductivity: if a border layer is already frozen, we have to extract heat from the specimen through this thermally insulating frozen layer and the cooling rates drop down rapidly. The result: we can only vitrify one cell layer of bulk samples (depth about 5 μm) and often also this cannot be achieved. Under these bad conditions any improvement by high pressure (or cryoprotection) is welcome. If we apply the optimum pressure of approximately 2 kbar (see Fig. 1a), cooling rates around 10,000°C/sec in most cases are sufficient for real vitrification. Such cooling rates are obtainable, if specimen thickness does not exceed 0.15 mm. If we cryoprotect our sample according to Tokuyasu (1973) with 2.3 M sucrose, even 1,000°C/sec vitrifies most structures and excludes I_h segregation despite the poor cooling by LN_2.

Rapid heat extraction from biological specimens ("*cryofixation*") results from direct contact between the specimen surface and a "*cryogen*". As already mentioned, both liquid and solid state cryogens are suited for this procedure. The most simple cryofixation is achieved by immersing the sample into a liquid cryogen: this procedure is called "*immersion cryofixation*" or "*plunge freezing*". It was observed a long-time ago (Hoerr, 1936, see also *e.g.*, Pearse, 1961), that boiling liquids (*e.g.*, LN_2 at its boiling temperature T_b, -196°C) are not well suited. These "*primary cryogens*", which maintain their low temperature by continuous boiling (evaporation heat), develop immediately a gas layer (bubble) around the immersed warm specimen ("*Leidenfrost phenomenon*"). Since gas layers (*e.g.*, gaseous nitrogen, GN_2) are very efficient thermal insulators, such layers prevent the desired rapid cooling. Suited cryogens are therefore only "*secondary cryogens*", that means liquids with boiling temperatures T_b around or slightly below ambient temperature and a very low melting / freezing temperature T_m, which results in a very big difference between their boiling- and freezing temperatures (T_b-T_m). Table 3 shows the data of some liquid cryogens proven for rapid freezing in comparison. To evaluate correctly the "*cooling efficiencies*, CE" it was necessary to cool small probes (microthermocouples) acting instead of specimens in the different cryogens under standardised conditions (see *e.g.*, Costello and Corless, 1978; Ryan, 1991, 1992; Ryan *et al.*, 1987; Schwabe and Terracio, 1980). By the same reason the "*cooling rate*" (cooling speed) was defined as the average speed measured during cooling such a probe from 0°C to -100°C (see Fig. 2). It is evident, that those "*cooling rates*" had practically nothing to do with the real cooling profiles within very small biological specimens. But they enabled elegantly to study the possibilities and limitations of different liquids. By similar measurements the importance of the different cooling parameters like the speed and the mount of the specimen or the influence of cold gas layers could be checked (see *e.g.*, Ryan, 1992; Ryan and Purse, 1984, 1985). It became clear by these experiments that ethane is the best suited liquid cryogen of all hitherto studied liquids. Its cooling efficiency CE is 1.3 times higher than CE of liquid propane under identical experimental conditions (see Table 3). CE itself is influenced by a confusing variety of physical parameters like *e.g.* specific heat, heat conductivity, heat transfer on the phase border between specimen and cryogen, and viscosity (for details see Bald, 1984, 1985). But all this is only of theoretical importance. The message for the application remains simply: ethane is not only the best, but for subsequent cryo-ultramicrotomy of vitrified specimens the only suited cryogen with respect to its fairly high vapour pressure at those low temperatures. LN2 is only suited and then excellent, if a high pressure (*e.g.*, 2 kbar) excludes the Leidenfrost phenomenon. Propane is a low cost and quite good coolant for daily work with hybrid technologies (freeze substitution and freeze drying) as far as the safety rules are strictly regarded (please read carefully the comments on safety *e.g.*, of Sitte *et al.*, 1987b or Robards and Sleytr, 1985, pp. 507-512). For subsequent cryo-ultramicrotomy at real low temperatures below -150°C propane is not applicable without special measures, since it does not evaporate sufficiently fast due to its low vapour pressure at those temperatures. It sticks therefore on all surfaces of an immersed specimen and disturbs sectioning by this liquid surface coat. Hyperbaric LN_2 should be an excellent cryogen also outside a high pressure freezer (Bald, 1984; Bald and Robards, 1978), but the results in practical use did not come up to the expectations.

Contrary to solid state cooling ("*metal mirror*" or "*impact freezing*") liquid cryogens extract heat in two spatial directions from a thin foil (*e.g.* "*bare grid*" according to Fig. 3a, see Adrian *et al.*, 1984; Dubochet *et al.*, 1982a, 1987; Lepault *et al.*, 1983a; or "*double jet*" M. Müller *et al.*, 1980; Th. Müller *et al.*, 1989; see also Taylor and Glaeser, 1974, 1976) and in all spatial directions from a small droplet (*e.g.*, "*spray freezing*" according to Bachmann and Schmitt, 1971; see also Bachmann and Schmitt-Fumian, 1973; Plattner *et al.*, 1973) as shown in Fig. 3c. Under such favorable conditions not only is heterogeneous nucleation minimised by minimum specimen (mostly suspension-) volumes, but also maximum cooling rates are achieved by multidirectional heat extraction. It is a severe disadvantage of most biological cryopreparations, that bulk specimens (tissue samples with diameters > 0.5 mm) have no profit from these favourable conditions: here extraction of heat in the border layers of some micrometers in depth during the deciding microseconds after the start of the freezing process is strictly unidirectional. This precondition is also given for the freezing in contact with precooled solid state surfaces. According to Fig. 3c multidirectional heat extraction from small droplets with diameters <50 μm creates cooling profiles (see lower part of the diagram), where the maximum cooling rates result in the center of the sphere. This behaviour is not only visible by the state of preservation of the ultrastructure but can also be demonstrated by ice crystal diameters and by calculations (see *e.g.*, Van Venrooij *et al.*, 1975; Ryan *et al.*, 1990). Similar profiles are estimated for double jet freezing, where heat is extracted only in two directions (Fig. 3b) : also in this case the diameter / thickness of the sheet is a deciding parameter. Only very thin layers (thicknesses _ 100 μm) show a useful improvement of cryopreservation in the center region. All specimens considerably thicker *e.g.*, with diameters in the order <0.5 mm have no noticeable profit from this phenomenon, because freezing of the thin border layers is already finished before heat extraction through the opposite surface region is able to contribute to the cooling rate. Both immersion of such a bulk specimen into a liquid cryogen and surface contact with a cold solid state surface (Fig. 3d) result therefore at their beginning in a unidirectional heat extraction through the already frozen and thermally insulating border layers of the specimen. It is understandable that under these preconditions only very small surface layers (*e.g.*, up to 5 μm) can be vitrified if an unprotected specimen is frozen under ambient pressure. Only high pressure or sufficient intra- and intercellular protection (*e.g.*, glycerol or sucrose) enable improvement in the situation.

Ambient Pressure Freezing

It has been shown several times, that also under ambient pressure well frozen specimens for subsequent cryosectioning are obtainable (see *e.g.*, Plattner and Knoll, 1983; Sitte *et al.*, 1987a). As far as liquid cryogens are used, this is at least valid for cell suspensions, which often can be cryoprotected partially, that means around the cells, by sugar or glycerol containing suspension media. As shown already by Dubochet and co-workers (Dubochet *et al.*, 1983; McDowall *et al.*, 1984; see also Moor, 1964; Moor and Mühlethaler, 1963), such extracellular protection does not affect different cells and is helpful for vitrification. Obviously this is true for some bacteria, yeast cells and a couple of other eucyte suspensions without the detrimental pre-treatment by chemical fixatives like aldehydes. The main advantage of this protection of the extracellular liquid is the more homogeneous freezing over larger areas, which results for sure in a much better overall sectioning consistency in comparison to completely unprotected samples.

The simplest cryofixation for subsequent cryo-ultramicrotomy results from immersion into propane or ethane, which is often done in home made laboratory set-ups (for details see *e.g.*, Ryan, 1991, 1992; Sitte, 1979, 1984; Sitte *et al.*, 1985, 1987a). Since these cryogens freeze above -196°C, direct cooling with LN_2 (T_b about -196°C) for longer periods of time is not possible. Either one needs thermostatic heating as shown in the schematic drawing (Fig. 4a) or additives, which reduce the freezing temperature, and affect the cooling properties accordingly. For freezing of bulk specimens in diameters exceeding 0.5 mm an element for convection of the liquid ethane is advantageous to prevent heating of the cryogen on the surface of the warm specimen: this may be for example realised by a magnetic stirrer (see *e.g.*, Robards and Sleytr, 1985, pp. 99 ff; see also Murray *et al.*, 1989) or more efficient by rotating the injector, which is equipped with a simple propeller (Fig. 4a). Specifically for very thin specimens with a minimal heat capacity another phenomenon has to be regarded, which was already noticed by Elder *et al.* (1982) and analysed in detail by Ryan and Purse (1984, 1985; see also Ryan, 1992): cold gas layers above the cryogen surface can freeze thin specimens. Since this initial freezing by GN_2 takes place with a very small cooling rate, ice segregation may result. A simple measure to prevent this precooling is, to lift up the cryogen container as recommended by Elder *et al.* (1982) just before immersion (see again Fig. 4a). Finally it is very important, to freeze the specimen already on a carrier, which has sufficient mechanical stability together with a minimum heat capacity and fits into the chuck of the cryo-ultramicrotome used. Standard carriers (*"pins"*, Fig. 4b) are not suited due to their relatively large heat capacity (see Ryan and Purse, 1984, or Ryan, 1992). A good compromise is given by the cone shaped hollow carrier developed for the FC-systems of Reichert-Leica (Fig. 4c). The wall thickness of this carrier is only 0.2 mm, the heat capacity according to the weight of 59 mg very small. Nevertheless the stability of the tube/cone design is absolutely sufficient for cryosectioning of very hard frozen specimens. Such carriers can be used for the application of microdroplets of suspensions or even for the mounting of small bulk samples with diameters less than 0.5 mm. By grinding with silica abrasive paper (finest grain size, control under the stereo microscope, see Fig. 4c) a small platform is produced for deposition of the microdroplet or the small specimen. Viscous cell suspensions or sticky tissue samples can be easily applied to such platforms on the tip of the cone under the stereomicroscope. Low viscosity cell suspensions in water or samples with smooth or rigid surfaces mostly do not adhere sufficiently to the ground platform. In such cases a coating of the rough aluminum surface with a viscous gelatine suspension is advantageous. As far as cell suspensions in water are concerned besides good adhesion an additional advantage of gelatine coating results, if a nearly dry gelatine coat is used : the gelatine in this case takes up liquid by swelling and enriches the cell concentration in the remaining surface layer without osmotic effects.

Within studies of immersion technology several times cooling columns with heights exceeding 20 cm were used (see *e.g.*, Ryan, 1991, 1992). In normal laboratory practice such columns do not offer advantages: mostly a height of 8 to 10 cm is absolutely sufficient if small specimens and light weight carriers are used as described above. Also the velocity of the specimen (entry speed and speed through the ethane) is not that important as one may conclude from some measurements or pictures (*e.g.*, Fig. 5). Of course an extremely *"forced convection"* at the specimen surface increases the initial cooling rate, as Handley *et al.* (1981) clearly demonstrated. But if cryogen columns with a height of 10 cm or less are used for practical reasons and specimens with a diameter > 0.5 mm are frozen, such higher entry speeds do not offer a real advantage (Robards and Crosby, 1983): after standstill the specimen is not completely frozen and heat exchange between the warm center of the specimen

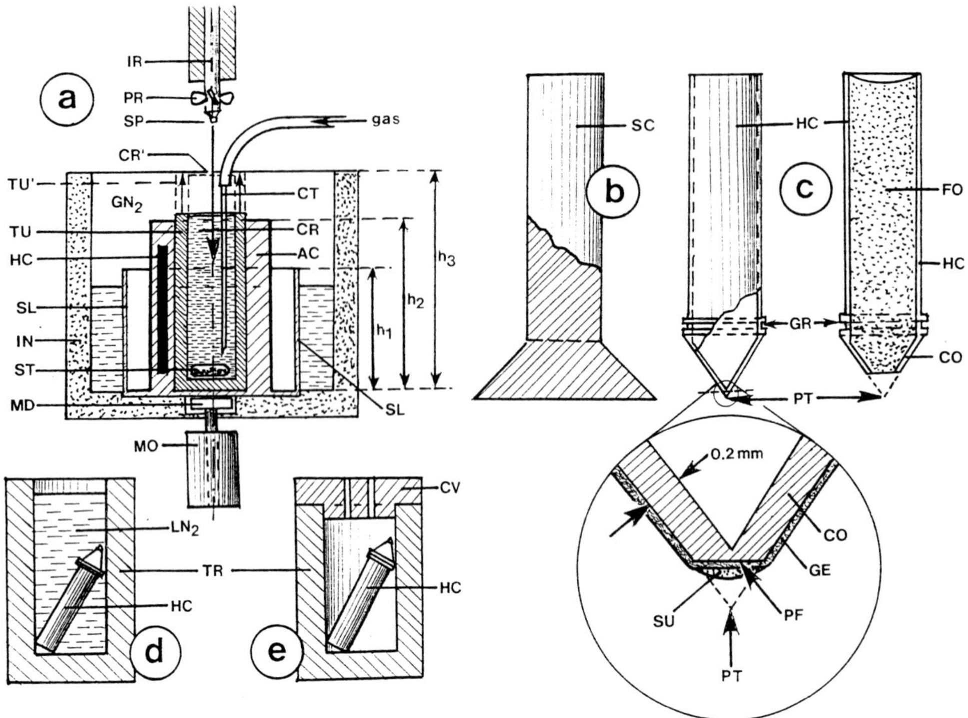

Figure 4. Essentials for immersion ("*plunge*") cryofixation and cryotransfer of plunge frozen specimens to the cryo-ultramicrotome. See Tables 3 and 4. Further explanations in the sections on "Ambient Pressure Feezing" and "Ultrathin Sectioning ...". **(a)** Plunging device (similar to Reichert KF 80 or CPC, see Sitte *et al.*, 1987a, Fig. 1b, p. 90): a thermally insulating container IN is divided into two compartments by a cylindrical sleeve SL. The aluminum container AC inside SL is thermostatically heated by the cartridge HC. In AC a tube TU is inserted for the secondary cryogen CR (*e.g.*, ethane), which is introduced through the capillary tube CT (connection TC to the container with pressurised propane or ethane). During liquefaction of the cryogen CR the level of LN_2 is lifted above h1 of SL, so that direct cooling of AC by LN_2 results. After liquefaction of a sufficient amount of CR (TU nearly completely filled) CT/TC are removed, LN_2-level again lowered below h1 and temperature of CR (AC/TU) set thermostatically slightly above its freezing temperature T_m (see Table 3). Immediately before immersion tube TU is lifted from h2 to h3, so that the specimen SP does not have to pass through cold GN_2. If larger bulk specimens with dimensions $\geq$ 0.5 mm have to be frozen, a convection of the cryogen after stand still at the end of the movement is recommendable. Such a convection is realised either by a magnetic stirrer ST driven by a magnetic dipole MD connected with motor MO, or more efficient with a small light weight propeller PR on the injector rod IR. IR starts speedy revolution after the superficially frozen specimen reaches its final lowest position in tube TU. **(b)** Standard carrier SC ("*pin*") of a cryo-ultramicrotome normally made of aluminium: this carrier is only suited for sugar protected specimens according to Tokuyasu (see Fig. 16) and for mounting a part of the slam frozen tissue slice (see Fig. 23), but not suited for plunge freezing of unprotected fresh samples. **(c)** Hollow cone carrier HC offered by Leica-Reichert for FC-systems FC4/FCS/FCR: the wall thickness of approx. 0.2 mm guarantees a very low thermal capacity and conductivity. A groove GR enables handling with the forceps inside the plunging device and cryochamber. The pointed tip PT of the cone CO is carefully flattened with fine grain abrasive paper as shown in the encircled inset below. If non viscous liquids (*e.g.*, suspensions in H_2O without protein content like blood plasm) have to be frozen, the cone must be coated with a "*sticky layer*" (*e.g.* gelatine GE) and only a very thin suspension layer has to be applied. Otherwise the suspension SU positioned on the ground platform PF is splashed away during speedy entry into CR. If larger bulk specimens with diameters exceeding 0.5 mm are to be frozen by immersion, a larger part of the cone has to be removed (see the right hand schematic diagram). In this case the hollow carrier HC' has to be filled with a low weight thermal insulator (*e.g.*, polyurethane foam FO) to avoid entry of water into the cavity which would increase heat capacity. **(d/e)** Cryotransfer of HC is either realised in a small transfer container TR below LN_2, or just "*dry*" within GN_2 in precooled TR sealed with a precooled cover CV.

Table 4. Details influencing the quality of immersion ("*plunge*") cryofixation

Details	Comments	References
Cryogen	**Ethane** for all cryosectioning below devitrification temperature T_d ~ -135°C and if real vitrification is needed (analytical grade ethane is expensive, handling more complicated because of high pressure) **Propane** is not recommendable for low temperature work < -135°C with respect to residual surface films on the specimens. Approx. 25 % lower cooling rates in comparison to ethane. There exists the risk of explosions due to incorrect handling (but less expensive and easier to handle) **Primary cryogens like LN2** are not suited under ambient pressure due to the "*Leidenfrost phenomenon*" (the same counts for partially frozen LN2 with respect to the small temperature difference between freezing and boiling temperature)	See Table 3 and Sections 2 and 3 (this review), Bald, 1994, Ryan, 1992, Sitte *et al.*, 1987a **Risks of propane** : see Sitte *et al.*, 1987b
Cryogen temperature	Liquid cryogens are normally used at a **temperature near the melting point** T_m. **Subcooled liquid cryogens** below the melting temperature Tm are more efficient, but mostly not recommendable (not stable: tendency to solidify)	
Cryogen container	Usually a **thermostatically heated container** of aluminium (wall thickness ≥ 5 mm) with a **height of 8 to 10 cm** and a **diameter of approximately 2 cm** guarantees reproducible temperatures without disturbing gradients. A height >> 10 cm does not pay. Stirring of the cryogen is not needed for small specimens (diameters ≤ 0.5 mm) on low weight carriers	See Figs. 4 and 5 (this review) and Handley *et al.*, 1981, Robards and Crosby, 1983, Robards and Sleytr, 1985, Ryan, 1992, Sitte *et al.*, 1987a
Specimen carrier	**Hollow cone carrier** or **4 μm-titanium foil sandwiches** deliver good results. Standard carriers ("*pins*" according to Fig. 4b) are not suited for unprotected highly hydrated biological specimens due to their high heat capacity (low cooling rates)	See Figs. 4c and 5a (this review). See also Handley *et al.*, 1981, Ryan and Purse, 1984, 1985, Ryan, 1992, Sitte *et al.*, 1987a
Immersion speed	For most specimens a **speed between 1.0 and 1.5 m/sec** is sufficient. Higher speeds up to 10 m/sec offer only an advantage, if an aerodynamically shaped and very pointed edge is formed and if the sample is protected by a fairly rigid metal foil (*e.g.* 4 μm-titanium) against the shock connected with the high-speed entry into the cryogen. Microdroplets on hollow cones according to Fig. 4c are mostly blown away at entry speeds approaching 10 m/sec. Similarly sandwiches between planchettes according to Fig. 3b are not stable enough to withstand this impact. For sandwiches double jet freezing is a helpful alternative in most cases	See Figs. 4c and 5a (this review). See also Handley *et al.*, 1981, Robards and Crosby, 1983, Robards and Sleytr, 1985, Ryan, 1991 and 1992

and the surrounding areas results. Usually such phenomena cannot be avoided completely by a magnetic stirrer, since the relatively slow convection of the liquid cryogen is not comparable with the "*forced convection*" occurring by an entry speed of *e.g.*, 10 m/sec as used by Handley and co-workers. Only in such exceptional cases remarkable differences can be observed between slow and rapid movements of the sample (see legend of Fig. 5). Summarising all discussed facts one can conclude, that immersion cryofixation of fresh specimens for subsequent cryosectioning should be made under the experimental conditions listed in Table 4 in order to get optimal results.

The elegant spray-freezing method according to Bachmann and Schmitt-Fumian (1973, see also Fig. 3c) is a special modification of plunge freezing of

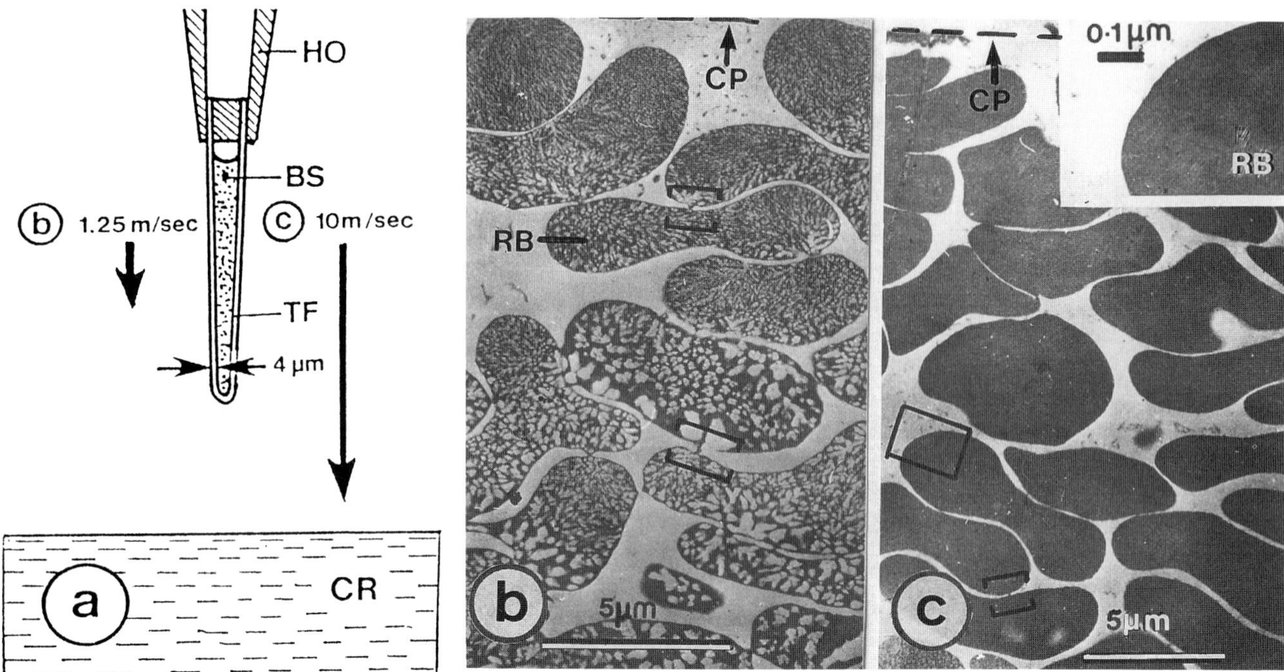

Figure 5. Under certain preconditions the results of immersion ("plunge") cryofixation depend on the speed of the specimen. See Handley *et al.* (1981), Sitte *et al.* (1987a). Further explanations in the section on "Ambient Pressure Freezing". Electron micrographs (b) and (c) reproduced with permission. **(a)** Specimen set-up of Handley *et al.* (1981): a blood sample BS is inserted between a V-shaped 4 μm thick titanium foil TF, mounted in a rigid low weight holder HO. By a spring loaded device HO/TF/BS is moved with adjustable speed into a secondary cryogen CR (Handley *et al.* used subcooled Freon 22 at -164°C). **(b)** Results obtainable with normal immersion speed of 1.25 m/sec. Arrow CP and dashed line show the position of the contact plane between CR and TF. Note the rapidly increasing ice segregation by insufficient cooling rate clearly visible in the red blood cells RB. Frozen suspension BS freeze substituted in OsO_4 acetone according to Van Harreveld and Crowell (1964) and embedded in epoxide. Ultrathin sections stained with uranyl acetate. **(c)** Cooling rate considerably enhanced by extremely high entry speed (10 m/sec) due to "*forced convection*". Note the absence of visible segregation artefacts in a border layer of approx. 20 μm. The conditions and results correspond closely to double jet cryofixation of sandwich specimens. It makes no sense to shoot larger bulk specimens with such a speed into a liquid cryogen of a height h2 $\leq$ 10 cm (see Fig. 4a), since without special measures after stand-still rediffusion of heat causes secondary phenomena and changes the well frozen layer.

Figure 6. (*on facing page*) Slam freezing followed by freeze substitution in OsO_4/acetone and epoxide embedding ("*hybrid method*") according to Van Harreveld and Crowell (1964) is a very fast procedure, which delivers results already over night. This hybrid method is very well suited for morphology and morphometry on resin blocks sectioned at ambient temperature on a standard ultramicrotome. It is often not suited for histochemistry since OsO_4 tends to damage a lot of antigens. One of the criteria for a good freezing quality (extremely fast cooling rate) is the dense appearance and the sharp delineation of heterochromatin (arrows) in the nuclei of slam frozen samples. In both cases (a) and (b) the nuclei are located in a distance of approx. 7 μm from the border CP of the samples (contact area between sample and metal mirror plane). Both samples are frozen on a simple LN2 cooled copper mirror without vacuum protection. Sections are stained as usual with uranyl acetate and lead citrate. Further explanation in the sections on "Freezing and the Frozen State of Water", "Ambient Pressure Freezing" and Discussion. See also description and discussion in Sitte *et al.* (1987a). **(a)** Frog sartorius muscle frozen on the inverted slammer according to Edelmann (1989b). Specimen and electron micrograph courtesy of L. Edelmann. **(b)** Neutrophil granulocyte frozen on Reichert MM80-system according to Fig. 7. Specimen and electron micrograph courtesy of E. Morgenstern.

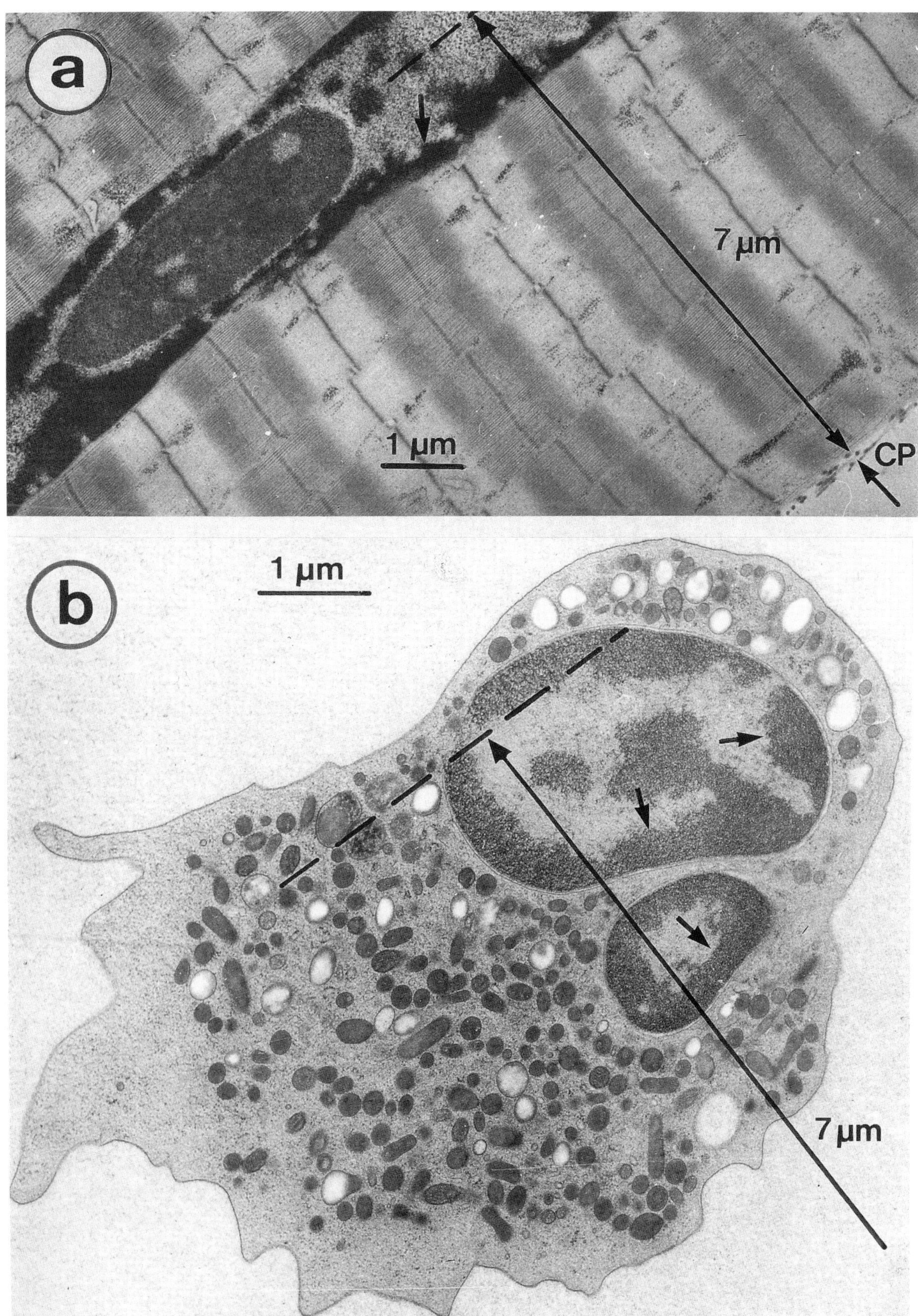
a
7 µm
1 µm
CP
b
1 µm
7 µm

Figure 7 (*on facing page*). Air damped slamming device with an LN_2 cooled metal mirror similar to the Leica-Reichert MM80. See also Figs. 3d, 7d, 8 and 9. Further explanations in the sections on "Freezing and the Frozen State of Water", and "Ambient Pressure Freezing". **(a)** Schematic diagram of the whole apparatus: the thermally insulated cryochamber CC within the base is partially filled with LN_2, which cools directly the aluminium cylinder AC with the metal block MB (mirror-like polished surface MM) to a temperature of approx. -190°C. After pushing the release button RB (arrow) the injector rod IR moves downward. The movement is accelerated by an adjustable helical spring HS to values between 1.5 and 2.5 m/sec at the impact of the specimen (*e.g.*, tissue slice TS) on the mirror plane MM. Bouncing is suppressed by minimum inertia of the hollow cylinder HC sliding in IR and by "*air bag*" AB between IR and HC (see "b" and Fig. 8). **(b)** Position of HC just before first contact between tissue slice TS and MM: since IR due to inertia continues the downward movement, air inside AB becomes increasingly compressed and starts to escape through the holes HO (arrows). The specimen carrier consists of a stainless steel plate SS with a foam support FS for the specimen (tissue slice TS). The extremely thin separation foil SF of polypropylene (the thinner the better) between FS and TS facilitates the separation of the specimen from the foam support after freezing. The loaded specimen carrier SS/FS/SF/TS is simply attached to the magnetic plate MP on the bottom of the injector IR. **(c)** View from top on the mirror plane MM of MB after removal of the frozen tissue TS: the borderline of TS is clearly visible by a grainy frost layer GF formed by the water vapour escaping from the side walls of the warm specimen immediately before freezing. Similar frost GF' is deposited, if a bubble of GN_2 was trapped between the specimen TS and MM. **(d)** The corresponding complementary contact plane to (c) after turn of the specimen TS by 180° around horizontal axis BB shows a grainy surface GS on the area formerly covered by the GN_2 bubble. The remaining surface MS is a mirror-like shiny replication of MM. **(e)** Specimen TS broken along the line BB in schematical drawing (d) in side view shows bright reflections BR of hexagonal ice crystals and the dark amorphous border layer AB on the well frozen mirror replication surface, if intense dark field illumination by fibre optics and sufficient magnification of the stereomicroscope are used. The dark border layer AB of the amorphously frozen (vitrified) contact area with MM disappears completely in the area GS formerly covered by the GN_2 bubble. **(f)** Compression phenomena according to Fig. 9 can be reduced by a carefully adjusted spacer ring SR around the specimen SP placed on propylene separation foil SF. The specimen SP or a suspension is either mounted inside SR or positioned on an intermediate layer IL (*e.g.*, liver homogenate, blood plasma, soft gelatine or β-gel according to Hanyu *et al.*, 1992). Suspensions can be directly filled into SR instead of SP/IL. The optimum height h and consistency of SR and the best intermedium IL have to be found empirically by trial and error for each specimen. Often thin (*e.g.*, h $\leq$ 0.5 mm) plastic foils with punched holes (diameter approximately 5 mm) are the best solution for slamming of suspensions (Warning: it is not easy to establish the best conditions, but this is the only way to obtain success reproducibly !).

---

microdroplets. To the best of my knowledge this method has not yet been used for subsequent cryosectioning: spray-freezing is therefore not included in this discussion, but probably remains an alternative worthy of consideration.

Immersion cryofixation is characterised by a travelling specimen which enters into a resting cryogen. In jet freezing (sometimes also called "*spray-freezing*", see *e.g.*, Bald, 1985) the conditions are just the inverse: the specimen rests in a firm position and a liquid cryogen is moved. Using one or two jet nozzles and an extrusion of the cryogen by a rather high pressure much higher relative speeds "*cryogen vs. specimen*" are possible, than with immersion systems. This thermodynamic advantage is coupled with the drawback that the specimens have to be protected from the mechanical impact of the cryogen jet by protection foils. Of course these foils enhance the heat capacity and diminish the cooling rates, if they are not extremely thin (Handley *et al.*, 1981; Ding *et al.*, 1992).

Jet freezing was used since 1978 mainly for freeze fracturing (M. Müller *et al.*, 1980; Plattner and Knoll, 1983) and sometimes in studies using hybrid methods like freeze substitution or freeze drying. According to my own knowledge it has never been used for subsequent cryo-ultramicrotomy. But it may be useful in future, since the recent generation of double jet systems (I refer exclusively to the Balzers JFD 030 - Jet-Freezer according to the principle of M. Müller *et al.*, 1980) work really reproducibly and reach evidently extremely high cooling rates (see Th. Müller *et al.*, 1989). Such double jet systems take advantage of the bi-directional heat extraction according to Fig. 3b. The precondition for this bi-directional heat extraction seems to be an exact synchronisation of both jet streams on a really thin specimen, which was according to obtainable second-hand information mostly not guaranteed by earlier or other double jet systems. According to Ding *et al.* (1992) the JFD 030 system "*vitrified*" plant material nicely within a thickness of

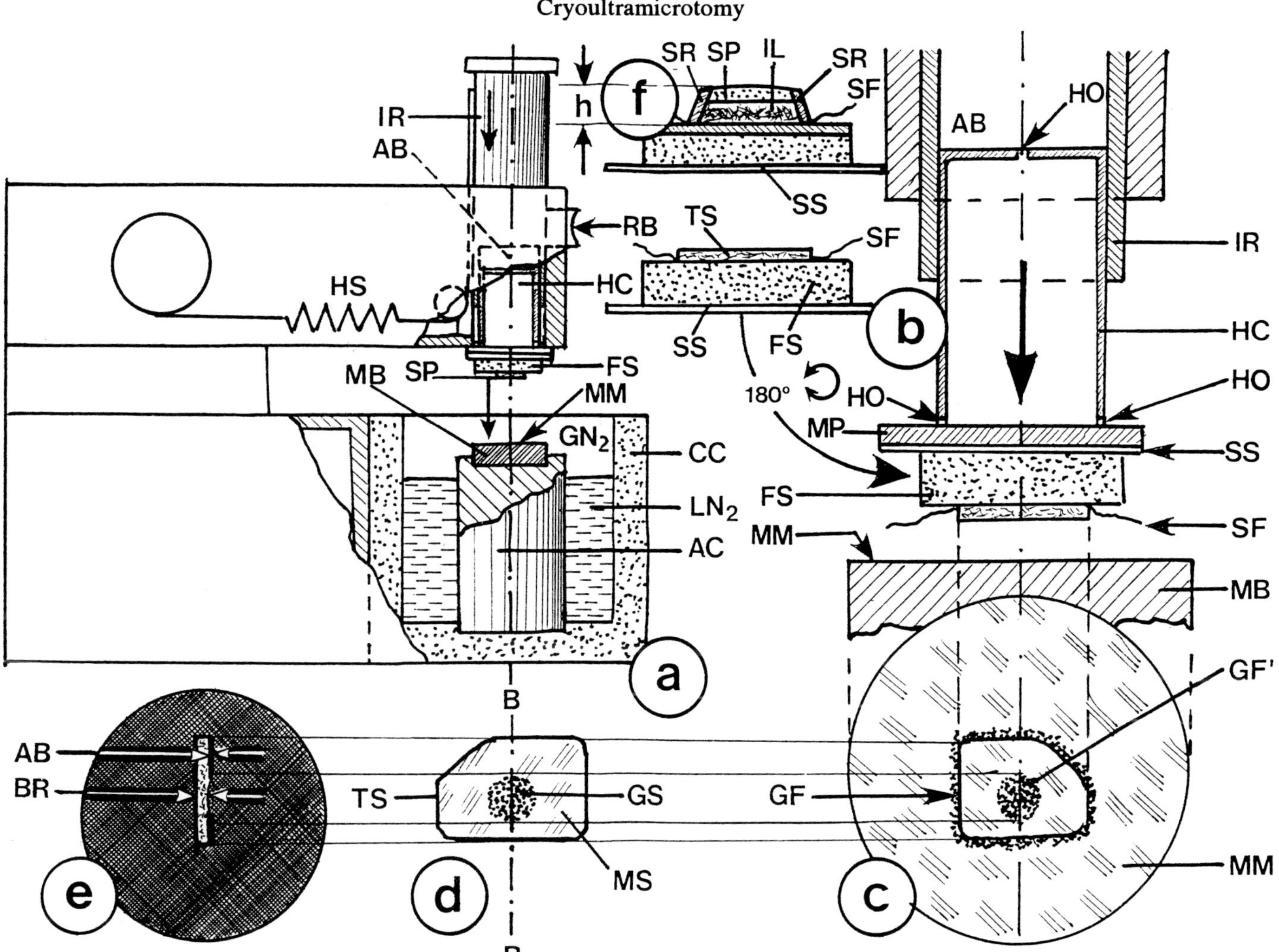

60 μm, if two titanium sandwich foils of minimum thickness (4 μm) in analogy to Handley *et al.* (1981) are used. Since Ding et al. used only freeze substitution, no evidence of real vitrification to I_v could be shown, but in agreement with other reports and results one can expect a proper freezing. Although the gluing of the frozen sandwich onto a specimen carrier for cryo-ultramicrotomy (see the section on "*cryo-ultramicrotomy*") may be tricky it seems to be a promising alternative to immersion or impact cryofixation if one enters into the development of a suitable follow-up technology. This follow-up procedure for cryosectioning must include at least the elimination of propane traces (*e.g.*, by liquid ethane) and a tight mounting of the thin frozen lamella (thickness approximately 60 μm) on a stable carrier similar to the gluing procedure for impact frozen layers on standard pins (see Fig. 23).

Probably the most powerful but also certainly the most tricky method for rapid freezing is the impact cryofixation, usually denominated as "*slamming*" or "*metal mirror cryofixation*". In this case heat extraction results from the contact between the specimen and a highly polished mirror-like solid state surface. This cooling method was already tried by Altmann in the last century without considerable success. It was applied by Eränkö 1954 for histochemistry and finally introduced by Van Harreveld and Crowell (1964) into electron microscopy. Most electron micrographs of slam frozen specimens were obtained with hybrid techniques (freeze substitution, freeze drying, resin embedding at ambient or low temperature, see *e.g.*, Armbruster *et al.* (1984), Carlemalm *et al.* (1982, 1986), Edelmann (1989a,b, 1991a,b, 1994a,b), Hobot *et al.* (1985), Humbel and Müller (1986), Menco (1986), Sitte *et al.* (1988, 1994), Steinbrecht and Müller (1987), but also successful application for subsequent cryo-ultramicrotomy was demonstrated in a couple of papers (*e.g.*, Edelmann, 1992; McDowall *et al.*, 1983). The results are often fascinating (see *e.g.*, Heuser *et al.*, 1979; Sitte *et al.*, 1987a, p. 96; Van Harreveld and Crowell, 1964), the drawbacks and difficulties mostly not known because they are not described: frustration of beginners reaches therefore a rather high level. Besides the power and the specific advantages therefore mainly the diversity of drawbacks of this elegant, absolutely recommendable and necessary method will be mentioned. Additionally some current misunderstandings concerning slam freezing will be explained and corrected.

Not to discourage all prospective applicants I want firstly to mention the striking advantages and the necessary instrumentation. Mirror freezing is really the only method which is suited for larger specimens. This concerns both the surface area for rapid freezing and the total volume of the sample. Areas around 1 cm^2 can be frozen without severe problems, *e.g.*, tissue slices in thicknesses between 0.1 and 1.0 mm (the thinner the better) sectioned either with a slicer (*e.g.*, a modified Stadie-Riggs slicer, see Sitte *et al.*, 1987a, p. 86, Fig. 4d) or with a vibratome (Bold, 1995). But also larger pieces of tissue *e.g.*, a segment of a kidney or liver with a volume of 1 ml or even more can be frozen on the metal mirror. The reason, why this is possible contrary to all other rapid freezing methods is simple: the heat conductivity of the just frozen border layer of the specimen is extremely low in comparison to the excellent heat conductivity of the mirror material (usually copper). Since the heat capacity of the copper block in comparison to the sample is extremely high, the temperature of the border region of the copper mirror remains practically unchanged during the whole freezing procedure, that means around -190°C for an LN_2 cooled mirror block. This is completely different from immersion cryofixation, jet and high pressure freezing, which all need very small specimen dimensions around 0.1 mm at least in one dimension. For some studies larger areas or volumes are of paramount importance: this concerns all cases, where with respect to changes of the object extremely speedy preparation is needed (*e.g.*, kidney cortex, central nervous tissue) or where dissection of a sensitive specimen is not possible without severe mechanical damage (*e.g.*, many soft tissues, which disintegrate if dissected into small pieces).

Similar to measurements during plunging, correct measurements of cooling rates at the border between specimen and mirror plane during slam freezing are not possible. But measurements in some distance from the border (Escaig, 1982, 1984; Escaig *et al.*, 1977) and results tell us, that slam freezing enables extremely high cooling rates - probably the highest obtainable. By immersion cryofixation in our laboratory (jet freezing was not available and is not considered) we could never obtain the same beautiful preservation of chromatin and other sensitive structures (see Sitte *et al.*, 1987a) as with slam freezing. These pictures (see Fig. 6) result obviously from extremely high cooling rates and could be obtained on LN_2 cooled copper mirrors with simple apparatuses (see Sitte *et al.*, 1987a; Edelmann, 1989b). Fig. 7a shows the principle of one of the used instruments, which is basically identical with the Reichert KF 80/MM 80 (new version CPC/MM 80). The copper block in those instruments is directly cooled by LN_2. The specimen is mounted onto an air damped low weight injector (Fig. 7b), which excludes or at least diminishes bouncing (see Boyne, 1979; Phillips and Boyne, 1984). Following release of the injector the specimen approaches finally with a speed between 1.5 and 2.5 m/sec (adjustable) against the mirror plane. A simple return stop element prevents a reverse upward motion of the injector after the first contact between specimen and mirror. Finally: the pressure of the injector system with the specimen against the mirror surface is increased automatically after the first contact and superficial freezing with a delay of approximately 5 msec. The border layer is therefore already frozen before this pressure is developed. This measure prevents distortion (bending) of thicker specimens by one-sided cooling, which can result in a partial separation of the specimen surface from the mirror plane and a subsequent recrystallisation of those areas.

In practical use, some specific peculiarities have to be regarded. For example: a gas bubble (GN_2) can be trapped between the mirror and the specimen surface. This concerns especially larger areas exceeding 50 mm^2. Areas of such gas bubbles are easily recognised by frost deposition on the mirror surface and a grainy appearance on the corresponding complementary surface of the frozen sample as shown in Fig. 7c and d. After complete freezing (finished after 15 sec or less for tissue slices in thicknesses < 1 mm) the frozen specimen is separated from the mirror and from its support and turned upside down. Both complementary areas on the specimen and on the mirror plane are inspected carefully with the stereomicroscope using a bright cold illumination. The former contact area on the mirror surface is surrounded by a grainy frost layer resulting from water evaporation from the side walls of the warm specimen to the cold mirror immediately after the first contact between specimen and mirror. If gas was trapped by the specimen surface, those areas show the same grainy frosting, too. The well frozen specimen surfaces represent a perfect mirror-like reflecting replica of the highly polished mirror surface. Only non-contacting areas due to gas inclusions have a grainy appearance which is completely different from the shiny replication and cannot be overlooked. Such areas have to be discarded since they are badly frozen and always show severe segregation artefacts. Another control of the quality of

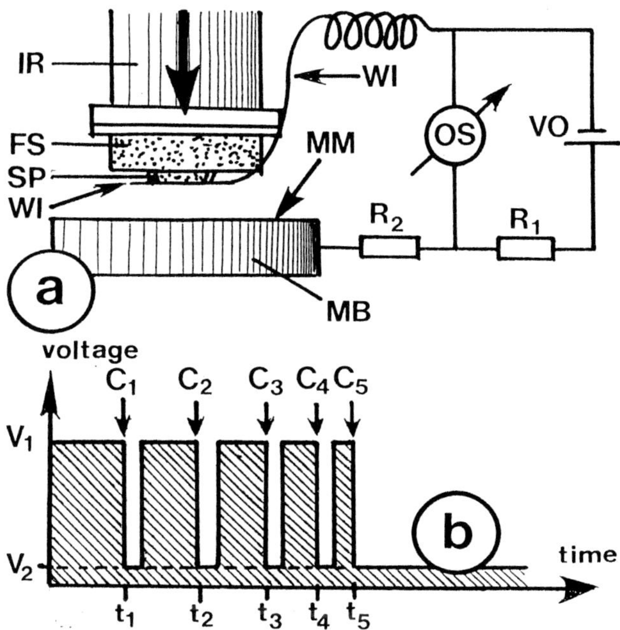

Figure 8. "*Bouncing*", i.e., repelling of a soft specimen SP from the rigid metal mirror surface MM is a common problem of slamming devices for impact cryofixation of biological samples. It can be recorded according to Boyne (1979). Further explanation in the section on "Ambient Pressure Freezing". **(a)** Set-up consisting of a voltage source VO connected by an extremely thin flexible wire WI to the surface of the specimen SP (see Fig. 7) and with the metal block MB via resistances R1 and R2. Recording of resistance or voltage with oscilloscope OS. **(b)** The oscilloscope OS registers a first contact C1 between wire WI and metal mirror MM at t1, followed by a separation due to bouncing (repelling) till C2 at t2 and so on ...C3 → C4 → ... Cn. Between the very short periods of contact there are periods of separation in the msec-range. It is understandable, that under such preconditions optimum freezing is not possible. It is therefore worth checking any slamming device with the specific set-up of specimen and specimen support and to change all influencing parameters (*e.g.*, speed of application, return stop, specimen support, spacer arrangement, specimen geometry) empirically by trial and error up to the moment that bouncing is reproducibly suppressed. One will find, that most of the slamming devices bounce and that suppression of bouncing is often an extremely frustrating and tough job. Sometimes it cannot be achieved at all.

----

cryofixation is possible by inspection after breaking the frozen specimen at -196°C in LN_2 filled styrofoam box (Fig. 7e). In side view the well frozen contact areas show in the dark field with a good stereomicroscope at higher magnification and fibre optic illumination no structure: they are just dark. In comparison to these vitrified border layers the deeper layers show clearly the reflections of larger hexagonal ice crystals. In areas of gas inclusions these bright crystalline reflections reach the free surface and no amorphous dark layer is visible. If these controls are made correctly, mistakes in the preparation are mostly excluded. Separation of the frozen specimen from the support (mostly a rubber foam) is simply obtained by an extremely thin film of an anti-adhesive material between sample and support. Mechanical compression of soft tissues can be reduced to some extent by spacer elements (*e.g.*, rings around the sample). Suspensions are easy to freeze using such spacer rings as shown in Fig. 7f (see also Gutensohn, 1993; Sitte *et al.*, 1987a, Fig. 7, p. 103).

The most severe drawbacks of slam freezing are bouncing and specimen compression. Both depend among other influences on the system used and the mounting of the specimen on the injection device. Boyne (1979) first observed and analysed the absolutely unexpected bouncing phenomenon with a simple set-up. He found, that the specimen after the first contact with the mirror is usually repelled, that means, that the contact between object and mirror plane is again interrupted. This procedure "*contact - interruption - contact - interruption...*" (Fig. 8) is repeated several times in the msec-range. Finally the specimen remains in a continuous contact with the mirror plane. Since freezing of the border layer occurs in fractions of a msec this bouncing phenomenon may provoke several "*freezing - thawing - freezing...*" or "*freezing - recrystallisation*" procedures. Structural preservation by such a bouncing slammer is bad - and many systems bounce against all precautions and expectations. It is therefore worth checking each system, that bouncing is properly suppressed before use according to Fig. 8 and to modify the parameters (speed of application, specimen support, specimen geometry, return stop, spacer elements etc.) before the start of experimental work. This check has to be repeated if one of the important parameters is changed (*e.g.*, freezing of another object with different consistency or geometry). According to my own experience and advice from literature and colleagues, pneumatic systems (see *e.g.*, Boyne, 1979; Sitte *et al.*, 1987a; Trachtenberg, 1993) have the best damping conditions and are to some extent bounce-proof.

In contrast to the completely unexpected bouncing phenomenon, everybody was aware that considerable compression or distortion artefacts will occur, if a soft specimen hits the rigid mirror plane. If the specimen

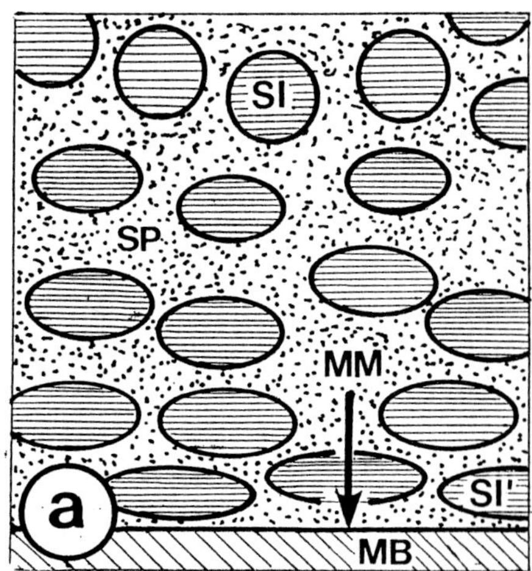

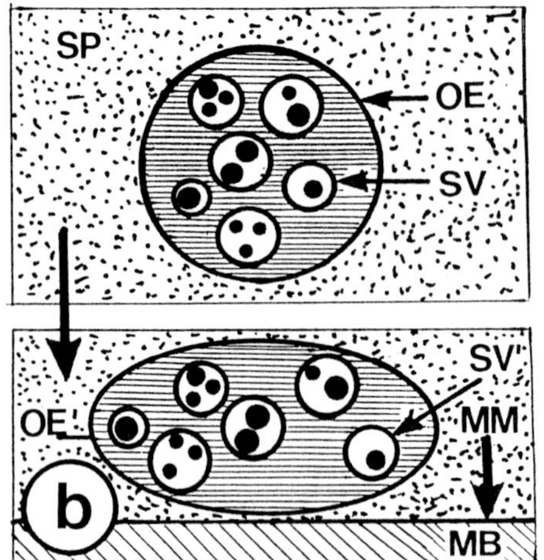

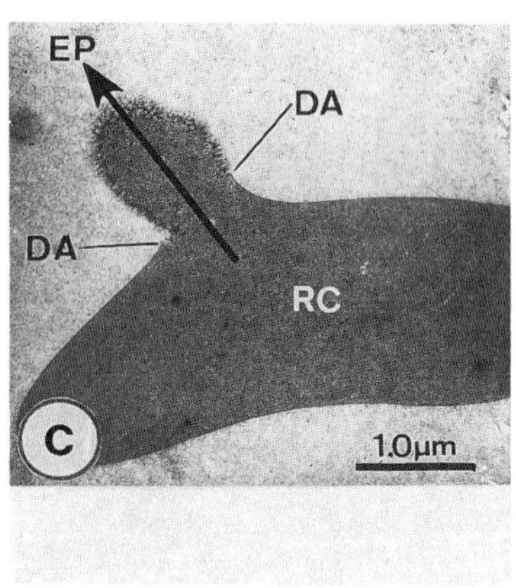

Figure 9. Compression (schematical drawings "a" and "b") and ruptures (electron micrograph "c") by slam freezing. Further explanations in the section on "Ambient Pressure Freezing". **(a)** Against former expectations according to Bold (1995) the most severe compression artefacts occur just in the well frozen border layer of the specimen SP, that means in the immediate vicinity of the contact plane between SP and metal mirror surface MM on the metal block MB. The deformation of originally spherical inclusions SI change these elements to rotational ellipsoids SI' of identical volume. The extent of compression depends on size and rigidity of the structures: smaller structures like mitochondria or tiny vesicles show less deformation than larger inclusions. Structures with some kind of a skeleton like nuclei or small cells are more resistant against compression than free membrane systems without stabilising skeletons. Deeper layers show less compression than the border layer near the contact plane between specimen SP and the metal mirror MM. **(b)** According to Bold (1995) different structures within the same area (*e.g.*, an outer envelope OE including small vesicles SV, original size before slamming in the upper half of the diagram) are compressed in different extents to OE' and SV' after slamming (lower half of diagram). They behave like a collection of rigid golf balls in a thin walled balloon. For example glomerula of the kidney cortex are deformed according to this schema. **(c)** Ruptures of the plasmalemma of single cells in suspension, which look like explosions in the direction of the pressure wave are not necessarily an artefact generated by the impact of the suspension on the metal mirror as argued formerly (see Sitte *et al.*, 1987b; DA = Damage of plasmalemma, EP = exit of cell plasma, RC = red blood cell). A recent study of Frei (1992) gave some support that such "*eruptions*" may be a result of the freezing process itself, which runs also in the direction of the arrow. The electron micrograph shows a slam frozen human red blood cell in human blood after freeze substitution in OsO_4-acetone according to Van Harreveld and Crowell (1964), followed by epoxide embedding. Ultrathin section stained with uranyl acetate and lead citrate. Electron micrograph courtesy of E. Morgenstern.

---

is not protected by a spacer (ring) element or is comparatively rigid as thin slices of muscle tissue, it looses completely its geometry: spatial relationships within tissues in such cases are completely changed in a way, which is not predictable. It was several times claimed or perhaps expected, that "*freezing is faster than compression*" (see *e.g.*, Escaig, 1982; Elder *et al.*, 1982). That should mean, that compression is not that important in the well frozen (vitrified) border layer of the specimen and that severe distortion artefacts occur only in the unusable badly frozen deeper layers of the sample. We have studied these phenomena on our system (S. Müller, 1993; Bold, 1995). Unfortunately it turned out during these investigations, that the worst case is reality: the most severe compressions mostly occur just in the well frozen border layer (Fig. 9a). The situation is somewhat confusing, since different structures are deformed to a different extent (see Fig. 9b). Generally smaller structures in the µm-range (*e.g.*, granules, vesicles, mitochondria, ER-profiles, protocytes) or very rigid structures with a skeleton (*e.g.*, nuclei, even some small cells) show less compression than larger structures. For example, the mainly spherical glomeruli in the kidney cortex are much more compressed than the smaller structures within or around those glomeruli. The situation corresponds to a collection of small rigid golf balls suspended in a thin walled balloon: if forces are applied, only the balloon is deformed - the small balls do not change their shape considerably. But the spatial relationships of the smaller spheres are completely altered. This should always be considered if mirror frozen specimens have to be investigated.

Finally, sometimes after slam freezing in cell suspensions defects on the outer cell membrane (plasmalemma) are observed, where cytoplasm exits (Fig. 9c). These openings or "*eruptions*" are always located on the cell surface opposite to the contact area between suspension and metal mirror. It was therefore argued, that this phenomenon is a result of the shock wave generated by the impact of the suspension onto the rigid metal mirror surface (see Sitte *et al.*, 1987a, Fig. 7 on p. 102). In the meantime this phenomenon was reinvestigated by Frei (1992). There are strong indications, that those ruptures are created by the freezing process itself, which has obviously the same direction as the pressure wave created by the impact shock. Further studies of this phenomenon including other freezing methods (*e.g.*, sandwich freezing by jet or high speed immersion according to Handley *et al.*, 1982) are needed, before definite statements concerning the generation of these artifactual ruptures are possible.

At least the discussed phenomena of bouncing and compression are inherent to slamming. Concerning bouncing there is only a gradual difference between different systems, *e.g.*, reduced bouncing tendency due to pneumatic damping or the application of β-gels according to Hanyu *et al.* (1992). The most important influence on compression phenomena is possible by a careful individual adaptation of spacer elements and support arrangements for each specimen: it seems that different specimens behave completely differently. It is also difficult to answer the question, which commercially available mirror freezer is the best or most convenient for a specific purpose. Probably the best will be to check different systems with one's own specimens of interest. In contradiction to several claims and calculations, our own practical experience and published pictures tell us that results obtainable with LHe cooled mirrors are not substantially better than results with LN_2 cooled mirrors [see in this respect the papers of Bald (1983, 1985), Escaig (1982, 1984) Heuser *et al.*, (1979), Kopstad and Elgsaeter (1982)]. Perhaps it was unfortunate, that Heuser and co-workers obtained their breathtaking pictures of synaptic vesicle release with a LHe slammer: practically everybody at that time drew the probably wrong conclusion, that such beautiful results are only obtainable with LHe cooled mirrors. But there is on the other hand some agreement between a considerable number of experienced scientists, that LHe does not offer a significant advantage in comparison to LN_2. From my own experience with LN_2 cooled systems a clear recommendation for the more complicated LHe technology is not possible today. But I must confess frankly, that a correct comparison up to now was not made, because the different authors were using completely different slamming devices for different objects. Following recommendations of Heuser practically exclusively copper mirrors were applied for LHe cooling. However, convincing calculations of Bald (1983) show, that copper mirrors cannot offer an advantage, if LHe is used instead of LN_2: only a silver mirror seems to be suited for the low temperature range around 15K. Finally, large differences in deeper layers of specimens (*e.g.*, zones in distances 20 μm and more from the contact area between specimen and mirror plane) cannot be expected with respect to the bad heat conductivity of the already frozen layer and the rapidly decreasing temperature gradients between mirror surface in the freezing frontier entering into the specimen. There may be a difference in the freezing of *e.g.*, a 1 to 5 μm border layer of the specimen, if a LHe cooled silver mirror is applied at 15K. Vacuum may also be important to eliminate surface coats from the mirror, which perhaps influence freezing at these very low temperatures of 15K and below [see Escaig (1982, 1984), and Wendt-Gallitelli *et al.* (1980)]. Recently Hanyu *et al.* (1992) reported about an improvement by a 2.45 GHz microwave irradiation of the sample starting 50 msec before the first contact between specimen and mirror plane. They believe, that this irradiation before and during the freezing procedure may crack H_2O-pentamers responsible for homogeneous nucleation. Also this report does not allow a correct comparison and a definite conclusion on the benefits of that somewhat complicated set-up. Up to now all the results presented with LHe, vacuum and microwave technology does not differ significantly from our own results obtained with the above mentioned simple LN_2 systems according to Sitte *et al.* (1987a), or Edelmann (1989b), working without LHe, without vacuum and without microwave irradiation. But there may be some potential to improve the situation for the already mentioned surface layer of 5 to 10 μm, which can be important, if true ambient pressure vitrification is wanted with respect to cryo-ultramicrotomy of unpretreated samples frozen with an extremely high cooling rate above 10^{6}°C/sec.

Concerning the other extreme: simplification of the slamming procedure by do-it-yourself instruments with a home made copper block according to Dempsey and Bullivant (1976) looks extremely convincing but is extremely frustrating in comparison with a commercial slammer. According to our own experience only in exceptional cases we could obtain useful results in this way, probably because a continuous sufficient contact between the mirror block and the specimen holder, operated with the right and left hand of the same person, was not possible (see Sitte *et al.*, 1987a, Fig. 4a, p. 96).

High Pressure Freezing

The basics of this method have already been discussed (see last part of the section "*Freezing and the Frozen State....*"). High pressure freezing was introduced by Moor and Riehle already in 1968 [see also Hunziker *et al.* (1984), Kaeser *et al.* (1989), Moor (1971, 1973, 1987), Moor and Hoechli (1970), M. Müller and Moor (1984), Riehle (1968), Studer *et al.* (1989)]. After long years of development and experiments the first commercial high pressure freezer of Balzers became available around 1985. In the meantime two slightly different high pressure systems are on the market [see M. Müller and Moor (1984), Sartori *et al.* (1993), and Studer *et al.* (1989, 1995)] and a lot of experience has been accumulated. Since 1985 this principle was successfully used to vitrify biological samples for subsequent cryo-ultramicrotomy and frozen-hydrated investigation in the TEM [see *e.g.*, Hohenberg *et al.* (1994), Michel *et al.* (1991, 1992), Studer *et al.* (1995)] as well as preparations with hybrid methods (see *e.g.*, Allenspach (1993), Hippe *et al.* (1989), Hippe-Sanwald (1993), Hohenberg *et al.* (1994), Hunzicker *et al.* (1994), Studer *et al.* (1989)]. In many respects high pressure freezing opened up a new era in frozen-hydrated work since for the first time homogeneously vitrified samples for cryosectioning were available. Within a short time it turned out, that comparatively large flat and astonishing perfect cryosections were much easier obtainable with such pressure frozen specimens (see Figs. 10 and 11).

The high pressure system is not as complicated as mostly expected (for physical and technical details see *e.g.*, Moor (1987), and M. Müller and Moor (1984). The hydraulic generation of the pressure of 2 kbar by pistons represents standard pressure technology. The most difficult problem was correct synchronisation of the pressurisation and the cooling procedure, since it became evident that such a high pressure produces within short times artefacts in living systems [see *e.g.*, Ding *et al.* (1992), Moor and Hoechli (1970), Moor *et al.* (1980), Studer *et al.* (1995)]. Firstly the pressure of approximately 2 kbar has to be developed by the hydraulic system to guarantee the full reduction of T_m and T_h (see Fig. 1a) for subsequent freezing. Immediately after the pressurisation the rapid cooling by LN_2-double jet must start. Heat extraction needs some time - more time in the center of the specimen than in a border region, even if the LN-double jet starts immediately after complete pressurisation. A proper synchronisation of pressurisation and cooling is therefore of great importance. This synchronisation is achieved in an elegant and very simple way by a certain amount of alcohol introduced into the tubing, which later on guides the LN_2 to both jet nozzles (see Fig. 12). If pressurisation starts, for 15 msec 2 ml of ethanol or isopropyl alcohol at ambient temperature is shot against the specimen container during the time that pressure is built up. With the delay of these 15 msec after the alcohol LN_2 enters through the nozzles into the pressure chamber and hits the specimen container from both sides. As mentioned before, LN_2 under this pressure does not boil and is therefore an excellent cryogen with a high specific heat and an extremely low viscosity (see Table 3).

Fig. 12c shows the sandwich elements for high

Figure 10 (*on facing page*). Martin Michel obtained the first beautiful flat cryosections free of artefacts from high pressure frozen discs (diameter 2 mm) of Golden Delicious apple leaves. Sometimes good cryosections were obtained with H2O-coated balanced broken glass knives (Michel *et al.*, 1991), but better reproducible cryosections were achieved with diamond knives (Michel *et al.*, 1992) using an ioniser. The electron micrographs (a) and (b) of the amorphous frozen-hydrated cryosections show representative areas of such cells: VA = vacuoles filled with a network of thick strands are existent besides other vacuoles VA' without this internal structures. CH = chloroplasts with the typical compound membranes of the thylacoids TH, NU = nucleus, MI = mitochondrium, ER = endoplasmic reticulum. The nuclear envelope NE shows the well known pore complexes (arrows). The cell walls CW are well sectioned. All sections show a severe compression up to 50 % perpendicular to the direction of sectioning (arrows in circles). The most astonishing fact is the nice contrast of the biomembranes, ribosomes and other components without any heavy metal staining. Sectioning with Reichert Ultracut/FC-system with automatic mode: cutting speed 0.4 mm/sec, maximum return speed, minimum sized cutting window and cycle length, advance setting 80 nm, chamber kept at the lowest possible temperature near -170°C. Electron micrographs: Zeiss EM 902 EFTEM with ESI, cold stage temperature approx. -170°C; Bar: 1 μm. Preparation and electron micrographs by Martin Michel, Electron Microscopy I, ETH Zürich. Reproduction with permission. Further explanation in the sections on "High-Pressure Freezing", Cryo-Ultramicrotomes", "Ultrathin Sectioning ..." and "... Frozen Hydrated Ultrathin Sections". See also Figs. 11, 12, 17, 24, and 25.

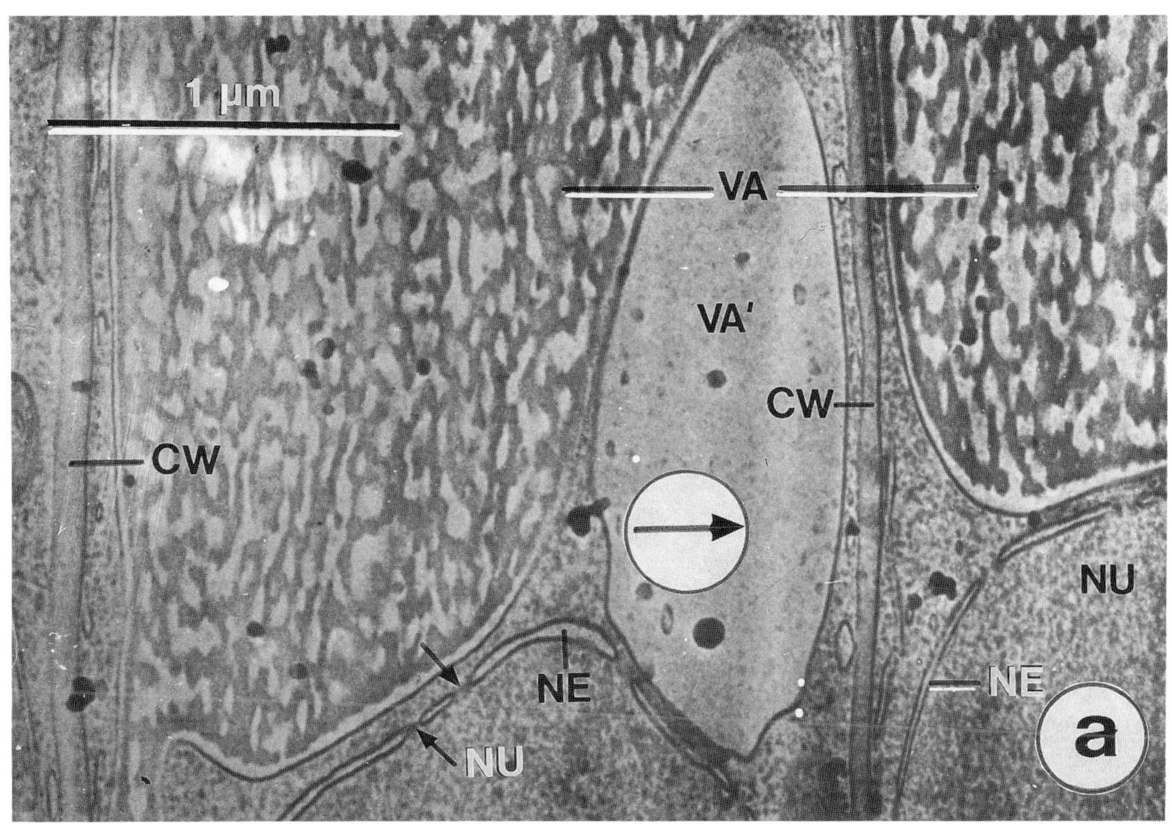
1 µm
VA
VA'
CW
CW
NU
NE
NE
NU
a

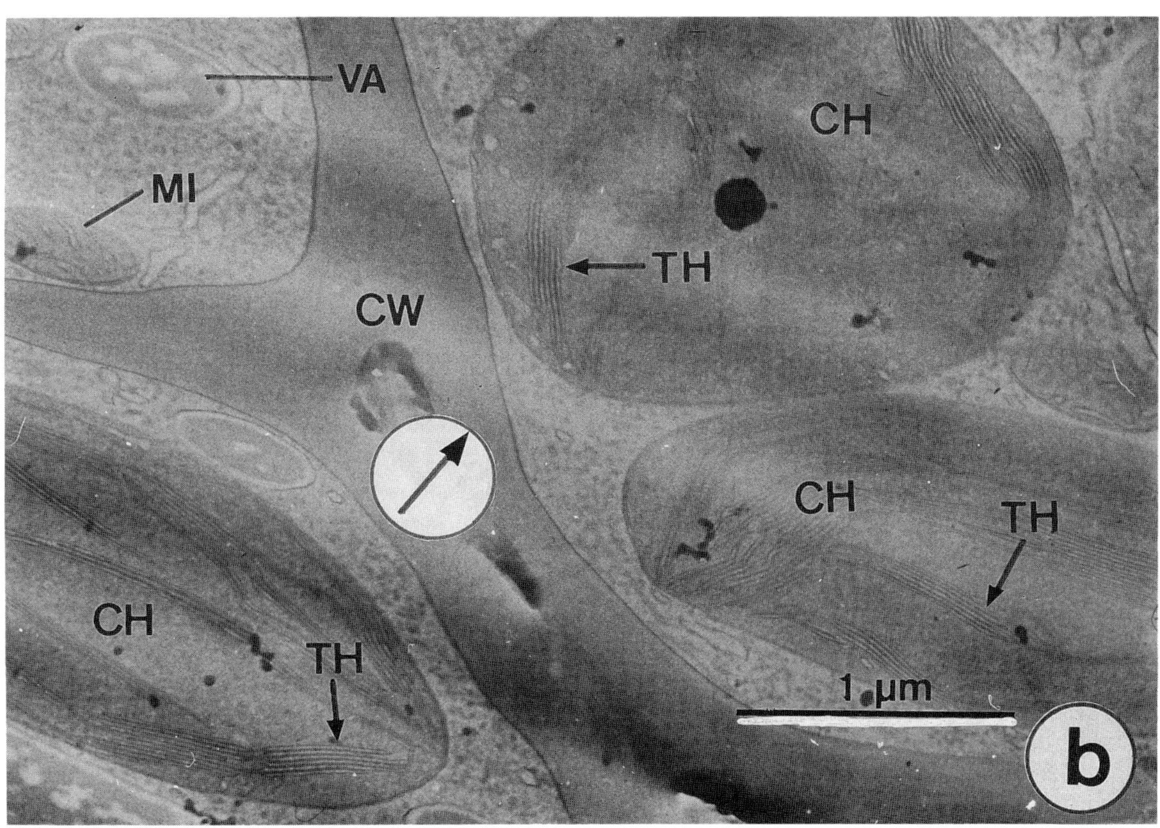
VA
MI
CH
TH
CW
CH
TH
CH
TH
1 µm
b

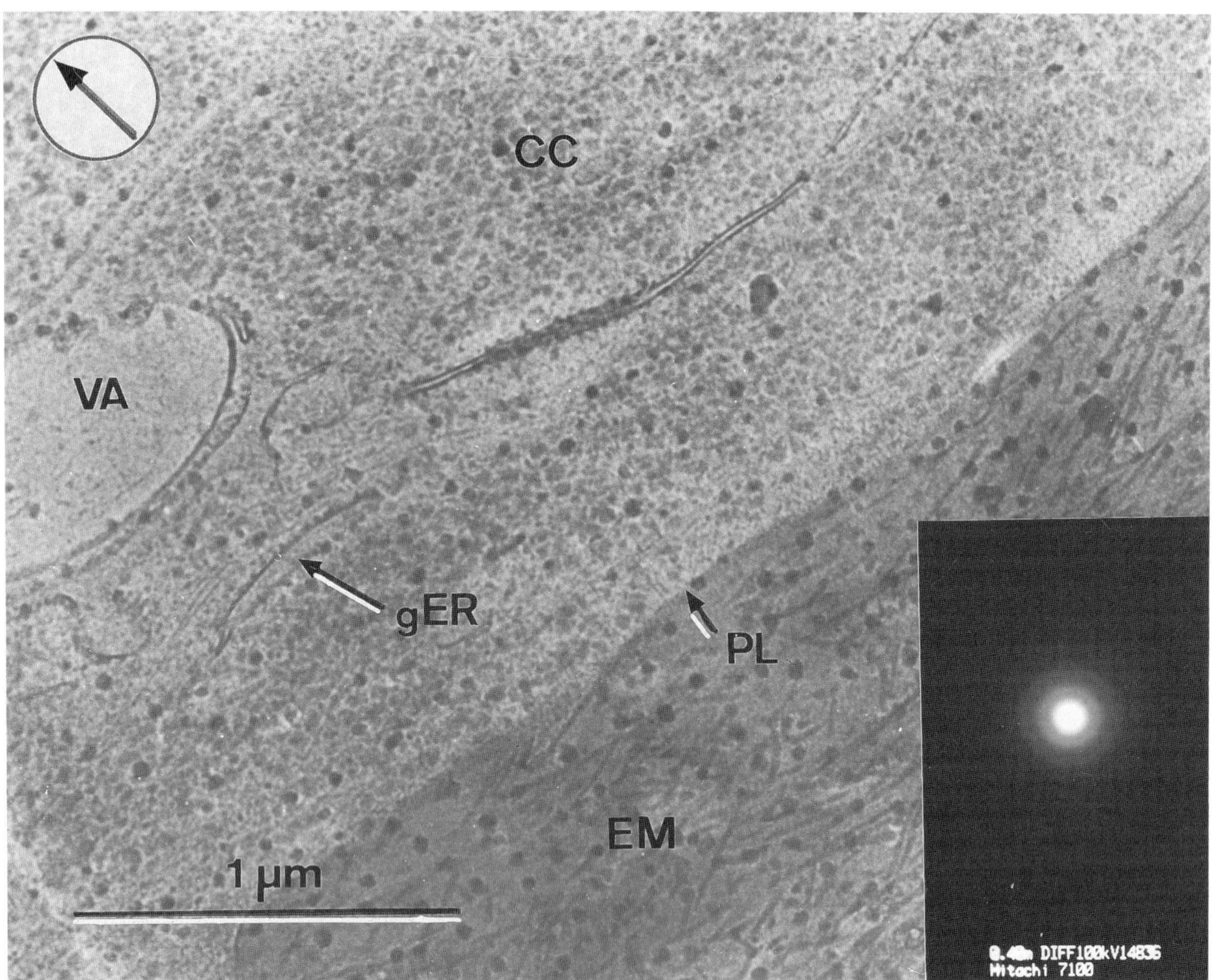

Figure 11. Frozen-hydrated cryosection of amorphously frozen bovine articular cartilage after high pressure freezing. Preparation and electron micrograph similar to Fig. 10. The diffraction pattern (insert) of this area shows the characteristic diffuse fringes of vitreous ice Iv. Chondrocyte CC with profiles of the granular endoplasmic reticulum gER (arrow) and a vacuole VA in the cytoplasmic matrix. Adjacent to the plasmalemma PL (arrow) of the chondrocyte CC the extracellular matrix EM with its typical ultrastructure only visible after perfect freezing to I_v. Sectioning direction: arrow in circle. Bar: 1 μm. Section preparation and electron micrograph by Martin Michel, M.E. Müller Institut für Biomechanik, Universität Bern, Switzerland. Reproduction with permission. Further explanation in the sections on "High-Pressure Feezing", Cryo-Ultramicrotomes", "Ultrathin Sectioning ..." and "... Frozen Hydrated Ultrathin Sections". See also Figs. 10, 12, 17, 24 and 25, and Studer *et al.* (1989, 1995).

---

pressure freezing in cross section diagrams. To guarantee an efficient heat extraction from the specimen to the outer wall, the space between those elements according to Studer *et al.* (1989) is completely filled with an inert medium of low viscosity (1-hexadecane). The same counts for cavities filled with gas inside the specimens [*e.g.*, plant leaves; see Michel *et al.* (1991)], since air bubbles, due to bad heat conductivity and compression during pressurisation, reduce the yield of useful preparations considerably. With respect to a perfect freezing of normal biological specimens, the thickness of the sample should not exceed 0.1 mm. Only specimens with a good internal cryoprotection or an unusual low water content can be frozen in thicknesses up to 0.6 mm. The high pressure systems therefore allow an adjustment for thicknesses between 0.1 and 0.6 mm in steps of 0.1 mm by combination of two different sandwich elements. The circular discs of

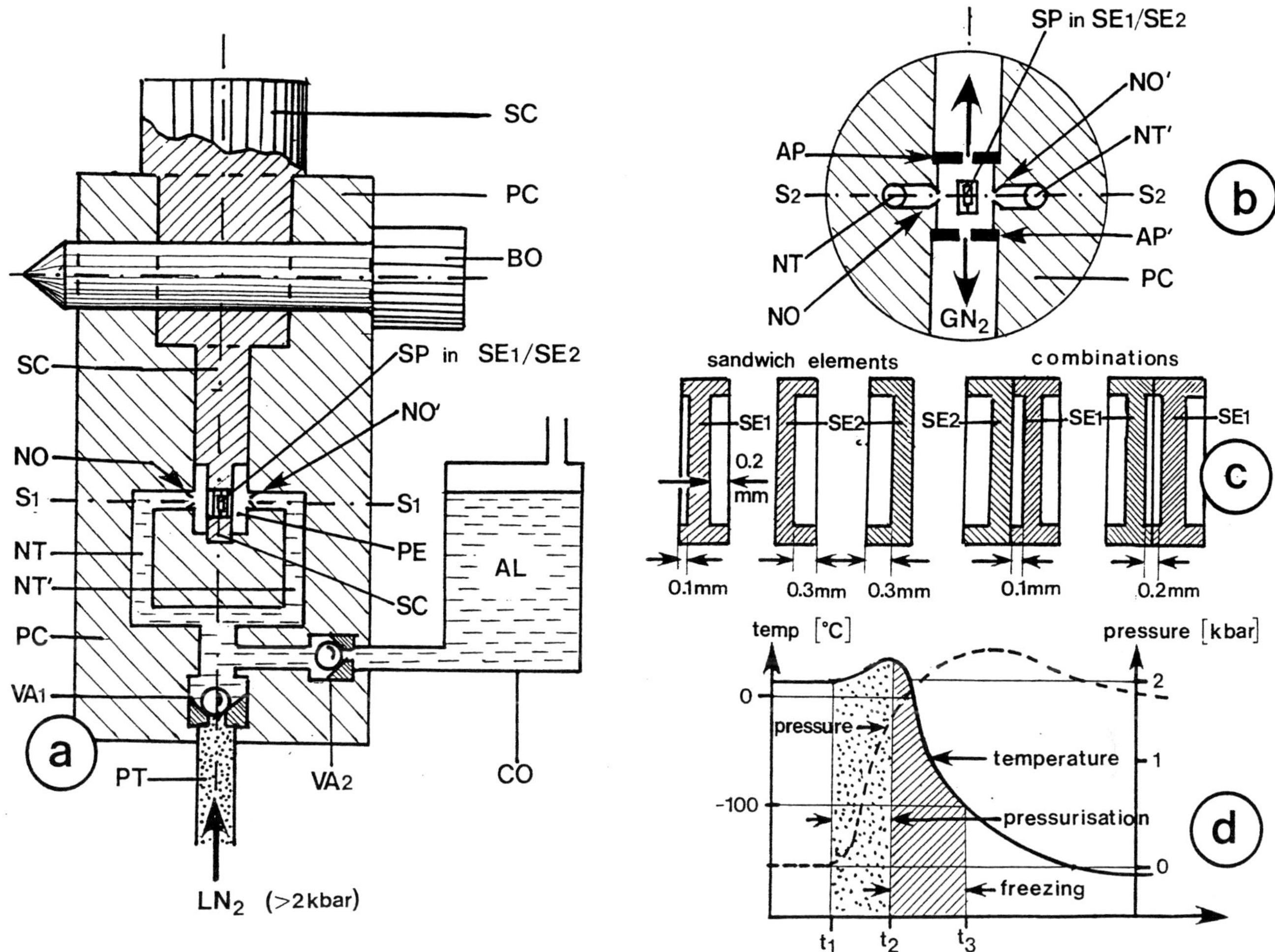

Figure 12. Simplified schematic diagrams of the basic design and function of the Balzers and Leica high pressure systems for cryofixation, which are basically very similar. See Moor (1987); M. Müller and Moor (1984); Studer *et al.* (1989). Further explanations in the sections on "Freezing and the Frozen State of Water..." and "High Pressure Freezing". See also Figs 10 and 11, and Table 5. **(a/b/c)** Design of the high pressure part for freezing (front view "a" sectioned in plane S2, top view "b", sectioned in plane S1, specimen holders "c"): the specimen carrier SC holds the sandwich elements SE1/SE2. SC is inserted into the pressure chamber PC and held by the transverse bolt BO. The LN_2 tubes NT/NT' above valve VA1 are filled with alcohol AL from container CO through valve VA2. In this condition the system is ready for a freezing cycle, which can be started by push button operation and runs then completely automatically in the following sequence: LN_2 is introduced with a pressure > 2.1 kbar through pressure tube PT and valve VA1 (valve VA2 closes automatically) → the alcohol AL in NT/NT' first hits the sandwich elements SE1/SE2 (initial period of approx. 40 msec) → LN_2 enters through the nozzles NO/NO' → the specimen is cooled to -100°C within 13 msec, if the preselected thickness of the specimen in the sandwich chamber SE1/SE2 is t = 0.1 mm (see "c" and Table 5) → the excess alcohol AL and nitrogen (GN_2) escape through the narrow apertures AP/AP' of the pressure chamber PC (see "b"). The whole procedure is finished within a period < 0.5 sec. The system allows approximately 40 freezing cycles per hour. Each cycle needs ≤ 0.2 liter LN_2. The thickness t of the specimen in the sandwich chamber (see "c") can be adjusted by combination of the different sandwich elements SE1 and SE2 between t = 0.1 mm and t = 0.6 mm in increments of 0.1 mm. Complete vitrification of unprotected highly hydrated specimens SP is only reproducibly possible with t = 0.1 mm (see Table 5). **(d)** Synchronisation of pressurisation and cooling: due to the small amount of alcohol AL at ambient temperature in the nitrogen tubes NT/NT' pressurisation from 0 to 2 kbar started at t1 is achieved without cooling. Cooling starts at t2 when subsequent LN2 hits the sandwich SE1/SE2. At t3 temperature of -100°C is reached.

Table 5. Cooling rates in the centre of samples during high pressure freezing depend on the thickness t of the specimen in the sandwich container according to Fig. 12c. The same counts for the time needed for complete freezing of the sample which may be responsible for pressure artefacts. Values according to Studer *et al.* (1995) and leaflet Leica EM HPF (6/94). See also Fig. 12.

Total Thickness of the Sample +)	Cooling Rate in the Centre of the Sample	Time Needed for Complete Freezing of the Sample to -100°C
100 µm	10,000 - 20,000 °C/sec	13 msec
200 µm	4,000 - 6,000 °C/sec	50 msec
600 µm	400 - 600 °C/sec	450 msec
+) Including intermedium (e.g. 1-hexadecane). Value corresponds to thickness "t" of the specimen in the sandwich chamber SE1/SE2 according to Fig. 12c.		

thin tissue slices or plant leaves are made with a special punch appropriate to the sandwich containers according to Fig. 12c. Fig. 12d shows the synchronisation of pressurisation and cooling under the preconditions mentioned above. For cell suspensions or small organisms, according to Hohenberg *et al.* (1994) dialysis tubes can be used to prevent blowing out during pressurisation: this is also a big advantage especially for handling tiny specimens during subsequent freeze substitution and embedding. Also cryosectioning after this preparation is possible [see Hohenberg *et al.* (1994), p. 37, Fig. 3). Under the conditions described above high pressure freezing is really an excellent tool for vitrification of biological material. But it should not be disregarded, that high pressure freezing is a relatively slow process: with respect to the stable sandwich container, the intermediate layer of 1-hexadecane and the dimension of the inserted specimen the obtainable cooling rates within the specimen are some orders of magnitude lower in comparison with the well frozen border layer of a specimen during slam freezing. The decrease of the cooling rates in the center region of the samples in the high pressure freezer is a function of the sample thickness, that means the preselected combination of the elements for the sandwich container (see Fig. 12c). With the overall thickness (specimen and intermedium, *e.g.*, 1-hexadecane) of 0.1 mm the cooling rate in the center of the sandwich chamber remains between 10.000 and 20.000°C/ sec. If the overall thickness is enhanced to 0.2 mm, this value drops already down to 4.000 to 6.000 °C/sec. With the formerly used value of 0.6 mm the rate in the center region is only around 400 to 600 °C/sec. Table 5 shows these values in comparison with the periods needed for complete freezing of samples in different thicknesses. True vitrification of normal biological samples is only possible with cooling rates between 10.000 and 20.000 °C/sec even under the optimum pressure of 2.1 kbar (Studer *et al.*, 1995). This has to be considered especially if hybrid follow-up procedures are used, which do not allow checking of the frozen-hydrated state by electron diffraction patterns.

Cryo-Ultramicrotomes

Probably Fernández-Morán in 1951 was the first who tried ultrathin sectioning at low temperatures for the study of cell structures and element analysis in the electron microscope with some success [see also *e.g.*, Robards and Sleytr (1985), Roomans *et al.* (1982), Roomans and Shelburne (1983), Sitte (1982), Sitte and Neumann (1983), Wendt-Gallitelli and Wolburg (1984), Zierold (1982, 1987)]. In the mid-sixties Wilhelm Bernhard used the cryostat technology of Linderström-Lang and Mogensen (1938) [cited by Pearse (1961) for studies on macromolecular cytochemistry with the TEM]: he simply installed a Porter-Blum ultramicrotome in a deep freeze box and sectioned cryoprotected material at temperatures around -40°C with remarkable results (Bernhard and Nancy, 1964; Bernhard, 1965, Bernhard and Leduc, 1967; Bernhard and Viron, 1971). On his recommendation LKB together with the group around Appleton at Cambridge UK developed a quite perfect cryostat system for cryosectioning at temperatures down to -80°C (Appleton, 1974). In the late sixties Dollhopf together with Hodson and Marshall in London (UK) and completely independently Christensen together with Blum in the USA introduced small cryochambers with a volume of approximately 1 liter [see Fig. 13 and Dollhopf (1968), Dollhopf and Sitte (1969), Christensen (1969, 1971), Blum (1970), Sitte (1982)]. Shortly afterwards LKB introduced a similar system [*"LKB CryoKit"*, Persson (1972)] and for the first time a set of special *"cryotools"* for a more convenient and efficient handling of dry cut frozen sections for element analysis [Sevéus, 1977; Barnard and Sevéus, 1977]. The new cryochambers surrounded only the area of knife and specimen. They made it possible to

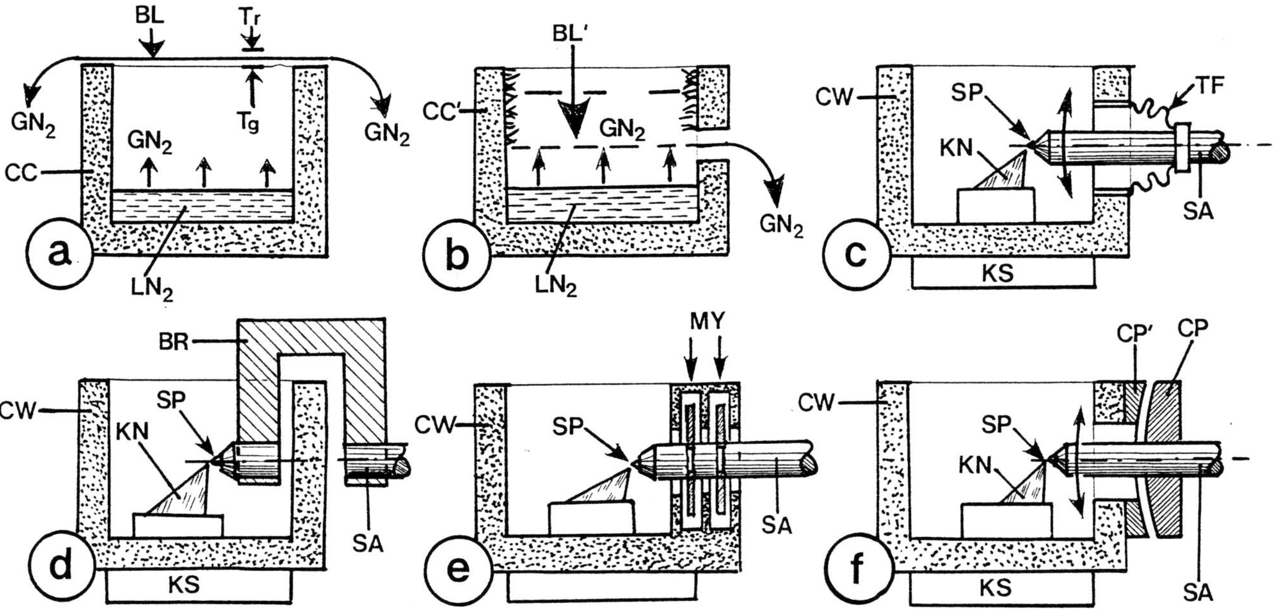

Figure 13. Physical basics and different designs of cryochambers for cryo-ultramicrotomes. Further explanations in the section on "Cryo-Ultrotomes". See also Figs. 14, 15 and 17. **(a)** Simple cryochamber CC with a thermally insulating wall, partially filled with LN_2. The boiling LN_2 delivers continuously GN_2, which fills the chamber CC completely and flows away over the wall (arrows). Since cold GN_2 is considerably denser than room atmosphere, air and GN_2 do not mix, but form a very stable border layer BL similar to a phase border liquid/gas ("*GN_2 lake*"). Temperature sensors show, that within a few millimetres of height the temperature changes from T_g < -100°C to T_r > +10°C. This physical behaviour is essential for cryo-ultramicrotomy in small chambers: since the temperature inside CC is very low and stable, room atmosphere cannot enter into CC and frost formation is completely prohibited similar to an open deep freeze in a supermarket. **(b)** Situation changes completely, if a hole HO is drilled in the wall CC' of the cryochamber: the dense LN_2 flows away through HO and the border layer between GN_2 and room atmosphere changes to a lower level BL'. The former cold wall of CC (see "a") now shows disturbing frost formation FF. The problem in the design of cryochambers is therefore to inhibit the outflow of cold GN2 through the chamber wall and to introduce the specimen arm of the ultramicrotome in a suited manner. **(c)** Dollhopf found the first suited solution: the specimen arm SA with the specimen SP was introduced through the chamber wall CW. The hole HO was sealed by an extremely thin flexible foil TF mounted on the rear wall of the chamber and on the specimen arm SA (CW mounted on knife support KS, knife KN mounted on the bottom of CW). **(d)** Christensen and Blum found an alternative with the bridge BR. **(e)** H. Hagler used a "*labyrinth sealing*" by some thin resin foils of Mylar MY, sliding in slits of the chamber wall CW (RMC-21 cryochamber). Through the labyrinth only small amounts of GN_2 escape from the chamber, which do not affect the correct function. **(f)** Leica FCS-cryochamber for Ultracut-S/R/UCT: R. Lihl developed a contact free labyrinth sealing by complementary cylinder surfaces CP/CP' mounted on SA and CW. The cylinder radius corresponds exactly with the radius around the bearing of the ultramicrotome. Without doubt this solution is both foolproof and simple and guarantees functioning of the cryo-ultramicrotome without any risk of disturbing frictions.

---

use in this smaller volume lower temperatures down to -150°C by cooling with cold GN_2 or even LN_2. The main advantage of those chambers besides the lower temperatures urgently wanted for element analyses was the better function of all moving parts of the ultramicrotome (motor drive, bearings, mechanical advance). But the draw-backs of those first generation chambers were very frustrating and practically nothing worked as expected. Besides cryostat systems and first generation cryochambers in the same decade some extremely simple freezing heads as accessories for standard ultramicrotomes (*e.g.*, Crudgington, 1966) were described. They were made and used exclusively for sectioning of rubberlike elastic or tough resins in material sciences and never worked successfully in biology.

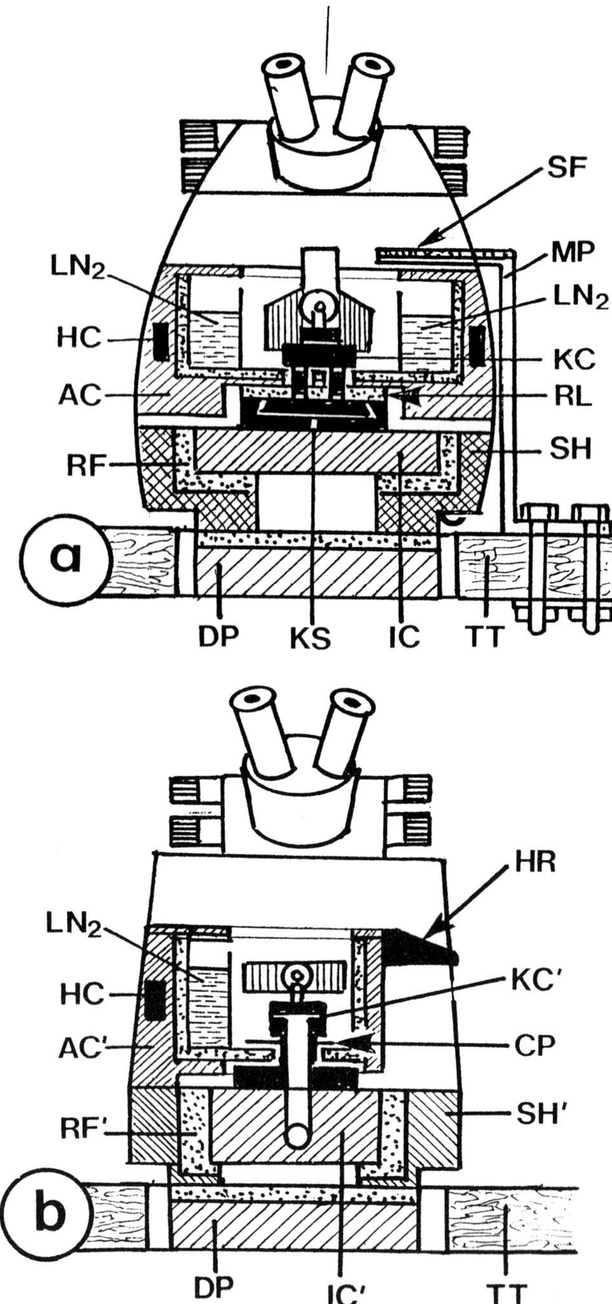

Figure 14. The mounting of the cryochamber decides the stability of a cryo-ultramicrotome: Most cryochambers are developed as an accessory for already existing ultramicrotomes and therefore mounted simply on the knife support KS (a). This makes the cryochamber sensitive against unavoidable manipulation forces. The only help in this case according to a recommendation of R. Ornberg (personal communication) is a stable metal profile MP mounted on the table top TT and covered by a 3 mm styrofoam sheet SF for handrest. Correct mounting is realised in the purpose designed Leica-Reichert cryochambers FCS/FCR on the ultramicrotomes Ultracut-S/R/UCT (b), where the aluminium cast AC' of the FC-system is mounted on the "*shell*" SH' of the ultramicrotome. Only the knife carrier KC' is directly connected to the "*iron core*" IC'. This design allows to use the handrest HR on the chamber AC' without annoying distortion of the system. Further explanations in Section 5. (a) FC-chamber on former Reichert-Ultracut models 1980/90 (front view cross section, roughly schematic): the base of the ultramicrotome rests on the damping plate DP and shows a "*shell construction*" according to Kenzian *et al.* (1975), consisting of an insulating shell SH, a flexible intermediate layer of rubber foam RF and an extremely rigid "*iron core*" IC. All precision elements influencing the regularity of sectioning (knife support KS, advance and bearing system) are mounted on IC. The thermostatically heated (heating cartridges HC) aluminium cast AC of the FC4-cryochamber was mounted on the considerably fortified knife support KS by a flexible rubber layer RL similar to RF. Only the knife carrier KC was mounted directly to the knife support. LN_2 was stored in two tanks (interconnected twin tanks) right and left-hand of the open top sectioning chamber with KC. (b) Improved chamber mount of FCS/FCR on Ultracut-S/UCT (models 1990 ff): AC' is mounted exclusively on shell SH'. Only the knife carrier KC' is directly connected with knife support KS'. A labyrinth between AC' and cover plate CP on KC' guarantees free x/y-translation of KC' with KS'. LN_2 is only stored in one tank left-hand. Right-hand handrest HR as mentioned above.

The first real break-through in cryo-ultramicrotomy was achieved with a system, where a purpose-designed ultramicrotome was developed together with a new type of a cryochamber (Sitte *et al.*, 1980; Sitte, 1982, 1984; Sitte and Neumann, 1983). This system (see Fig. 14a) worked with an open top sectioning chamber and direct LN_2 cooling. It reached a minimum temperature near -180°C and section preparation was simplified considerably even at lower temperatures. Nevertheless not all instrumental difficulties were properly eliminated. Like all former cryochambers also the heavy-weight FC4-chamber was mounted on the knife support, which was extremely rigid in comparison to former knife support constructions. But it turned out during practical work that this support was nevertheless sensitive in the nanometer-range. Also the specimen was mounted to a somewhat sensitive "*bridge*" according to the design of Christensen and Blum [see Fig. 13d and Christensen (1969, 1971), and Blum (1970)]. The automatic and electronically controlled refilling of LN_2 from a 35 l

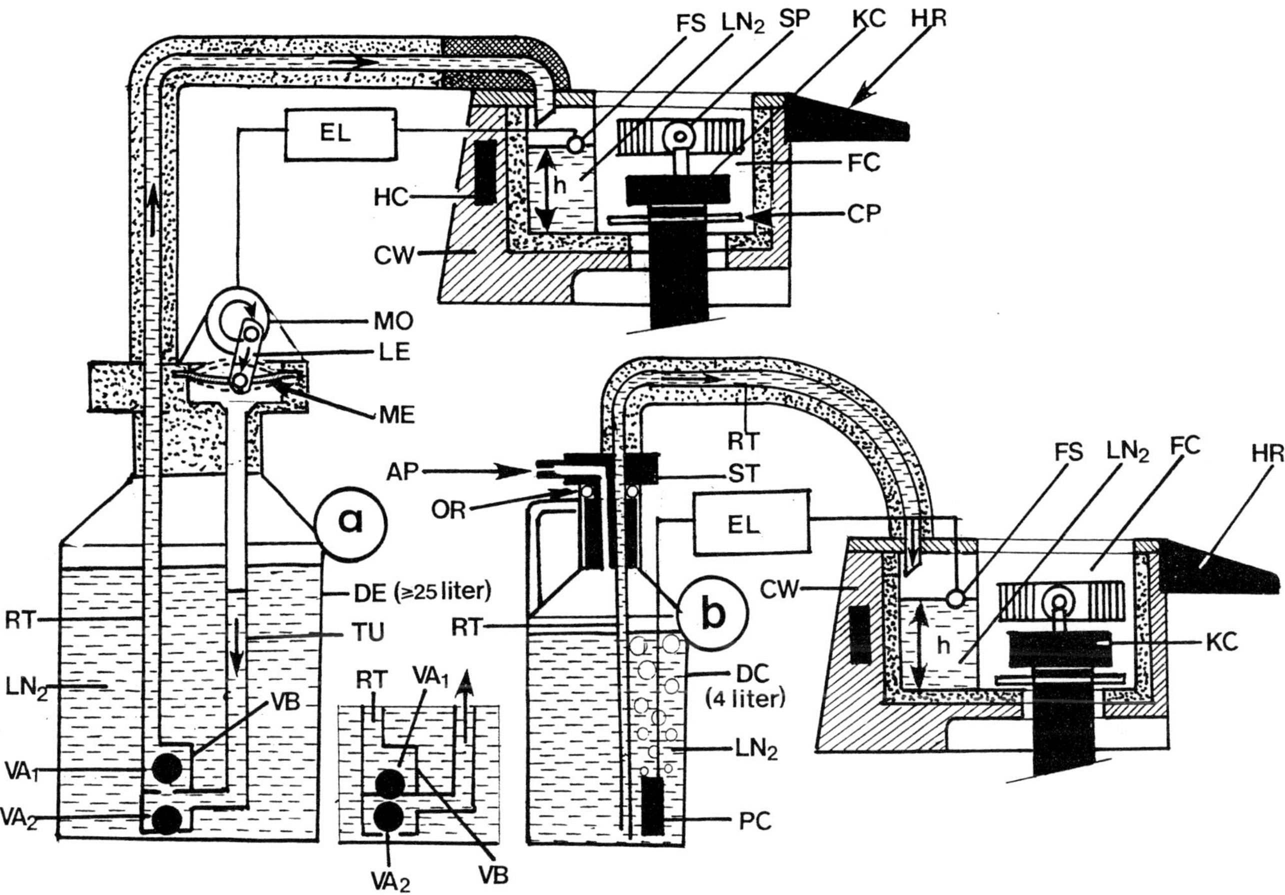

Figure 15. Electronically controlled continuous refilling of microlitre-amounts of LN_2 allows to maintain a height "h" of the LN_2-level in the LN_2 tank of a cryochamber with a precision better than $\pm$ 0.5 mm. This is important for absolutely regular cryosectioning of fresh frozen (unprotected) specimens for frozen-hydrated investigations in the cryo-TEM/STEM. Two different systems (a) and (b) are available for Leica-Reichert cryochambers FCS and FCR. Further explanations in the sections on cryo-ultramicrotomes and cryosectioning and in Figure 14b. (a) Refilling by membrane pump and two valves: a rubber membrane ME is bent by a lever LE linked to the eccentric disc on the driving wheel of the motor MO. This membrane pump causes an oscillating up- and down-movement of the LN_2-column in the tube TU connected with a valve box VB containing two spheres VA1 and VA2. A downward motion of LN_2 in TU closes VA2 and opens VA1. Opposite an upward motion of LN_2 in TU opens VA2 and closes VA1 (insert). The described automatic operation of both elements VA1 and VA2 results in a stepwise upward motion of LN_2 in the refilling tube RT inside the dewar vessel DE and finally in a transfer of LN_2 into the tank of the cryochamber. The LN_2-flux is precisely controlled by the electronic system EL connected with motor MO and filling sensor FS: if FS is wetted by LN_2, the motor drive is slowed down. If the LN_2 level drops below FS, the speed of the motor drive increases with increasing distance between LN2 level and FS (CW = chamber wall thermostatically heated by cartridge HC, FC = sectioning chamber with open top). (**b**) Refilling by slight pressurisation of the 4 liter Dewar can DC according to K. Neumann, H. Hässig and A. Kunz: DC is closed by the stopper ST (O-ring sealing OR), which contains an aperture AP adjusted for the escape of the small amount of GN_2, which evaporates from the LN_2 inside DC by static evaporation ("*defined leakage*"). Without heating of the pressurisation cartridge PC therefore no refilling of LN_2 results. If the LN_2 level in the cryochamber drops down below the sensor FS, the electronic control EL powers PC and additional GN_2 is evaporated. The pressure inside DC increases and LN_2 is transferred from DC to the cryochamber. The precision of LN_2 refilling operation is identical with system (a). Disadvantages of this simple system are the needed precooling of DC to reach a steady state for static evaporation before closing ST and the limited volume of DC (4 liter), but compare comments in the section on cryo-ultramicrotomes.

Dewar vessel into the twin tanks of the cryochamber was not continuous: refilling was started by a sensor, when the twin tanks of the chamber were nearly empty and interrupted as soon as a similar thermal sensitive diode sensor was wetted with LN_2. The consumption of LN_2 at higher temperatures above -100°C was increasing to 7 liter LN_2/h by counter heating, since this system was originally designed exclusively for extremely low temperatures and later on adapted for temperatures between -120 and -80°C needed for Tokuyasu work with sugar protected specimens. All these drawbacks became visible during some years of experience: they neither really inhibited sectioning in the low temperature range below -140°C for EDX and frozen-hydrated investigations [see *e.g.*, Chang *et al.* (1983), Dubochet and McDowall (1984), Dubochet *et al.* (1982b, 1983, 1985, 1987, 1988), Edelmann (1984, 1992), Hohenberg *et al.* (1994), McDowall *et al.* (1984, 1986), Michel *et al.* (1991, 1992), Ornberg (1985, 1989), Trus *et al.* (1989)], nor sectioning around -100°C with the Tokuyasu technique. But they made sectioning more sensitive, time consuming and by the high LN_2-consumption at temperatures above -120°C also more expensive.

Some years ago with the new RMC 21 cryochamber and with the Reichert-system Ultracut-S/FCS new standards were set for cryo-ultramicrotomy. Firstly the new RMC-cryochamber designed by Herbert Hagler used again a "*through-the-wall*" design for the specimen arm (see Fig. 13e). Such a system had already been used for the first time by Dollhopf (1968) [see also Dollhopf and Sitte (1969), Sitte (1982), and Fig. 13c), employing a thin membrane to close the cryochamber against the surrounding atmosphere and to prevent the cold dense GN2 inside the chamber from flowing out (see Fig. 13a and b). But this "*Dollhopf-sealing*" complicated the mounting and separation of the chamber on the ultramicrotome. Instead of the outside membrane of Dollhopf, Hagler used a set of Mylar sheets integrated in the rear wall of the RMC 21 cryochamber: rear wall and Mylar sheets formed a "*labyrinth sealing*" preventing sufficiently the escape of larger amounts of GN_2 from the chamber. A different solution was found by Reinhard Lihl from Leica AG at Vienna for the FCS-chamber (Fig. 13f): as "*labyrinth sealing*" he used two corresponding (complementary) cylinder surfaces, one on the rear wall of the cryochamber and one on a sealing profile mounted to the specimen arm. The two corresponding metal profiles are easily adjusted within a distance < 0.5 mm. The special advantage of this kind of "*cylinder labyrinth*" is that it works without any risk of a mechanical contact, from which frictions and vibrations may arise. This fact is of extreme importance for sectioning of fresh frozen material, as will be discussed later. The "*through-the-wall*" specimen arm enhances the stability of the cryo-ultramicrotome in comparison with any "*bridge design*" considerably. The stability is not only identical with the standard ultramicrotome for ambient temperature work but without doubt even better, since no self-locking adjustment similar to the segment arc is used in cryo-ultramicrotomy.

Another important detail concerns the "*shell-mounting*" of the cryochamber. Since all former cryochambers were accessories for already existing standard ultramicrotomes for ambient temperature sectioning, the cryochambers were mounted simply instead of the standard knife holder assembly on the knife support. For the FC4-chamber in 1980 (see Fig. 14a) the knife support was considerably fortified by roller guides and a dovetail mounting. Additionally the outer wall of the chamber was mechanically separated from the knife stage by an elastic intermediate rubber layer similar to the "*shell construction*" of the ultramicrotome (Kenzian *et al.*, 1975). Both measures reduced the sensitivity of the cryochamber against manipulation forces remarkably, but the chamber was still sensitive against touching and was for example not suited as handrest during the picking up of the sections. After this experience, it was understood that a complete separation of the outer chamber wall from the sensitive knife holder system was needed. A first successful attempt in this direction was made by LKB with the new model LKB Cryo-Nova: the molded poly-urethane cryochamber with incorporated cell-u-foam thermo-insulation was mounted directly on the heavy weight steelbase of the Ultrotome Nova. The cryo-knifeholder of the Cryo-Nova system was independently mounted. The new design reduced sensitivity considerably and allowed to use the outer chamber as handrest for cryosection preparation. Within the same development of LKB the "*cryotools*" were further improved and a possibility for a direct cryotransfer of loaded grids from the cryochamber to the cryo-TEM included. A real "shell-mounting" of a cryochamber with a strict complete mechanical separation of the cryochamber from the steel-core of an ultramicrotome was finally realised with the system Ultracut-S/FCS as shown in Fig. 14b. Within the new design the thermostatically heated aluminum wall of the FCS-cryochamber is now rigidly mounted to the resin shell of the Ultracut-S or Ultracut-UCT. Only the cryo-knifeholder inside the chamber is mounted to the knife support of the ultramicrotome by a metal-metal connection without intermediate plastic parts. The thermal insulation between ambient temperature and cryo elements is realised both at the object holder and at the knife holder system by intermediate tubes of stainless steel, which are a good compromise between heat resistance and mechanical stability. The backlight illumination works identically to the ambient temperature version through the tube of the knife holder. Besides the approach of the diamond knife against the mirror frozen and therefore perfectly reflecting specimen surface the special advantage of this backlight system for the cryowork is the simple identification of vitrified areas compared to segregation artefacts with hexagonal ice (see Fig. 16c). In comparison with the preceding FC4-chamber the FCS-chamber contains only one LN_2-tank instead of the FC4-twintank system. This results in a considerable reduction of LN_2 consumption, which remains considerably below 2 liter LN_2/h for temperatures down to -170°C. Instead of the right-hand tank there is now a handrest, which remarkably simplifies all preparation procedures as well as the mounting of the cryochamber on the ultramicrotome (see Fig. 14b).

A further improvement of cryo-ultramicrotomes relates to the refilling of the chamber tank with LN_2, which was formerly discontinuous as mentioned above. In the meantime two perfectly working systems for continuous LN_2 refilling are known. The more sophisticated FCS-refilling system (Fig. 15a) uses a motor driven membrane pump and two simple valves consisting of stainless steel balls with corresponding circular openings. If the eccentric piece on the driving wheel of the motor is rotating, the LN_2-column in the connective tube is stepwise moving upward and LN_2 is transferred from the Dewar vessel into the tank of the cryochamber. The motor is electronically controlled by a level sensor in the cryogen tank of the chamber. If the LN_2-level in the chamber drops down below the sensor, the motor drive is activated and LN_2 refilled. If the LN_2 level approaches to the sensor, the motor drive slows down and if the LN_2 wets the sensor itself, the motor comes almost to a stand-still. In this way the LN_2 level in the cryochamber during work is maintained within $\pm$ 0.5 mm. A much simpler set-up was developed for the FCR-cryochamber (Fig. 15b), primarily devoted to Tokuyasu preparation around -100°C. In comparison to the motor driven FCS-model this FCR-refilling system works with a simple electronic control of the heating cartridge used for pressurisation without any moving mechanical parts. The idea behind is a "*defined leak*" by an aperture just big enough to let the GN2 escape, which is continuously freed from the boiling LN_2 in the 4 liter Dewar can. If the sensor in the FCR-cryochamber (identical with the FCS-sensor) is wetted by LN_2, the power of the heating cartridge inside the Dewar can is reduced and the evaporation of GN_2 goes down to the minimum amount ("*static evaporation rate*"). If the LN_2-level in the chamber drops down below the sensor, the heating cartridge gets increasingly more power and evaporates additional GN_2. A pressure is built up in the can and LN_2 is transferred by this pressure to the tank of the cryochamber. Minimal pressures are sufficient for this transfer. The system is foolproof, since the head of the refilling system with the pressurisation device (heating cartridge) is not tightly mounted in the neck of the Dewar can: if by any failure (*e.g.*, aperture closed by contamination) a slightly higher pressure is developed, the head is lifted up and excess GN_2 released. This system works with the same precision as the motor driven unit according to Fig. 15a and delivers precisely μl-amounts of LN_2 to the chamber tank. It works absolutely silently and will have probably an unlimited life span. There are only two disadvantages: Precooling and limited LN_2-volume. Precooling of the 4 liter Dewar is needed, because the evaporation rate of the still warm and just filled Dewar lies considerably above the "*static evaporation*" of the cold container. It is a simple measure, first to fill the can with LN_2, then to mount the cryochamber and finally to close the can and to connect the LN_2 supply tube between refilling device and chamber. The second disadvantage (4 liter volume) turned out to be an advantage in reality, since it is in comparison to a big Dewar vessel so easy to exchange and to refill such a light can. Since the new cryochambers consume only small amounts of LN_2, the volume of 4 liter enables work for at least 3 h and that is sufficient for most cases. If not: refilling is easy. The continuous LN_2-refilling is an advantage mainly for cryosectioning at extreme conditions, that means if absolutely regular cryosections of minimum thicknesses are needed for frozen-hydrated investigation, where crevasses are the big problem. For Tokuyasu work I cannot believe, that continuous refilling makes a big difference in comparison to discontinuous refilling: only the start of refilling with the former FC4/FC4D/ FC4E-systems introduces a slight irregularity, that means a loss of 1 to 4 cryosections within a ribbon. But: better is better and we are happy that we reached the goal of continuous LN_2 refilling.

Cryosectioning According to Tokuyasu

The Tokuyasu method (Tokuyasu, 1973, 1978, 1980, 1986; Tokuyasu and Singer, 1976) is well introduced and some excellent reviews report on the technical and methodological details [see *e.g.*, Griffiths *et al.* (1983, 1984), Griffiths (1993), Leunissen and Verkleij (1989)]. Cryosectioning of biological material protected with 2.3 M sucrose according to Tokuyasu

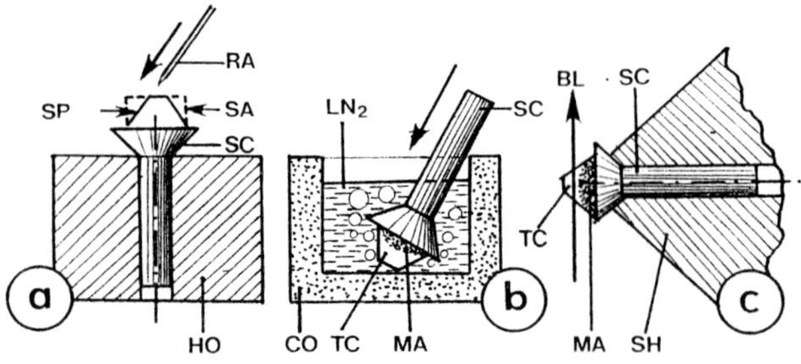

Figure 16. Compared to the mounting of most fresh frozen (unprotected) samples, mounting of fully (2.3 M sucrose) protected specimens according to Tokuyasu is simple. Compare the section on "Cryo-sectioning according to Tokuyasu" and cited reviews of Tokuyasu and Griffiths. **(a)** The protected sample (broken line SA) is presectioned with razor blade RB before mounting and freezing to a suited pyramid SP below a stereomicroscope. **(b)** The pyramid SP is positioned on the standard carrier SC ("*pin*") in holder HO. **(c)** Freezing in LN_2 (styrofoam container CO) is probably an advantage, since faster freezing in propane or ethane seems to provoke tension cracks and clefts inside the sample SP, which may impede regular sectioning. Usually fully protected samples are vitrified to an absolutely clear and transparent matrix CT by LN_2. Only in cases of insufficient sugar protection (*e.g.*, period of impregnation too short) or at sugar concentrations considerably below 2 M a milky area MA indicates lower cooling rates in centre regions at the border between carrier SC and the bottom of the pyramid SP. **(d)** Milky areas MA are easy to identify under the darkfield ("*backlight*") illumination BL in the cryochamber: the carrier SC is investigated with the stereomicroscope of the cryo-ultramicrotome in horizontal position after mounting in the specimen holder SH on the specimen arm (SA, see Fig. 13). Only the clear and transparent part CT of the sample should be sectioned, since the milky area MA contains severe hexagonal segregation artefacts.

(1973) as mentioned already in the "*Introduction*" (see Table 2) differs in most respects from cryosectioning of fresh frozen material described in the following sections of this review. This results mainly from "*sugar embedding*": the high amount of sugar reduces the amount of free water to an extent, that segregation of hexagonal ice is not possible even at the low cooling rates obtainable with immersion or dipping the samples into LN_2 (Fig. 16c). On the other hand, sugar changes the consistency of the sample considerably from a brittle to a somewhat jelly-like state. This enables sectioning of rather large areas with a width of 0.5 to 1.0 mm in most cases. The characteristics of protected materials change immediately, if the sugar content is reduced noticeably below 2.3 M, which is recommended for some special purposes [see *e.g.*, Tokuyasu (1973, 1980), or Tokuyasu and Singer (1976)]. The changes concern both sectioning consistency and ice segregation artefacts. Mostly in such preparations a more rapid freezing by propane or ethane is needed. Generally one has to check carefully by backlight illumination (see Fig. 16d), if a "*clear*" vitrification is obtained, since also for protected specimens proper freezing is a precondition for convenient sectioning and reproducible results.

Usually freezing is done after presectioning the chemically fixed and sugar protected sample to a pyramid with a razor blade (Fig. 16a) and mounting of these pyramids on a standard carrier ("*pin*", Fig. 16b) of the cryo-ultramicrotome used. Such a carrier has a weight of approximately 0.1 g. Such a "*heavy weight*" carrier is not suited for freezing of non pre-treated fresh material, but sufficient for a specimen with a good cryoprotection. Of course the heat extracted from the pin causes a profile of cooling with considerably smaller cooling rates inside: sometimes as a result of insufficient sucrose impregnation or lower sucrose concentration there is a contact region between specimen and pin, which looks "*milky*" and contains hexagonal ice. In such cases only the border regions directly exposed to the coolant (LN_2 or propane or ethane) are absolutely clear and properly vitrified (Fig. 16c): these well frozen (vitrified) regions are easy to identify by backlight illumination (Fig. 16d) as mentioned above. And only these regions of the specimen block should be sectioned and studied. If LN_2 cooling is not sufficient and the whole specimen looks milky, freezing in a secondary cryogen like propane or ethane (see Table 3) is necessary. As already mentioned above, slower freezing at lower cooling rates, that means dipping into LN_2 according to Fig. 16c in an advantage, if fully (2.3 M) sucrose protected specimens have to be frozen: faster cooling in a "*secondary cryogen*" like propane or ethane provokes often tension (stress) cracks. The resulting clefts inside the sample may affect the cryosectioning procedure considerably.

Formerly it was the general opinion, that diamond knives are not suited for dry sectioning of protected material at low temperatures. It was also claimed that glass knives have to be broken very slowly over several seconds and that both balanced break and tungsten coating are striking advantages (*e.g.*, Griffiths *et al.*, 1983, 1984; Roberts, 1975). We compared several times tungsten coated, slowly broken or balanced broken glass knives (Leica balanced break Knifemaker) with normal glass knives broken in the usual

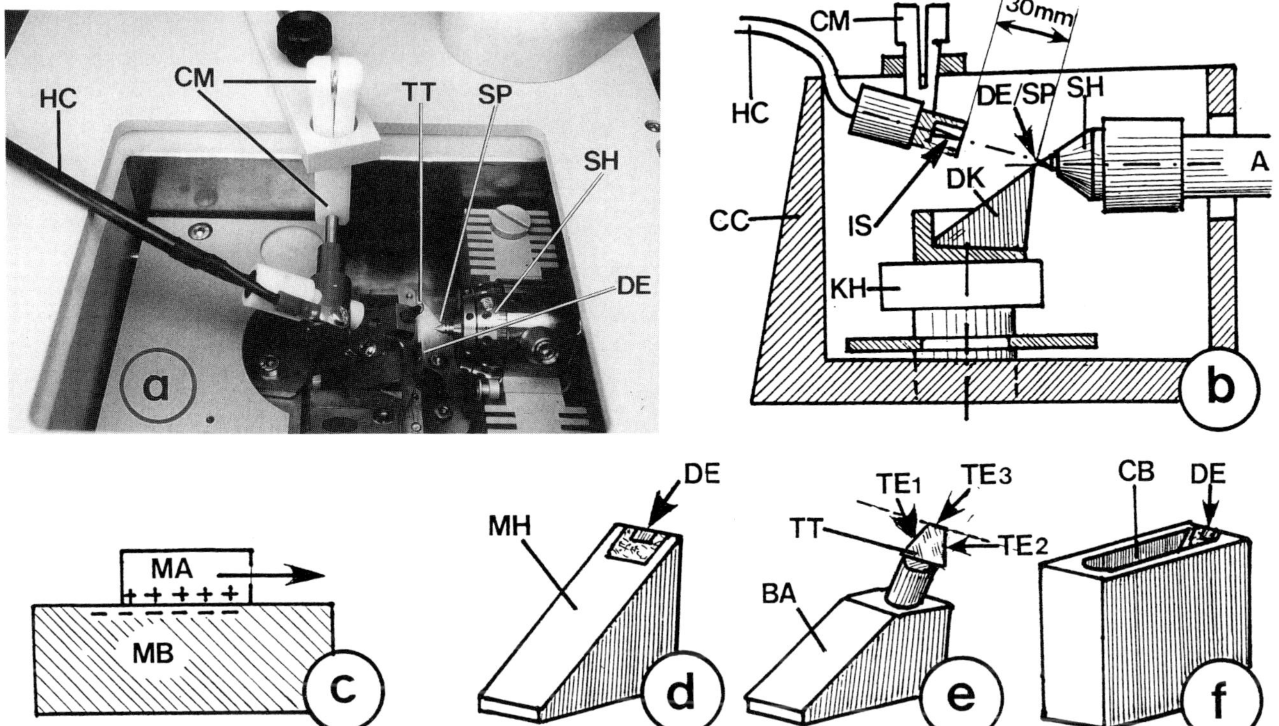

Figure 17. Cryosectioning with low angle cryodiamond knife, diamond trimming tool and an adjustable ioniser (*e.g.* Diatome static line II) simplifies the work both with protected and with fresh frozen samples considerably. Further comments in the sections on cryosectioning and in the Discussion, in Table 2 and the paper of Michel *et al.* (1992). **(a/b)** Professional set up in an FCS/FCR-cryochamber: near the specimen SP in holder SH the knife holder KH is visible. Usually in one of the two positions a triangular low angle diamond knife DK (see "d"), in the other a cryodiamond trimming tool TT (see "e") is mounted. The ioniser spike IS is positioned approx. 30 mm behind the knife edge (held by the cryomanipulator system CM). **(c)** Static positive and negative electricity is generated, if two non conductive materials MA and MB (*e.g.*, frozen water and diamond) slide in a good contact, which always creates mechanical friction. This corresponds to the well known production of static electricity by a rotating disc sliding over electrodes of non conductive material. Since opposite charges attract each other, electrostatic forces develop. Such forces must have annoying results, if one piece is very fragile like an ultrathin cryosection. Concerning the generation of static electricity, the couple "*diamond/ice*" seems to be much more efficient than the couple "*glass/ice*". **(d)** Triangular Diatome cryodiamond knife for "*dry* sectioning" with edge DE. The metal holder MH has the geometric shape of a triangular glass knife and fits therefore in all common knife holders. **(e)** Diatome diamond trimming tool TT with diamond edges TE1/TE2/TE3: the base BA corresponds to a triangular glass knife and fits in all knife holders like (d). The trimming tool TT has two oblique side edges T1/T2 for facing the pyramid and one normal edge T3 for presectioning the later cutting surface. **(f)** Diatome cryodiamond knife for "*wet sectioning*": the metal holder forms a section collection boat CB and is practical exclusively used in material sciences.

---

way with a well adjusted old LKB Knife Maker 7801A in our laboratory. In repeated experiments we could not find a reproducible difference between coated and uncoated, slowly and normally broken, or balanced and non-balanced broken glass knives: all delivered useful sections of the same quality as long as breaking was done with a well adjusted knife maker and care was taken not to damage the edge by improper handling. There was also no difference between freshly broken knives and edges stored up to several weeks in the laboratory, if care was taken, that the knife edges were not contaminated by dust or coating layers. But there was an evident difference between glass and diamond knives: sectioning with diamond knives in former times without special measures was frustrating because the sections seemed to stick to the dry diamond surface and it was practically impossible, to obtain good ribbons with the diamond knife. Especially at higher temperatures above -120°C and with fully protected samples after 2.3 M sucrose impregnation -

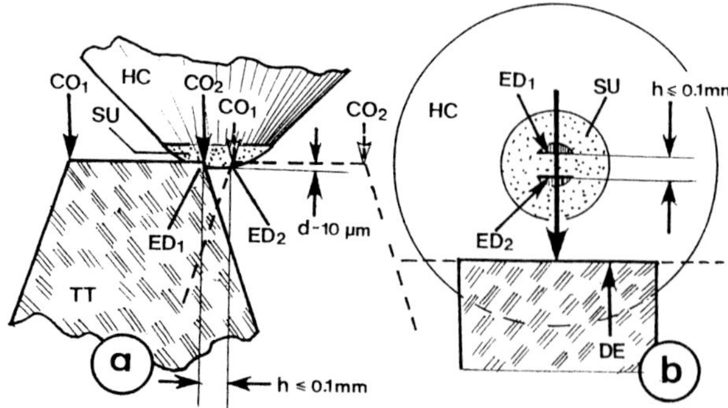

Figure 18. Trimming of a suspension droplet SU frozen on a hollow cone carrier HC after immersion cryofixation (see Fig. 4a and c) with a diamond trimming tool TT in the cryochamber (see Fig. 17a and e). Since suspension droplets usually have a regular convex surface and only a border layer ≤ 5 µm is properly vitrified by plunging, only two edges ED1 and ED2 have to be prepared. Additional remarks in Section 7. **(a)** Suspension droplet SU on carrier HC according to Fig. 4c in top view: sectioning of first edge ED1 with corner CO2 of trimming tool TT. Without changing the position of SU/HC the tool TT is retracted, laterally shifted to the right and again advanced against SU. Second edge ED2 is cut with corner CO1 (broken line). It makes no sense to enter deeper than d = 10 µm into droplet SU with respect to the depths ≤ 5 µm of the well frozen border layer. **(b)** Trimmed suspension droplet SU on holder HC with edges ED1/ED2 in front view above diamond edge DE. Holder HC with droplet SU is rotated by 90° around longitudinal axis of HC in comparison to schema (a). It is not needed to cut a complete "*pyramid*". After advance of the knife edge against the specimen or vice versa, sections with height "h" occur (arrow indicates cutting stroke). The width of these sections increases from cycle to cycle. It is recommendable to use the feed totaliser for control of the incision depth reached during the section preparation, since only a total depth of maximally 5 µm (maximum depth of vitrification) delivers useful results. This value corresponds to at most 50 sections of 0.1 µm or 100 sections of 0.05 µm thickness.

----

this phenomenon on diamond surfaces was evident: glass knives under these conditions made the job much better. That was a severe disadvantage, since the profit of the diamond edges in standard ultramicrotomy at ambient temperature was obvious, where only students and beginners in workshops are forced to use glass knives and practically all experienced users already switched over to the better, absolutely reproducible, and actually less expensive (if the costs for glass strips, knife maker, troughs and, not to forget, all wasted time of scientists and technicians are taken into consideration) diamond knives.

In the meantime the situation has changed completely and diamond edges are now usable with the same advantages as in standard ultramicrotomy for cryosectioning and trimming at low temperatures. Investigations of Helmut Gnägi (Diatome AG, Biel, Switzerland) around 1990 lead to the clear conclusion, that those detrimental phenomena result from charging phenomena of the specimen, the sections and the diamond surfaces, which can be eliminated by discharging ionisers. If such an ioniser is properly installed and used inside the chamber (see Fig. 17a and b), the diamond knife is clearly superior to the best glass knives. Discharging by antistatic pistols or ionisers was formerly recommended both for standard ultramicrotomy at ambient temperature (Nicholson, 1978; see also Mattheij and Dignum, 1975) and for cryowork with fresh frozen specimens (Sitte, 1982; Sitte and Neumann, 1983). But nobody had the idea, that sticking of the sections on the facet of a diamond knife could be caused by electric charges. It is the merit of Helmut Gnägi that this question is now clearly answered. Indeed: in retrospective this is easily understandable, since friction between two different nonconducting materials mostly provokes splitting of electrical charges (see Fig. 17c and Michel *et al.*, 1992). After this charging, attraction of those materials results from the electrical field, which develops between positive and negative loads of the corresponding non conductive materials. The proper solution for discharging is an ioniser with a spike within the GN_2-atmosphere of the cryochamber (Fig. 17a and b) connected to a transformer with an adjustable voltage in the range between 1 and 10 kV. This AC-spike produces negatively and positively charged nitrogen particles which are able to discharge both the sections and the knife surface. The spike of the ioniser has to be mounted inside the GN_2: outside mounting in the room atmosphere (*e.g.*, above the chamber) leads to a considerable turbulence at the border between ambient air and cold GN_2 at the open top of the chamber. Additionally such outside mounting results in severe frosting of both knife and specimen. Care has to be taken not to approach too close to temperature sensors (*e.g.*, Pt 100 sensors for object, knife or GN_2 temperature) or other electric elements of the cryochamber, since such ioniser spikes at higher voltages can easily damage IC elements of the electrical control unit. As far as protected material is sectioned, the ioniser spike is positioned as shown in Fig. 17b. The improvement is striking and easy to demonstrate by switching on and

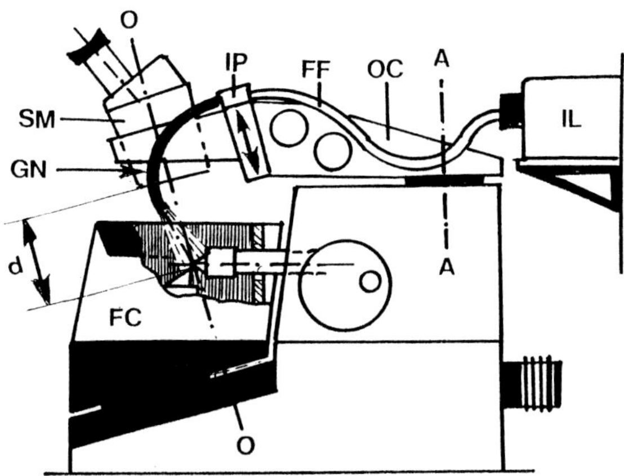

Figure 19. A high performance stereomicroscope SM and a bright cold fibre illumination are essential for cryosectioning of fresh-frozen biological samples, since surfaces below 0.1 x 0.1 mm^2 have to be trimmed and cut. According to our own experience a flexible fibre system FF between the illumination box IL with the light source (IL mounted separated from ultramicrotome table, *e.g.*, on the wall) and an intermediate piece IP mounted on a high performance stereomicroscope SM (for systems Ultracut-S/UCT the compatible Leica-Wild stereomicroscope M31 with wide field eyepieces 16 x /14B and 40 x /0B and mount WS and adapter lens 1.0 x /F = 89) or on the carrier of the stereomicroscope is the best solution. IP divides the light in two goose necks GN, which are bent right- and left-hand around the body of SM and remain always properly adjusted with respect to the working distance d and optical axis OO of SM. The advantage of the double system GN/FF is, that after turning away the optic holder OC around axis AA (*e.g.*, for exchange of specimen or knife) the illumination with respect to SM is again correctly centred, after SM is switched once more into working position. Additional remarks in the section on "Ultrathin Sectioning .." and in the Discussion.

----

off the ioniser during sectioning. With the ioniser long straight ribbons of incredible quality are obtainable from sugar protected samples with the dry diamond edge (I have never seen before such results !) and there is no question at all about the best suited methodology. In summary: the actual state-of-the-art is without doubt the sectioning with a triangular cryo diamond knife (best facet angle 35°) and an adjustable ioniser. One should forget all discussions about slow breaking, balanced breaking or coating of glass knives and change over to these considerably easier and better tools. Of course, glass knives remain the proper edges for all students, beginners and workshops for financial reasons (without tungsten coating and without all sophisticated breaking methods, but produced with a well adjusted knife maker and carefully checked).

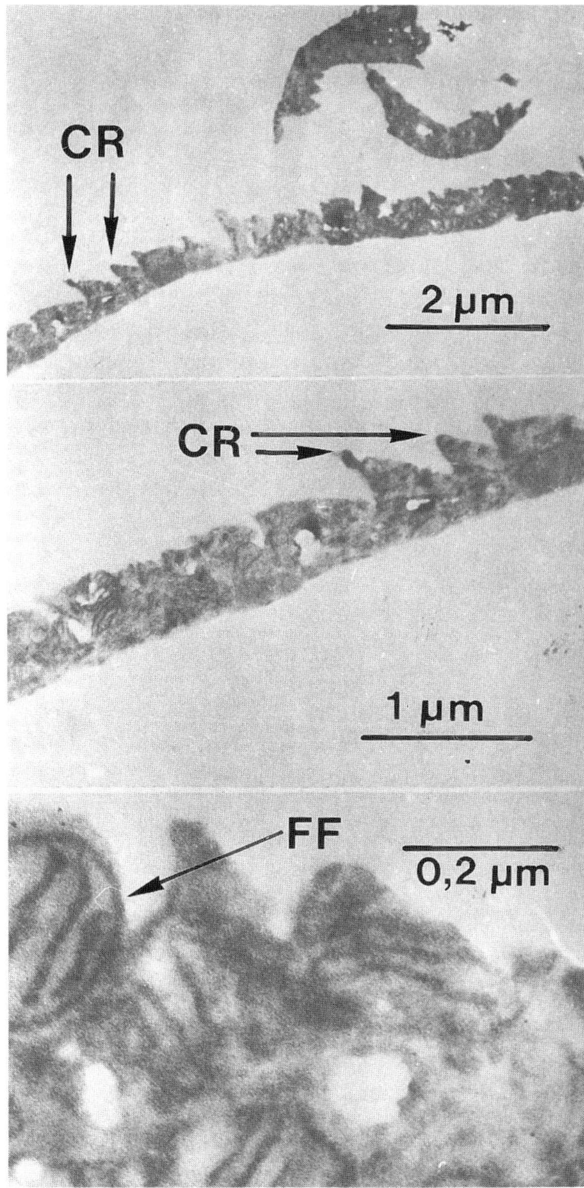

Figure 20. "*Crevasses*" (cold fractures) in cryosections of fresh-frozen material at different magnifications according to Frederik *et al.*, 1982 (reproduction with permission). The cryosections were freeze-dried, subsequently osmicated, reembedded in epoxide, cross sectioned and stained. It is clearly visible, that the crevasses form sharp edge like profiles. At higher magnification they look often like freeze fracture profiles FF of split membranes. See also Fig. 25b and further comments and discussion in the section on "Ultrathin Sectioning .." and in the Discussion.

All follow-up procedures remain unchanged as described within the already initially cited review articles.

Ultrathin Sectioning and Handling of Fresh Frozen Samples

As already stated in the "*Introduction*" and the preceding section "*Cryosectioning According to Tokuyasu*", there is a considerable difference between sectioning of samples impregnated in 2.3 M sugar solutions after a chemical prefixation by aldehyde and specimens frozen without chemical pre-treatment for element analysis or fresh frozen investigation or diffraction analysis in the TEM/STEM (see Table 2). If specimens have to be sectioned, which are protected by some reasons with a lower amount of sugar (< 2.3 M, *e.g.* 1 M, see Tokuyasu, 1973) or with another cryoprotectant (*e.g.*, according to Bernhard, 1965, see also Bernhard and Leduc, 1967), the methodology has to be changed and may have some similarities with fresh frozen sectioning described in this section. There may be also slight differences between the needs for EDX or EELS of freeze dried ultrathin sections and direct observation of frozen-hydrated ultrathin sections in TEM/STEM with respect to the solid state of water (I_h, I_c, I_v, see section "*Freezing and the Frozen State of Water .....*"). Nevertheless these pathways are now treated together, since it seems to be important to exercise also all freezing and cryosectioning for subsequent EDX or EELS at the best obtainable level of methodology in order not to run in the already mentioned pitfalls of misinterpretations. Finally it has to be mentioned, that ultrathin sectioning of fresh frozen material as described in this section and cutting of semithin sections for frozen-hydrated EDX in the STEM according to the methods of Hall and Gupta (see Gupta and Hall, 1981) are quite different: as mentioned already, I do not want to include this interesting but very special methodology needed for EDX of extracellular compartments not containing proteins (*e.g.*, urinary space) into this review, since we have no first hand experience in our laboratory with the methods of Hall and Gupta. I refer therefore to the previously cited literature.

Already freezing as the first step of cryopreparation deserves full attention, since proper vitrification is needed to exclude possible artefacts connected with any crystallisation process starting from a liquid phase. At the same time vitrification is the best precondition for an optimal sectioning consistency of frozen specimens. But remember: proper freezing makes no sense without a stable mounting of the frozen specimen to the carrier of the cryo-ultramicrotome, which will be described first. Sectioning also makes no sense without a diamond trimming tool and a low angle cryo diamond knife for sectioning together with an adjustable ioniser. Finally the mounting of the ultrathin sections on the grid and both the cryotransfer of the grids to the scope for direct studies in the frozen-hydrated state or to the freeze dryer for subsequent EDX are important steps which decide on success or failure of the whole effort. These steps are therefore also included here.

Proper vitrification at its best is achieved by high pressure freezing of sufficiently thin specimens. This means according to facts discussed in the previous section "*High Pressure Freezing*" mostly a thickness below 0.2 mm (see Table 5). In several cases vitrification is also possible by immersion or impact freezing (see preceding sections of this review). As far as immersion in ethane is used and the sample (cell suspensions or tiny bulks with diameters < 0.5 mm) is mounted on a gelatine coated hollow cone carrier (see Fig. 4c) no further mounting operation is needed. After immersion cryofixation the carrier has to be transferred to the precooled cryochamber (required minimum temperature around -170°C) in LN_2 following Fig. 4d. If this transfer has to bridge only a short distance in room atmosphere between the cryofixation unit and the cryo-ultramicrotome in the same laboratory a transfer in a LN_2-precooled but dry metal container closed with a similarly precooled metal cover is sufficient, because during this dry operation the container is warmed up only to a negligible extent while GN2 from inside escapes by expansion (Fig. 4e): there is no risk. Care should be taken, that the cone shaped carrier has the correct geometry and that heat capacity near the top is really small. After inserting of the carrier, trimming with the Diatome diamond trimming tool is easily carried out at minimum temperature with the motor drive and automatic feed according to Fig. 18. It makes no sense to speed up trimming by manual operation of the ultramicrotome with respect to the risk of loosing the sample from the carrier. With the motor driven cryo-ultramicrotome (feed: 0.2 to 0.5 μm, cutting speed approximately 0.5 mm/sec) trimming is also finished elegantly within a few minutes. It is mostly restricted in such cases to the upper and lower edges with-out presectioning the sectioning plane. Normally both edges are set in a distance of 0.1 mm or less. Splinters and section fragments on the trimming knife or the specimen are easily removed according to Herbert Hagler by splashing LN_2 from a small cotton ball mounted to a cocktail stick or in a forceps: the rapid evaporation of GN_2 normally takes off all fragments and splinters. An ioniser is also helpful during trimming, since fragments and splinters do not stick to the charged surfaces. Trimming is very difficult without a good stereo microscope and a bright cold light illumination. As far as a Reichert Ultracut-S or UCT is used, the compatible Wild stereo microscope M3C with wide field eyepieces 16x/14B and 40x/OB is recommended instead of the standard optics. The mount WS for this stereo microscope is available for the optic carrier of these Ultracuts. Of course the 40x eyepieces together with the lens 1.0x/F = 89 mm for the correct working distance on the Ultracut produce a magnification, which cannot increase the resolution compared to eyepieces with smaller magnification. But experience tells, that this high magnification is nevertheless an advantage for practical work in combination with a bright illumination. The brightness of the backlight illumination of the Ultracut-S/UCT together with an FCS cryochamber is absolutely sufficient for this high magnification. But the standard incident illumination (fluorescent tubes) of the Ultracuts has to be improved by a more powerful fibre optic illumination. In our laboratory we use the Schott cold light source KL 1500 with two flexible goose necks bent around the stereo microscope according to Fig. 20. For other ultramicrotomes a similar set-up is recommended. It makes no sense to enter into fresh frozen sectioning without this optimised optical equipment. It should be mentioned in this context, that the simpler cryosectioning system Ultracut R/FRC was developed mainly for Tokuyasu preparations (immuno-cytochemistry on sugar protected material), but is not suited for work below -120°C with fresh frozen samples directed to element analysis or frozen-hydrated investigations in the cryo-TEM.

Ultrathin sectioning is generally carried out for frozen-hydrated investigation in the TEM at the obtainable minimum temperature of the cryochamber. This temperature should be -165°C or below. If the temperature does not reach this value, check the temperature of the outside walls of the cryochamber and of the defroster at the specimen arm: the built-in heaters are meant to prevent condensation or frost. Therefore the temperature of the chamber walls and of the specimen arm defroster should be around room temperature, but not noticeable higher. When frozen suspension droplets according to Fig. 18 have to be sectioned, cutting is started at the original surface of the frozen droplet and continued up to the moment, that the first small ribbon is obtained. Proper setting of the cutting window and use of an adjustable ioniser (see preceding section and Fig. 17a and b) are preconditions for good sections. The cutting speed usually is set around 0.5 mm/sec, the advance as small as possible dependent on the specimen (*e.g.*, approx. 50 nm) with respect to the formation of "*crevasses*" (see Fig. 20 and 25b), which are much more obvious in thicker sections. The cycle length (time between two consecutive sections) is not that important with a new FCS cryochamber than with former systems (*e.g.*, FC4/FC4D/FC4E). The formerly often recommended speedy manual operation to shorten the cutting cycle is a disadvantage with respect to the regularity of the section thicknesses and the adjustment of the needed ioniser, since high spike voltages introduce risks with respect to the electronic system (see preceding section "*Cryo-Ultramicrotomes*"). For cutting a clean triangular low angle cryo diamond knife (facet angle 35° or less, the smaller the better with respect to crevasses and irreversible compression) with minimum clearance angle is recommended. The transfer of the sections to the grids [according to McDowall *et al.* (1983), a carbon coated, according to Michel *et al.* (1991) an uncoated 600-mesh copper grid, *e.g.* G 600 TT, SPI, West Chester, PA, USA) is carried out using a precooled eyelash. The grids are either mounted in an adjustable holder (Fig. 21a), available as an accessory for the FCS cryochamber or positioned on a polished metal plate according to R.L. Ornberg (personal communication), mounted on the knife holder of the cryo-ultramicrotome according to Fig. 21b. The Ornberg-device makes it very easy to have a couple of grids available and to shift single grids for loading with cryosections immediately behind the cutting edge of the diamond knife. This Ornberg-device is not offered commercially. But it may be possible to manufacture such a plate in one's own workshop and to adapt it carefully to the height of the knife used for cryosectioning. The device enables immediate first flattening and pressing of the cryosections with a precooled section flattening tool. Since a good thermal contact between the cryosections and the grid with respect to heat conduction under the electron beam is of greatest importance, efficient pressing of the sections onto the grid is an indispensable step even after a first pressing on the plate according to Fig. 21b. The best solution up to now remains the section press introduced by Dubochet and McDowall (see Fig. 21c), which is used in different versions by different groups and is available as accessory to the FCS-cryochamber. Pressing is carried out inside the chamber and the cold press may be used for a first cryotransfer operation to a loading station of a commercial cryotransfer system. An interesting home-made system was described by Tvedt *et al.* (1984) (see also Sitte, 1982, Fig. 2, p. 15). This system was especially designed for pressing and transfer of cryosections for subsequent freeze drying and EDX.

Cryo transfer units are offered for TEMs with side entry both by Gatan and Oxford Instruments. The

Gatan Model 626/626-DH (see Fig. 22a) is an improved version of the Philips transfer unit according to Hax and Lichtenegger (1982) with a protection shield ("*shutter*"), that can be moved from outside after the frost layer built up during the transfer operation through the humid room atmosphere has been evaporated in the vacuum of the scope. This evaporation process can be easily read on the vacuum indicator of the TEM. The new Oxford system Cryotrans CT 3500 according to R. Henderson uses an additional protection by inert GN_2 to avoid any frost formation during the transfer operation (see Fig. 22b). Both seem to work reproducibly according to second hand information. In our own laboratory we use successfully a Zeiss cryotransfer chamber for the top entry system of our Zeiss EM 902 filter lens scope (see Sitte, 1984), which was the only transfer system compatible to the former FC4/FC4D/FC4E-cryochambers. For the new FCS-cryochamber Ludwig Edelmann has designed an accessory (see Fig. 22c), which enables direct cryotransfer from the cryochamber to the EM 902: it consists of a polyurethane-foam profile mounted on the top of the FCS-chamber and is equipped with an entry port and holder for the Zeiss transfer chamber. Without doubt such a direct loading in the cryochamber of the ultramicrotome represents the simplest, most efficient and cheapest solution for this kind of cryo-work. But there is another very simple solution for side entry systems, which was used already in 1980 by Dubochet and co-workers [see McDowall *et al.* (1983) and Fig. 22d]. They transferred the sections with the old Philips cryo specimen holder PW 6591/100 for their Philips EM 400 without any modification, that means without protection shield (shutter) and workstation for loading. The grid with the sections was inserted under LN_2 into a small insulating container. Then the loaded cold stage with the grid below LN_2 was brought as close as possible to the opening of the side entry port on the column of the scope. Afterwards the rod of the cold stage was quickly moved through the ambient air and inserted. It is astonishing, that most of such simple transfer operations were successful: both the amorphous frozen state of the specimen is maintained and the contamination of the cryosections with hexagonal ice is within the usual limits - compared with much more sophisticated transfer operations and instruments. This fact is easily explained: since the rod of the cold stage with the grids is covered by a coat of LN_2 and has a temperature of -196°C, some seconds are necessary for evaporation of the LN_2 coat. Within these seconds ambient air cannot approach to the rod with the grids and evaporation of LN_2 acts as a heat sink providing a stable temperature even of the extremely thin grids with the cryosections. I agree with Jacques Dubochet that the most simple operation and set-up often is the best. Of course vitrified specimens have to be kept through the whole chain of cryopreparation according to the first idea of Heide and Grund (1974) always at a value below the devitrification temperature of ice, that means < -135°C (see section "*Freezing and the Frozen State*"). To be on the "*safe side*" temperature indication should never show values above -145°C in order to prevent devitrification to cubic ice. This precondition can be easily fulfilled with the cryochambers and the transfer systems mentioned above.

There remains only the question if the mechanical energy during the process of cryosectioning is transformed into heat to an extent, that may cause a "*through-the-section-melting*" of the knife edge as argued by Thornburg and Mengers (1957). Up to now a lot of serious evidence was collected against this hypothesis [see *e.g.*, Frederik (1982), Frederik and Busing (1981, 1982), Hodson and Marshall (1972), Karp *et al.* (1982)]. The most striking evidence against a melting process during cryosectioning under the conditions given above is the preservation of the amorphous-frozen state of vitrified specimens and vitrified water, shown by electron diffraction of the sections in the TEM [see McDowall *et al.* (1983), Michel *et al.* (1991, 1992)] and the formation of cold fracture profiles (crevasses) during that sectioning procedure (see Fig. 20). In comparison to freeze cleavage before freeze etching, where the whole splitting energy is freed instantaneously (Sleytr and Robards, 1977), cryosectioning is a very slow and smooth continuous process, where heat is generated and spread at rather small rates. Therefore it seems evident, that it is really possible with the vailable methodology and instrumentation to produce proper cryosections in thicknesses between 50 and 100 nm from vitrified biological samples, which show correctly cellular structures embedded in amorphous frozen water in the TEM/STEM.

The whole preparation procedure changes somewhat, if slam or high pressure frozen specimens have to be sectioned. Slam frozen material has the advantage, that the surface of the well frozen border layer of the specimen is a replica of the highly polished mirror plane. The most important step is the mounting of the mirror-frozen specimen on the carrier of the cryo-ultramicrotome. If tissue slices of thicknesses of 0.5 mm or less are used (Sitte *et al.*, 1987a, 1988), this mounting operation according to Fig. 23 is quite simple. Since the areas of such tissue slices mostly correspond to the cross section diameter of the sliced organ (*e.g.* a rat kidney with a diameter of approximately 1 cm), the frozen slice has to be divid-

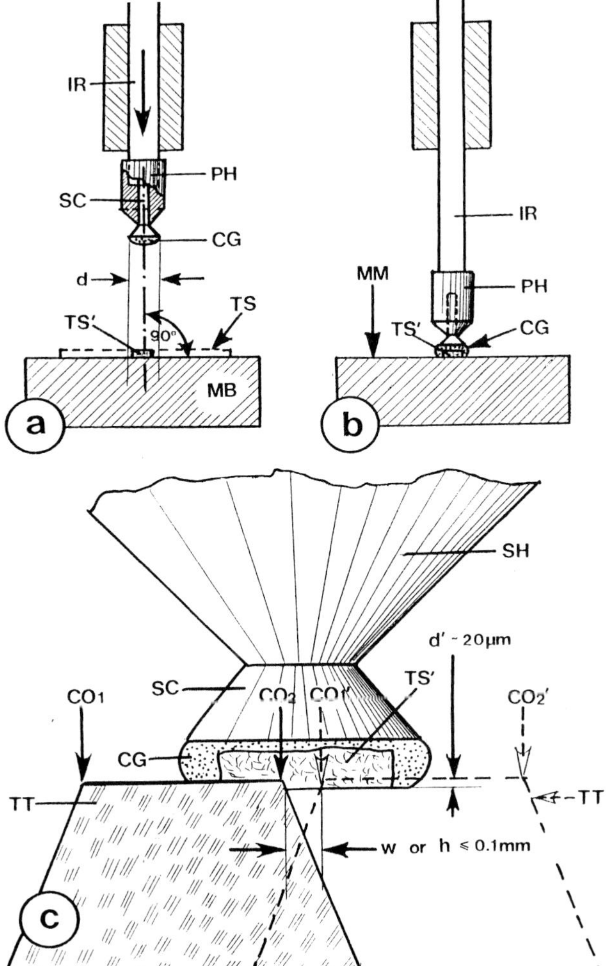

Figure 21. Manipulation of cryosections of fresh-frozen material in the cryochamber: the dry transfer of the sections SE from the edge DE of the diamond knife DK to the grid GR is mostly done with a cold eyelash-probe EP and considerably simplified by a properly adjusted ioniser (see Fig. 17a and b). The grids are either mounted on a holder GH (a) or positioned on a polished metal plate (b). All schematic drawings in side view (upper half) and top view (lower half). Further explanations in the section on "Ultrathin Sectioning ..". **(a)** Grid holder GH is available as cryotool for FC-systems of Leica-Reichert: it takes up to 5 grids and is mounted on the bolt BO of the cryomanipulator CM. By rotation around BO and x/y/z-translation of CM it is simply possible to adjust each grid GR exactly just below the edge DE of the knife DK or below the sections SE for the section transfer with EP. **(b)** Ornberg mounted a metal plate PL on the knife holder KH in the cryochamber: the highly polished surface PS of PL is adjusted approx. 0.5 to 1.0 mm below the edge of DK. It is easy to move a grid on the polished plate against the edge just below the ribbon of sections SE. This simplifies the transfer considerably. Afterwards provisional affixing with precooled stamp ST is easy. The set-up is not commercially available but highly recommendable, if manufacturing in ones own workshop is possible. **(c)** Up to the moment the section press according to Dubochet and McDowall is the best tool to obtain the needed thermal contact between cryosections and gridbars or cryosections and support film. In the commercially available Reichert-version the section press consists of a base BA, a hardened and polished steel ring RI, a glass plate GP and the screw SC. If SC is loosened, the glass plate is lifted up by spring SP and the grids GR can be inserted with a forceps FO (open corner OC). The grids are moved below glass plate GP by turning ring RI and pressed by tightening the screw SC between base BA and glass plate GP. If needed, the whole set up can be used after closing for a short cryotransfer into a loading station of a transfer unit.

----

ed into smaller pieces after the control described in the former section "Ambient Pressure Freezing" (see also Fig. 7). The piece used for cryosectioning should not exceed the diameter of the object table of a standard pin. It is positioned with the well frozen side downwards on the LN2 cooled mirror of the slam freezer (Fig. 23a and b). The pin with a droplet of cryoglue is mounted on a plunge injector device and guided vertically downwards to the frozen specimen. After contact the glue hardens due to heat extraction by the mirror block. Since this heat is extracted through the frozen specimen, one may believe that at least recrystallisation could occur due to this heat transfer. But this is certainly not possible at least in the well frozen border layer, since that layer remains cooled by the mirror. The situation is identical to the application of the warm specimen to the mirror plane : also during slam freezing the whole heat of the specimen definitely is extracted through this well frozen border layer without detrimental effects. But "*seeing is believing*" : it is easy to check the well frozen layer again by dark field inspection (Fig. 7e and 16d). The cryoglue used in our lab for this mounting is simply soft soap. But there are a lot of very useful cryoglues described in the literature, which may serve as "*more scientific*" alternatives [see *e.g.*, Bachmann and Schmitt-Fumian (1973), Karp *et al.* (1982), Michel *et al.* (1991), Steinbrecht and Zierold (1984)]. The special advantage of the described mounting operation is the exact perpendicular orientation of the mirror-like replicated surface of the slice with respect to the longitudinal axis of the pin (see again Fig. 23). A further advantage is the easy adjustment of the trimming diamond or the diamond

Figure 22 (*on facing page*). Special systems are needed to transfer frozen-hydrated sections from the cryo-ultramicrotome to the cryo-TEM. Such "*cryotransfer systems*" have to guarantee, that (1) the amorphous state of the vitrified sample does not change to crystalline ice due to temperature excursions above T_d = -135°C, (2) that the frozen specimen is not contaminated by frost particles during transfer operation, (3) that the evaporation of frost from the shutter sleeve SH and the side entry rod SE inside the vacuum of the TEM column does not cover the sections by an amorphous layer of ice, and finally (4) that drift phenomena or bubbling inside the mini-dewar MD do not offset the resolution of the TEM. The systems (a) to (c) fulfil these preconditions to a high degree. Systems (a) and (b) are designed for side entry ports, system (c) for the former Zeiss EM 10/ EM 902 top entry cryosystems. (a) and (b) represent the state-of-the-art and are both consisting of a "*work station*" WS for loading and the side-entry-rod SE with a mini-Dewar MD based on cryosorption vacuum (CS) as cryostage. All drawings are considerably simplified and roughly schematic. (a) and (b) show the loading set up (work station WS and transfer unit SE/MD) in side view (lower half) and the transfer unit alone in top view (upper half). Further explanations and comments in Sections 7 to 9. **(a)** Gatan Model 626/626-DH is a considerably improved version of the old Philips cryo specimen holder PW 6591/100 (see "d") and the Philips cryotransfer system according to Hax and Lichtenegger (1982). The most important new feature was the external control SC of the shutter sleeve SH. During transfer SH is moved over the grid GR (see circular insert, arrow) and remains closed up to the moment, that all frost deposited during transfer is evaporated inside the TEM column. In this way the above mentioned precondition (3) is fulfilled. **(b)** Oxford Cryotrans 3500 according to R. Henderson: the main difference to (a) is the flexible GN_2 filled bellows BE covering SE during transfer. Like (a) the grid during transfer operation is covered by shutter sleeve SH. Since dry GN_2 prevents frost formation, cryowork in TEM can start without delay. **(c)** Cryotransfer from Leica-Reichert cryochamber FCS to the top entry port of former Zeiss TEMscopes EM 10 CR/EM 902 CR: according to L. Edelmann the FCS-cryochamber is equipped with a "*chimney*" CH of styrofoam with an entry port EP for the Zeiss transfer chamber TC with mini-dewar MD. The transfer operation is carried out as usual (FC = cryochamber FCS, KN = knife on knife holder KH, KS = knife support, UM = Ultracut-S, VA1/VA2 = sliding valves to close CH/FC and TC; see Sitte, 1984). **(d)** Dubochet and McDowall used the former Philips cryo specimen holder PW 6591/100 without modification for cryotransfer: SE was loaded below LN_2 in foam container CT. Transfer was carried out through room atmosphere (EP = side entry port on TEM column CO, FO = forceps for grid GR). Further comments and references in the section on "Ultrathin Sectioning ...".

---

edge against the mirror-like surface which corresponds exactly the adjustment of a knife against a presectioned resin block with the aid of the backlight illumination reflection during standard ultramicrotomy at ambient temperature (see *e.g.*, Sitte and Neumann, 1983, Fig. 31 and 32, pp. 64 ff and Fig. 81, p. 132). Finally it is certainly nice never to be confronted with the compression artefacts according to Fig. 9a and b, which are connected which each slam freezing: of course they still exist, but are not visible, since the sectioning plane always shows circles if formerly spherical structures are sectioned. But one should remember, that the diameters of such circles or spheres do not show the original size, since all x/y-dimensions are enlarged by the compression artefact due to slamming.

Trimming in principle is similar to the "*droplet trimming*" described above (see Fig. 18). Since the slam frozen specimen has an absolutely plane surface, trimming is essentially a "*mesa trimming*". The trimming operation is made once more with a Diatome diamond trimming knife (see Fig. 23c) together with an ioniser (see Fig. 17a and b) at the motor driven ultramicrotome with automatic feed (speed: 0.5 mm/sec, feed: 0.5 μm). Also with such specimens this automatic operation at least is less time consuming than trimming by hand, because the risk of losing the specimen is minimised. Diamond trimming knives are superior to glass knives or the offered trimming tools of hard metal alloy, which are very brittle and therefore mostly defective. Trimming is done in four steps like "*mesa-trimming*" : firstly two edges right and left according to Fig. 23c are trimmed, the remaining two edges after a 90°-turn of the specimen around the longitudinal axis of the specimen holder. It is not necessary to enter more than 20 μm into the block surface. That makes 40 cutting strokes each with less than 10 sec or less than 7 min per edge of the pyramid. This time should be invested for proper trimming of a block face < 0.1 x 0.1 mm². Sectioning and section transfer are performed afterwards as described above.

Mounting of high pressure frozen specimens is described in detail by Michel *et al.* (1991). For the small disc-shaped specimens (diameter approximately 2 mm, thickness normally below 0.2 mm) either the use of a home made special holder compatible with the object holder system of the FCS cryochamber (see

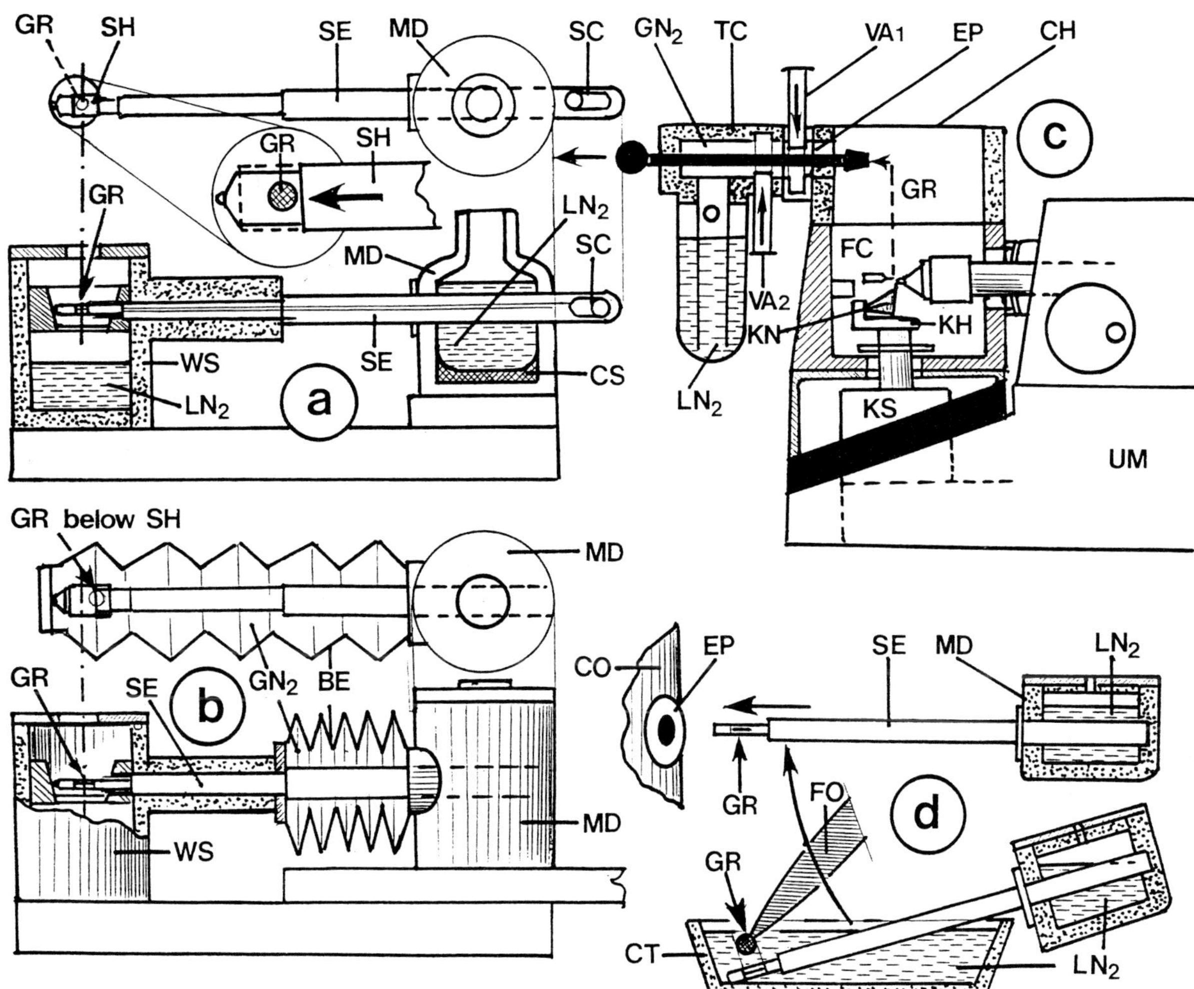

Fig. 24a) or the adaptation of the commercial Leica-Reichert cryo vise-type-holder (see Fig. 24b) is recommended. The holder according to Fig. 24a has a slit for the insertion of the tiny samples. The width of this slit can be varied by a screw inside the cryochamber. According to Michel *et al.* (1991) the high pressure frozen specimen is additionally held by a cryo-glue consisting of 30 % ethanol, 30 % methanol and 40 % tap water (v/v), which is highly viscous at -135°C. At sectioning temperatures of -165°C or below this mixture solidifies. An alternative for gluing in such arrangements may be an indium sheet as described by Dubochet *et al.* (1988). The pure metal indium is soft even at LN_2 temperature and therefore well suited as an intermediate layer which adapts excellently to each specimen profile. A suitable commercial vise type holder is available as accessory for Reichert-Leica FC-systems and can be adapted by a simple counter piece according to Fig. 24b.

Some sectioning artefacts have to be expected, if fresh-frozen specimens are cut. But also these artefacts - mostly "*crevasses*" and section compression - can be reduced considerably if all given possibilities are used together. "*Crevasses*", that means clefts or fractures within the sections (see Fig. 20), occur during most cryosectioning processes. There is a considerable discussion about the generation and the specific localisation of these crevasses [see *e.g.*, Dubochet *et al.* (1988), Fig. 21 on p. 164 or Richter *et al.* (1991), Fig. 7 on p. 26; detailed discussion in Richter (1992, 1994)], which is certainly of theoretical interest. But these fractures disturb the interpretation of the images of such frozen-hydrated sections in an extremely frustrating manner [see Chang *et al.* (1983) or Dubochet *et al.* (1988), Fig. 17, p. 161, or Fig. 25b in this review). The most important practical question is therefore, how this phenomenon can be avoided, suppressed or at least minimised. The following recommendations may help: regular thin sectioning (the thinner the better) would probably improve the results, since in really thin cryosections (thicknesses < 50 nm) this phenomenon mostly does not occur. Like metals or resins also films of ice seem to be flexible in sufficiently thin layers. As mentioned

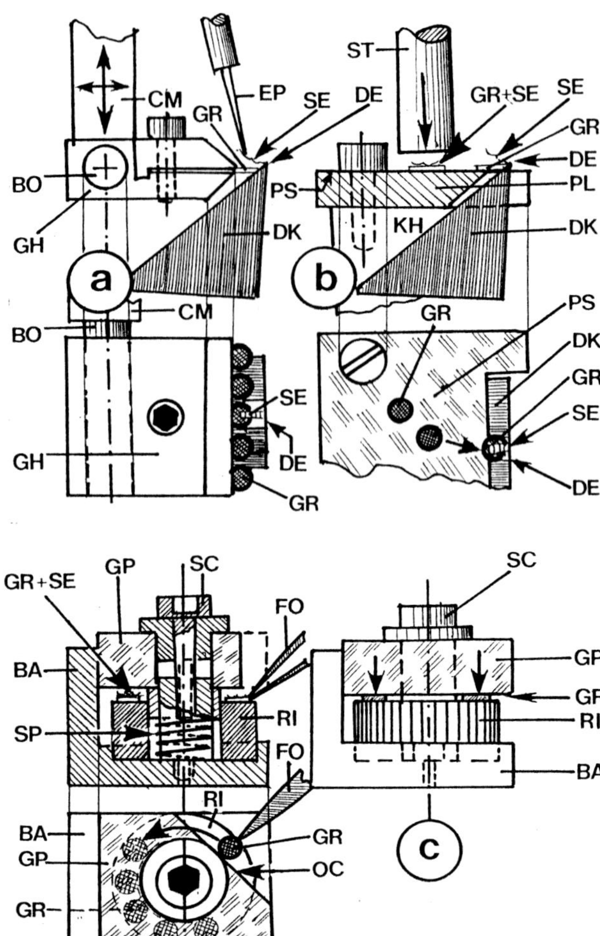

Figure 23. Mounting with a particular orientation of slam frozen material (*e.g.*, fraction TS' of a tissue slice TS, see Fig. 7) on a standard carrier SC ("*pin*", see Fig. 4b) of a cryo-ultramicrotome and subsequent trimming of a "*pyramid*" with a diamond trimming tool TT (see Figs 17e and 18). Mounting is easy on a Leica-Reichert KF 80/MM 80 or CPC/MM 80 system for plunge and impact freezing. The description of the procedure (a) to (c) refers to this apparatus. Further explanation in the section of "Ultrathin Sectioning...". **(a/b)** After careful inspection of the slam frozen tissue slice TS according to Fig. 7 a suited part TS' is selected, which corresponds to the diameter d of the object table on SC. TS' is deposited in the centre of metal mirror MM on MB exactly below SC mounted on pin-holder PH of injection rod IR. The well frozen original contact plane is oriented downwards and in good thermal contact with metal mirror surface MM. A droplet of a suited cryoglue CG (*e.g.* soft soap, but consider alternatives mentioned in the section on "Ultrathin Sectioning ...") is deposited on the table of SC, and moved with the plunge device IR/PH against TS'. It is important to work not too slow: the glue should not freeze but surround TS' to some extent (see "c"). There is no risk of recrystallisation of the vitrified border layer, which is in good thermal contact with MM. **(c)** Trimming with diamond trimming tool TT according to Figs. 17 and 18 is done like "*mesatrimming*": after facing the first sidewall of the pyramid with corner CO2 of TT (entry depth d' $\leq$ 20 μm) the corresponding side wall is faced after readjustment of the tool TT (new position TT': broken line in schema) with corner CO1'. The whole procedure is repeated after a 90° rotation of SC/TS' around longitudinal axis of SC. The width w and the height h of the final cutting surface should not exceed 0.1 mm. The orientation of the new sectioning surface is exactly 90° to the longitudinal axis AA of SC: this enables a perfect adjustment and approach of the diamond knife by darkfield-(backlight)-illumination due to the shiny mirror like replication of the metal mirror MM by the slamming process. Cryosectioning is done as described in the section on "Ultrathin Sectioning...".

----

several times above, homogeneous vitrification is also tremendously helpful and high pressure freezing a big advantage, if other cryofixation methods do not vitrify the specimen of interest properly. Finally, reduction of the dimension of the sectioning area parallel to the knife edge ("*width*" < 0.1 mm), reduction of the cutting speed and reduction of the facet and clearance angle of the knife are helpful. Of course thickness, proper freezing, reduced width, cutting speed and knife angles are not the only parameters of influence: more homogenous specimens with a lower content of membranes are less prone to this artefact. But homogeneity cannot be influenced in most cases. Also the other artefact, the severe irreversible compression of the cryosections (Dubochet *et al.*, 1988; Richter, 1992, 1994), can be reduced considerably by several measures: most of them are identical with the measures influencing the crevasses. The perhaps most important parameter also in this case is a reduction of the width of the specimen surface to values < 0.1 mm. Contrary to 2.3 M sugar protected specimens, which are so easy to section and to stretch after thawing, fresh-frozen specimens are extremely brittle below -150°C and tend to show considerable compression. This compression usually reaches values up to 50% of the original height of the sectioning area and is irreversible, since no stretching is possible. Therefore reduction of width is a precondition for good results and trimming under the optimum preconditions described above (see Figs. 19 and 23) is an urgent need. If severe compression occurs, reduction of sectioning speed may also help in some situations, but

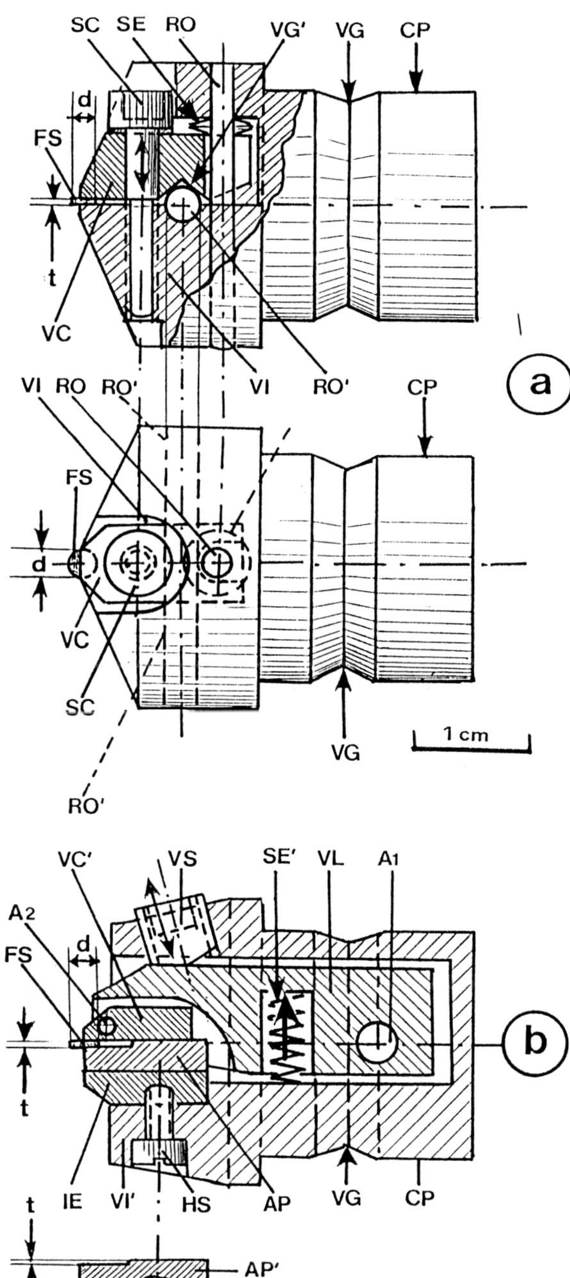

Figure 24. Vise-type holders for the extremely thin disc shaped high pressure frozen specimens FS (diameter d = 2 mm, thickness t = 0.1 to 0.6 mm) must guarantee proper clamping without breaking those small fragile samples. Further explanation and comments in the sections on "High Pressure Frezzing" and "Ultrathin Sectioning...". See also Figs. 12a, b and c. (a) If a suitable workshop exists, a home-made holder (similar to Michel *et al.*, 1991, Fig. 2, p. 6, modified) following the schematic drawings (side view above, top view below) is a suitable solution: the holder base of stainless steel consists of a cylinder part CP with V-shaped grove VG for mounting and of the lower part VI of the vise (dimensions of CP/VG identical with standard cryo-specimen holder of FCS-system). The corresponding vise clamp VC (also stainless steel) is orientated by the steel roll RO (diameter approx. 2 mm, similar to rolls used in roller bearings) and a second roll RO' inserted perpendicularly to RO between VI and VC. RO' acts as bearing for VC, which has a corresponding V-shaped grove VG'. Circular spring elements SE ("*Tellerfedern*") open the vise VI/VC automatically if screw SC is unlocked. With respect to the extremely thin frozen discs FS, the screw thread of SC should be as small as possible (*e.g.*, 0.5 mm per revolution). (**b**) Commercial vise-type holder (accessory for Leica-Reichert cryo-chambers FC4/FC4D/FC4E/FCS/FCR) is a useful alternative: the cylindrical part CP/VG corresponds exactly to (a). The vise lever VL turns around axis A1 mounted in CP. The small vise clamp VC' turns around axis A2 mounted in VL. With respect to the small diameter d and thickness t of the disc shaped frozen specimen FS an adjustment piece AP (upper diagram: profile corresponds to thickness t of FS) has to be inserted to guarantee proper clamping by screw VS. The vise VC'/VL opens automatically, if VS is unlocked (spring element SE'). Instead of a loose adjustment piece AP an element AP' with the same profile is recommended. AP' can be exchanged with insert element IE hold by screw HS (compare upper and lower diagram). An Indium plate between FS and one of the vise elements VC' or AP or AP' simplifies and ensures proper clamping of FS.

----

the facet angle of the cryo diamond knife used is much more important (Richter, 1994). Cryo diamonds with a facet angle of 35° produce far less compression than the usual standard cryo knives with a facet of 45°, which should not be used for fresh-frozen specimens. It is not difficult to draw the conclusion, that facet angles around 20°, which are technically possible will offer tremendous advantages in comparison to the "*low angle knife*" with a facet angle of 35° or less nowadays already commercially available (Jésior, 1986, 1989). If low angle knives are used, the clearance angle ϵ is an important additional parameter, because facet angle α and clearance angle ϵ influence compression together. It is therefore recommended to reduce also the clearance angle ϵ to the minimum value possible.

Ultrathin Cryosections for Element Analyses

There are some differences between section

preparation for frozen-hydrated investigation including electron diffraction and preparatory work on ultrathin cryosections for subsequent element analyses, since freeze-drying is an essential step preceding most of these analyses. If we exclude frozen-hydrated EDX according to Hall and Gupta not considered in this review article (see the excellent review of Gupta and Hall, 1981) we have to notice therefore those different preconditions.

Mostly the demanding criteria for frozen-hydrated work in the TEM/STEM with respect to vitrification and sectioning are not considered as strict guidelines for the variety of different analytical methods working on dry cryosections. Besides EDX and EELS these methods include nowadays microanalytical techniques like proton probe X-ray microanalysis, laser microprobe techniques (LAMMA), and even optical microscopy techniques concerned with ion and element localisation (see *e.g.*, Roomans *et al.*, 1994). In many cases (*e.g.*, EDX, LAMMA) somewhat thicker sections may be advantageous with respect to the total number of available signals. In other cases, for example for the very promising use of EELS in combination with a high performance field emission STEM system extremely thin cryosections in thicknesses below 100 nm are needed (Leapman *et al.*, 1994; Andrews *et al.*, 1994): only such ultrathin cryosections exclude disturbing multiple scattering processes and offer the wanted higher sensitivity of this set up for some physiologically interesting elements with low atomic numbers like Ca. But real vitrification also under these preconditions is not regarded as a precondition and as far as EDX analyses are concerned, even hexagonal ice segregation seems often to be disregarded (*e.g.*, Hagler and Buja, 1984; Roomans and Shelburne, 1983; Roomans *et al.*, 1982; Shuman *et al.*, 1976; Zierold, 1982, 1987; Zierold and Hagler, 1989). Fresh frozen specimens for EDX are mostly sectioned at temperatures considerably above the obtainable minimum temperature of cryochambers, *e.g.*, around -120°C. This makes a remarkable difference because brittleness of the material at -120°C is considerably reduced in comparison to minimum temperatures of cryochambers ranging between -165 and -180°C. Finally sectioning is much easier if the specimen is homogeneously frozen and coarse ice segregation is excluded. Microcrystalline cubic ice or generally homogeneous nucleation does not seem to be a severe drawback in cryosectioning.

As previously stated, element analyses on ultrathin sections by EDX with respect to the much higher electron doses needed for sufficient signals is only possible after freeze drying. Some new results in this respect have just been presented and discussed (Edelmann, 1994b). They lead to new conclusions. Formerly the considerable shrinkage of freeze dried ultrathin sections (see *e.g.*, Sitte *et al.*, 1994) was accepted like a physical law. There was agreement, that this shrinkage is a result of the "*thermal collapse phenomena*" described by Kellenberger and co-workers [see *e.g.*, Kellenberger (1987, 1991), Kellenberger *et al.* (1986)] and could not be influenced to a noticeable extent. In the meantime Ludwig Edelmann of our group has clearly demonstrated, that not only can shrinkage be reduced from the range of 50% to approximately 5% by elongated drying protocols, but also, that thermal collapse phenomena due to speedy drying may provoke redistribution of ions within the frozen tissue [Edelmann (1994b), see discussion with the reviewers, pp. 76 and following): mostly the time for sufficient drying, that means for minimising artefactual shrinkage is considerably underestimated and even ultrathin cryosections in thicknesses of 100 nm and less need drying times elongated to 12 h and over in the low temperature range for proper preservation. Such preparations show only a negligible shrinkage : besides the good preservation such freeze dried sections have the advantage, that all crevasses, which disturb the images of the frozen-hydrated sections, disappear nearly completely during freeze drying as shown in Fig. 25. Cryosorption freeze drying for such long-term freeze-drying is an elegant and efficient tool, since such instruments work absolutely silent and with negligible running costs (see *e.g.*, Sitte *et al.*, 1994). It does not seem to be necessary to carbon coat such freeze dried cryosections, since a cryotransfer system for the transfer of the cold dry cryosections on the grids from the cryosorption freeze dryer to the loading station of a cryotransfer system is available [grid holder of the dryer, see Fig. 7, p. 53, items GB/GC in Sitte *et al.* (1994)].

Investigation of Frozen-Hydrated Ultrathin Sections

TEM/STEM investigation and image recording of frozen-hydrated ultrathin sections suffer from two severe drawbacks: fresh frozen biological material is extremely sensitive to electron bombardment in the scope and reveals mostly a very faint contrast in thin layers additionally blurred by the large number of electrons inelastically scattered by the ice matrix and the elements with low atomic numbers. Low dose operation of the scope is an urgent need, since the sectioned material disintegrates rapidly, as already observed and clearly stated by Dubochet and co-workers in the early days of investigations on frozen-hydrated material [see *e.g.*, Chang *et al.* (1983), Dubochet and McDowall (1984), Dubochet *et al.*

(1983, 1986, 1987, 1988); see also Isaacson (1977), Talmon (1987), Talmon *et al.* (1986). In the eighties efficient low dose systems were not available. Only screening with quite simple TV-systems (*e.g.*, SIT-cameras) and focusing next to the preselected area ("*MDF-systems*" for Minimum Dose Focusing, mostly following the "*rocking beam*" principle) gave a possibility to reduce irradiation (electron doses). But just around 1980 some phenomena were observed, which supported the expectation that biological specimens at very low temperatures around 4K show a considerable (in favourable cases up to 300 times) higher stability under the electron beam in comparison to ambient temperature (Chiu *et al.*, 1981; Dietrich *et al.*, 1977, 1978, 1979, 1980; Dubochet, 1981; Dubochet *et al.*, 1981, 1982c; Freeman *et al.*, 1980; Knapek and Dubochet, 1980; Knapek *et al.*, 1982; K.H. Müller *et al.*, 1981). These data were obtained mainly on thin dry crystals of organic compounds by registration of the decay (fading) of diffraction patterns and showed a similar stabilising effect as measurements of the mass loss under the electron beam (Dubochet, 1975; Ramamurti, 1977; see also Isaacson, 1977; Salih and Cosslett, 1975; or Cosslett, 1978 and Fig. 26). That gave some hope, since such extremely low temperatures are obtainable by the use of super conducting LHe cryostat lenses in the TEM firstly designed by Fernández-Morán (1966; see also Fernández-Morán, 1985) and brought to a high degree of perfection by the Dietrich group in Munich (Dietrich *et al.*, 1977, 1979, 1980).

With respect to the importance of such a "*cryostabilisation*" the exciting values obtained in the above-mentioned first run were checked by several laboratories using different instrumentation but similar samples. The results obtained in this second run by an "*International Experimental Study Group*" and some others could in several respects not reproducibly confirm the data collected before [see "*International Experimental Study Group*" Chiu *et al.* (1986), Downing (1983), Wade (1984), Wade and Pelissier (1982), Lamvik *et al.* (1983); see also Lepault *et al.* (1983b), Dubochet *et al.* (1987, 1988). They confirmed more or less previous experiments which resulted in "*cryostabilisation effects*" between zero and x10 [see *e.g.*, Glaeser (1971, 1975), Glaeser and Hobbs (1975), Glaeser and Taylor (1978), Grubb and Groves (1971), Hayward and Glaeser (1979), Heide *et al.* (1982), Siegel (1972)]. Though in some details more material and time for additional experiments seemed to be necessary, there was at least in 1985 a general agreement on two important facts: (1) The "*cryostabilisation*" even at LHe temperature reaches in most studies only a factor of x3 to x10 in comparison to TEM work at room temperature, (2) There seems to be no noticeable difference between the "*cryostabilisation*" at LHe- and at LN2-temperature, if decay (fading) of diffraction patterns is considered. In other words: this stabilisation effect is mostly already available with LN2 cooling. The big effort for LHe-technology does not seem to be needed. Of course, most of these studies refer essentially to dry structures, which differ considerably from frozen-hydrated sections. But water-containing structures seem to behave similarly (Knapek and Dubochet, 1980). As far as a really similar "*cryostabilisation*" could be obtained on frozen-hydrated specimens, also the smaller stabilisation factor is highly welcome. But this kind of "*cryostabilisation*" was then obviously already reached and used by Dubochet's group during all investigations on frozen-hydrated specimens. They used the LN2 cold stage in the early eighties to maintain the amorphous frozen state both in ultrathin cryosections and in "*bare grid*" preparations of suspensions (*e.g.*, Dubochet *et al.*, 1988). It is important to remember that this kind of cryostabilisation could not prevent "*bubbling*" (immediate radiolysis effects) under the electron beam during investigation and diffraction analysis in the TEM. It is therefore of paramount importance for further work in this field to use real low dose systems, which effectively either reduce the irradiation needed for recording structures and diffraction patterns or lower the beam damage reproducibly.

As mentioned above, another parameter also influences the investigation of frozen-hydrated material in the TEM considerably: since only fresh frozen, not pre-treated and therefore unstained samples have to be investigated, there is a considerable lack in contrast. With respect to the "*phase contrast transfer function*" (see *e.g.*, Dubochet *et al.*, 1987, Fig. 3, p. 125) useful pictures with a sufficient contrast in the conventional TEM (CTEM) are mostly only obtainable considerably below the correct focus: usually the underfocus of such pictures ranges between 1 μm and 10 μm. It is difficult under those conditions to resolve small details and a big gap between the real resolving power of a modern CTEM in the atomic or molecular order (*e.g.*, approximately 0.3 nm) and the reality in the photographic picture (*e.g.*, 3 nm and over) results. At least partially this inconvenience can be compensated by "*Electron Spectroscopic Imaging*" (ESI) using an "*Energy Filtering TEM*" (EFTEM), since such systems allow the elimination of the relatively large amount of inelastic scattered electrons and their influence on the final picture (Bauer, 1988; Ottensmeyer and Andrew, 1980; Ottensmeyer *et al.*, 1981; Schröder, 1992; Schröder *et al.*, 1990). Since the ice matrix and the atoms with low atomic numbers of the specimen embedded in the ice matrix increase inelastic scattering processes dramatically, "*zero loss imaging*" and image recording with an underfocus below 0.1 μm close to the "*Scherzer focus*" (see Schröder *et al.*, 1990) is a considerable advantage for frozen-hydrated ultrathin sections. The combination of low dose systems with an ESI seems to be the most promising investigation technique in this field.

With respect to low dose investigation, there exists the already mentioned simple possibility to reduce the electron dose or the number of electrons needed for the formation of a photographic image by MDF-systems. This is possible by a lateral shift of the specimen or the electron beam between the first step of focusing and the subsequent recording of the image on a sensitive photographic emulsion. Similar to the general operation of the TEM usually an investigation of a frozen-hydrated specimen needs three steps. Screening, that means looking for a suitable section area for subsequent documentation is the first step: it is done at low magnification with a minimum beam intensity. Even this low intensity may be reduced by a suited image intensifying TV-system. The electron doses needed for this screening operation under these optimised conditions are negligible in comparison with the two following steps. These steps usually need considerably higher magnification and beam intensity for focusing and recording. According to experience in many laboratories, image intensifiers connected to a TV-screen are mostly not sufficient for focusing (even with special aids like Wobbler systems) and for the recording of images (Cosslett, 1978). The focusing step contributes considerably to the whole electron dose, if the selected area is used for focusing. Therefore this second step is often spatially separated from the definite third step, that means the electron exposure of the photographic emulsion: focusing takes place on an area nearby and not on the area selected for the picture. If the electron irradiation is limited to the field of the fluorescent screen observed with the light microscope (*e.g.*, by "*Köhler illumination*"), the electron dose could be reduced to the small dose needed for screening and the considerable higher dose needed for a sufficient exposure of the photographic emulsion. This irradiation could not be circumvented in the past as long as only photographic emulsions were available. But also this consideration is rather theoretical: the exposure is not only affected by drift phenomena often connected with a lateral x/y-movement of the specimen or a deviation of the beam, but suffers also from the irregular profiles typical for ultrathin cryosections, which influence the focus if focusing and recording are made in different areas of such a section. Real progress in low dose investigation is therefore only possible by replacing the photographic emulsion by another tool which is more sensitive for electron irradiation, and the application of all other means of reducing beam damage like higher voltages and better vacuum conditions (see *e.g.*, again Cosslett, 1978).

Nowadays two low dose systems fulfil the mentioned precondition, one of which (slow-scan charge-coupled devices, abbreviation: SS-CCD) is already used in many laboratories on different TEM models (see *e.g.*, De Ruijter, 1995; Krivanek and Mooney, 1993; Tietz, 1992). The other system ("*image plates*" with a photo and electron stimulable phosphor layer instead of a photographic emulsion) has been in experimental laboratory use for some time in different places and scopes (Ayato *et al.*, 1990; Burmester, 1992; Mori *et al.*, 1988, 1990; Oikawa, 1990; Shindo *et al.*, 1990, 1991). An image plate system of this kind for the TEM developed by Fuji has already been successfully tested in two Japanese EM-laboratories. This Fuji FDL-5000 system is just now introduced and commercially available for TEMs (1996). Besides other advantages both systems offer a real and remarkable reduction (> 10:1) of the electron dose needed for the digital recording of an electron image or an electron diffraction diagram in comparison with the most sensitive photographic emulsion available hitherto. The author of this review article has no first hand experience with SS-CCD and image plates. Therefore only second hand information from literature, colleagues and manufacturers was available. This has to be considered, if conclusions are drawn from the comments in this section and in the following discussion. But it seems necessary to include also these recent breakthroughs in the review with respect to their paramount importance for cryowork on the extremely beam sensitive frozen-hydrated sections.

The first Slow Scan CCD camera was invented and basically developed at Bell Laboratories in USA around 1970. It was in astronomy that SS-CCD systems were firstly used in scientific research for low-light-level imaging. Scientific grade CCDs were introduced by Janesick and co-workers (Janesick *et al.*, 1987). They are now used in different configurations for many purposes in scientific research. The first attempt for EM work was made by Mochel and Mochel (1986) using a Photometrics SS-CCD on a VG-HB5 SEM. The first purpose designed SS-CCD was built in co-operation between UCSF, Gatan and Photometrics in 1987 for a Philips EM 430. Similar systems are now commercially available from Gatan and compatible with most TEM models. In the following years a couple of SS-CCD devices were developed and described (De Ruijter, 1995). A new commercial-

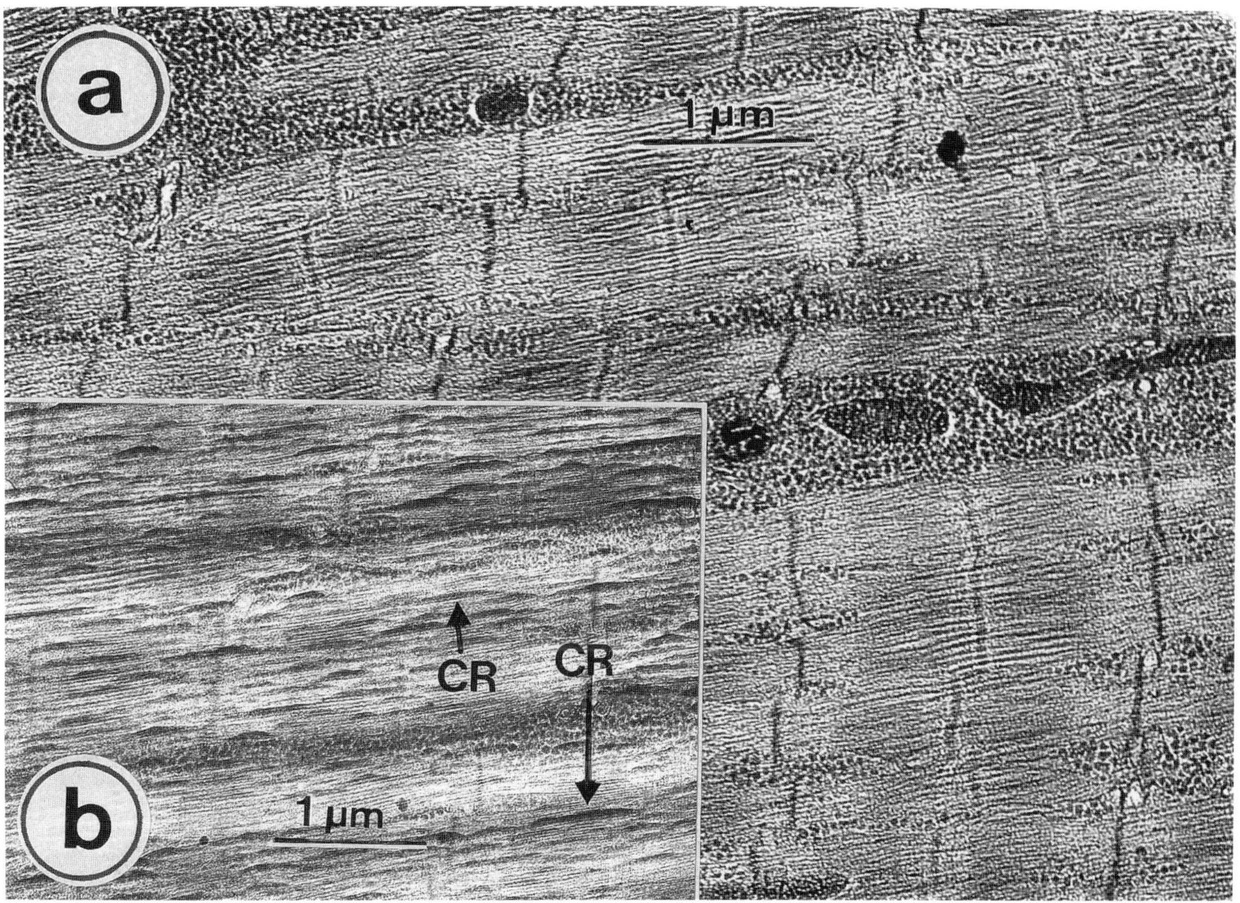

Figure 25. Freeze-drying of ultrathin cryosections is not necessarily connected with artefactual shrinking and cracks in the sections, if the drying procedure is extended over a sufficiently long period at low temperatures, as demonstrated by the electron micrographs (a) and (b). They show the same specimen at the same magnification before (b) and after freeze-drying (a). Preparation and electron micrographs by Edelmann (reproduced from Edelmann, 1994b, with permission): frog sartorius muscle, slam frozen on LN_2 cooled copper mirror (Edelmann, 1989b), cryosectioned at -160°C on Reichert Ultracut-S/FCS with Diatome diamond knife, Simco antistatic device, cutting speed 0.4 mm/sec. Sections placed on uncoated 600-mesh copper grids. Both pictures recorded at 4400 x in Zeiss EM 902 with energy filter (doses < 500 e^-/nm^2). Arrows in circles: direction of sectioning. Bar: 1 µm. Further explanation in the section on "Ultrathin Section Preparation..." and the Discussion. (**a**) Ultrathin section freeze-dried in Leica CFD cryosorption freeze dryer according to Edelmann and Sitte (see Sitte *et al.*, 1994) for 33 h at -100°C. (**b**) Frozen-hydrated section under identical conditions shows practically identical dimensions of A-bands. Ultrastructural details partially hidden by crevasses obliquely to the sectioning direction (arrows CR), which disappear during freeze-drying (see Figs 20 and 25).

SS-CCD was recently presented by Carl Zeiss, which developed a high resolution, variable speed camera for its models EM 906, EM 910 and EM 912 Omega. Finally Gatan has surpassed in 1995 the most severe limitation for SS-CCDs in electron microscopy by offering a new 2k x 2k SS-CCD "*MegaScan*" with a frame of approx. 40 x 40 mm^2 approaching towards the dimension of photographic films and image plates (Gatan, 1995). The principle of such scientific grade SS-CCDs for digital recording of TEM images or diffraction patterns according to Fig. 27 consists mostly of an electron sensitive scintillator, a fibre optic coupling plate and the CCD chip below. Both Yttrium-Aluminum Garnet-(YAG)-single crystal scintillators and Gadolinium Oxy-Sulphide (GOS) powder phosphorus (*e.g.*, P20 or P43) scintillators mounted on the top of the fibre optic plate were used successfully. Instead of the fibre optic coupling (*e.g.*, 1:1 or with a

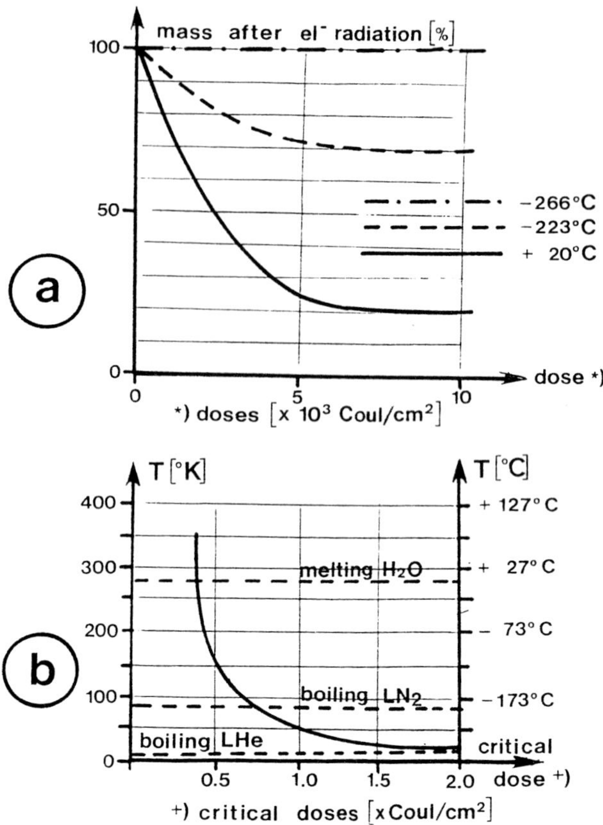

Figure 26. Electron beam damage: temperature dependence of mass loss (a) and "*critical dose*" (b) under electron irradiation. Diagrams according to Cosslett (1978), Figs. 3 (p. 120) and 5 (p. 125), redrawn and slightly modified, with permission. Both diagrams show, that beam damage is strongly temperature dependent and that near absolute zero (*e.g.* at LHe temperature) a considerable increase of the "*cryostabilisation effect*" seems to be possible. Nevertheless previous statements about stabilisation factors above x 100 had to be revised (see Chiu *et al.*, "*International Study Group*", 1986). The factor varies considerably from specimen to specimen and even from measurement to measurement (see Dubochet, 1975, Section 3 "*Irreproducibility*", pp. 284 ff) and remains probably generally in the range below x 10. See also further comments and additional references in the Introduction, the section on "..Frozen-Hydrated Ultrathin Sections" and the Discussion, and the review of Glaeser (1975). **(a)** Variations of the mass loss from a 40 nm thick film of phenylalanine with increasing electron doses at 15 kV at constant temperatures of -266°C (7°K), -223°C (50°K) and room temperature 20°C (293°K) according to preliminary data of Ramamurti (1977). The findings of Ramamurti agree very well with results of Dubochet (1975), which states after studies of the carbon loss of T4 phages and *E. coli* bacteria in different TEMs operated with 60 and 80 kV, that "*there is no perceptible carbon loss when the irradiation is made at LHe temperature*". **(b)** Variation of the critical irradiation dose for Coronene at 500 kV with temperature according to measurements of Salih and Cosslett (1975). The "*critical dose*" was measured by the disappearance of the diffraction patterns of the Coronene crystals. There occurs a considerable reduction in radiation sensitivity below LN_2 temperature (-196°C). But the variations in these measurements do not allow a definite conclusion about this "*stabilisation effect*" at LHe temperature.

----

reduction 2:1) a lens-mirror coupling is possible and was already used in practice (2:1 fibre optic reduction and lens-mirror-coupling are not included in the schematic diagram of Fig. 27). Under those circumstances one 100 keV electron frees 2000 visible photons on the YAG-scintillator or 7000 visible photons on the more sensitive GOS-scintillator. The most efficient channelling of those photons is achieved with a glass fibre coupling (10 to 15 % efficiency). Approximately 30 % of the channelled photons are transferred into "*well electrons*" for the digital read-out. Those electrons are confined into square adjacent pixels by "*potential wells*" generated by a conductive "*gate structure*" on the top of the insulating SiO_2-layer. The gate arrangement delivers typical square pixel sizes in ranges from 6 to 27 μm edge length [for technical details and additional literature see De Ruijter (1995), Section B and Figs. 1-3]. One 100 keV-electron corresponds under the mentioned preconditions to 60 (YAG) or 300 (GOS) well electrons. Since read-out noise at a read-out speed of 500 kHz ranges around 25 well electrons, one obtains a sufficient signal-to-noise ratio and a registration of single 100 keV-electrons [theoretical statistical value of detection: 0.3 electrons of 100 keV; all data and numbers according to Krivanek and Mooney (1993)].

The basic function of SS-CCD-systems is quite simple: the needed photons for the photon sensitive CCD are excited in the scintillator by the electrons of the beam and guided by the glass fibres of the fibre optic plate or collected by a lens to the polysilicon electrodes (CCD-chip; see Fig. 27a). The analogue output signals from the chip are further amplified and digitised by an intermediate electronic system and subsequently transferred to the memory of a computer. This "*read-out*" procedure is quite slow [approximately $2x10^4$ to $2x10^6$ pixels/sec according to De Ruijter (1995)] with respect to a sufficient charge transfer from relatively large areas (over 1024 x 1024 pixels) and to a low read-out noise. Read-out times are

Table 6. Influence of pixel size, detector surface area and SS-CCD-read-out time on low dose image acquisition: according to Burmester (1992) the limited usable polysilicon detector surface of slow scan CCD cameras influences the efficiency of the system considerably in comparison with photographic plates (films) or phosphor image plates. Burmester proposes and uses a "*Pixel Equivalent*" in the comparison: the real pixel size is calculated using the relation between the frame size of the common photographic materials (60 mm edge length) and the size of the polysilicon electrodes (CCD-chips, see Fig. 28), since smaller magnifications are necessary to image the same specimen area on the smaller CCD-chip. Different from Burmester (1992) as size of the photographic plate or film compared to the phosphorus image plate 80 mm edge length are calculated (Fuji image plate measures 99.6 x 80.9 mm^2, effective usable area 94.0 x 75.0 mm^2 identical to photographic plates for most scopes). The pixel equivalent (PE) according to Burmester is given by the equation "Pixel size x 80 mm divided by detector edge length", *e.g.* from Gatan 512 x 512 ..... PE = 19 x 80/9.7 = 157 μm). One has to remember, that during CCD read-out beam blanking is necessary, which either causes delay (deflection above specimen) or additional beam damage (deflection below the specimen with respect to drift phenomena). Read-out speed and mode therefore are important parameters especially for systems with high pixel numbers and special measures (*e.g.* "*subarea read-out*" as offered for Gatan MegaScan) are really necessary for low dose work. Calculation of minimum and maximum read-out times according to De Ruijter, 1995, with lower and upper read-out values of 2 x 10^4 and 2 x 10^6 pixels/sec. "*Subarea read-out*" and "*Modulation Transfer Function*" (MTF, see Krivanek and Mooney, 1993; De Ruijter, 1995; Weickenmeier *et al.*, 1995a,b) for the different SS-CCD systems are not considered in the calculation of the read-out times.

System	Array (Pixels)	Pixel Size	CCD-Chip Surface (mm^2)	Pixel Equivalent PE	Minimum Read-out Time	Maximum Read-out Time
Gatan	512 x 512	19 μm	9.7 x 9.7	157 μm	0.13 sec	6.5 sec
Gatan	1024 x 1024	27 μm	27.6 x 27.6	78 μm	0.52 sec	26.2 sec
Gatan Mega-Scan	2048 x 2048	24 μm	(49.2 x 49.2)	39 μm	2.10 sec	104.9 sec
Tietz	512 x 512	38 μm	19.5 x 19.5	155 μm	0.13 sec	6.5 sec
Tietz	1024 x 1024	19 μm	19.5 x 19.5	78 μm	0.52 sec	26.2 sec
Tietz	1024 x 1024	24 μm	24.6 x 24.6	78 μm	0.52 sec	26.2 sec
Zeiss Vario-speed	1024 x 1024	19 μm	19.5 x 19.5	78 μm	0.13 sec	500 sec

Most data concerning arrays, pixel sizes and CCD-chip surfaces according to Burmester (1992, Table 2.1, p. 6)

dependent on the frequency of the CCD-system: they are mostly situated between 500 kHz and 2 MHz. A further increase to 5 MHz seems to be possible and advantageous. Another possibility to cut down the read-out cycles are sub-area read-out modes. Contrary to broadcast compatible CCDs the whole chip surface is used as image detector. To avoid CCD-exposure during charge read-out a shutter is needed to avoid smearing of image information. This shutter is activated by computer control using an existing beam deflector of the scope. Mostly this "*beam blanking*" is realised above the specimen with one of the gun-shift controls. This procedure minimises the electron dose but is not always applicable if high resolution is needed and read-out times increase above a critical level. In such cases drift of the specimen, image and/or focus sometimes results from blanking by local heating or charging. In such cases continuous irradiation of the specimen during read-out is unavoidable and the deflection has to be arranged below the specimen. Fig. 27b shows the whole set up of a TEM equipped with an SS-CCD camera. The camera itself is mounted at the bottom of the column and is exchangeable against other detectors by a special construction. During venting the column of the scope the whole CCD-camera is retracted into a housing and in this way protected from frost layers on the cold surfaces.

Several details of the design of SS-CCD

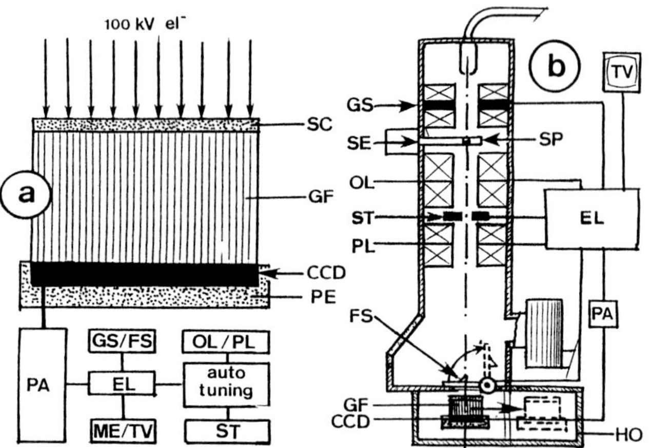

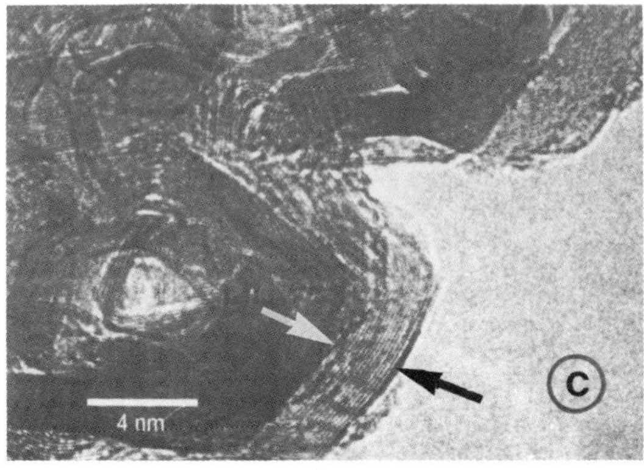

Figure 27. "*Slow-Scan Charge-Coupled Devices*" (SS-CCD systems) are one of the two promising technologies for low dose investigations on TEM/STEM-scopes. Further comments and references in the section on "..Frozen-Hydrated Ultrathin Sections" and the Discussion. See also Table 6 and Figs. 28 to 30. (**a**) Basic principle of an SS-CCD camera for electron microscopes according to Krivanek and Mooney (1993, Fig. 1, p. 96, modified): the fast high energy electrons (*e.g.*, 100 kV e^-) hit the scintillator SC (*e.g.*, YAG or GOS phosphorus) and produce visible photons (approx. 2000 to 7000 photons, dependent from the scintillator composition). Approximately 10% of these photons are "*channelled*" through glass fibres of the fibre plate GF to the polysilicon CCD-chip below. To reduce thermal noise the chip is cooled by the Peltier element PE. Approximately 30% of the channelled photons excite low energy "*well electrons*" suited as signals for digital "*read-out*" via pre-amplifier PA by the computer electronics EL. The read-out picture is transferred on-line to the memory ME of the computer and to the TV screen. During read-out "*beam blanking*" by shutter elements GS (gun-shift controls above specimen SP in side entry rod SE) or FS (fluorescent screen below object OB) is necessary to avoid continued CCD exposure and "*smearing*" of image information. "*Autotuning*" of the scope (*e.g.* autofocus and autostigmation) is possible by intersections with the lenses OL/PL and stigmator ST. (**b**) Set up of the scope with an SS-CCD system according to diagram (a): the SS-CCD camera is mounted on the base of the column CO below the fluorescent screen FS. EG = electron-gun, GP = glass plate for visual observation of picture on FS, HO = housing for SS-CCD camera retracted during venting of column. Computer control EL for read-out and autotuning according to (a). (**c**) The detector area of CCD-chips is limited for surfaces between 19 x 19 mm² (Standard with 1k x 1k Pixels) and 40 x 40 mm² (Gatan MegaScan with 2k x 2k Pixels). The CCD-picture of a carbon black particle shows the superb quality of such images (0.34 nm lattice planes of carbon: arrows) and represents less than 1% of the whole on-line image area of a Gatan MegaScan CCD (Specimen and electron micrograph: Gatan Inc., Pleasenton, CA, USA, with permission of Gatan-Munich).

----

systems are important for practical use. The scintillator has to be covered by a thin electrical conductive gold or aluminum layer to avoid charging phenomena. The SiO_2 layer of the CCD detector with its "*gates*" is usually cooled to a temperature around -40°C by a Peltier element or a cryogen to reduce thermal charging, which creates additional image noise. This cooling of the CCD detector once more makes some precautions (*e.g.*, by vacuum) necessary for the exchange of the scintillator or during venting of the TEM column to avoid frosting of the cold surface (see Fig. 27b). The amplifier and computer system are especially designed to compensate for the rather slow read-out of the charges for image areas sufficient for the interpretation of pictures with a useful resolution: actually there were till 1995 only image frames of about 20 x 20 mm² obtainable with a pixel size around 25 μm. At the end of 1995 Gatan introduced the already mentioned new 2k x 2k MegaScan system with a frame of approx. 40 x 40 mm². It is understandable, that both the scintillators, the glass fibre systems in the fibre plates and the CCD-chips for such extremely high pixel numbers cannot be manufactured without irregularities and that such defects will increase exponentially with the increase of the edge length of the detector area and with the decrease of the edge length of the pixel square. Besides defects due to manufacturing, dirt particles on the scintillator surface or "*burn spot damages*" by a concentrated electron beam of some nA have also a direct impact on the quality of the affected micro area of the CCD-picture. Since both manufacturing defects, contamination or burn spots on the scintillator cannot be avoided completely, an elec-

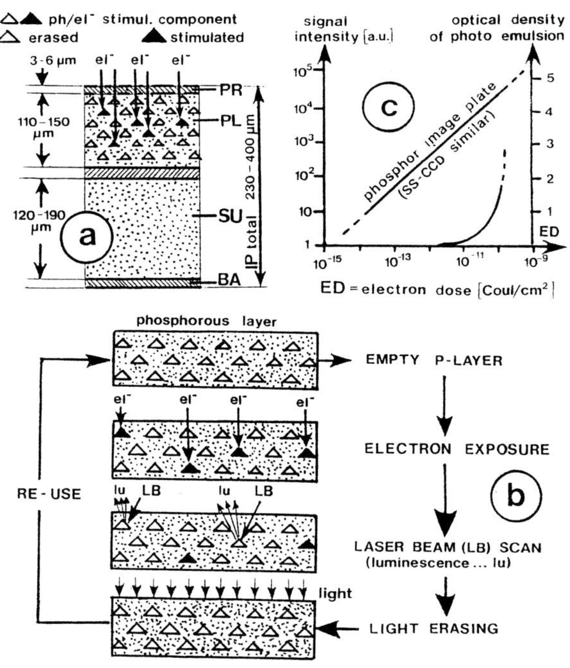

Figure 28. Photo- and electron-stimulable phosphor image plates (IP) are nowadays already commercially available for TEM work by Fuji (FDL 5000 system) as a substitute for photographic films. They have an identical size, similar thickness and consistency. The extremely high sensitivity to electrons of different energy levels (*e.g.*, 50 keV to 1000 keV) make these plates suited for low dose work in different TEM/-STEMscopes. Off-line read-out is accomplished by a laser beam IP-reader. Read-out information is mostly stored in DAT-cartridges. Final prints are made by Power PC and different compatible printers. After erasing the plates are re-usable. Further explanations and references in Sections 9 and 10. See also Fig. 29. Schematic diagrams according to Fuji leaflet 94.6 - SK 5-1 (AB), Mori et al. (1988, 1990) and Oikawa *et al.* (1990) with permission, **(a)** Schematic cross section diagram of an image plate IP: the IP consists essentially of the electron sensitive phosphor layer PL and a support SU. Both surfaces are covered by thin layers (PR = protection layer, BA = back layer). PL consists mainly of phosphorous with some barium halogenide compounds doped with Eu^{2+} to store the energy of incident electrons. **(b)** "*Empty*" IPs are inserted into the film magazine of the TEM (see Fig. 29). After exposure by electrons, off-line read-out by laser beam scanning and registration of luminescence photons by photomultiplier. Finally, after erasing by visible light they are re-usable. **(c)** Comparison between the ratio of signal intensity (arbitrary units / optical density of prints) and electron doses for phosphor IPs and standard photographic emulsions (Fuji FG film) used for TEM work. Note the excellent linearity of the IPs, which is identical to SS-CCD systems (Fig. 27) in comparison to photo emulsions.

----

tronic compensation of those defects by "*gain normalisation*" or "*flat field correction*" is a must, to obtain the wanted beautiful pictures. Of course: such cosmetic "*corrections*" or "*normalisations*" cannot restore the original information of the area affected by the above mentioned defects in the chain of information transfers, but they contribute probably considerably to a satisfying aesthetics and a good feeling, since easily recognisable irregularities like black dots or areas no more exist. But with respect to low dose work the sensitivity of the system and that means mainly the sensitivity of the scintillator (GOS-phosphor is 3.5x more sensitive than YAG) and the time required for all on-line operations (*e.g.*, read-out, autotuning of the TEM) are much more important than all cosmetics and there is now a lot of possibilities to improve the situation.

A considerable experience has already been acquired in the application of those CCD cameras and some remarkable features are well known. Among these features, besides a real dose reduction >10:1 in comparison to the most sensitive photographic emulsion there is strong linearity between electron dose and signal intensity (see Fig. 28c). This feature is especially advantageous for quantitative measurements and evaluations, since complicated corrections (linearisation) are not needed, which made quantitative work based on photographic images so complicated and frustrating. A further advantage of the SS-CCD system is the wide dynamic range, which enables the registration of a diffraction pattern within only one run. Since SS-CCD systems work "*on-line*", they are finally elegantly suited for different modes of automatic adjustments of the whole electron optical system of a TEM. This "*autotuning*" includes a couple of automatic adjustments like "*autofocusing*" (Krivanek, 1975 cited by Krivanek and Mooney, 1993; Fan and Krivanek, 1990; Krivanek and Mooney, 1993; Dierksen *et al.*, 1993; De Ruijter, 1995), automated electron tomography (Dierksen *et al.*, 1993; Koster and De Ruijter, 1992) and automatic astigmatism control ("*autostigmation*", see Krivanek and Mooney, 1993). Different to more complicated earlier autotuning procedures and to the "*Tilt Induced Displacement*" [TID, see Koster and De Ruijter (1992), Krivanek and Fan (1992)], the "*Automatic Diffractogram Analysis*" [ADA, see Fan and Krivanek (1990), Krivanek and Mooney (1993)] needs under optimum conditions only one diffractogram (exposure 0.5 to 1.0

sec) analysed by automatic computer routine, which takes additionally 8 sec of processing time on a Macintosh Quadra for complete autofocusing and autostigmation. According to Krivanek and Mooney (1993) this processing time will decrease again considerably, if new computer generations with increased power will be available. I suppose, that this is today (1996) certainly already reality. Gatan offers for these purposes a complete software package within its Digital Micrograph™ Model 700-0000. And this is probably just the start for further developments making the operation of a TEM more accurate and reproducible even under the extreme conditions of cryowork.

In some respects the image plate is a competitive system to the SS-CCD camera. In comparison to CCD the image plate does not need a special installation at the TEM itself, since this "*plate*" has the size and consistency of a photographic film and fits into the film magazine of practically any commercial TEM. Also the handling and exposure of the image plate has some similarity to the use of a photographic film. According to Fig. 28a the image plate consists basically of a flexible support film with a thickness between 0.1 and 0.2 mm. This support carries a photo- and electron-sensitive phosphorus layer in a thickness around 0.12 mm covered by a protection coat of several micrometers. The photo-stimulable phosphorus layer contains [BaFBr:Eu] which serves as detector for photons and electrons. The theory of the activation process seems to be still a matter of discussion [Burmester, 1992; Daberkow *et al.*, 1991; Hangleiter *et al.*, 1990 cited by Burmester (1992); von Seggern *et al.*, 1988; Takahashi *et al.*, 1988]. During exposure of the image plate photons or, in the TEM, electrons change charges of the detector complex (*e.g.*, [BaFBr:Eu] → [BaFBr:Eu^{2+}]). These charges are "*stored*" by the detector complexes. Digital read-out of these charges is possible by laser beam scanning, which causes an emission of visible light. After this read-out procedure any remaining TEM image on the plate is erased by visible light. After erasing the plate is usable again. The described cycle "*electron exposure → reading → erasing → electron exposure*" (see Fig. 28b) can be repeated several times. There is no actual experience available how often this cycle can be repeated after erasing, since delivery of the new Fuji FDL-5000 system just starts and exposure by electrons may be somewhat different to exposure by photons. Fig. 29 shows a schematic diagram of the new FDL-5000 image plate system of Fuji with all accessories necessary for TEM work. The image plates are mounted in TEM cassettes exactly identical to those for photographic films or plates and stored for use in the TEM magazine. After exposure they are inserted

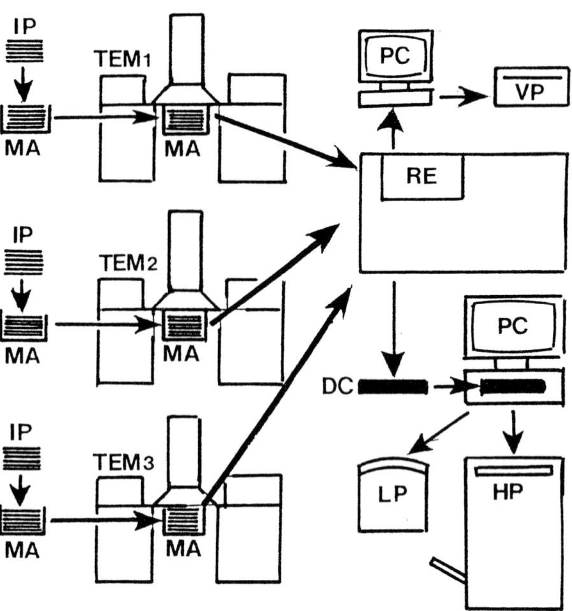

Figure 29: Schematic diagram of the new Fuji image plate hardware configuration (Fuji FDL 5000 system) for exposure, read-out and signal transfer to the prints. The image plates IP are loaded into the magazines MA for the cameras of the scopes. After normal exposure similar to photographic emulsion (only shorter times corresponding to the higher sensitivity) the exposed IPs are transferred in the laser beam IP-reader RE. A computer PC enables direct printing with videoprinter VP. High performance storage is accomplished on DAT-cartridges DC, which allow further processing by Power PC either using a laser printer LP or a more sophisticated high quality printer HP. Please note, that the whole equipment RE → PC → VP or RE → DC → PC → LP/HP may be used for different scopes TEM1/TEM2/TEM3 ff, since IPs are compatible with most modern TEM/STEM models. Read-out times are comparable with normal photographic dark room processes and no limiting parameter at all. After reading and erasing by visible light the IPs are usable again. Schema corresponds to Fuji Leaflet 94.6 - SK.5-1 (AB) in a modified version (with permission). See also Fig. 28. Further explanations and references in the section on "..Frozen-Hydrated Ultrathin Sections" and the Discussion.

---

in the "*reader*" (laser scanning device) connected to a Power PC terminal. All digital data are stored on DAT cartridges for printing either by a conventional laser printer or by a very sophisticated high quality printer operated by a second PC. Of course: this periphery like the processing hardware and software for an SS-CCD system is a considerable investment but in this case represents also the equivalent for the compatibility with practically any modern TEM.

According to the claims of Fuji the features of the new FDL-5000 system are really interesting: in comparison to commercial SS-CCD systems the image plates offer probably a slightly higher sensitivity and an even wider dynamic range together with a similar linearity of electron dose versus signal intensity (see Fig. 28c) and a similar pixel size of 25 μm. But similar pixel sizes count only under the precondition that the same magnification is used. The striking difference relates to the considerable larger area of the image plate (effective area 94 x 75 mm^2), which is identical to photographic films used for TEM work. This area surpasses CCD-frames 1k x 1k by a factor of 16 and even the new MegaScan 2k x 2k of Gatan by a factor of 4. If periodic crystalline structures of small sizes have to be recorded, at a first glance one may believe, that the area makes no important difference. But in practice this makes perhaps the point, since periodic structures in biology are exceptions and aperiodic ones are the rule. If we want to record a given surface, we must do this with the smaller detector size of the CCD-chip at a considerably smaller magnification: provided that we work at the correct focus in both cases, we loose accordingly a lot of pixels or resolution with the smaller CCD system. If we do not want to lose resolving power, we have another problem: at identical magnification, we lose a lot of surface area in our picture and therefore information about our aperiodically structured specimen. In comparison between the Fuji image plate and the largest SS-CCD detector for a TEM (Gatan MegaScan = 4k x 4k) we lose 3/4 of the picture. Vice versa the surface recorded on the image plate under comparable conditions (same resolution, same magnification, similar pixel size) is 4 x larger than the SS-CCD picture. This relates to the "*Pixel Equivalent*" defined by Burmester (1992, see also Table 6) and concerns certainly also future prospects of image plates and vice versa SS-CCD systems. From the technical point of view I see no severe problem, to increase the surface of image plates again above the value of 94 x 75 mm^2. One may for example use the whole circular field below the fluorescence screen of the scope to gain additional pixels. Read-out times do not hamper, since off-line recording is a must for image plates. In this respect the real potential of the image plate is not yet exploited. But I have the feeling, that some difficulties will arise, if the detector size of SS-CCD systems should be increased once more considerably above 40 x 40 mm^2. This relates mainly to the cost of manufacturing perfect CCD-chips and fibre plates in larger dimensions and to develop subarea read-out facilities working fast enough for a meaningful on-line operation. I guess that there will be no severe border of technical possibilities. But the question of the relation between manufacturing expenses and therefore prices and efficiency of the two different competitive systems will certainly be posed by financial reasons. Since read-out and autotuning facilities influence the value for low dose operation, nevertheless perfect large area SS-CCD systems will certainly gain increasing importance in our scientific field interested in frozen-hydrated cryo-sections.

To draw a conclusion from all the considerations mentioned above: since pixel size and sensitivity characteristics (linearity of electron dose versus signal intensity, see Fig. 28c) of the image plate are actually nearly identical to SS-CCD systems, the image plate seems to be superior to SS-CCDs in all respects, except one: an "*on-line use*" is absolutely impossible. The output of image plates is only "*off-line*" information. Of course this enables the reading-out of image plates exposed in different TEMs within one reader and the printing of the information of all DAT cartridges with the same equipment. But "*autotuning*" systems for tomography, autofocus, lens and stigmator and deflector calibration are not possible using image plates. It may be, that this severe disadvantage can perhaps be compensated by through focus series over the larger image areas offered by the more sensitive image plate. I cannot imagine this and I confess, that I for myself am in a somewhat "*schizophrenic*" situation and want to remind, that today (1996) all these considerations are situated in a risky field of speculation and that a clear prognosis is not possible. For sure, experience of the coming years will tell us more about the real advantages and disadvantages of both competitive systems in practical use for ultrathin frozen-hydrated cryosections, and also for sure, all scientists concerned with cryowork under extreme experimental conditions should be happy to have two fascinating alternatives for low dose operation of their scopes.

Considerable progress was also realised in "*Electron Spectroscopic Imaging*" (ESI) using TEM systems equipped with energy filters [see Bauer (1988) or Krivanek *et al.*, (1995)]: the first "*Energy Filter* TEM" (EFTEM) was an "*in column filter*" built and described by Castaing and Henry already in 1962 (see also Castaing, 1975). Different from "*magnetic prism spectrometers*" (Shuman and Somlyo, 1982), which are mounted as "*post column filters*" below the fluorescent screen of the scope (see again Krivanek *et al.*, 1995 and Fig. 30d), the "*Castaing-Henry-filter*" consists of two magnetic prisms and one electron mirror in-between (so called "*prism-mirror-prism*" or "*PMP-*

Figure 30. (*on facing page*) "Energy Filtering Transmission Electron Microscopes" (EFTEMs) with "Electron Spectroscopic Imaging" (ESI): different from "Conventional Transmission Electron Microscopes" (CTEMs) spectroscopic images in "*zero loss mode*" deliver better contrast of thin frozen-hydrated cryosections with underfocus values below 0.1μm. Further explanations and references in the section on "..Frozen-Hydrated Ultrathin Sections" and the Discussion. **(a)** Spectroscopy with photons (visible light, upper diagram): polychromatic light PL in glass prism GP behaves similar to an electron beam with different speeds ("*poly-energetic or polychromatic electron beam*" PE, lower diagram) in a magnetic field ("*magnetic prism*" MP). White light PL is split (spectrum SP). A monochromatic fraction ML can be selected by slit SS. Electrons PE in MP describe circles, whose radii depend on their speed (R1 = faster electrons FE, R2 = slower electrons SE, R1 > R2). A "*mono-energetic or monochromatic electron fraction*" ME of PE is selected by slit SS similar PL → ML. This SS-selection can be used for "*monochromatic imaging*" (elimination of chromatic defects of electron lenses). This is possible either by "*in column filters*" (b) and (c) or by "*post column filters*" (d). **(b)** "*Prism-Mirror-Prism-*(PMP)*-Filter*" according to Castaing-Henry and Ottensmeyer (Zeiss EFTEM EM902): the mainly monochromatic electron beam ME of the electron gun EG (CL = condenser lens) becomes polychromatic (PE) by inelastic scattering in the object OB. Between objective OL and projective lens PL the magnetic prism MP and the electrostatic 80kV mirror MI is installed. According to (a) the electrons describe circular pathways in the homogeneous magnetic field of MP, are rejected by MI and once more deflected by MP (opposite direction ... opposite deflection). The inelastically scattered slower electrons describe smaller radii and are eliminated by slit SS ("*zero loss imaging*"): for the image only unscattered and elastically scattered electrons are used, which correspond to ME. PMP-systems are limited to acceleration voltages ≤ 80 keV with respect to the electrostatic high tension mirror MI. **(c)** "*Omega*" in column filter (Zeiss EFTEM EM912): instead of PMP (b) four magnetic prisms MP1 → MP2 → MP3 → MP4 force the electron beam PE to a slope like the Greek letter Ω. Some correction lenses ("*sextupoles*" SP) are symmetrically arranged. In comparison to the PMP-systems (b) higher acceleration voltages are possible due to elimination of MI. Function and SS correspond closely to PMP (b). **(d)** "*Post column*" spectroscopic imaging (Shuman and Somlyo, 1982; Krivanek *et al.*, 1994, 1995; Gubbens *et al.*, 1995; modified diagram with permission of Gatan): the spectroscopic device uses only one magnetic prism MP and is mounted on the base of the TEM column CO. The TEM remains therefore unchanged and the spectrometer is suited as an accessory for conventional TEMs. Some correction lenses (8 quadrupoles QP and 7 sextupoles SP) are combined with MP and the selecting slit SS. Since the fluorescent screen (plane FS) cannot show the spectroscopic image, a TV system is used. After retraction of TV (arrow) on-line registration by CCD-camera is possible. Image plates (see Figs. 28 and 29) cannot be used with post-column spectrometers of this kind.

---

spectrometer", see Fig. 30b) incorporated in the column of the TEM between the magnetic lenses. Such PMP-arrangements allow all modes of observation, imaging and elemental mapping, but they have to be incorporated in a purpose-designed electron optical system and are not suited as an accessory for any TEMscope. After a period of complete silence Henkelmann and Ottensmeyer in 1974 had continued the developmental and application work with Castaing-Henry-filters (Henkelmann and Ottensmeyer, 1974; Ottensmeyer and Andrew, 1980; Ottensmeyer *et al.*, 1981). They inserted the PMP-filter and the energy selecting slit (see Fig. 30b) into commercial TEM-scopes above the projector lens. They were able to demonstrate, that ESI is possible without an annoying loss of resolution and that zero loss imaging, that means elimination of the inelastic background shows increased contrast and superbly clear pictures of biological structures [see Fig. 8, p. 92 in Henkelmann and Ottensmeyer (1974)]. Once more a period of silence was inserted up to the moment that Zeiss introduced with the EM 902 a PMP-filter lens system based on the developmental work of Henry, Castaing and Ottensmeyer which immediately was very successful (Bauer, 1988; Benner *et al.*, 1994). The main limitation of this system was the comparatively low beam voltage (80 kV or lower) given by the voltage limit of the electrostatic electron mirror. In the meantime this limit is eliminated by another "*in column filter*" of the Omega type working with magnetic prisms without an electron mirror [see Fig. 30c and Egerton (1986), Benner *et al.*, (1994), Bihr *et al.* (1991), Krahl *et al.* (1990), Lanio (1986), Lanio *et al.* (1986), Rose (1978)]. This new energy filter type already allows an accelerating voltage of 120 kV and is incorporated in the Zeiss EFTEM 912 Omega. A further increase of the voltage seems to be possible. There is now a considerable experience about ESI accumulated both with PMP and with the new Omega filter lens systems (see *e.g.*, Bauer, 1988; Benner *et al.*, 1994; Edelmann, 1992; Michel *et al.*, 1991, 1992; Murray, 1986; Schröder, 1992; Schröder *et al.*, 1990), which allows the conclusion, that frozen-hydrated specimens in comparison to CTEM show both better resolution and better contrast if ESI under "*zero-loss*" conditions, that means elimination of the inelastically scattered electrons is used. Actually the amount of underfocus needed for the recording of images can be reduced to minimum values below 0.1 μm far away from the need of any CTEM. The contrast of frozen-hydrated specimens available with ESI under these conditions is astonishing [see Schröder *et al.* (1990), see Fig. 3 on p. 32). The advantage of PMP and Omega systems is the normal use of the fluorescence screen and the possibility to work with image plates for ESI operation. According to Ottensmeyer and Andrew (1980) and Ottensmeyer *et al.* (1981) the resolution limit of such systems reaches values of 0.3 nm close to the CTEM. Resolutions down to 0.2 nm were demonstrated later on an experimental Omega filter by Lanio *et al.* (1986). Such a resolving power now really opens the entrance into molecular dimensions with EFTEMs including ESI with biological samples. And if all assumptions are correct, one can expect a general improvement of the resolving power of "Sub-Ångström Transmission Electron Microscopes" (SATEMs) down to approx. 0.6 Å (0.06 nm) by a strictly monochromatic imaging with new and more sophisticated types of magnetic filter lenses in future (Rose, 1994).

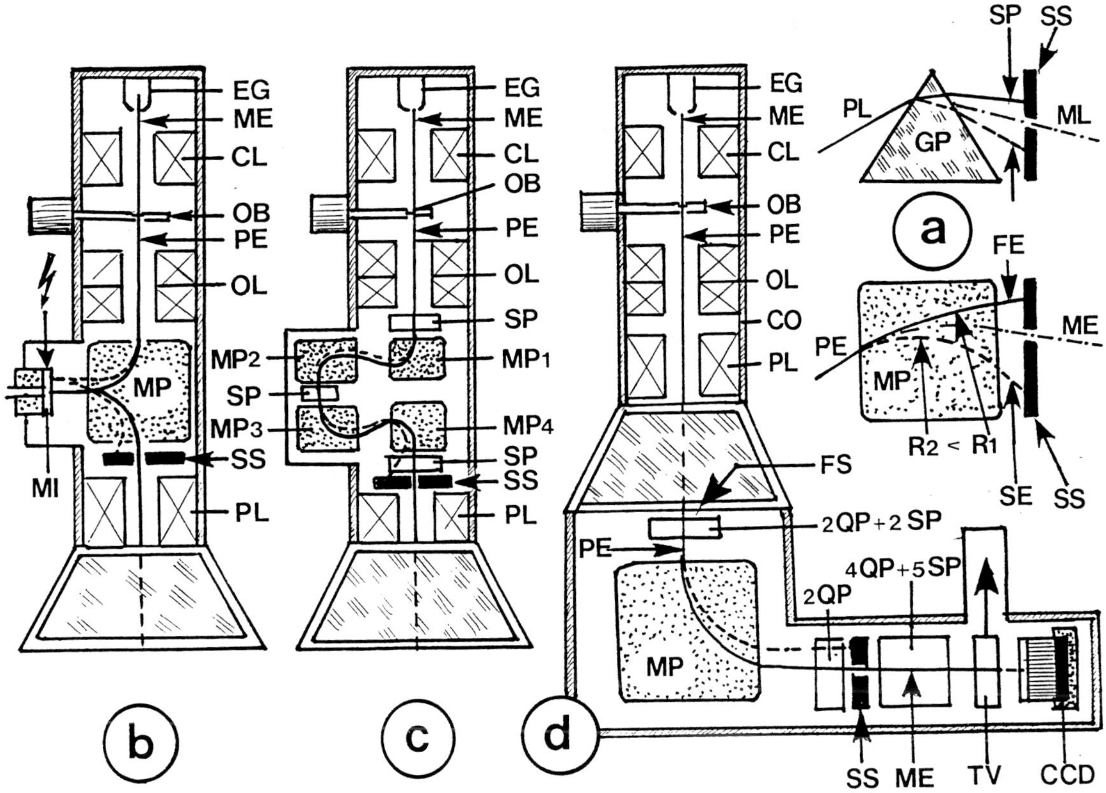

Discussion

No scientist interested in cryo-ultramicrotomy, including the author of this review article, can escape the basic question: does cryosectioning of fresh-frozen material really make sense? Or in other words: what is the gain from all this complicated work? There is no doubt, that cryosectioning according to Tokuyasu is a tremendously successful method. The same is true for other "*hybrid methods*" starting with a rapid cryofixation of the unpretreated fresh material and finishing with a resin block sectionable at room temperature. For example freeze substitution or freeze drying inserted between cryofixation and low temperature embedding are fully compatible with all known immuno techniques which offer in this respect much more information than frozen-hydrated sections. Both the Tokuyasu technique and the other mentioned hybrid methods open a fantastic opportunity to check, where important antigens (*e.g.*, receptors or enzymes) are located in cells and tissues. On the other hand X-ray microanalysis, EELS, LAMMA and all the methods focused on element analysis and mapping are well introduced and give a world of new information about cells and tissues. Cell biology would be a frustrating job without this direct information on the border between microscopic and molecular level respectively morphology and biochemistry. Only the precise knowledge of these *"hot spots"* in the architecture of cells open the view on what cellular engineering really means. It is my firm opinion, that all the fascinating new statistical and biochemical techniques of molecular biology like plotting methods, high performance liquid chromatography, ultracentrifugation and sequence analyses, which deliver actually plenty of absolutely new insights in cells and cell components cannot and will not substitute the direct investigation down to the molecular level only accessible with the superb resolution of the TEM and STEM. The same is also true for confocal laser scanning and even high resolution tunneling and atomic force microscopy: a complete three dimensional study of larger tissue complexes and even cells to demonstrate the arrangement of certain proteins or nucleic acids or lipids or carbohydrates *in situ*, will never happen with these new methods. And one should never forget again this dimension of life-important colloids (*"Die Welt der vernachlässigten Dimensionen"* - "the world of the neglected dimensions") already mentioned by the famous German chemist Wolfgang Ostwald in 1915. But returning to 1996: What can we really expect of fresh-frozen work from cryofixation to the final electron micrograph? That is the first and most important question.

Perhaps a retrospective view is helpful: in the era of light microscopy between the second half of the last and the first third of this century both the limited resolving power and all difficulties in observing specimens without staining resulted in a frustrating discussion around the questions about *"fact or artefact"* and about all cellular structures at the border of the resolution limit of the light microscope clearly defined by Ernst Abbe. In most of the issues no real progress resulted from this often extremely aggressive discussion. A possible approach was introduced by indirect methods like polarisation microscopy or X-ray diffraction, which allowed some access to the highly ordered structures of myelin or plant cell walls. But these methods required a good knowledge of physics not often found in the community of biologists and therefore did not get the general attention needed for real success. It is interesting nowadays to read again Albert Frey-Wysslings brilliant book *"Submicroscopic Morphology of Protoplasm"*. This second English edition was written in 1953 and includes just the first steps into electron microscopy. Nevertheless it describes clearly and correctly the painful situation in a field which missed the right tools. Some of this pain was eliminated by the invention of the phase contrast microscope by Frits Zernike (Zernike, 1945), making it possible to observe cells and tissues (at least in culture) in their living state - of course without chemical fixation, dehydration, embedding, extraction of the embedding medium, staining and all the other tortures needed for visualisation of cell and tissue structures with a light microscope. All the crazy optimists won, who always claimed with respect to stained paraffin sections, that *"such beautiful pictures are probably the correct ones"* after this first comparison between the living and the dead object: that was a message! Nevertheless all the unsolved problems and questions struggling on the resolution limit of the light microscope remained and neither phase contrast nor indirect methods (offering only a better *"resolution"* in one spatial direction for highly ordered periodic structures) could help.

If we take a look at the early days of electron microscopy (see for example Gabor, 1957 and E. Ruska, 1979) we gain the impression that most of the leading scientists in biology and medicine at that time between 1930 and 1950 had neither the confidence nor the good will to use this fascinating new tool. The strong aggression against this new technique mainly based on the feeling that most of the observed structures might be artefacts and that the electron beam probably destroys most organic matter of interest. But again: *"beautiful pictures are probably the correct ones"*. Already in 1966 Fawcett gave a solid support for this slogan with his brilliant view of the state-of-the-art in these early days of electron microscopy with his book *"The Cell: Its Organelles and Inclusions"*. Such crispy pictures were acting again like a tranquilliser. Finally *"freeze-cleavage"* together with *"freeze-etching"* [see *e.g.*, Steere (1957, 1973), Moor *et al.* (1961), Mühlethaler (1973), or in general Benedetti and Favard (1973)] as a strictly physical method demonstrated, that the beautiful pictures obtained after a chemical fixation, dehydration, embedding, ultrathin sectioning and staining with heavy metal compounds are not so far away from the reality of the living state. The introduction of the electron microscope with all auxiliary techniques into scientific research was again a powerful demonstration of the importance of instrumental and methodological progress for new insights into the secrets of nature (P. Sitte, 1973). It is now the question in our consideration, if freeze cleavage and freeze etching enable really clear and striking statements and if the conclusions drawn on the base of their results are correct in all respects. If yes, we would be in a splendid situation and have already reached our goal. If no, we have to continue our efforts to check morphology and chemical composition of our biological specimens in a more difficult terrain. Of course the answer is *"no"* .

Acoording to me there are two different serious arguments to continue research at the cellular

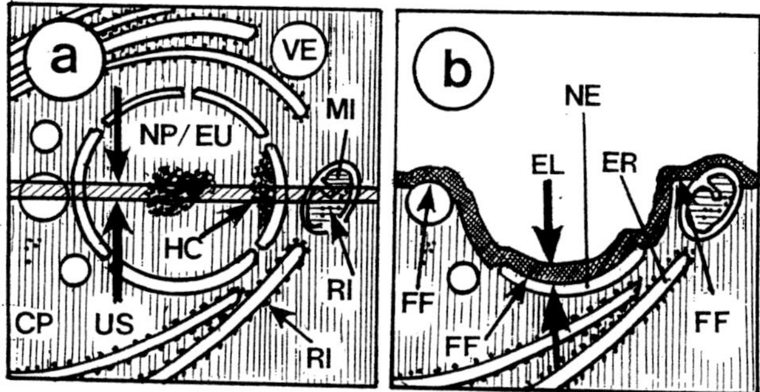

Figure 31: Comparison between the information available by ultrathin cryo-sectioning (a) and freeze-cleavage (b) followed by Pt/C freeze-replication (EL = evaporated layer for TEM investigation). Both schematic diagrams (a) and (b) show identical areas of a eucyte cell with some typical ultrastructures: CP = cytoplasm, ER = endoplasmic reticulum with ribosomes RI, HC = heterochromatin, MI = mitochondrial plasm with ribosomes RI, NE = nuclear envelope, NP = nucleoplasm with euchromatin EU, VE = vesicle. Further comments in the Discussion. **(a)** The ultrathin section US gives a complete survey over all structures including the content of the different plasmatic and extraplasmatic compartments of the sectioned cell (*e.g.*, CP with RI, mitochondrial plasma MI with RI, NP with HC and EU, content of VE). It shows particles sticking on membrane surfaces (*e.g.* RI on rough ER). For element analysis all different components of CP, NP, MI etc. are available. Three-dimensional reconstruction and mapping is possible both by serial sectioning or by stereological single section methods. **(b)** freeze cleavage or fracturing is something like a "*happening*" since the fractures FF follow preferentially the membrane structures as "*loci minoris resistentiae*". Most of the plasmatic (*e.g.*, CP, MI, NP) and extraplasmatic phases (*e.g.*, VE) are not visible. Element analysis is not possible after replication by layer EL and dissolving of cellular structures afterwards. Finally a meaningful three-dimensional reconstruction is impossible. For cryowork it is a severe disadvantage that identification of amorphous or crystalline freezing by diffraction patterns of I_v or I_h or I_c is not possible.

---

and molecular level with the new cryomethods described in this review article: (1) There is a general need to check and to re-evaluate all results obtained by standard procedures on a chemical basis using purely physical methods nowadays accessible, and (2) There were some limitations clearly visible which are connected with the chemical base of our standard procedures, which exclude water as the most important and major component of our biological samples from meaningful interpretations. With regard to the first argument: already the few examples given by the few groups around Jacques Dubochet, Martin Müller, Dick Ornberg and perhaps my own group really demonstrated, that nearly each object re-investigated by pure cryomethods offered at least interesting new aspects. For example: the ultrastructure of bacteria was in important respects different from that hitherto reported (distribution of nucleic acids, mesosomal structures; Dubochet *et al.*, 1983). The vacuoles of plant cells offered structures never observed before (Michel *et al.*, 1991, 1992). The extracellular matrix of cartilage was considerably different in comparison to the structures after chemical processing (Studer *et al.*, 1995). There was finally a lot of information by high resolution diffraction patterns not observable after chemical treatment. With regard to the second argument: It was already shown by van Harreveld and co-workers, that only rapid freezing offered a possibility to preserve the distribution of rapidly moving small molecules and ions (*e.g.*, H_2O, Na^+, K^+) correctly and to avoid redistribution not avoidable even after the most rapid chemically based perfusion fixation (Van Harreveld and Crowell, 1964). That means, that the volumes of extracellular and plasmatic compartments (*e.g.*, in central nervous tissue) cannot be correctly stabilised by chemical methods. Once more van Harreveld firstly demonstrated the possibility to observe fast running processes in cells and tissues by rapid freezing (Van Harreveld *et al.*, 1974). The most striking and "*fancy*" demonstration of this improvement (considerably better "*time resolution*" according to Plattner and Bachmann, 1982) came from John Heuser with his pictures of synaptic vesicle exocytosis [Heuser *et al.* (1979); see also Knoll *et al.* (1987, 1992), Morgenstern and Edelmann (1989), Plattner (1989), Ryan and Knoll (1994)]. Edelmann was able to demonstrate later on, that the correct analysis of ion distribution in muscle tissue depends not only on an optimum cryofixation procedure but also on the correct continuation of this purely physical approach by proper, that means, long enough, freeze drying [Edelmann (1994a,b; see also Sitte *et al.* (1994)]. I think, that these examples are sufficient to demonstrate the need for and the possibilities of physical procedures to solve some of the most important problems of cell and molecular biology correctly, that means the understanding of the distribution of the mixed aqueous phases in our biological samples, which probably gives us some insight into the secrets of life.

It is evident, that only purely physical processes starting with a true vitrification of the unpretreated sample are suited for the correct and meaningful re-evaluation claimed under (1) above. It remains the question, if perhaps freeze-fracturing followed by freeze-etching and Pt/C replication according to Steere (1957) or Moor and Mühlethaler (1963) may be a suited (and simpler) alternative to cryosectioning and low dose investigation of frozen-hydrated sections in the cryo-TEM. There is no doubt that also this alternative is a sequence of purely physical steps of preparation. As mentioned above, freeze-etch images really offered the first partial confirmation, that chemically based standard procedures delivered in most respects correct information about living cells. Freeze-etching brought (completely unexpectedly) a world of new important knowledge about the molecular architecture of biomembranes (see for example the beautiful demonstration by Orci and Perrelet, 1975) and gave a solid base for the current *"fluid-mosaic membrane"* model of Singer and Nicolson (1972). But by two reasons freeze-etching cannot be a real alternative to cryosectioning, since it does not fulfil some of the preconditions for a correct re-evaluation : firstly the frozen specimen must be completely removed from the Pt/C-replica before investigation in the TEM is possible. It therefore gives no correct information concerning the amorphous or crystalline state of the ice, which acts as an embedding matrix of the cellular ultrastructures, by electron diffraction. Since this diffraction pattern offers the only possibility to discern between amorphous (I_v), cubic (I_c) or hexagonal (I_h) ice and since we know, that already microcrystalline I_c segregation creates artefacts [see *e.g.*, McDowall *et al.* (1984) or Studer *et al.* (1995)] we cannot miss the frozen-hydrated investigation at least to check the quality of our cryofixation procedure. A second reason concerns geometry (see schematic diagrams in Fig. 31): only sections (Fig. 31a) have a properly defined geometry and show us all structural details incorporated in our sample. The normal *"freeze-fracture"* (Fig. 31b) splits the frozen specimens completely irreproducibly mostly along the *"loci minoris resistentiae"* (sites of least resistance), which are the contact areas of the apolar ends of the fatty acid chains inside the biomembranes. Most of the cytoplasmic or karyoplasmic matrix is hidden by this fracturing process. Of course: we have the possibility to produce an excellent surface for subsequent Pt/C-replication in the ultramicrotome using exactly the methodology for cryosectioning at temperatures of -165°C or lower, that means using a low angle diamond knife with $\alpha < 35°$, an ioniser, a cutting speed of 0.5 mm/sec, a width of the sectioned surface ≤ 0.1 mm and so on. Under these conditions we can expect a brilliant flat surface without fractured areas inside, but nobody would be happy to replicate these lousy 0.01 mm²: the whole procedure would be more complicated (cryosectioning plus replication plus preparation of the microreplica) than cryosectioning alone. Nevertheless we would miss the information about the amorphous or crystalline state of the frozen sample. The same problem occurs with element analysis by EDX: we have no possibility to analyse Na^+ or K^+ or any other element since they are gone. We can conclude, that only the frozen-hydrated cryosection gives us the desired answer, if true vitrification was obtained and - if "yes" - where the different fast moving ions are located *in situ* and *in vivo*. In other words: we cannot escape the cryo-ultramicrotome and the frozen-hydrated or freeze-dried investigation afterwards. Similar considerations may be necessary for a correct element analysis, since any crystallisation process during freezing without doubt is able to provoke redistribution of ions. I do not believe, that it makes a big difference for an experimental confirmation to section correctly a frozen specimen, to take a look in the cryo-TEM, to make a picture and (more important) a diffraction diagram and to move the still frozen-hydrated section on the grid by cryotransfer into a freeze dryer. Of course: freeze drying has to be realised at temperatures between -80°C and -100°C (Sitte *et al.*, 1994). The recrystallisation to cubic ice cannot be avoided under these circumstances. But big x/y-translations of particles, fibres, membranes and probably also ions in the still frozen matrix similar to the phase change *"liquid → crystalline"* do not seem to be possible, as discussed later in this review (Fig. 32). All knowledge presented in the preceding argumentation concerning arguments (1) and (2) was accumulated by very laborious and exhausting experimental work from a small group of cryofreaks. Neither Alisdair McDowall and Jacques Dubochet, nor Martin Müller together with Daniel Studer, Martin Michel and Heinrich Hohenberg, nor Richard Ornberg nor our group had full access to all instruments and techniques described in this review. Nevertheless it was possible to document the specific advantages of the new cryosectioning methods for fresh-frozen material by the already mentioned serious results. But it would be a nonsense to follow the description in those papers and to repeat all *"historical"* mistakes: this would mean to work nowadays with instruments of a former generation and with the methods cited. It is my firm opinion, that cryosectioning of fresh-frozen samples is such a tough job, that further progress on different objects can only be expected if the really best state-of-the-art instrumentation is available and if the best suited methods are used. That includes, that the best suited cryofixation device (often a high pressure freezer), the best cryo-ultramicrotome (at the moment the Ultracut-S or better UCT together with an FCS cryo-

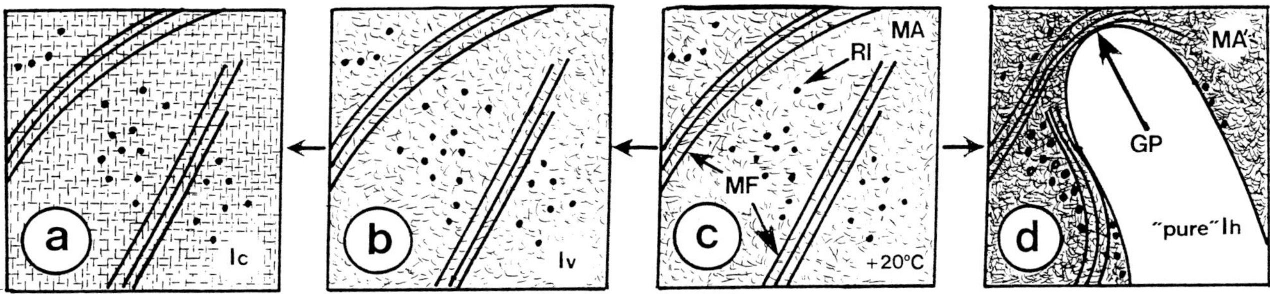

Figure 32. The schematical drawings (a) to (d) concern hypothetical assumptions about amorphous solidification ("*vitrification*") of a liquid plasmatic phase (c) by rapid freezing "c → b" in comparison to "*primary hexagonal crystallisation* c → d" (GD = growing direction of hexagonal ice I_h) occurring at cooling rates below 1 million °C/sec and "*secondary recrystallisation* b → a" after heating a vitrified sample (b) above devitrification temperature T_d of approx. -135°C. Since even the rapidly oscillating H_2O-dipoles of the plasmatic matrices MA during rapid cooling with rates above some million °C/sec have no chance to aggregate into a crystalline lattice, all included smaller molecules and especially the TEM-resolvable macromolecular components like microfilaments MF or ribosomes RI are encapsulated in vitreous ice (I_v) without noticeable x/y/z-translation during vitrification. Completely different from vitrification "c → b" during hexagonal crystallisation "c → d" the pure hexagonal crystalline phase I_h grows without incorporation of other constituents like dissolved smaller molecules, ions or larger macromolecular structures (*e.g.*, MF or RI), since such a crystallisation is an extremely precise selective procedure: we can suppose, that both small molecules, ions and larger molecules as well as particles RI or filaments MF are rapidly shifted from the Ih crystal borders by x/y/z-translations to other places they never occupied before. They are finally included in the dense eutectic mixture of the plasmatic (or extraplasmatic) components (enriched matrix MA'). Of special interest with respect to hybrid methods (*e.g.*, freeze-substitution or freeze-drying following a rapid vitrification) is the devitrification "b → a" of a primarily vitrified sample to a microcrystalline compound with a cubic Ic matrix observable by electron diffraction. One may speculate hypothetically upon the possibilities given in a still completely frozen solid state sample around T_d of approx. -135°C: there is certainly some probability, that during this "*secondary crystallisation event* b → a" x/y/z-translations of inclusions (*e.g.*, MF, RI) and maybe even ions and small molecules are not possible. This assumption is shown in the schematical drawings (b) and (a). But this hypothesis has to be checked by comparison of cryosections in the amorphous frozen state at *e.g.*, -170°C before increase of temperature above T_d and after devitrification to cubic I_c, *e.g.*, by heating above -135°C. This crucial experiment should be possible. Compare further remarks and comments in the Discussion.

---

chamber) equipped with the needed accessories (low angle cryo diamond knife with a facet angle < 35°, cryo diamond trimming tool, high performance M3C stereomicroscope for high magnification, high power cold light source, adjustable ioniser), a suited cryo-transfer system (Gatan 626-DH or Oxford CT 3500) and a cryo-TEM with a sufficient cold trap (decontamination system), one of the described low dose systems (SS-CCD or image plate) and an energy filter (EFTEM) for electron spectroscopic imaging (ESI) has to be used. If element analysis (*e.g.*, EDX or EELS) of ultrathin or semithin sections is the goal, a freeze dryer suited for proper drying - that means minimised artifactual shrinkage (*e.g.*, cryosorption drying according to Edelmann, see Sitte *et al.*, 1994) is necessary.

It makes no sense to start into fresh-frozen cryosectioning or element analysis without the whole chain mentioned above, since the exclusion of one of these tools reduces the efficiency considerably and makes it even more difficult to obtain useful results within a reasonable time. Of course starting without experience also makes no sense: a practical training in at least one of the leading laboratories is highly recommended to gain practice both in cryofixation, cryosectioning including cryotransfer and low dose operation of an EFTEM. If possible, this training should be made with one's own samples, since all methods need some specific adaptation on the material and it makes a big difference if a soft animal tissue or a rigid plant tissue or a cell suspension (probably the simplest case) has to be processed. Already the cryofixation as the first step is a crucial point and decides on success or failure of the whole procedure. For a first check of suitability of a special freezing method a subsequent freeze substitution and resin embedding according to Van Harreveld and Crowell (1964, see also Steinbrecht and M. Müller, 1987, or Fig. 6 in this review) of the specimen may be the most convenient way. If fast processes (*e.g.*, membrane fusions within exocytotic or endocytotic events) are considered, high pressure freezing may be too slow. Also this can be clarified on cryofixed, OsO_4/acetone substituted and resin embedded material. The same is true for an extracellular cryoprotection [*e.g.*, by sugar, see Dubochet *et al.* (1983), McDowall *et al.*, (1984, 1986) at least for cell suspensions. Cryo-ultramicrotomy should be started after this first check according to the rules given in the relevant preceding sections. Probably nobody will escape the somewhat time consuming process of trial and error in establishing the best individual sectioning parameters for his specific samples. And nobody will escape the painful and often frustrating transfer of the cryosections to the grid, simplified considerably by a well adjusted ioniser. As explained before, perfect vitrification is a must and already cryptocrystalline cubic ice may cause artefacts. One has to remember, that in the past many scientists were happy if severe damage by hexagonal ice could be avoided and often specimens without resolvable ice crystals were incorrectly described as "*vitrified*" (amorphous). Often cryptocrystalline cubic ice was considered as amorphously vitrified ice simply because it sections quite well and looks promising.

There is a lot of "*laboratory folklore*" around cryosectioning. Often it is claimed, that manual operation of the cryo-ultramicrotome delivers better (or even the only suitable) cryosections. That may be correct for prehistoric cryo-ultramicrotomes like MT2/FTS or OmU3/FC2. But this would be nonsense with a modern Ultracut-S/FCS-system with a perfectly adjustable cutting window, a fast return stroke and a very slow cutting speed: nobody is able to produce by manual operation of a cryo-ultramicrotome really regular serial sections below -160°C in the mostly wanted and needed thicknesses around 50 nm. Since irregularities due to manipulation forces cannot be excluded, manual operation does not work reproducibly and makes really no sense. Several steps in cryopreparation require some experience and special attention. Some adjustments have to be found by "*trial and error*" for each new specimen: this relates especially to the knife and specimen geometry and adjustments (clearance angle, width of sectioning surface) as well as for the best cutting speed, cycle length and ioniser adjustment (cycle length and ioniser adjustment depend on each other). As for ambient temperature work, the setting of the feed rarely corresponds exactly with the real advance or the section thickness finally obtained: it is understandable that cryo-ultramicrotomes react much more sensitively against environmental influences and that therefore thicknesses differ much more from the advance settings in comparison to standard ultramicrotomy. Slight irregularities in section thicknesses, which can be mostly ignored at ambient temperature work, severely disturb cryosectioning, since they provoke the formation of crevasses and are multiplied by compression phenomena too. An extremely smooth operation of the cryo-ultramicrotome is therefore of the same key importance as the careful exclusion of all outside influences (air draughts, thermal influences, building vibrations and manipulation forces). And again: generally the motor driven automatic ultramicrotome delivers the best results. If useful cryosections are on the knife edge, the transfer to the grid needs some experience and pressing of the cryosections onto the grid inside the cryochamber of the ultramicrotome is an essential step, since heat extraction under the electron beam is dependent on a good thermal contact between the solid state sections, the solid state grid and grid holder. If such sufficient thermal contact is obtained, the temperature difference between the cold stage and the cryosection will remain 10°C or less. Success or failure besides this pressing procedure depends on a correct cryotransfer and a sufficient cold trap (decontamination system) around the specimen in the column of the TEM. Success also depends on the low dose operation of the microscope and all measures, which allow to minimise beam damage on the extremely sensitive frozen-hydrated material. Besides the new SS-SSD or image plate systems for digital low-dose recording of images and an EFTEM for ESI at zero loss, both accelerating voltage and vacuum conditions influence the beam damage [see *e.g.*, Cosslett (1978), Glaeser (1971, 1975), Kobayashi and Sakaoku (1965)]. Simply: the better the vacuum and the higher the voltage the lower the beam damage. If the voltage is increased from 100 kV to 200 kV, the beam damage decreases already in the order of x0.5. In addition, an excellent vacuum seems to be helpful and will be available automatically by cryosorption, if an efficient cold trap decontamination system surrounds the specimen nearly completely.

The contradictory discussion about the "*cryostabilisation effect*" (misleadingly mostly termed "*cryoprotective effect*") of very low temperatures seems to be finished [see *e.g.*, Chiu *et al.* "*International Experimental Study Group*" (1986), Dubochet *et al.* (1987, 1988), but the real effect of the lower temperatures is not absolutely understood: the big effort invested in studies on the temperature dependence of beam damage resulted mainly in the simple conclusion, that this effect depends on the specific conformation of the compounds investigated, on some geometrical and material components of the set-up and of the frozen specimen: if the defined "*critical doses*" of different compounds are compared, aliphatic compounds show a higher sensitivity under the electron bombardment than aromatic ones. Additionally, the sensitivity is dependent on the support of the sample: carbon layers reduce, collodion films enhance the sensitivity. Similarly the thickness and the geometry of the sample influence the damage under comparable conditions. And most important under the point of view of this review article: these are mostly measurements based on the fading of diffraction patterns of crystals, which are in the strict sense not comparable with our frozen-hydrated specimens encapsulated in a matrix of amorphously vitrified ice. By similar reasons the results of radiochemistry based on measurements with ESR (Electron Spin Resonance) and ENDOR [Electron Nuclear Double Resonance; see *e.g.*, Box (1975, 1977)] are not really helpful, since the doses used for ESR and ENDOR are orders of magnitude lower than those used for imaging in the TEM. There are other indications about the benefit of low temperature on beam damage based on mass loss during electron irradiation [see *e.g.*, Dubochet (1975), Ramamurti (1977), Cosslett (1978), and Fig. 26 in this review): at 7°K only a very small mass loss occurs in comparison to room temperature. According to the studies of Grubb and Groves (1971) as well as Siegel (1972) there are at least two processes which are involved in beam damage and things are not so easy to reconstruct. In this context both the "*caging*" of fragments after chemical bond scission due to inelastic scattering of electrons, the re-unification of such fragments and cross-linking phenomena under the electron beam are discussed. Radiolysis of H_2O ["*beam etching*"; see *e.g.*, Draganic and Draganic (1971), Hochanadel (1960)] occurs also at LN_2 temperature and changes frozen-hydrated sections rapidly. Probably the observation of the rapid transformation of crystalline hexagonal ice (I_h) into amorphous I_v (Lepault *et al.*, 1983c; see also Heide, 1984; Heide and Zeitler, 1985; Talmon, 1987) is also a result of such H_2O radiolysis phenomena. It is not difficult from a thermodynamic point of view to expect a re-unification of the caged fragments of H_2O inside the frozen matrix. If this occurs, the apposition of the re-unified fragments at very low temperatures will certainly not follow the laws of crystallisation, but happen in a statistical manner similar to the production of amorphous Iv on a cold substrate (Burton and Oliver, 1935a,b; see also Eisenberg and Kauzman, 1969). Lepault *et al.* (1983c) report, that electron doses above 2000 el/nm² cause a complete transition from I_h to I_v. That means, that after rewarming above the devitrification temperature T_d (see the section "*Freezing and the Frozen State of Water ...* ") recrystallisation to cubic I_c occurs. After doses between 100 at 500 el/nm² Lepault and coworkers observed a change of the diffraction pattern from hexagonal I_c to amorphous I_v. They report, that under these conditions within the ice matrix after the complete decay of the hexagonal crystalline diffraction pattern there remains even at doses up to 2000 el/nm² "*some kind of memory*", which starts immediately hexagonal crystallisation again after warming up above the devitrification temperature T_d. This fits very well in the hypothetical concept: for sure there will still persist a number of small residual domains of crystalline I_h, far too small to create the characteristic diffraction pattern of crystalline hexagonal ice, but absolutely sufficient for nucleation and recrystallisation in the hexagonal lattice. The results of Lepault and coworkers give some idea, how fast H_2O radiolysis phenomena may change any frozen-hydrated specimen under the electron beam in the microscope.

How fast beam-induced H2O radiolysis acts is also visible by the "*bubbling phenomenon*" (*e.g.*, Frederik *et al.*, 1993), which starts immediately, if higher beam intensities are used for focusing or recording of images: the occurrence of gas bubbles within the frozen specimen has actually nothing to do with boiling processes, since the vapour pressure of ice below -150°C can be neglected. Also this event is based on beam-induced radiolysis phenomena, as perhaps another well known phenomenon too : it is evident and often reported (see *e.g.*, Michel *et al.*, 1991, Fig. 7, p. 13), that after electron irradiation the specimen contrast increases considerably. That may be a result of rapid beam etching due to H2O radiolysis. At the same time both knife marks and ripples (crevasses) lose contrast and therefore lose some of their annoying character. Since crevasses form sharp edges (see Fig. 20) it seems to be feasible, that radiolytic fragments disappear easily from the pointed edges and re-unification as well as deposition of those molecules occurs to a larger extent only in the grooves on the bottom of the crevasses. The same may occur on knife marks if they have the same expected irregular appearance as observed in resin sections as a result of a dull portion of a damaged knife edge, which creates superficial curling of the embedding resin. In both cases such irregularities will disappear immediately after start of irradiation by radiolysis and a flattening of the pointed surface profile will result both by apposition and etching, as demonstrated already by Zeitler and Bahr in 1965 (see also Isaacson, 1977).

Since the discussed welcome "*clearing up*" phenomenon is only the start of radiolytic beam damage and since further damage results within a very short time of irradiation respectively after a very small dose, a total destruction of the sample has to be expected immediately afterwards. Low dose operation of the scope is therefore perhaps the most important challenge related with cryo-ultramicrotomy of fresh-frozen samples. In this field one is in a conflicting compromise situation with respect to the signal-to-noise ratio : if an acceptable resolution is wanted, severe beam damage has to be expected in frozen-hydrated sections, since the signals result from scattering processes or a certain electron dose needed to reduce the noise of an image. Only in the very special cases of periodic structures noisy pictures can be used for "*image averaging*" to obtain a fairly good resolution ["*SNAP-shot method*" developed for "*Statistically Noisy Average Pictures*" by Kuo and Glaeser (1975); see also Unwin and Henderson (1975)]. As already mentioned, such samples are the exception in our biological collection and not the normal case and there is indeed more than a formal similarity with the indirect physical methods used in "*Submicroscopic Morphology*" in the sense of Albert Frey-Wyssling to circumvent the limit of the resolving power of the light microscope before the TEM was introduced. Perhaps the discussed low dose systems allow the making of a series of exposures at minimum dose at rather small magnification and to execute image averaging without periodic structures, just reproducing the same field to reduce the noise component. But the real success of such a procedure is still a matter of gambling and enterprise. More serious offers result both from reproducible experience of ESI (Fig. 30) and from direct dose reduction resulting from the application of SS-CCD systems or phosphor image plates read-out by laser scanning (Figs. 27 to 29). ESI is able to reduce the noise in pictures reproducibly by a considerable reduction of the underfocus desired to establish the needed contrast (see Bauer, 1988; Murray, 1986; Schröder, 1992; Schröder *et al.*, 1990). There is no doubt, that this improves the situation considerably and that EFTEM imaging (Fig. 30) is a powerful tool in low dose work with frozen-hydrated specimens. The observation of Michel *et al.* (1991) that zero-loss images of all sectioning artefacts (*e.g.*, ripples, crevasses, knife marks) are also recorded with better contrast are no real argument against ESI, but an additional challenge to improve the performance of our cryosectioning techniques in the way discussed above (*e.g.*, higher stability of the set up, lower α- and ϵ-angles of the knife, reduced width of the section surfaces). At least the minimising of the underfocus under zero loss conditions will correspond to a considerable reduction of noise and dose. This is a first step forward towards the desired improvement in resolution of beam sensitive structures already realised for amorphously vitrified suspension films by Schröder *et al.* (1990).

Both SS-CCD systems and electron stimulable phosphor image plates described in the preceding Section (see also Figs. 27 to 29) have a direct influence on the dose needed to record a digital image. In many respects (linear signal/dose ratio, dynamic range, exposure time) the characteristics of both systems are nearly identical. But there are also a lot of differences, which today impede a clear decision to acquire one or the other system. Burmester (1992, see also Table 6) has clearly demonstrated that the resolution is directly influenced by the effective surface (frame) of the recording area. The 19 x 19 mm² frame of the 1k x 1k polysilicon detector chip of the SS-CCD systems offered in the past covers only 1/16 of the area of the image plate (effectively 94 x 75 mm²) used in the Fuji FDL-5000 system. Since the pixel size of 25 μm is nearly identical, the image plate offers much better conditions at identical resolution with respect to needed magnification. Partially this disadvantage of the CCD chip is compensated by the new MegaScan system with 2k x 2k pixels offering the fourfold efficiency. That is already a lot (see Fig. 27c). But in comparison, the image plate has still an advantage of x 4 by the larger effective detector area. In practical work also the read-out time between two subsequent slow scan operations has to be regarded as limiting parameter for low dose work. This time must increase with respect to read-out noise linear with the amount of pixels. Read-out has an 1:1 influence on the periods of beam blanking and this may create shift problems, if the beam deflection above the specimen is used. But: beam blanking below the specimen recommended in such cases does not agree with low dose and larger read-out times due to increased pixel numbers may offset the gain in surface area under these preconditions. Both efficient and reproducible subarea read-out software as already announced by Gatan and the increase of read-out frequencies (*e.g.*, 500 kHz → 2 MHz → 5 MHz) are now important measures to reduce those inconveniences. Of course, one should not forget the most striking advantage of the on-line CCD-system: autotuning of focus and astigmatism etc. will never be possible with the off-line image plates. And just this autotuning according to Krivanek and Fan (1992, see also Krivanek and Mooney, 1993; Fan and Krivanek, 1990) by the above mentioned "*automatic diffractogram analysis* (ADA)" delivers results after only one pre-exposure of about 0.5 to 1.0 sec, followed by a computer calculation, which may be shortened considerably below 8 sec with a fast modern PC. It is no question at all, that a really fast autotuning routine will reduce the needed dose considerably since time consuming do-it-your-self focusing or dose intensive through-focus series are not needed. But also these considerations are only valid if autofocus as the most important automatism works really reproducibly on ultrathin fresh frozen-hydrated sections at the given faint contrast conditions. Today there is no practical experience with frozen-hydrated cryosections available and further comparisons between the offered low dose systems are needed to clarify the situation. If normal ambient temperature work on heavy metal-stained resin-embedded material would be the theme, probably the image plate would have real advantages because the whole off-line read-out hardware and software can be used for different TEM models without special modifications on these scopes: the image plate just substitutes for the photographic plates and eliminates the whole darkroom efforts and all costs for co-workers, materials and environment, since image plates are re-usable. But such nice arguments do not touch low dose. Probably a combination of all systems, that means autotuning by SS-CCD with a small frame (possibly considerably below 1k x 1k or 19 x 19 mm^2 detector surface with respect to extremely fast read-out and autofocus calculation), recording by large surface image plates (possibly even larger than the offered Fuji image plates of the FDL-5000 system with respect to maximum pixel numbers and minimum "*Pixel Equivalents*") and imaging with an energy filter at higher accelerating voltages in an ultrahigh vacuum would be the thrill - if all assumptions made are really correct and no new additional aspects arise in future.

I have described above, that in the past within an intermediate period starting after 1945 phase contrast optics in light microscopy had the important task to check, if after the chemical treatment of histoprocessing the specimens correspond correctly with the structures of the living object. After some years most of this job was done. It was possible to return to normality and to work again with (sometimes slightly improved) chemical methods which up to now remained the most important base in the wide fields of standard histology, microscopical anatomy, diagnostic histopathology and histochemistry. Only in the field of cell and tissue cultures were phase contrast and later on interference contrast indispensable tools for daily routine. Only in exceptional cases the question is again posed : do the standard procedures really deliver correct information? In such cases phase or interference contrast and nowadays confocal laser methods again serve as reference tools for checking. Cryo-ultramicrotomy and the related techniques of cryofixation and cryo-electron microscopy with ESI and low dose systems probably are going the same way. At the moment there is really the need to check, if and to what an extent all the beautiful pictures so easily obtainable with standard ultramicrotomy of resin blocks at ambient temperature correspond with the reality of living organisms, tissues and cells: this is the challenge of today. The re-investigation and new evaluation of the results obtained with standard ultramicrotomy in the recent decades certainly will yield an abundance of new results allowing us to improve standard techniques by comparison with the cryomethods described in this review. "*Hybrid techniques*" starting with rapid cryofixation at ambient or high pressure will probably also gain additional importance by such a comparison and re-evaluation. We can expect, that their results are close to reality. This certainly is true for cryofixation, freeze substitution, freeze drying and low temperature embedding. All these procedures have been improved considerably during the last years. For example Ludwig Edelmann could demonstrate that substitution without additives like OsO_4 or uranyl acetate gives excellent results in combination with Lowicryl low temperature embedding media (Edelmann, 1991a,b), if the frozen specimens are incubated for a sufficiently long time at low temperatures around -80°C and embedding in Lowicryls is performed really at temperatures of -60°C or below. Similarly freeze drying is now possible without severe artificial shrinkage ("*thermal collapses*") simply by long drying periods at low temperature (Edelmann, 1984a,b; Sitte *et al.*, 1994). Also freeze drying preparation gains importance by subsequent low temperature embedding in Lowicryls. First results indicate that ion distribution after these hybrid procedures agrees in principle with EDX results on freeze dried ultrathin sections. It is correct, that both freeze-substitution without additives like OsO_4 or uranyl acetate and freeze-drying at temperatures between -80°C and -100°C are time consuming procedures, but waiting really pays! There is no doubt, that a lot of investigations including three dimensional mapping of element distributions are easier possible by sectioning resin embedded samples with a dry knife, nowadays possible with purpose designed diamond knives without frustrating charging phenomena (Helmut Gnägi, personal communication, 1996). The application of cryo-ultramicrotomy for fresh-frozen (vitrified) biological samples will also in future certainly remain a tricky and time consuming job. But the proof by comparing the results obtained by cryo-ultramicrotomy is needed in all these areas to draw the correct conclusions.

The most important question for the meaningful use of any of the hybrid methods (cryofixation, freeze substitution or freeze drying followed by low temperature embedding) has to be repeated in this context: is true vitrification of a specimen really needed, when the amorphous sample is substituted or dried afterwards at a temperature around -80°C, where certainly devitrification and recrystallisation to cubic ice (I_c) occurs? I would say: "yes", and I want to repeat the already shortly presented opinion, that primary crystalline freezing of a liquid phase and secondary recrystallisation (*e.g.*, devitrification to cryptocrystalline I_c) within a solid state are completely different events. Solidification of a mixed liquid phase according to Fig. 32 is a very selective process as long as sufficient time for hexagonal crystallisation is available. That means, that the crystalline I_h formed by this "*liquid → solid transformation*" at ambient pressure (Fig. 32c and d) is certainly a perfect segregation into pure hexagonal crystalline I_h and the former suspended and dissolved components. In other terms: the mixed liquid cytoplasmic phases inside the cells or the mixed extraplasmatic phases around these cells (*e.g.*, lymphatic fluid, blood plasma or liquid phases in the big vacuoles of plant cells) during freezing segregate completely. With a very high probability in this I_h-lattice there is neither space for small ions with a hydration shell nor for macromolecular structures like globular or fibrous proteins or ribosomes. All these ions, molecules and macromolecular structures are shifted by the moving border of the growing I_h lattice to other places (Fig. 32d). As far as I_h-lattices are built up, such movements will be extremely fast over relatively large distances. The result must be a big difference to the original liquid suspension (Fig. 32c). Real vitrification according to Fig. 32b does not allow such shifts and redistributions: here water serves as an excellent amorphous embedding medium, which surrounds ("*encapsulates*") all these small components of the cytoplasmic or extracellular matrix. Since even the extremely fast oscillating H_2O dipoles (see *e.g.*, Bachmann and Mayer, 1987) have no chance to change their position, all larger particles [these are already ions with their hydration shells, see Kellenberger (1987, 1991), Kellenberger *et al.* (1986)] are captured by the vitrified H_2O matrix. If such a vitrified specimen is heated within a second thermodynamic event above the devitrification temperature T_d, H_2O molecules will reorganise to a cryptocrystalline cubic I_c-lattice (Fig. 32a). I cannot imagine that this could cause big movements in the still continuously solid state of the sample and I could expect the process shown in the diagrams of Fig. 32b and a, where all larger constituents of the frozen sample keep their original places and only the frozen matrix crystallises. Of course, this hypothesis has to be proven by experiments. Such proofs seem at least to be possible by freeze drying, if identical areas of a frozen specimen are shown before freeze drying in the amorphously frozen-hydrated state and after freeze drying without H_2O matrix. I suppose that results of such comparisons will support the general assumption made in Fig. 32 and that sufficient dehydration at a temperature around -80°C both by freeze drying (Edelmann, 1994a,b; Sitte *et al.*, 1994) and by freeze substitution (Edelmann, 1991a,b; Sitte *et al.*, 1994) can lead to quite correct pictures, which will in most respects correspond to the amorphously frozen-hydrated state. The mentioned hybrid methods will therefore probably offer the best alternatives to cryosectioning and frozen-hydrated investigation in future, if we have a solid base for meaningful conclusions by the mentioned comparisons between frozen-hydrated sections and resin sections of the same material after optimised cryodehydration (freeze-drying or freeze substitution) and optimal low temperature embedding.

Conclusion and Future Prospects

It was the goal of this review to show, that both instrumental and methodological progress in the recent years made cryosectioning of fresh-frozen samples much easier in comparison to former days. I wanted to show clearly, that these improvements count only if all the new possibilities including cryofixation (high pressure), cryo-ultramicrotomy and ESI/low dose investigation or diffraction analysis are used together and if all important accessories are at one's disposal (lowest angle cryo diamond knives, ioniser etc.). But I also wanted to point out, that "*easier*" does not mean "*easy*": cryosectioning at temperatures ≤ -165°C will always suffer from the brittleness of solid state water matrix of highly hydrated systems and will therefore remain more complicated in comparison with any hybrid procedure like freeze substitution, freeze drying or low temperature embedding, which were all also improved considerably during the last years. Cryosectioning of vitrified specimens still remains the proof, to what extent results obtainable with chemically based ambient temperature standard techniques (chemical fixation, polar solvent dehydration, resin embedding by heat polymerisation and sectioning at ambient temperature) or alternatively results obtainable with modern hybrid techniques (vitrification by rapid freezing, freeze substitution or freeze drying followed by low temperature embedding) agree with results from frozen-hydrated ultrathin sections in the cryo-TEM. I am absolutely confident that in many respects this comparison will give a clear recommendation, that standard procedures or at least hybrid methods deliver reliable results. Based on such comparative work some improvements of standard and hybrid procedures will certainly be possible. That is the challenge to realise this proof, which will render many new insights in the molecular structures of cells and tissues. It will stimulate and improve morphological work in many important respects. Of course there will be domains where purely physical cryopreparations will be needed also in the future: high performance cryowork of course will be needed also in the future for element analyses and for collecting pictures of our highly hydrated systems, not accessible with standard procedures. Besides these areas cryo-ultramicrotomy and frozen-hydrated work will remain the most important reference method in all cases of doubt, if the information obtained by standard or hybrid methods is really correct.

Acknowledgements

I would like to give my sincerest thanks to my colleagues of the Homburg FR 3.5 Medical Biology at the Saar-University for many helpful discussions, general support and astonishing patience. This concerns especially Dr. Klaus Neumann for careful reading of the manuscript and a lot of precious hints. The same counts for my friend Dr. Keith Ryan of the Marine Laboratory at Plymouth (UK). Both kindly eliminated a lot of mistakes resulting from my poor English. Many stimulating discussions and recommendations I owe to Dr. Ludwig Edelmann and Prof. Eberhard Morgenstern. Since I miss own first hand experience in high pressure freezing, the support of Dr. Daniel Studer and Dr. Martin Michel (both former co-workers of Dr. Martin Müller at ETH Zürich, later on members of Prof. Hunzickers staff at the University of Bern, Switzerland) was extremely helpful. Helmut Gnägi from Diatome Ltd. Biel, Switzerland, successfully engaged in the development of the diamond tools, that we use routinely, made a lot of extremely important contributions to this review article. Finally I would like to thank Dr. Wolfgang Probst, Erhard Zellmann and Volker Seybold from Carl Zeiss (now LEO Elektronenmikroskopie GmbH), Oberkochen, Dr. Heinrich Kolbe from Raytest, Straubenhardt FRG (European Representative of Fuji Systems) and Dr. Stephan Hiller (Gatan GmbH, München) for helpful comments on low dose operation of TEMscopes where I also lack first hand experience. The Royal Microscopical Society at Oxford kindly permitted the reproduction of numerous Figures within this review article from the Journal of Microscopy either in the original form or in a modified version. This related to Figs. 2, 5, 10, 11, 17, 20 and 26. Finally Mrs. Christa Rosinus carried out a lot of laboratory and photographic work and both Mrs. Barbara Reiland and Mrs. Christa Rosinus prepared this terribly voluminous manuscript. I sincerely thank them all.

List of Suppliers

Balzers Union AG (former Bal-Tec), Postfach 75, FL-9496 Balzers, Fürstentum Liechtenstein *(High pressure freezer HPM 10, systems for cryofixation and hybrid preparation of frozen specimens)*

Diatome AG, Postfach 557, CH-2501 Biel, Switzerland *(Cryo diamond knives and diamond tools for cryo-ultramicrotomy including trimming tools and ioniser "Static Line II")*

Fuji Photo Film Co., Ltd., 26-30 Minato-ku, Tokyo 106, Japan *(FDL 5000 Image Plate System for low dose image recording)*

Gatan Inc., 6678 Owens Drive, Pleasanton, CA 94588, USA *(SS-CCD camera systems for low dose image recording, post column electron spectroscopic systems for EFTEM with ESI, cryotransfer systems for side entries)*

Leica AG, Hernalser Hauptstraße 219, A-1171 Wien, Austria (Reichert-cryo-ultramictrotomes, high pressure freezer Leica EM HPF, automatic systems for freeze-substitution, freeze-drying and low temperature embedding, systems for ambient pressure freezing and cryopreparation)

LEO Elektronenmikroskopie GmbH, Carl Zeiss Straße 56, D-73446 Oberkochen, FRG and LEO Electron Microscopy Ltd, Clifton Road, Cambridge CB1 3QH England (former Carl Zeiss Elektronenoptik and Leica Cambridge) *(In column filter-lens Cryo-TEM Omega EM 912 for EFTEM and ESI including SS-CCD recording)*

LifeCell Corporation, 3606 Research Forest Drive, The Woodlands, TX 77381, USA (Slam freezer and systems for hybrid preparations of frozen specimens)

Oxford Instruments, Research Instruments, Scientific Research Division, Old Station Way, Eynsham Witney, Oxon OX8 1TL, England *(Side entry cryotransfer systems, anticontamination devices for TEMscopes and cryopreparation systems for TEM/STEM/SEM)*

RMC, 4400 South Santa Rita Avenue, Tucson, AZ 85714, USA *(Cryo-ultramicrotomes and cryochambers, systems for cryofixation and hybrid preparations of frozen specimens)*

Simco B.V., Aalsvoort 5, NL-7240 AA Lochem, Netherlands (German representative: Ziegener+Frick GmbH, Justinus-Kerner-Straße 8, D-71717 Beilstein, FRG) *(Adjustable ioniser for cryo-ultramicrotomy)*

References

Adrian M, Dubochet J, Lepault J, McDowall AW (1984) Cryo-electron microscopy of viruses. Nature **308**: 32-36.

Allenspach A (1993) Ultrastructure of early chick embryo tissues after high pressure freezing and freeze substitution. Microsc Res Techn **24**: 369-384.

Andrews SB, Buchanan RA, Leapman RD (1994) Quantitative dark-field mass analysis of ultrathin cryosections in the field-emission scanning transmission electron microscopy. Scanning Microsc Suppl **8**: 13-24.

Appleton TC (1974) A cryostat approach to ultrathin "dry" frozen sections for electron microscopy: a morphological and X-ray analytical study. J Microsc **100**: 49-74.

Armbruster BC, Kellenberger E, Carlemalm E, Villiger W, Garavito RM, Hobot JA, Chiovetti R, Acetarin JD (1984) Lowicryl resins - Present and future applications. In: Science of Biological Specimen Preparation for Microscopy and Microanalysis (Revel JP, Barnard T, Haggis GH, eds). Scanning Electron Microscopy Inc, AMF O'Hare, IL 60666, pp 77-81.

Ayato H, Mori N, Miyahara J, Oikawa T (1990) Application of the imaging plate to TEM observation. J Electron Microsc **39**: 444-448.

Bachmann L, Sitte P (1958) Dickenbestimmung nach Tolansky an Ultradünnschnitten (Thickness measurements according to Tolansky on ultrathin sections). Mikroskopie (Wien) **13**: 289-304.

Bachmann L, Sitte P (1960) Über Schnittdickenbestimmung nach dem Tolansky-Verfahren (Measurements of sections thicknesses using the Tolansky method). In: Verh 4ter Int Kongr Elektronen Mikroskopie 1958 (Bargmann W, Möllenstedt G, Niehrs H, Peters D, Ruska E, Wolpers C, eds.). Band II Biologisch-Medizinischer Teil. Springer Verlag, Berlin-Göttingen-Heidelberg, pp 75-79.

Bachmann L, Schmitt WW (1971) Improved cryofixation applicable to freeze-etching. Proc Nat Acad Sci USA **68**: 2149-2152.

Bachmann L, Schmitt-Fumian WW (1973) Spray-freezing and freeze-etching. In: Freeze-Etching - Techniques and Applications (Benedetti EL, Favard P, eds.). Société Française de Microscopie Électronique, Paris. pp 73-80.

Bachmann L, Mayer E (1987) Physics of water and ice: Implications for cryofixation. In: Cryotechniques in Biological Electron Microscopy (Steinbrecht RA, Zierold K, eds.). Springer-Verlag, Berlin-Heidelberg, pp 258-271.

Bald WB (1983) Optimizing the cooling block for the quick freeze method. J Microsc **131**: 11-23.

Bald WB (1984) The relative efficiency of cryogenic fluids used in the rapid quench cooling of biological samples. J Microsc **134**: 261-270.

Bald WB (1985) The relative merits of the various cooling methods. J Microsc **140**: 17-40.

Bald WB, Robards AW (1978) A device for the rapid freezing of biological specimens under precisely controlled and reproducible conditions. J Microsc **112**: 3-15.

Barnard T, Sevéus L (1977) Preparation of biological material for X-ray microanalysis of diffusible elements. II. Comparison of different methods of drying ultrathin cryosections cut without a trough liquid. J Microsc **112**: 281-291.

Bauer R (1988) Electron spectroscopic imaging: an advanced technique for imaging and analysis in transmission electron microscopy. Methods in Microbiology **20**: 113-146

Benedetti EL, Favard P (eds) (1973) Freeze-Etching - Techniques and Application. Société Française de Microscopie Électronique, Paris.

Benner G, Probst W, Martin JP (1994) Das Energiefilter-Transmissions-Elektronenmikroskop EM 912 OMEGA - Prinzip und Anwendung (The energy-filter transmission electron microscope EM 912 OMEGA - Principles and application). Zeiss Information mit Jenaer Rundschau **3**: 4-8.

Bernhard W (1965) Ultramicrotomie à basse température (Ultramicrotomy at low temperature). Ann Biol **4**: 5-19.

Bernhard W, Nancy MT (1964) Coupes a congélation ultrafines de tissu inclus dans la gélatine (Ultrathin cryosections of tissue embedded in gelatin). J Microscopie (Paris) **3**: 579-588.

Bernhard W, Leduc EH (1967) Ultrathin frozen sections. I. Methods and ultrastructural preservation. J Cell Biol **34**: 757-771.

Bernhard W, Viron A (1971) Improved techniques for the preparation of ultrathin frozen sections. J Cell Biol **49** : 731-746.

Bihr J, Benner G, Krahl D, Rilk A, Weimer E (1991) Design of an analytical TEM with integrated imaging Ω-spectrometer. In: Proc 49th Meeting EMSA (Bailey GW, ed.). San Francisco Press Inc, San Francisco, pp 354-355.

Blum J (1970) Cryogenic microtome apparatus (US-Patent application for the "bridge-design" of the object holder in Sorvall FTS cryo-ultramicrotome). Definite US-Patent No. 3,680,420 (1972)

Bold MA (1995) Vergleichende Studien zum Einsatz des Tissue Slicers und des Vibratomes vor einer Metallspiegel-Kryofixation biologischer Objekte (Comparative Studies on the Preparation of Biological Specimens for Metal Mirror Cryofixation with a Tissue Slicer and a Vibratome) Doctoral Thesis, Universität des Saarlandes, Homburg, Germany.

Box HC (1975) Cryoprotection of irradiated specimens. In: Physical Aspects of Electron Microscopy and Microbeam Analysis (Siegel BM, Beaman DR, eds.). Wiley & Sons, New York, pp 279-285.

Box HC (1977) Radiation Effects: ESR and ENDOR Analysis. Academic Press, New York.

Boyne AF (1979) A gentle, bounce-free assembly for quick-freezing tissues for electron microscopy: application to isolated torpedine ray electrocyte stacks. J Neurosci Meth **1**: 353-364.

Burmester C (1992) Entwicklung eines hochauflösenden Phosphor-Speicherplatten-Scanners (Design of a High Resolution Scanning Device for Readout of Phosphor Image Plates). Diplomarbeit aus dem Max-Planck-Institut für Medizinische Forschung. Ruprecht-Karls-Universität Heidelberg, Fakultät für Physik und Astronomie (Master's Thesis, Max-Planck Institute for Medical Research, University of Heidelberg). pp. 1-109.

Burton EF, Oliver WF (1935a) The crystal structure of ice at low temperature. Proc R Soc Lond A **153**: 166-171.

Burton EF, Oliver WF (1935b) X-ray diffraction patterns of ice. Nature **135**: 505.

Carlemalm E, Garavito RM, Villiger W (1982) Resin development for electron microscopy and an analysis of embedding at low temperature. J Microsc **126**: 123-143.

Carlemalm E, Villiger W, Acetarin JD, Kellenberger E (1986) Low temperature embedding. In: The Science of Biological Specimen Preparation 1985 (Müller M, Becker RP, Boyde A, Wolosewick JJ, eds.). Scanning Electron Microscopy Inc, AMF O'Hare, IL 60666, pp 147-154.

Castaing R (1975) Energy filtering in electron microscopy and electron diffraction. In: Physical Aspects of Electron Microscopy and Microbeam Analysis (Siegel BM, Beaman DR, eds.). John Wiley & Sons, New York. pp 287-301.

Castaing R, Henry L (1962) Filtrage magnétique des vitesses en microscope électronique (Magnetic filtering of velocities in an electron microscope). C R Acad Sci Paris **B 255**: 76-78.

Chang JJ, McDowall AW, Leupault J, Freeman R, Walter CA, Dubochet J (1983) Freezing, sectioning and observation artefacts of frozen hydrated sections for electron microscopy. J Microsc **132**: 109-123.

Chiu W, Knapek E, Jeng TW, Dietrich I (1981) Electron radiation damage of a thin protein crystal at 4K. Ultramicroscopy **6**: 291-296.

Chiu W, Downing KH, Dubochet J, Glaeser RM, Heide HG, Knapek E, Kopf DA, Lamvik K, Lepault J, Robertson JD, Zeitler E, Zemlin F ("*International Experimental Study Group*") (1986) Cryoprotection in electron microscopy. J Microsc **141**: 385-391.

Christensen AK (1969) A way to prepare frozen thin sections of fresh tissue for electron microscopy. In: Autoradiography of Diffusible Substances (Roth LJ, Stumpf WE, eds.). Academic Press, New York. pp 349-362.

Christensen AK (1971) Frozen thin sections of fresh tissue for electron microscopy, with a description of pancreas and liver. J Cell Biol **51**: 772-804.

Cosslett VE (1978) Radiation damage in the high resolution electron microscopy of biological materials: a review. J Microsc **113**: 113-129

Costello MJ, Corless JM (1978) The direct measurements of temperature changes within freeze-fracture specimens during quenching in liquid coolants. J Microsc **112**: 17-37.

Crudgington JR (1966) A freezing head for the ultrathin-sectioning of rubber-like materials. Science Tools (LKB Instrum J) **13**: 23-27.

Daberkow I, Herrmann KH, Liu L, Rau WD (1991) Performance of electron image converters with YAG single-crystal screen and CCD sensor. Ultramicroscopy **38**: 215-223.

Dempsey GP, Bullivant S (1976) A copper block method for freezing non-cryoprotected tissue to produce ice-crystal-free regions for electron microscopy. J Microsc **106**: 251-260.

De Ruijter WJ (1995) Imaging properties and applications of slow-scan charge-coupled device cameras suitable for electron microscopy. Micron **26**: 247-275.

Dierksen K, Typke D, Hegerl R, Baumeister W (1993) Towards automatic electron tomography II. Implementation of autofocus and low-dose procedures. Ultramicroscopy **49**: 109-120.

Dietrich I, Fox F, Knapek E, Lefranc G, Nachtrieb K, Weyl R, Zerbst H (1977) Improvements in electron microscopy by application of superconductivity. Ultramicroscopy **2**: 241-249.

Dietrich I, Fox F, Heide HG, Knapek E, Weyl R (1978) Radiation damage due to knock-on processes on carbon foils cooled to liquid helium temperature. Ultramicroscopy **3**: 185-189.

Dietrich I, Formanek H, Fox F, Knapek E, Weyl R (1979) Reduction of radiation damage in an electron microscope with a superconducting lens system. Nature **277**: 380-381.

Dietrich I, Dubochet J, Fox F, Knapek E, Weyl R (1980) Reduction of radiation damage by imaging with a superconducting lens system. In: Electron Microscopy at Molecular Dimensions (Baumeister W, Vogell W, eds.). Springer Verlag, Berlin-Heidelberg-New York, pp. 234-244.

Ding B, Turgeon R, Parthasarathy MV (1992) Effect of high-pressure freezing on plant microfilament bundles. J Microsc **165**: 367-376.

Dollhopf FL (1968) Kühlvorrichtung zum Herstellen von Ultradünnschnitten bei Temperaturen zwischen 0 und -150°C (Cryogenic apparatus for production of ultrathin sections at temperatures between 0 and -150°C). In: Electron Microscopy 1968 (Bocchiarelli StD, ed). Tipografia Poliglotta Vaticana, Roma. **2**: 39, 46.

Dollhopf FL, Sitte H (1969) Die Shandon-Reichert-Kühleinrichtung FC-150 zum Herstellen von Ultradünn- und Feinschnitten bei extrem niederen Temperaturen. I. Gerätetechnik (The Shandon-Reichert cryosystem for production of ultrathin and semithin sections at extremely low temperatures. I. Design). Mikroskopie (Wien) **25**: 17-32.

Downing KH (1983) Temperature dependence of the critical electron exposure for hydrocarbon monolayers. Ultramicroscopy **11**: 229-238.

Draganic IG, Draganic ZD (1971) The Radiation Chemistry of Water. Vol 26 in the Series: Physical Chemistry (Loebl EM, ed.)Academic Press, New York. pp. 1-242.

Dubochet J (1975) Carbon-loss during irradiation of T4 bacteriophages and *E. coli* bacteria in electron microscopes. J Ultrastruct Res **52**: 276-288.

Dubochet J (1981) Beam damage at low temperature (Abstract). Proc Royal Microsc Soc **16**: 170.

Dubochet J, McDowall AW (1984) Frozen hydrated sections. In: The Science of Biological Specimen Preparation (Revel JP, Barnard T, Haggis HG, eds) Scanning Electron Microscopy Inc. AMF O'Hare, pp 147-152.

Dubochet J, Booy FP, Freeman R, Jones AV, Walter CA (1981) Low temperature electron microscopy. Ann Rev Biophys Bioeng **10**: 133-149.

Dubochet J, Chang JJ, Freemann R, Lepault J, McDowall A (1982a) Frozen aqueous suspensions. Ultramicroscopy **10**: 55-62.

Dubochet J, Lepault J, Freemann R, Berriman JA, Homo JC (1982b) Electron microscopy of frozen water and aqueous solutions. J Microsc **128**: 219-237.

Dubochet J, McDowall A, Freeman R, Lepault J (1982c) Cryoprotection on organic specimens. In: Electron Microscopy 1982, 10th Int Congr Electr Micr, Hamburg 1982 (The Congress Organizing Committee, ed.). Vol 1: pp 19-23.

Dubochet J, McDowall AW, Menge B, Schmid EN, Lickfeld KG (1983) Electron microscopy of frozen-hydrated bacteria. J Bacteriol **155**: 381-390.

Dubochet J, Adrian M, Lepault J, McDowall AW (1985) Cryo-electron microscopy of vitrified biological specimens. Trends Biochem Sci **10**: 143-146.

Dubochet J, Adrian M, Schultz P, Oudet P (1986) Cryo-electron microscopy of vitrified SV40 minichromosomes: the liquid drop model. EMBO J **5**: 519-528.

Dubochet J, Adrian M, Chang JJ, Lepault J, McDowall AW (1987) Cryo-electron microscopy of vitrified specimens. In: Cryotechniques in Biological Electron Microscopy (Steinbrecht RA, Zierold K, eds.). Springer-Verlag, Berlin-Heidelberg, pp 112-131.

Dubochet J, Adrian M, Chang JJ, Homo JC, Lepault J, McDowall AW, Schultz P (1988) Cryo-electron microscopy of vitrified specimens. Quart Rev Biophys **21**: 129-228.

Echlin P (1992) Low Temperature Microscopy and Analysis. Plenum Press, New York, London, pp. 101-140.

Edelmann L (1984) Frozen-hydrated cryosections of thallium loaded muscle reveal subcellular potassium binding sites. Physiol Chem Phys and Med NMR **16**: 499-501.

Edelmann L (1989a) Freeze-substitution and low temperature embedding for analytical electron microscopy. In: Electron Probe Microanalysis, Applications in Biology and Medicine (Zierold K, Hagler HK, eds.). Springer Verlag, Berlin. pp 33-46.

Edelmann L (1989b) The contracting muscle: A challenge for freeze-substitution and low temperature embedding. Scanning Microscopy Suppl **3**: 241-252.

Edelmann L (1991a) Freeze-substitution and low temperature embedding. In: EMAG-Micro 89 (Elder HY, Goodhew PJ, eds), Vol. 2: Biological, Proc. Inst. EM Phys. Conf., pp 763-768.

Edelmann L (1991b) Freeze-substitution and the preservation of diffusible ions. J Microsc **161**: 217-228.

Edelmann L (1992) Biological X-ray microanalysis of ions and water: artefacts and future strategies. In: Electron Microscopy (Megias-Megias L, Rodriguez-Garcia MI, Rios A, Arias JM, eds.). Proc Eur Congr Electron Microscopy, Secretariado de Publicaciones de la Universidad de Granada, Vol. 1, pp 333-336.

Edelmann L (1994a) Low temperature embedding of chemically unfixed biological material after cryosorption freeze-drying. Scanning Microsc **8**: 551-562.

Edelmann L (1994b) Optimal freeze-drying of cryosections and bulk specimens for X-ray microanalysis. Scanning Microscopy Suppl **8**: 67-81.

Egerton RF (1986) Electron Energy-Loss Spectroscopy in the Transmission Electron Microscope. Plenum Press, New York.

Eisenberg D, Kauzmann W (1969) The Structure and Properties of Water. Clarendon Press, Oxford.

Elder HY, Gray CC, Jardine AG, Chapman JN, Biddlecombe WH (1982) Optimum conditions for cryoquenching of small tissue blocks in liquid coolants. J Microsc **126**: 45-61.

Eränkö O (1954) Quenching of tissues for freezing. Acta Anat **22**: 331-336.

Escaig J (1982) New instruments which facilitate rapid freezing at 83K and 6K. J Microsc **126**: 221-229.

Escaig J (1984) Control of different parameters for optimal freezing conditions. In: The Science of Biological Specimen Preparation (Revel JP, Barnard T, Haggis GH, eds.). Scanning Electron Microscopy Inc, AMF O'Hare, IL 60666, pp 117-122.

Escaig J, Geraud G, Nicolas G (1977) Congélation rapide de tissus biologiques. Mesure des températures et des vitesses de congélation par thermocouple en couche mince (Rapid freezing of biological tissue. Measurement of temperatures and freezing velocities by thermocouple in thin sections). C R Acad Sc Paris Serie D **284:** 2289-2292.

Fan GY, Krivanek OL (1990) Computer-controlled HREM alignment using automated diffractogram analysis. In: Proceedings XIIth International Congress Electron Microscopy (Peachey LD, Williams DB, eds) Vol. 1, San Francisco Press, San Francisco, pp. 532-533.

Fawcett DW (1966) An Atlas of Fine Structure: the Cell, its Organelles and Inclusions. 1st edition. B Saunders Co., Philadelphia-London.

Fernández-Morán H (1951) Application of ultrathin freezing-sectioning technique to the study of cell structures with the electron microscope. Arkiv Fysik **4:** 471-483 (with plates I-VIII without pagination).

Fernández-Morán H (1966) High-resolution electron microscopy with superconducting lenses at liquid helium temperatures. Proc Nat Acad Sci USA **56**: 801-808.

Fernández-Morán H (1985) Cryo-electron microscopy and ultramicrotomy: reminiscences and reflections. In: Advances in Electronics and Electron Physics, Academic Press, New York, Suppl **16**, pp 167-223.

Franks F (1972 - 1982) Water - a Comprehensive Treatise. Plenum, New York.

Frederik PM, Busing WM (1981) Strong evidence against section thawing whilst cutting on the cryo-ultratome. J Microsc **122**: 217-220.

Frederik PM, Busing WM, Persson A (1982) Concerning the nature of the cryosectioning process. J Microsc **125**: 167-175.

Frederik PM, Bomans PHH, Stuart MCA (1993) Matrix effects and the induction of mass loss or bubbling by the electon beam in vitrified hydrated specimens. Ultramicroscopy **48**: 107-119.

Freeman R, Leonard KR, Dubochet J (1980) The temperature dependence of beam damage to biological samples in the scanning transmission electron microscope (STEM). In: Electron Microscopy 1980, Proc 7th Europ Conf Electr Micr Den Haag, 1980 (Brederoo P, de Priester, W, eds) Vol **2**: 640-641.

Frei M (1992) Untersuchungen über einen Artefakt (Membranrupturen) bei der Kryofixation von Zellen (Investigations about an artefact, membrane ruptures, during cryofixation of cells). Doctoral Thesis, Universität des Saarlandes, Homburg, Germany.

Frey-Wyssling A (1953) Submicroscopic Morphology of Protoplasm. 2nd English Edition. Elsevier, Amsterdam.

Gabor D (1957) Die Entwicklungsgeschichte des Elektronenmikroskopes (The history of the electron microscope). Elektrotechn Zschr **78**: 522-543.

Gatan Inc. (1995) Mega Scan 2k x 2k Multi-Scan CCD Camera for TEM. Leaflet, 2 pages (see List of Suppliers).

Glaeser RM (1971) Limitations to significant information in biological electron microscopy as a result of radiation damage. J Ultrastruct Res **36**: 466-482.

Glaeser RM (1975) Radiation damage and biological electron microscopy. In: Physical Aspects of Electron Microscopy and Microbeam Analysis (Siegel BM, Beaman DR, eds.). Wiley & Sons, New York, pp 205-229.

Glaeser RM, Hobbs LW (1975) Radiation damage in stained catalase at low temperature. J Microsc **103**: 209-214.

Glaeser RM, Taylor KA (1978) Radiation damage relative to transmission electron microscopy of biological specimens at low temperature: a review. J Microsc **112**: 127-138.

Griffiths G (1993) Fine Structure Immunocytochemistry. Springer Verlag, Berlin-Heidelberg.

Griffiths G, Simons K, Warren G, Tokuyasu KT (1983) Immunoelectron microscopy using thin, frozen sections: application to studies of the intracellular transport of semliki forest virus spike glycoproteins. Meth Enzymol **96**: 466-485.

Griffiths G, McDowall A, Back R, Dubochet J (1984) On the preparation of cryosections for immunocytochemistry. J Ultrastruct Res **89**: 65-78.

Grubb DT, Groves GW (1971) Rate of damage of polymer crystals in the electron microscope: Dependence on temperature and beam voltage. Phil Mag **24**: 815-828.

Gubbens AJ, Kraus B, Krivanek OL, Mooney PE (1995) An imaging filter for high voltage electron microscopy. Ultramicroscopy **59**: 255-265.

Gupta BL, Hall TA (1981) The X-ray microanalysis of frozen-hydrated sections in scanning electron microscopy: an evaluation. Tissue & Cell **13**: 623-643.

Gutensohn J (1993) Kryofixation von Zellsuspensionen am Metallspiegel am Beispiel der Hefe *Rhodotorula glutinis* und *Saccharomyces cerevisiae* (Cryofixation of cell suspensions on the metal mirror demonstrated with yeast cells of *Rhodotorula glutinis* and *Saccharomyces cerevisiae*). Doctoral Thesis, Universität Bonn, Germany.

Hagler HK, Buja LM (1984) New techniques for the preparation of thin freeze dried cryosections for X-ray microanalysis. In: The Science of Biological Specimen Preparation for Microscopy and Microanalysis (Revel JP, Barnard T, Haggis GH eds.). Scanning Electron Microscopy Inc., AMF O'Hare, pp 161-166.

Handley DA, Alexander JT, Chien S (1981) The design and use of a simple device for rapid quench-freezing of biological samples. J Microsc **121**: 273-282.

Hanyu Y, Ishikawa M, Matsumoto G (1992) An improved cryofixation method: cryoquenching of small tissue blocks during microwave irradiation. J Microsc **165**: 255-271.

Hax WMA, Lichtenegger S (1982) Transfer, observation and analysis of frozen hydrated specimens. J Microsc **126** 275-284.

Hayward SB, Glaeser RM (1979) Radiation damage of purple membrane at low temperature. Ultramicroscopy **4**: 201-210.

Heide HG (1984) Observations on ice layers. Ultramicroscopy **14**: 271-278.

Heide HG, Grund S (1974) Eine Tiefkühlkette zum Überführen von wasserhaltigen biologischen Objekten ins Elektronenmikroskop (A "cryo-chain" for the transfer of frozen-hydrated biological samples to the TEM). J Ultrastruct Res **48**: 259-268.

Heide HG, Zeitler E (1985) Physical behaviour of solid water at low temperatures and the embedding of electron microscopical specimens. Ultramicroscopy **16**: 151-160.

Heide HG, Hermann KH, Jäger J (1982) Radiation damage of L-valine between 8 and 300 K. Electron Microscopy 1982, Proc 10th Int Congr Electr Micr Hamburg (The Congress Organizing Comitee, ed.) Vol 2, pp 457-458.

Henkelman RM, Ottensmeyer FP (1974) An energy filter for biological electron microscopy. J Microsc **102**: 79-94.

Heuser JE, Reese TS, Dennis MJ, Jan Y, Jan L, Evans L (1979) Synaptic vesicle exocytosis captured by quick freezing and correlated quantal transmitter release. J Cell Biol **81**: 275-300.

Hippe S, Düring K, Kreuzaler F (1989) *In situ* localization of a foreign protein in transgenic plants by immunoelectron microscopy following high pressure freezing, freeze substitution and low temperature embedding. Eur J Cell Biol **50**: 230-234.

Hippe-Sanwald S (1993) Impact of freeze substitution on biological electron microscopy. Microsc Res Techn **24**: 400-422.

Hobot JA, Villiger W, Escaig J, Maeder M, Ryter A, Kellenberger E (1985) Shape and fine structure of nucleoids observed on sections of ultrarapidly frozen and cryo-substituted bacteria. J Bacteriol **162**: 960-971.

Hochanadel CJ (1960) Radiation chemistry of water. In: Comparative Effects of Radiation (Burton M, Kirby-Smith JS, Magee JL, eds.). Report Conf San Juan 1960. John Wiley & Sons Inc, New York-London.

Hodson S, Marshall J (1972) Evidence against the through-section thawing whilst cutting on the ultracryotome. J Microsc **95**: 459-465.

Hoerr NL (1936) Cytological studies by the Altmann-Gersh freeze-drying method. I. Recent advances of the technique. Anat Rec **65**: 293-317.

Hohenberg H, Mannweiler K, Müller M (1994) High-pressure freezing of cell suspensions in cellulose capillary tubes. J Microsc **175**: 34-43.

Humbel BM, Müller M (1986) Freeze substitution and low temperature embedding. In: The Science of Biological Specimen Preparation (Müller M, Becker RP, Boyde A, Wolosewick JJ, eds). Scanning Electron Microscopy, AMF O'Hare, pp 175-183.

Hunziker EB, Herrmann W, Schenk RK, Müller M, Moor H (1984) Cartilage ultrastructure after high pressure freezing, freeze substitution, and low temperature embedding. 1. Chondrocyte ultrastructure - implications for the theories of mineralization and vascular invasion. J Cell Biol **98**: 267-276.

International Experimental Study Group (see Chiu W *et al.*, 1986) Cryoprotection in electron microscopy. J Microsc **141**: 385-391.

Isaacson MS (1977) Specimen damage in the electron microscope. In: Principles and Techniques of Electron Microscopy - Biological Applications (Hayat MA, ed.). Van Nostrand Reinhold, New York. pp 1-77.

Janesick JR, Elliot T, Collins S, Blouke MM,

Freeman J (1987) Scientific charge-coupled devices. Opt Eng **26**: 692-714.

Jésior JC (1986) How to avoid compression II. The influence of sectioning conditions. J Ultrastruct Molec Res **95**: 210-217.

Jésior JC (1989) Use of low angle diamond knives leads to improved ultrastructural preservation of ultrathin sections. Scanning Microsc Suppl **3**: 147-153.

Kaeser W, Koyro HW, Moor H (1989) Cryofixation of plant tissues without pretreatment. J Microsc **154**: 279-288.

Karp RD, Silcox JC, Somlyo AV (1982) Cryoultramicrotomy: evidence against melting and the use of a low temperature cement for specimen orientation. J Microsc **126**: 157-165.

Kellenberger E (1987) The response of biological macromolecules and supra molecular structures to the physics of specimen preparation. In: Cryotechniques in Biological Electron Microscopy (Steinbrecht RA, Zierold K, eds) Springer-Verlag, Berlin-Heidelberg, pp 35-63.

Kellenberger E (1991) The potential of cryofixation and freeze substitution: observations and theoretical considerations. J Microsc **161**: 183-203.

Kellenberger E, Carlemalm E, Villiger W (1986) Physics of the preparation and observation of specimens that involve cryoprocedures. In: The Science of Biological Specimen Preparation for Microscopy and Microanalysis 1985 (Müller M, Becker RP, Boyde A, Wolosewick JJ, eds.). Scanning Electron Microscopy Inc., AMF O'Hare, pp 1-20.

Kenzian A, Kleber H, Sitte H (1975) Schalenbauweise für Ultramikrotome (Shell construction for ultramicrotomes). Öst Patent Nr 320 308, Klasse 42 h (application 9622).

Knapek E, Dubochet J (1980) Beam damage to organic material is considerably reduced in cryo-electron microscopy. J Mol Biol **141**: 147-161.

Knapek E, Lefranc G, Heide HG, Dietrich I (1982) Electron microscopical results on cryoprotection of organic materials obtained with cold stages. Ultramicroscopy **10**: 105-110.

Knoll G, Verkleij AJ, Plattner H (1987) Cryofixation of dynamic processes in cells and organelles. In: Cryotechniques in Biological Electron Microscopy (Steinbrecht FA, Zierold K, eds.). Springer-Verlag, Berlin-Heidelberg, pp 258-271.

Knoll G, Braun C, Müller Th, Plattner H (1992) Time resolved analysis of rapid events in intact cells. In: Electron Microscopy 1992 (Megías-Megías L, Rodríguez-García MI, Ríos A, Arias JM, eds.). Proc Eur Congr Electron Microscopy, Secretariado de Publicaciones de la Universidad de Granada, Vol 3: 37-41.

Kobayashi K, Sakaoku K (1965) Irradiation changes in organic polymers at various accelerating voltages. Lab Invest **14**: 359-376.

Kopstad G, Elgsaeter A (1982) Theoretical analysis of specimen cooling rate during impact freezing and liquid-jet freezing of freeze-etch specimens. Biophys J **40**: 163-170.

Koster AJ, De Ruijter WJ (1992) Practical autoalignment of transmission electron microscopes. Ultramicroscopy **40**: 89-107.

Krahl D, Pätzold H, Swoboda M (1990) An aberration-minimised imaging energy filter of simple design. In: Proceedings XIIth International Congress Electron Microscopy (Peachey LD, Williams DB, eds) Vol. 1, San Francisco Press, San Francisco, USA.

Krivanek OL, Fan GY (1992) Complete HREM autotuning using automated diffractogram analysis. In: Proceedings 50th Annual EMSA Meeting (Bailey GW, Bentley J, Small JA, eds.). San Francisco Press, San Francisco. Part I, pp. 96-97.

Krivanek OL, Mooney PE (1993) Applications of slow-scan CCD cameras in transmission electron microscopy. Ultramicroscopy **49**: 95-108.

Krivanek OL, Kundmann MK, Bourrat X (1994) Elemental mapping by energy-filtered electron microscopy. Mat Res Soc Symp Proc **332**: 341-350.

Krivanek OL, Kundmann MK, Kimoto K (1995) Spatial resolution in EFTEM elemental maps. J Microsc **180**: 277-287.

Kuo IAM, Glaeser RM (1975) Development of methodology for low exposure, high resolution electron microscopy of biological specimens. Ultramicroscopy **1**: 53-66.

Lamvik, MK, Kopf DA, Robertson JD (1983) Radiation damage in L-valine at liquid helium temperature. Nature **301**: 332-334.

Lanio S (1986) High-resolution imaging magnetic energy filters with simple structure. Optik **73**: 99-107.

Lanio S, Rose H, Krahl D (1986) Test and improved design of a corrected imaging magnetic energy filter. Optik **78**: 56-68.

Leapman RD, Sun SQ, Hunt JA, Andrews SB (1994) Biological electron energy loss spectroscopy in the field-emission scanning transmission electron microscope. Scanning Microsc Suppl **8**: 245-259.

Leduc EH, Bernhard W, Holt SJ, Tranzer JP (1967) Ultrathin frozen sections. II. Demonstration of enzymic activity. J Cell Biol **34**: 773-786.

Lepault J, Booy FP, Dubochet J (1983a) Electron microscopy of frozen biological suspensions. J Microsc **129**: 89-102.

Lepault J, Dubochet J, Dietrich I, Knapek E,

Zeitler E (1983b) Amendment to: electron beam damage to organic specimens at liquid helium temperature. J Mol Biol **163**: 511.

Lepault J, Freeman R, Dubochet J (1983c) Electron beam induced "vitrified ice". J Microsc **132**: RP3-RP4.

Leunissen JLM, Verkleij AJ (1989) Cryo-ultramicrotomy and immuno-gold labeling. In: Immuno-Gold Labeling in Cell Biology (Verkleij AJ, Leunissen JLM, eds.). CRC Press, Boca Raton, FL, pp 95-114.

Mattheij JAM, Dignum PH (1975) A device for the complete elimination of static electricity in paraffin sectioning. Stain Techn **50**: 157-159.

Mayer E (1985) Vitrification of pure liquid water. J Microsc **140**: 3-15.

McDowall AW, Chang JJ, Freeman R, Lepault J, Walter CA, Dubochet J (1983) Electron microscopy of frozen hydrated sections of vitreous ice and vitrified biological samples. J Microsc **131**: 1-9.

McDowall AW, Hofmann W, Lepault J, Adrian M, Dubochet J (1984) Cryo-electron microscopy of vitrified insect flight muscle. J Mol Biol **178**: 105-111.

McDowall AW, Smith JM, Dubochet J (1986) Cryo-electron microscopy of vitrified chromosomes *in situ*. EMBO J **5**: 1395-1402.

Menco BPM (1986) A survey of ultra-rapid cryofixation methods with particular emphasis on application to freeze-fracturing, freeze-drying, and freeze-substitution. J Electr Microsc Tech **4**: 144-240.

Meryman HT (1971) Cryoprotective agents. Cryobiology **8**: 173-183.

Michel M, Hillmann T, Müller M (1991) Cryo-sectioning of plant material frozen at high pressure. J Microsc **163**: 3-18.

Michel M, Gnägi H, Müller M (1992) Diamonds are a cryosectioner's best friend. J Microsc **166**: 43-56.

Mochel ME, Mochel JM (1986) A CCD imaging and analysis system for the VG HB5 STEM. In: Proc 44th Annual Meeting EMSA (Bailey GW, ed.). San Francisco Press, San Francisco, pp 616-617.

Moor H (1964) Die Gefrier-Fixation lebender Zellen und ihre Anwendung in der Elektronenmikroskopie (The cryofixation of living cells and their application in electron microscopy). Z Zellforsch **62**: 546-580.

Moor H (1971) Recent progress in the freeze-etching technique. Philos Trans Roy Soc London Ser B **261**: 121-131.

Moor H (1973) Cryotechnology for the structural analysis of biological material. In: Freeze-etching - Techniques and Applications (Benedetti EL, Favard P, eds.). Société Française de Microscopie Électronique, Paris. pp. 11-20.

Moor H (1987) Theory and practice of high pressure freezing. In: Cryotechniques in Biological Electron Microscopy (Steinbrecht RA, Zierold K, eds.). Springer-Verlag, Berlin. pp. 175-191.

Moor H, Mühlethaler K (1963) Fine structure of frozen-etched yeast cells. J Cell Biol **17**: 609-628.

Moor H, Riehle U (1968) Snap-freezing under high pressure: A new fixation technique for freeze-etching. In: Electron Microscopy 1968, Proc 4th Europ Reg Conf Electron Microcopy, Roma (Bocciarelli StD, ed.). Tipografia Poliglotta Vaticana, Roma **2**: 33-34.

Moor H, Hoechli M (1970) The influence of high pressure freezing on living cells. In: Proc. 7th Int Congr EM, Grenoble (Favard P, ed.). Société Française de Microscopie Électronique, Paris, **1**: 445-446.

Moor H, Mühlethaler K, Waldner H, Frey-Wyssling A (1961) A new freezing-ultramicrotome. J Biophys Biochem Cytol **10**: 1-13. (*Author's note*: The title is misleading: the paper refers to the first prototype of the Balzers freeze-cleavage/freeze-etch system).

Moor H, Bellin G, Sandris C, Akert K (1980) The influence of high pressure freezing on mammalian tissue. Cell Tissue Res **209**: 201-216.

Morgan AJ (1985) X-Ray Microanalysis in Electron Microscopy for Biologists. Oxford University Press-Royal Microscopical Society.

Morgenstern E, Edelmann L (1989) Analysis of dynamic cell processes by rapid freezing and freeze substitution. In: Electron Microscopy of Subcellullar Dynamics (Plattner H, eds.) CRC Press Inc. Boca Raton, pp 119-140.

Mori N, Oikawa T, Katoh T, Miyahara J, Harada Y (1988) Application of the "Imaging Plate" to TEM image recording. Ultramicroscopy **25**: 195-202.

Mori N. Oikawa T, Harada Y, Mijahara J (1990) Development of the imaging plate for the transmission electron microscope and its characteristics. J Electron Microsc **39**: 433-436.

Mühlethaler K (1973) History of freeze-etching. In: Freeze-Etching - Techniques and Applications (Benedetti EL, Favard P, eds.). Soc Tranc Microscopie Électronique, Paris, pp 1-10.

Müller KH, Zemlin F, Zeitler E (1981) Cryo-protection of electron-irradiated organic crystals. In: Proc 39th EMSA Meeting Atlanta 1981, pp 26-27.

Müller M, Moor H (1984) Cryofixation of thick specimens by high pressure freezing. In: The Science of Biological Specimen Preparation for Microscopy and Microanalysis (Revel JP, Barnard T, Haggis GH, eds.) Scanning Electron Microscopy Inc,

AMF O'Hare, pp. 131-138.

Müller M, Meister N, Moor H (1980) Freezing in a propane jet and its application in freeze-fracturing. Mikroskopie (Wien) **36:** 129-140.

Müller S (1994) Vergleichende Studien zur Metallspiegel-Kryofixation, Kryosubstitution und Kunstharz-Einbettung für lichtmikroskopische Untersuchungen. (Comparative Studies Concerning Metal Mirror Cryofixation, Freeze Substitution and Resin embedding for Light Microscopy) Doctoral Thesis, Universität des Saarlandes, Homburg, Germany.

Müller Th, Hakert H, Eckert Th (1989) Rheological and electron microscopic characterization of aqueous carboxymethyl cellulose gels. Part II: Visualization of the gel structure by freeze-fracturing. Colloid Polym Sci **267**: 230-236.

Murray JM (1986) Application of electron spectroscopic imaging to frozen hydrated specimens. Proc 44th EMSA Meeting (Bailey GW, ed.). San Francisco Press Inc, San Francisco, pp 92-93.

Murray PWLeR, Robards AW, Waites PR (1989) Countercurrent plunge cooling: a new approach to increase reproducibility in the quick freezing of biological tissue. J Microsc **156**: 173-182.

Nicholson PW (1978) A device for static elimination in ultramicrotomy. Stain Technol **53:** 237-239.

Oikawa T, Mori N, Takano N, Ohnishi M (1990) The development of an image recording system using the imaging plate in a TEM. J Electron Microsc **39**: 437-443.

Orci L, Perrelet A (1975) Freeze Etch Histology, a Comparison between Thin Sections and Freeze-Etch Replicas. Springer-Verlag, Berlin-Heidelberg-New York.

Ornberg RL (1985) Cryoultramicrotomy and electron microanalysis of isolated bovine adrenal chromaffin cells. In: The Science of Biological Specimen Preparation for Microscopy and Microanalysis 1985 (Müller M, Becker RP, Boyde A, Wolosewick JJ, eds.). Scanning Electron Microscopy Inc, AMF O'Hare, pp 135-139.

Ornberg RL (1989) Cryoultramicrotomy of ultra-rapidly frozen specimens. Scanning Microscopy Suppl **3**: 227-230.

Ostwald W (1915) Die Welt der vernachlässigten Dimensionen - Eine Einführung in die Kolloidchemie mit besonderer Berücksichtigung ihrer Anwendungen, 1. Auflage (The World of Ignored Dimensions - An Introduction to Colloid Chemistry with Special Consideration to Application, 1st Edition). Verlag von Theodor Steinkopff, Dresden-Leipzig.[1]

Ottensmeyer FP, Andrew JW (1980) High-resolution microanalysis of biological specimens by electron energy loss spectroscopy and by electron spectroscopic imaging. J Ultrastruct Res **72**: 336-348.

Ottensmeyer F, Bazett-Jones D, Adamson-Sharp K (1981) Electron energy-loss microanalysis with high spatial resolution, and sensitivity. In: Microprobe Analysis of Biological Systems (Hutchinson E, Somlyo AP, eds.). Academic Press Inc, New York, pp 309-324.

Peachey LD (1958) Thin sections. I. A study of section thickness and physical distortion produced during microtomy. J Biophys Biochem Cytol **4** 233-242.

Pearse AGE (1961) Histochemistry - Theoretical and Applied. 2nd edition. Churchill, London, pp. 28-31.

Persson A (1972) Equipment for cryoultramicrotomy. The LKB CryoKit. J de Microscopie (Paris) **13**: 162.

Phillips TE, Boyne AF (1984) Liquid nitrogen-based quick freezing: experiences with bounce-free delivery of cholinergic nerve terminals to a metal surface. J Electron Microscopy Techn **1**: 9-29.

Plattner H (ed) (1989) Electron Microscopy of Subcellular Dynamics. CRC Press, Inc. Boca Raton, FL.

Plattner H, Bachmann L (1982) Cryofixation: A tool in biological ultrastructural research, Int Rev Cytol **79**: 237-304.

Plattner H, Knoll G (1983) Cryofixation of biological material for electron microscopy by the methods of spray-, sandwich-, cryogen-jet- and sandwich-cryogen-jet-freezing: a comparison of techniques. In: The Science of Biological Preparation for Microscopy and Microanalysis (Revel J-P, Barnard T, Haggis GH, eds.). Scanning Electron Microscopy Inc., AMF O'Hare, pp.139-146.

Plattner H, Zingsheim HP (1987) Elektronenmikroskopische Methodik in der Zell- und Molekularbiologie - Ein kritischer Leitfaden zur biologischen Ultrastrukturforschung für Biologen und Mediziner. (Methods in Electron Microscopy in Cell and Molecular Biology. A Critical Guide for Biological Ultrastructure Research for Biologists and Physicians). Gustav Fischer Verlag, Stuttgart-New York.

Plattner H, Schmitt-Fumian WW, Bachmann L (1973) Cryofixation of single cells by spray-freezing. In: Freeze-Etching - Techniques and Applications

[1] From 1916 on follow numerous editions of this well known book about colloid chemistry; see also from the same author: Kolloidchemie (Colloid Chemistry), 1st ed. 1909, and: Handbuch der Kolloidwissenschaft (Handbook of Colloid Science), 1924.

(Benedetti EL, Favard P, eds). Société Française de Microscopie Électronique, Paris, pp. 81-100.

Polge C, Smith AU, Parkes AS (1949) Revival of spermatozoa after vitrification at low temperatures. Nature **164**: 666.

Porter KR, Blum J (1953) A study in microtomy for electron microscopy. Anat Rec **117**: 685-710.

Ramamurti K (1977) Temperature dependence of beam induced mass loss of phenylalanine. In: Proc 35th Ann. Meeting EMSA, pp 560-561 (available from San Francisco Press, San Francisco).

Reid N, Beesley JE (1991) Sectioning and cryosectiong for electron microscopy. In: Practical Methods in Electron Microscopy (Glauert AM, ed) Vol 13, Elsevier, Amsterdam, pp. 249-289.

Richter K (1992) Cryoelectron Microscopy of Frozen-Hydrated Ultrathin Sections of Biological Material. Doctoral Thesis, University of Lausanne.

Richter K (1994) Cutting artefacts on ultrathin cryosections of biological bulk specimens. Micron **25**: 297-308.

Richter K, Gnägi H, Dubochet J (1991) A model for cryosectioning based on the morphology of vitrified ultrathin sections. J Microsc **163**: 19-28.

Riehle U (1968) Schnellgefrieren organischer Präparate für die Elektronen-Mikroskopie - Die Vitrifizierung verdünnter wäßriger Lösungen (Rapid freezing of organic samples for electron microscopy - the vitrification of diluted solutions in water). Chem Ing Techn **40**: 213-218.

Riehle U, Hoechli M (1973) The theory and technique of high pressure freezing. In: Freeze-Etching - Techniques and Applications (Benedetti EL, Favard P, eds). Société Française de Microscopie Électronique, Paris, pp 31-62.

Robards AW, Crosby P (1983) Optimisation of plunge freezing: linear relationship between cooling rate and entry velocity into liquid propane. Cryo-Letters **4**: 23-32.

Robards AW, Sleytr UB (1985) Low temperature methods in biological electron microscopy. In: Practical Methods in Electron Microscopy (Glauert AM, ed) Vol **10**, Elsevier, Amsterdam. pp. 201-241, p. 504.

Roberts IM (1975) Tungsten coating - a method of improving glass microtome knives for cutting ultrathin sections. J Microsc **103**: 113-119.

Roomans GM, Shelburne JD (1983) Basic Methods in Biological X-ray Microanalysis. Scanning Electron Microscopy Inc., AMF O'Hare.

Roomans GM, Wei X, Sevéus L (1982) Cryoultramicrotomy as a preparation method for X-ray microanalysis in pathology. J Ultrastruct Res **3**: 65-84.

Roomans GM, Gupta BL, Leapman RD, von Zglinicki T (eds) (1994) The Science of Biological Microanalysis. Scanning Microscopy, Supplement **8**.

Roos N, Morgan AJ (1990) Cryopreparation of Thin Biological Specimens for Electron Microscopy: Methods and Applications. In: Royal Microscopical Society Microscopy Handbooks, Vol. 21. Oxford University Press - Royal Microscopical Society.

Rose H (1978) Aberration correction of homogeneous magnetic deflection systems. Optik **51**: 15-38.

Rose H (1994) Correction of aberrations, a promising means for improving the spatial and energy resolution of energy-filtering electron microscopes. Ultramicroscopy **56** 11-25.

Roth E (1977) Gesamtausgabe: Heitere Verse (Collected Works: Cheerful Poems). Karl Hanser Verlag, München-Wien.[2]

Ruska E (1979) Die frühe Entwicklung der Elektronenlinsen und der Elektronenmikroskopie (The Early History of Electron Lenses and Electron Microscopy). In: Acta historica Leopoldina (Uschmann G, ed) Nr. **12**. Deutsche Akademie der Naturforscher Leopoldina, Halle/Saale.

Ryan KP (1991) Rapid Cryogenic Fixation of Biological Specimens for Electron Microscopy. Doctoral Thesis, Polytechnic South West, Plymouth (UK). Microfiche DSC DX96010 from The British Library

---

[2]The original version has the title "Gute Vorsätze" (Good Intentions) and the precise wording in German is:

Den guten Vorsatz, sich zu bessern,
Muß mancher manchmal arg verwässern
Die so erzielte Wasserkraft
Treibt dann den Alltag fabelhaft"

Like other poems of this category it is difficult to translate correctly, since some puns are included, for example "*verwässern*" ("*dilute*") or "*Wasserkraft*", which in English means "*impetus*" rather than "*water power*". Nevertheless, my friend Dr. Keith Ryan from the Marine Laboratory at Plymouth (UK) made the successful attempt to translate these four lines

"*The great intent to improve oneself*
must often be much diluted,
the water power thus achieved
to drive the day is aptly suited"

I want just to add, that Eugen Roth was born 100 years ago at Munich in 1895. I like him very much and I recommend all colleagues who (a) are frustrated about my tough review and (b) have some knowledge of the German language, to escape into this exciting poetry. I found also some psychological support during my exhausting literature studies about SS-CCD, ESI, and EFTEM for this review article in those extremely warm hearted short poems full of real humanity.

Document Supply Centre, Boston Spa, Wetherby, West Yorkshire LS23 7BQ, UK.

Ryan KP (1992) Cryofixation of tissues for electron microscopy: a review of plunge cooling methods. Scanning Microsc **6**: 715-743.

Ryan KP, Purse DH (1984) Rapid freezing: specimen supports and cold gas layers. J Microsc **136**: RP5-RP6

Ryan KP, Purse DH (1985) Plunge-cooling of tissue blocks: determinants of cooling rates. J Microsc **140**: 47-54.

Ryan KP, Knoll G (1994) Time-resolved cryofixation methods for the study of dynamic cellular events by electron microscopy: a review. Scanning Microsc **8**: 259-288.

Ryan KP, Purse DH, Robinson SG, Wood JW (1987) The relative efficiency of cryogens used for plunge-cooling biological specimens. J Microsc **145**: 89-96.

Ryan KP, Bald WB, Neumann K, Simonsberger P, Purse DH, Nicholson DN (1990) Cooling rate and ice-crystal measurement in biological specimens plunged into liquid ethane, propane and Freon 22. J Microsc **158**: 365-378.

Salih SM, Cosslett VE (1975) Radiation damage in electron microscopy of organic materials: effect of low temperatures. J Microsc **105**: 269-276.

Sartori N, Richter K, Dubochet J (1993) Vitrification depth can be increased more than 10-fold by high-pressure freezing. J Microsc **172**: 55-61.

Schröder RR (1992) Zero-loss energy filtered imaging of frozen hydrated proteins: model calculations and implications for future developments. J Microsc **166**: 389-400.

Schröder RR, Hofmann W, Menetret JF (1990) Zero-loss energy filtering as improved imaging mode in cryoelectronmicroscopy of frozen-hydrated specimens. J Struct Biol **105**: 28-34.

Schwabe KG, Terracio L (1980) Ultrastructural and thermocouple evaluation of rapid freezing techniques. Cryobiology **17**: 571-584.

Sevéus L (1978) Preparation of biological material for X-ray microanalysis of diffusible elements. I. Rapid freezing of biological tissue in nitrogen slush and preparation of ultrathin frozen sections in the absence of trough liquid. J Microsc **112**: 269-279.

Shindo D, Hiraga K, Oikawa T, Mori N (1990) Quantification of electron diffraction with imaging plate. J Electron Microscopy **39**: 449-453.

Shindo D, Hiraga K, Oku T (1991) Quantification in high-resolution electron microscopy with the imaging plate. Ultramicroscopy **39**: 50-57.

Shuman H, Somlyo AV, Somlyo AP (1976) Quantitative electron probe microanalysis of biological thin sections: method and validity. Ultramicroscopy **1**: 317-339.

Shuman H, Somlyo A (1982) Energy-filtered transmission electron microscopy of ferritin. Proc Natl Acad Sci USA **79**: 106-107.

Siegel G (1972) Der Einfluß tiefer Temperaturen auf die Strahlenschädigung von organischen Kristallen durch 100 keV-Elektronen (The influence of low temperatures on beam damage of organic crystals by 100 kV-electrons). Z Naturforsch **27a**: 325-332.

Singer SJ, Nicolson GL (1972) The fluid mosaic model of the structure of cell membranes. Science **175**: 720-731.

Sitte H (1955) Ein einfaches Ultramikrotom für hochauflösende elektronen-mikroskopische Untersuchungen (A simple ultramicrotome for high resolution investigations in the TEM). Mikroskopie (Wien) **10**: 365-396.

Sitte H (1979). Cryofixation of biological material without pretreatment - A review. Mikroskopie (Wien) **35**: 14-20.

Sitte H (1982) Instrumentation for cryosectioning. In: Electron Microscopy 1982, 10th Int Congr Electr Microsc (The Congress Organizing Committee, ed.). Vol. 1, pp 9-18.

Sitte H (1984) Instruments for cryofixation, cryoultramicrotomy and cryosubstitution for biomedical TEM. Zeiss MEM **3**: 25-31.

Sitte H, Neumann K (1983) Ultramikrotome und apparative Hilfsmittel für die Ultramikrotomie (Ultramicrotomes, Design - Function - Accesories). In: Methodensammlung der Elektronenmikroskopie (Methods of Electron Microscopy with English Summaries) (Schimmel G, Vogell W, eds.). Wiss. Verlagsges. Stuttgart, Beitrag (Contribution) 1.1.2.

Sitte H Neumann K Hässig H Kleber H Kappl G (1980) FC4-cryochamber for Reichert-OM U 4-ultramicrotome Ultracut. In: Electron Microscopy 1980, Proc 7th Eur Conf Electr Microsc Den Haag 1980 (Brederoo P, Boom G, eds.). Electron Microscopy 1980, Vol. I, pp. 540-541.

Sitte H, Neumann K, Edelmann L (1985) Cryofixation and cryosubstitution for routine work in transmission electron microscopy. In: The Science of Biological Specimen Preparation for Microscopy and Microanalysis 1985. (Müller M, Becker RP, Boyde A, Wolosewick JJ, eds.). pp 103-118.

Sitte H, Edelmann L, Neumann K (1987a) Cryofixation without pretreatment at ambient pressure. In: Cryotechniques in Biological Electron Microscopy (Steinbrecht RA, Zierold K, eds.). Springer-Verlag, Berlin. pp 87-113.

Sitte H, Neumann K, Edelmann L (1987b)

Safety rules for cryopreparation. In: Cryotechniques in Biological Electron Microscopy (Steinbrecht RA, Zierold K, eds.). Springer-Verlag, Berlin. pp 285-289.

Sitte H, Neumann K, Edelmann L (1988) Cryosectioning according to Tokuyasu versus rapid-freezing, freeze-substitution and resin embedding. In: Immuno-Gold Labeling in Cell Biology (Verkleij AJ, Leunissen JLM, eds.). CRC Press, Boca Raton, FL, pp. 63-93.

Sitte H, Edelmann L, Hässig H, Kleber H, Lang A (1994) A new versatile system for freeze-substitution, freeze-drying and low temperature embedding of biological specimens. Scanning Microscopy, Suppl **8**: 47-66.

Sitte P (1973) Methodengefüge und Erkenntnisfortschritt (The influence of methodology on the progress in Natural Sciences). Naturwissenschaften **60**: 333-339.

Sjöstrand FS (1953) A new microtome for ultrathin sectioning for high resolution electron microscopy. Experientia **9**: 114-117.

Sjöstrand FS (1954) Die routinemäßige Herstellung von ultradünnen Gewebeschnitten (The routine production of utrathin tissue sections). Z wiss Mikrosk **62**: 65-86.

Skaer H (1982) Chemical cryoprotection for structural studies. J Microsc **125**: 137-147.

Sleytr UB, Robards AW (1977) Plastic deformation during freeze-cleavage: a review. J Microsc **110**: 1-25.

Steere RL (1957) Electron microscopy of structural detail in frozen biological specimens. J Biophys Biochem Cytol **3**: 45-60.

Steere RL (1973) Preparation of high-resolution freeze-etch, freeze-fracture, frozen surface, and freeze-dried replicas in a single freeze-etch module, and the use of stereo electron microscopy to obtain maximum information from them. In: Freeze-Etching Techniques and Applications (Benedetti EL, Favard P, eds.). Societé Française de Microscopie Électronique Paris, pp 223-256.

Steinbrecht RA, Zierold K (1984) A cryoembedding method for cutting ultrathin cryosections from small frozen specimens. J Microsc **136**: 69-75.

Steinbrecht RA, Müller M (1987) Freeze-substitution and freeze-drying. In: Cryotechniques in Biological Electron Microscopy (Steinbrecht RA, Zierold K, eds). Springer-Verlag, Berlin. pp. 149-172.

Studer D, Michel M, Müller M (1989) High pressure freezing comes of age. Scanning Microsc Suppl **3**: 253-269.

Studer D, Michel M, Wohlwend M, Hunziker EB, Buschmann MD (1995) Vitrification of articular cartilage by high-pressure freezing. J Microsc **179**: 321-332.

Takahashi K, Kohda K, Mijahara J, Kanemitzu Y, Amitani K, Shionoya S (1988) Mechanism of photo-stimulated luminescence in BaFX: Eu (X = Cl, Br) phosphors. J Luminescence **31&32**: 266-288.

Talmon Y (1987) Electron beam radiation damage to organic and biological cryo specimens. In: Cryotechniques in Biological Electron Microscopy (Steinbrecht RA, Zierold K, eds.). Springer-Verlag, Berlin, pp 64-84.

Talmon Y, Adrian M, Dubochet J (1986) Electron beam radiation damage to organic inclusions in vitreous, cubic, and hexagonal ice, J Microsc **141**: 375-384.

Taylor KA, Glaeser RM (1974) Electron diffraction of frozen, hydrated protein crystals. Science **186**: 1036-1037.

Taylor KA, Glaeser RM (1976) Electron microscopy of frozen hydrated biological specimens. J Ultrastruct Res **55**: 448-456.

Thornburg W, Mengers PE (1957) An analysis of frozen section techniques I. Sectioning of fresh-frozen tissues. J Histochem Cytochem **5**: 47-52.

Tietz HR (1992) On-line processing and analysis of TEM slow-scan CCD images: theory and practical aspects. In: Electron Microscopy 1992, Proc 10th Eur Congr Electr Microsc (Megías-Megías L, Rodríguez-García MI, Ríos A, Arias JM, eds.) Secretariado de Publicaciones de la Universidad de Granada, Vol. 1, pp 131-135.

Tokuyasu KT (1973) A technique for ultracryotomy of cell suspensions and tissues. J Cell Biol **57**: 551-565.

Tokuyasu KT (1978) A study of positive staining of ultrathin frozen sections. J Ultrastruct Res **63**: 287-307.

Tokuyasu KT (1980) Immunocytochemistry on ultrathin frozen sections. Histochem J **12**: 381-403.

Tokuyasu KT (1986) Application of cryo-ultramicrotomy to immunocytochemistry. J Microsc **143**: 139-149.

Tokuyasu KT, Singer SJ (1976) Improved procedures for immunoferritin labeling of ultrathin frozen sections. J Cell Biol **71**: 894-906.

Trachtenberg S (1993) Fast-freezing devices for cryo-electron microscopy. Micron **24**: 1-12.

Trus BL, Steven AC, McDowall AW, Unser M, Dubochet J, Podolsky RJ (1989) Interactions between actin and myosin filaments in skeletal muscle visualized in frozen-hydrated thin sections. Biophys J **55**: 713-724.

Tvedt KE, Kopstad G, Haugen OA (1984) A section press and low elemental support for enhanced preparation of freeze-dried cryosections. J Microsc

133: 285-290.

Unwin PNT, Henderson R (1975) Molecular structure determination by electron microscopy of unstained crystalline specimens. J Mol Biol **94**: 425-440.

Van Harreveld A, Crowell J (1964) Electron microscopy after rapid freezing on a metal surface and substitution fixation. Anat Rec **149**: 381-385.

Van Harreveld A, Trubatch J, Steiner J (1974) Rapid freezing and electron microscopy for the arrest of physiological processes. J Microsc **100**: 189-198.

Van Venrooij GEPM, Aertsen AMHJ, Hax WMA, Ververgaert PHJT, Verhoeven JJ, Van der Vorst HA (1975) Freeze-etching: freezing velocity and crystal size at different locations in samples. Cryobiology **12**: 46-61.

von Seggern H, Voigt T, Knüpfer W, Lange G (1988) Physical model of photostimulated luminescence of x-ray irradiated BaFBr:EU^{2+}. J Appl Phys **64**: 1405-1412.

Wade RH (1984) The temperature dependence of radiation damage in organic and biological materials. Ultramicroscopy **14**: 265-270.

Wade RH, Pelissier J (1982) The temperature dependence of the electron irradiation resistance of crystalline paraffin. Ultramicroscopy **10**: 285-290.

Weickenmeier AL, Nüchter W, Mayer J (1995a) Quantitative determination of modulation transfer function and detection quantum efficiency for standard and antireflection YAG scintillator slow-scan CCD-camera. In: Proc Microscopy and Microanalysis 1995 (Bailey GW, Ellisman MH, Hennigar RA, Zaluzec NJ, eds) MSA Jones and Begell Publ, pp. 34-35.

Weickenmeier AL, Nüchter W, Mayer J (1995b) Quantitative characterisation of point spread function and detection quantum efficiency for a YAG scintillator slow scan CCD camera. Optik **99**: 147-154.

Wendt-Gallitelli MF, Stöhr P, Wolburg H, Schlote W (1980) Cryoultramicrotomy, electron probe microanalysis and STEM of myocardial tissue. Scanning Electron Microscopy 1980; II: 499-510.

Wendt-Gallitelli MF, Wolburg H (1984) Rapid freezing, cryosectioning, and X-ray microanalysis on cardiac muscle preparations in defined functional states. J Electron Microsc Techn **1**: 151-174.

Zeitler E, Bahr GF (1965) Determination of dry mass in populations of isolated particles. In: Quantitative Electron Microscopy (Bahr GF, Zeitler E, eds). Williams and Wilkins Co, Baltimore, pp. 217-239 (Identical paper in: Lab Invest **14**: 955-977).

Zernike F (1945) Phase contrast, a new method for the microscopic observations of transparent objects. Physica **9**: 686-698 and 974-986.

Zierold K (1982) Preparation of biological cryosections for analytical electron microscopy. Ultramicroscopy **10**: 45-54.

Zierold K (1987) Cryoultramicrotomy. In: Cryotechniques in Biological Electron Microscopy (Steinbrecht RA, Zierold K, eds) Springer Verlag, Berlin, pp. 132-148

Zierold K, Hagler HK (eds) (1989) Electron Probe Microanalysis. Applications in Biology and Medicine. Springer-Verlag, Berlin-Heidelberg.

Discussion with Reviewers

R. Wróblewski: I think that the most important breakthrough in cryo-ultramicrotomy was the method of tissue preparation for immunocytochemical purposes by Tokuyasu, and the way to pick up sections using a sucrose droplet in a loop. Using Tokuyasu's methods, sections can be obtained on all cryo-ultramicrotomes, even the oldest ones. My view is supported by a number of papers produced using different cryomethods for different purposes - ranging from pure morphology through immunocytochemistry and ending with microanalysis. All other methods described by the author have some applications but are still not accepted in daily routine research.

Author: As already stated in the Abstract "The article is mainly directed to the investigation of frozen-hydrated sections in the low dose cryo-TEM and -EFTEM" and "both cryoultramicrotomy for macromolecular cytochemistry (Tokuyasu technique) and cryo-ultramicrotomy for element analysis are only shortly mentioned". It was not the intention to describe again well introduced techniques, which were already several times excellently reviewed (see literature cited at the begin of the section 6 "Cryosectioning According to Tokuyasu") and "accepted in daily routine research". I agree that "Tokuyasu's work was the most important breakthrough in cryo-ultramicrotomy" and it is true that "using Tokuyasu's method, sections can be obtained even with the oldest cryo-ultramicrotomes". But nevertheless I would not like to prepare material according to Tokuyasu protected with 2.3 M sucrose using one of the old systems like the Reichert OmU3/-FC2 or the LKB-CryoKit or the Sorvall MT2/FTS: the differences in ergonomy and efficiency are too large. For frozen-hydrated work below -160°C those oldies are really not suited and the progress shown in Table 1 is really important for further studies in the field, which I wanted to stimulate.

K.P. Ryan: There is no mention of Freons as coolants - are they recommended for any aspect of rapid cooling?

Author: According to Table 6 there is no real need to work with coolants, which will be not so easy obtainable in the future (ozone layer) and which have an inferior cooling efficiency CE in comparison with propane and ethane. This is especially true for the often used Freon 12.

K.P. Ryan: Electron microscopy is full of hazards which include high voltages, toxic chemicals, vacuum equipment involving high temperatures with heavy electrical currents during coating activities, and critical point drying which involves high pressures. In cryo-methodology we encounter further hazards which might be worth detailing in this review - these are low temperatures which can *"burn"* and flammable gases which can form explosive mixtures, although these gases are normally kept below their flash point and under an inert atmosphere of gaseous nitrogen. The gaseous nitrogen is perhaps the most insidious hazard because it is unseen, has no odour and is generally not considered. Can you comment on the amount of nitrogen gas produced in a typical cryo-ultramicrotomy session ?

Author: The question related to hazards and risks is indeed of highest importance and should not be neglected ! I want again to draw the attention to our article "Safety Rules for Cryopreparation" (Sitte *et al.*, 1987b), which gives answers to all important questions. The mentioned risks were mostly overlooked or at least underestimated, especially the dangers of propane (explosions and burning) and the evaporation of LN_2 (unconsciousness and possibly death by asphyxiation). 1 liter of LN_2 develops nearly 800 liters of GN_2 (see Ryan and Liddicoat, 1987). The most severe danger results, if cryosystems with a high consumption of LN_2 run in small cabinets: this is often the case with cryo-ultramicrotomes. For example older FC-models of Reichert-Jung (FC4/FC4D/FC4E) consume up to 7 liters LN_2/h which correspond to approx. 5 m^3 GN_2/h. If a cryosystem of this kind is operated in a small cabinet (often with respect to undisturbed work with closed door, switched-off air condition and in the late evening) the risk is deadly: unconsciousness results, if the O_2-content drops from a normal level of 21% to below 18% ! Since unconsciousness arrives without any preceding signal like dizziness and reanimation is only possible within approximately 30 min, this is a potentially life-threatening situation. To answer in this context the question directed to the "amount of GN_2 produced in a typical cryo-ultramicrotomy session" correctly: if the session takes 2 h and work is done on one of the old FC-models 1980/90 at sectioning temperatures above -100°C, then up to 11 m^3 GN_2 are produced. In a cabinet of (3 x 3 x 2) = 12 m^3 volume with bad ventilation, danger for life exists obviously already within these 2 hours. Cryo-ultramicrotomy and similar cryowork therefore should only be done in well ventilated and larger laboratories. Risk is completely excluded if at least a second person is present in the room. Newer cryo-ultramicrotomes like Leica-Reichert FCS and FCR consume only approx. 1 to 2 liters LN_2/h - but risk still exists in the worst case discussed above.

Also splashing of liquid cryogens is really dangerous and it is often overlooked, that splashes of a *"primary cryogen"* like LN_2 and splashes of a *"secondary cryogen"* like liquefied ethane or propane behave completely different. Splashes of LN_2 are terribly dangerous, since they enter due to the superfluidity (low viscosity) of LN_2 through narrow slits easily into closed compartments like shoes (especially boots), protection gloves (often used for handling Dewar vessels) or safety spectacles: one can loose foot or hand by necessary amputation or the eyesight by LN_2 splashes. Just opposite is the risk of splashes from the viscous secondary cryogens which do not evaporate immediately and do not enter through narrow slits: here protection of skin and eyes by gloves and glasses is recommended.

I do not want to enumerate all risks of cryowork, but I recommend everybody responsible for co-workers concerned with cryo systems to refer to a paper on hazards and risks (*e.g.* Sitte *et al.*, 1987b) and to inform all co-workers carefully and completely about the danger. Probably it is the best, that those persons are forced to subscribe a declaration, that they have knowledge about the danger and that they will observe all safety regulations.

G.M. Roomans: Can you give a comment to the statements of Saubermann and co-workers in the late seventies and early eighties, who recommended cryosectioning of fresh-frozen material for element analyses at temperatures around -30 to -50°C ?

Author: Sectioning at higher subzero temperatures in the range between -50 and -30°C is of special interest, if semithin and thicker cryosections (e.g. thicknesses between 0.5 and 2.0 μm) are cut. This is well known for sugar protected samples: higher temperatures around -80°C simplify the production of thicker sections considerably. Saubermann *et al.* (1977) analysed the cutting forces (cutting work) in thick section cryomicrotomy of fresh frozen specimens by strain gauges and found, that forces and work depend sometimes on the clearance angle of the knife and on the sectioning temperature. Based on these measurements and continued experiments they recommended cryosectioning at -

30° for scanning electron microscopy and subsequent EDX (Saubermann *et al.*, 1981). There was some controversial discussion about the pros and cons of section preparation at these temperatures (see *e.g.* Saubermann, 1981, discussion on pp. 391-396). Even if this discussion does not fit within the theme of this review article ("*cryo-ultramicrotomy*" with section thicknesses $\leq$ 0.1 μm), a re-evaluation and re-investigation of this work would be of interest in several respects: up to now it is not absolutely clear, what happens if perfectly frozen, that means correctly vitrified fresh samples, according to Fig. 32b are warmed up to higher subzero temperatures. There are investigations of Ryan (1991, see especially pp. 200-210 and Figs. 44-46) on red blood cells and spleen of the flounder, which after cryofixation were warmed up to temperatures between -80 and -10°C for some hours to several days before standard freeze substitution according to Van Harreveld and Crowell (1964). The results were really astonishing: the expected severe segregation artefacts were only observable at temperatures $\geq$ -20°C. Similar results were reported by Steinbrecht (1985), who "*heated*" fresh-frozen silk moth sensory hairs for 45 min to -43°C without visible damage (segregation of ice). These findings seem to be in good agreement with earlier experiments and considerations of Meryman (1957), Luyet (1960), Nei (1971, 1973) and MacKenzie (1981). They are of considerable importance for all hybrid methods (see Fig. 32 and Section 10 "*Discussion*", this review). But obviously different samples will react in a different manner if temperature is increased after rapid freezing above -80°C and a lot of experimental work will be needed to understand, what really occurs in this case.

G.M. Roomans: One should realise that most of the freezing methods recommended in this paper make it necessary to dissect the tissue (unless one works with cultured cells). The dissection procedure itself may cause artefacts: mounting of the tissue (slammer, plunging) and building up the pressure (high pressure freezing) introduce a time period during which the supply of oxygen and nutrients to the tissue is reduced, and where also drying of the tissue through evaporation of water from the outer layers is a risk. In conventional fixation, perfusion fixation is used to reduce or eliminate this problem. For X-ray microanalysis *e.g.*, *in situ* freezing techniques have been proposed, even though these techniques do not give optimal results of freezing. Do you not think that in pursuing optimal freezing techniques. there is a risk of losing sight of possible artefacts occurring just before the freezing step? How should we attack this problem?

K. Ryan: You describe in several places the sizes of specimens suitable for various specimen holders or supports. Would you agree that the minimal possible size is always of utmost importance, bearing in mind the damage that can be done to the specimen during dissection or vibratoming?

Author: "*Pre-freezing artefacts*" are a very serious limitation of all freezing procedures, which need very small samples for efficient freezing, if studies on specific parts of larger organs or organisms are the goal and dissection is unavoidable. In this respect, already the loss of H2O from the freed surfaces is a problem: but this problem can be settled by continuous work in simple "*humid chambers*", which are relatively easy to realise (*e.g.*, glove boxes with a wet tissue on the bottom develop rapidly approx. 100 % relative humidity, see also Sitte *et al.*, 1987a, Fig. 4d4, p. 96). Much more difficult is the exclusion of pre-freezing artefacts due to dividing of larger organs by dissection or vibratoming: this is a severe limitation for high pressure, double jet or plunge freezing, which need extremely small specimens. Only direct *in situ* freezing of free surfaces with a propane jet (Green and Walsh, 1994) or with a slammer (*e.g.*, an "*inverted slammer*": Edelmann, 1989b) enable with a minimum delay after dissection a proper freezing of such surfaces with minimum damage and changes. Often there is no other possibility, for example if deeper layers of organs like brain or kidney are of interest: they change morphology and certainly also physico-chemistry immediately after interruption of blood supply and speedy work is necessary. There already exist examples for such procedures: Bernard and Krigman (1974) have studied slam frozen deeper layers of the brain after fast slicing immediately followed by impact cryofixation. These experiments were repeated and confirmed by Van Harreveld and Fifkova (1975). They demonstrated, that the structures after this fast procedure were intact, if the delay between slicing and freezing is very small (*e.g.*, below 10 sec). But this rapid work is only possible, if a suited cryogen-jet- or metal-mirror- or another purpose-designed-system is available, which is either not sensitive against large areas or volumes like single-jet or slamming devices (see discussion in the section on "*Ambient Pressure Freezing*") or excise and freeze a small fraction of the sample within a very fast event (*e.g.*, Gatan cryo-snapper).

Additional References

Bernard SA, Krigman MR (1974) Ultrastuctural analysis of the deeper layers of freeze-substituted guinea pig cortex. Brain Res **76**: 325-329.

Green WB, Walsh LG (1994) Cryo-jet preservation of calcium in the rat spinal cord. Scanning Microsc **8**:587-600.

Luyet B (1980) On various phase transitions occurring in aqueous solutions at low temperatures. Ann NY Acad Sci **85**: 549-569.

MacKenzie AP (1981) Modelling the ultrarapid freezing of cells and tissues. In: Microprobe Analysis of Biological Specimens (Hutchinson TE, Somlyo AP, eds). Academic Press, New York, pp 397-421.

Meryman HT (1957) Physical limitations of the rapid freezing methods. Proc R Soc London B **147**: 452-459.

Nei T (1971) Hemolysis during the rewarming process of frozen erythrocytes. Proc. XIIIth Int Congr Refrig **10:** 907-911.

Nei T (1973) Growth of ice crystals in frozen specimens. J Microsc **99**: 227-233.

Ryan KP, Liddicoat MC (1987) Safety considerations regarding the use of propane and other liquefied gases as coolants for rapid freezing purposes. J Microsc **147**: 337-340.

Saubermann A (1981) Cryosectioning of biological tissue for X-ray microanalysis of diffusible elements. In: *Microprobe Analysis of Biological Systems* (Hutchinson TE, Somlyo AP, eds). Academic Press, New York and London, pp 377-396.

Saubermann A, Riley WD, Beeuwkes III R (1977) Cutting work in thick section cryomicrotomy. J Microsc **111**: 39-49.

Saubermann A, Echlin P, Peters PD, Beeuwkes III R (1981) Application of scanning electron microscopy to X-ray analysis of frozen-hydrated sections. I. Specimen handling techniques. J Cell Biol **88**: 257-267.

Steinbrecht RA (1985) Recrystallisation and ice crystal growth in a biological specimen, as shown by a simple freeze substitution method. J Microsc **140**: 41-46.

Van Harreveld A, Fifkova E (1975) Rapid freezing of deep cerebral structures for electron microscopy. Anat Rec **182**: 377-386.

Scanning Microscopy Supplement 10, 1996 ISSN: 0892-953X
Scanning Microscopy International, Chicago (AMF O'Hare), IL 60666, USA

LIST of REVIEWERS

The conference organizers and publisher are grateful to the following reviewers for their help with the papers included in this Supplement.

Albrecht, R.M.	University of Wisconsin, Madison
Allen, T.D.	Paterson Inst. Cancer Res., Manchester, U.K.
Allison, D.P.	Oak Ridge Natl. Lab., TN
Amrein, M.	Univ. Muenster, Germany
Andrews, S.B.	National Institutes of Health, Bethesda, MD
Beesley, J.	Glaxo Wellcome Med. Res. Ctr., Stevenage, U.K.
Bell, P.B.	Univ. Oklahoma, Norman, OK
Castejón, O.J.	Univ. Maracaibo, Venezuela
Childs, G.V.	Univ. Texas Medical Branch, Galveston
DeBault, L.E.	Univ. Oklahoma Health Sci Ctr., Oklahoma City
Elder, H.Y.	University of Glasgow, Scotland, U.K.
Fine, A.	Dalhousie Univ. Sch. Med., Halifax, NS, Canada
Franzini-Armstrong, C.	Univ. Pennsylvania, Philadelphia, PA
Frederik, P.M.	Univ. of Limburg, Maastricht, The Netherlands
Fritzsche, W.	Techn. Univ. Jena, Germany
Gu, J.	Inst. Molec. Morphology, Mount Laurel, N.J.
Hainfeld, J.	Brookhaven Natl. Lab., Upton, NY
Hamkalo, B.	Univ. California, Irvine
Hörber, H.J.K.	Eur. Mol. Biol. Lab., Heidelberg, Germany
Hozak, P.	Inst. Exptl. Medicine, Prague, Czech Republic
Izutsu, K.	Univ. Washington Sch. Med., Seattle
Jones, N.	Bowman Gray Sch. Med., Winston-Salem, NC
Konat, G.	W. Virginia Univ. Sch. Med., Morgantown, WV
Lakowicz, J.R.	Univ. Maryland Sch. Med., Baltimore, MD
Lal, R.	Univ. California, Santa Barbara
Low, F.	Louisiana State Univ., New Orleans
Marchant, R.	Case Western Res. Univ., Cleveland, OH
Mestres, P.	Univ. Saarland, Homburg, Germany
Nuovo, G.J.	MGN Medical Research Lab., Setauket, NY
Ottensmeyer, F.P.	Univ. Toronto, Canada
Pasquinelli, G.	Univ. Bologna, Italy
Peachey, L.	Univ. Pennsylvania, Philadelphia
Robinson, K.A.	Emory Univ. Hosp., Atlanta, GA
Quist, A.	Univ. Uppsala, Sweden
Ralston, E.	National Institutes of Health, Bethesda, MD
Ryan, K.P.	Plymouth Marine Biol. Lab., U.K.
Sevéus, L.	Pharmacia-Upjohn Diagnostics, Uppsala, Sweden
Small, J.V.	Austrian Acad. Sci., Salzburg
Terasaki, M.	National Institutes of Health, Bethesda, MD
Thiery, M.	Univ. Liège, Belgium
Thornell, L.-E.	Univ. Umeå, Sweden
Turner, J.E.	Univ. North Texas Health Sci. Ctr., Fort Worth
Vesely, P.	Czech Acad. Sci., Prague, Czech Republic
Wroblewski, J.	Karolinska Institute, Stockholm, Sweden
Wróblewski, R.	Karolinska Institute, Stockholm, Sweden
Zasadzinski, J.A.N.	Univ. California, Santa Barbara
Zierold, K.	Max Planck Inst., Dortmund, Germany

Scanning Microscopy Supplement 10, 1996 ISSN: 0892-953X
Scanning Microscopy International, Chicago (AMF O'Hare), IL 60666, USA

Subject Index

Scanning Microscopy Supplement 10, 1996 ISSN: 0892-953X
Scanning Microscopy International, Chicago (AMF O'Hare), IL 60666, USA

Author Index

Scanning Microscopy Supplement 8, 1994 ISSN: 0892-953X
Proceedings of the Twelfth Pfefferkorn Conference, Univ. Cambridge, U.K., September 1993
Scanning Microscopy International, AMF O'Hare (Chicago), IL 60666, USA

The Science of Biological Microanalysis

Table of Contents

Price: US delivery: $73 (by book rate; add $5 for insured UPS, if desired); outside US: $78 (by surface mail); $88 (by surface air-lift); or $100 (by regular airmail). Reprints of individual articles are available at $5 each for upto 10 pages, and $10 each for 11+ pages.

Scanning Microscopy Supplement 5, 1991 **ISSN: 0892-953X**
Proceedings of the Ninth Pfefferkorn Conference, Univ. California, Santa Cruz, August 1990
Scanning Microscopy International, AMF O'Hare (Chicago), IL 60666, USA

The Science of Biological Specimen Preparation for Microscopy and Microanalysis, 1990

Table of Contents

Bound in same issue as Scanning Microscopy vol. 5, no. 4, 1991; $52 (US Delivery), $58 (Foreign).

Scanning Microscopy Supplement 3, 1989 ISSN: 0892-953X
Proceedings of the Seventh Pfefferkorn Conference, Univ. Surrey, Guildford, U.K., Sept. 1988
Scanning Microscopy International, AMF O'Hare (Chicago), IL 60666, USA

The Science of Biological Specimen Preparation for Microscopy and Microanalysis, 1988

Table of Contents

Price: $57 (US delivery); $62 (elsewhere).

The Science of Biological Specimen Preparation for Microscopy and Microanalysis, 1985

Proceedings of the 4th Pfefferkorn Conference held March 1985 at Grand Canyon, Arizona
Edited by: M. Mueller, R.P. Becker, A. Boyde, J.J. Wolosewick   Publ. by: Scanning Microscopy Intl.

TABLE of CONTENTS

Price: $46 (US delivery); $50 (elsewhere).